Identifying Marine Phytoplankton

Contributors

Grethe R. Hasle

Department of Biology
Marine Botany Section
University of Oslo
Oslo, Norway

Karen A. Steidinger

Florida Department of
Environmental Protection
Florida Marine Research Institute
St. Petersburg, Florida

Erik E. Syvertsen

Norwegian Pollution Control Board
Division of Offshore Activity and Monitoring
Oslo, Norway

Karl Tangen

OCEANOR
Trondheim, Norway

Jahn Throndsen

Department of Biology
Marine Botany Section
University of Oslo
Oslo, Norway

Berit R. Heimdal

Department of Fisheries and
  Marine Biology
University of Bergen
Bergen, Norway

Identifying
Marine
Phytoplankton

Editor

Carmelo R. Tomas

Florida Department of Environmental Protection
Florida Marine Research Institute
St. Petersburg, Florida

Academic Press
An Imprint of Elsevier

San Diego   New York   Boston   London   Sydney   Tokyo   Toronto

Copyright © 1997 by ACADEMIC PRESS

Academic Press
An Imprint of Elsevier
525 B Street, Suite 1900, San Diego, California 92101-4495, USA
http://www.academicpress.com

Academic Press
Harcourt Place, 32 Jamestown Road, London NW1 7BY, UK
http://www.academicpress.com

Library of Congress Cataloging-in-Publication Data

Identifying marine phytoplankton / edited by Carmelo R. Tomas.
       p.   cm.
    Includes index.
      ISBN-13: 978-0-12-693018-4   ISBN-10: 0-12-693018-X (pbk.: alk.paper)
      1.   Marine phytoplankton—Identification. 1. Tomas, Carmelo R.
   QK934.I44   1997   97-13861
   579.8'1776—dc21   CIP
      ISBN-13: 978-0-12-693018-4
      ISBN-10: 0-12-693018-X

Printed and bound in Great Britain by
CPI Antony Rowe, Chippenham and Eastbourne

Transferred to Digital Printing, 2010

Contents

Editor's Foreword   ix
Contributors' Forewords   xi
Acknowledgments   xv

Chapter 1   *Introduction and Historical Background*   1

Grethe R. Hasle and Carmelo R. Tomas

Chapter 2   *Marine Diatoms*   5

Grethe R. Hasle and Erik E. Syvertsen

Introduction   5
   General Characteristics   6
   Life Cycles   7
   Morphology and Terminology   13
   Classification   23
Genera Represented in Marine Plankton   23
   Centric Diatoms   23
   Pennate Diatoms   26
   Identification   27
   Content   27
Description of Taxa   28
   Centric Diatoms   28
   Pennate Diatoms   240
Methodology   331
   Collection and Concentration   331

Unialgal Cultures as a Means for Species Identification   333
Preservation and Storage   333
Preparation for Light Microscopy   334
Preparation for Electron Microscopy   335
Microscopy   336
What to Look for—General Hints for Identification
    and Preparation   337
Taxonomic Appendix   338
    New Genus   338
    New Names   339
    Validation of Names   339
    New Nomenclatural Combinations   340
Common Diatom Synonyms   341
Index of Diatom Taxa   351
References   361

Chapter 3 *Dinoflagellates*   387

Karen A. Steidinger in Collaboration with Karl Tangen

Introduction   387
    General Characteristics   387
    Dinoflagellates: Eukaryotic Unicells   390
Terminology and Morphology   391
    Morphological Types   391
    General Cell Terms   391
    Microanatomy   401
    Characters Used in Identifying Prorocentroid Desmokont
        Cells   402
    Characters Used in Identifying Dinokont Cells   403
Identification of Species   416
Techniques for Preparation of Dinoflagellates for
    Identification   551
Common Dinoflagellate Synonyms   554
Index of Dinoflagellate Taxa   565
References   570

Chapter 4   *Introduction*   585
Carmelo R. Tomas

Chapter 5   *The Planktonic Marine Flagellates*   591
Jahn Throndsen

Introduction   591
   General Considerations   592
   Algal Flagellate Characteristics   594
   Flagellate Terminology   600
   Phytoflagellate Taxanomy   602
   Phytoflagellate Systematics   602
Chromophyta   605
   Cryptophyceae   605
   Raphidophyceae   612
   Chrysophyceae   616
   Dictyochophyceae   626
   Prymnesiophyceae–Haptophyceae (Exclusive of
      Coccolithophorids)   633
Chlorophyta   645
   Euglenophyceae   645
   Prasinophyceae   651
   Chlorophyeae   664
Zooflagellates (Phylum Zoomastigophora)   671
   Choanoflagellidea   671
   Kinetoplastidea   680
   Ebriidea   685
Techniques   686
   Preparing Samples for Observation   688
   Cultivation for Identification   694
   Preparation of Samples for Further Studies   697
   Specific Problems to Avoid   698
Common Flagellate Synonyms   699
Glossary   702
Index of Flagellate Taxa   708
References   715

Chapter 6 *Modern Coccolithophorids* *731*

Berit R. Heimdal

Introduction   731
  General Characteristics   742
  Terminology and Morphology   744
  Problems in Studying Recent Coccolithophorids   749
  Classification   750
  Outline for Classification and Arrangement of Genera   752
Systematic Descriptions   755
  Holococcolithophorids   755
  Heterococcolithophorids   773
Common Coccolithophorid Synonyms   816
Index of Coccolithophorid Taxa   819
References   821
Glossary   832

General Index   835

Editor's Foreword

This volume is a reprinted version of the two volume series previously published as "Marine Phytoplankton: A Guide to Naked Flagellates and Coccolithophorids" Academic Press, 1993, and "Identifying Marine Diatoms and Dinoflagellates" Academic Press, 1996. It achieves one of the original project objectives of providing in one volume, updated information required for the identification of marine phytoplankton. The response to both volumes was gratifying and confirmed our belief that there was a need for information of this kind. The reprinting of this literature allowed us an opportunity to combine both book texts, to correct errors, and to offer the complete text in a softbound format. While the discipline has not stood still since the first publishing, a complete revision was not possible at this time. Therefore, we urge users of this literature to review the latest journal publications for recent changes. The historical literature is well represented in the reference lists following each chapter and should suffice for returning to original descriptions.

The organization of this book was made with the broadest of outlines to accommodate the specific needs of the contributors—each had their own preference as to how the chapters should be designed. As a result, the reader will recognize that certain elements are common to all chapters. A general introduction, terminology list (in one case illustrated), numerous line drawings as well as SEM and TEM photomicrographs, cryptic descriptions giving essential characters, a synonymy list, index taxa, and extensive reference citations accompany each chapter. Since the material of each chapter varied, a flexible approach was required. Thus readers will find differences in format from chapter to chapter accommodating these differences. I hope this will not be too distracting and that the users will freely move from chapter to chapter with little difficulty.

For the most part, the illustrations are unique and have been produced for this publication. Photographs loaned for inclusion here are noted as are figures taken after published drawings. All original authors were contacted for approval prior to publication of the book. In some instances there are unlabeled illustrations as in a plate series in the Dinoflagellate chapter. These figures, appearing in silhouette, were purposely unlabeled forcing the users to match the silhouette with the corresponding line drawings appearing in the other plates of that chapter. In this manner, the authors of this chapter hoped to encourage researchers to use cell shape and outline as one of the important diagnostic tools in accurate identification of species. This was a deliberate challenge and not an oversight in labeling illustrations.

All chapters have special guides for identification. In the Diatom chapter, there are numerous keys, list of characters, and tables comparing species to aid the researcher in distinguishing between similar species. In the Coccolithophorid chapter, there are comparisons of similar or problem taxa, which is of particular help with closely related confusing species. In the Flagellate chapter, there is a diagnostic "What to Look For" outline which can serve as a quick reference in narrowing the appropriate class, while in the Dinoflagellate chapter, the illustrated terminology section is oriented at showing the various features used in identifying species. Again, this reflects the preference of the authors. Of particular note, all readers should be aware of the Taxonomic Index that appears near the end of the Diatom Chapter. In this section, a new genus, new names and new taxa as well as novel nomenclatural combinations, are presented. All nomenclatural novelties were validated in the original publications in, Tomas, C. (ed.), "Identifying Marine Diatoms and Dinoflagellates," Academic Press, 1996 and Tomas, C. (ed.), "Marine Phytoplankton: A Guide to Naked Flagellates and Coccolithophorids," Academic Press, 1993. Additionally, common synonyms are presented at the end of each chapter. The most recently presented valid name is given with the others as equal synonyms including the full author citation.

A word of caution should be mentioned regarding distributions of species. The common feeling was that distributional information was to be given along broad climatic zones such as temperate waters, arctic or antarctic, subtropical, tropical oceans. In some instances, specific locations are mentioned, primarily because the species illustrated may have first been described from that region. The lack of a location reference for any species should not be construed as denoting the absence of that species from a region. Also, it would be impossible to cover all species of marine phytoplankton. The species chosen for description were selected to give a good representation of the commonly important species as well as enough different species as to give a full spectrum of characters to be found in the phytoplankton. The definitive book covering all phytoplankton species in the sea will probably never be realized.

As a final word, I would like again to restate my gratitude to my author/colleagues who devoted considerable effort in creating, revising, and modifying this literature. Without their full commitment, support, and patience this work could not have been completed. Numerous other colleagues also have assisted including those at the Zoological Station of Naples, Italy, the Florida Marine Research Institute, and the students of the International Phytoplankton Course, who have suggested modifications and corrections. Dr. Paul Silva gave invaluable assistance in matters regarding nomenclature. A special word of thanks is due to the editorial staff at Academic Press that has endured numerous modifications, delays, and revisions yet was supportive of the combined book project. Finally, I wish to acknowledge the continued effort and encouragement of my wife Cele, who throughout the various versions of these books has been the constant support without which this work would not have been completed.

Carmelo R. Tomas

Contributors' Forewords

Chapter 2

The diatoms have been studied for almost 300 years. A multitude of monographs and floras covering smaller and larger areas has been published, and the exact number of thousands of species distributed can hardly be given. Although the marine planktonic diatoms probably constitute a smaller fraction of the total number of species described, we are still dealing with some thousands of species. The elaborately and intricately ornamented siliceous diatom frustule was a challenge to the first transmission electron microscopist in the 1940s, and in the 1960s scanning electron microscopy was introduced in diatom studies providing even better insight into the structure of the diatom cell. This information led to new combinations of species, rejection of species, and description of taxa of all taxonomic categories. The thousands of species, the hundreds of years of studies, the clarification of intricate structures and relationships between taxa obtained by electron microscopy, and the confusion caused by introduction of new names may explain the length of the present chapter.

The history and development of the diatom chapter coincide with the rest of the project, starting with a simple text in 1976, mainly based on the authors' own research. The basis for a manuscript was therefore at hand when the possibility to publish the course notes as a book started to materialize in 1989. The first draft for a complete text was ready for the editor's corrections at the end of 1991 and was returned to the authors at the end of 1992. This version went back to the editor in April–May 1993, to be returned to the authors 1 year later. In April 1994, the editor and the senior author sat together for a short week to finally prepare a manuscript ready to submit to Academic Press.

Diatom research fortunately did not stand still between the start and the final step of the preparation of the diatom chapter. Efforts were made to incorporate, although to a limited extent, literature published in 1992–1994, but with the qualification that time and space did not permit a detailed treatment. During the last years of preparation nomenclatural problems related to the diatoms under study came to our notice. Thanks to Dr. Paul C. Silva as the nomenclature specialist on algae, most of the problems have been solved. New taxa and nomenclatural combinations having their first appearance in this chapter will hopefully be dealt with in detail in future publications.

The authors are grateful to Tyge Christensen for correction of the latin, to Paul Silva for his patience with the senior author's numerous questions, to Greta Fryxell for comments on *Pseudo-nitzschia* and *Thalassiosira,* to Frithjof Sterrenburg for comments on *Pleurosigma,* and to Bo Sundström for letting us copy his *Rhizosolenia* drawings. Carmelo Tomas is especially thanked for his editorial assistance; his initiative and sustained effort fulfilled the senior author's long-dreamt dream to get literature prepared for the International Phytoplankton Courses formally available to a greater audience. E. Paasche and Carina Lange carefully read and commented on parts of the manuscript; Berit Rytter Hasle assisted with the preparation of the line drawings, and the electron micrographs were made at the Electron Microscopical Unit for Biological Sciences at the University of Oslo.

The project was supported by grants from the Norwegian Fisheries Research Council (1202-203.075 to E.E.S.), and from the Norwegian Research Council for Science and the Humanities (457.90/027 to E.E.S., 457.91/001 and 456.92/006 to G.R.H.). The senior author expresses gratitude to the Department of Biology, University of Oslo, for financial support and also for continued working facilities after retirement.

Grethe R. Hasle

Chapter 3

Advances in microscopy have furthered our ability to differentiate genera and species based on morphology and cytology. Concurrent with these advances in equipment and technique were individual studies that clarified useful characters; for example, E. Balech's recognition and characterization of sulcal and cingular plates; D. Wall's, B. Dale's, and L. Pfiester's characterization of life-cycle stages; H. Takayama's characterization of apical grooves or what B. Biecheler described as acrobases; J. Dodge's characterization of apical pore complexes; and F. J. R. Taylor's synthesis and interpretations on dinoflagellate taxonomy, biology, and evolution. These scientists are counted among my heroes. In the future, there will be more heroes who will have worked on optical pattern recognition, biochemical systematics and molecular probes, and other new avenues to identify species and relatedness among species.

My deepest respect and appreciation go to my Norwegian colleagues to whom I am indebted for inviting me to be an instructor and for sharing their knowledge, wisdom, kindness, and sense of humor with me. To Dr. Karl Tangen of OCEANOR, my collaborator, I offer special thanks. To my friend and mentor, Dr. Enrique Balech of Argentina, I offer my sincerest appreciation for teaching me to see beyond what is obvious and to interpret plate patterns and species differences. To Dr. Jan Landsberg (Florida Department of Environmental Protection, Florida Marine Research Institute) and Julie Garrett (Louisiana State University) I offer my gratitude for encouraging and helping me to complete this project. To the editor of this series, Dr. Carmelo Tomas, I express my gratitude for his patience, resolve, and continued friendship. I also thank and acknowledge Dr. Earnest Truby (Florida Department of Environmental Protection, Florida Marine Research Institute) and Dr. Elenor Cox and Clarence Reed (Texas A&M University) for the loan of their exceptional, unpublished scanning electron micrographs of armored species that were used to draw some of the composite illustrations in the plates. Julie Garrett provided most of the scanning electron micrographs of apical pore complexes. Consuelo Carbonell-Moore (Oregon State University) shared her knowledge of the Podolampaceae with me and is credited for photographs in Plate 7. Llyn French (Florida Department of Environmental Protection, Florida Marine Research Institute) assisted in preparation of the plates and provided artistic advice. Diane Pebbles, a biological illustrator and artist, provided 80% of the species illustrations, many of them original drawings based on scanning electron micrograph images. Her work increases the value of this chapter. Dr. Haruyoshi Takayama (Hiroshima Fisheries Experimental Station) provided all the photographs of apical grooves in Plates 1 and 2.

Karen A. Steidinger

Chapter 5

Among today's experts in flagellate taxonomy, the electron microscope has become an indispensable tool for identification of the species. However, the light microscope still remains the main instrument for routine use, and in many cases electron microscopy is merely used to verify identifications already made with conventional light microscopy. For most phytoplankton ecologists, however, the light microscope remains the only accessible equipment for taxonomic identification.

My account of marine planktonic flagellates (excluding dinoflagellates) is an introductory guide to this group and is based primarily on observations using the light microscope. The number of species illustrated for each genus is limited to presumably the most characteristic ones. Within genera such as *Chrysochromulina* (Prymnesiophyceae) and *Pyramimonas* (Prasinophyceae) large numbers of species need electron microscopy for reliable identification. In these cases, reference to features observed using both light and electron microscopy is given. Whenever possible the original description should be consulted, as all later ones depend on the interpretation of the first. Emended diagnoses and descriptions, however, are very important for an up-to-date identification. The variation in morphology common within one flagellate species makes it desirable to give several illustrations, but for practical reasons only one illustration for each species is included here.

When dealing with flagellates, as with other types of plankton, personal experience with each taxon is most important for practical work. The variation within the species required for determining the typical cell shape can be observed by using culture techniques or studying the species during boom conditions. It should be noted, however, that some species vary in appearance with growth condition (such as the number of cells in the *Oltmannsiella* colonies; Carmelo Tomas, personal communication).

The systematics of marine flagellates is presently in a dynamic state. The contents of the plates presented here were fixed to a classification system that becomes further modified as time goes on. The contents of this chapter have, within practical limits, been updated to accommodate recent systematic revisions.

I am indebted to colleagues at the Institut for Sporeplanter, Copenhagen University; Tyge Christensen for providing Latin diagnoses and critically commenting upon the etymology of terms in the glossary, and Øjvind Moestrup, Helge A. Thomsen, and Jacob Larsen for comments on the manuscript. David Hill, School of Botany, University of Melbourne, Australia, commented on the cryptophycean systematics.

J. Throndsen

Chapter 6

Coccolithophorids, characterized by an outer covering of calcified scales or coccoliths, present special problems in identification due to the small cellular and coccolith size, often requiring observations with electron microscopes for reliable identification. However, species that are larger and/or have characteristic gross morphologies, like *Discosphaera tubifer* and *Scyphosphaera apsteinii*, can be readily identified during routine analysis of water samples with light microscopy. Morphological details of coccolith structure of smaller cells are discernible only under the best optical conditions. Thus, difficulties may be encountered if species like *Emiliania huxleyi* and smaller cells of *Gephyrocapsa oceanica* are present in the same sample. Use of an oil immersion objective and a total magnification of 800–1000 times should allow clear distinction between these species.

My chapter provides an introductory guide to the coccolithophorids. It presents the key literature, which is scattered in numerous publications. The text includes systematic descriptions and line drawings of a number of species as examples of the presumably most common ones encountered in marine samples. The chapter also includes species-related information such as synonyms, characteristics for identification, and distribution.

Several colleagues have kindly read parts of this chapter in manuscript form. This has been a great help. Among the many individuals to whom personal thanks are due, I wish to especially mention the late K. R. Gaarder, who introduced me to the coccolithophorids and so greatly enriched our knowledge of extant species. She has given a solid foundation for future work. Grateful thanks are also due to C. R. Tomas for editorial assistance as well as to R. Heimdal and E. Holm, who assisted with the preparation of the line drawings and other technical aspects of this chapter.

B. Heimdal

Acknowledgments

In addition to the acknowledgments given in the Forewords, the following agencies and institutions have substantially contributed to this work. Without their financial and logistical support, this work could not have been completed. The editor and authors are indebted to the support of the following:

The Norwegian Research Council for Science and the Humanities
The Norwegian Fisheries Research Council
Stazione Zoologica "Anton Dohrn" di Napoli
UNESCO
Department of Biology, University of Oslo
University of Bergen, Department of Fisheries and Marine Biology
Florida Department of Environmental Protection, Florida Marine
     Research Institute

For supporting the International Phytoplankton Course from which this literature was developed, contributions were also made by:

NORAD
Italian National Research Council
Italian Ministry of Foreign Affairs
U.S. Office of Naval Research
Carl Zeiss Company, Oberkochen

Numerous individuals helped during various phases of the preparation of this text and illustrations. Special thanks are given to Cele Tomas for her invaluable help, countless hours of labor, continued support, and infinite patience. She not only gave editorial assistance but remained a staunch supporter over the long duration of the project. Ms. Llyn French freely gave of her expertise in illustration, graphics, and the many aspects of page design. Colleagues at Stazione Zoologica have been especially supportive. These include Donato Marino, Marina Montresor, and Adriana Zingone. Dr. Dirk Troost from UNESCO and Professor Gaetano Salvatore, President of Stazione Zoologica, provided financial support from their respective institutions.

Chapter 1

Introduction and Historical Background

Grethe R. Hasle and Carmelo R. Tomas

The content of this book as well as the earlier companion volume "Marine Phytoplankton: A Guide to Naked Flagellates and Coccolithophorids" had its origins as teaching and "handout" literature developed for the Advanced International Phytoplankton Course. Since the original course in 1976, the literature has been updated, improved, and tested on the talented selected participants for each course. With each course offering, requests were made to have the literature presented in a more permanent format as a published book(s). The urgency for the need of such literature was seen as photocopies of the handouts began to appear in various laboratories around the world. Prior to the 1990 course, an attempt to finalize this goal was realized with the agreement to write one book containing this literature. Here we will briefly present the steps of the procedure leading to the publication of this volume.

The idea of an International Course in Phytoplankton had its origins with Professor Trygve Braarud at the University of Oslo, Norway. Within his archived files are notes where Professor Braarud considered a course to teach young students of phytoplankton. The faculty would consist of Professor F.

Hustedt (Diatoms), Professor J. Schiller (Dinoflagellates), and Professor E. Kamptner (Coccolithophorids). These names, gurus of the phytoplankton studies of the first half of this century, would have truly made an all-star teaching team. This dream was realized but not with the cast originally designed, as by the time the course was ready to be taught, most of these mentors were deceased.

In January 1969, a working Group of Phytoplankton Methods (WG 33) was established during the executive meeting of the Scientific Committee in Oceanic Research (SCOR). During this meeting, Professor Braarud pointed out the urgent need for considering phytoplankton methods other than those involving pigment and other chemical analyses. The IOC Working Group on Training and Education also commented on the need for modern textbooks and manuals (Unesco technical papers in marine science no. 18, Paris, 1974).

In Item 4 of the Terms of Reference to WG 33, the Working Group was asked to prepare a report including reference to literature in taxonomy of the main groups and on methods for using quantitative phytoplankton data in ecological studies. To fulfill this request, the Working Group suggested a list of the contents of such a manual and a tentative plan for a "Phytoplankton Course for Experienced Participants." The University of Oslo was chosen as the place for the course and the Marine Botany Section as responsible for the teaching program.

The preparation of a Phytoplankton Manual of Methodology started with a meeting at the University of Oslo under the auspices of SCOR in 1974. The "phytoplankton manual" was published in 1978 by Unesco as "Monographs on Oceanographic Methodology 6" with A. Sournia as the editor. No further steps were taken to prepare a corresponding manual on phytoplankton taxonomy although a need had been expressed by some members of WG 33.

The first "Phytoplankton Course for Experienced Participants" was held at the University of Oslo during 4 weeks in August–September 1976 with 17 participants from 13 different countries. After the first offering, the length of the course was cut to 3 weeks, and the next two courses, in the autumns of 1980 and 1983, were held at the Biological station in Drøbak, belonging to the University of Oslo. Stazione Zoologica "Anton Dohrn," Naples, Italy, hosted and organized the courses now called "Advanced Phytoplankton Courses, Taxonomy and Systematics" in 1985, 1990, and 1995. Another session of this course is presently being planned for Spring 1998 to be held in the Naples area.

From the very beginning interest in the courses was considerable and increased with each offering. In 1995, more than 170 applications were received for the 15–17 places available. The apparent need for a course dealing with identification of phytoplankton species became more evident with the increased activity in mariculture, the recurrence of harmful phytoplankton blooms, the documented toxicity of certain species, the apparent increased pollution of the sea, and global atmospheric changes.

A total number of 99 participants, representing 38 countries, participated in the five courses to date. The instructors in 1976 were the late Karen Ringdal Gaarder (coccolithophorids, dinoflagellates), Grethe Rytter Hasle (diatoms, dinoflagellates), E. Paasche (algal physiology, cultures), Karl Tangen (dinoflagellates), Jahn Throndsen (naked flagellates), and Berit Riddervold. All the instructors with the exception of Berit Riddervold Heimdal (coccolithophorids), from the University of Bergen, were from the University of Oslo. In 1983, Karen A. Steidinger, Florida Marine Research Institute, and Karl Tangen, now Oceanor, Trondheim, Norway, taught dinoflagellates and Barrie Dale, University of Oslo, lectured on dinoflagellates cysts. Erik E. Syvertsen, University of Oslo, assisted G. R. Hasle with the diatoms. From 1985 the staff of Stazione Zoologica also participated in the teaching.

The courses were sponsored by SCOR and IABO, and financially by UNESCO, NORAD (Norwegian Agency for International Development), the Norwegian Ministry of Foreign Affairs, the Italian Ministry of Foreign Affairs, the Italian National Research Council, the U.S. Office of Naval Research, Stazione Zoologica "A. Dohrn" di Napoli, and the University of Oslo.

Despite the unique collection of reprints and identification literature available during the course at the University of Oslo, and later at the Stazione Zoologica, class notes and handouts had to be prepared. They started out with a few pages on each group and increased gradually with additional information from the literature and the respective instructor's own research. In 1983, mainly by Karen Steidinger's initiative, contacts were made with publishing companies to formalize an officially published text. These attempts failed, but in 1989 Carmelo R. Tomas (participant of the 1983 course) started successful negotiations with publishing companies for a text to be used in the 1990 course. Again this deadline was not accomplished but a firm commitment from the authors, editor, and publishing company was definitely made. Consequently, the course notes changed in format and increased in content to form the basis of a manuscript for publication. It became evident that the flagellate and coccolithophorid texts would be completed ahead of those on the diatoms and dinoflagellates. This plus the fact that the newly expanded version of the diatom and dinoflagellate sections exceeded the original project would make a book containing all parts too large for a handy volume. After renegotiation between Academic Press and Carmelo Tomas as the editor, it was decided that a volume on flagellates and cocclithophorids would be published first to be followed by the present one on diatoms and dinoflagellates.

Running expenses inside Norway, related to the manual project, were covered by grants from the Department of Biology, University of Oslo; the Norwegian Research Council for Science and the Humanities (NAVF 457.90/041); and from the Norwegian Fisheries Research Council (project 66170).

Planning funds for the literature were also awarded to the editor from UNESCO while funds for illustrations, technical assistance, postage, and communications were given by Stazione Zoologica of Naples. Since no member of this team was funded to work full-time on this project, each person gave of their personal time and effort to accomplish the goal of completing these manuals. The respective institutions gave support, as was possible, affording each author and editor the opportunity to work on this project. The support notwithstanding, each member of this team worked on this literature while assuming full duties of their permanent work assignments.

Chapter 2

Marine Diatoms

Grethe R. Hasle and Erik E. Syvertsen

INTRODUCTION

The study of diatoms began in the 18th century. The name of the class Bacillariophyceae was derived from the genus *Bacillaria* Gmelin 1791, whereas "Diatom" refers to the genus *Diatoma* De Candolle 1805. Despite more than a century of devoted morphological and taxonomic investigations, electron microscopy, introduced to diatom research in the middle of the 20th century, revealed additional information. A reevaluation of the established classification systems and the current ideas and information on biogeography was required, and a new era of diatom investigations began.

Simonsen (1979) introduced a diatom system based on results from light and electron microscopy and constructed a key to the diatom families. Other ideas on classification, evolution, and critical evaluations at the higher taxonomic levels followed, based on the increasing amount of information (Cox, 1979; Round & Crawford, 1981, 1984; Fryxell, 1983; Glezer, 1983; Nikolaev, 1984; Williams & Round, 1986, 1987), resulting in two partially diverging diatom systems (Glezer et al., 1988; Round et al., 1990).

Identifying Marine Phytoplankton
Copyright © 1996 by Academic Press, Inc. All rights of reproduction in any form reserved.

Publications summarizing the new information on diatom morphology as well as a revision of the classical identification literature were needed. To meet this requirement several diatom atlases, floras and handbooks were published during the past decade or so, most of them concentrating on a particular geographical region. Ricard (1987) constructed keys to families and genera with genus as the lowest rank, the genera being illustrated with light and electron micrographs of one or a few species of each. The diatom handbooks by Priddle & Fryxell (1985) and Medlin & Priddle (1990) both deal with polar species. The focus of the former is on some planktonic diatoms commonly recorded in the Southern Ocean. The latter, a more comprehensive handbook, includes the two polar regions and has an ecological as well as a taxonomic part with keys to species. The diatom atlas from India and the Indian Ocean region (Desikachary, 1986–1989) contains only light micrographs of the diatoms recorded in the area with no additional text, and the phytoplankton atlas by Delgado & Fortuño (1991) has text as well as line drawings and scanning electron micrographs of diatoms from the Mediterranean.

The publications by Rivera (1981), Makarova (1988) and Rines & Hargraves (1988) have the character of monographs of the marine planktonic genera *Thalassiosira* (the former two publications) and *Chaetoceros* (the latter publication), although based on material from specific geographical areas. The investigation of *Rhizosolenia,* a third important marine planktonic genus, by Sundström (1986) is based on material from almost all oceans, and the Unesco Manual on Harmful Microalgae has a chapter on this category of diatoms (Hasle & Fryxell, 1995).

The monumental diatom volume by Round et al. (1990) differs from all the publications mentioned previously in content as well as size; it consists of sections on the biology of the diatoms, a summary of the introduced classification, and a generic atlas. Linnaeus, a catalogue and expert system for the identification of protistan species (Estep et al., 1992), includes diatoms, and the catalogue by Gaul et al. (1993) lists papers containing electron micrographs of diatoms and is thus useful to those studying the fine structure of the diatom frustule.

Despite these recent publications, teaching experience tells us that there is still a need to fill in respect to the global aspect of the identification of marine planktonic diatoms at the specific level. We hope to fill a part of this need with this chapter.

GENERAL CHARACTERISTICS

Systematics: Class Bacillariophyceae in the division Chromophyta.

Closest relatives: Chrysophyceae and Xanthophyceae. (See Round et al., 1990, p. 122.)

Number of species: 10,000–12,000, approx 50,000 (Round & Crawford, 1984, p. 169), or in excess of 100,000 (Round & Crawford, 1989, p. 574); or in marine plankton approx 1400–1800 (Sournia et al., 1991, p. 1085).

Size: ca. 2 μm–ca. 2 mm.

Level of organization: Unicellular, often in colonies.

Cell covering: Siliceous wall and organic layer.

Flagella: Male gametes with one flagellum with stiff hairs.

Chloroplasts: Lamellae with three thylakoids, girdle lamella, and four membranes around the chloroplast.

Pigments: Chlorophylls *a* and *c,* betacarotene, fucoxanthin, diatoxanthin, and diadinoxanthin.

Mitochondria: Tubular type.

Storage products: Chrysolaminarin and oil.

Motility: Present in pennate diatoms with a raphe.

Biotopes: Marine and freshwater, plankton, benthos, epiphytic, epizoic (e.g., on whales and crustaceans), endozoic (e.g., in foraminifera), endophytic (e.g., in seaweed), on and in sea ice, and "air diatoms."

Geological age: Centrics: Jurassic (a few species) and Early Cretaceous (Gersonde & Harwood, 1990). Araphid pennates: Late Cretaceous (Medlin et al., 1993, with references). Raphid pennates: Middle Eocene (Medlin et al., 1993, with references).

LIFE CYCLES

Reproduction (Figs. 1a and 1b)

Diatoms reproduce vegetatively by binary fission, and two new individuals are formed within the parent cell frustule. Each daughter cell receives one parent cell theca as epitheca, and the cell division is terminated by the formation of a new hypotheca for each of the daughter cells. This type of division, with formation of new siliceous components inside the parent cell, leads to size reduction of the offspring. The possible size range of the diatom cells seems to be species dependent, and the specific variation may be as large as 8 to 10 times the length of the apical axis or the diameter.

The considerable size variation is often accompanied by a pronounced size dependent change in cell proportions, normally in the form of an increase in the ratio between the length of the pervalvar axis and the apical axis or diameter. In addition, size variation often causes changes in valve ornamentation, like a reduction in the number of central clustered processes in *Thalassiosira* spp. (E. Syvertsen, personal observations), a loss of special structures like the pili of certain species of the Cymatosiraceae (Hasle et al., 1983), and an alteration of the valve outline of morphologically bipolar species from elongate toward almost circular, e.g., *Fragilaria* spp. (Hustedt, 1959) and Cymatosiraceae.

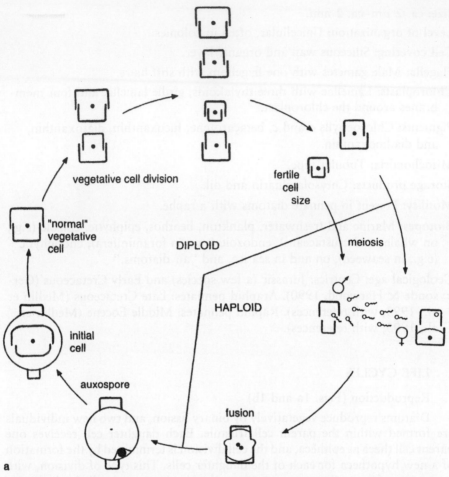

FIGURE 1   (a) Sexual reproduction of a centric diatom (oogamy) and (b) of a pennate diatom (morphological isogamy, physiological anisogamy). ●, Zygote; •, nucleus; ○, pycnotic nucleus.

The decrease in the average cell size of a diatom population during vegetative growth implies a need for a means of restoring the cell size. This is made possible by auxospore formation, in which a cell sheds its siliceous theca, thereafter forming a large sphere surrounded by an organic membrane. Within this sphere, a new diatom frustule of maximal size is formed, and the cycle starts anew. The first cell formed inside the auxospore, the initial cell, may have a morphology deviating in girdle structure, valve outline, and process pattern from that of a "normal" vegetative cell (*vide, Thalassiosira decipiens*, Hasle, 1979, Fig. 41; *Cymatosira lorenziana*, Hasle et al., 1983, Fig. 19).

Auxospore formation is size dependent and normally takes place when the cell has reached about one-third of its maximal size (Drebes, 1977). Below

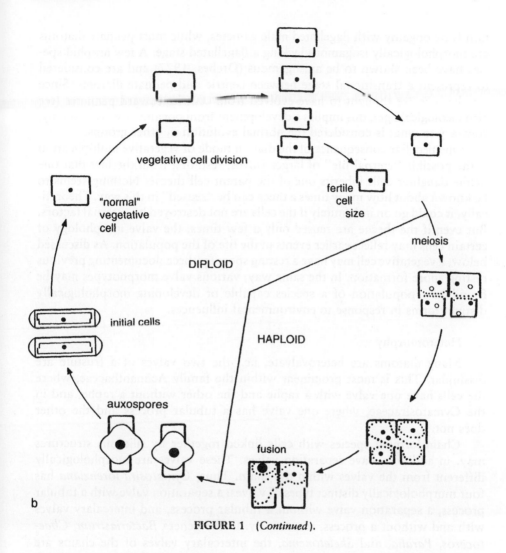

vegetative cell division

"normal" vegetative cell

fertile cell size

meiosis

DIPLOID

initial cells

HAPLOID

auxospores

fusion

b

FIGURE 1 *(Continued)*.

this limit the diatoms seem unable to rejuvenate themselves, and they continue to divide until they reach a stage at which cell division is no longer possible. There are reports in the literature of taxa which do divide without a simultaneous size reduction (Drebes, 1977), and some species seem to be able to multiply at their lower size limit without further size reduction for an extended length of time (E. Syvertsen, personal observations). For most species, however, auxospore formation is a necessary and normal occurrence in their life cycle. It may take place as a vegetative event or as the result of sexual reproduction.

All diatoms are diplonts with a meiosis at the end of the gametogenesis. The zygote develops into an auxospore. In the centric diatoms, sexual reproduction is by oogamy with flagellated male gametes, while most pennate diatoms are morphologically isogamous lacking a flagellated stage. A few araphid species have been shown to be anisogamous (Drebes, 1977) and are considered to represent a transitional stage between centric and pennate diatoms. Since the diatoms are thought to have evolved from centrics toward pennates (see also Geological age), this implies a development from oogamy toward isogamy, contrary to what is considered the normal evolution in other groups.

One peculiar consequence of the diatom mode of vegetative multiplication is the possible "eternal life" of larger valves, resulting from the fact that one of the daughter cells inherits one of the parent cell thecae. Nothing seems to be known about how many times a theca can be "reused" in this way. Theoretically, it could go on indefinitely if the cells are not destroyed by external factors. But even if the thecae are reused only a few times, the valve morphology of certain cells may reflect earlier events in the life of the population. As discussed below, a vegetative cell may have a resting spore epitheca documenting previous resting spore formation. In the same way, various valve morphotypes may be found in a population of a species capable of developing morphologically distinct forms in response to environmental influences.

Heteromorphy

Many diatoms are heterovalvate, i.e., the two valves of a frustule are dissimilar. This is most prominent within the family Achnanthaceae, where the cells have one valve with a raphe and the other without a raphe, and in the Cymatosiraceae, where one valve has a tubular process and the other does not.

Chain-forming species with cells linked together by siliceous structures may, in addition, have separation valves. These valves are morphologically different from the valves within the chain. Thus, *Cymatosira lorenziana* has four morphologically distinct types of valves: a separation valve with a tubular process, a separation valve without a tubular process, and intercalary valves with and without a process, respectively. In the genera *Bacteriastrum*, *Chaetoceros*, *Paralia*, and *Skeletonema*, the intercalary valves of the chains are all alike and different from the separation valves (Fryxell, 1976; Crawford, 1979).

Another type of heteromorphy may be found with species in which the morphology varies in response to changes in the environment. These morphotypes are generally considered to be forms of the species. During and after environmental changes specimens may be found which have two different valves reflecting different environmental conditions. This type of morphological adaptation has been found in *Thalassiosira rotula*. In this species, the valve morphology changes in response to variations in temperature and the girdle morphology changes in response to available nutrients (Syvertsen, 1977).

Resting Spore Formation (Fig. 2)

The diatom resting spores are first and foremost recognized by their heavily silicified frustules. The resting spore morphology of some species is similar to that of the corresponding vegetative cells, whereas in other species, the resting spores and the vegetative cells differ drastically (Syvertsen, 1979, 1985).

Diatom resting spores are normally formed as a response to unfavorable environmental conditions, and germination occurs when the conditions improve (see Hargraves & French, 1983, for a review). Resting spore formation is common in centric, but rare in pennate marine planktonic diatoms. Whereas resting spores of several centric marine planktonic diatoms germinate in culture within a few days, the freshwater benthic pennate species, *Eunotia soleirolii* (Kützing) Rabenhorst, requires a dormancy of several weeks (von Stosch & Fecher, 1979) before germination. *Achnanthes taeniata* and *Fragilariopsis oceanica* are pennate marine planktonic diatoms known to form resting spores; whether a dormancy period is present in these species is unknown.

Three types of resting spores can be distinguished: exogenous resting spore—the mature resting spore is not physically in contact with a parent cell theca; semiendogenous resting spore—the spore hypovalve is enclosed within one of the parent cell thecae; and endogenous resting spore—the whole spore is enclosed within the parent cell frustule. Normally two or more exogenous resting spores [e.g., a chain of 13 resting spores of *Detonula confervacae* (Syvertsen, 1979)], two semiendogenous resting spores [e.g., *Thalassiosira australis* (Syvertsen, 1985)], and one endogenous resting spore [e.g., *Chaetoceros* spp. (Hargraves, 1979)] are formed. All three types were found in clonal cultures of *Thalassiosira nordenskioldii* and *T. antarctica* with the semiendogenous type as the most common (Syvertsen, 1979).

Resting spore morphology is a more constant, specific feature than the type and mode of formation and, thus, is of greater taxonomic value. Until disproven by Syvertsen (1979) for centric diatoms and by von Stosch & Fecher (1979) for pennate diatoms, it was generally believed that resting spores had no girdle and thus differed from vegetative cells. Among the centric diatoms, the general trend seems to be that resting spores of species, possibly early in the phylogenetic diatom system (e.g., *Thalassiosira* and *Stellarima*), have a girdle and are often morphologically similar to the vegetative cells. Resting spores of species in the possibly more advanced part of the system (e.g., *Bacteriastrum* and *Chaetoceros*) are usually very different from the vegetative frustules and often lack a girdle. This seems to coincide with a suggested development from exogenous or semiendogenous toward endogenous resting spores (Syvertsen, 1979). On the other hand, phylogenetically advanced pennate diatoms, e.g., *Achnanthes taeniata* and *Fragilariopsis oceanica,* form resting spores with a girdle. A special case occurs when resting spores are formed within auxospores. This takes place, for instance, in *Chaetoceros eibenii* (von Stosch et

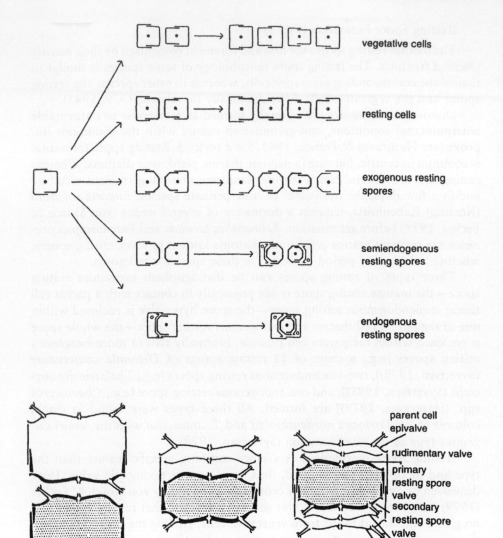

FIGURE 2 Formation of vegetative cells, resting cells, and resting spores from a vegetative parent cell. *Thalassiosira nordenskioeldii* is an example of a species forming all three types of resting spores.

al., 1973) and in *Leptocylindrus danicus* and *L. minimus* (Hargraves, 1990). Unlike other diatoms known so far, the resting spores are an obligate part of the life cycle of *L. danicus* (French & Hargraves, 1985).

Resting spore formation includes two cytokinetic mitoses (von Stosch et al., 1973), where one or both may be unequal. Depending on the degree of dissimilarity, the rudimentary cells may or may not be visible (Syvertsen, 1979). In terms of morphology, the rudimentary valves are often intermediate between vegetative and resting spore valves, but may be sufficiently different to risk being described as separate species unless their origin is known [*vide Thalassiosira australis* (Syvertsen, 1985)].

The two valves of a resting spore may be similar or distinctly different. Often the first valve formed (primary resting spore valve) is more similar to the valves of the vegetative cells than is the second valve (secondary resting spore valve). Thus, during resting spore formation at least four morphologically different valve types may be found which can easily be and probably often have been identified as belonging to different species. These valve types are (1) normal vegetative valves, (2) rudimentary valves, (3) primary resting spore valves, and (4) secondary resting spore valves. In addition, intermediate valve types between those mentioned and representing various degrees of development are often seen (E. Syvertsen, personal observations). This diversity of valve types belonging to one and the same species calls for caution in identification work using cleaned diatom material.

Resting spores germinate in two ways, according to whether or not they have a girdle. Spores with a girdle germinate to form two new vegetative cells where the resting spore thecae serve as epithecae [e.g., *Thalassiosira* (E. Syvertsen, personal observations)], while spores lacking a girdle shed the spore valves in the process of vegetative cell formation, as with *Bacteriastrum* and *Chaetoceros* (von Stosch et al., 1973). In the first case, chains formed after resting spore germination have the resting spore valves as epivalves on the end cells, and these cells are thus heterovalvate.

MORPHOLOGY AND TERMINOLOGY

With an increasing amount of information on details of the siliceous diatom cell wall, especially that obtained with electron microscopy, a need for a generally accepted terminology became evident in the early 1970s. The first attempt along this line was published in 1975 as "Proposals for a Standardization of Diatom Terminology" (Anonymous, 1975; von Stosch, 1975) followed by "An Amended Terminology for the Siliceous Components of the Diatom Cell Wall" (Ross et al., 1979). These publications contain glossaries in Latin, English, German, and French. A Russian translation of Anonymous (1975) was published by Makarova (1977).

When the fine structure of pennate diatoms became more extensively studied, new terms were introduced (Mann, 1978, 1981; Cox & Ross, 1981; Williams, 1985, 1986). Terms specific to certain centric diatom families or genera, partly applicable to light microscopy, were also suggested (Hasle et al., 1983; Sundström, 1986; Rines & Hargraves, 1988).

The text of this chapter follows the current terminology, including, in part, that of Barber & Haworth (1981). The gross morphology of the diatom frustule and structures more generally distributed within the class are defined in this chapter. Terms specific to particular taxa are defined in the introductory text to these taxa. The definitions of the terms may include elements not readily revealed by light microscopy. This does not exclude the possibility to recognize the presence of a particular structure. For example, the tubular parts of strutted processes may be visible in the light microscope, while the satellite pores are usually not observable.

Gross Morphology (Figs. 3 and 4)

Apical axis—long axis of a bilateral diatom—axis between the poles of a frustule.

Pervalvar axis—axis through the center point of the two valves.

Transapical axis—third axis of a bilateral diatom.

Valvar plane—parallel to the valves—plane of division.

Apical plane—perpendicular to the transapical axis.

Transapical plane—perpendicular to the apical axis. (If more specified terms are required, see Round et al., 1990, p. 23, Fig. 18.)

Valve view—frustule seen from top or bottom.

Broad girdle view—frustule seen from broad side.

Narrow girdle view—frustule seen from narrow side.

Frustule—the whole diatom box.

Epitheca—upper overlapping part of frustule.

Hypotheca—lower part of frustule.

Valve—epi-, hypo-.

Valve mantle—marginal part of valve, set off from valve face at an angle.

Valve face—part of valve surrounded by mantle.

Girdle—part of frustule between epi- and hypovalves consisting of epi- and hypocingula.

Cingulum—portion of the girdle associated with a single valve.

Band or segment—a single element of the girdle.

Intercalary band(s)—copula(e)—element(s) nearest to the valves, different in structure from elements farther away from the valves.

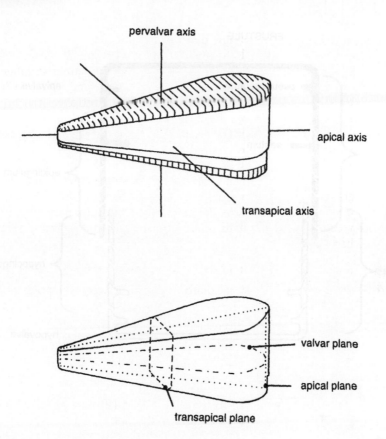

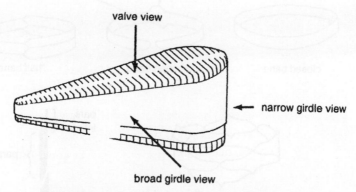

FIGURE 3 Axes and planes of a diatom frustule.

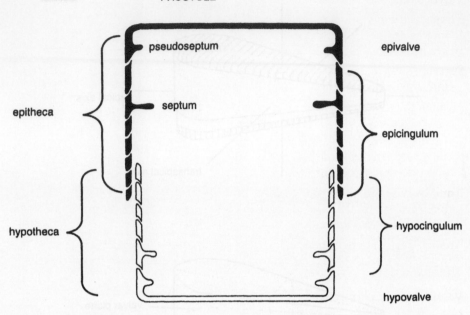

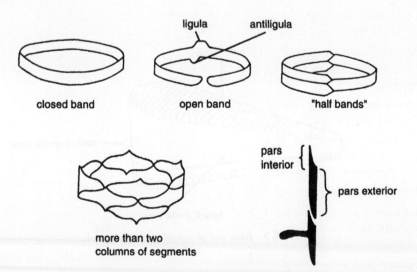

FIGURE 4 Gross morphology of the frustule, types of girdle bands and segments, and overlapping of bands.

Valvocopula—band adjacent to a valve.

Connecting band(s)—pleura(e)—element(s) in the middle of the girdle when intercalary bands are present or any element when no intercalary bands are present.

Septum—a sheet or ridge in the valvar plane projecting from a girdle band into the interior of the frustule, often with several openings.

Hyaline band—element of girdle with no perforations (see Hemidiscaceae).

Fine Structure of the Siliceous Cell Wall (Figs. 5 and 7)

Basal siliceous layer—the layer that forms the basic structure of the various components of the frustule.

Annulus (von Stosch, 1977)—a ring of costal thickness, often surrounding one or more processes and with a structure different from that of the rest of the valve (see *Porosira* and *Actinocyclus*).

Areola—regularly repeated perforation through the valve wall, often marked by more or less elaborate multiangular walls or ribs (definition slightly deviating from Ross et al., 1979, p. 527).

Velum—a thin perforated layer of silica across an areola.

Cribrum—a velum perforated by regularly arranged pores.

Foramen—the passage through the constriction at the surface opposite the velum.

Poroid areola or poroid—an areola not markedly constricted at one surface of the valve.

Loculate areola or loculus—an areola markedly constricted at one surface of the valve and occluded by a velum at the other.

Alveolus—an elongated chamber running from the central part of the valve to margin, open to the inside and covered by a perforate layer on the outside.

Stria—one or more rows of areolae or pores, or an alveolus. Uniseriate, one row; biseriate, two rows; multiseriate, many rows.

Interstria—the nonperforate siliceous strip between two striae.

Processes (Figs. 6–8)

Process—projection with homogeneously silicified walls.

Labiate process—rimoportula—a tube or an opening through the valve wall with an internal flattened tube or longitudinal slit surrounded by two lips.

Spine—a closed or solid structure projecting out from the surface of the frustule.

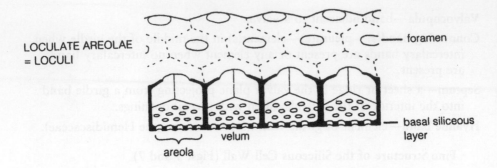

LOCULATE AREOLAE
= LOCULI

— foramen

— basal siliceous
layer

velum

areola

POROID AREOLAE
= POROIDS

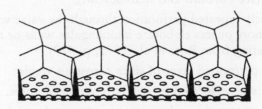

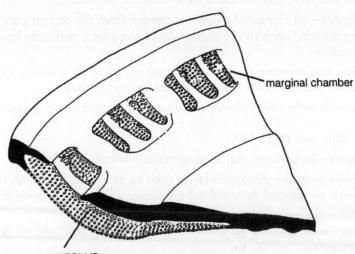

marginal chamber

ALVEOLUS

FIGURE 5 Fine structure of the siliceous cell wall.

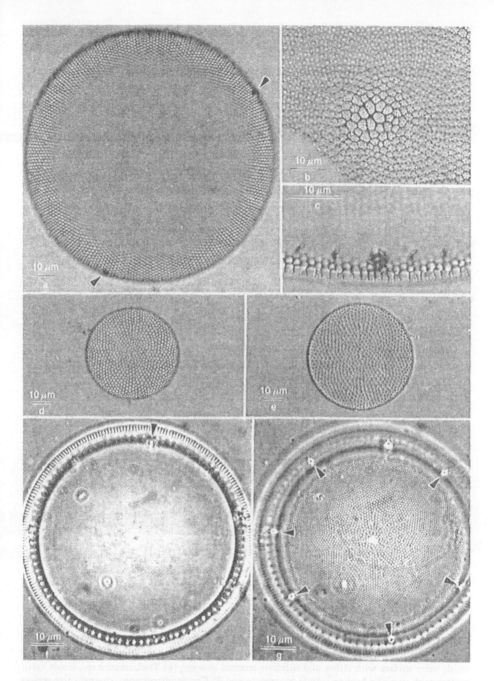

FIGURE 6 Light micrographs showing areola and process patterns. Scale bars = 10 μm. (a) *Coscinodiscus centralis,* radial areolation with striae incerted from valve margin, one marginal ring of smaller and two larger (arrows) labiate processes; (b) *C. centralis,* central rosette of larger areolae, cribra discernible, decussating arcs of areolae; (c) *C. centralis,* four smaller long-necked labiate processes and one larger process; (d and e) *Coscinodiscus radiatus,* radial areolation, indistinct decussating rows and fasciculation with striae parallel to the edge row; (f) *Thalassiosira punctigera,* ribbed margin, one marginal ring of small, densely spaced strutted processes, one larger labiate process (arrow); (g) *T. punctigera,* bases of occluded processes (arrows), fasciculate areolation with striae parallel to the middle row.

Marginal ridge—a ridge between the valve face and the valve mantle, continuous or interrupted, perforated or solid especially in Lithodesmiaceae.

Types of Colonies

Separable colonies (von Stosch, 1977)—cells connected by organic substances, separable into smaller units under appropriate conditions (e.g., in Lithodesmiaceae).

Inseparable colonies (von Stosch, 1977)—cells joined by fusion of, or by inseparable interlockings of, silica (e.g., in *Skeletonema*, *Chaetoceros*, and *Cymatosira*).

Separation valve (Florin, 1970)—valves where the separation of "inseparable" chains takes place with a structure and process pattern different from that of the other valves.

Chains
Thalassiosira—by threads from strutted processes.
Skeletonema—by external parts of strutted processes.
Leptocylindrus—by abutting valve faces.
Rhizosolenia—by external part of the single labiate structure (process), contiguous area, and claspers.
Eucampia—by bipolar elevations.
Cerataulina—by bipolar elevations with spines.
Chaetoceros—by setae.
Lithodesmium—by marginal ridge.

Ribbons
Fragilariopsis—by abutting valve faces.
Cymatosira—by marginal linking spines.

Stepped chains
Pseudo-nitzschia—overlapping of cell ends.

Zig zag or star-shaped chains
Thalassionema—mucilage pads.

FIGURE 7 Fine structure of the siliceous valve wall and processes as seen with the scanning electron microscope. Scale bars for a, b, and d–g = 1 μm; for c = 10 μm. (a) *Thalassiosira* sp., inside valve surface with cribra and trifultate strutted process; (b) *Thalassiosira* sp., inside valve surface with partially broken labiate process; (c and d) *Coscinodiscus* spp., central valve region seen from inside valve, internal foramina and external cribra and central unperforated (hyaline) area; (e) *Coscinodiscus* sp., marginal part of valve seen from the inside with foramina, six smaller and one larger labiate processes; (f and g) *Coscinodiscus* sp., marginal part of valve seen from the inside, foramina, smaller labiate processes and a larger process of a type different from the larger labiate process in 7e.

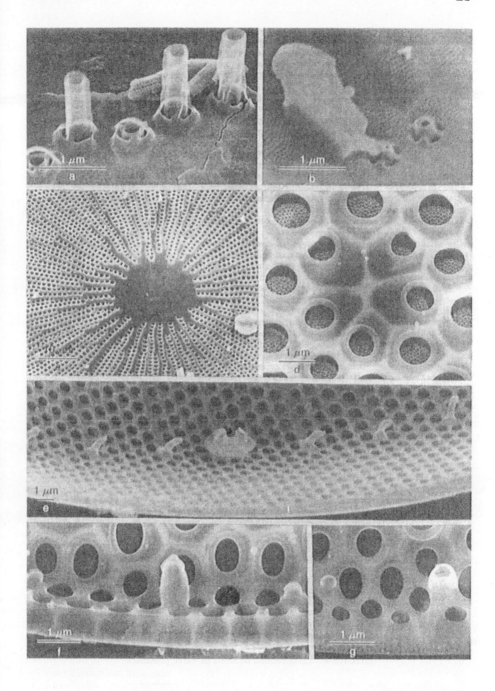

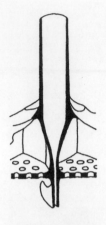
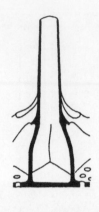
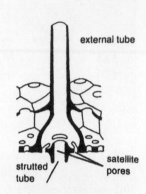

LABIATE PROCESS
= RIMOPORTULA

OCCLUDED PROCESS

STRUTTED PROCESS
= FULTOPORTULA

(see Thalassiosiraceae)

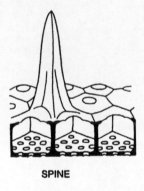

TUBULAR PROCESS
(see Cymatosiraceae)

BILABIATE PROCESS
(see Lithodesmiaceae)

SPINE

FIGURE 8   Longitudinal sections of valve processes.

CLASSIFICATION (Fig. 9)

Past and current classification systems for diatoms are mainly based on phenetic data confined to the structure and shape of the siliceous parts, especially the valves (see Round et al., 1990, pp. 117–121, for a discussion on diatom systematics). However, in recent schemes other kinds of characters have been added for consideration. For instance, the type of sexual reproduction and the structure of the auxospore envelope are considered distinctive features separating the centric and pennate diatoms (Simonsen, 1979; von Stosch, 1982). Round et al. (1990) also included the type of habitat (e.g., planktonic, epiphytic, marine, freshwater) as important in ordinal descriptions.

This chapter follows Simonsen (1979, p. 45) in dividing the centric diatoms into three suborders, each characterized by the shape of the cells, the polarity, and the arrangement of the processes (process pattern). The pennate diatoms are divided into two suborders, one including diatoms without a raphe, the other one encompassing taxa with a raphe. The further differentiation of the centric diatoms into families follows Simonsen (1979) in gross features, whereas the classification of the pennate diatoms is partially from Round et al. (1990).

GENERA REPRESENTED IN MARINE PLANKTON

CENTRIC DIATOMS

Order Biddulphiales
Valve striae arranged basically in relation to a point, an annulus, or a central areola.

Suborder Coscinodiscineae
Valves generally with a marginal ring of processes; symmetry primarily with no polarities.

Family Thalassiosiraceae Lebour 1930 emend. Hasle 1973
Genera: *Bacterosira* Gran; *Cyclotella* (Kützing) Brébisson; *Detonula* Schütt ex De Toni; *Lauderia* Cleve; *Minidiscus* Hasle; *Planktoniella* Schütt; *Porosira* Jørgensen; *Skeletonema* Greville; *Thalassiosira* Cleve.

Family Melosiraceae Kützing 1844
Genera: *Melosira* C. A. Agardh; *Paralia* Heiberg; *Stephanopyxis* (Ehrenberg) Ehrenberg.

Family Leptocylindraceae Lebour 1930
Genera: *Leptocylindrus* Cleve; *Corethron* Castracane.

Family Coscinodiscaceae Kützing 1844
Genera: *Coscinodiscus* Ehrenberg; *Ethmodiscus* Castracane; *Palmeria* Greville.

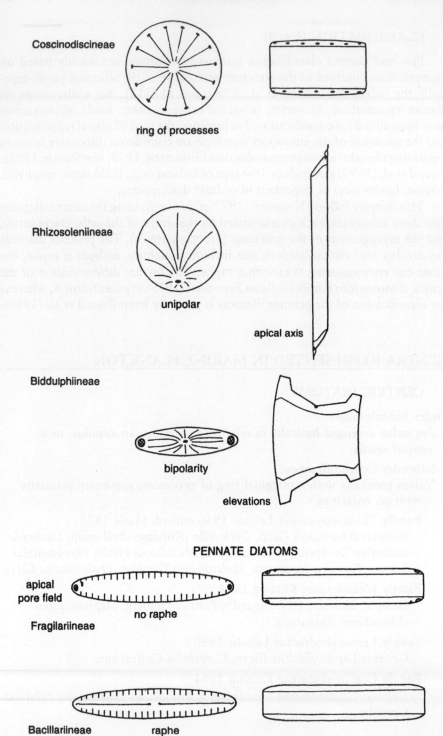

Coscinodiscineae

ring of processes

Rhizosoleniineae

unipolar

apical axis

Biddulphiineae

bipolarity

elevations

PENNATE DIATOMS

apical
pore field

no raphe

Fragilariineae

Bacillariineae   raphe

Family Stellarimaceae Nikolaev 1988 ex Sims & Hasle 1990
  Genera: *Stellarima* Hasle & Sims.

Incertae sedis: Gossleriella Schütt

Family Hemidiscaceae Hendey 1937 emend. Simonsen 1975 ex Hasle
  (this publication)
  Genera: *Actinocyclus* Ehrenberg; *Azpeitia* M. Peragallo in
  Tempère & Peragallo, *Hemidiscus* Wallich; *Roperia* Grunow ex
  Pelletan.

Incertae sedis: Pseudoguinardia von Stosch

Family Asterolampraceae H. L. Smith 1872 emend. Gombos 1980
  Genera: *Asterolampra* Ehrenberg; *Asteromphalus* Ehrenberg.

Family Heliopeltaceae H. L. Smith 1872
  Genera: *Actinoptychus* Ehrenberg; *Aulacodiscus* Ehrenberg.

Suborder Rhizosoleniineae
Valves primarily unipolar; no marginal ring of processes.

Family Rhizosoleniaceae De Toni 1890
  Genera: *Rhizosolenia* Brightwell; *Proboscia* Sundström;
  Pseudosolenia Sundström; *Guinardia* H. Peragallo; *Dactyliosolen*
  Castracane.

Suborder Biddulphiineae
Valves primarily bipolar; no marginal ring of processes.

Family Hemiaulaceae Heiberg 1863
  Genera: *Cerataulina* H. Peragallo ex Schütt; *Climacodium* Grunow;
  Eucampia Ehrenberg; *Hemiaulus* Heiberg.

Family Cymatosiraceae Hasle, von Stosch, & Syvertsen 1983
  Genera: *Arcocellulus* Hasle, von Stosch, & Syvertsen; *Brockmanniella*
  Hasle, von Stosch, & Syvertsen; *Cymatosira* Grunow; *Minutocellus*
  Hasle, von Stosch, & Syvertsen; *Plagiogrammopsis* Hasle, von Stosch,
  & Syvertsen.

Incertae sedis: Lennoxia Thomsen & Buck

Family Chaetocerotaceae Ralfs in Pritchard 1861
  Genera: *Bacteriastrum* Shadbolt; *Chaetoceros* Ehrenberg; *Attheya* T.
  West.

FIGURE 9 Schematic diagrams of centric and pennate diatom suborders with main features. Coscinodiscineae, valves with a marginal ring of processes, symmetry with no polarities; Rhizosoleniineae, valves unipolar, no marginal ring of processes; Fragilariineae, valves with sternum and no raphes; Bacillariineae, valve with sternum and raphe.

Family Lithodesmiaceae H. & M. Peragallo 1897–1908 emend.
Simonsen 1979
Genera: *Bellerochea* Van Heurck emend. von Stosch; *Ditylum* J. W.
Bailey ex L. W. Bailey; *Lithodesmioides* von Stosch; *Lithodesmium*
Ehrenberg, *Helicotheca* Ricard; *Neostreptotheca* von Stosch emend.
von Stosch.

Family Eupodiscaceae Kützing 1849
Genus: *Odontella* C. A. Agardh.

PENNATE DIATOMS

Order Bacillariales
Valve striae arranged basically in relation to a line

Suborder Fragilariineae—Araphid pennate diatoms
Sternum present (indistinct in one family)—Raphe absent.
Fragilariineae is used here for families belonging to several orders of the
class Fragilariophyceae F. E. Round.

Family Fragilariaceae Greville 1833
Genera: *Asterionellopsis* F. E. Round in Round et al; *Bleakeleya*
F. E. Round in Round et al.; *Striatella* C. A. Agardh; *Synedropsis*
Hasle, Medlin, & Syvertsen + the species *Fragilaria striatula*
Lyngbye.

Family Rhaphoneidaceae Forti 1912
Genera: *Adoneis* G. W. Andrews & P. Rivera; *Delphineis* G. W.
Andrews; *Neodelphineis* Takano; *Rhaphoneis* Ehrenberg.

Family Toxariaceae F. E. Round in Round et al., 1990
Genus: *Toxarium* J. W. Bailey.

Family Thalassionemataceae F. E. Round in Round et al., 1990
Genera: *Lioloma* Hasle gen. nov., *Thalassionema* Grunow ex
Mereschkowsky; *Thalassiothrix* Cleve & Grunow; *Trichotoxon* F.
Reid & Round.

Suborder Bacillariineae—Raphid pennate diatoms: sternum and raphe
present.
Bacillariineae is used here for families belonging to Achnanthales,
Naviculales, and Bacillariales of the subclass Bacillariophycidae
D. G. Mann.

Family Achnanthaceae Kützing 1844
Genus: *Achnanthes* Bory de St.-Vincent.

Family Phaeodactylaceae J. Lewin 1958
Genus: *Phaeodactylum* Bohlin.

Incertae sedis: Nanoneis R. E. Norris

Family Naviculaceae Kützing 1844
Genera: *Meuniera* P. C. Silva nom. nov., *Navicula* Bory de
St. Vincent; *Haslea* Simonsen; *Pleurosigma* W. Smith; *Ephemera*
Paddock; *Banquisia* Paddock; *Membraneis* Paddock; *Manguinea*
Paddock; *Plagiotropis* Pfitzer emend. Paddock; *Pachyneis* Simonsen.

Family Bacillariaceae Ehrenberg 1831
Genera: *Bacillaria* Gmelin; *Cylindrotheca* Rabenhorst; *Fragilariopsis*
Hustedt in A. Schmidt, 1990; *Neodenticula* Akiba & Yanagisawa;
Pseudo-nitzschia H. Peragallo in H. & M. Peragallo, 1897–1908;
Nitzschia Hassall.

IDENTIFICATION

The emphasis of this chapter in general is oriented toward routine identification of genera and species by investigators primarily using light microscopy.
The species dealt with are a selection of those which, according to the literature
and our own experience, are regularly encountered in marine plankton.

Identification of diatoms usually must rely on the siliceous frustule. Shape,
size, number, and arrangement of chloroplasts and the presence or absence of
pyrenoids may, however, be used for identification on the generic and specific
level. This is especially true in pennate diatoms (Cox, 1981) and for the centric
genera *Chaetoceros* and *Leptocylindrus*. For example, many species of *Chaetoceros* and *Rhizosolenia* may be identified by their gross morphology as seen in
water mounts or when embedded in a medium of higher refractive index than
that of silica. The proper identification of other species, such as *Thalassiosira*,
Coscinodiscus, and *Pseudo-nitzschia*, may require permanent mounts of
cleaned single valves (see Methodology). Phase and/or differential interference
contrast optics are especially helpful when working with weakly silicified diatoms.

CONTENT

Some families consist of more genera than those dealt with in this chapter.
In these cases, only characters common to the genera presented are listed.
Keys to genera and/or a list of generic characters (for monotypic genera the
description of the generitype) characterize each genus. Keys to species and/or
a list of characters distinguishing the species of a certain genus describe each
species. For the larger genera the species are grouped (A, B, C, etc.) according
to a few common characters.

When up to date information from the literature is lacking, morphometric
and distributional data from Hustedt (1930, 1959, 1961), Cupp (1943), Hen-

dey (1964), and the authors personal observations are given. Few planktonic species and localities have been regularly investigated over a longer period of time. It is therefore seldom worthwhile to go into detail concerning species distribution. We accordingly confine ourselves to the simplest system, viz. three main biogeographical provinces as distinguished for marine plankton communities by Zeitzschel (1982). These are the circumglobal warm water region, a northern cold water region, and a southern cold water region. This classification is fairly consistent with the results of a detailed investigation of 26 marine planktonic diatom species (Hasle, 1976a). The diatoms classified as "cold water" species are often associated with sea ice. Those classified as "warm water" species may have a much wider latitudinal range than that of the warm water region due to transport by currents out of their reproduction areas.

The catalogue by VanLandingham (1967–1979), although not always followed, was consulted for references to publications and synonyms. The generitypes are from Farr et al. (1979, 1986), Greuter et al. (1993), and from more recent literature with some modifications (P. C. Silva, personal communication). The types are cited as they were originally named. In cases of synonomy, the names now regarded as the correct names (the names recommended to be used) are also listed.

Taxa, names, and combinations previously unpublished and used under Description of Taxa are formally described in the Taxonomic Appendix.

The synonymy list is meant as an aid to avoid confusion concerning names commonly used in the literature. Principally, one or more common synonyms, not necessarily including the basionym, are presented. In general, synonymy information too recent to be found in Hustedt (1930, 1959, 1961), Cupp (1943), and Hendey (1964) has been updated. For further information the reader is referred to these publications as well as to synonyms and literature references in the present chapter.

As for the plates presented in this chapter, the sources are given for figures redrawn from the literature; if no source is indicated, the figure is original. In the Thalassiosiraceae, strutted processes are illustrated by a dot, sometimes with a short line attached to indicate the longer extension of the process, and labiate processes are illustrated by a line. In other diatoms, the labiate processes are illustrated by dots, triangles, etc. depending on their relative size and shape. In a series of figures with the same magnification, the scale bar is marked only for the first figure.

DESCRIPTION OF TAXA

CENTRIC DIATOMS

Order Biddulphiales

Terminology specific to centric diatoms (**Fig. 10**)

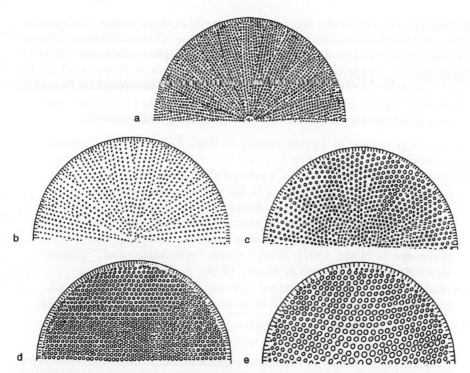

FIGURE 10 Valve striation in centric diatoms; (a) fasciculation ("curvatulus" type), striae parallel with long edge stria of the bundle; (b) fasciculation, striae parallel with long central stria; (c) radial striae, running from the center toward the margin of the valve with shorter striae inserted; (d) tangential straight striae; (e) tangential curved striae.

Striae

Radial striae run from the center of the valve toward the margin (Fig. 10c).

Decussating arcs—spiraling rows of areolae.

Fasciculate striae are grouped in bundles (sectors, fascicles) and are parallel to a radial stria, either the edge stria of the bundle (Fig. 10a) or the central stria (Fig. 10b).

Tangential striae are either straight (Fig. 10d) or curved (Fig. 10e).

Suborder Coscinodiscineae

Family Thalassiosiraceae Lebour 1930 emend. Hasle 1973

The family, as delineated by Simonsen (1979), comprises marine as well as freshwater planktonic diatoms, all having strutted processes as the main morphological, taxonomic character. The genera dealt with in this chapter were classified into two families by Glezer et al. (1988)—Thalassiosiraceae

Lebour emend. Hasle and Stephanodiscaceae Makarova, order Thalassiosirales—and into four families by Round et al. (1990)—Thalassiosiraceae Lebour, Skeletonemataceae Lebour emend. Round et al., Stephanodiscaceae Glezer & Makarova and Lauderiaceae (Schütt) Lemmermann emend. Round et al., order Thalassiosirales Glezer & Makarova, subclass Thalassiosirophycidae Round & Crawford. Silva & Hasle (1994) proposed conservation of Thalassiosiraceae against Lauderiaceae and Planktoniellaceae (Schütt) Lemmermann.

Terminology specific to Thalassiosiraceae: (**Figs. 6–8,** scanning electron microscopy)

Strutted process—fultoportula—a process through which a thread of organic material (mostly chitan, as far as is known) is extruded, consisting of (1) a narrow tube through the basal siliceous layer ("strutted tube") surrounded by struts and satellite pores and (2) "the external tube" which may be missing except for the basal chamber (Syvertsen & Hasle, 1982). Special terms: "operculate" and "trifultate" strutted processes (Fryxell & Hasle, 1979a, p. 378).

Occluded process—hollow external tube not penetrating the valve wall, sometimes (always?) at the top of an areola (Syvertsen & Hasle, 1982).

The genera treated here are characterized by:
Cells in chains or embedded in mucilage.

Cells in chains linked by threads of organic matter from **strutted processes** or by external tubes of marginal strutted or **occluded processes.**

Valve outline circular.

Valve surface thin radial costae, rows of poroids or loculate areolae, or alveoli.

External parts of processes usually more conspicuous than the internal parts.

One or a few labiate processes.

Internal cribra and external foramina (SEM).

Chloroplasts small, rounded bodies.

Resting spores/cells present.

KEY TO GENERA

1a. Valves with one to three marginal rings of strutted processes 2
1b. Processes away from valve margin. . . . . *Minidiscus,* p. 37
2a. Girdle with organic extrusions . . . . *Planktoniella,* p. 39
2b. Girdle with no such extrusions 3
3a. Valve wall alveolate . . . . *Cyclotella,* p. 33
3b. Valve.surface with loculate areolae or radial ribs; not alveolate 4

4a. Chain formation by external tubes of marginal strutted processes . . . 5
4b. Chain formation by threads from strutted processes 6
5a. Central process present . . . . *Detonula*, p. 34
5b. No central strutted process . . . . *Skeletonema*, p. 43
6a. Strutted processes organized in a pattern on valve face 7
6b. Strutted processes scattered on valve face, no particular central processes
    or central processes rudimentary (EM) 8
7a. Adjacent cells in chains abutting . . . . *Bacterosira*, p. 31
7b. Cells in chains separated by shorter or longer distance, or cells solitary or
    embedded in mucilage . . . . *Thalassiosira*, p. 45
8a. Long occluded processes, central processes rudimentary or missing, valve
    structure mainly consisting of radial ribs . . . . *Lauderia*, p. 36
8b. Occluded processes absent, valve surface areolated . . . . *Porosira*, p. 41

Remarks: Strutted processes from which linking threads are extruded may
be situated (1) exactly in valve center inside an areola (seldom) or an
annulus, (2) next to a central areola (often) or an annulus, or (3)
somewhere between valve center and margin, in some species in a more or
less modified ring. Central and subcentral processes thus stand for "strutted
processes through which interconnected threads instrumental in chain
formation may be extruded."

Genus *Bacterosira* Gran 1900
Type: *Bacterosira fragilis* (Gran) Gran.
Correct name: *Bacterosira bathyomphala* (Cleve) Syvertsen & Hasle (*vide*
Hasle & Syvertsen, 1993, p. 298).
Monospecific genus.

Bacterosira bathyomphala* (Cleve) Syvertsen & Hasle (Plate 1)
Basionym: *Coscinodiscus bathyomphalus* Cleve.
Synonyms: *Lauderia fragilis* Gran; *Bacterosira fragilis* (Gran) Gran.
References: Cleve, 1883, p. 489, Plate 38, Fig. 81; Gran, 1897a, p. 18, Plate
1, Figs. 12–14; Gran, 1900, p. 114; Hustedt, 1930, p. 544, Fig. 310;
Hendey, 1964, p. 141, Plate 7, Fig. 5; Hasle, 1973a, p. 27, Fig. 88; Hasle &
Syvertsen, 1993, p. 298.
  Girdle view: Tight chains by abutting valve faces. Pervalvar axis usually
  longer than cell diameter. Apparent lens-shaped structure between cells.
  Cell wall weakly silicified.
  Valve view: Cluster of central processes. One marginal ring of small
  strutted processes, one marginal labiate process. Radial ribs from valve
  center toward valve margin.
  Resting spores: Semiendogenous, heterovalvate, primary valve with
  flattened valve face, secondary valve highly elevated in the center, first
  described as *Coscinodiscus bathyomphalus* (Cleve, 1883).

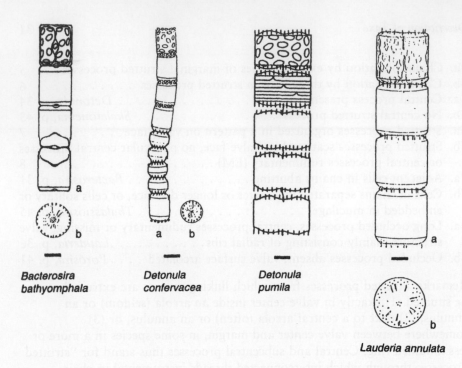

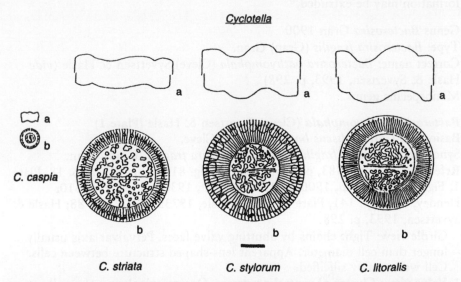

PLATE 1 *Bacterosira bathyomphala:* (a) chain in girdle view including two semiendogenous resting spores, lenticular openings between cells; (b) single valve of vegetative cell. *Detonula confervacea:* (a) chain in girdle view with four exogenous resting spores; (b) single valve of vegetative cell. *Detonula pumila:* chain in girdle view showing central process and thread between cells and marginal processes. *Lauderia annulata:* (a) chain in girdle view; (b) valve with many strutted and occluded processes and one labiate process. *Cyclotella caspia, C. striata, C. stylorum,* and *C. litoralis:* (a) single cells in girdle view showing undulations of valves; (b) single valves with structure. Scale bars = 10 μm.

Morphometric data: Diameter 18–24 μm, more than 30 radial ribs in 10 μm, five to seven marginal strutted processes in 10 μm.
Distribution: Northern cold water region.
How to identify: *Bacterosira* and *Thalassiosira* may be confused as single valves but are readily distinguished in water mounts in chains.

Genus *Cyclotella* (Kützing 1833) Brébisson 1838 (Plate 1, Table 1)
Type: *Cyclotella tecta* Håkansson & Ross.
Correct name: *Cyclotella distinguenda* Hustedt (*vide* Håkansson, 1989, p. 266).
References: Kützing, 1844, p. 131, Plate 1, Fig. 8; Brightwell, 1860, p. 96, Plate 6, Fig. 16; Grunow, 1878, p. 126, Plate 4, Fig. 19; Cleve & Grunow, 1880, p. 119; Hustedt, 1930, p. 334, Figs. 176, 177, and 179; Håkansson & Ross, 1984; Nagumo & Ando, 1985; Nagumo & Kobayasi, 1985; Håkansson, 1989; Lange & Syvertsen, 1989; Takano, 1990, pp. 166–167; Sancetta, 1990, Plate 1, Figs. 1–3.

Most *Cyclotella* species belong to freshwater. Detailed morphological, taxonomic studies of the species found in brackish water/marine environments are sparse (Lange & Syvertsen, 1989). Species often recorded from marine plankton are *C. caspia*, *C. litoralis* (as *C. striata* and/or *C. stylorum*), *C. meneghiniana* Kützing (often as *C. cryptica* Reimann, Lewin, & Guillard), *C. striata*, and *C. stylorum*.

Generic characters:
   Cells usually solitary.
   Valves tangentially undulated.
   Valve wall alveolate.
   A central field distinctly different from the rest of the valve.
   One to many strutted processes within the central field.
   Central field reticulate rugose or with warts or granules.
   One marginal ring of strutted processes.
   One marginal labiate process.

Characters showing differences between species:
   The presence or absence of marginal chambers [= marginal spaces on the inside of the valve encompassing two or more alveolus openings (see Fig. 5)].
   Location of marginal strutted processes versus interstriae.
   Degree of undulation.

TABLE 1   Morphometric Data of *Cyclotella* spp.

Species	Diameter (μm)	Striae in 10 μm
C. caspia	3.5–22	20–28
C. litoralis	10–60	9–14
C. striata	25–48	8–11
C. stylorum	35–67	9–12

KEY TO SPECIES

1a. Undulation of valve pronounced 2
1b. Undulation of valve evident but not pronounced. 3
2a. A marginal strutted process on every second, occasionally on every interstria; valve without marginal chambers. . *C. litoralis* Lange & Syvertsen
2b. Marginal strutted processes grouped in pairs or triplets; valve with prominent marginal chambers covering three or four alveolus openings . . . . .
. . . . *C. stylorum* Brightwell
3a. A marginal strutted process on every second to fourth interstria; valve
with marginal chambers covering two alveolus openings. . . . .
. . . . *C. striata*[1] (Kützing) Grunow in Cleve & Grunow
3b. A marginal strutted process on every third or fourth interstria; valve
without marginal chambers . . . . *C. caspia* Grunow

Distribution:
C. caspia—northern temperate region, euryhaline.
C. litoralis—southern and northern temperate region, coastal, marine
(Lange & Syvertsen, 1989; Sancetta, 1990).
C. striata—northern temperate region, littoral.
C. stylorum—warm water region to southern temperate region, littoral.
How to identify: The species may be identified with light microscopy (LM)
as cleaned valves mounted in a medium of a high refractive index.
Remarks: Nagumo & Ando (1985) made a comparative study of *C. stylorum*
and a species identified as *Cyclotella* sp. The latter is most likely *C. litoralis*.

Genus *Detonula* Schütt ex De Toni 1894 (Plate 1, Table 2)
Lectotype: *Detonula pumila* (Castracane) Gran (*vide* Round et al., 1990, p.
142).
Basionym: *Lauderia pumila* Castracane.
Synonyms: *Schroederella delicatula* Pavillard, *Thalassiosira condensata* Cleve
(for other synonyms see Hasle, 1973a, p. 18).

---

[1] Basinoym: *Coscinodiscus striatus* Kützing.

TABLE 2   Morphometric Data of *Detonula* spp.

Species	Pervalvar axis (μm)	Diameter (μm)	Marginal processes in 10 μm	Areolae in 10 μm
D. confervacea	15–30	6–20	10	30–40
D. pumila	15–120	16–30	6–8	ca. 20
D. moseleyana	60–212	28–120	8–12	21–28

References: Castracane, 1886, pp. 89, 90, Plate 9, Fig. 8, Plate 24, Fig. 9; Cleve, 1896a, p. 11, Plate 2, Fig. 21 (no number on the plate); Schütt, 1896, p. 83, Fig. 135; Gran, 1900, p. 113, Plate 9, Figs. 15–20; Pavillard, 1913, p. 126, Fig. 1a; Hustedt, 1930, pp. 551 and 554, Figs. 314 and 315; Cupp, 1943, p. 76, Fig. 36; Hendey, 1964, p. 142, Plate 5, Fig. 4, Plate 7, Fig. 6, p. 143, Plate 7, Fig. 7; Hasle, 1973a, pp. 15–27, Figs. 44–86; Syvertsen, 1979, p. 55, Figs. 63–69; Round et al., 1990, pp. 142–143; Takano, 1990, pp. 168–169.

Generic characters:
Tight chains.
Cylindrical cells.
Weakly silicified vegetative valves.
Valve surface with radial ribs and few well-developed areolae.
One central strutted process.
One marginal ring of strutted processes.
One marginal labiate process.

Characters showing differences between species:
Size of pervalvar axis and diameter.
Shape of external tubes of marginal strutted processes (SEM).
Size of labiate process.
Presence or absence of resting spore formation.

Hasle (1973a) distinguished between three *Detonula* species. *Detonula confervacea* (Cleve) Gran (basionym: *Lauderia confervacea*) and *D. pumila* are common in marine plankton, while there are few records in the literature of the much larger *D. moseleyana* (Castracane) Gran (basionym: *Lauderia moseleyana* Castracane). The single connecting thread from a strutted process in a central valve depression is usually conspicuous in *D. pumila* and less so in *D. confervacea*. *Detonula pumila* is also distinguished from *D. confervacea* by the longer external tubes of the marginal strutted processes linked in a distinct zig zag pattern. The external tubes of D. *confervacea* are laterally expanded into a T shape, whereas those of D. *pumila* and D. *moseleyana* have the shape of half tubes. The marginal labiate process of D. *pumila* is smaller than that in the two other species, and D. *moseleyana* differs from D. *pumila* mainly by its greater diameter.

Distribution:
   D. *confervacea*—northern cold water region to northern temperate.
   D. *pumila* —probably cosmopolitan with a preference for warmer waters.
   D. *moseleyana*—Indian Ocean.
How to identify: The species may be distinguished in girdle view in water mounts especially by their size (**Table 2**).
Remarks: Based on Cleve's (1900a, p. 22, Plate 8, Figs. 12 and 13) original description of *Thalassiosira condensata* as well as on the description and illustrations of the species in Lebour (1930, p. 63, Fig. 35) and Hendey (1937, p. 238, Plate 11, Fig. 11) we suggest that this species also should be put into synonomy with D. *pumila*. Resting spores (exogenous) are common in D. *confervacea* and not reported for the two other species. The resting spore valves are coarsely areolated and smoothly curved, similar to the valves of *Thalassiosira* spp. with one central process. The external tubes of the marginal strutted processes lack the lateral expansions linking vegetative cells together. Chains of D. *confervacea* of maximum size and *Bacterosira bathyomphala* are similar, both having many chloroplasts and a short or no distance between adjacent cells. The distinction is seen by paying attention to the external parts of the marginal strutted processes of end valves of D. *confervacea* chains, visible with LM in this species but not in B. *bathyomphala*.

Genus *Lauderia* Cleve 1873
Type: *Lauderia annulata* Cleve.
Monospecific genus.

Lauderia annulata Cleve (Plate 1)
Synonym: *Lauderia borealis* Gran.
References: Cleve, 1873a, p. 8, Plate 1, Fig. 7; Gran, 1900, p. 109, Plate 9, Figs. 1–8; Hustedt, 1930, p. 549, Fig. 313; Cupp, 1943, p. 74, Fig. 35; Hasle, 1973a, p. 3, Figs. 1–3; Syvertsen & Hasle, 1982; Takano, 1990, pp. 170–171.
   Girdle view: Cells in chains fairly close (separated by occluded processes on marginal part of valve). Pervalvar axis slightly longer than diameter.
   Valve view: Valve surface with faint radial ribs. Prominent central annulus, sometimes with a few processes. A large marginal labiate process. Numerous strutted processes on valve face and margin. Long

occluded processes in marginal zone (types of processes not differentiated with LM).

Morphometric data: Cell diameter 24–75 μm, pervalvar axis 26–96 μm, more than 30 radial ribs in 10 μm on valve face.

Distribution: Warm water region to temperate.

How to identify: Whole cells of *B. bathyomphala*, *D. pumila* and *Lauderia annulata* are distinguished in water mounts by the way the cells are linked together in chains. At a certain focus, adjacent cells of *B. bathyomphala* seem to be separated by a central lenticular opening. The external structures of the marginal strutted processes of *D. pumila* are linked midway between adjacent valves, and the tubular occluded processes of *L. annulata* run from one adjacent valve to the next. In critical cases cleaned valves mounted in a medium of a high refractive index may be examined to show the differences in process patterns.

Genus *Minidiscus* Hasle 1973 (Plate 2, Table 3)
Type: *Minidiscus trioculatus* (F. J. R. Taylor) Hasle.
Basionym: *Coscinodiscus trioculatus* F. J. R. Taylor.
References: Taylor, 1967, p. 437, Plate 5, Fig. 43; Hasle, 1973a, p. 29, Figs. 101–108; Takano, 1981; Rivera & Koch, 1984; Takano, 1990, pp. 172–175; Sancetta, 1990, Plate 1, Figs. 5 and 6.

Generic characters:

Usually observed as single cells.

Valves with a more or less prominent hyaline margin.

Mantle usually high.

Processes more or less concentrated in valve center.

KEY TO SPECIES

1a. Hyaline valve margin prominent 2
1b. Hyaline valve margin missing or extremely narrow . . . .
. . . .*M. comicus* Takano
2a. Processes close together in a nonareolated, undulated central part. . . . .
. . . . *M. chilensis* Rivera
2b. Processes separated by one to several areolae . . . .
. . . . *M. trioculatus* (F. J. R. Taylor) Hasle

Distribution:

M. comicus—described from Japanese waters (Takano, 1981) and recorded from Argentine waters (Lange, 1985), the English Channel, the Adriatic, and the Gulf of Mexico (G. Hasle and E. Syvertsen, unpublished observations).

<u>*Minidiscus*</u>

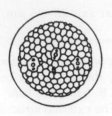

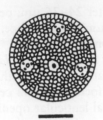

M. trioculatus

M. comicus

M. chilensis

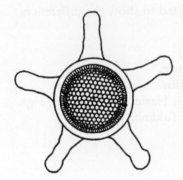

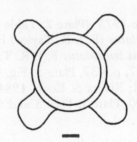

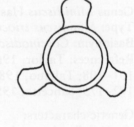

Planktoniella blanda

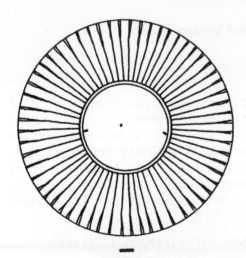

Planktoniella sol

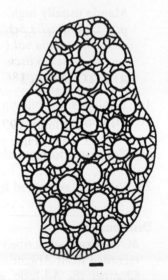

Planktoniella muriformis

TABLE 3 Morphometric Data of
Minidiscus spp.

Species	Pervalvar axis (µm)	Diameter (µm)
M. comicus	2–3?	1.9–7.0
M. chilensis	1.4–3.0	3.0–7.5
M. trioculatus	3.0–5.5	2.0–5.0

M. chilensis—described from Chile (Rivera & Koch, 1984) and recorded from localities in the Pacific as well as the Atlantic Oceans between ca. 60°N and 50°S (Sancetta, 1990, and Hasle and Syvertsen, unpublished observations) and off the Argentine coast (Ferrario, 1988).
M. trioculatus—described from the Indian Ocean and regarded as cosmopolitan in distribution (Hasle, 1973a).
How to identify: *Minidiscus* spp. belong to the smallest known planktonic, centric diatoms. Due to the small size they can hardly be identified to species with LM. With electron microscopy (EM) they are distinguished by differences in areolation and location of the processes.
Remarks: Since these species are numerically important in marine nanoplankton but easily overlooked, their worldwide distribution is still uncertain.

Genus *Planktoniella* Schütt 1892 (Plate 2, Table 4)
Type: *Planktoniella sol* (Wallich) Schütt.
Basionym: *Coscinodiscus sol* Wallich.
References: Wallich, 1860, p. 38, Plate 2, Fig. 1; Schmidt, 1878, Plate 59, Figs. 35–37; Schütt, 1892, p. 258, Fig. 64; Hustedt, 1930, p. 464, Fig. 259; Cupp, 1943, p. 63, Fig. 27; Loeblich et al., 1968; Fryxell & Hasle, 1972, Figs. 34–36; Round, 1972; Desikachary, 1989, p. 9, Plates 742–744; Hallegraeff, 1992, Figs. 1–12; Hasle & Syvertsen, 1993, p. 303, Figs. 17–31.

PLATE 2 *Minidiscus trioculatus, M. comicus,* and *M. chilensis:* Single valves. Structure seen with EM. Scale bar = 1 µm. *Planktoniella blanda:* cells in valve view having varying number of lobes; cell to the left with areolation and process pattern. Scale bar = 10 µm. *Planktoniella sol:* cell in valve view with wing, central strutted process, and two marginal labiate processes. Scale bar = 10 µm. *Planktoniella muriformis:* colony of many cells in valve view connected by an organic matrix. Scale bar = 10 µm.

TABLE 4   Morphometric Data of *Planktoniella* spp.

Species	Diameter (μm)	Areolae in 10 μm	Marginal proc. in 10 μm	No. of labiate proc.
P. blanda	25–55	3–4	3–4	2
P. muriformis	11–15	18–24	7–9	1
P. sol	10–60	5–9	4–5	2

Note. Proc., processes.

Generic characters:
   Cells discoid.
   Organic extrusions from the girdle.
   Radial or tangential areolation.
   One central strutted process.
   One marginal ring of processes.
   One or two labiate processes.

KEY TO SPECIES

1a. Cells usually solitary, occasionally connected by a thread from a central
    strutted process. 2
1b. Cells in a matrix extruded from the girdle forming sheet like colonies . .
    . . . . *P. muriformis*[2] (Loeblich, Wight, & Darley) Round
2a. Girdle with a continuous wing . . . . *P. sol* (Wallich) Schütt
2b. Girdle with lobes. . . . . *P. blanda*[2] (A. Schmidt) Syvertsen & Hasle

Distribution: Warm water region (*P. sol* has also been recorded in Atlantic
waters in the Norwegian Sea and along the Norwegian west coast).
How to identify: Since the organic material attached to the girdle disappears
during acid cleaning, *Planktoniella* spp. are easily misidentified as
Thalassiosira spp. when cleaned, mounted valves are examined.
Examination of water mounts is therefore more reliable.
Remarks: *Coscinodiscus blandus* Schmidt (1878) as well as *C.*
latimarginatus Guo (1981) were described with lobes attached to the girdle.
The combination *Thalassiosira blandus* (sic!) Desikachary & Gowthaman

---

[2] Basionyms: *Coenobiodiscus muriformis* Loeblich, Wight, & Darley and *Coscinodiscus blandus*
   A. Schmidt, respectively.

appeared in Desikachary (1989, p. 9), evidently based on LM observations. The valve gross morphology of *C. blandus* as resolved with SEM is much the same as those in *Thalassiosira* and *Planktoniella*. The justification for the combination *Planktoniella blanda* is the fact that organic extrusions from the girdle (in this case the lobes) are present in *Planktoniella* but not in *Thalassiosira*. Hallegraeff (1992) apparently regarded *C. blandus* A. Schmidt and *C. bipartitus* Rattray as two separate taxa and made the new combination, *Thalassiosira bipartita* (Rattray) Hallegraeff. *Thalassiosira simonsenii* Hasle & G. Fryxell has occluded processes but may even so belong to this genus, possibly as conspecific with *P. blanda*.

Genus *Porosira* Jørgensen 1905 (Plate 3, Table 5)
Type: *Porosira glacialis* (Grunow) Jørgensen.
Basionym: *Podosira hormoides* var. *glacialis* Grunow.
References: Grunow, 1884, p. 108, Plate 5, Fig. 32; Jørgensen, 1905, p. 97, Plate 6, Fig. 7; Hustedt, 1930, p. 314, Fig. 153; Hustedt, 1958a, p. 117, Figs. 20 and 21; Jousé et al., 1962, p. 66, Plate 4, Figs. 15 and 17; Simonsen, 1974, p. 11, Plates 7 and 8; Hasle, 1973a, pp. 6–15, Figs. 4, 5, and 13–43; Takano, 1990, pp. 176–177; Syvertsen & Lange, 1990.

Generic characters:
Cells single or in loose chains.

Cells discoid.

A central annulus.

Radial areolation.

One large labiate process in the marginal zone.

Numerous (strutted) processes all over the valve face.

KEY TO SPECIES

1a. Valve areolae distinct, central annulus mostly indistinct 2
1b. Valve areolae indistinct, striae wavy, annulus distinct 3
2a. Radial and spiraling striae, external part of labiate process short . . . . .
. . . . *P. pseudodenticulata*[3] (Hustedt) Jousé
2b. Areolation fasciculate, external part of labiate process long . . . .
. . . . *P. denticulata* Simonsen
3a. Labiate process close to the valve margin with two small strutted processes on a radial line on each side. . . . . . *P. pentaportula* Syvertsen & Lange
3b. Labiate process at some distance from the valve margin, no regular arrangement of strutted processes close to the labiate process . . . .
. . . . *P. glacialis* (Grunow) Jørgensen

[3] Basionym: *Coscinodiscus pseudodenticulatus* Hustedt.

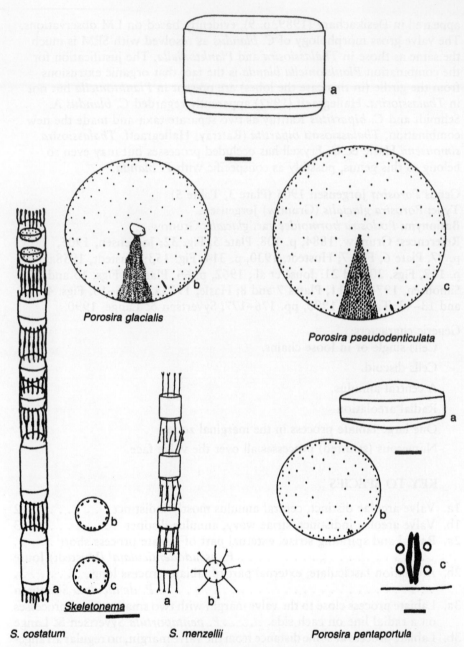

Porosira glacialis

Porosira pseudodenticulata

Skeletonema

S. costatum

S. menzellii

Porosira pentaportula

PLATE 3 *Porosira glacialis* and *P. pseudodenticulata*: (a) single cell in girdle view; (b) *P. glacialis* valve with central annulus, processes, and sector with wavy areolation; (c) *P. pseudodenticulata* with straight rows of areolae and no annulus. *Porosira pentaportula*: (a) single cell in girdle view; (b) valve with process pattern; (c) detail, labiate process with two strutted processes on each side.

TABLE 5 Morphometric Data of *Porosira* spp.

Species	Pervalvar axis (μm)	Diameter (μm)	Areolae in 10 μm
P. denticulata	—[a]	50–170	14–20
P. glacialis	36–64	30–40	25–26
P. pentaportula	20–25	25–50	ca. 30
P. pseudodenticulata	—	60–80	10–12

[a] —, No data.

Distribution:

P. denticulata—Indian Ocean, Equatorial Atlantic Ocean (?) (Simonsen, 1974, probably the only records).

P. glacialis—northern cold water region to temperate, southern cold water region.

P. pentaportula—off Uruguayan and Brazilian coasts, off Norwegian south coast (Syvertsen & Lange, 1990), Rhode Island (Hargraves, personal communication).

P. pseudodenticulata—southern cold water region.

How to identify: The organic threads extruded from the processes scattered on the valve surface, connecting the cells in chains, are usually destroyed by fixation and preservation. *Porosira* spp. are therefore mainly present as single cells in preserved samples. The larger specimens may be misidentified as *Coscinodicus* spp. when observed in girdle view in water mounts. In valve view in water mounts, *Porosira* spp. are recognized by the numerous strutted processes, seen as dark spots. The large labiate process is also usually discernible, especially if the numerous small chloroplasts are destroyed. Processes and areolation are easily discernible on permanent mounts with LM.

Genus *Skeletonema* Greville 1865 (Plate 3, Table 6)
Type: *Skeletonema barbadense* Greville.
References: Greville, 1866, p. 77, Plate 8, Figs. 3–6; Cleve, 1873a, p. 7; Cleve, 1900a, p. 22, Plate 8, Figs. 30 and 31; Cleve-Euler, 1912, p. 509,

---

Scale bar = 1 μm. *Skeletonema costatum:* (a) chain in girdle view, uppermost cell with two chloroplasts (note that the linking structures may be much shorter and less distinct); (b) intercalary valve with labiate process in the ring of marginal strutted processes; (c) separation valve (terminal/ end valve) with labiate processe close to valve center. *Skeletonema menzellii:* (a) chain in girdle view; (b) valve with long, thin external parts of strutted processes and labiate process close to the valve center. Scale bars = 10 μm except detail in *P. pentaportula.*

TABLE 6   Morphometric Data of *Skeletonema* spp.

Species	Pervalvar axis (μm)	Diameter (μm)	Marginal processes
S. costatum	2–61	2–21	6–30 (Total)
S. memzelii	3–10[a]	2.7–7	5–10 (Total)[a]
S. subsalsum	7.5–17.5	3–6	10 in 10 μm
S. tropicum	10	4.5–38	7–9 in 10 μm

[a] Unpublished observations (Australian clone E.E.S.).

Fig. 1; Bethge, 1928, p. 343, Plate 11, Figs. 1–12; Hustedt, 1930, p. 310, Fig. 149; Cupp, 1943, p. 43, Fig. 6; Hendey,1964, p. 91, Plate 7, Fig. 3; Hasle, 1973b; Round, 1973; Guillard et al., 1974; Hasle & Evensen, 1975; Takano, 1981, p. 46, Figs. 1–3; Medlin et al., 1991.

   Skeletonema barbadense has been transferred to a new genus *Skeletonemopsis* P. A. Sims and a proposal is being made that *Skeletonema costatum* should be conserved as the type of *Skeletonema*, by replacing *S. barbadense* (Sims, 1994).

Generic characters:
   Cells in chains united by external tubes of strutted processes (complete or
      split longitudinally) arranged in one marginal ring.
   One labiate process inside the ring of strutted processes or close to valve
      center.
   Valve structure (barely seen with LM): radial areolation, central annulus
      more or less developed.

KEY TO SPECIES

1a. Many chloroplasts per cell, diameter usually greater than 20 μm . . . . .
    . . . . *S. tropicum* Cleve
1b. One or two chloroplasts per cell, diameter smaller 2
2a. External tubes of marginal strutted processes threadlike . . . .
    . . . . *S. menzelii* Guillard, Carpenter, & Reimann
2b. External tubes of marginal strutted processes more robust 3
3a. External tubes of marginal strutted processes not split longitudinally, usu-
    ally extremely short . . . . *S. subsalsum*[4] (A. Cleve) Bethge
3b. External tubes of marginal strutted processes trough shaped with a wide
    longitudinal slit facing the valve margin . . *S. costatum*[4] (Greville) Cleve

[4] Basionyms: *Melosira subsalsa* A. Cleve and *Melosira costata* Greville, respectively.

Distribution:

S. costatum—cosmopolitan (absent from the high Arctic and Antarctic).

S. tropicum—warm water region.

S. subsalsum—brackish water.

S. menzelii—probably warm water region.

How to identify: Due to the unique appearance of the external tubes of the marginal strutted processes and of the colonies *Skeletonema* spp. will scarcely be confused with any other diatoms as long as typical chains or sibling valves are observed. However, a slight similarity between *Stephanopyxis turris* and large, coarsely silicified specimens of *Skeletonema costatum* and *S. tropicum* does exist. The linking structures of *S. costatum*, especially of cultured specimens, may be extremely short. In such cases, single cells may easily be misidentified as small *Thalassiosira* spp. In the same way, single valves, especially terminal valves with a central labiate process, may easily be confused with *Thalassiosira* spp. even when seen with EM. Whereas *S. costatum* and *S. tropicum* are differentiated on characters seen with LM, and the poorly known *S. menzellii* is distinguished with LM by the generally weak silicification and the thin connecting structures, SEM may be necessary to recognize the tubular connecting structures of *S. subsalsum*. *Skeletonema costatum* as presented here includes *S. pseudocostatum* Medlin in Medlin et al. (1991). With LM *S. pseudocostatum* has the appearance of an extremely small *S. costatum* (diameter 2–4 μm). *Skeletonema costatum* forms inseparable colonies, i.e., the external tubes of the marginal strutted processes of sibling valves are permanently attached also when organic material is removed by acid cleaning. *Skeletonema pseudocostatum* occurs in separable colonies, i.e., permanent attachments are absent. As a consequence of this difference *S. pseudocostatum* may in general appear in shorter chains than *S. costatum*. As shown by Medlin et al. (1991, Figs. 1–4), the distinction between the two species, i.e., the presence or absence of permanent attachment, is discernible with LM.

Remarks: The length of the linking structures (external tubes of strutted processes) of *S. subsalsum* varies with salinity (Hasle & Evensen, 1975; Paasche et al., 1975). The presence of resting spores/resting cells in *Skeletonema costatum* is still disputed.

Genus *Thalassiosira* Cleve 1873 emend. Hasle 1973
Type: *Thalassiosira nordenskioeldii* Cleve.

Thalassiosira, with its more than 100 species, is probably the marine planktonic genus most thoroughly examined by modern methods. Regional investigations, some of them including descriptions of new species, have been published in a high number since the first transmission electron microscopy (TEM) examinations started to appear in the 1950s (Helmcke & Krieger, 1953, 1954) and the first examinations taking with SEM in the 1960s (Hasle, 1968a). Attempts have been made to impose structure on this genus which seems to increase in number of species parallel to the number of localities investigated. Location of labiate process linked with the presence or absence of external process tubes seemed for some time to be a promising distinctive character (Hasle, 1968a), and may still be, although several exceptions do exist. Makarova (1988) grouped 53 *Thalassiosira* species into the sections *Tangentales*, *Fasciculigera*, *Thalassiosira*, and *Inconspicuae* with keys to species for each section. Rivera (1981), Johansen & Fryxell (1985), Fryxell & Johansen (1990), and Hasle & Syvertsen (1990a) also constructed keys to species, none of them starting with the same morphological characters.

Generic characters:
   Chains or cells embedded in mucilage.

   Cells in chains connected by organic thread(s) extruded from strutted process(es).

   Cells usually discoid.

   Valve wall with loculate areolae in various patterns or with faint radial ribs.

Characters (LM) showing differences between species:
   Girdle view—water mounts
      Curvature/undulation of valve face.

      Shape and height of valve mantle.

      Connecting thread(s)—Length, thickness (indicating number of central strutted processes).

      Threads extruded from the margin of valve face and/or mantle (indicating location and number of marginal rings of processes).

      Length (and shape) of external part of processes.

      The presence or absence of occluded processes.

   Valve view—mostly cleaned, mounted valves
      Length and location (external/internal) of process tubes.

      Number and arrangement of strutted processes in or near valve center (see Central processes, *Remarks*, pg. 32).

      The presence or absence of strutted processes on the rest of the valve face.

      Number of marginal rings of strutted processes.

      Distance between marginal strutted processes (number in 10 μm).

      Distance of marginal strutted processes from margin (number of areolae).

      Number and location of labiate process(es).

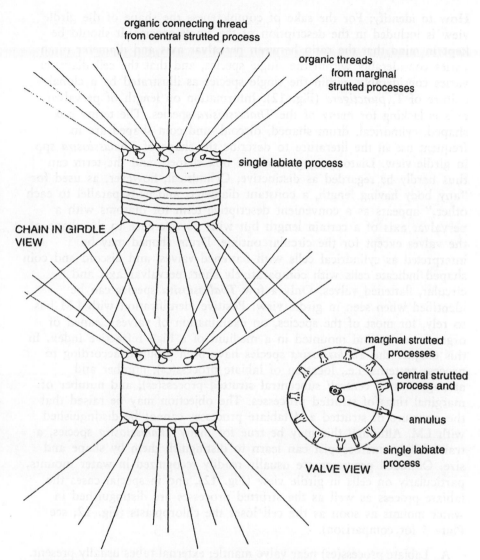

THALASSIOSIRA

organic connecting thread
from central strutted process

organic threads
from marginal
strutted processes

single labiate process

CHAIN IN GIRDLE
VIEW

marginal strutted
processes

central strutted
process and
annulus

single labiate
process

VALVE VIEW

FIGURE 11 Schematic illustrations of chain formation and processes in *Thalassiosira*.

Areolae size and array.

Occurrence of occluded processes.

The presence of a central areola or an annulus.

The presence of marginal ribs.

How to identify: For the sake of completeness the shape of the girdle view is included in the description of the single species. It should be kept in mind that the ratio between pervalvar axis and diameter often varies considerably within the single species, and that the cell diameter varies considerably within the single species as illustrated by a clonal culture of *T. punctigera* (**Fig. 12**). Information on length of pervalvar axis is lacking for many of the *Thalassiosira* species. The terms box shaped, cylindrical, drum shaped, discoid, and coin shaped are in frequent use in the literature to describe the shape of *Thalassiosira* spp. in girdle view. Diatoms in general are box shaped, and the term can thus hardly be regarded as distinctive. Cylindrical, however, as used for "any body having length, a constant diameter, and ends parallel to each other," appears as a convenient descriptive term for diatoms with a pervalvar axis of a certain length but with no reference to the shape of the valves except for the circular outline. Drum shaped may be interpreted as cylindrical cells with flattened valves, and discoid and coin shaped indicate cells with comparatively short pervalvar axis and circular, flattened valves. Only a few *Thalassiosira* species can be identified when seen in girdle view. Positive identification with LM has to rely, for most of the species, on examination of valves cleaned of organic matter and mounted in a medium of a high refractive index. In this chapter the *Thalassiosira* species have been grouped according to process patterns, i.e., location of labiate process(es), number and arrangement of central/ subcentral strutted process(es), and number of marginal rings of strutted processes. The objection may be raised that the structure of strutted and labiate processes cannot be distinguished with LM. Although this may be true for many *Thalassiosira* species, a trained light microscopist can learn to distinguish them by shape and size. Occluded processes are usually readily recognized in water mounts, particularly on cells in girdle view (**Fig. 12**), and in special cases the labiate process as well as the strutted processes are distinguished in water mounts as soon as the cell loses the chloroplasts (**Fig. 12**; see **Plate 5** for comparison).

A. Labiate process(es) near valve mantle; external tubes usually present.
　　1. One central or subcentral strutted process.
　　　　a. One marginal ring of strutted processes: *T. aestivalis*, *T. allenii*, *T. angulata*, *T. binata*, *T. bulbosa*, *T. conferta*, *T. decipiens*, *T. ferelineata*, *T. hispida*, *T. licea*, *T. mala*, *T. minuscula*, *T. nordenskioeldii*, *T. oceanica*, *T. pacifica*, *T. partheneia*, *T. punctigera*, *T. subtilis*, and *T. tenera* (**Table 7**).

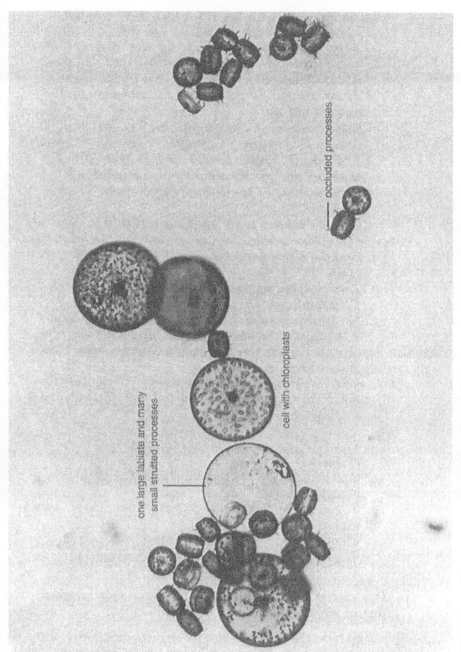

one large labiate and many
small strutted processes

cell with chloroplasts

occluded processes

FIGURE 12 A light micrograph of a clonal culture of *Thalassiosira punctigera* shortly after auxospore formation, showing the size variation.

      b. More than one marginal ring of strutted processes:
         T. delicatula, T. diporocyclus, T. eccentrica, T. fragilis,
         T. hendeyi, T. mendiolana, and *T. simonsenii* (Table 8).
  2. Usually more than one central or subcentral strutted process.
      a. One or a few (2–6) subcentral strutted processes:
         T. curviseriata, T. dichotomica, T. minima, and *T.*
         tealata (Table 9).
      b. Cluster of central strutted processes: *T. antarctica, T.*
         baltica, T. constricta, T. gerloffii, T. gravida, T. hyalina,
         T. karenae, T. rotula, and *T. tumida* (Table 10).
      c. Modified ring of subcentral strutted processes: *T.*
         anguste-lineata, T. australis, T. hyperborea, T.
         kushirensis,
         T. mediterranea, and *T. weissflogii* (Table 11).
  3. No or a variable number of strutted processes on valve face.
      a. No strutted processes on valve face: *T. leptopus*
         (Table 12).
      b. Zero to three subcentral strutted processes:
         T. guillardii and *T. pseudonana* (Table 12).
      c. No particular central or subcentral strutted process:
         T. lentiginosa, T. lineata, and *T. lineoides* (Table 13).
B. Labiate process on valve face; external process tubes absent, one
  ring of marginal strutted processes.
  1. One central or subcentral strutted process: *T. bioculata,*
      T. gracilis, T. oestrupii, T. perpusilla, T. proschkinae, and *T.*
      rosulata (Table 14).
  2. More than one central strutted process.
      a. Cluster of central strutted processes: *T. ritscheri*
         (Table 15).
      b. One or two rows of central/subcentral strutted
         processes:
         T. confusa, T. frenguellii, T. poroseriata, and *T. trifulta*
         (Table 15).
      c. Modified ring of subcentral strutted processes:
         T. endoseriata and *T. poro-irregulata* (Table 15).

A. Labiate process(es) near valve mantle.
    1a. One central or subcentral strutted process. One marginal
       ring of strutted processes.

    This is the most numerous *Thalassiosira* group containing species that are
the most difficult to distinguish. Except for the few species with an invariable
lineatus structure (areol. in straight tangential rows), the areolation pattern
varies with the size of the cell diameter from straight or tangential rows in the

smaller specimens to distinct fasciculation in the larger ones, or from radial rows to fasciculation. Dependening on the degree of silicification the valve structure of one and the same species may vary from merely faint radial ribs to a fully developed areolation.

Thalassiosira allenii Takano (Plate 4, Table 7)
References: Takano, 1965, p. 4, Fig. 2, Plate 1, Figs. 9–11; Hasle, 1978a, p. 101, Figs. 100–128; Rivera, 1981, Figs. 432 and 433; Takano, 1990, pp. 180–181.
  Girdle view: Cells quadrangular; valve face flat; valve mantle oblique. Connecting thread slightly longer than pervalvar axis.
  Valve view: Hexagonal areolae in straight or curved tangential rows, or in radial rows with a tendency of fasciculation; no distinct central areola; mantle areolae about half the size of those on valve face. Labiate process taking the place of a marginal strutted process; long, coarse external tubes of marginal processes.
  Distinctive features: Similar to *T. nordenskioeldii* in girdle view but with lower mantle, marginal processes closer together, and much smaller mantle areolae.

Thalassiosira angulata (Gregory) Hasle (Plate 4, Table 7)
Basionym: *Orthosira angulata* Gregory.
Synonym: *Thalassiosira decipiens* (Grunow) Jørgensen *non Thalassiosira decipiens* (Grunow) Jørgensen in Hasle 1979.
References: Gregory, 1857, p. 498, Plate 10, Figs. 43 and 43b; Jørgensen, 1905, p. 96 *pro parte,* Plate 5, Fig. 3a–e; Hustedt, 1930, p. 322, Fig. 158; Cupp, 1943, p. 48, Fig. 10, in both as *Thalassiosira decipiens* (Grunow) Jørgensen; Hasle, 1978a, p. 93, Figs. 4 and 70–99; Rivera, 1981, Fig. 436; Makarova, 1988, p. 56, Plate 27, Figs. 1–4 and 7–10.
  Girdle view: Pervalvar axis usually shorter than the diameter. Valve face flat; mantle smoothly curved. Connecting thread distinctly longer than pervalvar axis. External tubes of marginal processes usually readily seen.
  Valve view: Hexagonal areolae in curved tangential rows (eccentric structure), sometimes in straight rows or in sectors; no distinct central areola. Marginal processes with long external tubes; wide apart.
  Distinctive features: Large labiate process with long external tube located close to a marginal strutted process; sharp distinction between valve mantle and valvocopula.
  Remarks: *Thalassiosira decipiens* (Grunow) Jørgensen differs morphologically from *T. angulata* by more closely spaced marginal processes and a lower valve mantle. *Thalassiosira decipiens* is unlike *T. angulata,* littoral more than planktonic, and is found in great inland seas, estuaries, bays, shallow coastal waters, and rivers influenced by the tide (Hasle, 1979).

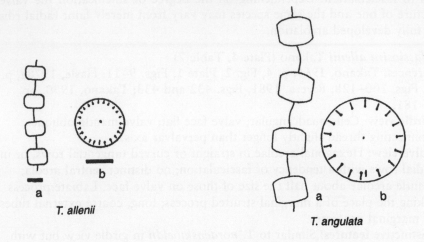

T. allenii

T. angulata

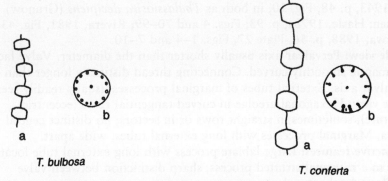

T. bulbosa

T. conferta

PLATE 4 *Thalassiosira allenii, T. angulata, T. bulbosa,* and *T. conferta.* (a) Chains in girdle view; (b) valves with process pattern. Scale bars = 10 μm.

TABLE 7 Morphometric Data of *Thalassiosira* spp. with One Central or Subcentral Strutted Process and One Marginal Ring of Strutted Processes

Species	Diameter (μm)	Valve face	Areolae in 10 μm	
			Valve mantle	Marginal
T. aestivalis[a]	14–56	18	>20	4
T. allenii	5–20	18–24	30–40	6
T. angulata	12–39	8–18	14–24	3
T. binata[a]	4–18	30–40	30–40	5–13 Total
T. bulbosa, veg.	2–16	36–42	36–42	ca. 4.5
Rest. cells		20–26	20–26	—[b]
T. conferta	3.5–23	25–27	25–27	3–5 μm apart
T. decipiens[a]	9–40	8–12	10–15	4–6
T. ferelineata[a]	20–43	6–7	6–7	3–4
T. hispida[a]	15–25	18	24–26	5
T. liceae[a]	16–36	11–12	14–15	2
T. mala	4–10	25–30	ca. 50	4–10
T. minuscula	10–27	32–48	32–48	3–4 μm apart
T. nordenskioeldii	10—50	14–18	14–18	3
T. oceanica	3–12	40–60	40–60	3–4
T. pacifica	7–46	10–19	20–28	4–7
T. partheneia	6–14	40–60	40–60	3–5
T. punctigera	40–186	10–23	10–23	4–5 (Mostly)
T. subtilis	15–32	ca. 30	ca. 30	ca. 3–4 μm apart
T. tenera	10–29	10–16	10–16	3–5

[a] Mentioned in the text for comparison.
[b] —, No data.

Thalassiosira bulbosa Syvertsen (Plate 4, Table 7)
Reference: Syvertsen & Hasle, 1984.

Girdle view: Cells discoid; valve face slightly convex or flat; valve mantle rounded. Solitary or less frequently in short chains with connecting thread slightly longer than pervalvar axis.

Valve view: Dimorphic, lightly silicified vegetative cells and heavily silicified resting cells or semiendogenous resting spores. The valve areolation of the latter two just visible with LM. Central strutted process close to a prominent annulus. Labiate process between two marginal strutted processes, the latter with low bulb-shaped outer parts.

Distinctive feature: Shape of marginal strutted processes.

Thalassiosira conferta Hasle (Plate 4, Table 7)
References: Hasle & Fryxell, 1977a, p. 239, Figs. 1–23; Makarova et al.,
1979, p. 922, Plate 1, Figs. 8 and 9; Rivera, 1981, p. 50, Figs. 48–62;
Takano, 1990, pp. 190–191.
   Girdle view: Cells octagonal; valve mantle highly vaulted. Pervalvar axis
   from less than to twice the cell diameter. Long external tubes of marginal
   processes, two of them longer and thicker than the others and extruding
   longer and thicker threads.
   Valve view: Valve areolae in radial rows, in sectors in larger specimens,
   central areola or annulus more or less prominent. Labiate process
   extremely small, located between the two larger marginal strutted
   processes.
   Distinctive features: Size and location of labiate process and adjacent
   strutted processes.

Thalassiosira mala Takano (Plate 13, Table 7)
References: Takano, 1965, p. 1. Fig. 1, Plate 1, Figs. 1–8; Takano, 1976;
Hasle, 1976a, Figs. 42 and 43; Hallegraeff, 1984, p. 497, Fig. 2; Takano,
1990, pp. 210–211.
   Girdle view: Cells discoid; valve face flat; mantle low and rounded, and
   embedded in cloud-like gelatinous masses of various shapes.
   Valve view: Areolation visible with LM by focusing on central part of
   valve face. Central strutted process off center; marginal strutted processes
   with inconspicuous external tubes, labiate process midway between two
   of them.
   Distinctive features: In water mounts the mucilage colonies appear
   cloudlike. On cleaned, mounted valves the location of the central process
   and the coarsely structured central part of valve face.
Remarks: *Thalassiosira mala* is one of the first or probably the first marine
planktonic diatom to be reported as being harmful to shellfish. A bloom of
this species discoloring the water of Tokyo Bay in September 1951 was
considered responsible for damages amounting to 57,958,000 Yen (Takano,
1956). "Mechanical closing of the respiration by a gelatinous substance
densely attached to the gills" (Takano, 1956, p. 65) exuded from the
diatom, together with poor quality water, was regarded as responsible for
the death of the bivalves.

Thalassiosira minuscula Krasske (Plate 5, Table 7)
Synonym: *Thalassiosira monoporocyclus* Hasle.
References: Krasske, 1941, p. 262, Plate 5, Figs. 4–6; Hasle, 1972a,
p. 129, Figs. 46–60; Hasle, 1976b, p. 104, Figs. 6–10; Rivera, 1981,
p. 95, Figs. 246–262.
   Girdle view: Pervalvar axis equal to or shorter than valve diameter, valve
   surface evenly curved. One or more cells surrounded by capsules and
   embedded in mucilage to form colonies of various shape and size.

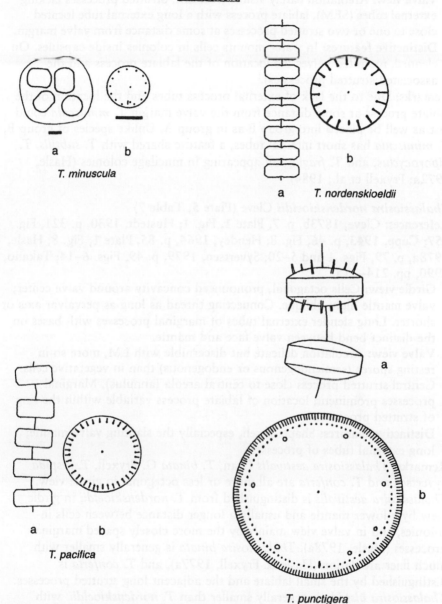

Thalassiosira

a b
T. minuscula

a T. nordenskioeldii

a b
T. pacifica

T. punctigera

PLATE 5 *Thalassiosira minuscula*: (a) mucilage colony of four cells, three in girdle and one in valve view; valve with processes, labiate process with adjacent strutted process. *Thalassiosira nordenskioeldii*: (a) chain in girdle view; (b) valve view showing processes. From Hasle & Syvertsen (1990a). *Thalassiosira pacifica*: (a) chain in girdle view; (b) valve with process pattern. *Thalassiosira punctigera*: (a) chain in girdle view with evident occluded processes on one cell; (b) valve with marginal ribs, ring of close set strutted processes, and occluded processes (open rings). Scale bars = 10 μm.

Valve view: Areolation barely visible with LM. Strutted processes lacking external tubes (SEM), labiate process with a long external tube located close to one or two strutted processes at some distance from valve margin. **Distinctive features:** In water mounts cells in colonies inside capsules. On cleaned, mounted valves the location of the labiate process and the associated strutted process(es).

Remarks: Due to the lack of external process tubes and the location of the labiate process at some distance from the valve margin *T. minuscula* could just as well be placed into group B as in group A. Unlike species of group B, *T. minuscula* has short internal tubes, a feature shared with *T. subtilis, T. diporocyclus,* and *T. fragilis,* all appearing in mucilage colonies (Hasle, 1972a; Fryxell et al., 1984)

Thalassiosira nordenskioeldii Cleve (Plate 5, Table 7)
References: Cleve, 1873b, p. 7, Plate 1, Fig. 1; Hustedt, 1930, p. 321, Fig. 157; Cupp, 1943, p. 46, Fig. 8; Hendey, 1964, p. 85, Plate 1, Fig. 8; Hasle, 1978a, p. 79, Figs. 1 and 5–20; Syvertsen, 1979, p. 49, Figs. 6–14; Takano, 1990, pp. 214–215.
Girdle view: Cells octagonal, pronounced concavity around valve center; valve mantle high; oblique. Connecting thread as long as pervalvar axis or shorter. Long slender external tubes of marginal processes with bases on the distinct bend between valve face and mantle.
Valve view: Areolation delicate but discernible with LM, more so in resting spores (semiendogenous or endogenous) than in vegetative cells. Central strutted process close to central areola (annulus). Marginal processes prominent; location of labiate process variable within the ring of strutted processes.
Distinctive features: Shape of cell, especially the slanting valve mantle; long external tubes of processes.
Remarks: *Thalassiosira aestivalis* Gran, *T. binata* G. Fryxell, *T. hispida* Syvertsen and *T. conferta* are all more or less octagonal in girdle view. *Thalassiosira aestivalis* is distinguished from *T. nordenskioeldii* in girdle view by a lower mantle and usually a longer distance between cells in colonies, and in valve view mainly by the more closely spaced marginal processes (Hasle, 1978a). *Thalassiosira binata* is generally smaller with much finer areolation (Hasle & Fryxell, 1977a), and *T. conferta* is distinguished by the small labiate and the adjacent long strutted processes. *Thalassiosira hispida* is generally smaller than *T. nordenskioeldii,* with smaller mantle areolae than on valve face, and as conspicuous with EM, "knobs and hairs on valve surface" (Syvertsen, 1986, Table 1).

Thalassiosira oceanica Hasle (Plate 13, Table 7)
Synonym: *Cyclotella nana* Guillard clone 13-1 in Guillard & Ryther.

References: Guillard & Ryther, 1962, Plate 1, Fig. 1D; Hasle, 1983a, p. 220, Figs. 1–18.

Girdle view: Cells rectangular; pervalvar axis somewhat shorter than diameter.

Valve view: Valve structure (radial ribs or poroid areolae, EM) not resolved with LM. Processes distinct; marginal processes with short external tubes located close to valve margin; labiate process close to one strutted process.

Distinctive features: On cleaned mounted valves the widely spaced marginal processes and the location of labiate process.

Remarks: An undulated marginal ridge, spinules and sometimes a finely perforated top layer covering the radial ribs are revealed with EM (Plate 13; Hasle, 1983a; Hallegraeff, 1984).

Thalassiosira pacifica Gran & Angst (Plate 5, Table 7)

References: Gran & Angst, 1931, p. 437, Fig. 12; Hasle, 1978a, p. 88, Figs. 42–69; Rivera, 1981, p. 105, Figs. 281–307; Makarova, 1988, p. 57, Plate 28, Figs. 1–9; Takano, 1990, pp. 218–219.

Girdle view: Cells rectangular with rounded low mantle; valve face flat or slightly concave. Connecting thread about as long as the pervalvar axis.

Valve view: Loculate areolae in linear, eccentric or fasciculate patterns depending on the diameter of the cell. Central process adjacent to a central areola (annulus), marginal strutted processes with distinct coarse external tubes, labiate process positioned as for a marginal strutted process.

Distinctive features: Mantle areolae smaller than those on valve face, valve margin ribbed. Distinguished from *T. angulata* by the ribbed margin; the more closely spaced marginal processes with shorter external tubes and the location of the labiate process.

Thalassiosira partheneia Schrader (Table 7)

References: Schrader, 1972; Hasle, 1983a, p. 223, Figs. 19–36; Fryxell et al., 1984, p. 143, Figs. 2–16.

Girdle view: Pervalvar axis equal to or larger than cell diameter; valve evenly curved; occurring in large gelatinous masses containing numerous threads extruded from strutted processes.

Valve view: Valve structure (radial ribs or poroid areolae in radial rows, EM) not revealed with LM. Processes distinct; internal parts longer than inconspicuous external parts; labiate process almost equidistant between two strutted processes.

Distinctive features: In water mounts the colonies with cells are entangled in threads. On cleaned mounted valves the shape of the strutted processes and the location of labiate process (as distinct from *Thalassiosira oceanica*).

Thalassiosira punctigera (Castracane) Hasle (Figs. 6f and 6g and 12, Plate 5, Table 7)
Basionym: *Ethmodiscus punctiger* Castracane.
Synonym: *Thalassiosira angstii* (Gran) Makarova; for further synonomy see Hasle, 1983b, p. 602.
References: Castracane, 1886, p. 167, Plate 3, Fig. 1; Makarova, 1970, p. 13; Fryxell, 1978a, p. 133, Figs. 9–20; Hasle, 1983b; Makarova, 1988, p. 67, Plate 38, Figs. 1–10; Takano, 1990, pp. 224–225.
   Girdle view: Cells low cylindrical; valve surface convex, evenly curved, usually with long tubes (occluded processes). Connecting thread thin.
   Valve view: Areolation fasciculate, valve margin ribbed. Small, densely spaced strutted processes with short external tubes close to valve margin; larger widely spaced occluded processes more away from valve margin.
   Distinctive features: Occluded processes; structure of valve margin.
Remarks: *Thalassiosira puntigera* is recognized by the ribbed margin and the presence of occluded processes. *Thalassiosira licea* G. Fryxell and *T. lundiana* G. Fryxell are closely related to *T. punctigera*; *T. licea* has more widely spaced marginal strutted processes, and *T. lundiana* has strutted processes scattered on the valve face in two marginal rings and a finer areolation (Fryxell, 1978a). Occluded processes may or may not be present in the three species.

Thalassiosira subtilis (Ostenfeld) Gran (Table 7)
Basionym: *Podosira* (?) *subtilis* Ostenfeld.
References: Ostenfeld, 1899, p. 55; Ostenfeld, 1900, p. 54; Gran, 1900, p. 117; Ostenfeld, 1903, p. 563, Fig. 119; Hasle, 1972a, p. 112, Figs. 1–20; Rivera, 1981, p. 123, Figs. 359–377.
   Girdle view: Pervalvar axis and diameter of about the same size, valve surface curved; marginal zone usually more than valve center. Embedded in mucilage without any particular order.
   Valve view: Areolation fasciculate; areola rows parallel to central row. Strutted processes scattered on valve face (between one subcentral process and one marginal ring of processes). Strutted and labiate processes lacking external tubes (SEM). Labiate process located at some distance from valve margin.
   Distinctive features: On cleaned mounted valves, the scattered strutted processes and the labiate process are from valve margin. See remarks to *T. minuscula.*
Remarks: *Thalassiosira subtilis* was the first species of this genus to be described as being embedded in mucilage (Ostenfeld, 1899, p. 55). Other, more recently described species occurring in this type of colony (Hasle, 1972a; Fryxell et al., 1984) may therefore have been misidentified as *T. subtilis.*

Thalassiosira tenera Proschkina-Lavrenko (Table 7)
References: Proschkina-Lavrenko, 1961, p. 33, Plate 1, Figs. 1–4, Plate 2,
Figs. 5–7; Hasle & Fryxell, 1977b, p. 28, Figs. 54–65; Takano, 1990, pp.
232–233.

Girdle view: Cells rectangular; valve face and mantle forming an almost
right angle; pervalvar axis about the same length or longer than cell
diameter. Chloroplasts 5–9.

Valve view: Hexagonal areolae in straight tangential rows (lineatus
structure); a slightly larger central areola (encompassing the central
strutted process, SEM). External parts of marginal strutted processes
wedge shaped.

Distinctive features: The linear areola array and the shape of marginal
strutted processes.

Remarks: *Thalassiosira ferelineata* Hasle & Fryxell is slightly larger and
coarser in structure but with the same areola array and process pattern as
T. tenera (Hasle & Fryxell, 1977b). The main difference is the lack of
external process tubes in *T. ferelineata*.

Distribution:

T. bulbosa—northern cold water region.

T. hispida and *T. nordenskioeldii*—northern cold water region to
temperate.

T. ferelineata, T. licea, T. oceanica, and *T. partheneia*—mainly warm
water region.

T. aestivalis, T. allenii, T. binata, T. mala, T. minuscula, T. punctigera,
and *T. subtilis*—warm water region to temperate?

T. conferta, T. pacifica, and *T. tenera*—cosmopolitan, exclusive polar
regions.

T. angulata—North Atlantic Ocean (see Hasle, 1978a). It should be
noted that *T. hibernalis* Gayoso, described from Argentina, is
morphologically very close to *T. angulata*.

1b. One central or subcentral strutted process. More than one marginal
ring of strutted processes.

Thalassiosira delicatula Ostenfeld in Borgert, 1908 (Plate 6, Table 8), *non
Thalassiosira delicatula* Hustedt.
Synonym: *Thalassiosira coronata* Gaarder.
References: Borgert, 1908, p. 16, Figs. A and B; Gaarder, 1951, p. 30, Fig.
17; Hasle, 1980, p. 170, Figs. 18–34; Rivera, 1981, p. 53, Figs. 63–107.

Girdle view: Pervalvar axis as long as or longer than cell diameter; valve
face with a central cavity; valve mantle high and smoothly rounded.
Connecting thread thin and as long as the pervalvar axis. Single cells or

Thalassiosira

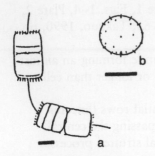

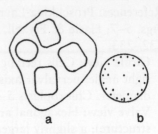

T. delicatula

T. diporocyclus

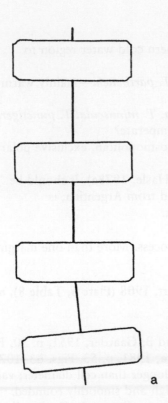

T. eccentrica

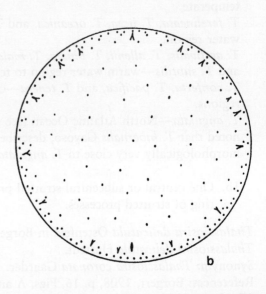

TABLE 8   Morphometric Data of *Thalassiosira* spp. with One Central or Subcentral Strutted Process and More Than One Marginal Ring of Strutted Processes

Species	Diameter (μm)	Valve areolae in 10 μm	Marginal proc. in 10 μm
T. delicatula	9–30	22–26	4–5
T. diporocyclus	12–24	24–31	2.7–3.8 μm apart
T. eccentrica	15–110	5–11	2–5
T. fragilis[a]	19–30.5	40–50	3–5 μm apart
T. hendeyi	42–110	5–6	5–6
T. mendiolana[a]	26–45	11–17	6–8
T. simonsenii	30–59	4–5	5–6

[a] Mentioned in the text for comparison.

chains sometimes embedded in mucilage forming large colonies. Long tubular (occluded) processes.

Valve view: Areolae in radial rows with slight fasciculation in larger specimens. Strutted processes on valve alternating, located in three rings, scattered strutted processes on valve face of larger specimens; external parts of strutted processes conspicuous but much shorter than those of occluded processes located in a ring between valve face and mantle.

Distinctive features: High pervalvar axis; occluded processes.
Remarks: Rivera (1981) put *T. chilensis* Krasske into synonymy with *T. delicatula* Ostenfeld.

Thalassiosira diporocyclus Hasle (Plate 6, Table 8)
References: Hasle, 1972a, p. 113, Figs. 25–45; Rivera, 1981, p. 60, Figs. 108–128; Takano, 1990, pp. 196–197.

Girdle view: Pervalvar axis of about the same size as cell diameter; central part of valve flat; mantle extremely curved. Cells in irregular mucilage colonies containing vast numbers of threads from the cells.
Valve view: Areolation fasciculate. Two marginal rings of strutted processes lacking external tubes; the single labiate process in peripheral ring.

PLATE 6   *Thalassiosira delicatula:* (a) short chain; (b) valve with processes. From Gaarder (1951). *Thalassiosira diporocyclus:* (a) mucilage colony; (b) valve with processes. *Thalassiosira eccentrica:* (a) chain in girdle view; (b) valve with process pattern. Scale bars = 10 μm.

Distinctive features: In water mounts, colonies formed by numerous distinct threads. On cleaned mounted material the two rings of processes with the labiate process in the peripheral ring.

Remarks: *Thalassiosira fragilis* G. Fryxell is similar to *T. diporocyclus* in valve morphology and colony formation but differs by having two labiate processes and the strutted processes, especially those in the inner ring, wider apart (Fryxell et al., 1984).

Thalassiosira eccentrica (Ehrenberg) Cleve (**Plate 6, Table 8**)
Basionym: *Coscinodiscus eccentricus* Ehrenberg.
References: Ehrenberg, 1841a, p. 146; Ehrenberg, 1843, Plate 1/3, Fig. 20, Plate 3/7, Fig. 5; Cleve, 1904, p. 216; Hustedt, 1930, p. 388, Fig. 201; Cupp, 1943, p. 52, Fig. 14; Fryxell & Hasle, 1972, p. 300, Figs. 1–18; Rivera, 1981, p. 64, Figs. 129–140; Makarova, 1988, p. 48, Plates 20 and 21; Takano, 1990, pp. 198–199.

Girdle view: Depending on cell diameter, pervalvar axis from one-sixth to longer than the length of the diameter. Valve face flat; mantle low and rounded. Connecting thread about twice the cell diameter.

Valve view: Areolae in curved tangential rows (eccentric structure) with a tendency of fasciculation. Central strutted process adjacent to a central areola surrounded by a ring of seven areolae; scattered strutted processes on valve face; two marginal rings of strutted processes with short external tubes; one ring of pointed spines further away from valve margin.

Distinctive features: Areola pattern of valve center; marginal spines.

Remarks: *Thalassiosira angulata* and *T. pacifica* with eccentric structure lack the central areola and process arrangement, the marginal spines and the processes scattered on the valve face. *Thalassiosira mendiolana* Hasle & Heimdal has the same process pattern as *T. eccentrica* but is usually more weakly silicified with smaller areolae arranged in sectors (Fryxell & Hasle, 1972).

Thalassiosira hendeyi Hasle & G. Fryxell (**Table 8**)
Synonym: *Coscinodiscus hustedtii* Müller-Melchers.
References: Müller-Melchers, 1953, p. 2, Plate 1, Figs. 2-5; Hasle & Fryxell, 1977b, p. 25, Figs. 35-45.

Girdle view: Pervalvar axis somewhat shorter than cell diameter. Valvocopula coarsely structured. Low, wavy marginal ridge. Two labiate processes with long external tubes.

Valve view: Hexagonal areolae in straight tangential rows (lineatus structure). Small strutted processes in three marginal rings (not resolved with LM).

Distinctive features: External tubes of two labiate processes; lineatus type of areolation; marginal ridge.

Thalassiosira simonsenii Hasle & G. Fryxell (Table 8)
Reference: Hasle & Fryxell, 1977b, p. 23, Figs. 26–34.
  Girdle view: Cells discoid; mantle low and slanting.
  Valve view: Hexagonal areolae in straight tangential rows (lineatus structure). Central strutted process adjacent to a small areola. Ribbed margin. Small strutted processes in two alternating marginal rings (not readily visible with LM); two labiate processes with long external tubes; occluded processes.
  Distinctive features: Lineatus type of areolation, ribbed margin, occluded processes.
Distribution
  T. delicatula and *T. eccentrica*—cosmopolitan, exclusive polar regions.
  T. diporocyclus, T. fragilis, T. hendeyi and *T. simonsenii*—warm-water region to temperate.
  T. mendiolana—South Pacific Ocean only.

2a. One or a few (two to six) subcentral strutted processes.

Thalassiosira curviseriata Takano (Plate 7, Table 9)
References: Takano, 1981, p. 34, Figs. 26–38; Hallegraeff, 1984, p. 498, Fig. 8; Takano, 1990, pp. 192–193.
  Girdle view: Cells discoid; valve face with central concavity; mantle high and slanting. Connecting thread long. Cells also found embedded in mucilage to form large colonies. Several large disc-shaped chloroplasts.
  Valve view: Areolation radial; just discernible with LM (cleaned, embedded material). One or probably more often, two strutted processes adjacent to a more or less off centered annulus. Marginal strutted processes conspicuous being equipped with two opposing wings; labiate process close to a marginal strutted process.
  Distinctive features: Off-centered valve structure, winged marginal strutted processes.
Remarks: *Thalassiosira tealata* Takano has one central process and marginal processes with longer wings than those of *T. curviseriata* (Takano, 1980).

Thalassiosira dichotomica (Kozlova) G. Fryxell & Hasle (Table 9)
Basionym: *Porosira dichotomica* Kozlova.
References: Kozlova, 1967, p. 56, Fig. 11; Fryxell & Hasle, 1983, p. 54, Figs. 2–14; Johansen & Fryxell, 1985, p. 161, Figs. 21–23 and 33–35.
  Girdle view: Valve face flat; mantle extremely low.
  Valve view: Areolae in radial; partly wavy rows; areolae walls thickened in pervalvar direction (SEM), visible with LM as irregular ribs. One to six central processes; labiate process midway between two marginal strutted processes, all with external tubes.
  Distinctive features: Irregular ribs superimposed on the areolation although variable in shape and presence.

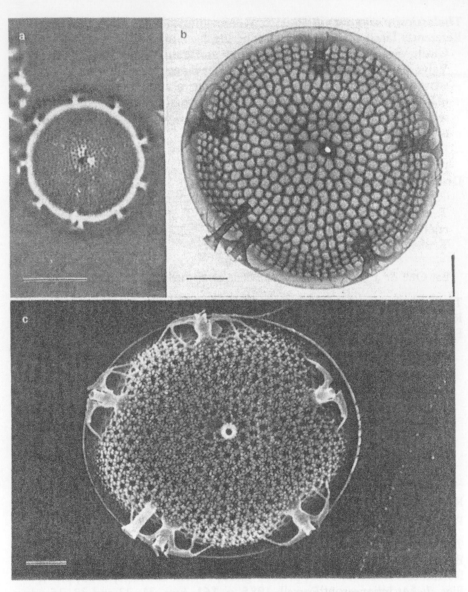

PLATE 7   Micrographs: single valves. *Thalassiosira curviseriata:* (a) areolae, process pattern, and characteristic outer parts of strutted processes, LM. Scale bar = 5 μm; (b) as described in a, TEM. Scale bar = 1 μm; (c) external valve surface, SEM. Scale bar = 1 μm.

TABLE 9   Morphometric Data of *Thalassiosira* spp. Having One to Six Subcentral Strutted Processes

Species	Diameter (μm)	Valve areolae in 10 μm	Marginal proc. in 10 μm
T. curviseriata	5–14	26–30	2–3
T. dichotomica	15–26	ca. 30	3–7
T. minima	5–15	30–40	3–6
T. tealata[a]	6.3–9.6	30–40	3

[a] Mentioned in the text for comparison.

Thalassiosira minima Gaarder (Table 9)

Synonyms: *Coscinosira floridana* Cooper; *Thalassiosira floridana* (Cooper) Hasle.

References: Gaarder, 1951, p. 31, Fig. 18; Cooper, 1958; Hasle, 1972b, p. 544; Hasle, 1976a, Figs. 44 and 45; Hasle, 1980, p. 167, Figs. 1–17; Rivera, 1981, p. 90, Figs. 226–245.

Girdle view: Cells rectangular; pervalvar axis less than half to slightly shorter than cell diameter; valve face flat and slightly depressed in the center; mantle low and beveled. Connecting thread as long as or longer than pervalvar axis.

Valve view: Areolation barely visible with LM (cleaned, embedded material). Processes distinct although with short external tubes; two central processes, sometimes one, seldom three; labiate process close to one marginal strutted process.

Distinctive feature: The two central processes.

Remarks: Cells in girdle view are similar to small specimens of *T. nordenskioeldii* as well as to *T. conferta*, *T. binata*, and *T. curviseriata*. The specific distinctions are evident especially in the process shape and pattern when examined as cleaned, embedded material.

Distribution:

T. curviseriata, and *T. minima*—cosmopolitan, exclusive polar regions.

T. dichotomica—southern cold water region.

T. tealata—Japanese warm water (Takano, 1980), English waters, Oslofjord, Norway (unpublished observations).

2b. Cluster of central strutted processes.

These species are very similar in girdle view, and despite a fairly large size and mostly coarse valve structure, reliable identification requires cleaned mounted valves.

Thalassiosira antarctica Comber (Plate 8, Table 10)
Synonyms: *Thalassiosira antarctica* var. *borealis* G. Fryxell, Doucette, &
Hubbard: *Thalassiosira fallax* Meunier (see Syvertsen, 1979, p. 59); Takano,
1990, pp. 184–185.
References: Comber, 1896; Meunier, 1910, p. 268, Plate 30, Figs. 1–4;
Hustedt, 1930, p. 327, Fig. 162; Hasle & Heimdal, 1968; Syvertsen, 1979,
p. 52, Figs. 15–59; Fryxell et al., 1981; Hasle & Syvertsen, 1990b, p. 289,
Figs. 25–27.

> **Girdle view:** Cells rectangular; valve face flat or slightly convex; mantle
> slightly sloping. Connecting thread thin, usually as long as or longer than
> pervalvar axis.
> **Valve view:** Areolation radial, bifurcate, occasionally lightly fasciculate.
> Two to several central processes; two or three rings of marginal strutted
> processes, located in valve mantle and hardly distinguishable with LM in
> intact valves; occluded processes present (usually not seen with LM); large
> labiate process located near valve mantle.
> **Distinctive feature:** Valve mantle with more than one ring of processes.
> **Remarks:** Fryxell et al. (1981) distinguished between *T. antarctica* var.
> *antarctica* from the southern hemisphere and *T. antarctica* var. *borealis*
> from the northern hemisphere. The valves of the vegetative cells are
> essentially the same but there are some differences in the bands (EM). Both
> varieties form heavily silicified resting spores, morphologically different from
> the vegetative cells. Three types of resting spores, endogenous,
> semiendogenous, and exogenous, have been found together in clonal culture
> of the northern variety (Oslofjord, Norway; Syvertsen, 1979).
> Semiendogenous resting spores were observed by Hasle & Heimdal (1968,
> Fig. 4) on a slide prepared of Comber's original material of *T. antarctica*
> (British Museum mount No. 59677). The primary valve of the exogenous/
> semiendogenous resting spores is flattened and the secondary valve is highly
> vaulted in both varieties. The main differences between the resting spore of
> var. *antarctica* and var. *borealis* seem to be the overall coarseness of the
> areolation and the more stable presence of strong, prominent marginal
> occluded processes, visible with LM, in the nominate variety. The flattened
> primary resting spore valve of var. *borealis* may easily be confused with

---

PLATE 8 *Thalassiosira antarctica* var. *borealis:* (a) chain in girdle view; (b) valve with process
pattern, open circles indicate occluded processes. From Hasle & Syvertsen (1990a). *Thalassiosira
gravida:* (a) chain in girdle view; (b) valve with process pattern. From Hasle & Syvertsen (1990a).
Thalassiosira hyalina: (a) chain in girdle view; (b) valve with process pattern. From Hasle &
Syvertsen (1990a). *Thalassiosira baltica:* (a) chain in girdle view; (b) valve with process pattern.
From Hasle & Syvertsen (1990a). *Thalassiosira rotula:* (a) chain in girdle view; (b) valve with
process pattern. Scale bars = 10 μm.

Thalassiosira

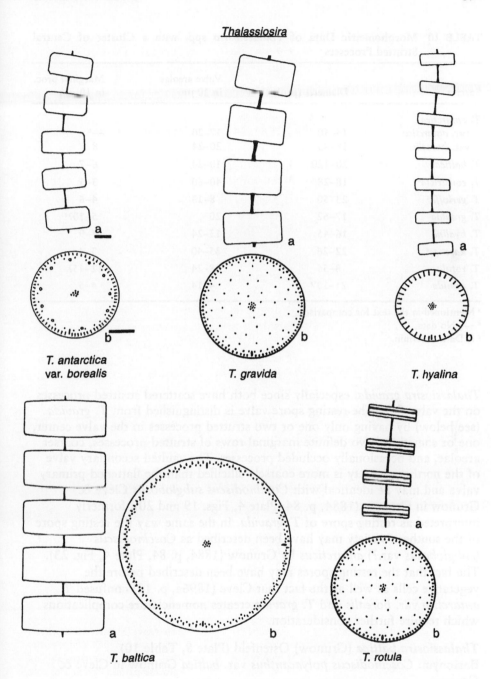

T. antarctica
var. borealis

T. gravida

T. hyalina

T. baltica

T. rotula

TABLE 10 Morphometric Data of *Thalassiosira* spp. with a Cluster of Central Strutted Processes

Species	Diameter (μm)	Valve areolae in 10 μm	Marginal proc. in 10 μm
T. antarctica			
var. *antarctica*	14–50	17–20	—[b]
var. *borealis*	18–43	20–24	8
T. baltica	20–120	10–20	6–7
T. constricta[a]	18–28	40–60	3–5
T. gerloffii[a]	23–50	8–13	4–6
T. gravida	17–62	20	5–10?[c]
T. hyalina	16–45	13–24	5–9
T. karenae[a]	22–26	35–40	3–4
T. rotula	8–55	18–24	12–15?
T. tumida	21–137	4–14	4–5

[a] Mentioned in the text for comparison.
[b] —, No data.
[c] ?, Data uncertain.

Thalassiosira gravida, especially since both have scattered strutted processes on the valve face. The resting spore valve is distinguished from *T. gravida* (see below) by having only one or two strutted processes in the valve center, one or sometimes two definite marginal rows of strutted processes, coarser areolae, and occasionally occluded processes. The vaulted secondary valve of the northern variety is more coarsely silicified than the flattened primary valve and may be identical with *Coscinodiscus subglobosus* Cleve & Grunow in Grunow (1884, p. 84, Plate 4, Figs. 19 and 20), formerly interpreted as resting spore of *T. gravida.* In the same way the resting spore of the southern variety may have been described as *Coscinodiscus* (*subglobosus* var.?) *antarcticus* by Grunow (1884, p. 84, Plate 4, Fig. 23). The fact that the resting spores may have been described before the vegetative cells as well as the fact that Cleve (1896a, p. 13) confused *T. antarctica* var. *borealis* and *T. gravida* creates nomenclature complications which require further consideration.

Thalassiosira baltica (Grunow) Ostenfeld (Plate 8, Table 10)
Basionym: *Coscinodiscus polyacanthus* var. *baltica* Grunow in Cleve & Grunow.
References: Cleve & Grunow, 1880, p. 112; Ostenfeld, 1901a, p. 290, Fig. 3; Hustedt, 1930, p. 328, Fig. 164; Hasle, 1978b, p. 266, Figs. 5–11;

Makarova, 1988, p. 63, Plate 33, Figs. 1–13; Hasle & Syvertsen, 1990b, p. 290, Figs. 28–30.

Girdle view: Cells almost rectangular; pervalvar axis about one-third to one-half of cell diameter; valve face flat; mantle low and slightly slanting. Connecting thread about twice as long as pervalvar axis.

Valve view: Areolation fasciculate. Two to nine central strutted processes; two rings of marginal strutted processes situated on valve mantle together with three or four labiate processes.

Distinctive features: Fasciculation and number of labiate processes.

Thalassiosira gravida Cleve (Plate 8, Table 10)

References: Cleve, 1896a, p. 12, Plate 2, Figs. 14–16; Fryxell, 1975, pp. 133–138; Syvertsen, 1977; Fryxell, 1989, Figs. 19–21.

Girdle view: Cells rectangular; valve face flat; mantle low and slightly sloping. Connecting thread thick, especially close to valve surface.

Valve view: Hexagonal areolae in radial rows. Numerous central strutted processes and scattered strutted processes on valve face and mantle; one large marginal labiate process.

Distinctive features: Connecting thread; hexagonal areolae; scattered strutted processes.

Remarks: It should also be noted that *T. gravida* has not been found to form resting spores (Syvertsen, 1977), whereas resting spores corresponding in structure to *C. subglobosus*, regarded by Cleve (1896a) as endocysts of *T. gravida*, have been found as resting spores of *T. antarctica* (Syvertsen, 1979). *Thalassiosira gravida* may appear in mucilage colonies in the Antarctic (Fryxell, 1989).

Thalassiosira hyalina (Grunow) Gran (Plate 8, Table 10)

Basionym: *Coscinodiscus hyalinus* Grunow in Cleve & Grunow.

References: Cleve & Grunow, 1880, p. 113, Plate 7, Fig. 128; Gran, 1897a, p. 16, Plate 1, Figs. 17 and 18; Hustedt, 1930, p. 325, Fig. 159; Hendey, 1964, p. 86, Plate 1, Fig. 6.

Girdle view: Pervalvar axis about one-third of cell diameter; valve face flat or slightly convex; mantle low and rounded. Connecting thread thick.

Valve view: Valve face with radial areolae rows or ribs; mantle always areolated. Two to 15 central strutted processes; one marginal ring of strutted processes with conspicuous external tubes; labiate process taking the place of a marginal strutted process.

Distinctive features: Low cells; one ring of marginal processes.

Remarks: The well-areolated valves belong most likely to resting spores (or resting cells) which occur in groups of two with no or very short connecting thread (E. Syvertsen, unpublished observations). The Arctic *Thalassiosira constricta* Gaarder, the Antarctic *T. karenae* Semina, as well as *T. gerloffii*

Rivera from South American Antarctic and Subantarctic waters have many central strutted processes and one marginal ring. In contrast to *T. hyalina*, *T. constricta* has resting spores drastically different from the vegetative cells, has higher cells, the marginal strutted processes are wider apart, and the labiate process is midway between two strutted processes (Heimdal, 1971). Except for the distribution and lack of observations of resting spores, *Thalassiosira karenae* is almost identical with *T. constricta* (Semina, 1981a). *Thalassiosira gerloffii* is generally coarsely silicified with a fairly high valve mantle and areolation more similar to *T. angulata* and *T. pacifica* but with marginal processes closer together (G. Hasle, unpublished observations; Rivera, 1981).

Thalassiosira rotula Meunier (Plate 8, Table 10)
References: Meunier, 1910, p. 264, Plate 29, Figs. 67–70; Hustedt, 1930, p. 326, Fig. 328; Cupp, 1943, p. 49, Fig. 12; Fryxell, 1975, pp. 95–100; Syvertsen, 1977; Takano, 1990, pp. 226–227.

> Girdle view: Cells flattened and discoid. The band (copula) next to the valvocopula sometimes broadened, septate ("unevenly thickened"); otherwise as *T. gravida*.
>
> Valve view: Valve face with radial ribs and few tangential areola walls (i.e., poorly developed areolae) except on valve mantle; otherwise as *T. gravida*.
>
> Distinctive features: Structure of valve face and the band next to valvocopula.

Remarks: Experiments indicated that the special copula was formed by nutrient deficiency. The difference between *T. rotula* and *T. gravida* in valve structure seemed to be temperature dependent, thus indicating a conspecificity or close relationship between the two taxa (Syvertsen, 1977).

Thalassiosira tumida (Janisch) Hasle in Hasle et al. (Table 10)
Basionym: *Coscinodiscus tumidus* Janisch in A. Schmidt.
References: Schmidt, 1878, Plate 59, Figs. 38 and 39; Hasle et al., 1971; Johansen & Fryxell, 1985, p. 176, Figs. 28–32; Fryxell, 1989, Fig. 18.

> Girdle view: Cells rectangular; valve face flat with slight, central depression or lightly convex; mantle rounded. Several connecting threads twinned to form one thick thread.
>
> Valve view: Areolation varying from linear to eccentric to fasciculate. Numerous strutted processes in valve center in an irregular ring, sometimes with processes also inside the ring; strutted processes scattered on valve face; one regular marginal ring of strutted processes including three to nine labiate processes.
>
> Distinctive features: Arrangement of central processes; number of labiate processes.

Remarks: *Thalassiosira tumida* has been regarded as an extremely variable species. Fryxell et al. (1986a) hypothesized the existence of two varieties, and Fryxell (1988) found a certain relationship between temperature and variation in areola pattern.

Distribution:

T. antarctica var. *antarctica*, *T. karenae*, and *T. tumida*—southern cold water region.

T. antarctica var. *borealis*, *T. constricta*, and *T. hyalina*—northern cold water region to temperate.

T. gerloffii—Chilean coastal waters, southern Atlantic Ocean (Rivera, 1981, p. 80; Lange, 1985), southern cold water region (Johansen & Fryxell, 1985).

T. gravida and *T. rotula*—cosmopolitan if regarded as one species. Diatoms identified as *T. gravida* from the Antarctic may belong to a separate taxon.

T. baltica—brackish water (e.g., Baltic Sea and the Arctic).

2c. +/− Modified ring of subcentral strutted processes.

Thalassiosira anguste-lineata (A. Schmidt) G. Fryxell & Hasle (Plate 9, Table 11)

Basionym: *Coscinodiscus anguste-lineatus* A. Schmidt.

Synonyms: *Coscinodiscus polychordus* Gran; *Thalassiosira polychorda* (Gran) Jørgensen; *Coscinosira polychorda* (Gran) Gran.

References: Schmidt, 1878, Plate 59, Fig. 34; Gran, 1897b, p. 30, Plate 2, Fig. 33, Plate 4, Fig. 56; Jørgensen, 1899, p. 15; Gran, 1900, p. 115; Hustedt, 1930, p. 317, Fig. 154; Cupp, 1943, p. 44, Fig. 7; Hendey, 1964, p. 89, Plate 1, Fig. 11; Hasle, 1972b; Fryxell & Hasle, 1977, p. 73, Figs. 22–34; Rivera, 1981, p. 45, Figs. 29–47; Makarova, 1988, p. 55, Plate 25, Figs. 10–13, Plate 26, Figs. 1–14; Takano, 1990, pp. 182–183.

Girdle view: Cells rectangular; valve face flat or slightly curved; mantle rounded. Many connecting threads in groups at some distance from valve center.

Valve view: Areolation varying from linear to eccentric to fasciculate. Central strutted processes in arcs in a ring at some distance from valve center, each arc with one to nine processes. Strutted processes with conspicuous external tubes in one marginal ring including one labiate process.

Distinctive features: Arrangement of central processes, coarse marginal processes.

Remarks: The subcentral strutted processes of coarsely silicified *T. anguste-lineata* specimens are often difficult to recognize, and these diatoms may

Thalassiosira

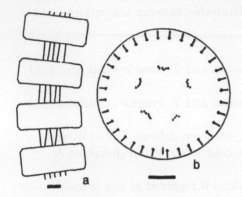

T. anguste-lineata

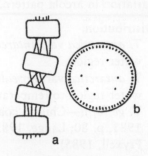

T. kushirensis

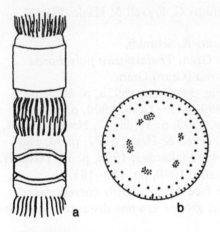

T. australis

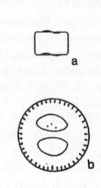

T. hyperborea

PLATE 9 *Thalassiosira anguste-lineata*: (a) chain in girdle view; (b) valve, groups of subcentral processes in a modified ring. From Hasle & Syvertsen (1990a). *Thalassiosira kushirensis*: (a) chain in girdle view; (b) valve, single subcentral processes in a modified ring. From Hasle & Syvertsen (1990a). *Thalassiosira australis*: (a) two cells in girdle view, lower one with endogenous resting spore and rudimentary valves. Marginal processes evident; (b) valve with irregular groups of strutted processes in a modified ring. *Thalassiosira hyperborea*: (a) cell in girdle view with valve undulation; (b) valve with process pattern. From Hasle & Syvertsen (1990a). Scale bars = 10 μm.

TABLE 11 Morphometric Data of *Thalassiosira* spp. with a Modified Ring of Subcentral Strutted Processes

Species	Diameter (μm)	Valve areolae in 10 μm	Marginal proc. in 10 μm
T. anguste-lineata	14–78	8–18	3–6
T. australis	23–56	11–16	4–9
T. hyperborea	16–70	8–18	3–5
T. kushirensis[a]	8–35	15–24	7–8
T. mediterranea	6–20	ca. 30	3–6
T. weissflogii	5–32	30–40?[b]	9–16

[a] Mentioned in the text for comparison.
[b] ?, Data uncertain.

have been reported as "*Coscinodiscus lineatus*" or "*Coscinodiscus sublineatus*." *Thalassiosira kushirensis* Takano (**Plate 9, Table 11**) is a smaller species similar to *T. anguste-lineata*. It occurs in short chains, as solitary cells or in mucilage, has single central strutted processes in an irregular ring, and areolae in radially clustered rows (Takano, 1985).

Thalassiosira australis M. Peragallo (**Plate 9, Table 11**)
References: Peragallo, 1921, p. 84, Plate 4, Fig. 17; Fryxell, 1977, p. 96, Figs. 1–12; Johansen & Fryxell, 1985, p. 158, Figs. 19, 20, and 46–48; Syvertsen, 1985, p. 116, Figs. 15–27; Fryxell, 1989, Fig. 22.
   Girdle view: Cells cylindrical; valve face flat or slightly concave; mantle high, steep, and slightly slanting; distance between cells in chains about half as long as pervalvar axis. Long occluded processes.
   Valve view: Areolation fasciculate. Subcentral strutted processes in three to seven clusters in a ring midway between valve center and margin; one marginal ring of strutted processes not easily seen with LM because of the high steep mantle; one ring of occluded processes on valve mantle and one labiate process displaced toward valve center.
   Distinctive features: Arrangement of central strutted processes; size, shape, and number of occluded processes.
Remarks: The endogenous resting spores of *T. australis* differ distinctly from the vegetative cells, having a central cluster of strutted processes, one marginal ring of coarse, trumpet-shaped strutted processes situated close together, and a labiate process with long, thick external tube, thus being similar to *T. gerloffii* (Syvertsen, 1985).

Thalassiosira hyperborea (Grunow) Hasle (Plate 9, Table 11)
Basionym: *Coscinodiscus (lacustris* var.?) *hyperboreus* Grunow.
References: Grunow, 1884, p. 85, Plate 4, Fig. 26; Hasle & Lange, 1989, p.
125, Figs. 20–22, 28–33.
  Girdle view: Cells almost rectangular; valve face tangentially undulated;
  mantle low and fairly steep. Chains not observed.
  Valve view: Hexagonal areolae in radial rows. The deep or the raised part
  of the undulated valve surface, or both, sigmoid in outline. One to eight
  (sometimes more) separated strutted processes in no particular pattern,
  mostly on the externally raised part of the valve (occasionally discernible
  with LM); one ring of coarse strutted processes close to valve margin; one
  labiate process. Coarsely silicified interstriae extending from the marginal
  processes, visible at a certain focus.
  Distinctive features: Shape of valve; distinct ribs associated with marginal
  processes.
Remarks: Four varieties of *T. hyperborea* were distinguished based on
areola size and slight differences in process pattern and structure. The
diatoms from the Baltic Sea and adjacent waters known as *Coscinodiscus
lacustris* Grunow, *Coscinodiscus lacustris* var. *pelagica* Cleve-Euler,
Thalassiosira lacustris (Grunow) Hasle, and *Thalassiosira bramaputrae*
(Ehrenberg) Håkansson & Locker are identical with *T. hyperborea* var.
pelagica (Cleve-Euler) Hasle (Hasle & Lange, 1989, p. 129). The nominate
variety, *T. hyperborea* var. *hyperborea* (Grunow) Hasle belongs to the
Arctic, probably brackish water. The fresh- or brackish water species *T.
lacustris* (Grunow) Hasle, *T. gessneri* Hustedt, and *T. australiensis*
(Grunow) Hasle have tangentially undulated valves but differ from *T.
hyperborea*, especially in internal valve structure (Hasle & Lange, 1989).

Thalassiosira mediterranea (Schröder) Hasle (Table 11)
Basionym: *Coscinosira mediterranea* Schröder.
Synonym: *Thalassiosira stellaris* Hasle & Guillard in Fryxell & Hasle.
References: Schröder, 1911, p. 28, Fig. 5; Hustedt, 1930, p. 318, Fig. 156;
Hasle, 1972b, p. 544; Fryxell & Hasle, 1977, p. 72, Figs. 16–21; Rivera,
1981, p. 120, Figs. 353–358; Hasle, 1990.
  Girdle view: Pervalvar axis usually shorter than diameter; valve surface
  convex. Several connecting threads approximately as long as or slightly
  longer than pervalvar axis.
  Valve view: Areolation not resolved with LM; two to eight radial rays
  extending from a central annulus to about midway between valve center
  and margin, each ending at a strutted process. One ring of strutted
  processes close to valve margin including a labiate process located about
  midway between two strutted ones.

Distinctive features: Radial rays and regular ring of single strutted processes midway between valve center and margin.

Thalassiosira weissflogii (Grunow) G. Fryxell & Hasle (Table 11)
Basionym: *Micropodiscus weissflogii* Grunow in Van Heurck.
Synonym: *Thalassiosira fluviatilis* Hustedt.
References: Van Heurck, 1880–1885, p. 210; Hustedt, 1926; Hustedt, 1930, p. 329, Fig. 165; Fryxell & Hasle, 1977, p. 68, Figs. 1–15; Takano, 1990, pp. 234–235.
Girdle view: Cells rectangular; pervalvar axis usually less than the diameter; valve flat in the center, valve mantle fairly steep or rounded.
Valve view: Irregular areolation; areolae small or ill defined (LM). Irregular ring of strutted processes (2–28) at some distance from valve center; one marginal ring of strutted processes including one very large labiate process taking the place of a strutted process.
Distinctive features: Meshwork on valve surface; irregular ring of central strutted processes.
Remarks: *Thalassiosira weissflogii* is often found in bloom proportions in brackish waters, apparently without chain formation although chains have been observed in natural samples (G. Hasle, personal observations).

Distribution:
  T. australis—southern cold water region.
  T. mediterranea—warm water region to temperate.
  T. anguste-lineata—cosmopolitan.
  T. hyperborea—brackish water, e.g., Baltic Sea and the Arctic.
  T. weissflogii—fresh- to brackish water, cosmopolitan?
  T. kushirensis—Japanese waters (Takano, 1985), North Atlantic waters (G. Hasle and E. Syvertsen, personal observations) including Skagerrak (Lange et al., 1992).

3a. No strutted processes on valve face.

Thalassiosira leptopus (Grunow) Hasle & G. Fryxell (Plate 10, Table 12)
Basionym: *Coscinodiscus* (*lineatus* var.?) *leptopus* Grunow in Van Heurck.
Synonyms: *Coscinodiscus lineatus* Ehrenberg; *Coscinodiscus pseudolineatus* Pantocsek; *Coscinodiscus praelineatus* Jousé.
References: Ehrenberg, 1839, p. 129; Ehrenberg, 1841a, p. 146, Plate 3, Fig. 4; Van Heurck, 1880–1885, Plate 131, Figs. 5 and 6; Hustedt, 1930, p. 392, Fig. 204; Jousé, 1968, p. 15, Plate 2, Fig. 1; Hasle & Fryxell, 1977b, p. 20, Figs. 1–14; Hasle & Syvertsen, 1984.
Girdle view: Valve face flat; mantle low and evenly slanting.

Thalassiosira

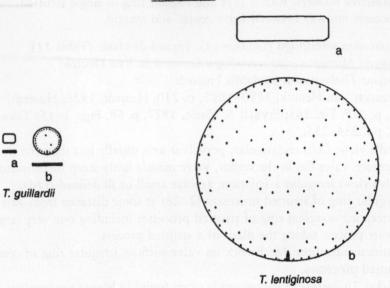

T. guillardii

T. lentiginosa

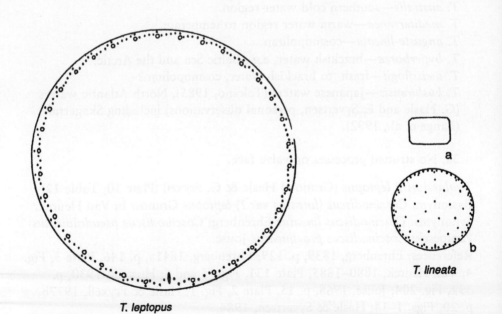

T. lineata

T. leptopus

TABLE 12   Morphometric Data of *Thalassiosira* spp. with Zero to Three Subcentral Strutted Processes

Species	Diameter (μm)	Valve areolae or ribs in 10 μm	Marginal proc. in 10 μm
T. guillardii	4–14	30–40[a] : 70–80[b]	7–8
T. leptopus	26–165	4–7	3–8
T. pseudonana	2.3–5.5	50–70	6–14

[a] Valve face.
[b] Valve mantle.

Valve view: Areolae on valve face in straight tangential rows (lineatus structure), those on the mantle smaller and irregularly arranged; a central areola larger than the others. Small strutted processes in two to three marginal rings not revealed with LM; one ring of coarser processes further away from margin composed of one large labiate process and evenly spaced coarse, short occluded processes.
Distinctive features: Large labiate process and occluded processes; larger central areola.

3b. Zero to three subcentral strutted processes.

Thalassiosira guillardii Hasle (Plates 10 and 11, Table 12)
Synonym: *Cyclotella nana* Guillard clone 7–15 in Guillard & Ryther.
References: Guillard & Ryther, 1962, Plate 2, Figs. 2A and 2B; Hasle, 1978b, p. 274, Figs. 28–50; Makarova et al., 1979, p. 924, Plate 3, Figs. 8–10.
Girdle view: Cells almost rectangular; pervalvar axis from about one-third to as long as cell diameter. Colonies not observed.
Valve view: Areolation not resolved with LM; slightly raised siliceous ribs radiating from an annulus; less distinct in outer half of valve face (EM). Subcentral strutted processes located about one-third to one-half radius away from valve center; one marginal ring of regularly spaced strutted processes; one labiate process taking the place of a strutted process.
Distinctive features: When compared with *T. pseudonana,* the location of the labiate process and an apparent lack of areolation.

PLATE 10   *Thalassiosira guillardii:* (a) cell in girdle view; (b) valve with two subcentral processes. *Thalassiosira lentiginosa:* (a) cell in girdle view; (b) valve with process pattern. *Thalassiosira leptopus:* valve with pattern of marginal strutted and occluded (open circles) processes. *Thalassiosira lineata:* (a) cell in girdle view; (b) valve with process pattern. Scale bars = 10 μm.

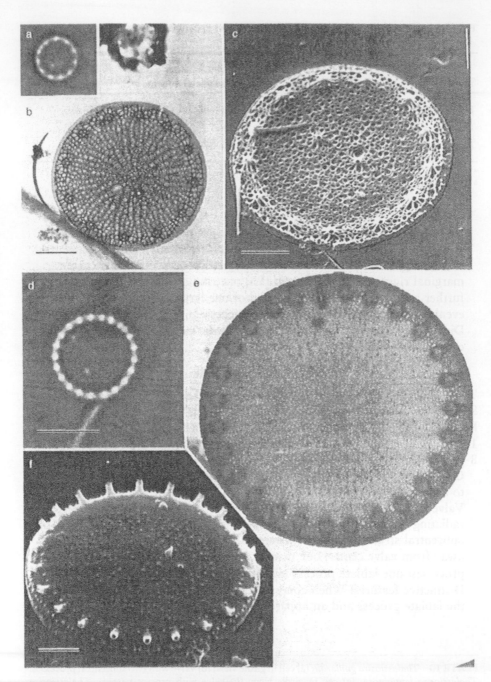

PLATE 11   Micrographs: single valves. *Thalassiosira pseudonana* (a–c): (a) valve with process pattern, LM. Scale bar = 5 μm; (b) typical valve structure, TEM. Scale bar = 1 μm; (c) external valve structure, SEM. Scale bar = 1 μm. *Thalassiosira guillardii* (d–f): (d) valve with process pattern, LM. Scale bar = 5 μm; (e) valve process pattern, TEM. Scale bar = 1 μm; (f) external valve surface, SEM. Scale bar = 1 μm.

Thalassiosira pseudonana Hasle & Heimdal (Plate 11, Table 12)
Synonym: *Cyclotella nana* Guillard clone 3H in Guillard & Ryther.
References: Guillard & Ryther, 1962, Plate 1, Fig. 1A; Hasle & Heimdal,
1970, p. 565, Figs. 27–38; Hasle, 1976b, p. 105, Figs. 11–16; Makarova et
al., 1979, p. 923, Plate 2, Figs. 1–8, Plate 3, Figs. 1–7; Hasle, 1983a, Table
1; Simonsen, 1987, Plate 657, Figs. 20 and 21 (?), but probably not Figs.
13–19.

Girdle view: Cells rectangular; pervalvar axis shorter than or as long as
cell diameter; valve face flat or slightly convex; mantle rounded. Colonies
not observed.
Valve view: Valve structure not revealed with LM. Process pattern distinct
in cleaned, mounted material. One marginal ring of strutted processes
including one labiate process midway between two strutted ones.
Distinctive features: When compared with *Thalassiosira oceanica*, the
location and number of central strutted processes.

Remarks: *Thalassiosira pseudonana* is an extremely variable species, and a
precise description based on LM observations can hardly be given. The
diatom here named *T. pseudonana* is identical with specimens in Hustedt's
samples from Wümme (the type locality of *Cyclotella nana*) and with clonal
cultures established from Wümme (Hasle, 1976b). The kind of valve
structure, radial ribs, a more or less developed areolation as well as a
central annulus, or a network of cross venation (as in *T. weissflogii*) are
apparently dependent on the amount of silica available (Paasche, 1973).
Chang & Steinberg (1989) "ordered *T. pseudonana* back to *Cyclotella*."
Although *T. pseudonana* has characters not exactly consistent with the
Thalassiosira type, it is hard to see that it is any closer to the *Cyclotella*
type (e.g., Håkansson, 1989).

3c. Strutted processes on the whole valve face; no particular central or
subcentral process.

Thalassiosira lentiginosa (Janisch) G. Fryxell (Plate 10, Table 13)
Basionym: *Coscinodiscus lentiginosus* Janisch in A. Schmidt.
References: Schmidt, 1878, Plate 58, Fig. 11; Fryxell, 1977, p. 100, Figs. 13
and 14; Johansen & Fryxell, 1985, p. 170, Figs. 7, 49, and 50.

Girdle view: Cells narrowly rectangular; pervalvar axis about one-fourth
diameter; valve flat; mantle low. Not observed in chains (Fryxell, 1977).
Valve view: Areolation usually fasciculated. Strutted processes evenly
scattered over valve face, resembling small areolae (not easily seen with
LM); one marginal ring of inconspicuous strutted processes lacking
external tubes; a large easily visible marginal labiate process.
Distinctive features: Labiate process; fasciculate valve structure.

TABLE 13   Morphometric Data of *Thalassiosira* spp. with Strutted Processes on the Whole Valve Face and No Central or Subcentral Process

Species	Diameter (μm)	Valve areolae or ribs in 10 μm	Marginal proc. in 10 μm
T. lentiginosa	47–95	7–9	—[b]
T. lineata	9–45	8–16	5–6
T. lineoides[a]	38.5–50	9–11	3–5

[a] Mentioned in the text for comparison.
[b] —, No data.

Thalassiosira lineata Jousé (Plate 10, Table 13)

References: Jousé, 1968, p. 13, Plate 1, Figs. 1 and 2; Hasle, 1976a, Figs. 22–28; Hasle & Fryxell, 1977b, p. 22, Figs. 15–25.

    Girdle view: Valve face flat; mantle low. Observed as single cells.

    Valve view: Hexagonal areolae in strictly straight rows throughout the whole valve face. Strutted processes scattered over the whole valve; two marginal rings of strutted processes (not revealed with LM); one large labiate process.

    Distinctive features: Straight areola rows; distinct strutted processes on the valve face (LM).

Remarks: The scattered processes are readily seen in water mounts.

Thalassiosira lineoides Herzig & G. Fryxell is distinct from *T. lineata* by certain differences in the location and structure of the scattered processes, mainly discernible with SEM (Herzig & Fryxell, 1986).

Distribution:

    T. lentiginosa—southern cold water region.

    T. lineata—warm water region.

    T. lineoides—warm water region to temperate(?).

    B. Labiate process on valve face.

    1. One central or subcentral strutted process.

Thalassiosira bioculata (Grunow) Ostenfeld (Plate 12, Table 14)

Basionym: *Coscinodiscus bioculatus* Grunow.

References: Grunow, 1884, pp. 107 and 108, Plate 3, Fig. 30, Plate 4, Fig. 2; Ostenfeld, 1903, p. 564, Fig. 120; Meunier, 1910, p. 266, Plate 29, Figs. 45–62; Hustedt, 1930, p. 331, Fig. 168.

    Girdle view: Cells cylindrical; pervalvar axis up to twice as long as cell diameter; valve face flat or slightly convex; mantle low and rounded. Connecting thread thin and short.

Thalassiosira

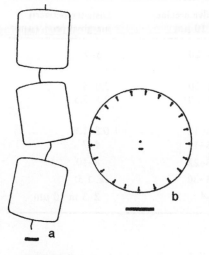

T. gracilis

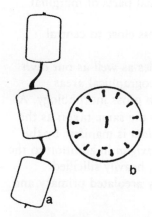

T. bioculata

T. oestrupii

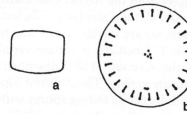

T. poroseriata

T. trifulta

PLATE 12 *Thalassiosira bioculata:* (a) chain in girdle view; (b) valve with one central strutted and one subcentral labiate process. From Hasle & Syvertsen (1990a). *Thalassiosira gracilis:* (a) cell in girdle view; (b) valve with submarginal labiate process. *Thalassiosira oestrupii:* (a) cell in girdle view, striated valvocopula; (b) valve with process pattern. *Thalassiosira poroseriata:* (a) chain in girdle view; (b) valve with irregular line of central strutted processes. From Hasle & Syvertsen (1990a). *Thalassiosira trifulta:* (a) cell in girdle view; (b) valve with V-shaped configuration of central strutted processes. From Hasle & Syvertsen (1990a). Scale bars = 10 μm.

TABLE 14   Morphometric Data of *Thalassiosira* spp. with One Central or Subcentral
Strutted Process

Species	Diameter (μm)	Valve areolae in 10 μm	Distance between marginal proc. (μm)[a]
T. bioculata	20–60	16–20	5–7
T. gracilis			
var. *gracilis*	5–28	8–20?[b]	3.0–3.5
var. *expecta*	7–15	14–20?	2.5–3.5
T. oestrupii			
var. *oestrupii*	7–60	6–12	0.8–1.9
var. *venrickae*	5.5–39	6–11	4–7
T. perpusilla	6–15	18–22	8–10
T. proschkinae	3–11.5	25–30	ca. 1.5
T. rosulata[c]	7–22	—[d]	2–3 in 10 μm

[a] Proc., processes.
[b] ?, Data uncertain.
[c] Mentioned in the text for comparison.
[d] —, No data.

Valve view: Areolation fasciculated, made up of rows of hexagonal
areolae or radial ribs. A single strutted process precisely in valve center, a
labiate process a little off centered, and long internal parts of marginal
strutted processes readily seen at a certain focus.
Distinctive features: Cylindrical cells; labiate process close to central
strutted process; long internal process tubes.
Remarks: Our study material includes Grunow's slides as well as our own
samples from the Arctic. Our material from other geographical areas
includes diatoms similar to *Thalassiosira bioculata* in gross morphology. At
present we are unable to tell whether they belong to the same taxon as the
Arctic *T. bioculata*. A certain variability of *T. bioculata* is manifest by the
fact that Grunow (1884) described the smaller var. *exigua* in addition to the
nominate variety. *Thalassiosira bioculata* forms low, heavily silicified
semiendogenous resting spores with convex, coarsely areolated primary and
secondary valves (Meunier, 1910).

Thalassiosira gracilis (Karsten) Hustedt var. *gracilis* (Plate 12, Table 14)
Basionym: *Coscinodiscus gracilis* Karsten.
References: Karsten, 1905, p. 78, Plate 3, Fig. 4; Hustedt, 1958a, p. 109,
Figs. 4–7; Fryxell & Hasle, 1979a, p. 382, Figs. 12–22; Johansen &
Fryxell, 1985, p. 168, Figs. 8, 58 and 59.

Thalassiosira gracilis var. *expecta* (VanLandingham) G. Fryxell & Hasle
(Plate 12, Table 14)
Basionym: *Thalassiosira expecta* VanLandingham.
Synonym: *Thalassiosira delicatula* Hustedt *non Thalassiosira delicatula*
Ostenfeld in Borgert.
References: Hustedt, 1958a, pp. 110, Figs. 8–10; VanLandingham, 1978, p.
3994; Fryxell & Hasle, 1979a, p. 384, Figs. 23–28; Johansen & Fryxell,
1985, p. 170, Figs. 8 and 60–63.
   Girdle view: Cells discoid; valve center mainly flat; rest of valve evenly
   rounded. Connecting thread fairly thin.
   Valve view: Areolae in central part large, widely spaced and irregularly
   arranged, smaller and less spaced on outer part of valve face and on
   mantle. Central strutted process a little off centered; single labiate process
   on border between valve face and mantle (EM).
   Distinctive features: Heavily silicified, rounded cells; larger areolae in
   central part of valve.
Remarks: *Thalassiosira gracilis* var. *expecta* differs from *T. gracilis* var.
gracilis mainly by smaller areolae on the central part of valve face. Whereas
the processes of heavily silicified specimens of var. *gracilis* may be difficult
to see with LM, those of var. *expecta* are readily observed when focusing on
their longer internal parts (Hustedt, 1958a; Fryxell & Hasle, 1979a;
Johansen & Fryxell 1985).

Thalassiosira oestrupii (Ostenfeld) Hasle (Plate 12, Table 14)
Basionym: *Coscinosira oestrupii* Ostenfeld.
References: Ostenfeld, 1900, p. 52; Hustedt, 1930, p. 318, Fig. 155; Hasle,
1972b, p. 544; Fryxell & Hasle, 1980.
   Girdle view: Pervalvar axis half to twice the diameter; valve face flat or
   slightly convex; mantle low and rounded; valvocopula striated. Cells in
   chains united by a thread from the central strutted process.
   Valve view: Areolae usually larger in central part of valve than closer to
   the margin, sometimes in sublinear array. One nearly central strutted
   process; labiate process usually one or two areolae distant.
   Distinctive features: Striated valvocopula; coarse areolation.
Remarks: *Thalassiosira oestrupii* var. *venrickae* G. Fryxell & Hasle is
distinguished from *T. oestrupii* var. *oestrupii* by a distinct eccentric areola
pattern and by more widely separated marginal processes. The labiate as
well as the strutted processes are discernible with LM (Fryxell & Hasle,
1980).

Thalassiosira perpusilla Kozlova (Table 14)
References: Kozlova, 1967, p. 60, Figs. 12 and 13; Fryxell & Hasle, 1979a,
p. 380, Figs. 1–11; Johansen & Fryxell, 1985, p. 173, Figs. 5 and 53–55.

Girdle view: Valve evenly rounded. No observations on chain formation.
Valve view: Areolae in irregular radial rows. Nearly central strutted
process; labiate process eccentric; widely separated marginal processes.
Distinctive feature: Few marginal strutted processes.
Remarks: The processes are readily seen with LM due to the light
silicification of the valves. *Thalassiosira rosulata* Takano (Takano, 1985, p.
3) is similar to *T. perpusilla* as well as to small specimens which may belong
to *T. bioculata*. The labiate process of *T. rosulata* is closer to the valve
margin than those of *T. perpusilla* and *T. bioculata,* and *T. rosulata* is
similar to *T. bioculata* regarding valve structure and distance between
marginal strutted processes.

Thalassiosira proschkinae Makarova in Makarova et al. (Plate 13,
Table 14)
References: Makarova et al., 1979, p. 922, Plate 1, Figs. 1–7; Makarova,
1988, p. 80, Plate 51, Figs. 13–22; Feibicke et al., 1990; Takano, 1990, pp.
220–221.

Girdle view: Cells low cylindrical; valve face flat; mantle low. A few large
chloroplasts. Cells in cultures as well as in natural samples found
entangled with threads in large colonies.
Valve view: Areolae in curved tangential rows (eccentric structure).
Central strutted process and labiate process close together and marginal
strutted processes closely spaced, all readily seen with LM on cleaned
valves mounted in a medium of a high refractive index.
Distinctive features: Marginal strutted processes not as widely spaced,
central strutted and labiate processes closer and areolae smaller than
those in *T. perpusilla,* which is the most similar species.
Remarks: Makarova (1988) reduced *Thalassiosira spinulata* Takano in rank
to *T. proschkinae* var. *spinulata* (Takano) Makarova. Our observations on
cultures and natural samples showed that *T. spinulata* is certainly no
distinct species. The two varieties can hardly be distinguished with LM.

PLATE 13   Micrographs: single valves. *Thalassiosira proschkinae* (a–c): (a) valves with areola
and process patterns distinct, LM. Scale bar = 5 μm; (b) valves with processes, areolae, and cribra,
TEM. Scale bar = 1 μm; (c) external valve surface, SEM. Scale bar = 1 μm. *Thalassiosira mala*
(d–f): (d) valves with processes, LM. Scale bar = 5 μm; (e) valve with process pattern, areolae,
and cribra, TEM. Scale bar = 1 μm; (f) external valve structure, areolae with foramina, SEM.
Scale bar = 1 μm. *Thalassiosira oceanica* (g–i): (g) valves with process pattern, LM. Scale bar =
5 μm; (h) areolation and processes, TEM. Scale bar = 1 μm; (i) external view of areolation and
processes, SEM. Scale bar = 1 μm.

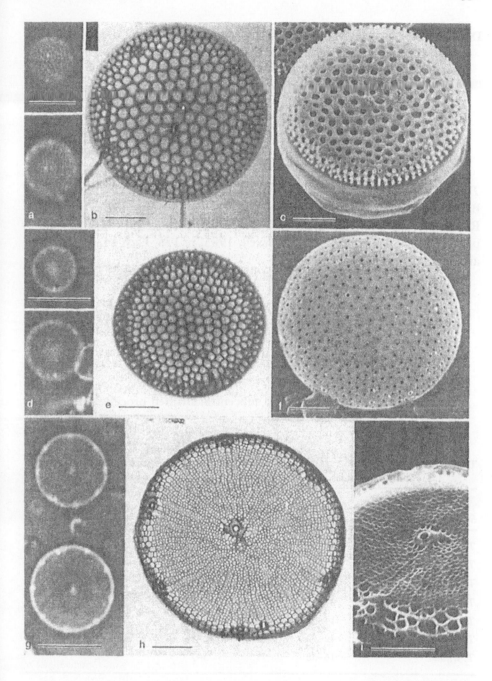

Distribution:
T. gracilis, and T. perpusilla—southern cold water region.
T. oestrupii—warm water region to temperate.
T. proschkinae—cosmopolitan (found as far north as 79°N).
T. bioculata—northern cold water region?
T. rosulata—Japanese waters.

2a. Cluster of central strutted processes.

Thalassiosira ritscheri (Hustedt) Hasle in Hasle & Heimdal (Table 15)
Basionym: *Coscinodiscus ritscheri* Hustedt.
References: Hustedt, 1958a, p. 19, Figs. 44–46; Hasle & Heimdal, 1970,
p. 569, Figs. 39–43; Johansen & Fryxell, 1985, p. 176, Figs. 14, 56,
and 57.
Girdle view: Valve face slightly convex. Chain formation by numerous
central threads.
Valve view: Areolation fasciculated, sublinear or linear. Labiate process a
few (two or three?) areolae away from the central cluster of strutted
processes.
Distinctive feature: Location of strutted and labiate processes.

2b. One or two rows of central to subcentral strutted processes.

TABLE 15 Morphometric Data of *Thalassiosira* spp. with More Than One Central
Strutted Process

Species	Diameter (μm)	Valve areolae in 10 μm	Marginal proc.[a] in 10 μm
T. confusa[b]	22–28	10–11	3
T. endoseriata	20–60	11–18	5–6
T. frenguellii[b]	12–52?[c]	9–13	1–2
T. poro-irregulata[b]	21–30	10–15	4–6
T. poroseriata	14–38	11–16	1–2
T. ritscheri	42–72	12–16	3–4
T. trifulta	16–58	5–7	2

[a] Proc., processes.
[b] Mentioned in the text for comparison.
[c] ?, Data uncertain.

Thalassiosira poroseriata (Ramsfjell) Hasle (Plate 12, Table 15)
Basionym: *Coscinosira poroseriata* Ramsfjell.
References: Ramsfjell, 1959, p. 175, Plates 1g and 2a; Hasle, 1972b, p. 544;
Fryxell & Hasle, 1979b, p. 20, Figs. 31–36; Rivera, 1981, p. 117, Figs.
340–352; Johansen & Fryxell, 1985, p. 175, Figs. 10, 51, and 52.
  Girdle view: Pervalvar axis often longer than cell diameter; valve face flat;
  mantle low and gently rounded. Several connecting threads about as long
  as pervalvar axis.
  Valve view: Valves usually weakly silicified; areolation radial or
  fasciculated. One to eight central strutted processes on a straight, curved,
  or zigzag line; a central areola or annulus at the end of the line; labiate
  process away from margin almost in a line with the central processes.
  Distinctive feature: Configuration of central processes.
Remarks: *Thalassiosira confusa* Makarova has central strutted processes in
two rows, a labiate process closer to valve margin, and marginal processes
more densely spaced than those of *T. poroseriata*.

Thalassiosira trifulta G. Fryxell in Fryxell & Hasle (Plate 12, Table 15)
Reference: Fryxell & Hasle, 1979b, p. 16, Figs. 1–24.
  Girdle view: Pervalvar axis shorter than cell diameter; valve face flat;
  mantle evenly rounded. No observations on chain formation.
  Valve view: Valves often coarsely silicified; areolae in straight or slightly
  curved tangential rows. One to eight strutted processes in valve center in
  one or two lines; labiate process mostly closer to valve mantle than to
  valve center(?); eight or nine areolae from central process on large valves.
  Distinctive features: Areola pattern; arrangement of central processes.
Remarks: The process pattern is difficult to see with LM because of the
heavy silicification of the valves. *Thalassiosira frenguellii* Kozlova is similar
to *T. trifulta* but it is more weakly silicified; the areolae are smaller, and
the marginal strutted processes are more widely spaced. A possible
conspecificity of *T. trifulta* and the Arctic *T. latimarginata* Makarova was
discussed by Fryxell & Hasle (1979b) who did not find sufficient evidence
to put them into synonymy at that time. This was done by Makarova (1988).

2c. +/− Modified ring of subcentral strutted processes.

Thalassiosira endoseriata Hasle & G. Fryxell (Table 15)
References: Hasle & Fryxell, 1977b, p. 78, Figs. 45–49; Rivera, 1981, p.
68, Figs. 145–157.
  Girdle view: Valve face flat; mantle low and rounded. No observations on
  chain formation.
  Valve view: Areolation usually fasciculated. Four to 14 central strutted
  processes in an irregular ring, each process taking the place of an areola
  inside the middle row of a sector; labiate process about one-fourth the
  distance from margin toward center.

Distinctive feature: Location of central processes.

Remarks: *Thalassiosira poro-irregulata* Hasle & Heimdal (Hasle & Heimdal, 1970, p. 573) is another species with an irregular ring of subcentral strutted processes and fasciculated areolation. The internal parts of the processes are much longer and also built differently (SEM) from those of *T. endoseriata,* and the labiate process is farther away from the margin.

Distribution:

T. confusa—northern cold water region.

T. endoseriata—warm water region.

T. frenguellii, T. ritscheri, and *T. trifulta*—southern cold water region.

T. poroseriata—cosmopolitan.

T. poro-irregulata—Chilean coastal waters.

Family Melosiraceae Kützing 1844

Melosiraceae is characterized by:

Strongly developed pervalvar axes.

A marginal ring of labiate processes (sometimes reduced).

Cells in chains.

Primarily circular valve outline.

Simonsen (1979), who listed these characters, found the family circumscription unsatisfactory and suggested that Melosiraceae should be split into a number of families without doing so himself.

Melosira and *Stephanopyxis* share the characters mentioned previously; *Paralia* differs, however, by having a low mantle and a short pervalvar axis. Round et al. (1990) placed *Paralia* into the family Paraliaceae Crawford in the order Paraliales and Glezer et al. (1988) placed it into the Pseudopodosiraceae (Sheshukova) Glezer. *Stephanopyxis,* considered synonymous with *Pyxidicula* Ehrenberg, was placed into Pyxidiculaceae Nikolaev by Glezer et al., whereas Round et al. retained the name *Stephanopyxis* and placed it in the Stephanopyxidaceae Nikolaev.

Terminology specific to Melosiraceae

Carina—Collar—a circular membranous costa on the outer side of valve.

Corona—a ring of larger irregular spines at valve apex.

Pseudoloculus—a chamber formed on the outer side of the valve surface by expansion of the distal parts of anastomosing costae.

KEY TO GENERA

1a. Cells in chains close together 2
1b. Cells in chains united by long external extensions of labiate processes . .
. . . . *Stephanopyxis,* p. 91

2a. Cells in chains united by mucilage pads, sometimes also by a corona consisting of larger irregular spines, valve mantle high and strongly curved. . . . .*Melosira,* p. 89

2b. Cells in chains united by interlocking ridges and grooves and marginal spines, valve mantle low and straight . . . .*Paralia,* p. 89

Genus *Melosira* C. A. Agardh 1824 (Plate 14, Table 16)
Type: *Melosira nummuloides* C. A. Agardh.
References: Dickie, 1852, p. cxcvi; Van Heurck, 1880–1885, Plate 85, Figs. 3 and 4; Hustedt, 1930, pp. 231–234, Figs. 95 and 96; Hendey, 1964, p. 72, Plate 1, Fig. 1: Crawford, 1975; Syvertsen & Hasle, 1988.

Melosira, as typified by *M. nummuloides* (Crawford, 1975), is mainly a marine, nonplanktonic genus. The chloroplasts are numerous, small, and plate like lying in the peripheral cytoplasm.

KEY TO SPECIES

1a. Collar away from valve apex, corona present. . . . .
. . . . *M. nummuloides* C. A. Agardh

1b. Collar close to valve apex, corona absent . . . . *M. arctica*[5] Dickie

Distribution:
   M. arctica—northern cold water region (also in the Baltic Sea and in the Oslofjord, Norway).
   M. nummuloides—cosmopolitan?
How to identify: The two species may be identified as whole cells in water mounts. With EM preparations they are also distinguished by the structure of the striae and the labiate process pattern.
Remarks: Formation of resting spores (semiendogenous) is common in *M. arctica* and not known in *M. nummuloides. Melosira nummuloides* often appears in coastal plankton, detached from the substratum, and *M. arctica* and the Baltic form or variety (Syvertsen & Hasle, 1988) may be just as common attached to sea ice as in the plankton under or in the vicinity of the ice. According to the literature (e.g., Van Heurck, 1909; Manguin, 1960) *Melosira* is represented in the Antarctic as a tychopelagic species but never abundant. The taxonomy of the Antarctic *Melosira* species is confusing, and a thorough reexamination is needed.

Genus *Paralia* Heiberg 1863
Type: *Paralia marina* (W. Smith) Heiberg.
Basionym: *Orthosira marina* W. Smith.

---

[5] Synonym: *Melosira hyperborea* Grunow in Van Heurck, 1880–1885, Plate 85, Figs. 3 and 4.

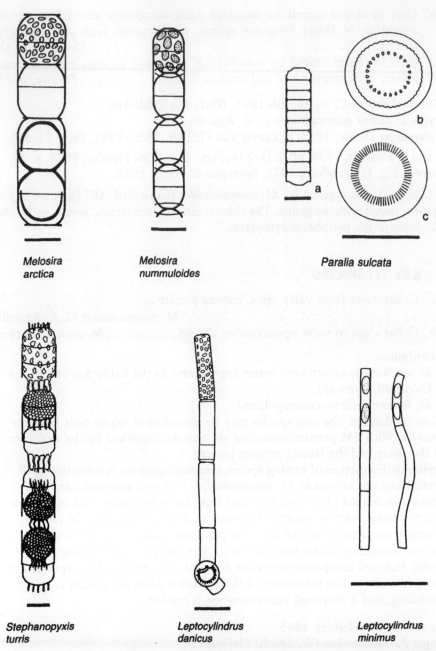

Melosira
arctica

Melosira
nummuloides

Paralia sulcata

Stephanopyxis
turris

Leptocylindrus
danicus

Leptocylindrus
minimus

PLATE 14 *Melosira arctica:* chain with semiendogenous resting spores. *Melosira nummuloides:* chain, collars evident. *Paralia sulcata:* (a) chain in girdle view; (b) intercalary valve; (c) separation valve. *Stephanopyxis turris:* chain with two semiendogenous resting spores. *Leptocylindrus danicus:* chain, cells with numerous chloroplasts, one cell with resting spore. *Leptocylindrus minimus:* two chains, one slightly helical, two chloroplasts. Scale bars = 20 μm.

TABLE 16   Morphometric Data of
Melosira spp.

Species	Valve diameter (μm)	Valve height (μm)
M. arctica	10–40	11–17
M. nummuloides	10–40	10–14

Correct name: *Paralia sulcata* (Ehrenberg) Cleve (*vide* Crawford, 1979, p. 209).
References: Crawford, 1979; Crawford, 1988, p. 422; Crawford et al., 1990.

Paralia sulcata (Ehrenberg) Cleve (**Plate 14**)
Basionym: *Gaillonella sulcata* Ehrenberg.
Synonym: *Melosira sulcata* (Ehrenberg) Kützing.
References: Ehrenberg, 1838, p. 170, Plate 21, Fig. 5; Ehrenberg, 1841a, p. 152, Plate 3, Fig. 5; Kützing, 1844, p. 55, Plate 2, Fig. 7; Cleve, 1873b, p. 7; Hustedt, 1930, p. 276, Fig. 119; Cupp, 1943, p. 39, Fig. 2; Hendey, 1964, p. 73, Plate 23, Fig. 5; Crawford, 1979.
   Girdle view: Cell wall extremely heavily silicified. Cells more wide than long. Valve face flattened. Chloroplasts several per cell; small discoid.
   Valve view: Separation valves at ends of chains without marginal spines and reduced ridges (heterovalvy).
Morphometric data: Pervalvar axis 3–45 μm; diameter 8–130 μm.
Distribution: *Paralia sulcata* is a bottom form but fairly common in coastal plankton, probably cosmopolitan.
How to identify: *Paralia sulcata* may be identified as whole cells in water mounts and as cleaned material on permanent mounts either as sibling or single valves in girdle as well as valve views. The difference in structure of the separation and intercalary valves has to be taken into consideration. If not, this heterovalvy may cause confusion.

Genus *Stephanopyxis* (Ehrenberg) Ehrenberg 1845 (Plate 14, Table 17)
Type: *Pyxidicula aculeata* Ehrenberg.
References: Ehrenberg, 1844c, p. 264; Greville, in Gregory, 1857, pp. 538 and 540, Plate 14, Fig. 109; Pritchard, 1861, p. 826, Plate 4, Fig. 74; Greville, 1865a, p. 2, Plate 1, Fig. 9; Grunow, 1884, p. 90; Gran & Yendo, 1914, p. 27, Fig. 16; Hustedt, 1930, p. 302, Figs. 140 and 147; Cupp, 1943, p. 40, Figs. 3–5; Hasle, 1973b, Figs. 91–112; Round, 1973, Figs. 1–14; Glezer et al., 1988, pp. 43 and 46, text Fig. 2: 3–7, Plate 25, Figs. 8

TABLE 17 Morphometric Data of *Stephanopyxis* spp.

Species	Diameter (μm)	Valve areolae in 10 μm	Mantle areolae in 10 μm
S. nipponica	24–36	—[a]	5–6
S. palmeriana	27–71	1.5–2.5	5–5.5
S. turris	36–57	3.5–5	3.5–5

[a] —, No data.

and 9, Plate 27, Figs. 1–11, Plate 28, Figs. 1–10; Takano, 1990, pp. 240–241.

Stephanopyxis has usually been distinguished from *Pyxidicula* by its long siliceous external extensions of labiate processes, uniting cells in chains. Strelnikova & Nikolaev (1986) did not accept this distinction and made the point that this type of linking is also found in *Pyxidicula*. The two genera are similar in valve structure and, despite the information in the older literature, they have girdles consisting of many bands (for *Pyxidicula mediterranea*, Hasle, unpublished observation made on a von Stosch culture). For the sake of convenience the name *Stephanopyxis* is retained for the planktonic diatoms dealt with in this chapter.

Electron microscopy revealed external valve wall structures which have been interpreted in various ways and which have a certain similarity to pseudoloculi (Hasle, 1973b, p. 131; Round et al., 1990, p. 158). Here they are called loculate areolae with wide foramina for the sake of simplicity.

Generic characters:

Valve mantle high, more or less curved.

Valve wall with large hexagonal areolae with large external foramina.

Resting spores present (semiendogenous).

KEY TO SPECIES

1a. External structures of labiate processes joined midway between cells in chains. 2

1b. Not so . . . . *S. nipponica* Gran & Yendo

2a. Areolae of same size on whole valve . . . .
. . . . *S. turris*[6] (Arnott in Greville) Ralfs in Pritchard

2b. Areolae larger on valve face than on valve mantle. . . . .
. . . . *S. palmeriana*[6] (Greville) Grunow

[6] Basionyms: *Creswellia turris* Arnott in Greville; *Creswellia palmeriana* Greville, respectively.

Distribution:
   S. nipponica—temperate to northern cold water region (Cupp, 1943).
   S. palmeriana—temperate to warm water region (Cupp, 1943).
   S. turris—temperate to warm water region (?).
How to identify: The specific distinctive characters are readily seen in water mounts.
Remarks: Resting spores are known for *S. turris* and *S. palmeriana*.

Family Leptocylindraceae Lebour 1930

   Simonsen (1979, p. 17) tentatively considered *Leptocylindrus* to belong to the family Melosiraceae, whereas Round et al. (1990) retained the family Leptocylindraceae and suggested a new order, Leptocylindrales, in the subclass Chaetocerotophycidae. Glezer et al. (1988) classified the genus under Pyxillaceae Schütt in the order Rhizosoleniales.
   Lebour's (1930, p. 75) family description is as follows: "Cells cylindrical, living singly or united in chains by the flat valve faces. Many collar-like intercalary bands." This diagnosis has to be emended, independent of whether Leptocylindraceae is regarded as monotypic, as done by Round et al. (1990), or as to encompass two genera as suggested in this chapter. The character to be added is: the presence of a marginal ring of spines, small, flap like, or triangular in *Leptocylindrus* and long and unique in shape in *Corethron*.

Genus *Leptocylindrus* Cleve 1889 (Plates 14 and 15, Table 18)
Type: *Leptocylindrus danicus* Cleve.
References: Cleve, 1889, p. 54; Peragallo, 1888, p. 81, Plate 6, Fig. 45; Gran, 1915, p. 72, Fig. 5; Hustedt, 1930, pp. 556–557, Figs. 317–321; Cupp, 1943, p. 77, Fig. 38; French & Hargraves, 1986; Fryxell, 1989, p. 4, Figs. 1–5; Takano, 1990, pp. 236–237; Hargraves, 1990; Delgado & Fortuño, 1991, Plate 61, Figs. c and d.

Generic characters:
   Tight chains by abutting valve faces.

   Cells cylindrical.

   Numerous half bands, trapezoidal in outline (TEM).

   Central parts of valves slightly convex or concave.

   Short flap-like spines on the border between valve face and mantle.

   Resting spores inside an auxospore-like sphere, distinctly different from vegetative cells.

KEY TO SPECIES

1a. Cell wall weakly silicified, neither valve nor girdle structure resolved with
    LM 2

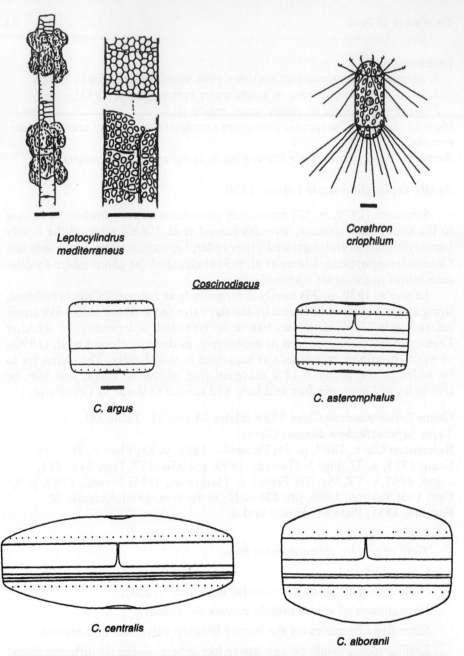

PLATE 15. *Leptocylindrus mediterraneus:* (a) chain with epiphyte; (b) detail of girdle. Scale bar = 10 μm. After Cupp (1943). *Corethron criophilum:* single cell with characteristic spines. Scale bar = 10 μm. *Coscinodiscus argus:* Girdle view, processes away from margin. Scale bar = 20 μm. *Coscinodiscus asteromphalus:* girdle view, processes close to margin. Scale bar = 20 μm. *Coscinodiscus centralis:* girdle view, evenly vaulted valves with a central depression. Scale bar = 20 μm. *Coscinodiscus alboranii:* girdle view, steep mantle and evenly vaulted valve face. Scale bar = 20 μm.

TABLE 18 Morphometric Data of
Leptocylindrus spp.

Species	Diameter (μm)	Bands in 10 μm
L. danicus	5–16	—[a]
L. mediterraneus	7–35	1–5
L. minimus	1.5–4.5	—

[a] —, No data.

1b. Cell wall coarsely structured .. *L. mediterraneus*[7] (H. Peragallo) Hasle
2a Two (seldom one) elongate chloroplasts, chains sometimes slightly undulated . . . . *L. minimus* Gran
2b. Numerous small rounded chloroplasts . . . . *L. danicus* Cleve

Distribution:

L. danicus—cosmopolitan—absent or scarce in the subantarctic/Antarctic (recorded from the Drake Passage Rivera, 1983); recorded as dominant species at 81°43'N (Heimdal, 1983); one of the dominant summer diatoms in Norwegian fjords.

L. mediterraneus—cosmopolitan—recorded from the Antarctic to the Arctic, seldom or never as a dominant species (Hasle, 1976a); scarce in the Arctic.

L. minimus—cosmopolitan—apparently absent from the Subantarctic/Antarctic (Hargraves, 1990).

How to identify: All three species may be identified in water mounts, *L. mediterraneus* especially by the presence of the epiphytic flagellate *Rhizomonas setigera* (Pavillard) Patterson, Nygaard, Steinberg, & Turley (= *Solenicola setigera* in Fryxell, 1989, Figs. 1–5). Single valves of *L. mediterraneus* mounted in a medium of a high refractive index are recognized by its coarse, open structure (Hasle, 1976a, Fig. 34), whereas those of *L. danicus* may be recognized under the best optical conditions by the small spines in the bordering zone between valve face and mantle and the sometimes more coarsely structured valve center.

Remarks: It should be noted that of two adjacent *L. danicus* valves in a chain one is slightly convex and the other concave, and neither are exactly flattened. *Leptocylindrus minimus* and *L. danicus* are distinguished not only by size and chloroplasts but also by the shape of the resting spores. The *L. minimus* resting spore is globular with a cylindrical neck, whereas that of

[7] Basionym: *Dactyliosolen mediterraneus* (H. Peragallo) H. Peragallo (see Hasle, 1975).

L. danicus consists of two unequal valves (Hargraves, 1990). *Leptocylindrus mediterraneus* is seldom found with chloroplasts and the cell wall has a double-layered structure (Hasle, 1975). Its taxonomic position is therefore questionable. *Leptocylindrus* includes two taxa in addition to those dealt with here, viz. *L. danicus* var. *adriaticus* (Schröder) Schiller from the Adriatic Sea and *L. curvatulus* Skvortzov from the Sea of Japan. The main character of the former is the ratio between pervalvar axis and diameter, which is greater than that in *L. danicus,* whereas the latter is distinguished by undulated chains. Their identity as separate taxa remains to be proven.

Genus *Corethron* Castracane 1886 (Plate 15, Table 19)
Lectotype: *Corethron criophilum* Castracane (*vide* Boyer, 1927, p.114).
References: Castracane, 1886, p. 85, Plate 21, Figs. 3–6, 12, 14, and 15; Karsten, 1905, p. 100, Plates 12–14; Hendey, 1964, p. 144, Plate 7, Fig. 4; Fryxell & Hasle, 1971; Fryxell, 1989, p. 9, Figs. 27–32; Thomas & Bonham, 1990.

Corethron was placed in Melosiraceae by Simonsen (1979) and in Chaetocerotaceae by Glezer et al. (1988), whereas Round et al. (1990) retained the family Corethraceae Lebour 1930 and introduced Corethrophycidae and Corethrales as a new subclass and a new order, respectively. The position in Chaetocerotaceae probably reflects an emphasis on the presence of long siliceous outgrowths, called spines for lack of a better term, from the border of the valve face. The unique shape of the marginal spines may be a reasonable justification for establishing a separate subclass and order for the genus.

Hendey (1937) concluded that *Corethron* was a monotypic genus, with *C. criophilum* as the only species, but appeared in different phases. Thomas & Bonham (1990) claimed the existence of two *Corethron* species in the Antarctic.

Generic characters:
 Cells cylindrical with more or less dome-shaped valves.
 Girdle composed of many bands.
 Valves with marginal long (barbed) and short hooked (clawed) spines.
 Chloroplasts numerous rounded or oval bodies.

KEY TO SPECIES (mainly after Thomas & Bonham, 1990)

1a. Cells solitary, robust, heterovalvate, valves with both hooked and long spines or long spines only, bands open, ligulate . . . .
. . . . *C. criophilum* Castracane
1b. Cells usually in colonies, weakly silicified, only end cells of colonies heterovalvate, intercalary valves with short spines interlocking sibling valves, cingulum composed of half bands . . . . *C. inerme* Karsten

TABLE 19 Morphometric Data of *Corethron* spp. (Hendey, 1937)

Species	Pervalvar axis (μm)	Diameter (μm)
C. criophilum	20–200	5–20
C. inerme	40–350	30–40

Distribution: If Hendey's concept of *Corethron* as a monotypic genus is followed, *C. criophilum* is a cosmopolitan species recorded as far north as ca. 80°N (Heimdal, 1983) and occurring in its greatest abundances in Antarctic waters (Fryxell & Hasle, 1971). As far as known, *C. inerme* has not been recorded outside the Antarctic.

How to identify: The species may be identified to genus from specimens in water mounts either of whole cells or of a single valve.

Remarks: The structure of the spines was described by Karsten (1905) and verified with EM by Fryxell & Hasle (1971).

Family Coscinodiscaceae Kützing 1844

Palmeria was classified under this family by Simonsen (1979) as well as by Round et al. (1990). *Ethmodiscus* was tentatively placed in Stictodiscoideae Simonsen, subfamily in Biddulphiaceae, by Simonsen (1979), in Ethmodiscaceae Round in the order Ethmodiscales Round by Round et al. (1990), and in Coscinodiscaceae by Glezer et al. (1988).

The genera dealt with here are characterized by:

Solitary cells (exception, *C. bouvet*).

No external tubes of processes.

Marginal labiate processes, sometimes in more than one ring [may be absent in *Ethmodiscus* (Rivera et al., 1989)].

Labiate processes are sometimes also between valve center and margin, irregularly positioned if present in central part of valve.

Labiate processes usually of two types (shape and/or size).

Areolae loculate; cribra external; foramina internal (SEM).

KEY TO GENERA

1a. Valves circular 2
1b. Valves semicircular . . . . *Palmeria*, p. 111

2a. Pervalvar axis **high**, cell diameter up to 2 mm . . *Ethmodiscus,* p. 110
2b. Pervalvar axis and diameter smaller. . . . . *Coscinodiscus,* p. 98

Genus *Coscinodiscus* Ehrenberg 1839 emend. Hasle & Sims 1986
Lectotype: *Coscinodiscus argus* Ehrenberg (proposed by Ross & Sims, 1973, conservation proposed by Fryxell, 1978b).
References: Ehrenberg, 1839, p. 128; Fryxell, 1978b, p. 122; Hasle & Sims, 1986a, p. 316.

Coscinodiscus is usually regarded as one of the largest marine planktonic diatom genera [400–500 validly described taxa (VanLandingham, 1968)]. A great number of the most frequently recorded *Coscinodiscus* species have been transferred to *Thalassiosira, Azpeitia,* and *Actinocyclus,* or to new genera as illustrated by the fact that of the approximately 20 *Coscinodiscus* species recorded in the Arctic literature between 1853 and 1911 only 4 are now regarded as belonging to the genus.

The *Coscinodiscus* species dealt with in this chapter are frequently recorded in the literature and/or they have been critically examined with LM and EM. Some of them are readily distinguished by special, pronounced morphological features, whereas others are easily confused. Our interpretation of the species is based on the often incomplete original diagnoses, the descriptions in Hustedt (1930), as well as recent LM and EM investigations referred to under the various species. It should be noted that the data on cell diameter of these species vary considerably in the literature. We have tried to use a compilation of data, and the sources are not given specifically in all cases.

All the species dealt with here have (see **Fig. 13**):
  Radial areola pattern.
  Two larger marginal labiate processes (macro-
    rimoportulae).
  Marginal ring of smaller labiate processes.
  Numerous, usually disc-shaped, chloroplasts.

Comments on terminology (Figs. 6a–6e, 7c–7g, 10, and 13)
  The radial rows of areolae (striae) may be grouped into more or less distinct bundles (fasciculation), and/or spiraling rows (decussating arcs) may be present (**Figs. 6a–6e and 10b and 10c**). Radial rows inserted from the margin (inserted striae, incomplete striae) are necessarily present in this type of areola pattern (**Figs. 6d and 6e and 13**). The bundles are separated by more or less distinct unperforated radial areas (wide interstriae, hyaline spaces, hyaline lines), on valve face, as well as on valve mantle (**Plate 16**). Valve center is usually occupied by either a hyaline (nonareolate) area or by a rosette of larger areolae (**Figs. 6b, 7c and 7d, and 13**). The structure of this central area may vary parallel to the cell diminution by the the vegetative multiplication (Schmid, 1990).

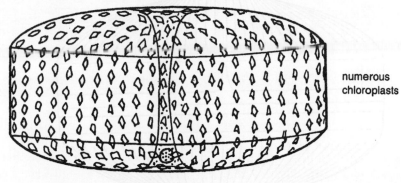

COSCINODISCUS

numerous
chloroplasts

GIRDLE VIEW

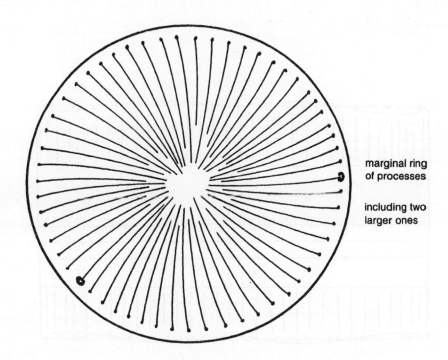

marginal ring
of processes

including two
larger ones

VALVE VIEW

FIGURE 13 Schematic illustration of *Coscinodiscus* sp. showing numerous chloroplasts, marginal ring of smaller labiate processes and the two larger processes (macrorimoportulae).

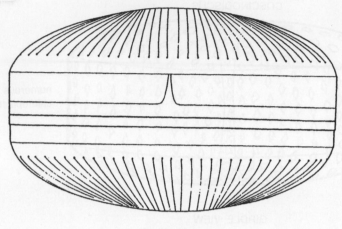

C. concinnus

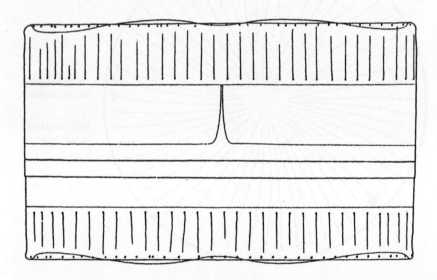

C. wailesii

PLATE 16 *Coscinodiscus concinnus:* girdle view. Broader interstriae distinct, one marginal ring of processes at ends of interstriae. *Coscinodiscus wailesii:* girdle view. High, steep valve mantle with two rings of labiate processes, one ring in the junction zone between valve face and mantle, and one ring close to valve margin at end of wide interstriae. Scale bar = 20 μm.

The two larger processes (**Figs. 7e–7g**) are usually observed with LM, in the larger species in water mounts, and in most species in acid cleaned material mounted in a medium of a high refractive index (**Figs. 6a and 6c and 13**). When noticed in the past, they were termed "unsymmetrische Prozesse" (Hustedt, 1930), or "asymmetrical processes or apiculi" (Cupp, 1943; Hendey, 1964). In some species the internal part and the shape of the processes are not revealed with LM; their position may be revealed, however, by indentations of the valve margin. When discernible with LM, the smaller labiate processes (**Figs. 6a and 6c and 13**) were termed in the past "Randdornen" (Hustedt, 1930) or "spinulae" (Cupp, 1943; Hendey, 1964). Smaller labiate processes may be present in a second marginal ring or on the valve face, usually associated with wide interstriae, or at the points of origin of incomplete striae.

"Interstitialmaschen—interstitial meshes" (Hustedt, 1930; Cupp, 1943) are located at the points of origin of incomplete striae and are identical with labiate processes, small pentagonal areolae or larger adjacent areolae, or unperforated areas.

Characters showing differences between species (partly based on Fryxell & Ashworth, 1988):

Cell shape in girdle view.

Cell diameter.

Valve shape.

Areolae (size) in 10 μm near center and peripheral.

Height of mantle, measured as number of areolae.

Areola pattern on valve face.

Central area, shape (rosette of larger areolae or nonareolated = hyaline).

The presence/absence of hyaline lines (interspaces, wide interstriae).

Shape of the larger processes.

Distance between the larger processes.

Distance between the smaller marginal processes (measured as number of areolae).

The presence or absence of processes on valve face.

Chloroplast outline.

See Schmid (1990) for further details on variation of some of these characters through the life cycle.

The species dealt with in this chapter are grouped according to the shape of the frustules in girdle view with the same qualifications as made for *Thalassiosira* spp.

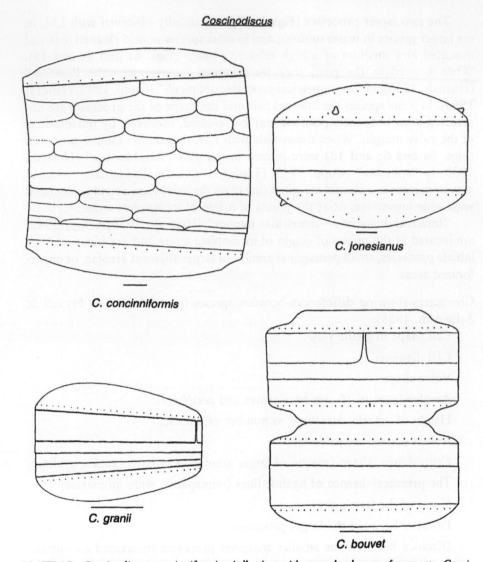

Coscinodiscus

C. concinniformis

C. jonesianus

C. granii

C. bouvet

PLATE 17 *Coscinodiscus concinniformis*: girdle view with several columns of segments. *Coscinodiscus jonesianus*: girdle view with evident outer parts of larger labiate processes. *Coscinodiscus granii*: girdle view, showing valve mantle and bands of uneven width. *Coscinodiscus bouvet*: girdle view, short chain. Scale bar = 20 μm.

A.   Frustules discoid to cylindrical: *C. alboranii*, *C. argus*, *C. asteromphalus*, *C. centralis*, *C. concinniformis*, *C. concinnus*, *C. jonesianus*, *C. marginatus*, and *C. wailesii* (**Tables 20a and 20b**).

B. Frustules coin shaped: *C. radiatus* (Tables 20a and 20b).
C. Frustules with a stepped elevated valve face: *C. bouvet* (Tables 20a and 20b).
D. Frustules wedge shaped: *C. granii* (Tables 20a and 20b).

A. Frustules discoid.

Coscinodiscus argus Ehrenberg (Plate 15, Tables 20a and 20b)
References: Ehrenberg, 1839, p. 129; Ehrenberg, 1841a, p. 145; Ehrenberg, 1854, Plate 21, Fig. 2, Plate 22, Fig. 5; Hustedt, 1930, p. 422, Fig. 226; Hasle & Sims, 1986a, Figs. 1–7, 33, and 34.
  Girdle view: Valve face flat or slightly depressed in the center; rounded margins; valve mantle steep and high.
  Valve view: Small central hyaline area just discernible with LM; areolae in the center differ slightly from the others in shape, but not in size. Decussating arcs present but not prominent. Hyaline lines (ribs) visible internally (SEM) but not perceived with LM, neither are the marginal processes. Foramina and cribra just discernible with LM.
  Distinctive features: Areolae increase in size from the center of the valve toward the middle of the radius, then decrease in size toward the margin.

TABLE 20a Morphometric Data of *Coscinodiscus* spp.

Species	Diameter (μm)	Valve areolae in 10 μm	No. of bands/theca	Valvocopula width (μm)
Group A				
C. alboranii	38–215	7–9.5	2	17 ?
C. argus	31–110	2–6	—	—
C. asteromphalus	80–400	3–5	4	13–14
C. centralis	100–300	4–6	3	20–24
C. concinniformis	150–500	ca. 10	Segmented	—
C. concinnus	110–500	7–9	5 ?	20–26
C. jonesianus	140–280	5–9	—	—
C. marginatus	35–200	1–4	—	12–24
C. wailesii	280–500[a]	5–6	3	44–60
Group B				
C. radiatus	30–180	2–9	3	—
Group C				
C. bouvet	100–290	6–10	2	23–28
Group D				
C. granii	40–200	8–11	2 ?	17–20

Note. ?, Data uncertain; —, no data available.
[a] 550–50 μm under laboratory conditions (Schmid, 1990, p. 109).

TABLE 20b   Processes in *Coscinodiscus* spp.

Species	Margin rings	Areolae from margin	Areolae apart	Scattered on valve face
Group A				
C. alboranii	1 + 1	4 + 2–3	7–16	–
C. argus	1	3–4	1–3	–
C. asteromphalus	1	1–2	3–6	–
C. centralis	1	3–4	2–5	–
C. concinnus	1	4–9	3–9	–
C. concinniformis	1	2 ?	9–11	+
C. jonesianus	1	2 ?	10–20	–
C. marginatus	1	2 ?	1–2	–
C. wailesii	2ᵃ	2–3ᵇ	2–10	+
Group B				
C. radiatus	1	1–2	4–5	+
Group C				
C. bouvet	1	2–3	3–4	+
Group D				
C. granii	1	2–4	5–8	–

Note. +, Present; –, absent; ?, data uncertain.

ᵃ One ring at the junction of valve face and mantle.

ᵇ The second ring including the two macrolabiate processes 2–3 areolae from margin.

Coscinodicus asteromphalus Ehrenberg (Plate 15, Tables 20a and 20b)

References: Ehrenberg, 1844a, p. 77; Ehrenberg, 1854, Plate 18, Fig. 45; Hustedt, 1930, p. 452, Figs. 250 and 251; Hasle & Lange, 1992, p. 42, Figs. 1–14.

Girdle view: Cells discoid with slightly convex valves; valve center depressed and valve mantle gently sloping. Numerous large rounded plate-like chloroplasts.

Valve view: Central rosette of somewhat larger areolae present, usually not prominent. Decussating arcs fairly distinct. No hyaline lines. Cribra just visible with LM. Ring of processes close to valve margin visible with LM. Interstitial meshes (probably smaller areolae) present at the points of origin of short incomplete striae. Larger labiate processes about 120–135° apart, comparatively small.

Distinctive features: Compared to *C. argus* and *C. centralis,* which are the most similar species, *C. asteromphalus* has the ring of processes closer to the margin.

Coscinodiscus centralis Ehrenberg (Figs. 6a–6c, Plate 15, Tables 20a and 20b)
References: Ehrenberg, 1844a, p. 78; Ehrenberg, 1854, Plate 18, Fig. 39, Plate 21, Fig. 3, Plate 22, Fig. 1; Hustedt, 1930, p. 444, Fig. 243; Semina & Sergeeva, 1980; Hasle & Lange, 1992, p. 45, Figs. 15–30.
  Girdle view: Cells discoid; valves gently convex. Numerous small plate-like chloroplasts.
  Valve view: Distinct rosette of large areolae (Fig. 6b), sometimes with a small hyaline area in the middle. Areola rows grouped into narrow bundles bordered by hyaline lines, distinct in weakly silicified valves and indistinct in coarsely silicified valves. Decussating arcs in the central part of the valve. Hyaline lines associated with labiate processes at the valve margin. Interstitial meshes present, identical with the pentagonal areola at the point of origin of an incomplete stria or an adjacent larger areola. Cribra resolved with LM (Fig. 6b), consisting of one central pore and a marginal ring of pores. Marginal processes readily resolved with LM; the smaller processes are long necked and slightly curved, the two larger processes, ca. 135° apart, have two "horns" (Fig. 6c).
  Distinctive features: Distinguished from *C. argus* and *C. asteromphalus* by the central areolae distinctly set off from the much smaller surrounding areolae, and the cribra and processes readily resolved with LM; and distinguished from *C. concinnus* by the shorter pervalvar axis, the more coarsely structured valves, and the marginal processes closer to the valve margin (Plates 15 and 16).

Coscinodiscus concinnus Wm. Smith (Plate 16, Tables 20a and 20b)
References: Smith, 1856, p. 85; Hustedt, 1930, p. 442, Figs. 241 and 242; Boalch, 1971, Plate 2; Brooks, 1975; Hasle & Lange, 1992, p. 50, Figs. 31–48.
  Girdle view: Cells thin walled; valves convex. Numerous small chloroplasts, fimbriate in outline, markedly H shaped when dividing. Marginal processes discernible in water mounts.
  Valve view: Central rosette large with a star-shaped hyaline area in the middle. The size of the hyaline area and the development of areola walls variable. Areola rows in bundles bordered by distinct hyaline lines running from marginal labiate processes toward valve center (Plate 16). Interstitial meshes, sometimes visible, probably identical with the pentagonal areola at the point of origin of each incomplete stria. The cribra barely seen with LM. Two larger marginal processes, ca. 135° apart, with raised parts as well as the smaller marginal processes, readily seen with LM.
  Distinctive features: Greater distance between valve margin and processes than in *C. centralis, C. concinniformis,* and *C. wailesii;* species with which *C. concinnus* may be confused.

Coscinodiscus wailesii Gran & Angst (Plate 16, Tables 20a and 20b)
References: Gran & Angst, 1931, p. 448, Fig. 26; Schmid & Volcani, 1983; Takano, 1990, pp. 250–251; Schmid, 1990; Hasle & Lange, 1992, p. 55, Figs. 49–62.

Girdle view: Cells low to tall cylindrical, often as high as wide, and rectangular in outline at a certain focus. Valves flattened with a concentric depression near the steep high mantle. Numerous chloroplasts irregular in outline. Marginal processes discernible in water mounts.
Valve view: Prominent central hyaline (unperforated) area; wide interstriae (hyaline lines) radiating from the central area. Irregular fasciculation formed, partly by wider interstriae, partly by distinct incomplete striae, originating near the valve center at a labiate process or a small hyaline area (area of solid silica). Cribra visible with LM. One ring of smaller processes in the junction zone between valve face and mantle, and one ring including two larger processes, 120–180° apart, close to the valve margin. Processes in the first ring more densely spaced than those in the second ring. Hyaline lines more conspicuous and regular on valve mantle than on valve face, associated with processes.
Distinctive features: High, steep mantle with hyaline lines and two marginal rings of processes.

Coscinodiscus concinniformis Simonsen (Plate 17, Tables 20a and 20b)
References: Simonsen, 1974, p. 14, Plates 10 and 11; von Stosch, 1986, p. 303, Figs. 5 and 6.

Girdle view: Cells cylindrical, as high as wide or higher. Girdle consisting of several columns of segments, curved with the two valves inclined toward each other. Valves uniformly and moderately convex; mantle extremely low. Weakly silicified. Numerous small chloroplasts.
Valve view: Prominent central hyaline (unperforated) area from which wide interstriae (hyaline lines) bordering bundles of radial areola rows radiate and partly branch. Labiate processes at the points of origin of incomplete striae. One ring of processes close to the valve margin, each process at the end of a hyaline line. Processes discernible with LM. The two larger processes ca. 125° apart.
Distinctive features: Distinguished in valve view from *C. wailesii* by one ring of processes on valve face and from *C. concinnus* by scattered processes on valve face and the ring of processes closer to valve margin.

Coscinodiscus alboranii Pavillard (Plate 15, Tables 20a and 20b)
References: Pavillard, 1925, p. 13, Fig. 16; von Stosch, 1986, pp. 295–303.

Girdle view: Cells cylindrical, up to 1.5 times as long as wide. Valves convex, flattened in the center, and nearly semiglobular in small cells. Four lobed or oval chloroplasts. Cell wall extremely weakly silicified.

Valve view: (structures seen in air mounts with phase contrast) A circular central area without areolae, distinct hyaline lines between the central area and one marginal ring of processes. Two larger processes, ca. 125° apart, closer to the margin than the smaller processes. Valve face with two types of areolae, the rarer of which appears darker in phase contrast. Valve margin usually with a special type of areolae, endochiastic areolae, appearing as dark circlets with an internal structure similar to a cross.
Distinctive features: Two types of valve face areolae; endochiastic areolae on valve mantle.

Coscinodiscus jonesianus (Greville) Ostenfeld (Plate 17, Tables 20a and 20b)

Basionym: *Eupodiscus jonesianus* Greville.
References: Greville, 1862, p. 22, Plate 2, Fig. 3; Ostenfeld, 1915, p. 13, Fig. 7; Hustedt, 1930, p. 438, Fig. 239; Hendey, 1964, p. 79; Makarova, 1985, p. 52, Plate 2; Takano, 1990, pp. 246–247.
Girdle view: Cells about as high as wide. Valves convex, slightly concave in the center.
Valve view: Central rosette of larger areolae more or less distinct. Areolae in radial and spiraling rows. Cribra visible with LM. At about half radius an irregular ring of interstitial meshes (probably labiate processes). One marginal ring of processes, visible with LM, with hyaline lines bordered by one row of slightly larger areolae on either side toward the valve center. Inside the marginal ring two larger processes with prominent external areolated protuberances, ca. 100° apart.
Distinctive features: Large external protuberances of the two larger marginal processes.

Coscinodicus marginatus Ehrenberg (Plate 18, Tables 20a and 20b)

References: Ehrenberg, 1844a, p. 78; Ehrenberg, 1854, Plate 18, Fig. 44, Plate 33/12, Fig. 13, Plate 38b/22, Fig. 8; Hustedt, 1930, p. 416, Fig. 223; Hendey, 1964, p. 78, Plate 22, Fig. 2; Sancetta, 1987, p. 231, Plate 1, Figs. 1–13.
Girdle view: Cells discoid, thick walled. Valves flat or nearly so with steeply sloping or straight mantle. Several small rounded chloroplasts.
Valve view: No central rosette. Coarse areolae in irregular radial rows. Cribra discernible with LM, processes not.
Distinctive feature: The coarse areolation.

B.   Frustules coin shaped.

Coscinodiscus radiatus Ehrenberg (Figs. 6d and 6e, Plate 18, Tables 20a and 20b)

References: Ehrenberg, 1841a, p. 148, Plate 3, Fig. 1; Hustedt, 1930, p. 420, Fig. 225; Hendey, 1964, p. 76, Plate 22, Fig. 7; Hasle & Sims, 1986a,

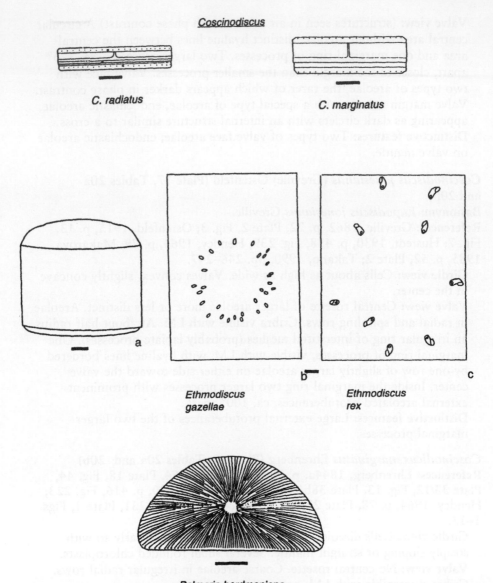

Coscinodiscus

C. radiatus

C. marginatus

a

b

c

Ethmodiscus gazellae

Ethmodiscus rex

Palmeria hardmaniana

PLATE 18 _Coscinodiscus radiatus:_ girdle view, flattened valves. Ring of processes close to margin. Scale bar = 20 μm. _Coscinodiscus marginatus:_ girdle view. Flattened valve, ring of processes close to junction between valve mantle and valve face. Scale bar = 20 μm. _Ethmodiscus_ spp.: (a) girdle view. Scale bar = 200 μm.; (b) _E. gazellae_ and (c) _E. rex,_ valve center with processes. After Kolbe (1957). Scale bar = 10 μm. _Palmeria hardmaniana:_ valve view. From Allen & Cupp (1935). Scale bar = 10 μm.

pp. 310 and 312, Figs. 8–32 and 35–39; Sancetta, 1987, p. 234, Plate 2, Figs. 1–10; Fryxell & Ashworth, 1988.

Girdle view: Cells discoid, flat, coin shaped. Numerous cocciform chloroplasts.

Valve view: Indistinct central rosette of slightly larger areolae. Areolae in radial rows, sometimes in indistinct decussating arcs (Figs. 6d and 6e). Small labiate processes at the points of origin of incomplete striae (interstitial meshes). No hyaline lines. One marginal ring of processes, including two slightly larger processes visible with LM as indentations of the valve margin, ca. 135° apart. The smaller marginal processes barely discernible.

Distinctive features: Flat cells, large areolae. Distinguished from *C. argus* by areolae of uniform size throughout the entire valve face and from *C. marginatus* by smaller areolae in more regular radial rows.

C. Frustules with a stepped elevated valve face.

Coscinodiscus bouvet Karsten (Plate 17, Tables 20a and 20b)
References: Karsten, 1905, p. 83, Plate 3, Fig. 9; Hendey, 1937, p. 244, Plate 13, Figs. 3 and 4; Priddle & Thomas, 1989; E. Syvertsen, unpublished observations.

Girdle view: Cells about as high as wide, octagonal in outline due to an elevated area of the valve face, about one-half to two-thirds of the valve diameter, and occasionally with no elevation. Sometimes in chains (up to 18 cells). Numerous irregular stellate chloroplasts.

Valve view: Central rosette of larger areolae. Radial rows of areolae. One ring of processes, including two larger processes ca. 125° apart, close to the valve margin, visible with LM (water mounts).

Distinctive features: The shape of the cell in girdle view. In chains.

D. Frustules wedge shaped.

Coscinodiscus granii Gough (Plate 17, Tables 20a and 20b)
References: Gough, 1905, p. 338, Fig. 3B; Hustedt, 1930, p. 436, Fig. 437; Boalch, 1971, Plate 1; Brooks, 1975; Karayeva & Dzhafarova, 1984; Takano, 1990, pp. 244–245; Hasle & Lange, 1992, p. 60, Figs. 63–76.

Girdle view: Cells asymmetric, one side much higher than the other. Valvocopula wedge shaped, widest opposite to the opening. Greatest convexity of the valve not in the center of the valve but nearest the widest part of the valvocopula. Chloroplasts discoid and smooth in outline.

Valve view: Central rosette of larger areolae. Radial areolation, incomplete striae, and decussating arcs in the central part of the valve. Cribra barely discernible with LM. One ring of marginal processes including two larger processes ca. 135° apart, readily seen with LM; the

larger processes seen as deep indentations of the valve mantle. Hyaline
lines from the marginal processes toward the valve center more or less
distinct.

Distinctive features: Cells wedge shaped in girdle view. Eccentric
convexity of the valve, manifest with LM in valve view by only a part of
the valve in focus. Valvocopula of uneven width, a second narrow band
with rectangular, tall ligula positioned ca. 90° apart from the band
opening (Hasle & Lange, 1992, Figs. 63 and 67).

It should be noted that the illustrations (**Plates 15–18**) show cells consisting
of epitheca and incomplete hypotheca. The epithecae have, when known, the
number of bands listed (**Table 20a**) with valvocopula included. The formation
of all the bands of a hypotheca is usually first completed when the cell starts
to divide.

Distribution:
  C. bouvet—southern cold water region.
  C. alboranii, C. concinniformis, and *C. jonesianus*—warm water region.
  C. wailesii—warm water to temperate (recently introduced to North
  Atlantic waters).
  The remaining species have a wide distribution according to the literature
  and may be cosmopolitan with a wide temperature tolerance.
How to identify: When not specifically mentioned, the morphological details
are discernible with LM. *Coscinodiscus wailesii, C. bouvet,* and *C. granii*
can be identified in water mounts in girdle view because of their special
shape. To ensure a correct identification the remaining species should, in
addition to water mounts, be examined as valves cleaned of organic matter
and mounted in a medium of a high refractive index.

Genus *Ethmodiscus* Castracane 1886 (Plate 18, Table 21)
Lectotype: *Ethmodiscus gigas* Castracane selected by F. E. Round (*vide*
Round et al., 1990, p. 206).
Correct name: *Ethmodiscus gazellae* (Janisch ex Grunow) Hustedt (*vide*
Hustedt, 1930, p. 375).

TABLE 21   Morphometric Data of *Ethmodiscus* spp.

Species	Diameter (μm)	Areolae in 10 μm	Size of processes (μm)[a]
E. gazellae	1280–2000	6–9	2–4
E. rex	600–1900	6–8 (3)[b]	4–9

[a] Measured with the light microscope.
[b] No. in parentheses occasionally found.

References: Grunow, 1879, p. 688; Castracane, 1886, p. 169, Plate 14, Fig. 5; Rattray, 1890a; Hustedt, 1930, p. 374, Fig. 196; Wiseman & Hendey, 1953, p. 49, Plates 1 and 2; Kolbe, 1957, p. 33, Figs. 5 and 6, Plate 4, Figs. 46–49; Round, 1980; Rivera et al., 1989; Round et al., 1990, p. 206.

Ethmodiscus rex and *E. gazellae* are the only two of the 15 taxa recognized by VanLandingham (1969) that are regularly referred to. They are usually regarded as the largest single celled members of the marine phytoplankton having the shape of large boxes, with approximately equally sized diameters and pervalvar axes.

Generic characters:

Cell wall weakly silicified.

Valve areolae small, distinctly separated, in radial rows outside a central area.

Each theca with one wide band with wide hyaline margins and areolae in straight parallel rows.

Processes (revealed as labiate processes with SEM) in central part of the valve.

Heterovalvy, one strongly convex valve and the other with flattened valve face and a broad and flat mantle (Rivera et al., 1989).

KEY TO SPECIES

1a. Valve center with large (up to 50 μm in diameter) nonareolated area bordered by an irregular ring of processes . . . .
. . . .*E. gazellae*[8] (Janisch ex Grunow) Hustedt
1b. Valve center irregularly areolated with nonareolated areas interspersed, central processes larger and scattered . . . .
. . . . : *E. rex*[8] (Rattray) Hendey in Wiseman & Hendey

Distribution: Warm water region to temperate.

How to identify: Although the unusually large size may distinguish the genus in water mounts, examination of cleaned material on permanent mounts is needed to distinguish between the two species.

Remarks: Marginal labiate processes are present in *E. gazellae* (Round, 1980), but were not seen in *E. rex* (Rivera et al., 1989).

Genus *Palmeria*[8a] Greville 1865 (Plate 18, Table 22)
Type: *Palmeria hardmaniana* Greville.

---

[8] Basionyms: *Coscinodiscus gazellae* Janisch ex Grunow and *Coscinodiscus rex* Rattray, respectively.

[8a] *Palmeria* Greville 1865 is a later homonym of *Palmeria* F. von Müller 1864 and was given the new name Palmerina Hasle 1995 (Diatom Research 10, 357–358).

Synonym: *Hemidiscus hardmanianus* (Greville) Mann.
References: Greville, 1865b, p. 2, Plate 5, Figs. 1–4; Ostenfeld, 1902, p. 222, Figs. 1 and 2; Mann, 1907, p. 316; Allen & Cupp, 1935, p. 152, Figs. 91 and 91a–91e; Simonsen, 1972, p. 270, Fig. 12; von Stosch, 1987, pp. 31–41, Figs. 1–45.

Generic characters:
> Valves semicircular.
>
> Hyaline central area.
>
> Radial areolation with incomplete striae inserted from the margin.
>
> Distinct hyaline lines associated with marginal processes.
>
> Two processes larger than the others.

KEY TO SPECIES (von Stosch, 1987, p. 32)

1a. Valve face approximately plane . . . . *P. hardmaniana* Greville
1b. Valve face excavated by deep narrow fold parallel to dorsal circumference
     of valve face and usually settled by epiphytic ciliates . . . .
     . . . . *P. ostenfeldii*[9] (Ostenfeld) von Stosch

Distribution:
> *P. hardmaniana*—warm water region.
>
> *P. ostenfeldii*—Gulf of Thailand and Townsville, Australia (von Stosch, 1987), but may have a much wider distribution because it was confused with the former species in the past.

How to identify: The valve morphological structures are seen with LM by examination of material mounted in air or Pleurax (von Stosch, 1987). In *P. hardmaniana*, the ventral line of small labiate processes (those along the straight valve margin) is ca. 10 areolae away from the margin. In *P. ostenfeldii*, it is ca. 20 areolae from the straight margin in its middle part.

TABLE 22   Morphometric Data of *Palmeria* spp.

Species	Length (μm)	Width (μm)	Areolae in 10 μm
P. hardmaniana	303–534	162–270	12–14
P. ostenfeldii	352–650	189–300	13–15

Note. Length is measured along the straight margin of the valve; width is the greatest distance from the curved to the straight margin of the valve.

[9] = *Palmeria hardmaniana* in Ostenfeld, 1902.

Family Stellarimaceae Nikolaev 1983 ex Sims & Hasle 1990

To date the family includes *Stellarima* Hasle & Sims with living and fossil members and the fossil genus *Fenestrella* Greville (Sims, 1990). The family is based on *Stellarima* Hasle & Sims.

Genus *Stellarima* Hasle & Sims 1986 (Plate 19, Table 23)
Type: *Stellarima microtrias* (Ehrenberg) Hasle & Sims.
Basionym: *Symbolophora? microtrias* Ehrenberg.
Synonyms: *Coscinodiscus symbolophorus* Grunow; *C. furcatus* Karsten; *Symbolophora furcata* (Karsten) Nikolaev.
References: Ehrenberg, 1844b, p. 205; Roper, 1858, p. 21, Plate 3, Fig. 3; Grunow, 1884, p. 82, Plate 4, Figs. 3–6; Karsten, 1905, p. 82, Plate 4, Fig. 7; Hustedt, 1930, p. 396, Figs. 207 and 208; Cupp, 1943, p. 53, Fig. 16; Hustedt, 1958a, pp. 113 and 118, Figs. 18, 19, 36–39, and 47; Nikolaev, 1983; Syvertsen, 1985, p. 113, Figs. 1–14; Hasle & Sims, 1986b; Hasle et al., 1988; Fryxell, 1989, p. 4, Figs. 6–17; Hasle & Syvertsen, 1990c.

Generic characters:

Cells drum shaped, discoid, or lenticular.

Valves more or less convex depending on size of diameter.

Areolae in radial rows.

Areolae loculate, cribra external, foramina internal (SEM).

"Specialized areolae" with a special type of velum (SEM).

Center of valve with small hyaline area with one or a single group of (two to eight) labiate processes (SEM).

No marginal processes.

Many small chloroplasts, irregular (angular?) in outline.

KEY TO LIVING SPECIES

1a. Valves distinctly convex, areolation fasciculate 2
1b. Valves flat to slightly convex, areolation furcate . . . .
. . . . *S. microtrias* (Ehrenberg) Hasle & Sims, vegetative cells
2a. Many areola rows per sector, areolae small . . . .
. . . . *S. stellaris*[10] (Roper) Hasle & Sims
2b. Sectors narrower, areolae larger . . . .
. . . . *S. microtrias* (Ehrenberg) Hasle & Sims, resting spores

---

[10] Basionym: *Coscinodiscus stellaris* Roper.

Stellarima

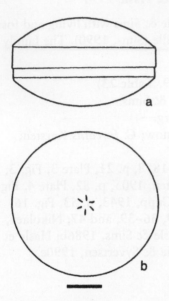

a

b

S. stellaris

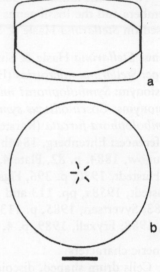

a

b

S. microtrias

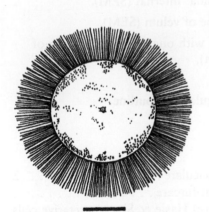

Gossleriella tropica

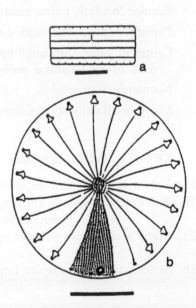

a

b

Actinocyclus curvatulus

TABLE 23   Morphometric Data of *Stellarima* spp. (Hasle et al., 1988)

Species	Diameter (μm)	Areolae in 10 μm	Striae in 10 μm	Rows/sector
S. microtrias, veg. cells	35–105	11–15	12–16	No sectors
S. microtrias, rest. spores	40–100	9–11	10–12	10–17
S. stellaris	40–115	15–16	18–22	30–40

Distribution:

S. microtrias—southern cold water region—particularly common on and in fast sea ice and in the surrounding plankton.

S. stellaris—warm water region to temperate, planktonic.

How to identify: The species may be identified in valve view in water mounts although examination of valves cleaned of organic matter and mounted in a medium of a high refractive index may be needed in most cases, especially to identify *S. stellaris*.

Remarks: *Stellarima microtrias* forms heavily silicified endogenous resting spores, usually identified as *C. symbolophorus,* inside weakly silicified cells previously identified as *C. furcatus* (Syvertsen, 1985). *Stellarima stellaris* has not been found to form resting spores. The labiate processes are readily observed with LM in the lightly silicified *S. stellaris* (e.g., Cupp, 1943, Fig. 16) and are less easily seen in the *S. microtrias* resting spores. *Stellarima microtrias* was found in long chains in living natural material (Fryxell, 1989) as well as in cultures (G. Hasle, personal observations on a M. Elbrächter isolate). The attachment point was central to eccentric on sibling valves.

Incertae sedis (Stellarimaceae)

Genus *Gossleriella* Schütt 1892
   Type: *Gossleriella tropica* Schütt.
   Monospecific genus.
      The taxonomic position of this genus is disputed. Hargraves (1976) suggested Rhizosoleniaceae as a possible family; Simonsen (1979)

PLATE 19   *Stellarima stellaris:* (a) girdle view; (b) valve with central labiate processes. Scale bar = 20 μm. *Stellarima microtrias:* (a) girdle view with endogenous resting spore; (b) valve with central labiate processes. From Hasle & Syvertsen (1990c). Scale bar = 20 μm. *Gossleriella tropica:* valve with girdle spines. From Hustedt (after Schütt) (1930). Scale bar = 100 μm. *Actinocyclus curvatulus:* (a) cell in girdle view; (b) valve with curvatulus structure indicated in two sectors with the pseudonodulus located in one. Scale bars = 20 μm.

retained it in Coscinodiscaceae, and Round et al. (1990) introduced the monotypic family Gossleriellaceae Round. Nikolaev (1983) placed it in the family he described as Symbolophoraceae which was later renamed Stellarimaceae (Sims & Hasle 1990). The justification for this classification is the absence of a marginal ring of processes and a single central labiate (SEM) process similar in shape to those of *Stellarima*. The genus most probably comprises only the generitype.

Gossleriella tropica Schütt (Plate 19)
References: Schütt, 1892, p. 258, Fig. 63; Hustedt, 1930, p. 500, Fig. 280; Hargraves, 1976; Delgado & Fortuño, 1991, Plate 44, Fig. b.
  Girdle view: Discoid with flat or slightly convex
  valves and a ring of siliceous spines attached to the cingulum.
  Chloroplasts numerous small plates, each with a pyrenoid.
  Valve view: Valve surface poroid. A single process in or near valve center.
Morphometric data: Pervalvar axis ca. 5–7.5 μm; diameter of a valve without spines 96–250 μm, with spines 162–293 μm (Hargraves, 1976).
Distribution: Warm water region—lower photic zone (Hargraves, 1976).
How to identify: Whole cells are easily identified in water mounts; the valve structure is visible with LM by examination of cleaned material in permanent mounts.

Family Hemidiscaceae Hendey 1937 emend. Simonsen 1975 ex Hasle 1995 (this publication)
Terminology specific to Hemidiscaceae
Pseudonodulus—a marginal to submarginal structure, always only one per valve, with LM evident as an open hole or an area covered by densely packed smaller areolae.

The pseudonodulus is the principle diagnostic feature of the Hemidiscaceae as emended by Simonsen (1975) and encompassing *Actinocyclus*, *Hemidiscus* and *Roperia*. Watkins & Fryxell (1986) questioned the taxonomic preeminence of the pseudonodulus, and Fryxell et al., 1986b, p. 33) found that the best place for *Azpeitia*, which has no pseudonodulus, was within Hemidiscaceae. This placement was "based in part on the fact that the Coscinodiscaceae is ill-defined at the present dynamic stage of diatom systematics."

  Glezer et al. (1988) placed *Actinocyclus, Hemidiscus,* and *Roperia* in Hemidiscaceae whereas the monotypic family Azpeitiaceae Glezer & Makarova was introduced for *Azpeitia*. Round et al. (1990) followed Fryxell et al. (1986b), placing *Azpeitia* and the three genera mentioned in Hemidiscaceae.

Characters common to *Actinocyclus, Azpeitia, Hemidiscus,* and *Roperia:*
  Cells low cylindrical to discoid.
  Few bands, unperforate (hyaline).
  Areolae loculate (SEM)
  Cribra external, foramina internal (SEM).
  Valve areolation radial, often fasciculate.
  Valve face and mantle more or less different in areolation.
  One marginal ring of large labiate processes, similar in shape and generally
    also in size.

KEY TO GENERA

1a. Pseudonodulus present (sometimes revealed only with EM). 2
1b. Pseudonodulus absent, central labiate process present (sometimes revealed
    only with EM) . . . . *Azpeitia*, p. 123
2a. Valves circular or elliptical. 3
2b. Valves semicircular . . . . *Hemidiscus*, p. 128
3a. Areolation radial, usually fasciculate . . . . *Actinocyclus*, p. 117
3b. Areolation basically linear in central portion of the valve, fasciculate or
    radial in marginal part . . . . *Roperia*, p. 130

Genus *Actinocyclus* Ehrenberg 1837
  Lectotype: *Actinocyclus octonarius* Ehrenberg (*vide* Boyer, 1927, p. 80).

VanLandingham (1967) listed 80–90 validly described *Actinocyclus* species of which only 4 or 5 are commonly recorded from marine or brackish water plankton. This chapter includes these 5 taxa and other marine planktonic species newly described or transferred to *Actinocyclus*.

Generic characters:
  Valvocopula wide, a narrow second band with ligula, and a probable
    narrow third band.
  Valve mantle deep and straight.
  Valve outline circular.
  Areola pattern usually fasciculate.
  Marginal zone often denser in areolation and/or different in direction of
    the striae.
  Central annulus, variable in size and presence.
  No labiate process in valve center.
  Marginal ring of labiate processes.
  Pseudonodulus, variable in size.

Characters showing differences between species:
> Type of fasciculation.
>
> Position of labiate processes: at the end of edge and/or central areola row of a fascicle.
>
> Distinct or indistinct annulus.
>
> Distinct or indistinct pseudonodulus.
>
> Position of pseudonodulus.

The 11 *Actinocyclus* species dealt with here exhibit three different types of fasciculation:

A. Radial areola rows parallel to central row: *A. actinochilus*, *A. circellus*, *A. kützingii*, *A. normanii*, *A. octonarius*, *A. sagittulus*, and *A. subtilis* (Table 24).

B. Radial areola rows parallel to edge (side) row: *A. curvatulus* and *A. exiguus* (Table 25).

C. Areola rows parallel to central and/or edge row: *A. spiritus* and *A. vestigulus* (Table 26).

TABLE 24 Morphometric Data of *Actinocyclus* spp. with Areola Rows Parallel to the Central Row of the Fascicles

Species	Pervalvar axis (μm)	Diameter (μm)	Valve areolae in 10 μm	Marginal striae in 10 μm	μm Between marginal proc.
A. actinochilus	7–17	20–112	5–11	13–21	9.5–15
A. circellus	28	88–140	8–9	17	7–9
A. kützingii	—	30–70	7–8	10–12	—
A. normanii f. normanii	—	30–110	ca. 9	—	—
A. normanii f. subsalsa	—	16–44	8–11	18–22	—
A. octonarius var. octonarius	—	50–300	6–8	15	8–11
A. octonarius var. tenellus	—	18–48	8.5–11.5	17–23	10–16
A. sagittulus	—	41–110	8.5–10.5	15–18	7–9
A. subtilis	—	35–160	12–18	—	7

Note. Proc., processes; —, no data.

A. Radial areola rows parallel to central row.

Actinocyclus actinochilus (Ehrenberg) Simonsen (Table 24)
Basionym: *Coscinodiscus actinochilus* Ehrenberg.
Synonym: *Charcotia actinochila* (Ehrenberg) Hustedt.
References: Ehrenberg, 1844b, p. 200; Ehrenberg, 1854, Plate 35A/21, Fig.
5; Hustedt, 1958a, p. 126, Figs. 57–80; Hasle, 1968b, p. 7, Plate 9, Fig. 6,
Map 2; Simonsen, 1982; Villareal & Fryxell, 1983, p. 461, Figs. 21–32.
Large labiate processes, clearly seen with LM, positioned in an apparently
hyaline band between valve areola rows and margin, at end of edge rows
of fascicles. The presence and structure of a central annulus and the
presence and width of hyaline spaces between radial areola rows variable.
Pseudonodulus of about the same size as an areola, located in the hyaline
band, evident in LM although difficult to recognize in many specimens,
ranges internally (SEM) from a large circular depression to a small hole
with a central depression (Villareal & Fryxell, 1983).

Actinocyclus circellus T. P. Watkins in Watkins & Fryxell (Table 24)
Reference: Watkins & Fryxell, 1986, p. 294, Figs. 1–8.
Pronounced central annulus enclosing areolae usually smaller than those
on the rest of the valve and more or less linearly arranged. Areolae on the
rest of the valve arranged in irregular, obscure fascicles. Labiate processes,
discernible with LM, positioned at the end of the central and edge row of
each fascicle. Pseudonodulus slightly away from the marginal ring of
processes toward the valve center, easily discernible with LM.

Actinocyclus kützingii (A. Schmidt) Simonsen (Table 24)
Basionym: *Coscinodiscus kützingii* A. Schmidt.
References: Schmidt, 1878, Plate 57, Figs. 17 and 18; Hustedt, 1930,
Fig. 209; Simonsen, 1975, p. 92.
Pseudonodulus, very small, obscure. Areolation fasciculated with
secondary, curved rows. Marginal processes discernible with LM,
positioned at the end of the central row of each fascicle. Fairly wide
marginal zone, structure consisting of small areolae in two crossing
systems. Small central annulus.

Actinocyclus normanii (Gregory) Hustedt f. *normanii* (Table 24)
Basionym: *Coscinodiscus normanii* Gregory in Greville.
References: Greville, 1859a, p. 80, Plate 6, Fig. 3; Hustedt, 1957, p. 218;
Hasle, 1977, Figs. 2, 11–17, 20 and 21.

Actinocyclus normanii f. *subsalsus* (Juhlin-Dannfelt) Hustedt (Table 24)
Basionym: *Coscinodiscus subsalsus* Juhlin-Dannfelt.
References: Juhlin-Dannfelt, 1882, p. 47, Plate 3, Fig. 33; Hustedt, 1957, p.
219; Hasle, 1977, Figs. 1, 3–10, 15–19, 22, and 23.

Width of sectors, number of complete areola rows and relationship
between number of fascicles and marginal labiate processes variable.
Position of processes at the end of central rows of fascicles more stable.
Processes readily discernible in water mounts. Pseudonodulus positioned
in the bend between valve face and mantle and thus more difficult to
discover. Central annulus indistinct or missing.

Remarks: The two taxa especially differ in size and also slightly in ecology
but not in morphology, and are regarded here as forms.

Actinocyclus octonarius Ehrenberg (Table 24)
Synonym: *Actinocyclus ehrenbergii* Ralfs in Pritchard.
References: Ehrenberg, 1838, p. 172, Plate 21, Fig. 7; Pritchard, 1861, p.
834; Hustedt, 1930, p. 525, Figs. 298–302; Hendey, 1964, p. 83, Plate 24,
Fig. 3; Villareal & Fryxell, 1983, p. 453, Figs. 1–14.

Areolation distinctly fasciculated giving the impression of the valve face
being divided into "compartments." Fascicles separated by pronounced
complete striae running from the margin to a more or less well developed
central annulus. Fasciculation accentuated by hyaline areas filling out
spaces left open by shorter rows adjacent to the complete striae. A wide
marginal zone with areolae smaller than those on the valve face. Processes
positioned at the end of edge rows and a large pseudonodulus in the bend
between valve face and mantle are readily seen with LM.

Remarks: Four varieties, in addition to the nominate variety dealt with by
Hustedt (1930), differ in size and development of the central annulus, the
amount of hyaline spaces and the width of the marginal zone. The central
annulus was shown to be highly variable in cultured material of *A.
octonarius* var. *tenellus,* whereas the valve areolation was stable
(Villareal & Fryxell, 1983).

Actinocyclus sagittulus Villareal in Villareal & Fryxell (Table 24)
Reference: Villareal & Fryxell, 1983, p. 458, Figs. 15–20.

Areolation weakly fasciculated, hyaline spaces and a marginal zone less
pronounced than in *A. octonarius,* otherwise similar to this species by the
presence of a central annulus and in morphometric data. Marginal
processes and a large pseudonodulus discernible with LM. A special
feature, located close to the pseudonodulus and considered diagnostic, is
evident with LM as a small heavily silicified arrowhead shaped region
pointing to the valve center (Villareal & Fryxell, 1983).

Actinocyclus subtilis (Gregory) Ralfs in Pritchard (Table 24)
Basionym: *Eupodiscus subtilis* Gregory.
References: Gregory, 1857, p. 501, Plate 11, Fig. 50; Pritchard, 1861,
p. 835; Hustedt, 1930, p. 534, Fig. 304.

Distinct irregularly delimited central annulus filled by closely spaced areolae. Narrow, inconspicuous bundles of more or less wavy, radial areolae rows with processes at the ends of edge rows. Valve face with incomplete striae and hyaline spaces resulting in a spotted appearance, absent on valve mantle. Large pseudonodulus positioned on valve face and labiate processes on valve mantle easily discerned with LM.

Remarks: The central annulus, the pseudonodulus and the densely areolated valve structure are the features emphasized in the original description of the species.

Distribution:

A. actinochilus—southern cold water region.

A. circellus, A. sagittulus—warm-water region?

A. octonarius and *A. subtilis*—cosmopolitan.

A. normanii—brackish water, probably cosmopolitan.

A. kützingii—known from North Atlantic coastal waters.

B. Radial areola rows parallel to edge (side) row.

Actinocyclus curvatulus Janisch in A. Schmidt **(Plate 19, Table 25)**
Synonyms: *Coscinodiscus curvatulus* var. *subocellatus* Grunow, *Actinocyclus subocellatus* (Grunow) Rattray.
References: Schmidt, 1878, Plate 57, Fig. 31; Grunow, 1884, p. 82, Plate 4, Figs. 8–16; Rattray, 1890b, p. 145; Hustedt, 1930, p. 538, Fig. 307.

Areola rows slightly curved. A process at the end of each side row of the fascicles. Central annulus irregular in shape. Areolae decreasing in size close to the margin. Small irregular pseudonodulus located close to valve mantle.

Remarks: *Actinocyclus curvatulus* was described by means of Plate 57, Fig. 31 in Schmidt's Atlas (1878), which illustrates a part of a valve ca. 110 μm in diameter. The prominent feature is the type of fasciculation.
Coscinodiscus curvatulus Grunow (Schmidt's Atlas 1878; Plate 57, Fig. 33) with several varieties has the same fasciculation as *Actinocyclus curvatulus*.

TABLE 25 Morphometric Data of *Actinocyclus* spp. with Areola Rows Parallel to Edge (Side) Row of the Fascicles

Species	Diameter (μm)	Valve areolae in 10 μm	Marginal processes in 10 μm	μm between marginal proc.
A. curvatulus	13–160	8–18	12–18	7–9
A. exiguus	6.5–13	15–18	21–24	—

Note. Proc. processes. —, no data.

According to the literature, *C. curvatulus* is smaller than *A. curvatulus* and has no pseudonodulus, except for *C. curvatulus* var. *subocellatus*, which has been transferred to *Actinocyclus* (Rattray, 1890b). Grunow (1884) recorded *C. curvatulus* var. *inermis*, var. *genuina*, var. *karianus*, and var. *minor* from the Kara Sea and Franz Josef Land. Clonal cultures isolated from ice in the Barents Sea showed the whole range of variation indicated by these varieties within one clone, and all clones had a more or less obscure pseudonodulus (E. Syvertsen, personal observations). Our proposal is therefore that *C. curvatulus* should be put into synonomy with *A. curvatulus* as a species with a much wider range of size of cells and areolae than is evident from the current literature. Other diatomists would regard *A. curvatulus* as defined here as more than one single species (Fryxell, 1990).

Actinocyclus exiguus G. Fryxell & Semina (Table 25)
Reference: Fryxell & Semina, 1981.
Small species. Areolae in irregular rows, sometimes with a tendency toward fasciculation. Valve usually heavily silicified and the labiate processes not easily seen with LM. Pseudonodulus obscure with irregular outline as in *A. curvatulus*.

Distribution:
A. exiguus—southern cold water region.
A. curvatulus—cosmopolitan (it should be noted that our circumscription of *A. curvatulus* includes diatoms common on Arctic sea ice).

C. Areola rows parallel to central and/or edge row.

Actinocyclus spiritus T. P. Watkins in Watkins & Fryxell (Table 26)
Reference: Watkins & Fryxell, 1986, p. 302, Figs. 17–23.
Similar to *A. actinochilus* but differs by the pseudonodulus being clearly visible with LM, by the lack of a hyaline band between valve areola rows

TABLE 26   Morphometric Data of *Actinocyclus* spp. with Mixture of Areola Rows Parallel to Central and Edge Rows of Fascicles

Species	Pervalvar axis (μm)	Diameter (μm)	Valve areolae in 10 μm	μm Between marginal proc.
A. spiritus	—	27–36	13–15	10–12
A. vestigulus	15	46–81	6–10	7–9

Note. Proc., processes; —, no data.

and the margin, and by the lack of isolated rows of areolae typical for *A. actinochilus*.

Actinocyclus vestigulus T. P. Watkins in Watkins & Fryxell (Table 26)
Reference: Watkins & Fryxell, 1986, p. 296, Figs. 9–16.

Sometimes with a not very distinct, more or less central annulus with the smaller areolae almost linearly arranged. Position of labiate processes in relation to the irregular fascicles obscure. Pseudonodulus difficult to see with LM.

Distribution:

A. *spiritus*—southern cold water region
A. *vestigulus*—warm water region?

How to identify: The *Actinocyclus* species presented here can hardly be identified as whole cells in water mounts since, as far as known, they all have numerous small chloroplasts lying against the cell wall and more or less the same cell shape. *Actinocyclus* spp. may easily be confused with *Coscinodiscus* spp. and *Azpeitia* spp. when seen in water mounts, and examination of cleaned material mounted in a medium of a high refractive index is necessary to ensure correct identification.

Remarks: The **pseudonodulus** was not mentioned in the original description of the genus, but was first noted by Smith (1853) according to Simonsen (1975), and is now regarded as typical for the genus. In some species the pseudonodulus is similar to a valve areola in size and structure, and/or it is positioned in the bend between valve face and mantle. This explains why it has so often been overlooked (Simonsen, 1982). The morphology of the pseudonodulus varies considerably within the genus as well as within one species. Andersen et al. (1986) showed that each labiate process of the *Actinocyclus* sp. that they examined contained material which they proposed is related to movement noted in this centric diatom. Their investigation thus throws light upon the long-speculated function of the labiate process.

Genus *Azpeitia* M. Peragallo in Tempère & Peragallo
Type: *Azpeitia temperi* M. Peragallo.
Correct name: *Azpeitia antiqua* (Pantocseck) Sims (*vide* Fryxell et al., 1986b, p. 6).
References: Fryxell et al., 1986b; Sims et al., 1989.

The genus was described from fossil material and encompasses living marine planktonic species as well. The living ones are better known as *Coscinodiscus* spp. and were transferred to *Azpeitia* by Fryxell et al. (1986b).

Generic characters:
  Marked difference in the areola patterns on valve face and mantle.

  Valves generally flat.

  Valve center generally with an annulus.

  Large nearly central labiate process on the edge of the annulus.

  A ring of large labiate processes, similar in shape, at the edge of valve
  mantle.

Characters showing differences between species:
  Valve face areolation: pattern and areola size.

  Position and shape of annulus.

  Size of labiate processes.

  Shape and structure of bordering zone between valve face and mantle
  (hyaline, marginal ridge).

Azpeitia africana (Janisch ex A. Schmidt) G. Fryxell & T. P. Watkins in
Fryxell et al. (**Plate 20, Table 27**)
Basionym: *Coscinodiscus africanus* Janisch ex A. Schmidt.
References: Schmidt, 1878, Plate 59, Figs. 24 and 25; Hustedt, 1930,
p. 428, Fig. 231; Fryxell et al. 1986b, p. 22, Figs. 22, 23, 32-1, and 32-2.
  Valves circular, sometimes slightly elliptical. Distinct external marginal
  slits leading into labiate processes. Central labiate process on edge of an
  eccentric circle of linearly arranged areolae. Areola rows radiating from
  the annulus and, in larger specimens, in spiraling rows.

Azpeitia barronii G. Fryxell & T. P. Watkins in Fryxell et al. (**Plate 20,
Table 27**)
Reference: Fryxell et al. 1986b, p. 20, Figs. 18-3, 18-5, 19–21, and 31.
  Sublinear areolation. Areolae of same size over most of the valve. Curved
  marginal ridge; some rounding between valve mantle and face. Central
  labiate process large, noticeable with LM. Central annulus possibly
  missing.

Azpeitia neocrenulata (VanLandingham) G. Fryxell & T. P. Watkins in
Fryxell et al. (**Plate 20, Table 27**)
Basionym: *Coscinodiscus neocrenulatus* VanLandingham.
Synonym: *Coscinodiscus crenulatus* Grunow *non C. crenulatus* Castracane.
References: Grunow, 1884, p. 83, Plate 4, Fig. 17; Hustedt, 1930, p. 411,
Fig. 219; VanLandingham, 1968, p. 930; Fryxell et al. 1986b, p. 18, Figs.
16, and 30-2.
  Areolation fasciculated with radial rows of areolae usually parallel to an
  edge row. Marginal labiate process and a depression on the mantle at the

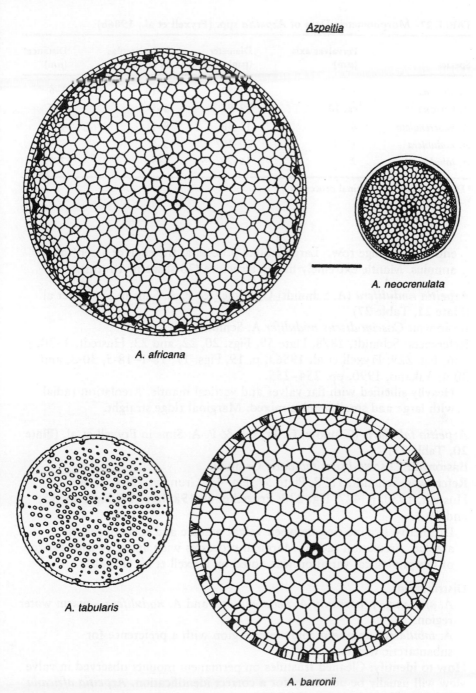

A. africana

A. neocrenulata

A. tabularis

A. barronii

PLATE 20 *Azpeitia africana, A. neocrenulata, A. tabularis,* and *A. barronii.* Valves with areolation. Scale bar = 10 μm.

TABLE 27   Morphometric Data of *Azpeitia* spp. (Fryxell et al., 1986b)

Species	Pervalvar axis (μm)	Diameter (μm)	Valve areolae in 10 μm	Distance[a] (μm)
A. africana	—[b]	30–90	5–10	3–6
A. barronii	ca. 10	39–95	3–5	4–7
A. neocrenulata	6	13–48	9–11	4–7
A. nodulifera	17–18	20–102	3–8	5–9
A. tabularis	7	16–70	5–9	7–12

[a] Distance between marginal processes.
[b] —, No data.

end of each edge row. Large labiate process at the edge of a central annulus. Mantle extremely fine in structure (20 striae in 10 μm).

Azpeitia nodulifera (A. Schmidt) G. Fryxell & P. A. Sims in Fryxell et al. (Plate 21, Table 27)
Basionym: *Coscinodiscus nodulifer* A. Schmidt.
References: Schmidt, 1878, Plate 59, Figs. 20, 22, and 23; Hustedt, 1930, p. 426, Fig. 229; Fryxell et al. 1986a, p. 19, Figs. 17, 18-1–18-5, 30-3, and 30-4; Takano, 1990, pp. 254–255.
  Heavily silicified with flat valves and vertical mantle. Areolation radial with large and small areolae mixed. Marginal ridge straight.

Azpeitia tabularis (Grunow) G. Fryxell & P. A. Sims in Fryxell et al. (Plate 20, Table 27)
Basionym: *Coscinodiscus tabularis* Grunow.
References: Schmidt, 1878, Plate 57, Fig. 43; Grunow, 1884, p. 86; Hustedt, 1930, p. 427, Fig. 230; Fryxell et al. 1986b, p. 16, Figs. 14, 15, and 30-1.
  Distinctive hyaline ring between valvar areolae and rows of mantle areolae. Areolation radial. Marginal processes wide apart. A wide range of morphological variation was noted by Fryxell et al. (1986b).

Distribution:
  A. africana, A. barronii, A. neocrenulata, and *A. nodulifera*—warm water region.
  A. tabularis—southern cold water region with a preference for subantarctic waters.
How to identify: Cleaned frustules on permanent mounts observed in valve view will usually be necessary for a correct identification. *Azpeitia africana* will be readily recognized with LM by the central area and the marginal

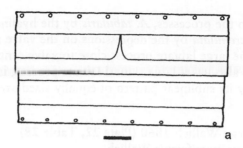

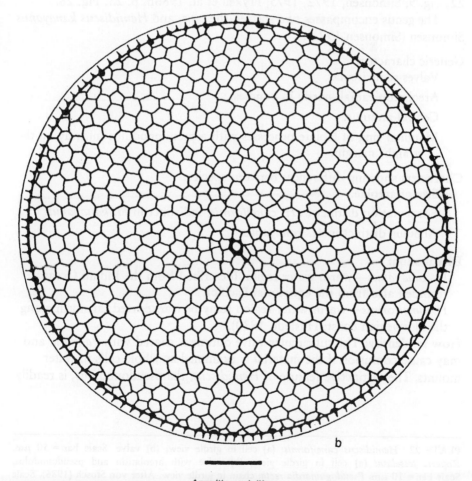

Azpeitia nodulifera

PLATE 21 *Azpeitia nodulifera*: (a) girdle view, steep mantle; (b) valve with areolation. Scale bar = 10 μm.

slits leading into the processes, *A. tabularis* by the hyaline ring close to the margin, *A. neocrenulata* by the depressions on the valve mantle, and *A. nodulifera* by the large labiate process close to valve center. *Azpeitia barronii,* which also has a large central labiate process, is distinguished from *A. nodulifera* by its sublinear pattern of equally sized areolae over most of the valve.

Genus *Hemidiscus* Wallich 1860 (Plate 22, Table 28)
Type: *Hemidiscus cuneiformis* Wallich.
References: Wallich, 1860, p. 42, Plate 2, Figs. 3 and 4; Hustedt, 1930, p. 904, Fig. 542; Cupp, 1943, p. 170, Fig. 121; Hendey, 1964, p. 94, Plate 22, Fig. 9; Simonsen, 1972, 1975; Fryxell et al. 1986b, p. 25, Fig. 26.

The genus encompasses two species, the type, and *Hemidiscus kanayanus* Simonsen (Simonsen 1972).

Generic characters:
Valves semicircular.

Areolation radial, partly in bundles.

Central annulus.

Marginal ring of labiate processes with a pseudonodulus midway on the straight margin.

Characters showing differences between species:
Size of areolae.

Fasciculation distinct in *H. cuneiformis,* indistinct in *H. kanayanus.*

Annulus distinct in *H. kanayanus.*

Distribution:
H. kanayanus—warm water region (described from the Indian Ocean, recorded also from the equatorial Pacific).
H. cuneiformis—warm water region (may be transported far north along the Norwegian coast; Hustedt, 1930).
How to identify: The subcircular valve outline is shared with *Palmeria* and may cause confusion if the material is examined as whole cells in water mounts. The distinctive feature of *Hemidiscus,* the pseudonodulus, is readily

---

PLATE 22 *Hemidiscus cuneiformis:* (a) cell in girdle view; (b) valve. Scale bar = 50 μm. *Roperia tesselata:* (a) cell in girdle view; (b) valve with areolation and pseudonodulus. Scale bar = 10 μm. *Pseudoguinardia recta:* chain in girdle view. After von Stosch (1986). Scale bar = 50 μm. *Actinoptychus senarius:* (a) girdle view with undulated valve; (b) valve with sectors. Scale bar = 10 μm.

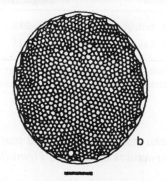

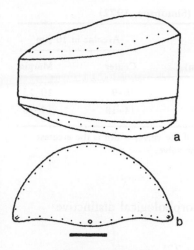

Hemidiscus cuneiformis

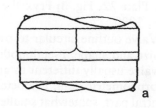

Roperia tesselata

Pseudoguinardia recta

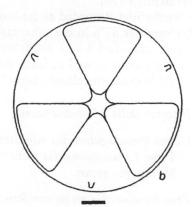

Actinoptychus senarius

TABLE 28   Morphometric Data of *Hemidiscus* spp. (Simonsen, 1972)

Species	Length (μm)	Width (μm)	Areolae in 10 μm Center	Areolae in 10 μm Margin
H. cuneiformis	58–288	32.5–158	6–9	10–13
H. kanayanus	111–225	76–120	14–18	14–18

Note. Length is measured along the more straight margin of the valve; width is the greatest distance from the curved to the more straight margin of the valve.

seen with LM in cleaned material as are the morphological distinctive structures of the two *Hemidiscus* species.

Genus *Roperia* Grunow ex Pelletan 1889 (Plate 22)
Type: *Roperia tesselata* (Roper) Grunow ex Pelletan.
Basionym: *Eupodiscus tesselatus* Roper.
References: Roper, 1858, p. 19, Plate 3, Fig. 1; Van Heurck, 1880–1885, Plate 118, Figs. 6 and 7; Hustedt, 1930, p. 523, Fig. 297; Hendey, 1964, p. 85, Plate 22, Fig. 3; Fryxell et al. 1986b, p. 24, Figs. 25, 32-3, 32-4; Lee & Lee, 1990.
    Valve outline circular to ovate. Marginal ring of processes—all similar in size and shape. Pseudonodulus prominent. Central and marginal part of valve usually different in areolation.
Morphometric data: Diameter, 40–70 μm; valve areolae, six in 10 μm in central part, somewhat smaller near margin; marginal processes, two in 10 μm (Hustedt, 1930).
Distribution: Recorded as far north as 66°N in the Norwegian Sea and as far south as 57°S in the subantarctic Pacific by Hasle (1976a) who, with doubt, classified it as a warm water species. The paucity of records from the Pacific Ocean as well as from the western part of the Atlantic Ocean north of 30°N made the classification "cosmopolitan" just as doubtful.

Incertae sedis (Hemidiscaceae)

Genus *Pseudoguinardia* von Stosch 1986
    Type: *Pseudoguinardia recta* von Stosch.
    Monotypic genus.

Pseudoguinardia recta von Stosch (Plate 22)
Reference: von Stosch, 1986, p. 307, Figs. 7–11.
    Girdle view: Cells cylindrical in long straight colonies. Bands with poroids (TEM) numerous, open, not or barely discernible with LM. Chloroplasts numerous, subglobular with a large globular pyrenoid.

Valve view: Valve outline circular. Valve face flat; mantle low and rounded. Pseudonodulus close to valve margin, discernible (LM) in valve view as a circular hyaline area. Marginal ring of small labiate processes (SEM) and one larger labiate process discernible (LM) as a small nodule in girdle view. Valve face structure composed of faint, branched lines oriented toward a diameter (a line) combining pseudonodulus and the large labiate process (LM). Scattered poroids on valve face and mantle (TEM).

Morphometric data: Pervalvar axis, 93–270 μm; diameter, 26–83 μm; ratios of pervalvar axis to diameter, 3.0–6.7 in smaller and 1.6–3.9 in larger cells.

Distribution: Warm water region—listed by von Stosch (1986) from Australian waters, the Indian Ocean, coastal waters off Northwest Africa, off Portugal, both coasts of Florida and the Mediterranean Sea. Other records are from Columbian and Californian coastal waters (G. Hasle, personal observations) and from the Gulf Stream warm core rings (Fryxell, personal communication).

How to identify: *Pseudoguinardia recta* is easily recognized in chains and as single whole cells in girdle view in water mounts; single valves are also recognizable with LM on permanent mounts.

Remarks: Von Stosch (1986) assigned *Pseudoguinardia* a provisional place in Hemidiscaceae on the assumption that the structures interpreted as a pseudonodulus and labiate processes were correctly named. As evident from the generic characters, particularly the numerous structured bands and the valve structure, it differs from the four genera assigned to Hemidiscaceae in this chapter on most or all other characters. No other described family can be pointed out as the right place for this species. In the past *P. recta* has most probably been identified as *Guinardia flaccida* or *Leptocylindrus* sp. It differs (LM) from *G. flaccida* by the straight chains, the indistinct bands, the presence of the pseudonodulus, the shape and number of labiate processes, and the orientation of the valve structure, and from *Leptocylindrus* spp. by the two latter features as well as by the greater size. A TEM picture of *P. recta*, identified as *Lauderiopsis costata*, is provided in Sundström (1986, Plate 39, Fig. 291). *Lauderiopsis costata* Ostenfeld was described with the ends of the numerous, very distinct bands forming "a slowly twined line" (Ostenfeld & Schmidt, 1901, p. 159) in contrast to the bands of *P. recta* being "hardly visible in water mounts" and with openings and ligulae forming "two opposite straight rows" (von Stosch, 1986, p. 313).

Family Asterolampraceae H. L. Smith 1872 emend. Gombos 1980

In the classification of Round et al. (1990) this family constitutes a separate order, Asterolamprales Round & Crawford, whereas Simonsen (1979) as well as Glezer et al. (1988) placed it in the order Centrales.

The family is characterized (**Fig. 14**) by "a partially areolated valve surface, which exhibits varying modes of development of hollow, hyaline rays that open to the interior of the valve by way of elongate slit-like openings and open to the exterior of the valve through holes at the marginal ends of the rays" (Gombos, 1980, p. 227). The hyaline rays extend from a hyaline (nonareolate) central area and terminate short of the valve margin in a labiate process (spine; apiculus in the older literature). The central area is traversed by a number of straight, zig-zaged, or branched lines termed umbilical lines (Greville, 1860) or separating lines (Gombos, 1980). Due to the raised hyaline rays the valve surface is radially undulated, and the cell is otherwise discoid with flat or slightly convex valves.

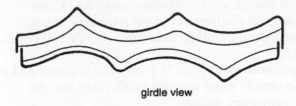

girdle view

ASTEROMPHALUS

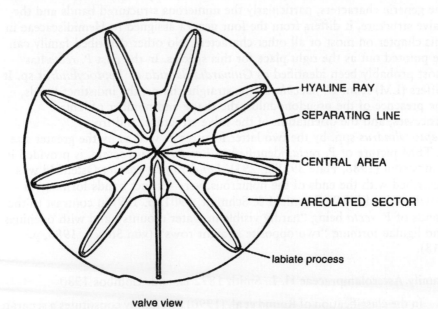

valve view

FIGURE 14 Schematic illustration of *Asteromphalus* sp. with terminology.

Two genera of this family have species represented in recent marine plankton.

KEY TO GENERA

1a. All hyaline rays of the same shape and width. . . *Asterolampra*, p. 133
1b. One of the hyaline rays narrower than the others . . . .
. . . . *Asteromphalus*, p. 133

Genus *Asterolampra* Ehrenberg 1844 (Plate 23, Table 29)
Lectotype: *Asterolampra marylandica* Ehrenberg (*vide* Boyer, 1927, p. 71).
Synonym: *Asterolampra vanheurckii* Brun (*vide* Simonsen, 1974, p. 24).
References: Ehrenberg, 1844a, pp. 73 and 76; Ehrenberg, 1844c, Fig. 10; Greville, 1860, p. 113; Wallich, 1860, p. 47, Plate 2, Fig. 5; Brun, 1891, p. 10, Plate 14, Fig. 1; Hustedt, 1930, pp. 485 and 489, Figs. 270, 271 and 274; Cupp, 1943, p. 68, Fig. 31.
  Asterolampra grevillei (Wallich) Greville and *A. marylandica* Ehrenberg
  are distinguished by the size of areolae and number of hyaline rays.
Distribution: Warm water region (e.g., Mediterranean, Indian Ocean, Gulf of California).
How to identify: The species may be identified in valve view in water mounts.

Genus *Asteromphalus* Ehrenberg 1844
Lectotype: *Asteromphalus darwinii* Ehrenberg (*vide* Boyer 1927, p. 72).
References: Ehrenberg, 1844b, pp. 198 and 200; Ehrenberg, 1844c, Figs. 1–7; Hernández-Becerril, 1991a.
The genus encompasses 10 or more species commonly recorded from marine plankton. This chapter deals with a few distinct species and describes them as they are seen with LM.

Characters showing differences between species:
  Valve outline (variable in some species).

  Number of hyaline rays (variable to a certain extent).

  Position of central area (centric or eccentric).

  Size of central area compared to valve diameter.

  Shape of separating lines.

  Shape of narrow hyaline ray within central area.

  Shape of areolated sectors.

  Size of areolae.

A. Valve outline oblong to elliptical: *A. flabellatus* and *A. sarcophagus* (Table 30).

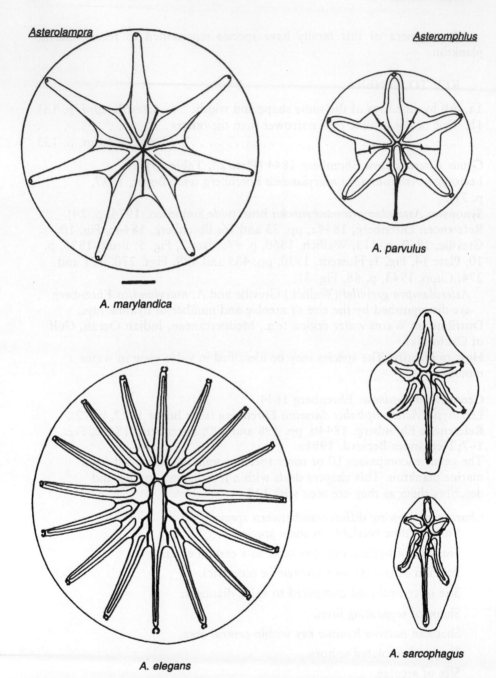

Asterolampra

Asteromphlus

A. parvulus

A. marylandica

A. elegans

A. sarcophagus

PLATE 23   *Asterolampra marylandica, Asteromphalus parvulus, A. elegans,* and *A. sarcophagus.* Valves with hyaline rays. Scale bar = 10 μm.

TABLE 29   Morphometric Data of *Asterolampra* spp. (Hustedt, 1930)

Species	Diameter (μm)	Valve areolae in 10 μm	No. of hyaline rays	Size of central area
A. grevillei	70–125	20–22	7–17 (13)	ca. 0.25 × Diameter
A. marylandica	50–150	13–17	4–12 (7)	0.17–0.33 × Diameter

Note. Numbers in parentheses are most common.

B. Valve outline broadly oval to circular: *A. arachne, A. elegans, A. heptactis, A. hookeri, A. hyalinus, A. parvulus,* and *A. roperianus* (Table 31).

A. Valve outline oblong to elliptical.

Asteromphalus flabellatus (Brébisson) Greville   (Plate 24, Table 30)
Basionym: *Spatangidium flabellatum* Brébisson.
References: Brébisson, 1857, p. 297, Plate 3, Fig. 3; Greville, 1859b, p. 160, Plate 7, Fig. 4; Hustedt, 1930, p. 498, Fig. 279; Hernández-Becerril, 1991a, p. 14, Plates 12 and 13.
   Central area slightly eccentric. Separating lines straight. Extension of the narrow ray inside the central area rectangular to club shaped. The remaining hyaline rays straight or slightly curved and narrow. Areolated sectors narrow and curved toward valve center.

Asteromphalus sarcophagus Wallich (Plate 23, Table 30)
References: Wallich, 1860, p. 47, Fig. 12; Taylor, 1967, p. 443, Plate 1, Fig. 6; Simonsen, 1974, p. 26, Plate 22, Figs. 3–6; Hernández-Becerril, 1991a, p. 30, Plates 32 and 33.
   Valve outline varying from having slightly convex margins and one apex narrower than the other to having a lateral inflation more toward the end opposite the narrow hyaline ray giving the valve a pyriform shape. Central area often extremely eccentric with a narrow hyaline ray occupying about half the length of the valve. Two rays pointing to the same end of the valve as the narrow ray and curved with the concave side facing the valve margin. Areolae decreasing in size toward the narrow apex.

Distribution: *A. flabellatus,* and *A. sarcophagus*—warm water region (Simonsen, 1974).

Asteromphalus

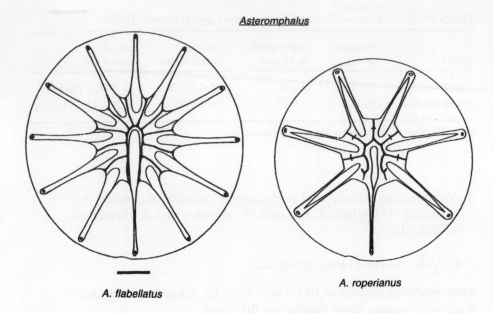

A. flabellatus

A. roperianus

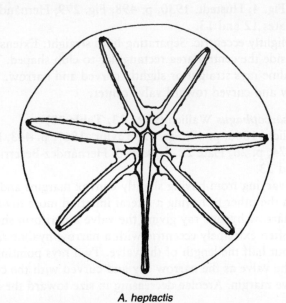

A. heptactis

PLATE 24 *Asteromphalus flabellatus, A. roperianus,* and *A. heptactis.* Valves with hyaline rays. Scale bar = 10 μm.

TABLE 30 Morphometric Data of *Asteromphalus* spp. with Oblong to Elliptical Valve Outline

Species	Diameter (μm)[a]		Sector areolae in 10 μm	No. of hyaline rays	Central area
	1	2			
A. sarcophagus	21–45	18–23	7–11	6	0.33–0.5 × diam.
A. flabellatus	40–60	24–50	16	7–11	0.4–0.6 × diam.

[a] Diameters 1 and 2, longer and shorter diameter of ellipse, respectively.

B. Valve outline broadly oval to circular.

Asteromphalus arachne (Brébisson) Ralfs in Pritchard (**Plate 25, Table 31**)
Basionym: *Spatangidium arachne* Brébisson.
References: Brébisson, 1857, p. 296, Plate 3, Fig. 1; Pritchard, 1861, p. 837, Plate 5, Fig. 66; Hustedt, 1930, p. 493, Fig. 276; Sournia, 1968, p. 25, Plate 9, Fig. 60; Hernández-Becerril, 1992a, p. 279, Figs. 1–14.
    Central area eccentric, extremely small. Hyaline rays narrow, the two farthest away from the narrow ray curve away from the valve center.

Asteromphalus elegans Greville (**Plate 23, Table 31**)
References: Greville, 1859b, p. 161, Plate 7, Fig. 6; Sournia, 1968, p. 24, Plate 9, Fig. 59; Hernández-Becerril, 1991a, p. 17, Plates 16 and 17.
    Separating lines genuflexed and sometimes once or twice forked. Extension of narrow hyaline ray inside the central area rectangular. Many hyaline rays. Finely areolated sectors, sharply truncated or somewhat pointed toward central area.

Asteromphalus heptactis (Brébisson) Ralfs in Pritchard (**Plate 24, Table 31**)
Basionym: *Spatangidium heptactis* Brébisson.
References: Brébisson, 1857, p. 296, Plate 3, Fig. 2; Pritchard, 1861, p. 838, Plate 8, Fig. 21; Hustedt, 1930, p. 494, Fig. 277; Cupp, 1943, p. 69, Fig. 32; Hendey, 1964, p. 96, Plate 24, Fig. 5; Sournia, 1968, p. 25, Plate 9, Fig. 58; Hernández-Becerril, 1991a, p. 26, Plates 28 and 29.
    Central area small compared to similar species. Separating lines bent or branched. Extension of narrow hyaline line inside the central area almost rectangular. Hyaline rays broad. Areolated sectors coarsely structured, truncated, and narrow (the width of three or four areolae) toward the central area.

Asteromphalus

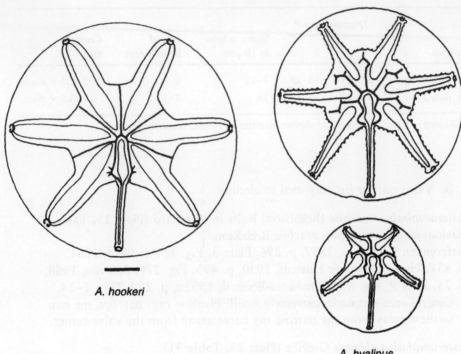

A. hookeri

A. hyalinus

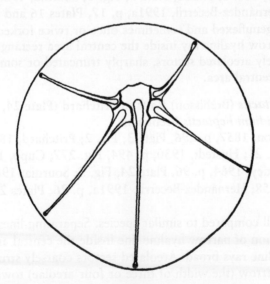

A. arachne

PLATE 25 *Asteromphalus hookeri*, *A. hyalinus*, and *A. arachne*. Valves with hyaline rays. Scale bar = 10 μm.

Asteromphalus hookeri Ehrenberg (Plate 25, Table 31)
References: Ehrenberg, 1844b, p. 200; Ehrenberg, 1844c, Fig. 3; Karsten, 1905, p. 90, Plate 8, Fig. 9; Hustedt, 1958a, p. 127, Figs. 88–90; Hernández-Becerril, 1991a, p. 23, Plates 24 and 25.
Separating lines straight. Extension of narrow hyaline ray rectangular or bell shaped. Areolated sectors wedge shaped and curved toward valve center.

Asteromphalus hyalinus Karsten (Plate 25, Table 31)
References: Karsten, 1905, p. 90, Plate 8, Fig. 15; Hustedt, 1958a, p. 128, Figs. 84–87.
Small. Central area large and more or less eccentric with convex margins toward wide areolated sectors (six to nine areolae between the hyaline rays). Separating lines genuflexed and branched. Extension of narrow hyaline ray inside central area wide at base with sudden constriction closer to valve center. Hyaline rays few and often slightly curved.

Asteromphalus parvulus Karsten (Plate 23, Table 31)
References: Karsten, 1905, p. 90, Plate 8, Fig. 14; Hustedt, 1958a, p. 128, Fig. 91.
Large central area. Separating lines broken. Areolated sectors coarsely structured and convex toward valve center.

Asteromphalus roperianus (Greville) Ralfs in Pritchard (Plate 24, Table 31)
Basionym: *Asterolampra roperiana* Greville.
References: Greville, 1860, p. 120, Plate 4, Fig. 14; Pritchard, 1861, p. 838; Sournia, 1968, p. 26, Plate 9, Fig. 61; Simonsen, 1974, p. 26, Plate 22, Fig. 2; Hernández-Becerril, 1991a, p. 21, Plates 22 and 23.

TABLE 31   Morphometric Data of *Asteromphalus* spp. with Broadly Oval to Circular Valve Outline

Species	Diameter (μm)	Sector areolae in 10 μm	No. of hyaline rays	Size of central area
A. arachne	40–60	7–8	5	0.17–0.14 × diam.
A. elegans	100–180	15–20 ?[a]	11–26	0.33 × diam.
A. heptactis	42–175	5–7	7	0.25–0.33 × diam.
A. hookeri	25–60	5–9 ?	6–9	0.33–0.50 × diam.
A. hyalinus	15–32	8–12	3–5	0.50 × diam.
A. parvulus	22–48	8–10	6	0.50–0.75 × diam.
A. roperianus	80–120	9–11	7	0.33 × diam.

[a]?, Data uncertain

Central area comparatively small. Separating lines obscurely bent and sometimes equipped with small branches in the middle. Extension of narrow hyaline ray inside central area conspicuous in shape being semicircular at the top, then contracted, and widely expanded. Areolated sectors moderately wide (six or seven areolae between two hyaline rays) and truncated toward valve center.

Distribution:

A. arachne, A. elegans, and *A. roperianus*—warm water region (Simonsen, 1974).

A. heptactis—temperate [Davis Strait—Cape of Good Hope (Hendey, 1964)].

A. hookeri, A. hyalinus, and *A. parvulus*—southern cold water region (Hustedt, 1958a).

How to identify: Examinations of cells in valve view, whole cells, or single valves are required for identification purposes. Water mounts may be sufficient in some cases; permanent mounts of cleaned material are recommended.

Remarks: Special distributional as well as taxonomic problems connected with the three species classified as Antarctic (southern cold water region) should be noted. *Asteromphalus hookeri* has been reported as common in north temperate waters (Hendey, 1964), and our own observations indicate that the species described by Karsten (1905) as *A. hyalinus* may be present in the Norwegian and the Barents Seas as well as in the North Pacific. Examination of Antarctic material showed that *A. parvulus* and larger specimens of *A. hyalinus* are not readily distinguished, and a comparison with Ehrenberg's (1844c) illustrations of *A. darwinii* and *A. rossii* indicates a possible conspecificity between these four species. Hernández-Becerril (1991a, 1992a) provided detailed information on the valve structure seen with SEM and on the taxonomy and biogeography of the genus (information that has not been included in this chapter). It should be noted, however, that *A. sarcophagus* was placed in the subgenus *Liriogramma* (Kolbe) Hernández-Becerril, and that reinstatement of the genus *Spatangidium* Brébisson with *S. arachne* as the type was suggested.

Family Heliopeltaceae H. L. Smith 1872

Glezer et al. (1988) and Simonsen (1979) regarded *Actinoptychus* as well as *Aulacodiscus* as members of this family. Round et al. (1990) put *Aulacodiscus* Ehrenberg in the family Aulacodiscaceae (Schütt) Lemmermann and retained *Actinoptychus* Ehrenberg in Heliopeltaceae. The two genera are primarily benthic. Two species are often recorded from plankton and are included here.

Actinoptychus senarius (Ehrenberg) Ehrenberg (Plate 22)
Basionym: *Actinocyclus senarius* Ehrenberg.
Synonym: *Actinoptychus undulatus* (Bailey) Ralfs in Pritchard.
References: Ehrenberg, 1838, p. 172, Plate 21, Fig. 6; Ehrenberg, 1843,
p. 400, Plate 1,1, Fig. 27, Plate 1,3, Fig. 21; Pritchard, 1861, p. 839, Plate 5,
Fig. 88; Hustedt, 1930, p. 475, Fig. 264; Cupp, 1943, p. 67, Fig. 29, Plate 5,
Fig. 1; Hendey, 1964, p. 95, Plate 23, Figs. 1 and 2; Takano, 1990, pp. 258–259.
   Cells disc shaped with valves divided into sectors, usually six, alternately
raised and depressed. A central nonareolated area, hexagonal in outline.
Each of the raised sectors with a labiate process with an external tube.
Areolation coarse and irregular. Chloroplasts large and numerous.
Morphometric data: Diameter, 20–150 μm; areolae, four to seven in
10 μm.
Distribution: Cosmopolitan?

Aulacodiscus argus (Ehrenberg) A. Schmidt
Basionym: *Tripodiscus argus* Ehrenberg.
References: Ehrenberg, 1844a, p. 73; Schmidt, 1886, Plate 107, Fig. 4; Hustedt,
1930, p. 503, Fig. 281.
   Cells box shaped with flat or slightly convex valves. Valve structure complex
and dense consisting of one layer with angular areolae overlying a layer with
finer pores. Three to six marginal labiate processes with prominent pear-
shaped extensions raised above valve surface. Chloroplasts fairly large
and roundish.
Morphometric data: Diameter, 80–260 μm; areolae, four to six in
10 μm.
Distribution: Cosmopolitan?
How to identify: The species may be identified in valve view in water mounts
or, in critical cases, as valves cleaned of organic matter and mounted in a
medium not necessarily of a high refractive index.

Suborder Rhizosoleniineae

Family Rhizosoleniaceae Petit 1888

   Dactyliosolen, Guinardia, and *Rhizosolenia* are the recent, marine, planktonic genera of this family as circumscribed by Simonsen (1979) and Glezer
et al. (1988); Round et al. (1990) included the more recently described genera
Proboscia, Pseudosolenia, and *Urosolenia.* In the latter classification Rhizosoleniaceae was placed in the order Rhizosoleniales P. Silva in the new subclass
Rhizosoleniophycidae Round & Crawford.
   Sundström (1986) stressed the necessity of splitting Rhizosoleniaceae and
establishing new families since he found that his circumscription of *Rhizosolenia,* the type of the family, comprised species that differed considerably from

those in *Guinardia* and *Dactyliosolen*. He suggested that Rhizosoleniaceae should be confined to *Rhizosolenia* Brightwell, *Pseudosolenia* Sundström, and possibly *Proboscia* Sundström. However, no new families were described.

For a lack of better alternatives this chapter will deal with the heterogeneous family Rhizosoleniaceae *sensu lato* comprising *Proboscia, Pseudosolenia, Rhizosolenia sensu lato, Guinardia,* and *Dactyliosolen.* Most of the data and information on *Rhizosolenia sensu stricto, Pseudosolenia,* and *Proboscia* are from Sundström (1986). *Rhizosolenia* and *Proboscia* species present in polar regions are dealt with specifically by Priddle et al. (1990).

Characters of Rhizosoleniaceae *sensu lato:*
　Cells in chains.
　Cells cylindrical.
　A single process with internal labiate, sometimes more tubular structure.
　Unipolar valve symmetry.
　Numerous small chloroplasts.
　Resting spores seldom.

Terminology specific to Rhizosoleniaceae (mainly after Sundström, 1986) (**Fig. 15**):

Contiguous area—part of the ventral side of the valve contiguous with the adjacent valve of linked cells, usually delimited by low marginal ridges.

Claspers—pair of membranous structures usually continuous with marginal ridges of contiguous area clasping otaria of the adjacent valve of linked cells. In *Proboscia* the structures clasping the distal part of the proboscis of linked cells.

Otarium—one of the pair of membranous costae that occur opposite each other at or near the base of the external process, previously called a wing.

Proboscis—elongated part of the valve with truncate tip; the distal part fits into a groove on the adjacent valve of linked cells.

Segment—band—copula—single element of the girdle.

KEY TO GENERA

1a. Valves conical to subconical, girdle segments (bands) generally with loculate areolae (SEM) 2
1b. Valves flat or rounded, girdle segments (bands) with poroid areolae (SEM) 4
2a. Valves with an external process 3
2b. Valves with a proboscis and no process . . . . *Proboscia,* p. 159

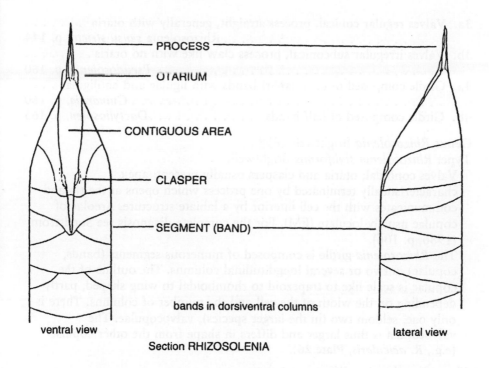

PROCESS

OTARIUM

CONTIGUOUS AREA

CLASPER

SEGMENT (BAND)

bands in dorsiventral columns

ventral view

lateral view

Section RHIZOSOLENIA

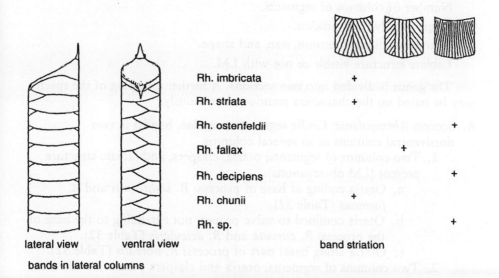

Rh. imbricata   +

Rh. striata   +

Rh. ostenfeldii   +

Rh. fallax   +

Rh. decipiens   +

Rh. chunii   +

Rh. sp.   +

lateral view   ventral view   band striation

bands in lateral columns

Section IMBRICATAE

FIGURE 15 Schematic illustrations of *Rhizosolenia* spp. with terminology and type of band striation for the section *Imbricatae*.

3a. Valves regular conical, process straight, generally with otaria . . . .
. . . . Rhizosolenia *sensu stricto,* p. 144
3b. Valves irregular subconical, process claw like with no otaria . . . .
. . . . *Pseudosolenia,* p. 160
4a. Girdle composed of open (split) bands with ligulae and antiligulae . . . .
. . . . *Guinardia,* p. 160
4b. Girdle composed of half bands . . . .*Dactyliosolen,* p. 165

Genus *Rhizosolenia* Brightwell 1858
Type: *Rhizosolenia styliformis* Brightwell.
Valves conoidal; otaria and claspers usually present; apex
characteristically terminated by one process which opens at the tip and
communicates with the cell interior by a labiate structure. Areolae of
copulae usually loculate (EM). For the complete diagnosis see Sundström
(1986, p. 106).
The *Rhizosolenia* girdle is composed of numerous segments (bands,
copulae) in two or several longditudinal columns. The outline of the
copulae is scale like to trapezoid to rhomboidal to wing shaped, partly
depending on the width of the cell and the number of columns. There is
only one, seldom two (in the larger species), valvocopulae. The
valvocopula is thus larger and differs in shape from the other copulae
(e.g., *R. acicularis,* **Plate 26**).

Characters showing differences between species:
Segments (bands) in dorsiventral or lateral columns.

Number of columns of segments.

Shape of valve and process.

Otaria: position, extension, size, and shape.

Labiate structure visible or not with LM.

The genus is divided into two sections. A further grouping of the species
may be based on the characters mentioned previously.

A. Section *Rhizosolenia:* Girdle segments (copulae, bands) in two
dorsiventral columns or in several columns.
    1. Two columns of segments; otaria, claspers, and labiate structure
present (LM observations).
        a. Otaria ending at base of process: *R. styliformis* and *R.
formosa* (**Table 32**).
        b. Otaria confined to valve proper, not extending to the base of
the process: *R. curvata* and *R. acicularis* (**Table 32**).
        c. Otaria along basal part of process: *R. borealis* (**Table 32**).
    2. Two columns of segments, otaria and claspers present or absent;
labiate structure (LM observations) present; dimorphism:
R. antennata, R. hebetata, R. polydactyla, and *R. sima* (**Table 33**).

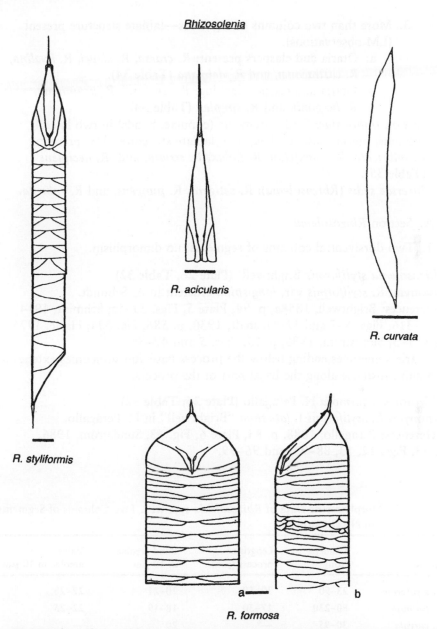

Rhizosolenia

R. aicularis

R. curvata

R. styliformis

a———— b

R. formosa

PLATE 26 *Rhizosolenia styliformis:* epitheca in ventral and hypotheca in lateral view. Otaria and claspers evident on epitheca. Scale bar = 20 μm. *Rhizosolenia aicularis:* valve with valvocopula, ventral view. After Sundström (1986). Scale bar = 50 μm. *Rhizosolenia curvata:* lateral view. From Hasle, 1968b. Scale bar = 50 μm. *Rhizosolenia formosa:* (a) ventral view; (b) lateral view. After Sundström (1986). Scale bar = 50 μm.

3. More than two columns of segments—labiate structure present (LM observations).
   a. Otaria and claspers present: *R. crassa, R. clevei, R. hyalina, R. castracanei,* and *R. debyana* (Table 34).
   b. Otaria and claspers lacking: *R. temperei, R. acuminata, R. bergonii,* and *R. simplex* (Table 34).

B. Section *Imbricatae:* Girdle segments (copulae, bands) in two lateral columns; otaria, claspers (LM), and labiate structure (EM) present: *R. imbricata, R. ostenfeldii, R. fallax, R. striata,* and *R. decipiens* (Table 35).

C. *Incertae sedis* (*Rhizosolenia*): *R. setigera, R. pungens,* and *R. robusta.*

A. Section *Rhizosolenia.*

1. Two dorsiventral columns of segments; no dimorphism.

Rhizosolenia styliformis Brightwell (Plate 26, Table 32)
Synonym: *R. styliformis* var. *longispina* Hustedt in A. Schmidt.
References: Brightwell, 1858a, p. 94, Plate 5, Figs. 5a–5e; Schmidt, 1914, Plate 316, Figs. 5–7 and 12; Hustedt, 1930, p. 586, Fig. 334; Hasle, 1975, Figs. 1–3; Sundström, 1986, p. 15, Figs. 5 and 47–56.
Otaria sometimes ending below the process base and sometimes extending a short distance along the basal part of the process.

Rhizosolenia formosa H. Peragallo (Plate 26, Table 32)
Synonym: *R. styliformis* f. *latissima* "Brightwell" in H. Peragallo.
References: Peragallo, 1888, p. 83, Plate 6, Fig. 43; Sundström, 1986, p. 33, Figs. 12, 13, 88–93, and 96–99.

TABLE 32   Morphometric Data of *Rhizosolenia* spp. with Two Columns of Segments and No Dimorphism

Species	Diameter (μm)	Length of process (μm)	Band areolae in 10 μm	Valve areolae in 10 μm[a]
R. styliformis	23–90	30–50	20–21	27–28
R. formosa	80–230	17–30	18–19	22–25
R. curvata	30–95	—[b]	20	—
R. acicularis	13–40	up to 50	20–21	—
R. borealis	13–65	15–28	15–17	22–24

[a] Measured with TEM.
[b] —, No data.

Bands "low." Apex shallow. Otaria sometimes ending slightly below the process base and sometimes extending a short distance along the basal part of the process.

Rhizosolenia curvata Zacharias (Plate 26, Table 32)
Synonym: *Rhizosolenia curva* Karsten.
References: Zacharias, 1905, p. 120; Karsten, 1905, p. 97, Plate 11, Fig. 2; Hasle, 1968b, p. 7, Plate 9, Fig. 8, Map 2; Sournia et al., 1979, p. 191, Fig. 20; Sundström, 1986, p. 20, Figs. 57–63; Priddle et al., 1990, p. 118, Plate 15.2, Fig. 3.
Cells curved. Otaria comparatively small, ending below valve apex.

Rhizosolenia acicularis Sundström (Plate 26, Table 32)
Reference: Sundström 1986, p. 22, Figs. 6 and 64–69.
Otaria pointed, narrow, ending below process base. Perhaps closely related to *R. styliformis* but differs in shape and position of otaria and in distribution.

Rhizosolenia borealis Sundström (Plate 27, Table 32)
Synonym: *Rhizosolenia styliformis* var. *oceanica* Wimpenny *pro parte*.
References: Wimpenny, 1946, p. 279, Text Fig. 1d; Hasle, 1975, p. 104, Figs. 8–20; Sundström, 1986, p. 30, Figs. 10, 11, and 80–87.
Otaria extending to about half the length of the thicker basal part of the process.
Remarks: The reasons for regarding *R. borealis* and *R. polydactyla* as two distinct species are the difference in the distribution and the fact that *R. polydactyla* is dimorphic and *R. borealis* is not (Sundström, 1986, p. 32).

Distribution:
  R. borealis—northern cold water region.
  R. styliformis—northern part of North Atlantic Ocean.
  R. acicularis and *R. formosa*—warm water region.
  R. curvata—Subantarctic/Antarctic convergence.

2. Two dorsiventral columns of segments: dimorphism.

Rhizosolenia antennata (Ehrenberg) Brown f. *antennata* (Plate 27, Table 33)
Basionym: *Dicladia antennata* Ehrenberg.
Synonyms: *Rhizosolenia bidens* Karsten *pro parte*; *Rhizosolenia hebetata* f. *bidens* Heiden in Heiden & Kolbe.
References: Ehrenberg, 1844b, p. 201; Karsten, 1905, p. 98, Plate 9, Fig. 13; Van Heurck, 1909, Plate 4, Fig. 64; Brown, 1920, p. 233, Fig. 8; Heiden & Kolbe, 1928, p. 519, Figs. 158–161; Sundström, 1986, p. 42, Figs. 115, 119, and 120; Priddle et al., 1990, p. 117, Plate 15.1, Figs. 5a–5c.
Two processes, no otaria.

Rhizosolenia

R. borealis

f. polydactyla

f. squamosa

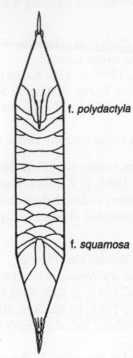

R. polydactyla

R. antennata
f. antennata

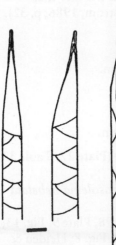

R. hebetata
f. hebetata

R. hebetata
f. semispina

R. sima
f. sima

R. sima
f. silicea

TABLE 33   Morphometric Data of *Rhizosolenia* spp. with Two Columns of Segments and Dimorphism

Species	Diameter (μm)	Length of process (μm)	Band areolae in 10 μm
R. antennata f. *antennata*	18–45	30–80	23–26
R. antennata f. *semispina*	6.5–42	—[a]	—
R. hebetata f. *hebetata*	15–44	15–25	—
R. hebetata f. *semispina*	4.5–25	—	28–30
R. polydactyla	15–105	18–34	20–23[b]
R. sima f. *sima*	12–39	—	31–36
R. sima f. *silicea*	18–40	—	30

[a] —, No data.
[b] *R. polydactyla* 26–28 valve areolae in 10μm, measured with TEM, information on the other species missing.

Rhizosolenia antennata f. *semispina* Sundström (Table 33)
Commonly identified as *Rhizosolenia hebetata* f. *semispina*.
Reference: Sundström, 1986, p. 44, Figs. 20, 114, and 116–118.
   Pointed otaria extending at least 3 μm along the basal part of the process.

Rhizosolenia hebetata Bailey f. *hebetata* (Plate 27, Table 33)
Synonym: *Rhizosolenia hebetata* f. *hiemalis* Gran.
References: Bailey, 1856, p. 5, Plate 1, Figs. 18 and 19; Gran, 1904, pp. 524–527, Plate 27, Figs. 9 and 10; Hustedt, 1930, p. 590, Fig. 337; Cupp, 1943, p. 88, Fig. 50a; Drebes, 1974, p. 56, Fig. 41; Sundström, 1986, p. 47, Figs. 18, 112, and 113.
   Process heavily silicified, no otaria.

Rhizosolenia hebetata f. *semispina* (Hensen) Gran (Plate 27, Table 33)
Basionym: *Rhizosolenia semispina* Hensen.

PLATE 27   *Rhizosolenia borealis*: ventral view. Scale bar = 50 μm. *Rhizosolenia polydactyla*: upper theca f. *polydactyla* and the lower theca f. *squamosa*. Scale bar = 50 μm. *Rhizosolenia antennata* f. *antennata*: Scale bar = 20 μm. *Rhizosolenia hebetata*: f. *hebetata*: (a) ventral and (b) lateral view. Scale bar = 20 μm. *Rhizosolenia hebetata* f. *semispina*: ventral view. Scale bar = 20 μm. *Rhizosolenia sima* f. *sima*: ventral view. Scale bar = 50 μm. *Rhizosolenia sima* f. *silicea*: ventral view. Scale bar = 50 μm. All figures after Sundström (1986).

References: Hensen, 1887, p. 84, Plate 5, Fig. 39; Gran, 1904, Plate 27, Figs. 11 and 12; Sundström, 1986, p. 48, Figs. 19, 20, 114, and 116–118.
Pointed otaria extending at least 3 μm along the basal part of the process.
Remarks: Although R. *antennata* f. *semispina* and R. *hebetata* f. *semispina* are almost indistinguishable with both LM and EM, Sundström (1986) regarded the species as separate taxa because of the widely different distribution and the great morphological dissimilarity between their nominate forms.

Rhizosolenia polydactyla Castracane f. *polydactyla* (Plate 27, Table 33)
Synonym: *Rhizosolenia styliformis* var. *oceanica* Wimpenny *pro parte*.
References: Castracane, 1886, p. 71, Plate 24, Fig. 2; Wimpenny, 1946, p. 279, Text Fig. 1e; Sundström, 1986, p. 24, Figs. 7–9, 70–73, 74, 76, 77, and 79.
Otaria extending to about half the length of the thicker basal part of the process.

Rhizosolenia polydactyla f. *squamosa* Sundström
Reference: Sundström, 1986, p. 26, Figs. 8, 9, 73, 75, and 78.
Bands in two or a multiple of two columns. Process coarsely structured; otaria lacking. "Exogenous resting spore" of f. *polydactyla?*

Rhizosolenia sima Castracane f. *sima* (Plate 27, Table 33)
References: Castracane, 1886, p. 71, Plate 24, Fig. 11; Van Heurck, 1909, Plate 4, Fig. 70; Sundström, 1986, p. 62, Figs. 28 and 144–149.
Process basal part bulbous; distal tube thin walled. Otaria large, extending along the basal part of the process.

Rhizosolenia sima f. *silicea* Sundström (Plate 27, Table 33)
References: Van Heurck, 1909, Plate 4, Fig. 71; Sundström, 1986, p. 63, Figs. 29 and 150–154.
Process heaviliy silicified, usually with uneven longitudinal ridges; no otaria. "Resting spore" of f. *sima?*

Distribution:
R. *antennata*, R. *polydactyla*, and R. *sima*—southern cold water region.
R. *hebetata*—northern cold water region.

3. More than two columns of segments.

Rhizosolenia crassa Schimper in Karsten (Plate 28, Table 34)
References: Karsten, 1905, p. 99, Plate 11, Fig. 6; Sundström, 1986, p. 59, Figs. 26, 27, and 139–143.
Two and four columns of bands within a single cell. Otaria prominent, bordering valve apex and basal part of process; distal margin concave.

Rhizosolenia clevei Ostenfeld var. *clevei* (Plate 28, Table 34)
References: Ostenfeld, 1902, p. 229, Fig. 6; Sournia, 1968, p. 78, Plate 10, Fig. 68; Sundström, 1984, p. 348, Figs. 1 and 4–9; Sundström, 1986, p. 53, Figs. 21, 121, 122, and 125.
Otaria arising at valve apex, extending along the basal part of the process. Host for the blue–green *Richelia intracellularis*.

Rhizosolenia clevei var. *communis* Sundström (Plate 28, Table 34)
References: Sundström, 1984, p. 348, Figs. 2, 3, and 10–15; Sundström, 1986, p. 54, Figs. 22, 23, 123, 124, 126, and 127.
Bands in two columns, otherwise as the nominate variety.

Rhizosolenia hyalina Ostenfeld in Ostenfeld & Schmidt.
(Plate 28, Table 34)
Synonym: *Rhizosolenia pellucida* Cleve.
References: Cleve, 1901a, p. 56, Plate 8, Fig. 4; Ostenfeld & Schmidt, 1901, p. 160, Fig. 11; Sournia, 1968, p. 79, Plate 3, Fig. 19; Sundström, 1986, p. 76, Figs. 34 and 190–194.
Process slightly bent near the end of the otaria. Otaria narrow, extending along process for ca. 4–6 μm.

Rhizosolenia castracanei H. Peragallo var. *castracanei* (Plate 28, Table 34)
References: Peragallo, 1888, p. 83, Plate 6, Fig. 42; Cupp, 1943, p. 94, Fig. 54; Sundström, 1986, p. 37, Figs. 15, 16, 102–104, and 108–111.
Bands coarsely structured. Valve apex shallow. Otaria confined to valve proper.

Rhizosolenia castracanei var. *neglecta* Sundström (Plate 28, Table 34)
Reference: Sundström, 1986, p. 39, Figs. 17 and 105–107.
Bands in two columns. Valves of narrow cells similar to those of R. *styliformis*.

Rhizosolenia debyana H. Peragallo (Plate 28, Table 34)
References: Peragallo, 1892, p. 111, Plate 2, Figs. 7 and 7a; Sundström, 1986, p. 57, Figs. 24, 25, and 128–138.
Valve and otaria similar in shape to those of R. *castracanei* var. *castracanei*; band areolae smaller.
Remarks: The number of columns of segments and the size of cell diameter are related to a certain extent. In *Rhizosolenia crassa* two columns of segments were observed in cells of all sizes and four columns in cells of 115–140 μm in diameter but not in the narrower ones.

Rhizosolenia temperei H. Peragallo (Plate 29, Table 34)
References: Peragallo, 1888, p. 83, Plate 5, Fig. 40; Hustedt, 1930, p. 605, Fig. 349; Sundström, 1986, p. 66, Figs. 30 and 155–163.
Valve contour sigmoid. Distal part of valve abruptly narrowed. No otaria.

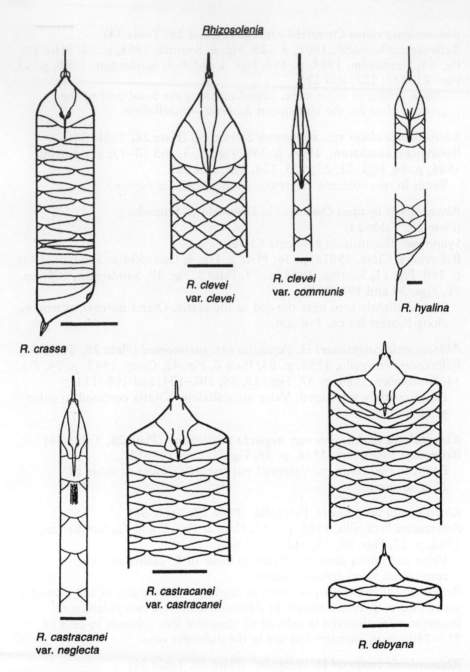

Rhizosolenia

R. crassa

R. clevei
var. clevei

R. clevei
var. communis

R. hyalina

R. castracanei
var. neglecta

R. castracanei
var. castracanei

R. debyana

TABLE 34   Morphometric Data of *Rhizosolenia* spp. with More Than Two Columns of Segments

Species	Diameter (μm)	Length of process (μm)	Band areolae in 10 μm[a]	Valve areolae in 10 μm[b]
Otaria present				
R. crassa	100–165	ca. 50	22–24	27
R. clevei				
var. clevei	80–250	—	—	—
var. communis	7–55	—	19–25	—
R. hyalina	9–60	up to 40	31–36	—
R. castracanei				
var. castracanei	108–250	25–35	9–12	ca. 24
var. neglecta	14–127	—	10–13	—
R. debyana	180–310	Short	19–22	20–26
Otaria absent				
R. temperi	125–278	10–30	20–21	16–18
R. acuminata	50–190	10–40	18–22	—
R. bergonii	9–115	10–20	19–24	—
R. simplex	5–48	15–20	30–33	24–28

[a,b] Measured with TEM, except those on the bands of *Rhizosolenia castracanei* and on the valves of *R. temperi*.

Rhizosolenia acuminata (H. Peragallo) H. Peragallo in H. & M. Peragallo (Plate 29, Table 34)
Basionym: *Rhizosolenia temperei* var. *acuminata* H. Peragallo.
References: Peragallo, 1892, p. 110, Plate 2, Fig. 4; H. & M. Peragallo, 1897–1908, p. 463, Plate 123, Figs. 7 and 8; Hustedt, 1930, p. 605, Fig. 350; Cupp, 1943, p. 94, Fig. 53; Sundström, 1986, p. 69, Figs. 31 and 165–176.
Valve and valvocopula deeply conical; distal part usually slightly narrowed. No otaria.

PLATE 28   *Rhizosolenia crassa:* epitheca in ventral view, hypotheca in lateral view. Scale bar = 100 μm. *Rhizosolenia clevei* var. *clevei:* part of theca with many rows of segments. Scale bar = 100 μm. *Rhizosolenia clevei* var. *communis:* part of theca with two columns of segments. Scale bar = 20 μm. *Rhizosolenia hyalina:* parts of thecae in ventral and lateral view. Scale bar = 20 μm. *Rhizosolenia castracanei* var. *neglecta:* two dorsiventral columns of segments. Scale bar = 20 μm. *Rhizosolenia castracanei* var. *castracanei:* many columns of segments. Scale bar = 100 μm. *Rhizosolenia debyana:* upper partial theca in ventral and lower part in dorsal view. Scale bar = 100 μm. All figures after Sundström (1986).

Rhizosolenia

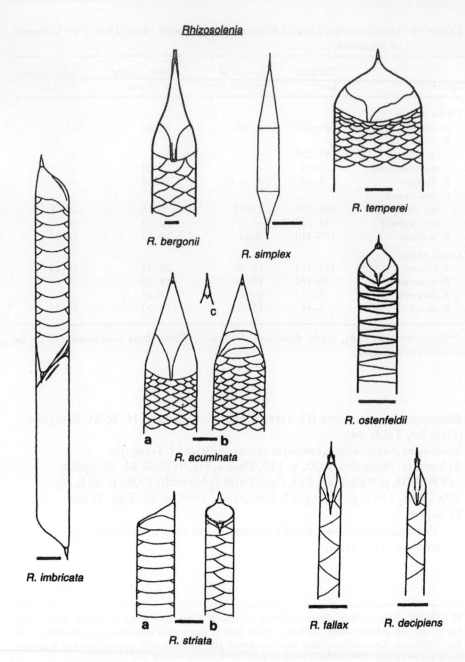

R. bergonii

R. simplex

R. temperei

R. imbricata

R. acuminata

a b

R. ostenfeldii

R. striata

a b

R. fallax

R. decipiens

Rhizosolenia bergonii H. Peragallo (Plate 29, Table 34)
References: Peragallo, 1892, p. 110, Plate 2, Fig. 5; Hustedt, 1930,
p. 575, Fig. 327; Cupp, 1943, p. 81, Fig. 43; Sundström, 1986, p. 72, Figs.
32, 33, and 177–189.

Narrow cells always with four columns of bands. Process tip appearing cleft in LM. Valve and valvocopula deeply conical and apex long and narrow, usually heavily silicified. No otaria.

Rhizosolenia simplex Karsten (Plate 29, Table 34)
References: Karsten, 1905, p. 95, Plate 10, Fig. 1; Sournia et al., 1979,
p. 191, Figs. 16 and 24; Sundström, 1986, p. 78, Figs. 35 and 195–199.

Bands indistinct. Process conical, narrowing abruptly and terminating in a short narrow tube. Valve and process a continuous cone. No otaria.

Distribution:

R. temperi—Mediterranean.

R. acuminata, R. bergonii, R. clevei, R. hyalina, R. castracanei, and *R. debyana*—warm water region.

R. simplex and *R. crassa*—southern cold water region.

B. Section *Imbricatae.* Two lateral columns of segments.

Rhizosolenia imbricata Brightwell (Plate 29, Table 35)
Synonyms: *Rhizosolenia shrubsolei* Cleve; *R. imbricata* var. *shrubsolei* (Cleve) Schröder.
References: Brightwell, 1858a, p. 94, Plate 5, Fig. 6; Cleve, 1881, p. 26; Schröder, 1906, p. 346; Sundström, 1986, p. 80, Figs. 200–208.

Cross section slightly elliptical. Valve obliquely conical. Process swollen basally, narrowing abruptly into a distal tube. Otaria small, extending along the swollen part of the process (not shown on Plate 29).

Rhizosolenia ostenfeldii Sundström (Plate 29, Table 35)
Reference: Sundström, 1986, p. 87, Figs. 37 and 218–226.

PLATE 29 *Rhizosolenia imbricata:* two cells in lateral view. After Brightwell (1858a). Scale bar = 20 μm. *Rhizosolenia bergonii:* stout process. After Sundström (1986). Scale bar = 20 μm. *Rhizosolenia simplex:* whole frustule. After Sundström (1986). Scale bar = 50 μm. *Rhizosolenia temperei:* ventral view. Many columns of segments. After Sundström (1986). Scale bar = 50 μm. *Rhizosolenia acuminata:* (a) ventral and (b) dorsal view; (c) detail of labiate structure. After Sundström (1986). Scale bar = 50 μm. *Rhizosolenia ostenfeldii:* ventral view. After Sundström (1986). Scale bar = 50 μm. *Rhizosolenia striata:* (a) lateral and (b) ventral view. After Sundström (1986). Scale bar = 20 μm. *Rhizosolenia fallax:* ventral view. After Sundström (1986). Scale bar = 20 μm. *Rhizosolenia decipiens:* ventral view. After Sundström (1986). Scale bar = 20 μm.

TABLE 35 Morphometric Data of *Rhizosolenia* spp. with Two Lateral Columns of Segments

Species	Diameter or apical axis (μm)	Length of processes (μm)	Band areolae in 10 μm
R. imbricata	2.5–57	8–18	10–18
R. ostenfeldii	32–56	11–15	21–25
R. fallax	3–23	8–12	17–25
R. striata	11–110	10–18	6–12
R. decipiens	3.5–20	9–14	22–30

Cross-section circular. Valve shallow. Process swollen basally; distal part tubular. Otaria extending along the basal part of the process; distal margin straight to weakly concave.

Rhizosolenia fallax Sundström (Plate 29, Table 35)
Reference: Sundström, 1986, p. 89, Figs. 38 and 227–233.
   Cross-section circular to slightly elliptical. Valve obliquely conical. Process swollen basally, narrowing abruptly into the distal tube. Otaria small, extending along the swollen part of the process.

Rhizosolenia striata Greville (Plate 29, Table 35)
References: Greville, 1864, p. 234, Plate 3, Fig. 4; Sundström, 1986, p. 84, Figs. 36 and 209–217.
   Cross-section elliptical. Valve shallow. Process triangular in outline with lateral edges usually weakly concave. Otaria small, extending along the lower part of the process.

Rhizosolenia decipiens Sundström (Plate 29, Table 35)
Reference: Sundström, 1986, p. 92, Figs. 39a, 39b, and 234–240.
   Cross-section usually circular. Valve obliquely conical. Basal part of the process conical, gradually narrowing into a distal tube of roughly equal length. Otaria narrow, extending along the conical part of the process.

Distribution:
   R. imbricata—widely distributed although not in polar regions.
   R. fallax—temperate to tropical waters.
   R. decipiens, R. ostenfeldii, and *R. striata*—warm water region.
Remarks: The bands of *R. imbricata* and *R. striata* have a similar striation pattern (Fig. 15) but differ in the shape of the valves and the basal part of the processes. The bands of *R. ostenfeldii* and *R. decipiens* have a similar striation pattern (Fig. 15) but these species also differ in the shape of the

valves and the basal part of the processes. The bands of R. *fallax* have a striation pattern different from that of the other species (Fig. 15), whereas the valve process is similar to those of R. *imbricata* and R. *ostenfeldii*. Although the structure may be too fine for the striae to be counted using LM, the striation pattern is visible with LM in all species of this section. *Rhizosolenia chunii* Karsten was also allocated to the section *Imbricatae* by Priddle et al. (1990, p. 118, Plate 15.3, Fig. 2).

C. *Incertae sedis* (*Rhizosolenia*)

Rhizosolenia setigera Brightwell (Plate 30)
References: Brightwell, 1858a, p. 95, Plate 5, Fig. 7; Hustedt, 1930, p. 588, Fig. 336; Cupp, 1943, p. 88, Fig. 49; Drebes, 1974, p. 52, Fig. 40; Sundström, 1986, p. 104, Figs. 286–288; Priddle et al., 1990, p. 120, Plate 15.4, Fig. 5.
Cell wall weakly silicified; structure not resolved with LM. Areolae poroid (SEM). Two dorsiventral columns of bands. Valves conical. External process long, almost straight along the whole length, slightly wider for some distance from the base, and gently tapering toward the tip. No otaria. Labiate structure present (G. Hasle, personal observations).

Rhizosolenia pungens Cleve-Euler (Plate 30)
References: Cleve-Euler, 1937, p. 43, Fig. 10; Drebes, 1974, p. 52, Fig. 34a.
Basal part of external process narrow, abruptly swollen for about half its length. Otherwise, same as R. *setigera*.
Morphometric data: *Rhizosolenia setigera*—4–25 μm in diameter
R. *pungens*—8–14 μm in diameter.

Distribution:
R. *pungens*—mainly brackish water (Swedish and Danish coastal waters, Kiel Bay, Brazil, Japan).
R. *setigera*—cosmopolitan, probably absent from polar waters.
Remarks: *Rhizosolenia setigera* and R. *pungens* would fit fairly well into *Pseudosolenia* except for the different shape of the valves and processes. Some species in *Rhizosolenia sensu stricto* have no otaria (Table 34) like R. *setigera* and R. *pungens*. Unlike R. *setigera* and R. *pungens* they have more than two columns of segments and also often have a large diameter. In addition, R. *setigera* forms resting spores in pairs distinctly different from the vegatative cells—a feature which, together with the poroid areolae of the bands (SEM), Sundström (1986) regarded as decisive for excluding this species from R. *sensu stricto*. The reason for listing R. *pungens* as a separate species and not a form or variety of R. *setigera* is the unique shape of the process.

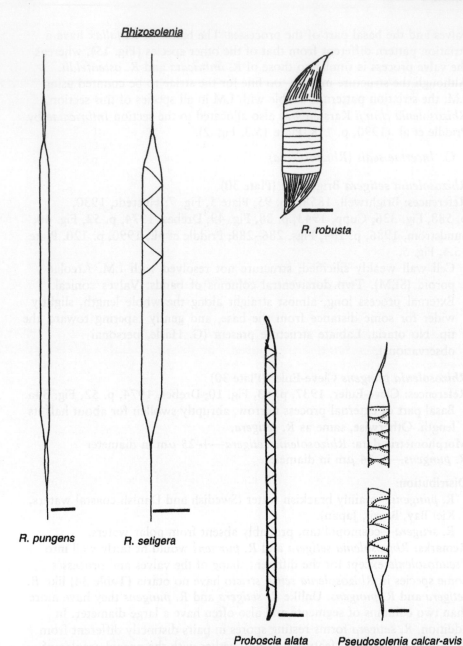

Rhizosolenia

R. robusta

R. pungens   R. setigera

Proboscia alata   Pseudosolenia calcar-avis

PLATE 30 _Rhizosolenia pungens:_ characteristic process, lower part swollen. After Cleve-Euler (1951). Scale bar = 20 μm. _Rhizosolenia setigera:_ long, tapering process. Scale bar = 20 μm. _Rhizosolenia robusta:_ frustule with characteristic valves and bands. After Cupp (1943). Scale bar = 100 μm. _Proboscia alata:_ proboscis. After Brightwell (1858a). Scale bar = 20 μm. _Pseudosolenia calcar-avis:_ parts of thecae. Scale bar = 20 μm.

Rhizosolenia robusta Norman in Pritchard (Plate 30)
References: Pritchard, 1961, p. 866, Plate 8, Fig. 42; Hustedt, 1930,
p. 578, Fig. 330; Cupp, 1943, p. 83, Fig. 46; Hasle, 1975, p. 110, Figs.
42–47; Sundström, 1986, p. 104, Figs 289 and 290.
  Cells cresent shaped or S shaped. Two columns of typically collar-shaped
  segments. Areolae loculate (SEM). Valves deeply convex or conical,
  curved, and with longitudinal lines. External process consisting of a
  needle-shaped part (?) extending from a short, wider tube (SEM).
Morphometric data: Diameter, 48–400 μm; cell length, 0.5–1 mm; valve
areolae, 19–20 in 10 μm; band areolae, 24–26 in 10 μm.

Distribution: Warm water region.
Remarks: Sundström's (1986) argument for not including this species in *R.*
sensu stricto, despite the presence of areolated segments, was the shape of
the valve apex and the tubular process.

Genus *Proboscia* Sundström 1986
Type: *Proboscia alata* (Brightwell) Sundström.
  Valves subconical, terminating in a proboscis. No process. Auxospores
  terminal. Claspers usually present.

Proboscia alata (Brightwell) Sundström (Plate 30)
Basionym: *Rhizosolenia alata* Brightwell.
References: Brightwell, 1858a, p. 95, Plate 5, Fig. 8; Drebes, 1974, p. 57,
Figs. 39c and 39d; Sundström, 1986, p. 99, Figs. 258–266; Jordan et al.,
1991, p. 65, Figs. 1–9.
  Bands in two columns, numerous, rhomboidal, with pores (LM) scattered
  between loculate (SEM) areolae. Proboscis, tip truncate, short longitudinal
  slit just below tip.
Morphometric data: Diameter, 2.5–13 μm.

Distribution: "The biogeographical limits cannot be determined without
further research" (Sundström, 1986, p. 101).
Remarks: Sundström (1986, p. 99) wrote: "*Proboscia* comprises the generic type
P. alata and an undetermined number of species commonly referred to in the
literature as *Rhizosolenia alata, Rh. arafurensis, Rh. indica, Rh. inermis, Rh.*
truncata, etc." Jordan et al. (1991) examined Antarctic phytoplankton and made
two new combinations, *Proboscia inermis* (Castracane) Jordan & Ligowski and
Proboscia truncata (Karsten) Nöthig & Ligowski. *Proboscia inermis* has
moderately prolonged valves with stout, wedge-shaped, markedly truncate
proboscis (Jordan et al., 1991, p. 66, Figs. 10–18). *Proboscia truncata* has either
somewhat rounded valves with short, truncate, oblique proboscis, or tapered
valves with long, straight or slightly oblique proboscis somewhat wider near its
tip (Jordan et al., 1991, p. 70, Figs. 19–29). *Proboscia eumorpha* Takahashi,
Jordan & Priddle and *P. subarctica* Takahashi, Jordan & Priddle were described
from subarctic waters (Takahashi et al., 1994), and the genus thus includes five
modern species.

Genus *Pseudosolenia* Sundström 1986
Type: *Pseudosolenia calcar-avis* (Schultze) Sundström.
Monospecific genus.

***Pseudosolenia calcar avis* (Schultze) Sundström (Plate 30)**
Basionym: *Rhizosolenia calcar avis* Schultze.
References: Schultze, 1858, p. 339; Schultze, 1859, p. 19, Figs. 5–8;
Hustedt, 1930, p. 592, Fig. 339; Cupp, 1943, p. 89, Fig. 51; Sundström,
1986, p. 95, Figs. 40–46 and 247–257.
   Bands in two, or a multiple of two, columns. Areolae poroid (SEM).
   Valves subconical. Process claw like. No otaria. Contiguous area roughly
   sigmoid. Labiate structure different from that in *Rhizosolenia* but similar
   to the two larger labiate processes in *Coscinodiscus* (SEM).
Morphometric data: Diameter, 4.5–190 μm; process, 28–52 μm; areolae on
bands 28–32 in 10 μm measured with TEM.
Distribution: Warm water region, occasionally in temperate waters.
Remarks: The shape of the valve and the external as well as the internal
parts of process and the poroid areolae distinguish the genus from
Rhizosolenia sensu stricto.

How to identify: Most of the *Rhizosolenia* species as well as *Proboscia* and
Pseudosolenia may be identified in girdle view in water mounts. In critical
cases in which information on the otaria is urgent, valves cleaned of organic
matter and mounted in a medium of a high refractive index may be
examined in valve view.

Genus *Guinardia* H. Peragallo 1892
Lectotype: *Guinardia flaccida* (Castracane) H. Peragallo (*vide* Round et al.,
1990, pp. 326 and 691).

Generic characters:
   Low, open ligulate bands usually distinct with LM.

   Band structure composed of regular rectangular poroids (EM).

   Valve structure composed of faint ribs radiating from the single process
      (occasionally revealed with LM).

   External process tube.

   External depression for process of sibling valve.

Characters showing differences between species:
   Shape of chains (loose or close set).

   Shape of cells (straight or curved).

   Shape of valves (flat or convex).

   Location of process (central or marginal).

   Shape of external part of process.

KEY TO SPECIES

1a. Process maringal 2
1b. Process central . . . . G. *cylindrus* (Cleve) Hasle comb. nov
2a. External part of process tubular 3
?h Enternal part of process low, inconspicuous. . . . . : 4
3a. External part of process coarse . . . .
. . . . G. *striata* (Stolterfoth) Hasle comb. nov.
3b. External part of process thin, oblique to pervalvar axis . . . .
. . . . G. *delicatula* (Cleve) Hasle comb. nov.
4a. Marginal process noticeable as a tooth and/or indent . . . .
. . . . .G. *flaccida* (Castracane) H. Peragallo
4b. External part of process short, stout (dome shaped). . . . .
. . . . G. *tubiformis* (Hasle) Hasle comb. nov.

A. Process central: G. *cylindrus* (Table 36).
B. Process marginal: G. *delicatula*, G. *flaccida*, G. *striata*, and G. *tubiformis* (Table 36).

A. Process central.

Guinardia cylindrus (Cleve) Hasle comb. nov. **(Plate 31, Table 36)**
Basionym: *Rhizosolenia cylindrus* Cleve.
References: Cleve, 1897a, p. 24, Plate 2, Fig. 12; Hustedt, 1930, p. 572, Fig. 325; Cupp, 1943, p. 80, Fig. 42; Hendey, 1964, p. 148, Plate 3, Fig. 3; Sundström, 1986, p. 103, Figs. 276–278; Priddle et al., 1990, p. 120, Plate 15.4, Figs. 1 and 2.
Solitary or loose chains adjacent cells in touch by the processes. External impression upon adjacent cell discernible with LM. Cells straight. Bands distinct with tapering ends. Valves convex or rounded. External process tube long and curved No observations on its internal part.

B. Process marginal.

Guinardia delicatula (Cleve) Hasle comb. nov. **(Plate 31, Table 36)**
Basionym: *Rhizosolenia delicatula* Cleve.
References: Cleve, 1900b, p. 28, Fig. 11; Hustedt, 1930, p. 577, Fig. 328; Cupp, 1943, p. 83, Fig. 44; Hendey, 1964, p. 147, Plate 4, Fig. 2; Drebes, 1974, p. 49, Fig. 35a; Sundström, 1986, p. 103, Figs. 272 and 273.
Cells in close set straight chains. Bands indistinct. Valves flat and only slightly rounded at the edges. External process thin, pointed, and oblique to the pervalvar axis. Internal part of process probably labiate shaped (G. Hasle, personal observations). External depression with claspers (EM; Sundström, 1986).

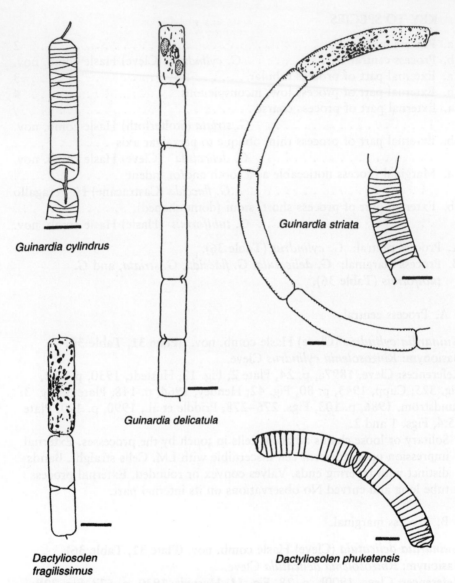

Guinardia cylindrus

Guinardia striata

Guinardia delicatula

Dactyliosolen
fragilissimus

Dactyliosolen phuketensis

PLATE 31   *Guinardia cylindrus:* Long processes. *Guinardia delicatula:* marginal process evident. After Cupp (1943). *Guinardia striata:* part of helical chain. Details of bands. After Cupp (1943. *Dactyliosolen fragilissimus:* Central process. After Cupp (1943). *Dactyliosolen phuketensis:* stout process and distinct bands. After Sundström (1986). Scale bars = 20 μm.

TABLE 36 Morphometric Data of *Guinardia* spp.

Species	Pervalvar axis (μm)	Diameter (μm)
Group A		
G. cylindrus	up to 300	8–50
Group B		
G. delicatula	3–5 × Diameter	9–22
G. flaccida	1.5 to Several × diameter	25–90
G. striata	up to 250	6–45
G. tubiformis	50–100	3–10

Guinardia flaccida (Castracane) H. Peragallo (Plate 32, Table 36)
Basionym: *Rhizosolenia* (?) *flaccida* Castracane.
References: Castracane, 1886, p. 74, Plate 29, Fig. 4; Peragallo, 1892, p. 107, Plate 1, Figs. 3–5; Hustedt, 1930, p. 562, Fig. 322; Cupp, 1943, p. 78, Fig. 40; Hendey, 1964, p. 141, Plate 5, Fig. 5; Drebes, 1974, p. 58, Fig. 43a; Hasle, 1975, p. 116, Figs. 64, 65, and 81–89; Takano, 1990, pp. 260–261.

Girdle view: Cells in close set chains; cells straight or slightly curved. Bands distinct and collar shaped. Valves flat or slightly concave. Chloroplasts rounded; more or less lobed or cleft plates each with a distinct pyrenoid.
Valve view: Valve structure visible with LM; process distinct; external part a curved tube, one end connected with a labiate-shaped internal part and the other end in touch with the external valve surface (EM). Indentation close to the process probably giving space for process of adjacent cell.

Guinardia striata (Stolterfoth) Hasle com. nov. (Plate 31, Table 36)
Basionym: *Eucampia striata* Stolterfoth.
Synonym: *Rhizosolenia stolterfothii* H. Peragallo.
References: Stolterfoth, 1879, p. 836, Figs. a and b; Peragallo, 1888, p. 82, Plate 6, Fig. 44; Hustedt, 1930, p. 578, Fig. 329; Cupp, 1943, p. 83, Fig. 45; Hendey, 1964, p. 148, Plate 4, Fig. 5; Drebes, 1974, p. 49, Fig. 35b; Hasle, 1975, p. 113, Figs. 66–73; von Stosch, 1986, p. 319, Figs. 13 and 14; Sundström, 1980, p. 580, Figs. 2 and 4; Sundström 1986, p. 103, Figs. 274 and 275.

Cells in close set curved, often spiraling chains. Bands with tapering ends discernible in water mounts. Valves flat and rounded at the edges.
External process fitting into a depression upon the adjacent valve (LM; Hendey, 1964).

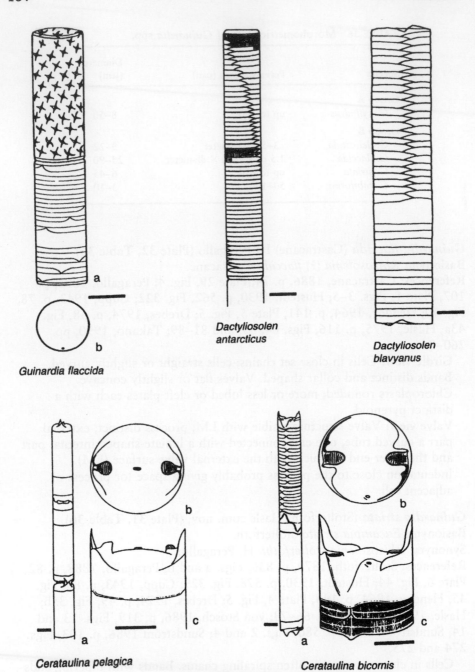

Guinardia flaccida

Dactyliosolen antarcticus

Dactyliosolen blavyanus

Cerataulina pelagica

Cerataulina bicornis

Guinardia tubiformis (Hasle) Hasle comb. nov. (Table 36)
Basionym: *Rhizosolenia tubiformis* Hasle.
References: Hasle, 1975, p. 115, Text Fig. 1, Fig. 80; Sundström, 1986, p. 104, Figs. 282–285.
Cells in close set straight chains. Bands with tapering ends discernible with LM. Valves flat. Depression corresponding to external part of the adjacent valve probably present (Sundström, 1986).

Distribution:
G. *cylindrus*—warm water region.
G. *delicatula*—warm water region (?) to temperate.
G. *flaccida* and G. *striata*—cosmopolitan but not seen in polar waters.
G. *tubiformis*—southern cold water region.
How to identify: The species may be identified in girdle view in water mounts or in a medium of higher refractive index or possibly as dried material mounted in air. Cleaned material mounted in a medium of a high refractive index is needed to reveal the valve structure and the internal shape of the process if visible at all with LM.

Genus *Dactyliosolen* Castracane 1886
Type: *Dactyliosolen antarcticus* Castracane.

Generic characters:
Ends of half of the bands wedge shaped.
Valve structure composed of branching ribs radiating from the single process (often too delicate to be revealed with LM).

Characters showing differences between species:
Band ends in an oblique or a straight line.
Bands ribbed or with regular rectangular poroid areolae.
Shape of chains (loose or close set).
Shape of cells (straight or curved).
Shape of valves (flat or weakly convex).

PLATE 32 *Guinardia flaccida:* (a) short chain with star-shaped chloroplasts; (b) valve with marginal process. Scale bar = 20 μm. *Dactyliosolen antarcticus:* partial chain in girdle view, band ends in an oblique line. Scale bar = 20 μm. *Dactyliosolen blavyanus:* cell with band ends in a straight line. After Hustedt (1930). Scale bar = 20 μm. *Cerataulina pelagica:* (a) partial chain in girdle view. Scale bar = 20 μm; (b) valve in valve view; (c) valve in girdle view. Costate ocelli, spines, and central labiate process. Scale bar = 10 μm. From Hasle & Syvertsen (1980). *Cerataulina bicornis:* (a) two cells in girdle view. Scale bar = 20 μm. (b) Valve in valve view with costate ocelli, spines (black), and marginal labiate process; (c) valve with two bands in girdle view. Scale bar = 10 μm. From Hasle & Syvertsen (1980).

Location of process (marginal or central).

Process with or without an external tube.

Shape of internal part of process (may be absent).

KEY TO SPECIES

1a. Band ends in a straight line   2
1b. Band ends in an oblique line   4
2a. Process marginal   3
2b. Process central with external tube fitting into a depression (pocket) in adjacent cell . . . . *D. fragilissimus* (Bergon) Hasle comb. nov.
3a. No external process tube . . . . *D. blavyanus* (H. Peragallo) Hasle
3b. Short external process tube fitting into a depression (pocket) in adjacent cell . . . . *D. phuketensis* (Sundström) Hasle comb. nov.
4a. Process central, no external tube . . . . *D. tenuijunctus* (Manguin) Hasle
4b. Process marginal, occasionally closer to valve center, no external tube . . . . *D. antarcticus* Castracane

A. Band ends in an oblique line: *D. antarcticus,* and *D. tenuijunctus* (Table 37).
B. Band ends in a straight line: *D. blavyanus, D. fragilissimus,* and *D. phuketensis* (Table 37).

A. Band ends in an oblique line.

Dactyliosolen antarcticus Castracane (Plate 32, Table 37)
References: Castracane, 1886, p. 75, Plate 9, Fig. 7; Hustedt, 1930, p. 556, Fig. 316; Cupp, 1943, p. 76, Fig. 37; Hasle, 1975, p. 119, Figs. 90–100, 109–112.

TABLE 37   Morphometric Data of *Dactyliosolen* spp.

Species	Pervalvar axis (μm)	Diameter (μm)	Bands in 10 μm	Band ribs in 10 μm
Group A				
D. antarcticus	Up to 140	13–90	2–3	4–15
D. tenuijunctus	18–55	6–12	5–9	28–36
Group B				
D. blavyanus	—	6–38	4–5	28–32
D. fragilissimus	42–300	8–70	—	—
D. phuketensis	76–236	4.5–54	—	—

Note. —, No data.

Girdle view: Close set chains; cells straight; valves flat. Bands coarsely ribbed (LM).

Valve view: Valve ribs visible with LM. Internal part of process a low tube with thickened rim (LM).

Dactyliosolen tenuijunctus (Manguin) Hasle (Table 37)
Basionym: *Rhizosolenia tenuijuncta* Manguin.
References: Manguin, 1957, p. 118, Plate 2, Fig. 15; Manguin, 1960, p. 270, Figs. 76 and 77; Hasle, 1975, p. 122, Figs. 114–120; Priddle & Fryxell, 1985, p. 60.

Cells straight, slightly curved, or undulated; valve face flat(?). Bands delicately ribbed. Internal part of process tubular (LM).

B. Band ends in a straight line.

Dactyliosolen blavyanus (H. Peragallo) Hasle (Plate 32, Table 37)
Basionym: *Guinardia blavyana* H. Peragallo.
References: Peragallo, 1892, p. 107, Plate 1, Figs. 1 and 2; Hustedt, 1930, p. 564, Fig. 323; Hasle, 1975, p. 121, Figs. 101–108 and 113; von Stosch, 1986, p. 317, Fig. 12.

Girdle view: Close set chains; cells straight; valve surface flat with a marginal indentation at the process. Bands delicately ribbed and discernible with LM.

Valve view: Faint valve ribs (LM). Process marginal; internal part tubular (LM).

Dactyliosolen fragilissimus (Bergon) Hasle comb. nov. (Plate 31, Table 37)
Basionym: *Rhizosolenia fragilissima* Bergon.
References: Bergon, 1903, p. 49, Plate 1, Figs. 9 and 10; Hustedt, 1930, p. 571, Fig. 324; Cupp, 1943, p. 80, Fig. 41; Drebes, 1974, p. 48, Figs. 34b and 34c; Hasle, 1975, p. 114, Figs. 61, 62, and 74–78; Sundström, 1986, p. 103, Figs. 268 and 269; Takano, 1990, pp. 262–263.

Girdle view: Cells straight and united in loose chains by the central part of valve surface. Bands difficult to see with LM; rectangular poroid areolae (EM). Valves weakly convex. External process tube narrow, slightly curved fitting into a depression (pocket) in adjacent cell, and visible with LM as an indentation near the tube.

Valve view: Valve structure too delicate to be resolved with LM. Internal part of process lacking (?).

Dactyliosolen phuketensis (Sundström) Hasle comb. nov. (Plate 31, Table 37)
Basionym: *Rhizosolenia phuketensis* Sundström.
References: Sundström, 1980, p. 579, Figs. 1 and 3; von Stosch, 1986, p. 323, Figs. 15–17; Sundström, 1986, p. 103, Figs. 270 and 271.

Cells curved; in close set curved or spiraling chains. Bands visible with
LM; rectangular poroid areolae (TEM). Internal part lacking (?).

Distribution:

  D. tenuijunctus—southern cold water region.

  D. blavyanus—warm water region to temperate including the Oslofjord,
  southern Norway in summer.

  D. phuketensis—warm water region to temperate including North Sea
  and the Skagerrak (Sundström, 1986).

  D. fragilissimus—cosmopolitan (?).

  D. antarcticus—cosmopolitan (?), especially important in the southern
  cold water region.

How to identify: The species may be identified in girdle view in water
mounts, in a medium of a higher refractive index, or possibly as dried
material mounted in air. Material cleaned of organic matter and mounted in
a medium of a high refractive index is needed in order to see the valve
structure and the shape of the internal part of the process.

Remarks: *Dactyliosolen blavyanus* differs from the other species of the
genus by being the only one known to form (endogenous) resting spores and
by having the protoplast concentrated to the middle of the cell (von Stosch,
1986). *Guinardia delicatula, D. fragilissimus, Leptocylindrus danicus,* and
Cerataulina pelagica often appear together as do *G. striata* and *G.
phuketensis.*

They are distinguished by:

  Shape of valve.

  The presence/absence, shape, and location of process.

  Sha

  Bands, shape, ends in straight/oblique line.

Suborder Biddulphiineae

Family Hemiaulaceae Jousé, Kisselev, & Poretsky 1949

The four recent planktonic genera dealt with here were all placed in this
family by Simonsen (1979) as well as by Round et al. (1990). In Round et al.
(1990), Hemiaulaceae was placed in the new order Hemiaulales Round &
Crawford. Glezer et al. (1988) placed *Eucampia* together with *Odontella* and
Biddulphia in Biddulphiaceae Kützing.

Terminology specific to Hemiaulaceae

Aperture—space between valves of adjacent cells in chains.

Elevation—raised portion of valve wall, not projecting laterally beyond
  valve margin, may bear some special structure; otherwise with the same
  structure as the valve.

Horn—long, narrow elevation.

Ocellus—plate of silica, pierced by closely packed holes, normally with a thickened structureless rim.

Costate ocellus—similar to an ocellus or pseudocellus (see Eupodiscaceae) but with siliceous ribs between rows of pores.

Common morphological characters:

Close set chains formed by apposition of tips of elevations.

Bipolar symmetry.

Bipolar elevations.

Apertures between cells in colonies formed by elevations.

A single labiate process.

Poroid areolae.

Numerous small disc-like chloroplasts.

KEY TO GENERA

1a. Elevations with spine or wing-like prolongations or with pointed ends. 2

1b. Elevations with obtuse ends 3

2a. Elevations short, top plates ribbed (costate ocellus) and with spine or wing-like prolongations, apertures between cells in chains narrow . . . . *Cerataulina*, p. 169

2b. Elevations usually long and slender with pointed ends and no ribbed top plate, apertures between cells in chains mostly wide . *Hemiaulus*, p. 176

3a. Pervalvar axis usually short, chains sometimes twisted . . . . *Climacodium*, p. 172

3b. Pervalvar axis longer, chains often helically curved, elevations with ribbed top plate (costate ocellus) . . . . *Eucampia*, p. 172

Genus *Cerataulina* H. Peragallo ex Schütt 1896 (**Plates 32 and 33, Table 38**)

Type: *Cerataulina bergonii* (H. Peragallo) Schütt.

TABLE 38 Morphometric Data of *Ceratulina* spp. (Hasle & Syvertsen, 1980)

Species	Pervalvar axis (μm)	Diameter (μm)	Valve striae in 10 μm
C. bicornis	87–200	5–75	18–30
C. dentata	26–88	5–12	ca. 20
C. pelagica	55–120	7–56	14–25

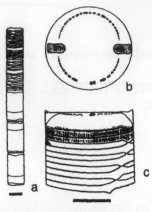

Cerataulina dentata

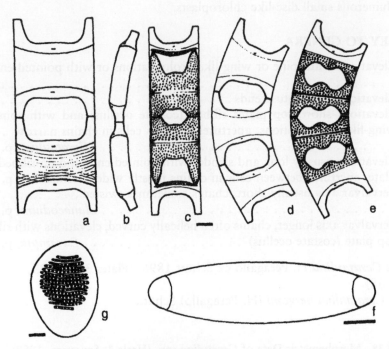

Climacodium frauenfeldianum

Eucampia antarctica

PLATE 33 *Cerataulina dentata:* (a) chain in girdle view. Scale bar = 20 μm; (b) valve with costate ocelli, marginal labiate process, and dentate margin; (c) valve in girdle view with bands. Scale bar = 10 μm. From Hasle & Syvertsen (1980). *Climacodium frauenfeldianum:* twisted chain. After Cupp (1943). Scale bar = 20 μm. *Eucampia antarctica:* (a and d) chains in broad and (b) narrow girdle view, (c) with semiendogenous resting spores and, (e) with "winter stages." Scale bar = 10 μm; (f) valve with marginal labiate process. Scale bar = 10 μm; (g) costate ocellus. Scale bar = 1 μm. From Syvertsen & Hasle (1983).

Basionym: *Cerataulus bergonii* H. Peragallo.
Correct name: *Cerataulina pelagica* (Cleve) Hendy.
References: Ehrenberg, 1845a, p. 365; Greville, 1866, p. 83, Figs. 22–28; Cleve, 1889, p. 54; Peragallo, 1892. p. 103, Plate 13, Figs. 15 and 16; Schütt, 1896, p. 95, Fig. 165; Ostenfeld & Schmidt, 1901, p. 153, Fig. 7; Hustedt, 1930, p. 869, Fig. 517 (vegetative cells of *C. pelagica* and resting spore of *C. bicornis*); Hendey, 1937, p. 279; Cupp, 1943, p. 167, Fig. 117; Hendey, 1964, p. 113, Plate 6, Fig. 4; Hasle & Syvertsen, 1980; Hasle, 1983c; Hasle & Sims, 1985; Takano, 1990, pp. 270–273.

Generic characters:

Cells twisted about pervalvar axis.

Valves circular to subcircular.

Elevations low with wing-like extensions.

KEY TO SPECIES

1a. Bands indistinct or not resolved in water mounts, conspicuous wing-like extensions of elevations. 2
1b. Bands distinct, elevations and wings inconspicuous, valve margin dentate, labiate process submarginal, areola array fan shaped and oriented toward the process. . . . .*C. dentata* Hasle in Hasle & Syvertsen
2a. Broad wing-like extensions of the elevations fitting into V-shaped deep furrows on mantle of adjacent valve, labiate process marginal, areola array irregular . . . .
. . . .*C. bicornis*[11] (Ehrenberg) Hasle in Hasle & Sims
2b. Wing-like extensions of elevations less conspicuous, labiate process central or subcentral, regular areolae rows oriented toward the process . . . . . .
. . . . *C. pelagica* (Cleve) Hendey

Distribution:

C. bicornis and *C. dentata*—coastal warm water region.

C. pelagica—cosmopolitan.

How to identify: The species may be identified in girdle view in water mounts. In critical cases examination of valve structure made on cleaned material mounted in a medium of a high refractive index may be necessary.

Remarks: *Cerataulina bicornis* is the only species of the genus that has been found with resting spores. These are morphologically distinctly different from the vegetative cells and were described as *Syringidium bicornis* (Ehrenberg, 1845a; Hasle & Sims, 1985).

[11] Basionym: *Syringidium bicorne* Ehrenberg. Synonyms: *Cerataulina compacta* Ostenfeld in Ostenfeld & Schmidt; *Cerataulina daemon* (Greville) Hasle in Hasle & Syvertsen.

TABLE 39   Morphometric Data of *Climacodium*
spp. (Hustedt, 1930)

Species	Pervalvar axis (μm)	Apical axis (μm)
C. biconcavum	ca. 60	35–65
C. frauenfeldianum	10–30	75–225

Genus *Climacodium* Grunow 1868 (Plate 33, Table 39)
Type: *Climacodium frauenfeldianum* Grunow.
References: Grunow, 1868, p. 102, Plate 1a, Fig. 24; Cleve, 1897a, p. 22,
Plate 2, Figs. 16 and 17; Hustedt, 1930, p. 776, Figs. 453 and 454; Cupp,
1943, p. 147, Fig. 105.

Generic characters:

Cells straight but usually forming somewhat twisted chains.

Valves elliptical.

Apertures between cells in chains large.

KEY TO SPECIES

1a. Valve surface between elevations flat, apertures right angled to oblong
and larger in pervalvar direction than in the cell proper . . . .
. . . . *C. frauenfeldianum* Grunow
1b. Valve surface between elevations concave, apertures elliptical lanceolate,
usually smaller than the cell proper . . . . *C. biconcavum* Cleve

Distribution: Warm water region.
How to identify: The species may be identified as colonies or whole cells in
water mounts.

Genus *Eucampia* Ehrenberg 1839 (Plates 33, 34, and 35, Table 40)
Type: *Eucampia zodiacus* Ehrenberg.
References: Ehrenberg, 1841a, p. 151, Plate 4, Fig. 8; Cleve, 1873a, p. 7, Plate
1, Fig. 6; Van Heurck, 1880–1885, Plate 95 bis, Fig. 5; Cleve, 1896a, p. 10,

---

PLATE 34   *Eucampia cornuta:* (a and b) two chains of unequal width in girdle view. Scale
bar = 10 μm; (c) valve in girdle view; (d) valve in valve view (central labiate process). Scale
bar = 10 μm; (e) costate ocellus. Scale bar = 1 μm. *Eucampia groenlandica:* (a–c) chains in broad
and   (d) narrow girdle views. Scale bar = 10 μm; (e) valve with central labiate process. Scale
bar = 10 μm; (f) costate ocellus. Scale bar = 1 μm.   From Syvertsen & Hasle (1983).

Eucampia

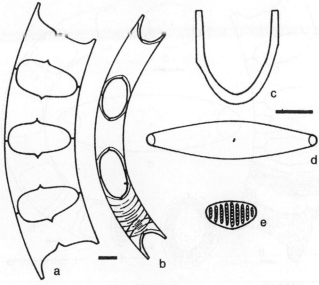

E. cornuta

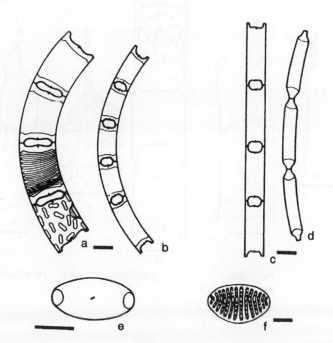

E. groenlandica

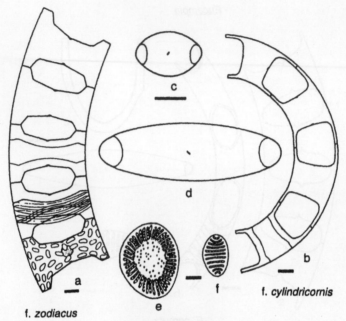

f. zodiacus

f. cylindricornis

Eucampia zodiacus

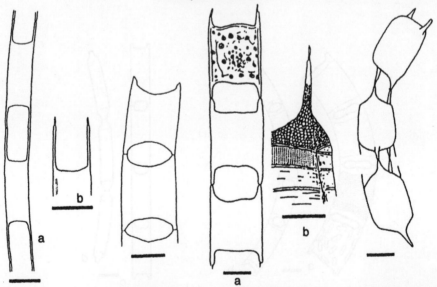

Hemialus
hauckii

Hemialus
membranaceus

Hemiaulus sinensis

Hemiaulus indicus

TABLE 40   Morphometric Data of *Eucampia*
spp. (Syvertsen & Hasle, 1983)

Species	Apical axis (μm)	Valve areolae in 10 μm
E. antarctica	18–92	3–10
E. cornuta	13–60	7–21
E. groenlandica	10–33	38–42 (SEM)
E. zodiacus	8–80	10–20

Plate 2, Fig. 10; Castracane, 1886, pp. 97 and 98, Plate 18, Figs. 5 and 8;
Mangin, 1915, p. 480, Figs. 7 and 8; Hustedt, 1930, p. 771, Figs. 451 and
452; Cupp, 1943, p. 145, Figs. 103 and 104; Drebes, 1974, p. 95, Fig. 79;
Syvertsen & Hasle, 1983; Fryxell, 1989, p. 4, Figs. 23–26; Fryxell et al., 1989;
Fryxell & Prasad, 1990; Takano, 1990, pp. 274–275.

Generic characters:

Valves elliptical.

Apertures fairly wide.

KEY TO SPECIES

1a. Valve face concave in broad girdle view 2
1b. Valve face convex or flat in broad girdle view 3
2a. Bands distinct in water mounts (ribbed), cells slightly curved in broad
girdle view, elevations (horns) long, narrow, apertures tall, elliptical, labiate process in a depression of valve center. . . . .
. . . . *E. cornuta* (Cleve) Grunow
2b. Bands not ribbed, scarcely visible, cells curved in broad girdle view, chains
helically coiled, horns low,[12] broad, apertures angular elliptical to square,
labiate process central. . . . . *E. zodiacus* Ehrenberg

---

[12] The nominate form. Basionym of *Eucampia cornuta: Mölleria cornuta* Cleve, 1873. Synonym
of *Eucampia antarctica: Eucampia balaustium* Castracane (see Fryxell et al., 1989).

PLATE 35 *Eucampia zodiacus:* (a and b) chains in broad girdle view. Scale bar = 10 μm;
(c and d) valves of different sizes with central labiate process and costate ocelli. Scale bar =
10 μm; (e and f) costate ocelli of f. *zodiacus* and f. *cylindricornis*, respectively. Scale bar =
1 μm. From Syvertsen & Hasle (1983). *Hemiaulus hauckii:* (a) chain in girdle view; (b) detail.
After Cupp (1943). Scale bar = 20 μm. *Hemiaulus membranaceus.* Chain in girdle view. After
Cupp (1943). Scale bar = 20 μm. *Hemiaulus sinensis:* chain, (a) in broad girdle view and (b)
detail of broad girdle view. After Cupp (1943). Scale bars = 20 μm. *Hemiaulus indicus:* chain
after Allen & Cupp (1935). Scale bar = 20 μm.

3a. Cells fairly coarsely silicified, cells curved or straight, horns fairly broad,
    apertures square to hexagonal, labiate process submarginal . . . .
    . . . . *E. antarctica* (Castracane) Mangin
3b. Cells lightly silicified, cells curved or straight, helical chains rare, horns
    usually low, apertures almost square or low elliptical, labiate process large,
    central . . . . *E. groenlandica* Cleve

Distribution:
   E. antarctica—southern cold water region.
   E. cornuta—warm water region.
   E. groenlandica—northern cold water region
   E. zodiacus—cosmopolitan, probably absent from polar waters.
How to identify: The species may be identified in girdle view in water
mounts. Location of labiate process is more easily seen in material
embedded in a medium of a high refractive index.
Remarks: Syvertsen & Hasle (1983) distinguished between two forms of *E.
zodiacus, E. zodiacus* f. *cylindricornis* Syvertsen and the nominate form
mainly based on the shape and length of the elevations. Fryxell & Prasad
(1990) distinguished between two varieties of *E. antarctica, E. antarctica*
var. *recta* (Mangin) Fryxell & Prasad and the nominate variety mainly
based on the shape of cells in broad girdle view. *Eucampia antarctica* var.
recta has a more polar distribution, and both taxa have heavily silicified
"winter stages," often termed resting spores.

Genus *Hemiaulus* Heiberg 1863 (Plate 35, Table 41)
Type: *Hemiaulus proteus* Heiberg.
References: Greville, 1865a, p. 5, Plate 5, Fig. 9; Cleve, 1873a, p. 6, Plate 1,
Fig. 5; Van Heurck, 1880–1885, Plate 103, Fig. 10; Karsten, 1907, p. 394,
Plate 46, Figs. 4 and 4a; Hustedt, 1930, p. 874, Figs. 518 and 519; Cupp,

TABLE 41 Morphometric Data of *Hemiaulus* spp.

Species	Apical axis (μm)	Areolae in 10 μm
H. hauckii	12–35	16–17
H. indicus	34–40	—
H. membranaceus	30–97	ca. 30 (TEM)
H. sinensis	15–36	7–9 (Valve center)
		11–13 (Valve mantle)

Note. —, No data.

1943, p. 168, Figs. 118–120; Sournia, 1968, p. 32, Fig. 29; Ross et al., 1977, p. 187, Figs. 20–41.

The genus has four recent marine plankton species and numerous fossil species.

Generic characters:

Cells straight in broad girdle view.

Chains sometimes curved or turned about the long axis.

Valves elliptical.

Elevations mainly long and ends claw like, flattened, or pointed.

KEY TO RECENT SPECIES

1a. Chains straight, curved or twisted, horns with claw like or narrow, flattened tips. 2
1b. Chains twisted, horns short, with a more or less sharp point, cell wall lightly silicified, areolation and labiate process indistinct. . . . .
. . . . *H. membranaceus* Cleve
2a. Chains often twisted, horns ending in claw like spines, areolation and labiate process indistinct 3
2b. Chains straight or curved, horns with flattened tips, areolae and labiate process distinct . . . . *H. sinensis* Greville
3a. Horns long, valve surface flat or slightly concave, apertures large and rectangular. . . . . *H. hauckii* Grunow in Van Heurck
3b. Horns shorter, valve surface convex. . . . . *H. indicus* Karsten

Distribution:

H. hauckii and *H. sinensis*—warm water region to temperate.

H. indicus—Indian Ocean, Sea of Java.

H. membranaceus—warm water region.

How to identify: The species may be identified as whole cells, especially in chains, in water mounts. The single process and the areolation are more distinct on permanent mounts of rinsed or cleaned specimens.

Remarks: EM investigations showed that the single labiate process is offset from center in *H. membranaceus* and *H. sinensis,* and apical in *H. hauckii* (Ross et al., 1977). The areolae of *H. sinensis* have well developed vela which seem to be missing in the other two species. *Hemiaulus indicus* has not been examined with EM.

Family Cymatosiraceae Hasle, von Stosch, & Syvertsen 1983

Round et al. (1990) established the order Cymatosirales Round & Crawford and a subclass Cymatosirophycidae Round & Crawford for this family, whereas Glezer et al. (1988) followed the authors of the family and retained it in Biddulphiales. Most of the species of this family are inhabitants of sand and mud flats. *Brockmanniella, Cymatosira,* and *Plagiogrammopsis* occur occasionally in plankton, and two genera, *Arcocellulus* and *Minutocellus,* belong to marine plankton and have been recorded as predominant in inshore nanoplankton (Hasle et al., 1983, p. 82).

Terminology partially specific to Cymatosiraceae:

Fascia (Fig. 18)—an extension of the central area forming a hyaline (unperforated) band extending the valve, i.e., transapically.

Linking spines—marginal spines that link sibling valves in chains.

Pilus—long hair flattened proximally.

Pilus valve—valve with two pili (the other valve of a cell has a process = heterovalvy).

Pseudoseptum—a membranous costa on the inner side of the valve (here in pervalvar direction).

Tubular process (Fig. 8)—a simple tube penetrating the valve wall; distinguished from a labiate process with EM.

Ocellulus—basically structured the same as an ocellus (see Hemiaulaceae), but with few porelli and a raised rim (EM).

The genera dealt with here are characterized by:
Cells single, in tight chains by linking spines, or in loose ribbons.

Low elevations, each with an ocellulus.

Bipolar symmetry.

One process per cell.

Heterovalvy.

One chloroplast.

KEY TO GENERA

1a. Pili present. 2
1b. Pili absent 4
2a. Cells curved in broad girdle view 3
2b. Cells straight in broad girdle view . . . . *Plagiogrammopsis,* p. 183
3a. Pilus valves concave . . . . *Arcocellulus,* p. 179
3b. Pilus valves convex or convex in the middle and concave closer to the
     elevations . . . . *Minutocellus,* p. 182
4a. Fascia present, linking spines absent. . . . . *Brockmanniella,* p. 179
4b. Fascia absent, linking spines present. . . . . *Cymatosira,* p. 181

Genus *Arcocellulus* Hasle, von Stosch, & Syvertsen 1983 (Plate 36, Table 42)
Type: *Arcocellulus mammifer* Hasle, von Stosch, & Syvertsen.
References: Hasle et al., 1983, p. 54, Text Figs. 10 and 11, Figs. 272–333 and 408–421.

Generic characters:
Solitary or in loose ribbons.

Valve outline narrowly lanceolate, elliptical to subcircular, smaller valves oblong to subcircular, often without pili.

No marginal linking spines.

Marginal ridge inconspicuous.

Cell wall weakly silicified; structure not resolved with LM.

Tubular process.

KEY TO SPECIES

1a. Cells in girdle view curved in the middle and nearly straight close to apices (genuflexed), the two pili of a valve crossing . . . .
. . . . *A. mammifer* Hasle, von Stosch, & Syvertsen
1b. Cells in girdle view evenly curved, the two pili of a valve not crossing . .
. . . . *A. cornucervis* Hasle, von Stosch, & Syvertsen

Distribution:
A. cornucervis—many records from polar and temperate waters of the northern hemisphere; two records from the South Island, New Zealand.
A. mammifer—temperate waters of the North Sea, subtropical habitats along the Texas coast of the Gulf of Mexico.

Genus *Brockmanniella* Hasle, von Stosch, & Syvertsen 1983
Type: *Brockmanniella brockmannii* (Hustedt) Hasle, von Stosch, & Syvertsen.
Monospecific genus.

Brockmanniella brockmannii (Hustedt) Hasle, von Stosch, & Syvertsen
(Plate 36)
Basionym: *Plagiogramma brockmannii* Hustedt.
References: Hustedt, 1939, p. 595, Figs. 11 and 12; Drebes, 1974, p. 101, Fig. 82; Hasle et al., 1983, p. 34, Text Fig. 5, Figs. 132–155; Gardner & Crawford, 1994, Figs. 30–37.
Girdle view: Long, loose, partly twisted ribbons. Cells rectangular, bulging in the middle. Sibling valves abutting in the middle and at the elevations.

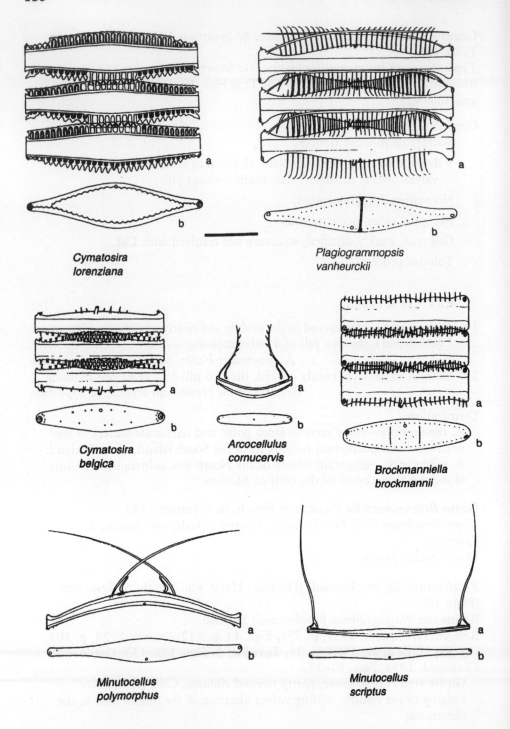

*Cymatosira
lorenziana*

*Plagiogrammopsis
vanheurckii*

*Cymatosira
belgica*

*Arcocellulus
cornucervis*

*Brockmanniella
brockmannii*

*Minutocellus
polymorphus*

*Minutocellus
scriptus*

TABLE 42   Morphometric Data of *Arcocellulus* spp.

Species	Pervalvar axis (μm)	Apical axis (μm)	Transapical axis (μm)
A. *cornucervis*	1–17	1.2–13	0.7–1.5
A. *mammifer*	2–4	2–12.5	1.1–1.4

Valve view: Valve outline narrowly lanceolate and oblong to subcircular. Process submarginal in the fascia. Coarse areolation. Labiate process.
Morphometric data: Apical axis, 4–36 μm, transapical axis, 3–5 μm; poroid areolae, 13–15 in 10 μm.
Distribution: North European coastal waters in sediment and in spring plankton (in bloom proportions).

Genus *Cymatosira* Grunow 1862 (Plate 36, Table 43)
Type: *Cymatosira lorenziana* Grunow.
References: Grunow, 1862, p. 378, Plate 7, Figs. 25a–25c; Van Heurck, 1880–1885, Plate 45, Figs. 38–42; Hustedt, 1959, p. 127, Figs. 648 and 649; Fryxell & Miller, 1978, p. 122, Figs. 20–31; Hasle et al., 1983, p. 17, Text Figs. 1 and 2, Figs. 1–71.

TABLE 43   Morphometric Data of *Cymatosira* spp.

Species	Pervalvar axis (μm)	Apical axis (μm)	Transapical axis (μm)	Valve areolae in 10 μm
C. *belgica*	ca. 4	10–40	3–5	9–12
C. *lorenziana*	—	12–60	6–11	8–10

Note. —, No data.

PLATE 36   *Cymatosira lorenziana:* (a) short chain in girdle view with central linking spines and separation valves on end cells; (b) valve with marginal ridge, labiate process, and ocelluli . *Cymatosira belgica:* (a) typical triplet with linking spines and separation valves; (b) separation valve with spines, labiate process, and ocelluli. *Plagiogrammopsis vanheurckii:* (a) short chain with marginal spines and pili; (b) valve with labiate process, pseudoseptum, and ocelluli. *Brockmanniella brockmannii:* (a) chain and (b) valve with fascia, labiate process, and ocelluli. *Arcocellulus cornucervis:* (a) cell in girdle view with branched pili; (b) process valve (tubular process). *Minutocellus polymorphus:* (a) cell in girdle view with pili; (b) process valve (tubular process). *Minutocellus scriptus:* (a) cell in girdle view; (b) Process valve. From Hasle et al. (1983). Scale bar = 10 μm.

Generic characters:
   Partly twisted ribbons.
   Marginal spines linking the middle part of the valves.
   Cell wall coarsely silicified.
   Labiate process.

KEY TO SPECIES

1a. Cells rectangular in girdle view, both valves slightly convex in the middle, valve outline broadly lanceolate with produced ends, nonareolated valve area absent. . . . . *C. lorenziana* Grunow
1b. Larger cells curved in girdle view, one valve convex in the middle, the other slightly concave or almost straight, valve outline lanceolate to linear oblong, nonareolated area present in the middle of the valve . . . .
   . . . . *C. belgica* Grunow

Distribution:
   C. belgica—cosmopolitan, mostly on sandy beaches.
   C. lorenziana—warm water region to temperate, coastal.

Genus *Minutocellus* Hasle, von Stosch, & Syvertsen 1983 **(Plate 36, Table 44)**
Type: *Minutocellus polymorphus* (Hargraves & Guillard) Hasle, von Stosch, & Syvertsen.
Basionym: *Bellerochea polymorpha* Hargraves & Guillard.
References: Hargraves & Guillard, 1974, p. 167, Figs. 1–8; Hasle et al., 1983, p. 38, Text Figs. 6 and 8, Figs. 156–189 and 220–242; Takano, 1990, pp. 278–279.

Generic characters:
   Solitary or in ribbons.
   Cell outline narrowly lanceolate, elliptical to subcircular.
   Cell wall weakly silicified, structure not resolved with LM.
   Tubular process.

TABLE 44 **Morphometric Data of *Minutocellus* spp.**

Species	Pervalvar axis (µm)	Apical axis (µm)	Transapical axis (µm)
M. polymorphus	2	2–30	2–3
M. scriptus	0.6–3	3–36	2–2.5

KEY TO SPECIES

1a. Both valves of cells in girdle view evenly curved, pili usually arising at some distance from the elevations on the convex valve, crossing each other in girdle view, smaller cells in ribbons . . . .

    M. polymorphus (Hargraves & Guillard) Hasle, von Stosch, & Syvertsen

1b. Pilus valve convex in the middle and concave distally, the other valve slightly convex, pili arising close to the elevations, more or less parallel in girdle view, diverging from the apical axis when seen in valve view, solitary . . . .*M. scriptus* Hasle, von Stosch, & Syvertsen

Distribution:

  M. polymorphus—probably cosmopolitan, marine, planktonic.

  M. scriptus—known only from Helgoland and Bremerhaven, Germany, probably planktonic.

Genus *Plagiogrammopsis* Hasle, von Stosch, & Syvertsen, 1983
Type: *Plagiogrammopsis vanheurckii* (Grunow) Hasle, von Stosch, & Syvertsen.
Monospecific genus.

Plagiogrammopsis vanheurckii (Grunow) Hasle, von Stosch, & Syvertsen
(Plate 36)
Basionym: *Plagiogramma vanheurckii* Grunow in Van Heurck.
References: Van Heurck, 1880–1885, Plate 36, Fig. 4; Hasle et al., 1983, p. 30, Text Fig. 4, Figs. 104–131; Gardner & Crawford, 1994, Figs. 20–29.

  Girdle view: Ribbons, often twisted around their long axis. Cells in girdle view convex in the middle and constricted near the elevations.

  Valve view: Valve outline narrowly lanceolate with rostrate apices or broadly lanceolate, rhombic or subcircular. Fascia with a pseudoseptum. Labiate process submarginal in the fascia. Coarse areolation.

Morphometric data: Apical axis 3–50 μm; transapical axis, 4 μm; valve areolae, 12 in 10 μm.

Distribution: Cosmopolitan.

How to identify: *Brockmanniella*, *Cymatosira*, and *Plagiogrammopsis* may be identified in water mounts in girdle as well as in valve view. *Arcocellulus* and *Minutocellus* may easily be overlooked in water mounts, especially the smaller specimens. The larger and medium sized specimens with pili are characteristic and readily observable. The extreme variation in valve outline depending on cell size, present in all genera, also complicates the identification when cleaned material mounted in a medium of a high refractive index or in EM is examined.

Remarks: The labiate process of *Brockmanniella, Cymatosira* and *Plagiogrammopsis* has a tubular external part (EM) and is discernible with

LM. The tubular process of *Arcocellulus* and *Minutocellus* is short externally and internally and is visible with EM.

Incertae sedis (Cymatosiraceae):

Genus *Lennoxia* Thomsen & Buck 1993
Type: *Lennoxia faveolata* Thomsen & Buck.
Monotypic genus.

Lennoxia faveolata Thomsen & Buck
Reference: Thomsen et al., 1993, p. 279, Figs. 1–16.
   Girdle view: Cells bipolar and rostrate. Approximately eight weakly silicified bands without ornmentation. One chloroplast.
   Valve view: Spindle shaped, apical axis curved; middle part roughly triangular. Valve face with honeycomb pattern of flat, hexagonal chambers (TEM). Reminiscent marginal tubular process on one valve (TEM).
Morphometric data: Apical axis, 10–22 μm; pervalvar axis, 1.5–2 μm; hexagonal chambers, 0.15–0.2 μm in diameter.

Distribution: South America, central California, West Greenland, and Denmark. Cell counts published by Thomsen et al. (1993) characterize the species as an important, frequently and abundantly occurring species in marine plankton.
How to identify: When examined with LM *L. faveolata* is confusingly similar to certain species of the subgenus *Nitzschiella*, and if ever observed, it has most probably been identified as *Nitzschia closterium/Cylindrotheca closterium*.

Family Chaetocerotaceae Ralfs in Prichard 1861

   Bacteriastrum and *Chaetoceros* belong in this family (Simonsen, 1979; Glezer et al., 1988; Round et al., 1990). For the sake of convenience *Attheya* is also mentioned here since two former *Chaetoceros* species, *C. armatus* and *C. septentrionalis,* are now transferred to *Attheya* (Crawford et al., 1994). *Attheya* is usually not classified under Chaetocerotaceae, and it is not primarily planktonic. Simonsen (1979) placed *Attheya* in Hemiauloideae, subfamily of Biddulphiaceae. The family Hemiaulaceae is, in this chapter, used for diatoms with valve outgrowths, called elevations or horns, structured like the rest of the valve and not projecting laterally beyond the valve margin. The valve outgrowths of Hemiaulaceae thus differ from the setae of Chaetocerotaceae in structure and orientation (see below). The valve outgrowths of *Attheya* spp. are structured very much like the rest of the valve, and they do not project outside the valve margin in the generitype *A. decora* but they do in the former *C. armatus* and *C. septentrionalis* (Plate 47; Round et al., 1990, p. 334). Round et al. (1990) solved the problem concerning the classification of *Attheya* by

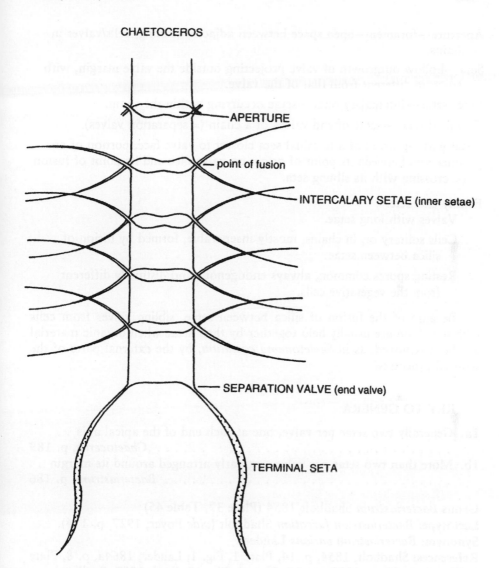

FIGURE 16 Schematic illustration of *Chaetoceros* sp. in broad girdle view with terminology.

describing a new monotypic family Attheyaceae Round & Crawford. The new family and Chaetocerotaceae were placed in the order Chaetocerotales Round & Crawford in the subclass Chaetocerotophycidae Round & Crawford.

Terminology specific to Chaetocerotaceae (partialy after Rines & Hargraves, 1988) (Fig. 16):

Aperture—foramen—open space between adjacent (sibling) cells/valves in chains.

Seta—hollow outgrowth of valve projecting outside the valve margin, with structure different from that of the valve.

Inner setae—intercalary setae—setae occurring within the chain.

Terminal setae—setae of end valves of a chain (= separation valves).

Basal part—portion of a terminal seta closest to valve face, portion of an inner seta between its point of origin on valve face and its point of fusion or crossing with its sibling seta.

Family characters:
>Valves with long setae.

>Cells solitary or, in chains, mostly inseparable, formed by fusion of silica between setae.

>Resting spores common, always endogenous and distinctly different from the vegetative cells.

Because of the fusion of silica between setae, sibling valves from cells within a chain are usually held together by their setae when organic material has been removed, as in *Skeletonema costatum,* by the external parts of the strutted processes.

KEY TO GENERA

1a. Generally two setae per valve, one at each end of the apical axis . . . . .
. . . .*Chaetoceros,* p. 189
1b. More than two setae per valve, regularly arranged around its margin . .
. . . . *Bacteriastrum,* p. 186

Genus *Bacteriastrum* Shadbolt 1854 (Plate 37, Table 45)
Lectotype: *Bacteriastrum furcatum* Shadbolt (*vide* Boyer, 1927, p. 114).
Synonym: *Bacteriastrum varians* Lauder.
References: Shadbolt, 1854, p. 14, Plate 1, Fig. 1; Lauder, 1864a, p. 8, Plate 3, Figs. 1–6; Cleve, 1897a, p. 19, Plate 1, Fig. 19; Ikari, 1927; Pavillard,

---

PLATE 37   *Bacteriastrum elongatum:* (a) part of chain in girdle view. After Cupp (1943). *Bacteriastrum comosum:* chain in girdle view. After Cupp (1943). *Bacteriastrum hylinum:* (a) terminal parts of chain, girdle view; (b) terminal valve with central process. After Cupp (1943). *Bacteriastrum delicatulum:* (a) part of chain in girdle view; (b) intercalary valve in valve view. After Cupp (1943). Bacteriastrum furcatum: (a) terminal parts of chain in girdle view; (b and d) terminal, and (c and e) intercalary valves in valve view. After Hustedt (1930). Scale bars = 20 μm.

Bacteriastrum

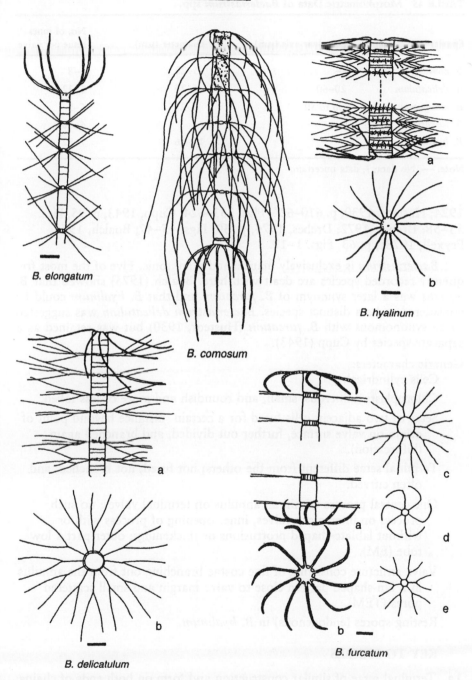

B. elongatum

B. comosum

B. hyalinum

B. delicatulum

B. furcatum

TABLE 45 Morphometric Data of *Bacteriastrum* spp.

Species	Pervalvar axis (μm)	Diameter (μm)	No. of inner setae per valve
B. comosum	—	5–22	6–8
B. delicatulum	20–60	6–20	6–12
B. elongatum	Up to 50	6–27	5–9
B. furcatum	—	—	6–10 ?
B. hyalinum	—	13–56	7–32

Note. —, No data; ?, data uncertain.

1924; Hustedt, 1930, p. 610–624, Figs. 353–624; Cupp, 1943, p. 95, Figs. 55–58; Drebes, 1972; Drebes, 1974, p. 60, Figs. 45–47; Boalch, 1975; Fryxell, 1978c, p. 63, Figs. 1–17.

Bacteriastrum is exclusively marine and planktonic. Five of the most frequently recorded species are dealt with here. Boalch (1975) showed that *B. varians* was a later synonym of *B. furcatum* and that *B. hyalinum* could be considered to be a distinct species. *Bacteriastrum delicatulum* was suggested to be synonomous with *B. furcatum* (Hustedt, 1930) but was retained as a separate species by Cupp (1943).

Generic characters:

Cells cylindrical.

Chloroplasts numerous, small, and roundish and more or less lobed.

Setae of two adjacent cells fused for a certain distance beyond point of origin on valve surface, further out divided, and branched again (bifurcation).

Terminal setae different from the others; not fused, not branched, and often curved.

One central process within an annulus on terminal valves; no such process on intercalary valves, inner opening of process with or without labiate-shaped protrusions or thickenings; outer part a low tube (EM).

Valve structure consisting of fine costae branching out from the annulus in a fan-shaped pattern close to valve margin and small scattered pores (TEM).

Resting spores (endogenous) in *B. hyalinum*.

KEY TO SPECIES

1a. Terminal setae of similar construction and form on both ends of chains
. . . . 2

1b. Terminal setae on either end of chain different in form and direction, inner setae very long and curved in the same direction along chain axis. . . . . *B. comosum* Pavillard

2a. Bifurcation in the apical plane (parallel to chain axis), fused part of inner setae short 3

2b. Bifurcation in the valvar plane (transverse to chain axis), fused part of inner setae long. 4

3a. Terminal setae with arched base, then running outward nearly parallel to chain axis, bell shaped, inner setae usually six on each valve . . . . *B. elongatum* Cleve

3b. Terminal setae umbrella shaped, inner setae numerous on each valve . . . . *B. hyalinum* Lauder

4a. Terminal setae of both ends directed toward the chain . . . . *B. delicatulum* Cleve

4b. Terminal setae first transverse to chain axis, then abruptly obliquely curved toward the chain in their outer part. . . . .*B. furcatum* Shadbolt

Distribution:

B. comosum—warm water region.

B. delicatulum—temperate waters.

B. elongatum—warm water region to temperate.

B. furcatum—poorly known, often recorded under the name *B. varians* and confused with *B. hyalinum* (Hustedt, 1930; Cupp, 1943).

B. hyalinum—common in temperate waters.

How to identify: The species may be identified in water mounts, especially if it is present as intact chains showing the shape of the terminal setae. Some other characters may be used for identification of single cells or part of chains, e.g., for *B. comosum,* the long, inner setae bent toward the one terminal valve with coarse setae; for *B. elongatum,* the long cells; for *B. hyalinum,* the large number of inner setae and the hairy appearance caused by bifurcation of the setae in the apical plane; for *B. delicatulum,* the slightly curved and wavy branches of the inner setae; and for *B. furcatum,* the inner part of the terminal setae perpendicular to the end valve surface.

Remarks: *Bacteriastrum furcatum* was found in "near-bloom numbers" in the northwest Gulf of Mexico (Fryxell, 1978c). *Bacteriastrum hyalinum* seems to be the most common species of the genus in the North Sea and the north Atlantic Ocean, especially in summer (Hendey, 1964; Drebes, 1972).

Genus *Chaetoceros* Ehrenberg 1844
Lectotype: *Chaetoceros tetrachaeta* Ehrenberg (*vide* Boyer, 1927, p. 104).

The genus, *Chaetoceros dichaeta* and *C. tetrachaeta,* were described in the same paper based on Antarctic material (Ehrenberg, 1844b, pp. 198 and 200, genus and species, respectively). *Chaetoceros tetrachaeta* is regarded as a taxonomic entry not recommended for use (VanLandingham, 1968), whereas *C. dichaeta* is frequently recorded from the Antarctic.

Chaetoceros is one of the largest, if not the largest genus of marine plank-tonic diatoms with approximately 400 species described. Although it is assumed that only a fraction, "no more than one-third to one-half," are still valid (Rines & Hargraves, 1988, p. 19), it leaves us with a high number of species, almost all of which are marine.

Several attempts have been made to impose structure on this large genus by a division into two subgenera and numerous sections (see Hustedt, 1930, p. 630; Cupp, 1943, p. 101; Rines & Hargraves, 1988, pp. 21, and 117). For the sake of simplicity only the names of the subgenera, *Phaeoceros* and *Hyalochaete*, will be used here, and species within both subgenera are grouped according to prominent morphological character(s). Hernández-Becerril (1993a) proposed a new subgenus *Bacteriastroidea* for *Chaetoceros bacterias-troides* Karsten and changed the subgenus name *Phaeoceros* to *Chaetoceros*. Hernández-Becerril (1992b; 1993a,b) and Hernandez-Becerril et al. (1993) examined seven warm water *Chaetoceros* species which are not dealt with in this chapter.

Generic characters:

Cells more or less rectangular in girdle view.

Cells elliptical to almost or rarely circular in valve view.

Opposite setae of adjacent cells (sibling setae) "touch" one another near their origin.

Characters showing differences between species (partially after Rines & Hargraves, 1988):

Chloroplasts

The presence/absence in setae.

Number.

Shape.

Size.

Setae

Coarseness (thickness, spines).

Direction.

Terminal setae different from the inner setae in shape and coarseness (e.g., *C. affinis*) and in direction (e.g., *C. atlanticus*).

Some inner setae different from the others (e.g., *C. compressus* and *C. diversus*).

Adjacent (sibling) setae fused for some distance (e.g., *C. decipiens*).

One seta of a cell longer than the three others (e.g., *C. socialis*).

Direction of basal part of setae.

Point of fusion of sibling setae: inside valve or chain margin, at the margin, external to the margin.

Shape and size of aperture determined by
> Point of origin of setae on valve surface.
> Point of fusion of sibling setae (e.g., basal part absent; i.e., apices of sibling valves touching: *C. curvisetus* and *C. constrictus;* basal part present and point of fusion of sibling setae at some distance away from the valves: *C. debilis* and *C. diadema*).

Height of girdle
> e.g., in *C. convolutus* > one-third of pervalvar axis; in *C. concavicornis* < one-third of pervalvar axis.

Direction of chain
> Straight (e.g., *C. laciniosus*).
> Curved or helical (e.g., *C. debilis*).
> Twisted (e.g., *C. tortissimus*).

Resting spores
> Primary valve (epivalve) and secondary valve (hypovalve) similar or dissimilar.
> Spiny or smooth.
> Protuberances with branches.
> Paired spores with fused setae of parent hypothecae.

For information on resting spores see Proschkina-Lavrenko (1955) who also includes species not mentioned in this chapter.

A. **Subgenus *Phaeoceros* Gran 1897**

Subgeneric characters:
> Chloroplasts numerous small granules throughout the whole cell, the setae included.
>
> Large robust forms.
>
> Setae strong, thick, often very long, striated, and armed with conspicuous spines.
>
> One, seldom many, central processes on every valve, often located closer to one side of the valve; no protrusion or thickening around the inner opening (EM).
>
> Valves irregularly perforated with simple holes; more weakly silicified valves with a weak pattern of costae branching out from an annulus (TEM).
>
> Mostly oceanic.
>
> Resting spores reported for one species.

1. Cells solitary or in short chains; external part of central process inconspicuous: *C. aequatorialis, C. criophilus, C. danicus, C. peruvianus,* and *C. rostratus* (Table 46).

2. Terminal setae differentiated from the others by length and direction, straight chains, wide apertures; external part of central process long and tubular: *C. atlanticus* and *C. dichaeta* (Table 47).

3. Terminal setae not distinctly differentiated from the others; setae often diverging in all directions; apertures smaller than cell body; external part of processes inconspicuous: *C. borealis, C. castracanei, C. coarctatus, C. concavicornis, C. convolutus, C. dadayi, C. densus, C. eibenii,* and *C. tetrastichon* (Table 48).

4. Setae not fused; apertures narrow in apical and pervalvar directions; setae delicate; external part of central process inconspicuous: *C. flexuosus*

B. Subgenus *Hyalochaetae* Gran 1897

Subgeneric characters:

Chloroplasts, one or a few plates or, more rarely, numerous small granules.

Setae thin, often hair-like; spines and structure seen with LM in some species; no chloroplasts.

One, seldom several, central processes on terminal valves and no such processes on intercalary valves; inner opening of processes with or without labiate-shaped protrusions (EM).

Valves with a more or less regular pattern of costae branching out from an annulus; often holes or poroids between the costae (TEM).

Mainly distributed in coastal and inshore waters.

Resting spores in many species.

1. Cells with more than two chloroplasts.
   a. Four to 10 chloroplasts; terminal setae more or less differentiated from the others by coarseness and orientation: *C. decipiens, C. lorenzianus,* and *C. mitra* (Table 49).
   b. Numerous small plate-like chloroplasts; terminal setae scarcely different from the others: *C. compressus, C. teres,* and *C. lauderi* (Table 49).

2. Cells with two chloroplasts.
   a. Cells with a hemispherical or conical protuberance: *C. didymus* (Table 50).
   b. Cells with deep constriction between valve and girdle band: *C. constrictus* (Table 50).
   c. Apertures high, elliptical, and square to rectangular: *C. laciniosus* (Table 50).
   d. Adjacent cells touching each other in the middle by a central raised region; apertures narrow and divided into two parts: *C. similis* (Table 50).

e. Setae of adjacent cells not crossing; connected by a pervalvarly directed bridge: *C. anastomomosans* (Table 50).
3. Cells with one chloroplast.
   a. Chains curved or helical, intercalary setae all bent in one direction: *C. curvisetus, C. debilis,* and *C. pseudocurvisetus* (Table 51).
   b. Chains mostly loose; resting spores united in pairs; resting spore parent cells with fused hypovalvar setae and no apertures: *C. cintus, C. furcellatus,* and *C. radicans* (Table 52).
   c. Inner setae of two kinds: *C. diversus* and *C. messanensis* (Table 53).
   d. Valves of adjacent cells touch: *C. affinis, C. costatus, C. karianus, C. subtilis,* and *C. wighamii* (Table 54).
   e. Valves of adjacent cells do not touch: *C. diadema, C. holsaticus,* and *C. seriacanthus* (Table 54).
   f. Chains curved and joined in irregular spherical colonies: *C. socialis* and *C. radians.*
   g. Unicellular species: *C. ceratosporus, C. simplex,* and *C. tenuissimus* (Table 55).
4. Cells with one chloroplast; unicellular; two, seldom three, setae per cell: *C. minimus* and *C. throndsenii* (Table 56).

A. Subgenus *Phaeoceros*

1. Cells solitary or in short chains.

Chaetoceros aequatorialis Cleve (Plate 38, Table 46)
Synonym:? *Chaetoceros pendulus* Karsten.
References: Cleve, 1873a, p. 10, Fig. 9; Karsten, 1905, p. 118, Plate 15, Fig. 7; Karsten, 1907, p. 389, Plate 45, Fig. 1; Schmidt, 1920, Plate 325, Figs. 1 and 2; Hendey, 1937, p. 294.

Cells heterovalvate; upper valve with central depression or smoothly concave and lower valve with projecting corners and sometimes short tubular extension of the central process. Setae long, very coarse, emerging well inside valve margin, not abutting at point of emergence, proceeding outwards almost at right angles to pervalvar axis, then curving downwards. Setae of upper and lower valves almost parallel and armed with spines.

Chaetoceros criophilus Castracane (Plate 38, Table 46)
References: Castracane, 1886, p. 78; Mangin, 1915, p. 34, Figs. 13 and 14; Hendey, 1937, p. 295, Plate 13, Fig. 7; Hasle, 1968b, p. 7, Plate 10, Fig. 12, Map 4; Fryxell, 1989, p. 10, Figs. 33–38.

Cells heterovalvate; upper valve almost flat and lower valve curved. Setae long, coarse, and curved. Upper setae emerge at valve margin giving the

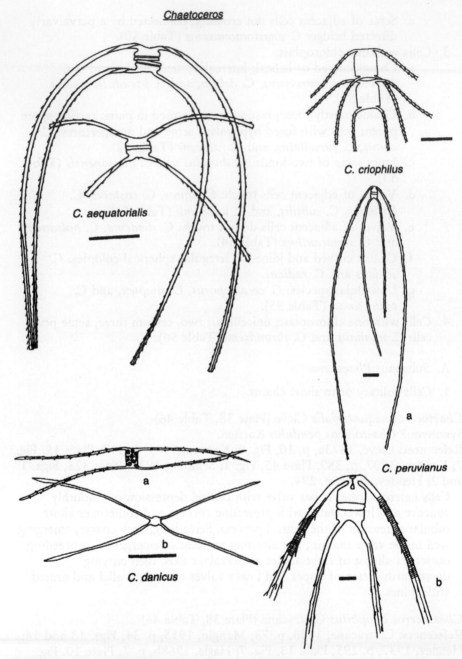

PLATE 38 *Chaetoceros aequatorialis:* cells in broad girdle view. After Schmidt (1920). Scale bar = 50 μm. *Chaetoceros criophilus:* chain in broad girdle view. Scale bar = 50 μm. *Chaetoceros danicus:* (a) cell in broad girdle view; (b) cell in valve view. Scale bar = 50 μm. *Chaetoceros peruvianus:* (a) cell in broad girdle view. Scale bar = 50 μm; (b) detail of cell in broad girdle view. Scale bar = 10 μm.

TABLE 46 Morphometric Data of *Chaetoceros*
spp. Appearing as Solitary Cells or
Short Chains

Species	Apical axis (μm)
C. aequatorialis	ca. 25
C. criophilus	16–50
C. danicus	8–20
C. peruvianus	10–32
C. rostratus	10–37

appearance of being a continuation of valve surface; lower setae emerge inside valve margin. Chains with lanceolate; narrow apertures.

Chaetoceros danicus Cleve (Plate 38, Table 46)

References: Cleve, 1889, p. 55; Hustedt, 1930, p. 659, Fig. 373; Cupp, 1943, p. 109, Fig. 65; Hendey, 1964, p. 122, Plate 10, Fig. 5; Drebes, 1974, p. 66, Fig. 50; Koch & Rivera, 1984, p. 71, Figs. 53–56; Rines & Hargraves, 1988, p. 49, Figs. 95–99.

Cells isovalvate; valve surface flat. Setae long, stiff, perpendicular to pervalvar or chain axis, and originating at valve margin. Chains with small apertures—terminal setae perpendicular to chain axis, intercalary setae basally directed toward one end of the chain then become parallel to the terminal setae. External part of central process a short, flattened tube (Koch & Rivera, 1984, Fig. 54, SEM).

Chaetoceros peruvianus Brightwell (Plate 38, Table 46)

Synonym: *Chaetoceros chilensis* Krasske.

References: Brightwell, 1856, p. 107, Plate 7, Figs. 16–18; Brightwell, 1858b, p. 155, Plate 8, Figs. 9 and 10; Hustedt, 1930, p. 671, Figs. 380 and 381; Krasske, 1941, p. 266, Plate 4, Fig. 3, Plate 6, Figs. 1 and 2; Cupp, 1943, p. 113, Fig. 68; Hendey, 1937, p. 296, Plate 13, Fig. 6; Hasle, 1960, p. 15, Fig. 2; Koch & Rivera, 1984, p. 69, Figs. 36–47.

Cells heterovalvate; upper valve rounded and lower valve flat. Setae of upper valve arising in pervalvar direction from near valve center, abutting with a groove between them, turning sharply, and running backward in more or less outwardly convex curves. Setae of lower valve originating inside valve margin and slightly convex toward outside. Central process between the bases of the setae, subcentral in location; external part conical (Koch & Rivera, 1984, EM).

Chaetoceros rostratus Lauder (Table 46)
References: Lauder, 1864b, p. 79, Fig. 10; Hustedt, 1930, p. 660, Fig. 374;
Rines & Hargraves, 1988, p. 55, Figs. 105–107; Guiffré & Ragusa, 1988.

Cells isovalvate; valve surface flat or slightly convex. Cells in chains (two
to six cells) held together by a central intervalvar connection [not a
labiate-like process (Rines & Hargraves, 1988)]. Setae perpendicular to
the pervalvar axis. Terminal cells heterovalvate i.e. terminal valve lacking
the protuberance forming the intervalvar connection. External part of
subcentrally located process a laterally compressed tube, slightly cone
shaped (Guiffré & Ragusa, 1988, SEM).

Distribution:

C. aequatorialis and *C. rostratus*—warm water region.

C. criophilus—southern cold water region.

C. danicus—cosmopolitan?

C. peruvianus—warm water region to temperate.

2. Terminal setae differentiated from the others by length and direction.

Chaetoceros atlanticus Cleve (Plate 39, Table 47)
References: Cleve, 1873b, p. 11, Plate 2, Fig. 8; Hustedt, 1930, p. 641, Figs.
363 and 364; Cupp, 1943, p. 103, Fig. 59; Evensen & Hasle, 1975, p. 157,
Figs. 6–11; Koch & Rivera, 1984, p. 63, Figs. 1–5; Takano, 1990, pp.
282–283.

Cells rectangular in broad girdle view. Apertures hexagonal and smaller
than the cells. Setae arising slightly within valve margin; basal part first
narrow, then widened, and diagonally oriented. Inner setae almost
straight. Terminal setae shorter than others, usually forming a V.

Chaetoceros dichaeta Ehrenberg (Plate 39, Table 47)
References: Ehrenberg, 1844b, p. 200; Mangin, 1922, p. 60, Fig. 6; Hendey,
1937, p. 291, Plate 6, Figs. 9 and 10; Evensen & Hasle, 1975, p. 157, Figs.
1–5; Koch & Rivera, 1984, p. 64, Figs. 6–12.

Cells rounded in broad girdle view. Apertures hexagonal to square,
usually larger in pervalvar direction than the cell proper. Setae arising far
inside valve margin; basal part long, more or less parallel to the chain
axis, then bent outward at nearly right angles to chain axis; terminal setae
later bent again so they are once more nearly parallel to chain axis.

PLATE 39   *Chaetoceros atlanticus:* (a) chain. Scale bar = 50 μm; (b) detail of end cell in broad
girdle view with external part of process. Scale bar = 10 μm. After Cupp (1943). *Chaetoceros
dichaeta:* partial chain in broad girdle view. After Mangin (1922). Scale bar = 50 μm. *Chaetoceros
borealis:* chain in broad girdle view. Scale bar = 50 μm. *Chaetoceros castracanei:* partial chain
in broad girdle view. After Karsten (1905). Scale bar = 50 μm.

Chaetoceros

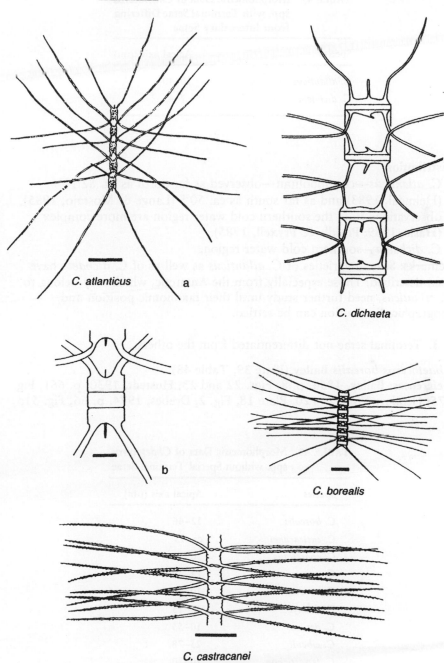

C. atlanticus a

C. dichaeta

b

C. borealis

C. castracanei

TABLE 47 Morphometric Data of *Chaetoceros*
spp. with Terminal Setae Differing
from Intercalary Setae

Species	Apical axis (μm)
C. atlanticus	10–40
C. dichaeta	7–45

Distribution:

C. atlanticus—cosmopolitan—observed as far north as ca. 82°N
(Heimdal, 1983) and as far south as ca. 50°S (Lange & Mostajo, 1985),
observations from the southern cold water region are more complex
(Hasle, 1969; Priddle & Fryxell, 1985).

C. dichaeta—southern cold water region.

Remarks: Several varieties of *C. atlanticus* as well as of *C. dichaeta* have
been described. Those especially from the Antarctic, which may belong to
C. atlanticus, need further study until their taxonomic position and
geographic distribution can be settled.

3. Terminal setae not differentiated from the others.

Chaetoceros borealis Bailey (Plate 39, Table 48)
References: Bailey, 1854, p. 8, Figs. 22 and 23; Hustedt, 1930, p. 661, Fig.
375; Hendey, 1964, p. 120, Plate 18, Fig. 2; Drebes, 1974, p. 66, Fig. 51;

TABLE 48 Morphometric Data of *Chaetoceros*
spp. without Special Terminal Setae

Species	Apical axis (μm)
C. borealis	12–46
C. castracanei	ca. 15
C. coarctatus	30–44
C. concavicornis	12–30
C. convolutus	10–27
C. dadayi	10–15
C. densus	10–55
C. eibenii	25–78
C. tetrastichon	10–20

Evensen & Hasle, 1975, p. 158, Figs. 12–14; Rines & Hargraves, 1988, p. 45, Fig. 93.
Chains straight, not twisted. Cells isovalvate; apertures elliptical to hexagonal. Setae arising well inside valve margin having a distinct basal part; point of fusion near chain edge, setae often crossing each other. External part of central process a short tube (Evensen & Hasle, 1975, EM).

Chaetoceros castracanei Karsten (Plate 39, Table 48)
References: Karsten, 1905, p. 116, Plate 15, Fig. 1; Priddle & Fryxell, 1985, p. 26.
Chains straight. Cells isovalvate. Apertures narrow, almost closed by a central valve protuberance. Setae emerging within valve margin, fused outside chain edge, and perpendicular to chain axis in their outer part.

Chaetoceros coarctatus Lauder (Plate 40, Table 48)
References: Lauder, 1864b, p. 79, Fig. 8; Hustedt, 1930, p. 655, Fig. 370; Hendey, 1937, p. 293, Plate 5, Figs. 7 and 8; Cupp, 1943, p. 107, Fig. 62; Hernández-Becerril, 1991b.
Chains long and robust. Cells isovalvate. Apertures almost absent. Valve surface flat. Posterior terminal setae large, strongly curved, and shorter than the others and anterior terminal setae less robust. Usually found with a species of *Vorticella* attached. Eighteen to 22 small, slit-like processes on each valve (Hernández-Becerril, 1991b, EM).

Chaetoceros concavicornis Mangin (Plate 40, Table 48)
References: Mangin, 1917, p. 9, Figs. 5–7; Hustedt, 1930, p. 665, Figs. 376 and 377; Cupp, 1943, p. 109, Fig. 66; Hendey, 1964, p. 122, Plate 9, Fig. 1; Evensen & Hasle, 1975, p. 158, Figs. 15–22.
Chains straight. Cells heterovalvate; upper valve rounded, setae arising from near center; lower valve flat, setae emerging from inside valve margin. Apertures trapezoid like. Girdle zone narrow (less than one-third of the length of pervalvar axis). Setae thin at base, wider outward, and all bent toward lower end of chain, and outside line concave. External part of central process a short tube (Evensen & Hasle, 1975, EM).

Chaetoceros convolutus Castracane (Plate 41, Table 48)
References: Castracane, 1886, p. 78; Hustedt, 1930, p. 668, Fig. 378; Cupp, 1943, p. 110, Fig. 67; Fryxell & Medlin, 1981, p. 9, Figs. 43-49; Koch & Rivera, 1984, p. 67, Figs. 23–35; Rines & Hargraves, 1988, p. 47, Fig. 94.
Chains sometimes twisted about pervalvar axis. Cells heterovalvate, upper valve vaulted, setae arising from near center; lower valve flat, setae emerging from well inside valve margin. Setae of the lower valves bent

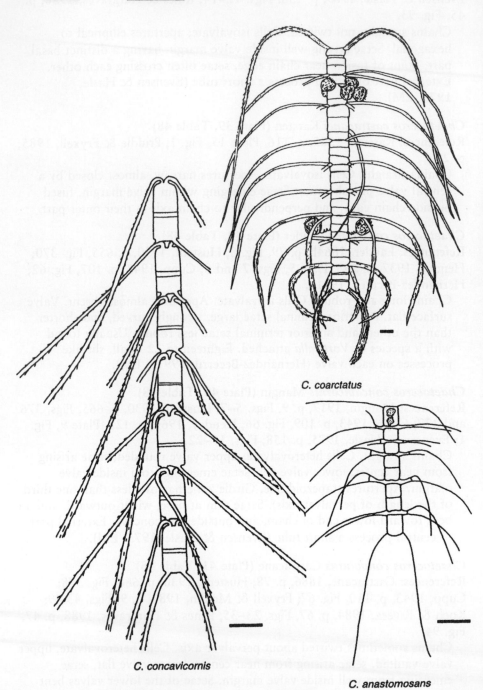

C. coarctatus

C. concavicornis

C. anastomosans

toward the same end of the chain. Sibling setae often twisted at their basal part thus partialy or wholly covering the apertures. Sibling valves held together by siliceous flaps on setae of upper valve (Fryxell & Medlin, 1981; Koch & Rivera, 1984, EM). Girdle none fairly broad (about one-third of the length of pervalvar axis). Setae not increasing in width from base and outward (cf. *C. concavicornis*). External part of central process short and tubular (Fryxell & Medlin, 1981; Koch & Rivera, 1984, EM).

Chaetoceros dadayi Pavillard (Plate 41, Table 48)
References: Pavillard, 1913, p. 131, Fig. 2b; Hustedt, 1930, p. 658, Fig. 372; Cupp, 1943, p. 109, Fig. 64; Hernández-Becerril, 1992c, p. 367, Figs. 9–22.
Recognized and differentiated from *C. tetrastichon* by formation and direction of setae (Cupp, 1943). Setae arising from valve margin without a basal part. Setae on one side of the chain short and on the other side very long with some running toward one end of the chain and some toward the other. Usually found with a tintinnid attached.

Chaetoceros densus (Cleve) Cleve (Table 48)
Basionym: *Chaetoceros borealis* var. *densus* Cleve.
References: Cleve, 1897a, p. 20, Plate 1, Figs. 3 and 4; Cleve, 1901a, p. 299; Hustedt, 1930, p. 651, Fig. 368; Drebes, 1974, p. 64, Fig. 48; Rines & Hargraves, 1988, p. 50, Figs. 100 and 101.
Cells tightly packed together in chains. Cells isovalvate; valve surface slightly convex. Apertures narrowly lanceolate. Setae perpendicular to chain axis or bent slightly toward chain ends, diverging strongly from apical plane.

Chaetoceros eibenii Grunow in Van Heurck (Plate 41, Table 48)
References: Van Heurck, 1880–1885, Plate 82, Figs. 9 and 10; Pavillard, 1921, p. 469, Figs. 1–11; Hustedt, 1930, p. 653, Fig. 369; Cupp, 1943, p. 106, Fig. 61; Drebes, 1974, p. 64, Fig. 49; Koch & Rivera, 1984, p. 65, Figs. 13–22; Rines & Hargraves, 1988, p. 52, Figs. 102–104.
Cells isovalvate; valve surface slightly concave with a small central process on each valve (LM). Apertures lanceolate to hexagonal. Setae generally bent toward nearest end of the chain. Resting spores formed inside lateral auxospores ("in need of further study" according to Rines & Hargraves, 1988). External part of central process laterally flattened; internally visible as a slit (Koch & Rivera, 1984, Figs. 17–20, SEM).

PLATE 40   *Chaetoceros coarctatus:* chain in broad girdle view with *Vorticella.* After Cupp (1943). *Chaetoceros concavicornis:* chain in broad girdle view. After Hustedt (1930). *Chaetoceros anastomosans:* chain in broad girdle view. After Hustedt (1930). Scale bars = 20 μm.

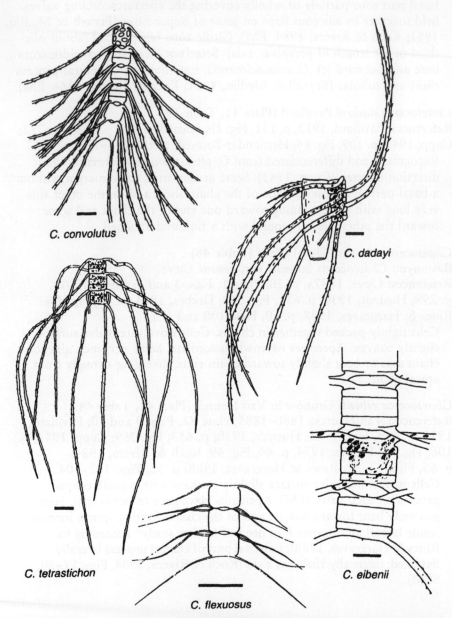

Chaetoceros

C. convolutus

C. dadayi

C. tetrastichon

C. flexuosus

C. eibenii

PLATE 41 *Chaetoceros convolutus:* chain in broad girdle view. *Chaetoceros dadayi:* with tintinnid. After Cupp (1943). *Chaetoceros tetrastichon:* chain. After Cupp (1943). *Chaetoceros flexuosus:* chain. *Chaetoceros eibenii:* Chain. After Cupp (1943). Scale bars = 20 μm.

Chaetoceros tetrastichon Cleve (Plate 41, Table 48)
References: Cleve, 1897a, p. 22, Plate 1, Fig. 7; Hustedt, 1930, p. 657, Fig. 371; Cupp, 1943, p. 108, Fig. 63; Hendey, 1964, p. 123, Plate 11, Fig. 1; Hernández-Becerril, 1992c, p. 367, Figs. 1-8.

Chains straight. Cells isovalvate; valve surface flat. Apertures almost absent. Setae emerging from valve margin, curved outward, almost at right angles to chain, toward ends turned nearly parallel to chain axis. One pair of setae generally curve more strongly and may be attached to a tintinnid (Hernández-Becerril, 1992c).

Distribution:
C. *borealis*—cosmopolitan?, probably most common in temperate and cold water.
C. *concavicornis*—cosmopolitan but absent from southern cold water region, common in northern cold water region and temperate waters.
C. *convolutus* and C. *densus*—cosmopolitan, probably most common in temperate waters.
C. *castracanei*—southern cold water region.
C. *coarctatus,* C. *dadayi,* and C.·*tetrastichon*—warm water region.
C. *eibenii*—warm water region to temperate.

Remarks: It may be questioned whether C. *hendeyi* Manguin is a separate species endemic to the southern cold water region or conspecific with C. *borealis*. It should be noted that specimens including at the same time features specific to C. *borealis* and C. *concavicornis,* are present in the northern cold water region (Gran, 1904; Braarud, 1935; Holmes, 1956). Based on these findings, Braarud (1935, p. 92) proposed C. *concavicornis* to be a form of C. *borealis*. *Chaetoceros convolutus* and C. *concavicornis* are similar in gross morphology as are C. *tetrastichon* and C. *dadayi*; distinctive characters have been pointed out previously.

4. Setae not fused.

Chaetoceros flexuosus Mangin (Plate 41)
References: Mangin, 1915, p. 45, Fig. 27; Hasle, 1968b, p. 7, Plate 10, Fig. 14, Map 4; Fryxell & Medlin, 1981, p. 6, Figs. 1–5 and 16–28.

Cells held together in colonies by sibling setae twisted around each other in one or two turns (no fusion of silica, Fryxell & Medlin, 1981, EM observations). Apertures narrow but conspicuous. A process located toward one margin from the valve center, externally forms a cone cut at the top (Fryxell & Medlin, 1981, SEM).

Morphometric data: Pervalvar axis, 9–20 μm; apical axis, 9–15 μm.

Distribution: Southern cold water region.

Remarks: *Chaetoceros flexuosus* has been found crowded with chloroplasts extending into the setae (Fryxell & Medlin, 1981). Despite the extremely delicate setae it thus belongs in *Phaeoceros*, which is a taxononmic position also supported by the presence of a central process on every valve.

B. Subgenus *Hyalochaete*

1a. Four to 10 chloroplasts.

Chaetoceros decipiens Cleve (Plate 42, Table 49)
References: Cleve, 1873b, p. 11, Plate 1, Figs. 5a and 5b; Hustedt, 1930, p. 675, Fig. 383; Cupp, 1943, p. 115, Fig. 70; Hendey, 1964, p. 123, Plate 12, Fig. 2; Drebes, 1974, p. 69, Fig. 52a; Evensen & Hasle, 1975, p. 161, Figs. 55–69; Rines & Hargraves, 1988, p. 75, Figs. 148, 149, and 152.

Chains straight and stiff. Valve corners touching those of adjacent cell. Apertures slit-like to broadly lanceolate. Sibling setae fused for a length several times their diameter and then diverging in pervalvar direction. Terminal setae thicker than the others, initially divergent, then bent, becoming parallel to chain axis. No resting spores. Central process with no distinct protrusion on the outside and a labiate structure on the inside (Evensen & Hasle, 1975, Figs. 63–66, EM).

Chaetoceros lorenzianus Grunow (Plate 42, Table 49)
References: Grunow, 1863, p. 157, Plate 5, Fig. 13; Hustedt, 1930, p. 679, Fig. 385; Cupp, 1943, p. 118, Fig. 71; Hendey, 1964, p. 124, Plate 16, Fig. 1; Rines & Hargraves, 1988, p. 85, Figs. 178–184.

Chains straight and stiff. Apertures elliptical to oval. Sibling setae fused only at the point of exit from margin; setae with distinct transverse rows of pores. Terminal setae divergent in broad girdle view for their whole length. Primary valve of resting spores with two conical protuberances which branch repeatedly and dichotomously at the tips; secondary valve smooth.

Chaetoceros mitra (Bailey) Cleve (Plate 42, Table 49)
Basionym: *Dicladia mitra* Bailey.
Synonym: *Dicladia groenlandica* Cleve.
References: Bailey, 1856, p. 4, Plate 1, Fig. 6; Cleve, 1873b, p. 12, Plate 2, Figs. 1 and 2; Cleve, 1896a, p. 8, Plate 2, Figs. 1 and 2; Hustedt, 1930, p. 677, Fig. 384; Hasle & Syvertsen, 1990b, p. 288, Figs. 5–9 and 11.

Chains straight. Apertures narrowly lanceolate and slightly constricted centrally. Setae crossing only at the point of exit from margin, setae without distinct structure (LM). Terminal setae parallel or convergent in broad girdle view and divergent in narrow girdle view. Primary valve of resting spores with two conical protuberances, terminating into a stalk which branches dichotomously at the tip; secondary valve smooth.

Remarks: Intermediate forms between *C. decipiens* and *C. lorenzianus* exist (Rines & Hargraves, 1988). Usually the two species are distinguished by the

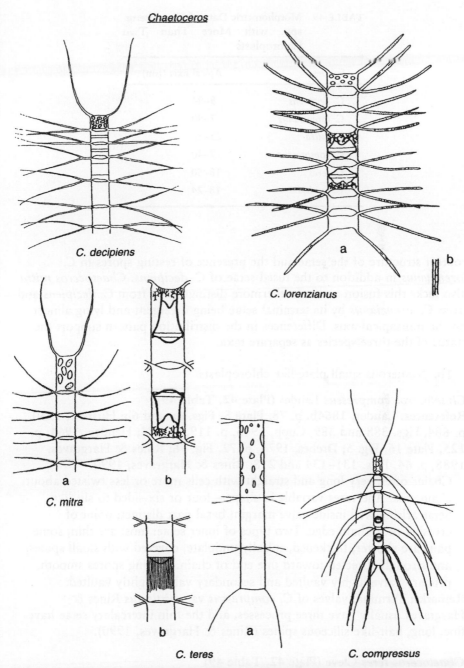

PLATE 42 *Chaetoceros decipiens:* setae fused at base. *Chaetoceros lorenzianus:* (a) chain; (b) detail of seta. Partly after Cupp (1943). *Chaetoceros mitra:* (a) partial chain with chloroplasts; (b) cells with resting spores. *Chaetoceros teres:* (a) chain with chloroplasts; (b) cell with resting spore. After Hustedt (1930). *Chaetoceros compressus:* chain with one pair of thickened intercalary setae and resting spores. Scale bars = 20 μm.

TABLE 49   Morphometric Data of *Chaetoceros*
spp. with More Than Two
Chloroplasts

Species	Apical axis (μm)
C. decipiens	9–84
C. lorenzianus	7–80
C. mitra	25–80
C. compressus	7–40
C. teres	18–50
C. lauderi	18–24

coarser structure of the setae and the presence of resting spores in *C. lorenzianus* in addition to the fused setae of *C. decipiens*. *Chaetoceros mitra* also lacks this fusion and is furthermore distinguished from *C. decipiens* and from *C. lorenzianus* by its terminal setae being divergent and lying almost on the transapical axis. Differences in the distribution pattern support the status of the three species as separate taxa.

1b. Numerous small plate-like chloroplasts.

Chaetoceros compressus Lauder (Plate 42, Table 49)
References: Lauder, 1864b, p. 78, Plate 8, Figs. 6a and 6b; Hustedt, 1930, p. 684, Figs. 388 and 389; Cupp, 1943, p. 119, Fig. 74; Hendey, 1964, p. 125, Plate 16, Fig. 5; Drebes, 1974, p. 72, Fig. 56; Rines & Hargraves, 1988, p. 64, Figs. 131–134 and 218; Rines & Hargraves, 1990.
    Chains often very long and straight with cells more or less twisted about chain axis. Apertures variable in shape—four or six-sided to slit like. Setae arising well inside valve margin; basal part distinct; point of crossing near chain edge. Two types of inner setae: most are thin, some pairs are shorter, thickened, spirally undulate, covered with small spines, and strongly directed toward one end of chain. Resting spores smooth, primary valve highly vaulted and secondary valve slightly vaulted.
Remarks: Terminal valves of *C. compressus* var. *hirtisetus* Rines & Hargraves usually have three processes, and the thin intercalary setae have fine, long, hair-like siliceous spines (Rines & Hargraves, 1990).

Chaetoceros teres Cleve (Plate 42, Table 49)
References: Cleve, 1896b, p. 30, Fig. 7; Hustedt, 1930, p. 681, Fig. 386; Cupp, 1943, p. 118, Fig. 72; Hendey, 1964, p. 124, Plate 10, Fig. 3;

Drebes, 1974, p. 70, Figs. 53 and 54; Rines & Hargraves, 1988, p. 102, Fig. 203.

Chains straight and tight. Cells cylindrical. Apertures narrow slits. Inner setae more or less perpendicular to chain axis; terminal setae widely divergent. Resting spores with evenly vaulted and smooth primary valve; secondary valve slightly vaulted, often with a ring of long hair-like siliceous spines.

Chaetoceros lauderi Ralfs in Lauder (Table 49)
References: Lauder, 1864b, p. 77, Plate 8, Figs. 4a and 4b; Hustedt, 1930, p. 683, Fig. 387; Cupp, 1943, p. 118, Fig. 73; Hendey, 1964, p. 125, Plate 13, Fig. 3; Drebes, 1974, p. 72, Fig. 55; Rines & Hargraves, 1988, p. 84, Figs. 170–173.

Chains somewhat twisted. Otherwise distinguished from *C. teres* by the shape of the primary valves of the resting spores, which are highly vaulted or capitate and spiny.

Distribution:
C. decipiens—cosmopolitan.
C. lorenzianus—warm water region.
C. mitra—northern cold water region.
C. compressus and *C. lauderi*—warm water region to temperate.
C. teres—northern cold water region to temperate.
Remarks: *Chaetoceros compressus* is a common species, often occurring in great abundances. *Chaetoceros teres* and *C. lauderi* are less frequently recorded, and they are differentiated by their resting spores and distribution.

Chaetoceros neglectus Karsten (Plate 44)
References: Karsten, 1905, p. 119, Plate 16, Fig. 5; Hasle, 1968b, p. 7, Plate 10, Fig. 13, Map 4; Priddle & Fryxell, 1985, p. 46.

Weakly silicified; one chloroplast, one type of inner setae, and one resting spore valve with setae extending into the setae of the parent cell (G. Hasle and E. Syvertsen, personal observations). Except for this difference, the description of *C. compressus* is also valid for *C. neglectus*.
Morphometric data: Apical axis, 10–15 μm.
Distribution: Southern cold water region.

2. Cells with two chloroplasts.

Chaetoceros didymus Ehrenberg (Plate 43, Table 50)
References: Ehrenberg, 1845b, p. 75; Hustedt, 1930, p. 688, Figs. 390 and 391; Cupp, 1943, p. 121, Fig. 75; Hendey, 1964, p. 125, Plate 17, Fig. 2; von Stosch et al., 1973; Drebes, 1974, p. 73, Fig. 57; Rines & Hargraves, 1988, p. 77, Figs. 154-163; Takano, 1990, pp. 288–289; Hernández-Becerril, 1991c.

Chaetoceros

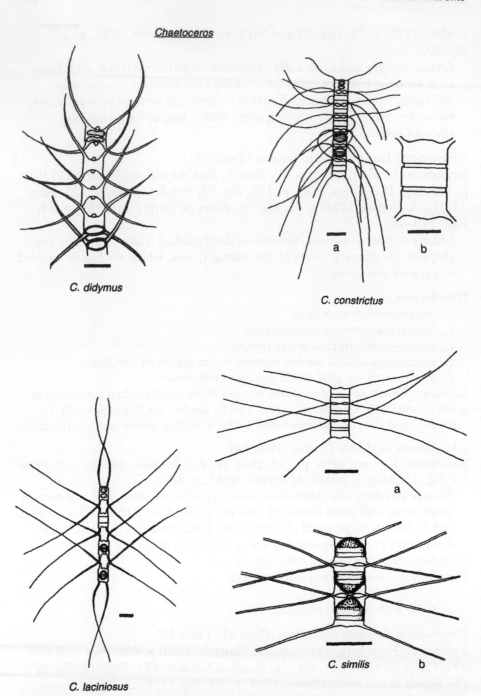

C. didymus

C. constrictus

a   b

a

b

C. laciniosus   C. similis

PLATE 43 *Chaetoceros didymus:* valve protrusions and resting spores. *Chaetoceros constrictus:* (a) chain with two resting spores; (b) detail of cell with strong constriction between valve mantle and band. *Chaetoceros laciniosus:* chain with two resting spores. *Chaetoceros similis:* (a) chain; (b) chain with partly developed resting spores. After Hustedt (1930). Scale bars = 20 μm.

TABLE 50   Morphometric Data of *Chaetoceros*
spp. with Two Chloroplasts

Species	Apical axis (μm)
C. anastomosans	8–20
C. constrictus	12–36
C. didymus	10–40
C. laciniosus	9–42
C. similis	7–17

Chains straight. Apertures large and partly filled by the valve
protuberance. Setae arising from corners of cells, crossing at their bases or
farther out, sometimes far outside chain edge. Each chloroplast with a
pyrenoid located in the protuberance. Resting spores smooth; within pairs
of resting spore parent cells with short, thick setae on hypovalve.
Remarks: Rines & Hargraves (1988) interpreted *C. didymus* as a complex
of taxa including a number of varieties. Based on EM investigations
Hernández-Becerril (1991c) suggested that *Chaetoceros protuberans* Lauder
should be regarded as a separate species and not as a variety of *C. didymus.*
According to the same investigation *C. didymus* has a centrally located
process consisting of a slit-like hollow with a very short projection to the
outside; a similar process is located at the center of the typical protuberance
of *C. protuberans.*

Chaetoceros constrictus Gran (Plate 43, Table 50)
References: Gran, 1897b, p. 17, Plate 1, Figs. 11-13, Plate 3, Fig. 42;
Hustedt, 1930, p. 694, Fig. 395; Cupp, 1943, p. 122, Fig. 76; Hendey,
1964, p. 126, Plate 9, Fig. 2; Rines & Hargraves, 1988, p. 67, Figs.
128-130.
Chains straight. Valve poles drawn up; corners of adjacent cells touching.
Apertures lanceolate, slightly narrowing in center. Terminal setae
diverging at an acute angle. Constriction between valve mantle and band
conspicuous. Resting spores with unequally vaulted, spiny valves.

Chaetoceros laciniosus Schütt (Plate 43, Table 50)
References: Schütt, 1895, p. 38, Plate 4, Fig. 5; Hustedt, 1930, p. 701, Fig.
401; Cupp, 1943, p. 128, Fig. 80; Hendey, 1964, p. 127, Plate 13, Fig. 2;
Evensen & Hasle, 1975, p. 160, Figs. 42–45; Rines & Hargraves, 1988, p.
83, Figs. 167–169.
Chains straight and loose. Setae thin; basal part parallel to chain axis and
then perpendicular to chain axis; far outer part usually bent toward one

chain end. Terminal setae different from the others—almost parallel in broad girdle view and more diverging in narrow girdle view. Apertures high, elliptical, and square to rectangular. Each chloroplast with a central pyrenoid. Resting spore valves smooth; primary valve more or less highly vaulted and secondary valve almost flat to highly vaulted. Central process built as that in *C. curvisetus* but located near the edge of valve face (Evensen & Hasle, 1975, EM).

Chaetoceros similis Cleve (Plate 43, Table 50)
References: Cleve, 1896b, p. 30, Fig. 1; Hustedt, 1930, p. 720, Fig. 411; Cupp, 1943, p. 135, Fig. 90; Hendey, 1964, p. 130, Plate 15, Fig. 2; Rines & Hargraves, 1988, p. 94, Figs. 222 and 223.
Short and straight chains. Apertures narrow and divided into two parts by a central, raised region of the valve. Setae arising from cell corners, directed diagonally toward chain ends, crossing outside chain margin. Terminal setae parallel to the others. Resting spores pear shaped with small spines.

Chaetoceros anastomosans Grunow in Van Heurck (Plate 40, Table 50)
References: Van Heurck, 1880-1885, Plate 82, Figs. 6–8; Hustedt, 1930, p. 743, Fig. 429; Cupp, 1943, p. 140, Fig. 96; Drebes, 1974, p. 81, Fig. 65 as *C. externus* Gran.
Chains straight or slightly curved, mostly loose. Setae thin, arising from cell corners, variously bent, and not crossing but connected by pervalvarly directed bridges. Apertures wide.

Distribution:
C. anastomosans and *C. didymus*—warm-water region to temperate.
C. constrictus, *C. laciniosus*, and *C. similis*—northern cold water region to temperate?
Remarks: Except for the two chloroplasts these species have few characters in common, whereas each of them is readily identified by their specific features. *Chaetoceros vanheurckii* Gran is very similar to *C. constrictus*, differing only by some minor features of the resting spores. *Chaetoceros pelagicus* Cleve and *C. brevis* Schütt have both one chloroplast but are otherwise similar to *C. laciniosus*. Resting spores are unknown for *C. pelagicus*, those of *C. brevis* and *C. laciniosus* are similar. There seems to be a general consensus that positive identification of *C. brevis* is problematic.

   3. Cells with one chloroplast.
     3a. Chains curved or helical; intercalary setae all bent in one direction.

Chaetoceros curvisetus Cleve (**Plate 44, Table 51**)
References: Cleve, 1889, p. 55; Hustedt, 1930, p. 737, Fig. 426; Cupp, 1943, p. 137, Fig. 93; Hendey, 1964, p. 133, Plate 17, Fig. 6; Drebes, 1974, p 79, Fig. 63; Evensen & Hasle, 1975, p. 159, Figs. 23–26; Rines & Hargraves, 1988, p. 71, Figs. 141 and 142.
    Adjacent cells in chains connected by drawn up poles of the concave valves; basal part of setae short or missing. All setae directed toward the outside of the chain spiral (best seen in narrow girdle view). Apertures a lanceolate slit, elliptical, or nearly circular. Resting spores smooth; primary valve evenly rounded and secondary valve less rounded to almost flat. Central process short, flattened with no protrusion or thickening on the inside (Evensen & Hasle, 1975, Figs. 25 and 26).

Chaetoceros debilis Cleve (**Plate 44, Table 51**)
References: Cleve, 1894a, p. 13, Plate 1, Fig. 2; Hustedt, 1930, p. 740, Fig. 428; Cupp, 1943, p. 138, Fig. 95; Hendey, 1964, p. 133, Plate 14, Fig. 7; Drebes, 1974, p. 81, Fig. 64; Evensen & Hasle, 1975, p. 159, Figs. 27–32; Rines & Hargraves, 1988, p. 72, Figs. 143–147; Takano, 1990, pp. 286–287.
    Valves flat or slightly convex. Valves of adjacent cells in chains do not touch. Basal part of setae conspicuous. Setae crossing slightly outside chain edge, extending outward from the spiral. Apertures almost rectangular or slightly constricted in the middle. Primary valve of resting spores with two humps and two setae extending into the corners of the parent cell. Secondary valve smooth or with setae.

Chaetoceros pseudocurvisetus Mangin (**Plate 44, Table 51**)
References: Mangin, 1910, p. 350, Fig. 3, II, Fig. 4, II; Hustedt, 1930, p. 739, Fig. 427; Cupp, 1943, p. 138, Fig. 94; Hendey, 1964, p. 134, Plate 18, Fig. 1; Fryxell, 1978c, p. 68, Figs. 22–26; Rines & Hargraves, 1988, p. 89, Figs. 185–191; Takano, 1990, pp. 290–291.
    Cells in colonies joined by fusion of sibling setae and at the edges of valves by four elevated projections leaving a large lenticular aperture in the center between cells in broad girdle view.

Distribution:
    C. curvisetus—cosmopolitan, mainly temperate and warm waters.
    C. debilis—cosmopolitan, mainly cooler waters.
    C. pseudocurvisetus—warm water region.

    3b. Resting spores in pairs; resting spore parent cells with fused hypovalvar setae and no apertures.

Chaetoceros cinctus Gran (**Plate 45, Table 52**)
References: Gran, 1897b, p. 24, Plate 2, Figs. 23–27; Hustedt, 1930, p. 748, Fig. 432; Cupp, 1943, p. 142, Fig. 98; Hendey, 1964, p. 135, Plate 11, Fig. 4.

Chaetoceros

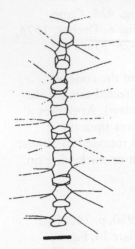

C. neglectus

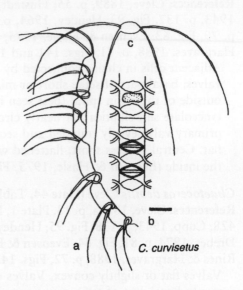

C. curvisetus

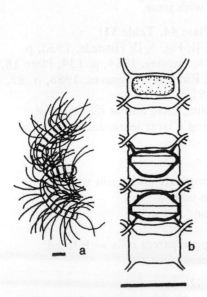

C. debilis

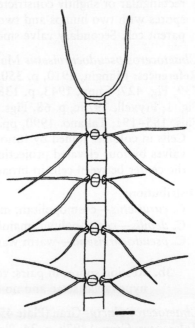

C. pseudocurvisetus

TABLE 51 Morphometric Data of *Chaetoceros* spp. with One Chloroplast and Curved or Helical Chains

Species	Apical axis (μm)
C. curvisetus	7–30
C. debilis	8–40
C. pseudocurvisetus	12–50

Chains straight or weakly curved. Setae thin, arising somewhat inside valve margin; basal part short, diagonal, and conspicuous. Setae crossing outside chain edge, perpendicular to chain axis. Apertures fairly large and rectangular. Resting spore setae curved to surround the cell like a girdle.

Chaetoceros furcellatus Bailey (Plate 45, Table 52)
References: Bailey, 1856, Plate 1, Fig. 4; Hustedt, 1930, p. 749, Fig. 433.
Chains straight or weakly curved. Setae thin, arising slightly inside valve margin; short and diagonal basal part. Setae of vegetative cells crossing slightly outside chain edge, irregularly oriented toward chain axis. Apertures rectangular, slightly compressed in center. Resting spores smooth, within resting spore parent cells with coarse hypovalvar setae (E. Syvertsen, personal observations), which are often fused for a fairly long distance, perpendicular to chain axis, twisting, branching, and diverging at a low angle.

Chaetoceros radicans Schütt (Plate 45, Table 52)
References: Schütt, 1895, p. 48, Fig. 27; Hustedt, 1930, p. 746, Fig. 431; Cupp, 1943, p. 141, Fig. 97; Hendey, 1964, p. 134, Plate 14, Fig. 4; Drebes, 1974, p. 82, Fig. 66; Fryxell & Medlin, 1981, p. 8, Figs. 9–15, and 29–35; Rines & Hargraves, 1988, p. 90, Figs. 192–198.
Chains straight or slightly curved, twisted about the chain axis. Setae arising from just inside the valve margin, all bent out transversely. Intercalary setae with hair-like siliceous spines; terminal setae without spines. Apertures narrow and elliptical with central constriction. Hypovalvar setae of resting spore parent cell thick and smooth and separate after the fused space to surround the cell like a girdle.

PLATE 44 *Chaetoceros neglectus:* part of chain with resting spores. *Chaetoceros curvisetus:* (a) narrow girdle view; (b) broad girdle view; (c) valve view. *Chaetoceros debilis:* (a) spiralled chain; (b) part of chain with resting spores. Partly after Cupp (1943). *Chaetoceros pseudocurvisetus:* partial chain. After Cupp (1943). Scale bars = 20 μm.

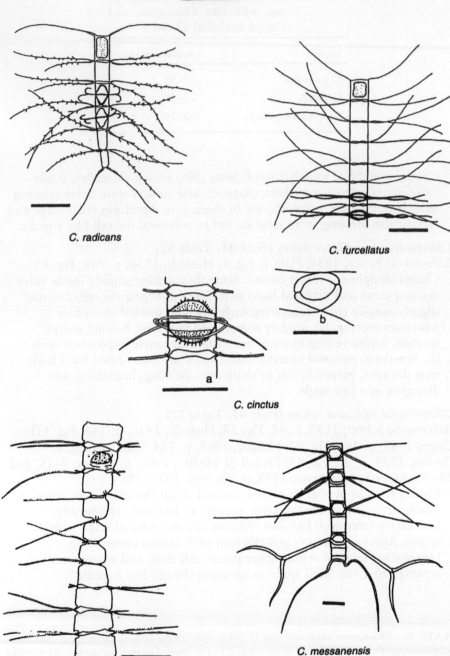

Chaetoceros

C. radicans

C. furcellatus

C. cinctus

C. tortissimus

C. messanensis

TABLE 52 Morphometric Data of *Chaetoceros* spp. with One Chloroplast and Paired Resting Spores

Species	Apical axis (μm)
C. cinctus	5–15
C. furcellatus	8–20
C. radicans	6–25
C. tortissimus	11–20

Chaetoceros tortissimus Gran (Plate 45, Table 52)
References: Gran, 1900, p. 122, Plate 9, Fig. 25; Hustedt, 1930, p. 751, Fig. 434; Cupp, 1943, p. 142, Fig. 99; Hendey, 1964, p. 135, Plate 11, Fig. 2; Drebes, 1974, p. 82, Fig. 67.

Chains straight or slightly bent, very strongly twisted about chain axis. Setae thin, arising somewhat inside the valve margin, perpendicular to chain axis but going in all directions. Apertures apparent only at corners; slightly convex valve surfaces of sibling cells touching in the middle. Gross morphology similar to that of the three species previously described. Resting spores unknown.

Distribution:
C. cinctus, C. radicans, and C. tortissimus—cosmopolitan.
C. furcellatus—northern cold water region.

Remarks: *Chaetoceros furcellatus* and *C. cinctus* can hardly be distinguished without resting spores. The typical *C. radicans* is conspicuous by the spiny setae and the twisted chains, but Rines & Hargraves (1988) found forms with few or no spines on the setae. Also, the shape of the *C. radicans* chains and apertures varied during the season, and a similarity with *C. cinctus* and *C. tortissimus* was therefore striking. *Chaetoceros furcellatus* is one of the most common species of this genus in the Arctic.

PLATE 45 *Chaetoceros radicans:* slightly twisted chain with two resting spores. *Chaetoceros furcellatus:* chain with two resting spores. *Chaetoceros cinctus:* (a) partial chain with resting spore; (b) resting spore in valve view. After Cupp (1943). *Chaetoceros tortissimus:* twisted chain. After Cupp (1943). *Chaetoceros messanensis:* part of chain with characteristic intercalary setae. After Hustedt (1930). Scale bars = 20 μm.

3c. Short rigid chains; two kinds of intercalary setae.

Chaetoceros diversus Cleve (Table 53)
References: Cleve, 1873a, p. 9, Plate 2, Fig. 12; Hustedt, 1930, p. 716, Fig.
409; Cupp, 1943, p. 132, Fig. 87; Hendey, 1964, p. 130, Plate 17, Fig. 4.
  Chains straight and usually short. Setae arising from cell corners, no basal
part. Apertures slit like. One type of intercalary setae thin, more or less
curved, and usually turned toward chain ends; and the other type heavy,
almost club shaped, first straight and at a sharp angle from chain axis,
then turning and running almost parallel to chain axis in outer part.
Terminal setae thin and differ from the intercalary setae in position, being
first U shaped, then nearly parallel to chain axis in outer part.

Chaetoceros messanensis Castracane (Plate 45, Table 53)
References: Castracane, 1875, p. 394, Plate 1, Fig. 1a; Hustedt, 1930, p.
718, Fig. 410; Cupp, 1943, p. 133, Fig. 89; Hendey, 1964, p. 129, Plate 12,
Fig. 3; Evensen & Hasle, 1975, p. 162, Figs. 70–74.
  Chains straight. Cells connected by drawn up poles of concave valves.
Apertures wide and linear six sided to almost round. Setae thin; no basal
part. Terminal setae strongly diverging, unlike, and usually one directed
backward from the chain. Some intercalary setae thicker than the others,
first fused, then forked. Central process as in *C. curvisetus* (Evensen &
Hasle, 1975, Figs. 71b).

Distribution: *C. diversus*, and *C. messanensis*—warm water region.
Remarks: *Chaetoceros laevis* Leuduger-Fortmorel is listed as a separate
species in Cupp (1943). Hustedt (1930) suggested that *C. laevis* and *C.
diversus* are conspecific.

3d. Valves of adjacent cells touch at valve poles.

Chaetoceros affinis Lauder (Plate 46, Table 54)
References: Lauder, 1864b, p. 78, Plate 8, Fig. 5; Hustedt, 1930, p. 695,
Figs. 396–398; Cupp, 1943, p. 125, Fig. 78; Hendey, 1964, p. 127, Plate

TABLE 53 Morphometric Data of *Chaetoceros*
spp. with One Chloroplast and
Two Types of Intercalary Setae

Species	Apical axis (μm)
C. diversus	8–12
C. messanensis	9–40

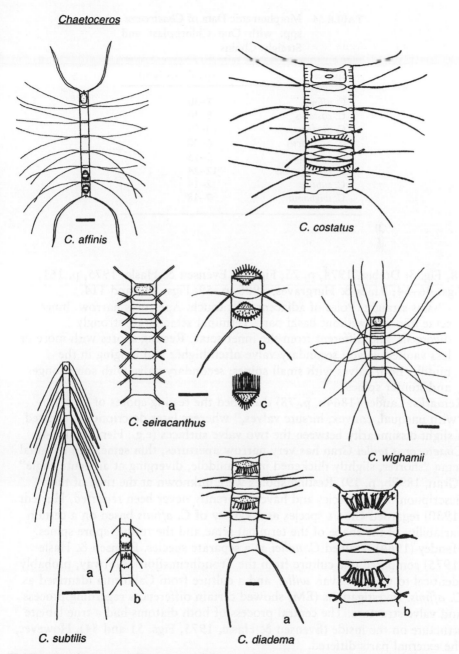

PLATE 46 *Chaetoceros affinis:* chain with two resting spores. *Chaetoceros costatus:* chain with resting spores. After Cupp (1943). *Chaetoceros seiracanthus:* (a) chain in broad girdle view; (b) part of chain with two resting spores; (c) single resting spore. After Hustedt (1930). *Chaetoceros wighamii:* chain with typical orientation of setae. After Hustedt (1930). *Chaetoceros subtilis:* (a) chain; (b) resting spores. After Hustedt (1930). *Chaetoceros diadema:* (a) partial chain; (b) two cells with resting spores. After Cupp (1943). Scale bars = 20 μm.

TABLE 54   Morphometric Data of *Chaetoceros*
spp. with One Chloroplast and
Straight Chains

Species	Apical axis (μm)
C. affinis	7–30
C. costatus	8–40
C. diadema	10–50
C. holsaticus	6–30
C. karianus	5–15
C. seiracanthus	12–24
C. subtilis	2–14
C. wighamii	7–18

18, Fig. 3; Drebes, 1974, p. 75, Fig. 58; Evensen & Hasle, 1975, p. 161,
Figs. 46–54; Rines & Hargraves, 1988, p. 59, Figs. 113 and 114.
   Chains straight. Poles of adjacent cells touch. Apertures narrow. Inner
   setae thin and without basal part. Terminal setae large, strongly
   divergent, and different from the inner setae. Resting spores with more or
   less vaulted valves; secondary valve often higher and bulging in the
   middle. Both valves with small spines; secondary valve with some longer
   and stouter spines.
Remarks: Lauder (1864b, p. 78) described the resting spores of *C. affinis*
"with unequal, convex, hirsute valves," whereas later descriptions reported
a slight dissimilarity between the two valve surfaces (e.g., Hendey, 1964).
Chaetoceros willei Gran has very narrow apertures, thin setae, and terminal
setae "shorter, slightly thickened in the middle, diverging at an acute angle"
(Gran, 1897b, p. 19). Resting spores were unknown at the time of the
description of the species and have apparently never been reported. Hustedt
(1930) regarded Gran's species as a variety of *C. affinis* based on a certain
variability of the shape of the terminal setae and the resting spore spines.
Hendey (1964) retained *C. willei* as a separate species. Evensen & Hasle
(1975) compared one culture from the Trondheimsfjord, Norway, probably
identical to *C. affinis* var. *willei,* and a culture from California identified as
C. affinis. Investigations (EM) showed certain differences regarding process
and valve structure. The central process of both diatoms had a true labiate
structure on the inside (Evensen & Hasle, 1975, Figs. 51 and 54). However,
the external parts differed.

Chaetoceros costatus Pavillard (Plate 46, Table 54)
References: Pavillard, 1911, p. 24, Fig. 1b; Hustedt, 1930, p. 699, Fig. 399;

Cupp, 1943, p. 127, Fig. 79; Hendey, 1964, p. 126, Plate 19, Fig. 3; Drebes, 1974, p. 77, Fig. 59; Rines & Hargraves, 1988, p. 69, Figs. 139 and 140.

Chains straight. Adjacent valves touch by two symmetrical valve protuberances at a short distance inside chain edge. Apertures small, elliptical, and shorter than apical axis. Setae thin and at nearly right angles to chain axis. Girdle bands conspicuous. Primary valve of resting spores evenly vaulted, with short spines; secondary valve smaller, centrally vaulted, and smooth.

Chaetoceros karianus Grunow (Table 54)
References: Grunow, in Cleve & Grunow, 1880, p. 120, Plate 7, Fig. 135; Hustedt, 1930, p. 736, Fig. 424.

Chains straight and short. Poles of adjacent valves touch. Apertures narrowly lanceolate to elliptical. Setae thin and without basal part. Some of the inner setae perpendicular to chain axis, others curved around the chain. Inner part of terminal setae U shaped and outer part bent and outward divergent. Resting spores not known.

Chaetoceros subtilis Cleve (Plate 46, Table 54)
References: Cleve, 1896b, p. 30, Fig. 8; Hustedt, 1930, p. 723, Fig. 413; Hendey, 1964, p. 130, Plate 10, Fig. 2; Rines & Hargraves, 1988, p. 96, Figs. 204–206.

Chains short. Valves flat adjacent valves fitting tightly together. Apertures missing. Setae thin, arising at valve edge, straight, and all directed toward one end of chain. Resting spores with spines; the two valves unequally vaulted.

Chaetoceros subtilis var. *abnormis* Proschkina-Lavrenko is characterized by having only one terminal seta (Proschkina-Lavrenko, 1955 as *Chaetoceros abnormis*).

Chaetoceros wighamii Brightwell (Plate 46, Table 54)
References: Brightwell, 1856, p. 108, Plate 7, Figs. 19–36; Hustedt, 1930, p. 724, Fig. 414; Cupp, 1943, p. 136, Fig. 91; Hendey, 1964, p. 131, Plate 11, Fig. 3.

Chains delicate and straight. Poles of adjacent valves touch. Apertures narrowly lanceolate to lanceolate. Setae thin, arising from valve margin, without basal part. Inner setae running very irregularly being perpendicular, bowed, or parallel to chain axis. Terminal setae not thicker than the others, often nearly parallel to chain axis. Primary valve of resting spores rounded, with fine spines, secondary valve constricted at base and blunt cone shaped in the middle.

3e. Valves of adjacent cells do not touch at the valve poles.

Chaetoceros diadema (Ehrenberg) Gran (**Plate 46, Table 54**)
Basionym: *Syndendrium diadema* Ehrenberg (= resting spore).
Synonyms: *Chaetoceros distans* var. *subsecunda* Grunow in Van Heurck,
Chaetoceros subsecundus (Grunow) Hustedt.
References: Ehrenberg, 1854, Plate 35a, Fig. 18/13; Van Heurck,
1880–1885, Plate 82bis, Figs. 6 and 7; Gran, 1897b, p. 20, Plate 2, Figs.
16–18; Hustedt, 1930, p. 709, Fig. 404; Cupp, 1943, p. 130, Fig. 83;
Hendey, 1964, p. 128, Plate 10, Fig. 1; Drebes, 1974, p. 77, Fig. 60;
Rines & Hargraves, 1988, p. 76, Figs. 150, 151, and 153.
  Chains slightly twisted about central axis. Setae arising inside valve
  margin; basal part extending outward in valvar plane, crossing at chain
  edge, generally perpendicular to chain axis but running fairly irregularly.
  Terminal setae diverging at an acute angle. Apertures rather wide and
  elliptical with a slight central constriction. Primary valve of resting spore
  topped with 4–12 dichotomously branching spines; secondary valve
  centrally inflated and smooth.

Chaetoceros holsaticus Schütt (**Table 54**)
References: Schütt, 1895, p. 40, Figs. 9a and 9b; Hustedt, 1930, p. 714,
Fig. 407; Cupp, 1943, p. 131, Fig. 85; Hendey, 1964, p. 128, Plate 15,
Fig. 4.
  Chains straight and sometimes slightly twisted. Setae thin, arising from
  inside valve margin, basal part running diagonally outward, crossing at
  chain edge; outer part of setae perpendicular to chain axis or bent toward
  chain ends. Terminal setae diverging at an acute angle. Apertures wide
  and hexagonal. Resting spores with small spines; primary valve larger
  than secondary valve.

Chaetoceros seiracanthus Gran (**Plate 46, Table 54**)
References: Gran, 1897b, p. 21, Plate 3, Figs. 39–41; Hustedt, 1930, p.
711, Fig. 405; Cupp, 1943, p. 131, Fig. 84; Hendey, 1964, p. 129, Plate 15,
Fig. 1; Rines & Hargraves, 1988, p. 92, Fig. 199.
  Setae thin. Otherwise as described for *C. diadema*. Primary valve of
  resting spores smoothly vaulted and covered with small spines, secondary
  valve rounded to capitate, also covered by spines.

Distribution:
  C. affinis and *C. diadema*—cosmopolitan.
  C. costatus and *C. seiracanthus*—warm water region to temperate.
  C. karianus—northern cold water region.
  C. holsaticus—cold water (Cupp, 1943), common in brackish water.
  C. subtilis and *C. wighamii*—probably restricted to brackish water.

3f. Chains curved and in irregular, spherical colonies.

Chaetoceros socialis Lauder (Plate 47)
References: Lauder, 1864b, p. 77, Plate 8, Fig. 1; Hustedt, 1930, p. 751,
Fig. 435; Cupp, 1943, p. 143, Fig. 100; Hendey, 1964, p. 136, Plate 15,
Fig. 3; Drebes, 1974, p. 82, Fig. 68; Evensen & Hasle, 1975, p. 160, Figs.
33–39; Rines & Hargraves, 1988, p. 95, Fig. 207; Takano, 1990, pp.
292–293.

Chains short. Poles of adjacent valves not touching one another. Three
setae of two adjacent valves short, the fourth one straight, elongated,
and serving in formation of the more or less spherical secondary colonies
by being entwined in the colony center with the elongated setae of
other chains. Resting spores, both valves rounded and smooth. The
central process is similar to that of *C. curvisetus* (Hasle & Evensen,
1975, EM).

Morphometric data: Apical axis, 2–14 μm.
Distribution: Probably cosmopolitan, very important in plankton close to
the ice in the northern cold water region and, according to Cupp (1943),
also one of the most prominent species in the Gulf of California.
Remarks: *Chaetoceros radians* Schütt, also appearing in spherical colony,
has usually been regarded as a separate species distinguished from *C.*
socialis by spiny resting spores (Plate 47) but was reduced in rank to a
variety of *C. socialis* by Proschkina-Lavrenko (1953).

3g. Unicellular species.

Rines & Hargraves (1988) mention seven unicellular *Chaetoceros* species
commonly reported in the literature, one of them has now been transferred to
Attheya and is discussed here. Of the other six Rines & Hargraves regarded *C.*
ceratosporus Ostenfeld, *C. muelleri* Lemmermann (inland waters), *C. simplex*
Ostenfeld, and *C. tenuissimus* Meunier as adequately described species possible
to recognize. According to the same authors "names such as *C. gracilis* Schütt
and *C. calcitrans* have most likely been applied to many different, not necessarily related forms which happen to have a similar appearance" (Rines & Hargraves, 1988, p. 99).

Chaetoceros ceratosporus Ostenfeld (Plate 47, Table 55)
References: Ostenfeld, 1910, p. 278; Hustedt, 1930, p. 760, Fig. 442;
Hendey, 1964, p. 138, Plate 17, Fig. 7; Rines & Hargraves, 1986, p. 104,
Figs. 1, 2, 22, and 23.

Cells most commonly single or in pairs. Valves drawn up at the poles,
each valve with one central process, usually visible with LM. Setae, thin,
originate at poles of apical axis and bend sharply outward. Primary valve

Chaetoceros

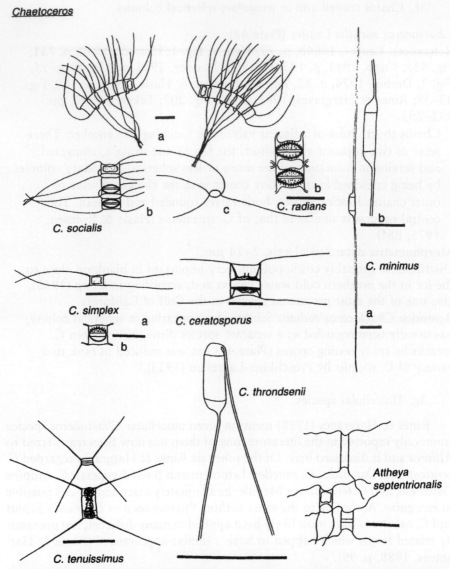

PLATE 47 *Chaetoceros socialis* and *Chaetoceros radians*: (a) chains and secondary colonies; (b) cells with resting spores; (c) cell in valve view. *Chaetoceros simplex*: (a) cell with chloroplast; (b) cell with resting spore. *Chaetoceros ceratosporus*: Cell with resting spore. After Hustedt (1930). *Chaetoceros minimus*: (a) whole cell; (b) detail. After Hustedt (1930). *Chaetoceros tenuissimus*: two cells. After Hustedt (1930). *Chaetoceros throndsenii*: part of cell with characteristic setae. After Marino et al. (1987). *Attheya septentrionalis*: three cells with characteristic setae. Scale bars = 20 µm.

TABLE 55   Morphometric Data of Unicellular
Chaetoceros spp.—Solitary Cells

Species	Apical axis (μm)
C. ceratosporus	4–20
C. simplex	4–30
C. tenuissimus	3–5

of resting spores more or less evenly vaulted with two large processes originating near valve edge, growing toward or into setae of vegetative valve, valve center covered with numerous small spines; secondary valve evenly vaulted to bluntly truncated, occasionally with processes and spines.

Chaetoceros ceratosporus var. brachysetus Rines & Hargraves
Reference: Rines & Hargraves, 1986, p. 105, Figs. 7–21 and 24–26; Rines & Hargraves, 1988, p. 63, Figs. 123–127.
   Setae thicker and shorter than in the nominate variety and constricted at the base.

Chaetoceros simplex Ostenfeld (Plate 47, Table 55)
References: Ostenfeld, 1901b, p. 137, Fig. 8; Hustedt, 1930, p. 755, Fig. 437; Hendey, 1964, p. 137, Plate 19, Fig. 2.
   Setae thin, long, straight and lying in the direction of apical axis of the cell. Resting spores with vaulted valves with spines.

Chaetoceros tenuissimus Meunier (Plate 47, Table 55)
Synonym: *Chaetoceros simplex* var. *calcitrans* Paulsen.
References: Paulsen, 1905, p. 6; Meunier, 1913, p. 49, Plate 7, Fig. 55; Hustedt, 1930, p. 756, Fig. 438; Rines & Hargraves, 1988, p. 97.
   Cells extremely small with apical and pervalvar axes approximately equal. Setae straight, sometimes scarcely longer than apical axis, emerging from poles of valve at a 45° angle to both apical and pervalvar axes. Resting spores unknown. Probably one chloroplast. See Rines & Hargraves (1988, p. 97) for discussion about *C. galvestonensis* Collier & Murphy and *C. calcitrans* f. *pumilus* Takano as synonyms of *C. tenuissimus*.
Remarks: In light of the evident identification problems the distribution of these species can scarcely be given. They seem to occur mainly in brackish and other inshore waters.

4. Two, seldom three, setae per cell.

Chaetoceros minimus (Levander) Marino, Giuffré, Montresor, & Zingone, 1991 **(Plate 47, Table 56)**
Basionym: *Rhizosolenia minima* Levander.
Synonym: *Monoceros isthmiiformis* Van Goor.
References: Levander, 1904, p. 115, Figs. 7 and 8; Van Goor, 1924, p. 303, Fig. 3; Hustedt, 1930, p. 598, Fig. 343; Marino et al., 1991, p. 318, Figs. 1–9.
Cells isovalvate. One seta per valve; the two setae running in opposite directions. Resting spores ellipsoidal; primary valve with scattered protuberances, knobs, and sometimes spines; secondary valve with regularly arranged numerous small knobs, a strong stud-like central protuberance, a wide flange, and a marginal collar.

Chaetoceros throndsenii (Marino, Montresor, & Zingone) Marino, Montresor, & Zingone var. *throndsenii* **(Plate 47, Table 56)**
Basionym: *Miraltia throndsenii* Marino, Montresor, & Zingone.
References: Marino et al., 1987; Marino et al., 1991.
Cells heterovalvate. One seta per valve; the two setae of a cell running in same direction. Primary valve of resting spores strongly convex with numerous small protuberances; secondary valve more flattened, with one or two central stud like protuberances, a wide flange and a marginal collar.

Chaetoceros throndsenii var. *trisetosus* Zingone (in Marino et al., 1991)
Reference: Marino et al., 1991, p. 319
Differs from the nominate variety by having a third seta.

Distribution:
  C. minimus—brackish water: northern part of the Baltic Sea, Dutch inshore waters, Tyrrhenian brackish water lagoons.
  C. throndsenii—Gulf of Naples.
Remarks: The small size and the weak silicification make the species difficult to recognize. Both species may appear in cell concentrations amounting to millions per liter (Marino et al., 1987). Whereas *C. minimus* is a typical

TABLE 56 Morphometric Data of *Chaetoceros* spp. with Two or Three Setae per Cell

Species	Pervalvar axis (μm)	Apical axis (μm)	Length of setae (μm)
C. minimus	6–32	2–7	10–220
C. throndsenii	8–15	1.5–5	30–40; 10–40

brackish water species, *C. throndsenii* made up 45% of the diatom
population at salinity of 35.9%.

Genus *Attheya* T. West
Type: *Attheya decora* T. West.
References: West, 1860, p. 152, Plate 7, Fig. 15; Hustedt, 1930, p. 768, Fig.
449; Crawford et al., 1994.

> *Attheya* is distinguished morphologically from *Chaetoceros* by structure
> of valve outgrowths or horns and by type of girdle bands as revealed with
> EM (Crawford et al., 1994). The frequent appearance of resting spores in
> *Chaetoceros* but not in *Attheya,* the planktonic habitat of *Chaetoceros,*
> and *Attheya* being attached to sand grains and other diatoms are other
> distinctive features.

Attheya septentrionalis (Østrup) Crawford (Plate 47)
Basionym: *Chaetoceros septentrionalis* Østrup.
Synonym: *Gonioceros septentrionalis* (Østrup) Round, Crawford, & Mann.
References: Østrup, 1895, p. 457, Plate 7, Fig. 88; Hustedt, 1930, p. 759,
Fig. 441; Hendey, 1964, p. 137, Plate 14, Fig. 5; Duke et al., 1973;
Evensen & Hasle, 1975, p. 164, Figs 79–82; Round et al., 1990, pp. 334
and 340; Crawford et al., 1994, p. 41, Figs. 42–49.

> Solitary or in pairs. Horns fairly long [three times cell length (Crawford et
> al., 1994, Table 2)], arising at the poles of the valves and projecting
> parallel to the valvar plane. Tips of the horns open and thickened.
> Chloroplasts one or two per cell.

Morphometric data: Apical axis, 4–6 μm.
Distribution: Northern cold water region to temperate(?).
Remarks: *Attheya longicornis* Crawford & Gardner (Crawford et al., 1994,
p. 38) has long (8–10 times cell length) and not markedly flexuous horns. It
has most likely been identified as *Chaetoceros septentrionalis* in the past,
and it may well be that specimens recorded as *C. septentrionalis* in
temperate waters belong to *A. longicornis.*
How to identify: *Bacteriastrum* spp., *Chaetoceros* spp., as well as
A. septentrionalis and *A. longicornis* may be identified in water mounts.
Phase contrast is recommended for the identification of the more delicately
structured, weakly silicified species especially for recognizing the setae.

**Family Lithodesmiaceae H. & M. Peragallo 1897–1908 emend.
Simonsen 1979**

> The circumscription of this family varies from including *Bellerochea,
> Ditylum,* and *Lithodesmium* (H. & M. Peragallo, 1897–1908; Glezer et al.,
> 1988) to including, in addition, *Streptotheca* and *Neostreptotheca* (Simonsen,
> 1979; Ricard, 1987) and, finally *Lithodesmioides* (von Stosch, 1987).

Lithodesmiaceae Round within the order Lithodesmiales Round & Crawford and subclass Lithodesmiophycidae Round & Crawford [all taxa described in Round et al. (1990)] includes *Lithodesmium, Lithodesmioides,* and *Ditylum.* Here, we follow von Stosch (1987) using the widest circumscription of the family.

We have taken into consideration the new name *Helicotheca* Ricard for the diatom genus described by Shrubsole (1890) as *Streptotheca,* the fungal genus *Streptotheca* Vuillemin being described in 1887 (Farr et al., 1979, p. 1692). The diatom genus *Streptotheca* was designated as the type of the new family Streptothecaceae Crawford (Round et al., 1990). Possible consequences regarding the name of the family should be taken into account by those using the classification by these authors.

Terminology specific to Lithodesmiaceae (after von Stosch, 1977, 1980, 1986, 1987):

Ansula—single element of the fringed marginal ridge of *Ditylum,* shaped as a ribbon longitudinally split in its medium part.

Bilabiate process—a process consisting of an external shorter or longer tube, sometimes reduced to a low ring (LM), and an internal part with a longer or shorter stalk and a trapezoid end piece closed at the tip but open at each of the two slanting sides by a longitudinal slit (EM, **Fig. 8**).

Fissiparity—split wall character, i.e., a localized *in vivo* separation of the siliceous and diatotepic layers of the cell wall, the diatotepic layer being the acidic layer rich in carbohydrates between the siliceous layer and the plasmalemma.

The two slits of the bilabiate process can hardly be seen with LM; however, the trapezoid shape of the internal part of a process in side view is discernible, e.g., in *Helicotheca.* The term bilabiate process was first introduced for *Bellerochea* and *Helicotheca* (von Stosch, 1977, p. 125). Von Stosch evidently considered fissiparity as an important descriptive character. We are not convinced of its usefulness for identification purposes, especially when dealing with preserved material, and it has therefore not been used here.

Family characters:

Cells solitary or in separable or inseparable ribbons.

Cells in girdle view rectangular, square, or shaped as a parallelogram.

Girdle consisting of several rows (columns) of bands (segments).

Valve outline biangular, triangular, quadrangular or quinqueangular.

Each valve with one bilabiate process.

Resting spores known in one genus.

KEY TO GENERA

1a. Marginal ridge present 2
1b. Marginal ridge absent, intercellular spaces missing 3
2a. Cells in inseparable or separable ribbons, seldom solitary 4
2b. Cells solitary 5
3a. Cells rectangular in broad girdle view . . . . *Helicotheca*, p. 234
3b. Cells like a parallelogram in broad girdle view . *Neostreptotheca*, p. 235
4a. Ribbons with conspicuous intercellular spaces . . . . *Bellerochea*, p. 227
4b. Cells in ribbons joined by a slight overlap of a conspicuous marginal ridge
. . . . *Lithodesmium*, p. 232
5a. Marginal ridge conspicuous, often fringed, no defined elevations at valve
corners . . . . *Ditylum*, p. 230
5b. Marginal ridge inconspicuous, well defined elevations at valve corners. .
. . . . *Lithodesmioides*, p. 231

Genus *Bellerochea* Van Heurck 1885 emend. von Stosch 1977 (Plate 48, Table 57)
Type: *Bellerochea malleus* (Brightwell) Van Heurck.
Basionym: *Triceratium malleus* Brightwell.
References: Brightwell, 1858b, p. 155, Plate 8, Figs. 6 and 7; Van Heurck, 1880–1885, p. 203, Plate 114, Fig. 1; Hustedt, 1930, p. 781, Fig. 456; Hendey, 1964, p. 122, Plate 6, Fig. 5; Drebes, 1974, p. 95, Figs. 80a and 80b; Hasle, 1975, Figs. 152–154; von Stosch, 1977, p. 128, Text Fig. 1, Figs. 1–70; von Stosch, 1987, p. 74, Figs. 204–208; Takano, 1990, pp. 294–295.

Generic characters:

Cells in ribbons.

Cells very weakly silicified.

Cells in broad girdle view roughly rectangular.

Cells in valve view biangular to quadrangular, rarely three or four armed.

Short elevation at each corner of the valve; those of adjacent cells in ribbons abutting.

Valves consisting of tracery of siliceous costae covering valve mantle, marginal ridge and most of valve face.

Bilabiate process with long external tube.

Chloroplasts numerous, oval and slightly constricted.

Characters showing differences between species:

Type of ribbons (separable or inseparable).

Shape of intercellular spaces.

Location of bilabiate process.

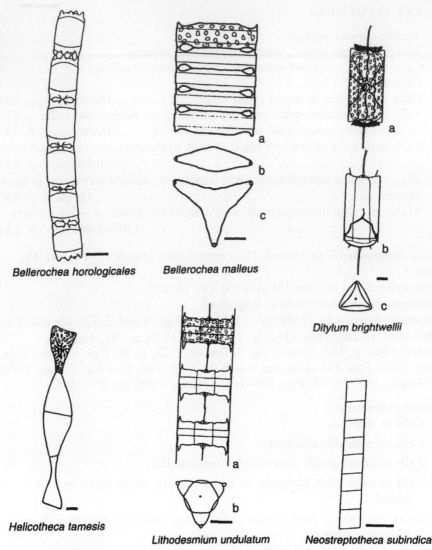

Bellerochea horologicales

Bellerochea malleus

Ditylum brightwellii

Helicotheca tamesis

Lithodesmium undulatum

Neostreptotheca subindica

PLATE 48 *Bellerochea horologicalis*: chain, short valve elevations and intercellular spaces. *Bellerochea malleus*: (a) short chain; (b and c) two types of valves. *Ditylum brightwellii*: (a) single cell with outer parts of bilabiate process; (b) cell with resting spore; (c) valve view. *Helicotheca tamesis*: twisted chain. *Lithodesmium undulatum*: (a) short chain; (b) valve with central bilabiate process and depressions across the corners. *Neostreptotheca subindica*: chain in girdle view. Scale bars = 20 μm.

TABLE 57   Morphometric Data of *Bellerochea* spp.

Species	Pervalvar axis (μm)	Apical axis (μm)	Transapical axis (μm)	Mantle costae in 10 μm
B. horologicalis	40–54	28–98	25–32	23–27
B. malleus[a]	13–34	22–210	20–27	10–22
B. yucatanensis	34–70	16–33	15.5–21	ca. 33 (EM)

[a] Side length of triangular cells: 52–180 μm.

KEY TO SPECIES

1a. Ribbons inseparable, cells tightly joined, bilabiate process marginal 2
1b. Ribbons separable, cells loosely joined, bilabiate process central . . . . .
. . . . *B. yucatanensis* von Stosch
2a. Ribbons usually straight, cells biangular, triangular, or quadrangular, intercellular spaces drop shaped, open only near elevations, costae interrupted in valve center. . . . .
. . . . *B. malleus* (Brightwell) Van Heurck emend. von Stosch
2b. Ribbons curved in transapical plane, or straight or nearly straight, cells biangular (quadrangular?), intercellular spaces dumbbell shaped, costae partly continuous . . . . *B. horologicalis* von Stosch

Distribution:
   B. horologicalis—known from Florida (Gulf of Mexico) and Melville Bay, Australia.
   B. malleus—known with certainty only from the North Sea, the English Channel, the French and Portuguese Atlantic coasts, Portuguese Guinea, and Leigh, New Zealand.
   B. yucatanensis—known from Australia and the type locality, Porto Progreso, Yucatan.
How to identify: The species are distinguished mainly by the shape of the ribbons and may thus be identified in water mounts. Details of valve structure may be recognized by phase contrast examination of material cleaned of organic matter as air mounts (von Stosch, 1977, 1987) or mounted in a medium of a high refractive index.
Remarks: The information on distribution is from von Stosch (1977, 1987), who emphasized that all three species might have been identified as *B. malleus* in the past. *Bellerochea yucatanensis* is characterized by the loosely connected cells in ribbons implying that only single valves will be present in cleaned material. The two other species occur mostly as pairs connected by the filaments of the marginal ridges (except for terminal valves). In valve view *B. malleus* differs from *B. horologicalis* by the median part of the valve being

without costae and by an unevenly ribbed marginal ridge. Von Stosch (1986) described *B. horologicalis* var. *recta* from Townsville, north Queensland, Australia, as distinguishable from the nominate variety by having straight or nearly straight chains.

Genus *Ditylum* J. W. Bailey ex L. W. Bailey 1861 (Plate 48, Table 58)
Lectotype: *Ditylum trigonum* J. W. Bailey ex L. W. Bailey (*vide* Round et al., 1990, pp. 292 and 689).
Correct name: *Ditylum brightwellii* (West) Grunow (*vide* Van Heurck, 1880–1885, plate 114).
References: West, 1860, p. 149, Plate 7, Figs. 6a and 6b; Bailey, 1861, p. 332, Plate 7; Van Heurck, 1880–1885, Plate 114, Figs. 3–9, Plate 115, Figs. 1 and 2; Schröder, 1906, p. 355, Fig. 24; Hustedt, 1930, p. 784, Figs. 457–460; Cupp, 1943, p. 148, Fig. 107; Hendey, 1964, p. 111, Plate 5, Fig. 1; Drebes, 1974, p. 59, Fig. 44; Hasle, 1975, Figs. 144–148; Hargraves, 1982; von Stosch, 1987, p. 57, Figs. 112–203; Takano, 1990, pp. 296–297; Delgado & Fortuño, 1991, Plate 60, Figs. b, c, and d.

Generic characters:
  Cells solitary.

  Cells in girdle view rectangular.

  Cells in valve view usually triangular.

  Marginal ridge fimbriate (with ansulae) or slotted (a basal membrane with entire margin but perforated by evenly spaced pervalvar slots).

  Valve structure consisting of radially arranged poroid areolae and/or ribs starting from a nonperforated area around a central bilabiate process.

  External part of process long.

  Chloroplasts numerous small granules.

TABLE 58 Morphometric Data of *Ditylum* spp.

Species	Pervalvar axis (μm)	Diameter (μm)	Valve areolae or ribs in 10 μm	Mantle areolae in 10 μm
D. brightwellii	80–130	25–100	10	18
D. buchananii	52–112	73–139[a]	14–15	23–24
D. pernodii	—[b]	—	27–28	ca. 36
D. sol	—	40–225	19–20	—

[a] Side length.
[b] —, No data.

Characters showing differences between species:
Shape and structure of marginal ridge.

Valve structure: areolae, interareolar costae, discernible or not
discernible with LM.

KEY TO SPECIES

1a. Areolae around valve central area elongated 2
1b. Areolae around valve central area not elongated 3
2a. Areolae on valve face conspicuously larger than those on valve mantle,
marginal ridge entire and slotted, or fimbriate with ansulae. . . . .
. . . .*D. brightwellii*[13] (West) Grunow in Van Heurck
2b. Areolae close to central area especially large, the rest on valve face like
those on valve mantle, the presence of ansulae variable . . . .
. . . .*D. sol*[13] Grunow in Van Heurck
3a. Ansulae restricted to apical fourth of marginal ridge. . . . .
. . . . *D. buchananii* von Stosch
3b. Ansulae along whole marginal ridge. . . . . *D. pernodii* Schröder

Distribution:
D. brightwellii—cosmopolitan although not recorded from polar regions.
D. buchananii—Gulf of Carpentaria, Australia, Gulf of Thailand (von
Stosch, 1987).
D. pernodii—Papua, New Guinea, Gulf of Carpentaria, Townsville and
Melville Bay, Australia (von Stosch, 1987).
D. sol—warm water region.

How to identify: The species may be discriminated by combined phase
contrast examinations of whole frustules, preferably in a medium of a high
refractive index (e.g., Pleurax) and of single valves cleaned of organic matter
and mounted in a medium of a high refractive index.

Remarks: *Ditylum brightwellii* and *D. buchananii* form resting spores
(Hargraves, 1982; von Stosch, 1987). *Ditylum buchananii* and *D. pernodii*
both have a very delicate ornamentation of the valve face. The triangular
marginal ridge of *D. buchananii* has rounded corners, whereas the marginal
ridges of *D. pernodii* and *D. sol* have subacute and narrow corners.

Genus *Lithodesmioides* von Stosch 1987 (Table 59)
Type: *Lithodesmioides polymorpha* von Stosch.
Reference: von Stosch, 1987, p. 46, Figs. 60–111.

---

[13] Basionyms: *Triceratium brightwellii* T. West and *Triceratium sol* Grunow in Van Heurck,
respectively.

TABLE 59   Morphometric Data of *Lithodesmioides* spp.

Species	Pervalvar axis (μm)	Width (μm)	Valve areolae in 10 μm	Length of external processes in μm
L. minuta	14–28	17–25	21–26	ca. 3
L. polymorpha	27–65	33–66	18	ca. 3

Generic characters:
   Cells solitary.
   Cells in valve view regularly quadrangular and in one species
      alternatively irregularly triangular to quinqueangular.
   Central bilabiate process with short external tube.
   Poroid areolae in radial rows from nonperforated central area.
   Circular concavity in middle of valve face.
   No depressions between valve corners and central area.

Characters showing differences between species:
   Cell size.
   Valve outline.
   Size of central depression.

KEY TO SPECIES

1a. Valves regularly quadrangular, valve face smooth . . . .
    . . . . *L. minuta*   von Stosch
1b. Valves regularly quadrangular, or irregularly triangular, quadrangular, or
    quinqueangular, spines around valve center . . . .
    . . . . *L. polymorpha* von Stosch

Distribution
   L. minuta—Townsville, north Queensland, Australia.
   L. polymorpha—mouth of Norman River, Australia, probably benthic.
How to identify: Examination in phase contrast or interference contrast of
whole frustules mounted in Pleurax is evidently the superior method for
identification (von Stosch, 1986, 1987).

Genus *Lithodesmium* Ehrenberg 1839 (Plate 48, Table 60)
Type: *Lithodesmium undulatum* Ehrenberg.
References: Ehrenberg, 1841a, p. 127, Plate 4, Figs. 13a–13c; West, 1860, p.
148, Plate 7, Fig. 5; H. & M. Peragallo, 1897–1908, p. 394, Plate 96, Figs. 4 and

TABLE 60 Morphometric Data of *Lithodesmium* spp.

Species	Pervalvar axis (μm)	Side length (μm)	Valve areolae in 10 μm	Length of external processes in μm
L. duckerae	17–43	22–61	9–10	18
L. intricatum	Up to 55	34–63	21–24	8
L. undulatum	Up to 74	37–93	12–13	—[b]
L. variabile	<Width	25–65[a]	30 (SEM)	3.5–4.5

[a] Width (Takano, 1979).
[b] —, No data.

5; Hustedt, 1930, p. 789, Fig. 461; Cupp, 1943, p. 150, Fig. 108; Hendey, 1964, p. 111, Plate 6, Fig. 6; Hasle, 1975, Figs. 149–151; Takano, 1979; von Stosch, 1980; von Stosch, 1987, p. 42, Figs. 46–59; Takano, 1990, pp. 298–299.

Generic characters:

Cells solitary or in ribbons.

Cells in girdle view rectangular to square.

Cells in valve view triangular, rarely biangular or quadrangular.

Marginal ridge membraneous.

Each valve with a central bilabiate process with long external tube.

Valve face with radial rows of poroid areolae starting from a nonperforated area around the process.

Conspicuous elevations at valve corners.

Proximal to each elevation a depression in valve face across the corners.

Chloroplasts, numerous small bodies.

Characters showing differences between species:

Solitary or in ribbons.

Direction of marginal ridge in relation to valve mantle.

Structure of marginal ridge.

Shape of valve mantle.

Height of elevations.

Size of valve areolae.

KEY TO SPECIES

1a. Cells solitary or in ribbons. 2
1b. Cells solitary . . . . *L. duckerae* von Stosch
2a. Marginal ridge only partly perforated. 3

2b. Marginal ridge with clear pattern of perforation. . . . .
. . . . *L. undulatum* Ehrenberg
3a. Valves triangular, perforation of marginal ridge without a clear pattern
and restricted to its advalvar half . . . .
. . . .*L. intricatum*[14] (T. West) H. & M. Peragallo
3b. Valves more or less irregularly biangular, triangular, or quadrangular,
marginal ridge sparsely or not perforated . . . .*L. variabile* Takano

Distribution:
L. duckerae—warm water region, probably benthic (von Stosch, 1987).
L. intricatum and *L. undulatum*—warm water region to temperate.
L. variabile—recorded in the literature only from Japanese coastal waters,
but probably also present in Gulf of Naples (G. Hasle, personal
observations).
How to identify: As for the other genera of this family a combination of
girdle and valve view examinations may be needed. The distinction between
L. undulatum and *L. intricatum* manifest in the perforation of the marginal
ridge may be seen in water mounts but more clearly in a medium of a high
refractive index.
Remarks: *Lithodesmium undulatum* and *L. intricatum* differ in the shape of
the valve mantle and the marginal ridge. The valve mantle of *L. intricatum* has
a widening in its transition to the girdle and the marginal ridge is inclined
outwards or inwards. The valve mantle of *L. undulatum* is at a right angle to
the valvar plane, and the marginal ridge is in the same plane as the mantle and
the girdle. The sides of *L. undulatum* valves are undulated, whereas those of
L. intricatum are straight or slightly concave. *Lithodesmium duckerae* is
similar to *L. intricatum* but has higher elevations; the depressions across the
corners are on an average deeper, the valve face areolation is much coarser,
and there are siliceous spines in the valve center. The taxonomic position
of *L. variabile* is dubious. Von Stosch (1987) suggested a similarity to
Lithodesmioides based on the presence of a nonperforated and low marginal
ridge. It has, however, depressions across the valve corners, and our own
observations indicate the presence of some perforation of the marginal ridge,
and, as also concluded by von Stosch (1987), its correct position therefore
seems to be in *Lithodesmium*.

Genus *Helicotheca* Ricard 1987
Synonym: *Streptotheca* Shrubsole 1890.
Type: *Helicotheca tamesis* (Shrubsole) Ricard; see Ricard (1987, p.75).
Monospecific genus.

***Helicotheca tamesis* (Shrubsole) Ricard (Plate 48)**
Basionym: *Streptotheca tamesis* Shrubsole.

[14] Basionym: *Triceratium intricatum* T. West.

References: Shrubsole, 1890; Hustedt, 1930, p. 778, Fig. 455; Cupp, 1943, p. 147, Fig. 106; Hendey, 1964, p. 113, Plate 7, Fig. 2; Drebes, 1974, p. 98, Fig. 80c; Hasle, 1975, p. 126, Figs. 131–141; von Stosch, 1977, p. 134, Figs. 78–84.

Girdle view: Chains ribbon shaped and separable. Chains and cells with a torsion in relation to pervalvar axis. Cells in broad girdle view rectangular apart from the torsion. Chloroplasts numerous and broadly oval with a slight constriction, each with a pyrenoid.

Valve view: Valves linear oblong and slightly inflated in the middle. Each valve with a subcentral bilabiate process, the external part a low ring, and the internal part with a long stalk. Valve structure not resolvable with LM.

Morphometric data: Pervalvar axis, 56–120 μm, apical axis, 26–160 μm; transapical axis, 9–11 μm.

Distribution: Warm water region to temperate.

How to identify: The species may be identified in chains in water mounts and as single cleaned valves mounted in a medium of a high refractive index by the valve outline and shape and position of the process.

Genus *Neostreptotheca* von Stosch 1977 emend. von Stosch 1987 (Plate 48, Table 61)
Type: *Neostreptotheca subindica* von Stosch.
References: von Stosch, 1977, p. 134, Figs. 85–94; von Stosch, 1987, p. 78, Figs. 209–234.

Generic characters:

Cells in straight separable ribbons.

Intercellular spaces missing.

Cells in broad girdle view shaped like a parallelogram.

Cells in valve view linear oblong and slightly inflated at the apices and in the middle.

TABLE 61 Morphometric Data of *Neostreptotheca* spp.

Specie	Pervalvar axis (μm)	Apical axis (μm)	Transapical axis (μm)
N. subindica	72–157	70–138	8–12
N. torta	—	28–77 (104)	—

Note: —, No data; number in parentheses occasionally found.

Each valve with a marginal bilabiate process, the external part a low
ring, and the internal part with a long stalk.

Valve structure not resolvable with LM.

Chloroplasts numerous and rounded.

KEY TO SPECIES (von Stosch, 1987, p. 82)

1a. Ribbons (or solitary), flat, valves oblique to girdle in broad girdle view
. . . . *N. subindica* von Stosch
1b. Chains (or solitary) twisted or spiraled, valves oblique or sometimes perpendicular to girdle in broad girdle view. . . . . . . *N. torta* von Stosch

Distribution: Warm water region.

Remarks: Von Stosch (1987) described *Neostreptotheca torta* with the form
triangularis from Australian plankton. *Neostreptotheca torta* f. *torta*
sometimes has rectangular cells and is thus habitually similar to *Helicotheca*
from which it differs by a special structure around the base of the process
which is revealed with TEM.

Family Eupodiscaceae Kützing 1849

Odontella is the only genus included in Eupodiscaceae *sensu* Simonsen, 1979
that has typically marine planktonic species. Round et al. (1990) placed *Odontella* in Triceratiaceae (Schütt) Lemmermann, order Triceratiales Round &
Crawford, subclass Biddulphiophycidae Round & Crawford, and Glezer et al.
(1988) placed *Odontella* in Biddulphiaceae Kützing, order Biddulphiales.

The species listed in this chapter as *Odontella* spp. were previously regarded
as *Biddulphia* spp. SEM investigations of genera of Biddulphiaceae showed,
however, that they belong in *Odontella* partly because they have ocelli and
labiate processes with long external tubes (Ross & Sims, 1971; Simonsen,
1974) whereas *Biddulphia* has pseudocelli.[15] Which species really belong to
Odontella is still disputed (Round et al., 1990); here we include those which
are usually regarded as planktonic.

Genus *Odontella* C. A. Agardh 1832 (Plate 49, Table 62)
Type: *Odontella aurita* (Lyngbye) C. A. Agardh.
Basionym: *Diatoma aurita* Lyngbye.
Synonym: *Biddulphia aurita*(Lyngbye) Brébisson.

---

[15] **Pseudocellus**—field of areolae decreasing in size from those on the main part of the valve.

PLATE 49 *Odontella aurita:* (a) short chain; (b) valve view. *Odontella sinensis:* single cell with
characteristic processes. *Odontella mobiliensis:* single cell, processes, and horns. After Cupp (1943).
Odontella longicruris: single cell, processes, and horns. After Cupp (1943). Scale bars = 20 μm.

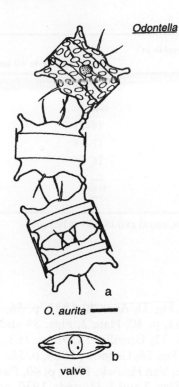

O. aurita

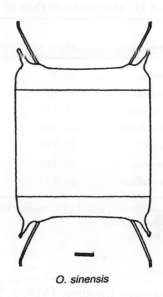

O. sinensis

b
valve

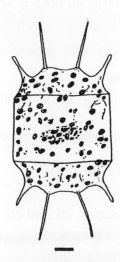

O. mobiliensis

O. longicruris

TABLE 62   Morphometric Data of *Odontella* spp.

Species	Apical axis (μm)	Valve areolae in 10 μm	Band areolae in 10 μm
O. *aurita*	10–97	8–11	8–14
O. *litigiosa*	30–80	—[b]	—
O. *longicruris*	15–110	12–17	18–21
O. *mobiliensis*[a]	45–200	14–16	17–18
O. *regia*[a]	90–200	14	16
O. *sinensis*	90–260	16–18	—
O. *weissflogii*	60–84	—	—

[a] O. *mobiliensis*, apical axis, usually 40–80 μm; O. *regia*, apical axis usually >100 μm (Hustedt, 1930).
[b] —, No data.

References: Lyngbye, 1819, p. 182, Plate 62, Fig. D; Agardh, 1832, p. 56; Brébisson & Godey, 1838, p. 12; Bailey, 1851, p. 40, Plate 2, Figs. 34 and 35; Schultze, 1858, p. 341; Schultze, 1859, p. 21; Greville, 1859b, p. 163, Plate 8, Fig. 10; Greville, 1866, p. 81, Plate 9, Fig. 16; Grunow, 1884, p. 58; Van Heurck, 1880–1885, Plate 101, Figs. 4–6; Van Heurck, 1909, p. 40, Plate 10, Fig. 141; Karsten, 1905, p. 122, Plate 17, Figs. 2 and 3; Hustedt, 1930, pp. 837, 840, and 846, Figs. 493, 495, and 501; Cupp, 1943, pp. 153, 154, and 161, Figs. 110–112; Hendey, 1964, pp. 103–105, Plate 20, Figs. 1 and 3, Plate 24, Fig. 6; Drebes, 1974, pp. 85, 90, and 91, Figs. 70, 73, and 76; Simonsen, 1974, p. 26; Hoban et al., 1980, p. 594, Figs. 15–38; Hoban, 1983, p. 283; Takano, 1984; Takano, 1990, pp. 300–303.

Generic characters:

Valves elliptical or lanceolate (bipolar).

An elevation (horn) with an ocellus at each pole.

Cells in straight (united by both elevations) or in zigzag chains (united by one elevation).

Two or more labiate processes per valve, usually with long external tubes.

Numerous small chloroplasts lying against valve wall.

Characters showing differences between species:

Planktonic species with weakly, littoral with coarsely silicified cell wall.

Curvature of valve face.

Position of processes.

Direction of external tubes of processes.

Shape and direction of elevations.

Valve wall spinose or not.

The presence or absence of resting spore formation.

KEY TO SPECIES

1a. Cell wall coarsely silicified, valve face between elevations evenly inflated 2

1b. Cell wall weakly silicified, middle part of valve face shaped in various ways. . . . . : 3

2a. Valve wall with shorter or longer spines, elevations robust, slightly divergent . . . . *O. litigiosa*[16] (Van Heurck) Hoban in Hoban et al.

2b. Valve wall coarsely areolated with no spines, elevations obtuse, inflated at base, divergent . . . . *O. aurita* (Lyngbye) C. A. Agardh

3a. Processes at a fair distance from the elevations. 4

3b. Processes close to slender elevations, valve face between processes flat or concave. . . . . *O. sinensis*[16] (Greville) Grunow

4a. Elevations prominent, valve face flat or concave or bulging in the middle 5

4b. Elevations inconspicuous, often more than two processes, divergent in direction, valve face flat or evenly convex . . . . *O. weissflogii*[16] (Janisch) Grunow

5a. Processes close together on a narrow, bulging middle part of valve face, external tubes diverging, elevations in pervalvar direction . . . . *O. longicruris*[16] (Greville) Hoban

5b. Middle part of valve face flat or slightly concave, external tubes of processes and elevations diverging . . . . *O. mobiliensis*[16] (Bailey) Grunow

Distribution:

O. aurita—cosmopolitan?, "usually in long chains attached to a substratum" (Hendey, 1964).

O. litigiosa and *O. weissflogii*—southern cold water region.

O. longicruris—warm water region to temperate.

O. mobiliensis, *O. sinensis*, and *O. regia*—cosmopolitan?

How to identify: The species may be identified in water mounts.

[16] Basionyms: *Biddulphia litigiosa* Van Heurck, *Biddulphia sinensis* Greville, *Biddulphia weissflogii* Janisch, *Biddulphia longicruris* Greville, *Zygoceros (Denticella?) mobiliensis* Bailey, respectively. Synonyms: *Biddulphia mobiliensis* (Bailey) Grunow in Van Heurck and *Biddulphia striata* Karsten (=*O. weissflogii*), respectively.

Remarks: *Odontella regia* (Schultze) Simonsen (basionym: *Denticella regia* Schultze) is close to *O. mobiliensis*. The Antarctic species have resting spores or appear in various phases (Hoban et al., 1980).

PENNATE DIATOMS

Order Bacillariales

Terminology specific to Pennate diatoms (Anonymous, 1975; Mann, 1978; Ross et al., 1979; Takano, 1983; Round et al., 1990): (Fig. 17)

Apical pore field—area of small pores at the apices of the valves; the pores are usually arranged in longitudinal rows, otherwise similar to the ocellus of the centric diatoms.

Apical slit field—area of elongate pores (slits) at the apices of the valves; the pores (slits) are separated laterally by long bars.

Foot pole (basal pole) and **head pole (apical pole)**—mainly used for cells in stellate, radiating, and bundle-shaped colonies in which neighboring cells are joined by valve faces of the foot poles; head pole is the other cell end.

Sternum—an elongate part of the valve where areolae are sparse (relative to the rest of the valve) or missing and which is often thickened pervalvarly (Mann 1978, p. 27); here used for the former pseudoraphe or axial area.

Striae—Parallel striae are perpendicular to the median line of the valve or to the raphe (Figs. 17a and 17c). Radiate striae are inclined from the valve

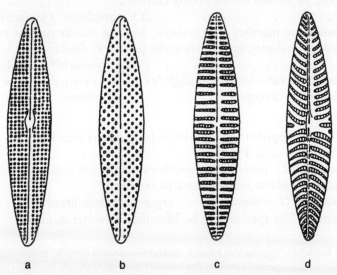

a   b   c   d

FIGURE 17   Valve striation in pennate diatoms. (a) Parallel striae; (b) radiate striae; (c) lineate, parallel striae; and (d) lineate, radiate striae.

margin toward the center of the valve (Figs. 17b and 17d). Convergent striae are inclined from the valve margin toward the apex. Lineate striae are crossed by finer longitudinal striation (Figs. 17c and 17d).

Whereas the pores and the slits will hardly be seen, the presence of an apical field is often recognized with LM as an open area lacking any resolvable structure.

Suborder Fragilariineae—Araphid pennate diatoms
Family Fragilariaceae Greville 1833

The family is poorly represented in marine phytoplankton; *Asterionellopsis glacialis* and *A. kariana* may be the only species dealt with in this chapter that may be characterized as being truly marine planktonic. It should be noted that *A. glacialis* also occurs in the surf zone communities (Lewin & Norris, 1970). *Striatella* is attached to a substratum by a mucilage stalk although it is often found in plankton collected close to the coast, as is also the only *Fragilaria* species mentioned in this chapter. *Synedropsis hyperborea* is fairly common in Arctic plankton but, like other *Synedropsis* species, it may be more common on ice and as an epiphyte on other diatoms, especially on *Melosira arctica* (Syvertsen, 1991, Table 1).

Common characters:

One labiate process near one or both apices.

One apical pore or slit field at each apex.

Uniseriate striae of poroid areolae.

Numerous(?) narrow bands.

Genus *Asterionellopsis* Round in Round et al., 1990 (Plate 50)
Type: *Asterionellopsis glacialis* (Castracane) Round in Round et al., 1990
Basionym: *Asterionella glacialis* Castracane.
Synonym: *Asterionella japonica* Cleve in Cleve & Möller.
References: Cleve & Möller, 1882, p. 3, No. 307; Cleve & Grunow, 1880, p. 110, Plate 6, Fig. 121; Castracane, 1886, p. 50, Plate 14, Fig. 1; Cupp, 1943, pp. 188–190, Figs. 138 and 139; Hustedt, 1959, pp. 254–256, Figs. 734 and 735; Körner, 1970, pp. 616–632, Figs. 36–41 and 108–128; Round et al., 1990, pp. 392 and 393; Takano, 1990, pp. 304–307.

Generic characters:

Cells with dissimilar ends in valve as well as in girdle views.

Cells joined by valve faces of expanded foot poles in star-like, spiral chains.

Foot poles angular in girdle view and more or less rounded in valve view.

Apical pore or slit fields (EM).

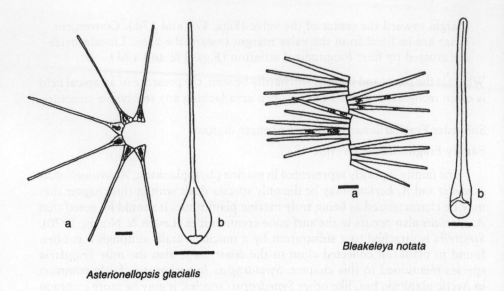

Asterionellopsis glacialis

Bleakeleya notata

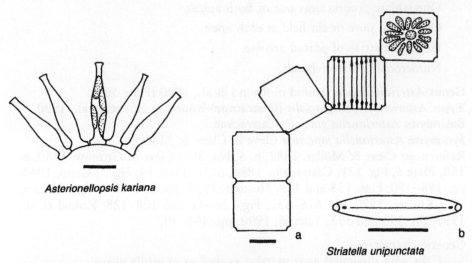

Asterionellopsis kariana

Striatella unipunctata

PLATE 50   *Asterionellopsis glacialis:* (a) chain in girdle view. Scale bar = 20 μm; (b) valve view, with apical field and sternum indicated in head pole. After Cupp (1943). Scale bar = 10 μm. *Asterionellopsis kariana:* chain in girdle view. Scale bar = 10 μm. *Bleakeleya notata:* (a) chain in girdle view. Scale bar = 20 μm; (b) valve view, with sternum and apical field. After Hustedt (1959). Scale bar = 10 μm. *Striatella unipunctata:* (a) chain in broad girdle view, with chloroplasts (top cell) and septa (second cell from top); (b) valve view, with sternum, apical labiate processes, and pore fields. Scale bars = 20 μm.

KEY TO SPECIES

1a. Cells in girdle view narrow with straight parallel sides and greatly expanded triangular foot pole; foot pole greatly widened and rounded in valve view, one or two chloroplasts in foot pole only . . . .
. . . . *A. glacialis* (Castracane) Round
1b. Cells in girdle view broad at foot pole, then suddenly constricted, then gradually widening again to the middle, then tapering toward head pole, two lobed, chloroplasts . . . . *A. kariana*[17] (Grunow) Round

Morphometric data:
  A. glacialis—apical axis, 30–150 μm; length of expanded part, 10–23 μm (ca. one-fourth of total length); transapical axis of expanded part, 8–12 μm; 28–34 transapical striae in 10 μm.
  A. kariana—Apical axis, 16–68 μm; transapical axis of foot pole, ca. 3 μm; at constriction, ca. 1 μm; median part, ca. 3–4 μm; head pole, ca. 0.5 μm (Körner, 1970).

Distribution:
  A. glacialis—cosmopolitan, sometimes abundant in plankton in cold to temperate coastal waters.
  A. kariana—northern cold water region to temperate (?).
How to identify: Whole cells in girdle view show sufficient details that the two species are distinguishable.
Remarks: *Asterionella socialis* Lewin & Norris (Lewin & Norris, 1970, p. 145), which was described from the surf zone in the State of Washington, should also be transferred to *Asterionellopsis* (Round et al., 1990, p. 392). *Asterionellopsis glacialis* cultured for some time regularly looses the narrow part of the cells and occurs as triangular or almost circular cells (G. Hasle and E. Syvertsen, personal observations; Körner, 1970, Figs. 119–122). Whereas both poles of *A. glacialis* have apical slit fields (Körner, 1970, Figs. 108–110; Hasle, 1973c, Figs. 22–25; Takano 1983, p. 27; Round et al., 1990, p. 393; Takano, 1990, p. 304), the broad foot pole of *A. kariana* carries pores and the narrow head pole slits (Körner, 1970, Fig. 126; Takano, 1990, pp. 306 and 307) and *A. socialis* seems to be like *A. kariana* in this respect (Lewin & Norris, 1970, Fig. 15).

Genus *Bleakeleya* Round in Round et al., 1990
Type: *Bleakeleya notata* (Grunow) Round in Round et al.
Monospecific genus.

[17] Basionym: *Asterionella kariana* Grunow in Cleve & Grunow

Bleakeleya notata (Grunow) Round in Round et al. (**Plate 50**)
Basionym: *Asterionella bleakeleyi* var. *notata* Grunow.
References: Grunow, 1867, p. 2; Hustedt, 1959, p. 254, Fig. 733; Round et al., 1990, p. 394.

Cells linear in girdle and valve views with dissimilar ends. Cells united by valve faces of expanded foot poles, in flat or twisted chains. Head pole rounded in valve view. Foot pole in valve view slightly inflated with a more or less angular or rounded outline, crossed by a transverse bar from which a narrow sternum rises. Basal part of foot pole with smaller areolae in radiating striae. Chloroplasts—numerous small granules scattered throughout the cell.

Morphometric data: Apical axis, 50–170 μm; transapical axis of foot pole, 4–10 μm; of head pole, 1.6–3 μm; 30–36 transapical striae in 10 μm (Körner, 1970).

Distribution: Warm water region.

How to identify: The shape of the colonies characterizes this species. Identification based on single cells or valves requires cleaned material mounted in a medium of a high refractive index.

Genus *Striatella* C. A. Agardh 1832
Lectotype: *Striatella unipunctata* (Lyngbye) C. A. Agardh (*vide* Ehrenberg, 1838, pp. 202 and 230).
Monospecific genus (as proposed by Round et al., 1990, p. 432).

Striatella unipunctata (Lyngbye) C. A. Agardh (**Plate 50**)
Basionym: *Fragilaria unipunctata* Lyngbye.
References: Lyngbye, 1819, p. 183, Plate 62, Fig. G; Agardh, 1832, p. 61; Cupp, 1943, p. 173, Fig. 122; Hustedt, 1959, p. 32, Fig. 560; Hendey, 1964, p. 161, Plate 26, Figs. 17 and 18.

Girdle view: Tabular with corners appearing as being cut off. Cells united to form ribbons or zigzag chains. Numerous open bands with narrow septa. Chloroplasts granular to oblong and radially arranged.

Valve view: Lanceolate with distinct apical pore fields, slightly sunk in and surrounded by a rim (SEM), thus the impression of cut off corners when seen with LM. One labiate process at each pole. Valve areolae in three self-crossing line systems. Sternum narrow.

Morphometric data: Apical axis, 35–125 μm; transapical axis, 6–20 μm; 6–10 bands in 10 μm; areolae in 18–25 oblique lines in 10 μm.

Distribution: Temperate species (Cupp, 1943).

How to identify: *Striatella unipunctata* is most easily identified as whole cells in girdle view in water mounts.

Remarks: *Tessella interrupta* Ehrenberg and *Hyalosira delicatula* Kützing, both listed as *Striatella* species by Hustedt (1959), have been transferred to *Microtabella* F. E. Round (Round et al., 1990).

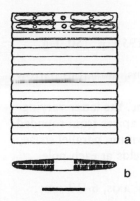

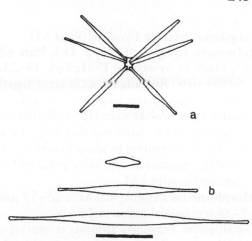

Fragilaria striatula

Synedropsis hyperborea

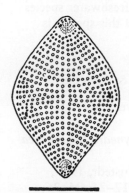

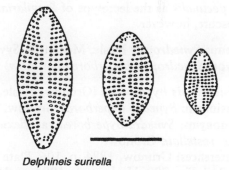

Delphineis surirella

Adoneis pacifica

PLATE 51 *Fragilaria striatula:* (a) ribbon in broad girdle view; (b) valve view with striation and sternum. Scale bar = 20 μm. *Synedropsis hyperborea:* (a) stellate colony, with cells in girdle and valve views; (b) valves, size variation. Scale bars = 20 μm. *Adoneis pacifica:* valve with areolation, apical pore fields and four labiate processes. After Andrews & Rivera (1987). Scale bar = 20 μm. *Delphineis surirella:* valves, size variation. Sternum broadened apically, apical pores, and labiate processes. After Hustedt (1959). Scale bar = 20 μm.

Fragilaria striatula Lyngbye (Plate 51)
References: Lyngbye, 1819, p. 183, Plate 63; Hustedt, 1959, p. 150, Fig. 663; Hasle & Syvertsen, 1981, Figs. 18–23.

Girdle view: Rectangular cells close together in ribbons, numerous narrow bands. Chloroplasts large, two per cell.

Valve view: Valves variable in silification and outline, the latter varying from broadly to narrowly elliptical, and the shape of apices from rounded to slightly attenuate to almost capitate. A narrow sternum, parallel striae, a labiate process at one valve pole, and apical pore fields at each pole discernible with LM.

Morphometric data: Apical axis, 25–53 μm; transapical axis, 6–10 μm; 17–28 transapical striae in 10 μm.

Distribution: *Fragilaria striatula* is benthic and a common epiphyte on larger algae but was also reported as a "neritic plankton species" by Hendey (1964). It was described from the Faeroe Is, North Atlantic, and is often recorded from other parts of the North Atlantic and from the Arctic. Due to the shape of the chains it may be confused with *Fragilariopsis* spp.

How to identify: *Fragilaria striatula* cannot be identified in girdle view and has to be examined in valve view as cleaned material mounted in a medium of a high refractive index.

Remarks: The typification of *Fragilaria* has been disputed. Williams & Round (1987) suggested that the genus name should be conserved for freshwater species. The marine *F. striatula* was placed into this genus by the author of the genus, and the type material has been examined with EM (Hasle & Syvertsen, 1981). Boyer (1927) designated the freshwater species *F. pectinalis* as the lectotype of *Fragilaria*. The identity of this species is obscure, however.

Genus *Synedropsis* Hasle, Medlin, & Syvertsen 1994
Type: *Synedropsis hyperborea* (Grunow) Hasle, Medlin, & Syvertsen.

Synedropsis hyperborea (Grunow) Hasle, Medlin, & Syvertsen (Plate 51)
Basionym: *Synedra hyperborea* Grunow.
Synonyms: *Synedra hyperborea* var. *flexuosa* Grunow; *Synedra hyperborea* var. *rostellata* Grunow.
References: Grunow, 1884, p. 106, Plate 2, Figs. 4–6; Hustedt, 1959, p. 217, Fig. 709; Hasle et al., 1994, p. 249, Figs. 1–12, 17–21, 24–26, 31–38, 45–47, and 142a.

Girdle view: Cells narrowly linear in stellate colonies.

Valve view: Specimens of maximum length rostrate; smaller valves with shorter prolongations; smallest specimens almost lanceolate; some valves with irregular inflations and indentations. A labiate process at one valve apex, a narrow sternum, and transapical striae discernible with LM. Apical slit fields (EM).

Morphometric data: Apical axis, 13–96 μm; transapical axis, 2.5–4 μm; 25–27 transapical striae in 10 μm.

Distribution: Northern cold water region, described from the undersurface of ice but often encountered in the plankton and as an epiphyte on *Melosira arctica.*

How to identify: *Synedropsis hyperborea* can probably be identified as whole cells in water mounts in Arctic material; in critical cases examination of many specimens mounted in a medium of a high refractive index using phase or interference contrast is needed to secure positive identification.

Remarks: The *Synedropsis* species are mainly living associated with sea ice, either attached to the ice itself or to the ice diatoms. The Arctic *S. hyperborea* and, probably more seldom, the Antarctic *S. recta* Hasle, Medlin, & Syvertsen and *S. hyperboreoides* Hasle, Syvertsen, & Medlin, are also found in the plankton close to the ice.

Family Rhaphoneidaceae Forti 1912

The habitat of the four genera dealt with here is shallow coastal water over sandy shores and mud and sand flats. They may be attached to sand grains or other particles [e.g., valves of other diatoms (Drebes, 1974)], but may be stirred up in turbulent water and thus become part of the plankton.

Characters common to *Adoneis, Delphineis, Neodelphineis,* and *Rhaphoneis:*

Cells solitary or in ribbons or zigzag or stellate colonies.

Cells rectangular in girdle view.

Valve outline linearly elliptical to broadly lanceolate, sometimes with produced apices or central inflation.

Large poroid areolae in uniseriate parallel or radiate striae.

Apical pore fields or one or two apical pores (EM).

One labiate process at each apex (one genus, *Adoneis,* usually with labiate processes also near the center of each lateral margin).

KEY TO GENERA

1a. Apical pore fields present 2
1b. Apical pore fields missing (one or two small apical pores present). . . 3
2a. One labiate process at each valve apex. . . . . *Rhaphoneis,* p. 251
2b. One labiate process at each valve apex and usually also near center of lateral margins . . . . *Adoneis,* p. 248
3a. Valve striae aligned across sternum . . . . *Delphineis,* p. 248
3b. Valve striae alternate, not aligned across sternum *Neodelphineis,* p. 249

Genus *Adoneis* G. W. Andrews & P. Rivera 1987
Type: *Adoneis pacifica* G. W. Andrews & P. Rivera.
Monospecific genus.

Adoneis pacifica G. W. Andrews & P. Rivera (Plate 51)
Reference: Andrews & Rivera, 1987.
   Valve view: Broadly lanceolate with smoothly rounded lateral margins
   and slightly produced rounded apices. Sternum narrow and distinct. Striae
   radiate. Single row of areolae on valve mantle continuous around apices.
   An apical pore field at each pole. One labiate process at each pole and
   usually one near the center of one or both lateral margins.
Morphometric data: Apical axis, 29–95 μm; transapical axis, 20–47 μm;
six to eight valve areolae in 10 μm.
Distribution: Chilean coastal waters, California coastal waters (Lange,
personal communication).

Genus *Delphineis* G. W. Andrews 1977, 1981 (Plate 51, Table 63)
Type: *Delphineis angustata* (Pantocsek) Andrews.
Basionym: *Rhaphoneis angustata* Pantocsek.
References: Ehrenberg, 1841a, p. 160, Plate 4, Fig. 12; Van Heurck,
1880–1885, Plate 36, Figs. 26 and 27; Boden, 1950, p. 406, Fig. 87;
Hustedt, 1959, p. 173, Fig. 679; Drebes, 1974, p. 103, Figs. 84a and 84b;
Simonsen, 1974, p. 35, Plate 23, Figs. 2–8; Andrews, 1977, 1981; Fryxell
& Miller, 1978, p. 116, Figs. 1–10; Round et al., 1990, p. 410.

   The genus comprises several fossil species (Andrews, 1977).

Characters common to recent species:
   Cells solitary or in shorter or longer ribbons.
   Valve outline linear or broadly elliptical to lanceolate.
   Striae parallel to slightly radiate.
   Rows of two or three areolae continue around valve apices.

TABLE 63   Morphometric Data of *Delphineis* spp.

Species	Pervalvar axis (μm)	Apical axis (μm)	Transapical axis (μm)	Areolae in 10 μm	Striae in 10 μm
D. karstenii	7–12.5	27–86	6–7	—	8–10
D. surirella	—	17–53	8–25	7–8	7–12
D. surirelloides	—	14–40	5.5–7.5	—	12–14

Note. —, No data.

Two small pores at each valve apex (usually not seen with LM).
Several chloroplasts.

KEY TO SPECIES

1a. Valves linear or linearly elliptical to oval. 2
1b. Valves broadly elliptical to lanceolate with slightly produced bluntly
   rounded apices, sternum distinct, narrow, and widening slightly near
   apices, cells solitary or in loose chains attached to particles (e. g. sand
   grains) . . . . *D. surirella*[18] (Ehrenberg) G. W. Andrews
2a. Valves linear with rounded apices to slightly inflated in center, wide sternum, long chains, planktonic . . . .
   . . . .*D. karstenii*[18] (Boden) G. Fryxell in Fryxell & Miller
2b. Valves linearly elliptical to broadly elliptical with broadly rounded apices,
   sternum variable in width, widening slightly near apices. . . . .
   . . . .*D. surirelloides*[18] (Simonsen) G. W. Andrews

Distribution:
D. karstenii—warm water region—off southwest coast of Africa
(Fryxell & Miller, 1981).
D. surirella—"cool to temperate seas" (Andrews, 1981), e.g., North Sea
(Drebes, 1974); Chile (Andrews & Rivera, 1987).
D. surirelloides—warm water region—Indian Ocean (Simonsen, 1974).

Genus *Neodelphineis* Takano 1982 (Plate 52, Table 64)
Type: *Neodelphineis pelagica* Takano.
References: Taylor, 1967, p. 440, Plate 3, Figs. 22–24; Simonsen, 1974, p.
36, Plate 23, Figs. 8–18; Takano, 1982; Hernández-Becerril, 1990; Takano,
1990, pp. 310–311; Round et al., 1990, p. 412; Hasle & Syvertsen, 1993,
p. 309, Figs. 32–41. Tanimura 1992, p. 136, Fig. 13.

Generic characters:
Cells solitary or in zigzag or stellate chains.

Valve outline linear to broadly elliptical with broadly rostrate apices or
   inflated at the center and sometimes at the apices.

Sternum narrow.

Striae parallel to slightly radiate.

Single row of valve mantle areolae continues around apices.

Usually one (occasionally two) fine pore near each apex (EM).

---

[18] Basionyms: *Zygoceros surirella* Ehrenberg, *Fragilaria karstenii* Boden, and *Rhaphoneis surirelloides* Simonsen, respectively.

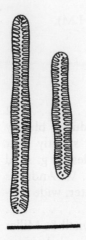

Neodelphineis indica

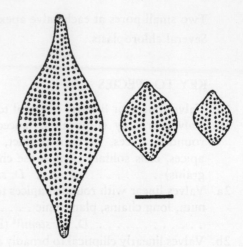

Rhaphoneis amphiceros

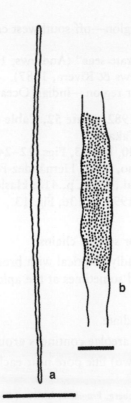

Toxarium undulatum

TABLE 64   Morphometric Data of *Neodelphineis* spp.

Species	Apical axis (μm)	Transapical axis (μm)	Striae in 10μm
N. indica	8–37	1.5–ca. 3	18–24
N. pelagica	5–23	2.8–5.8	14–18

KEY TO SPECIES

1a. Valve outline varying according to cell size: larger cells linearly elliptical, median cells broadly elliptical with broadly rostrate ends, smallest cells lanceolate . . . . *N. pelagica* Takano

1b. Valve outline of all cell sizes: linear, inflated at the center and less so at apices. . . . . *N. indica*[19] (F. J. R. Taylor) Tanimura

Distribution:

N. indica—warm water region to temperate (?)—Indian Ocean, Gulf of Mexico, Gulf of California, and Central and North Pacific.

N. pelagica—warm water region to temperate (?)—Japan, coasts of Texas and Florida (Round et al., 1990), and Pacific Ocean off Mexico (Hernández-Becerril, 1990).

Remarks: A further distinctive feature of *Neodelphineis* compared to *Delphineis* is the presence of pointed, raised single spines located on the interstriae on the edge of the valve face (SEM).The light micrographs of *Synedra indica* in Simonsen (1974, Plate 23, Figs. 9–18) illustrate the invariability of the valve outline of *N. indica* (11–37 μm long, Simonsen 1974, Figs. 9–14) and a part of the variability of the valve outline of *N. pelagica* (ca. 15–24 μm long; Simonsen 1974, Figs. 15–18).

Genus *Rhaphoneis* Ehrenberg 1844
Lectotype: *Rhaphoneis amphiceros* (Ehrenberg) Ehrenberg (*vide* Boyer, 1927, p. 190).

---

[19] Basionym: *Synedra indica* F. J. R. Taylor.

PLATE 52 *Neodelphineis indica:* valves showing outline, striation, apical pores and labiate processes. Scale bar = 10 μm. *Rhaphoneis amphiceros:* valves, size variation. Striation, sternum, apical pore fields, and labiate processes. After Hustedt (1959). Scale bar = 10 μm. *Toxarium undulatum:* (a) valve outline. Scale bar = 100 μm; (b) central part of valve, showing structure. Scale bar = 10 μm. After Cupp (1943).

The lectotype seems to be the only commonly recorded recent species left in this genus.

Rhaphoneis amphiceros (Ehrenberg) Ehrenberg (Plate 52)
Basionym: *Cocconeis amphiceros* Ehrenberg.
References: Ehrenberg, 1841b, p. 206; Ehrenberg, 1844a, pp. 74 and 87; Hustedt, 1959, p. 174, Fig. 680; Hendey, 1964, p. 154, Plate 26, Figs. 1–4; Drebes, 1974, p. 101, Fig. 83; Round et al., 1990, p. 406.
   Cells solitary and often attached to sand grains. Valve outline broadly elliptical or lanceolate with produced almost capitate apices to subcircular. Striae parallel or radiating. Sternum narrow and lanceolate. Small and distinct (LM) apical pore fields. Chloroplasts small and numerous.
   Morphometric data: Apical axis, 20–100 μm; transapical axis, 18–25 μm, six or seven striae in 10 μm.
Distribution: Probably cosmopolitan.

How to identify: Diatoms belonging to Rhaphoneidaceae can only be identified in valve view. The most coarsely silicified specimens may be identified in water mounts. Examination of cleaned valves mounted in a medium of a high refractive index is recommended.

Family Toxariaceae F. E. Round 1990

   Toxarium undulatum Bailey and *T. hennedyanum* Grunow are, at present, probably the only species in this family (Round et al., 1990). They are not typical plankton forms although *T. hennedyanum* was characterized as a "neritic plankton species" (Hendey 1964, p. 164). *Toxarium undulatum* was first found "attached in considerable numbers to *Sargassum vulgare,* in Narragansett Bay" (Bailey, 1854, p. 15), and Bailey's two other records of *T. undulatum* were also from *Sargassum*.

Genus *Toxarium* J. W. Bailey 1854 (Plate 52)
Type: *Toxarium undulatum* J. W. Bailey.
Synonym: *Synedra undulata* (J. W. Bailey) Gregory.
References: Bailey, 1854, p. 15, Figs. 24 and 25; Gregory, 1857, pp. 531–533, Plate 14, Figs. 107 and 108; Van Heurck, 1880–1885, Plate 42, Fig. 3; Hustedt, 1959, pp. 222–224, Figs. 713 and 714; Round et al., 1990, p. 422.

Generic characters:
   Needle like in valve and girdle views.
   Valves slightly expanded at both apices and at the center.
   No distinct sternum.
   Areolae scattered over valve face.
   Labiate processes and apical pore fields absent.

Toxarium undulatum has undulated valve margins; *T. hennedyanum* (Gregory) Grunow in Van Heurck (syn. *Synedra hennedyana* Gregory) has smooth valve margins.

Morphometric data:

T. hennedyanum—apical axis, 300–900 μm; transapical axis, 6–8 μm in valve center, 5–6 μm at the apices, and 2 μm in between; 9–11 striae in 10 μm (Hustedt, 1959).

T. undulatum—apical axis, up to 600 μm; 10–18 striae in 10 μm near apices (Hendey, 1964).

Distribution: Common, especially in tropical/subtropical waters (Round et al., 1990), but also occasionally in temperate waters.

How to identify: These two species are coarsely structured and may be identified in valve view in water mounts.

Family Thalassionemataceae Round 1990

In contrast to the other araphid families, Thalassionemataceae is exclusively marine and planktonic.

Family characters:

Cells solitary or in colonies of various types.

Cells needle shaped, often long, twisted, sometimes curved and expanded in the middle and at the apices.

Sternum usually wide and often varying in width along the cell length.

Areolae loculate with internal foramina and external vela (SEM).

Areolae circular to elongate transapically.

One labiate process at each end.

Apical spine(s) usually present at one or both ends.

Apical fields absent.

Marginal spines present or absent.

Numerous small chloroplasts scattered throughout the cell.

KEY TO GENERA

1a. Marginal spines lacking 2
1b. Marginal spines present . . . . *Thalassiothrix*, p. 263
2a. Cells solitary or in colonies of various shapes, cells usually straight. . 3
2b. Cells solitary or in bundles, bow shaped . . . . *Trichotoxon*, p. 267
3a. Cells in stellate, zigzag, or fan-shaped colonies, cells not twisted. . . . . . .
. . . . *Thalassionema*, p. 257
3b. Cells solitary or in stellate or fan-shaped colonies, cells twisted . . . . . .
. . . . *Lioloma*, p. 254

Additional characters to distinguish between genera (EM)
    Size of vela compared to that of foramina.
    Velum reticulate or consisting of struts (bars).

Genus *Lioloma* Hasle gen. nov. (Plates 53 and 54, Table 65)
Type: *Lioloma elongatum* (Grunow) Hasle comb. nov.
Basionym: *Thalassiothrix elongata* Grunow in Van Heurck.
Synonym: *Thalassiothrix vanhoeffenii* Heiden in Heiden & Kolbe.
References: Van Heurck, 1880–1885, Plate 37, Fig. 9; Karsten 1907, p.
397, Plate 46, Fig. 11; Pavillard, 1916, p. 39, Plate 2, Fig. 3; Heiden &
Kolbe, 1928, p. 566, Plate 6, Figs. 124 and 125; Allen & Cupp, 1935,
p. 153, Fig. 95; Hustedt, 1959, p. 248, Fig. 728; Cupp, 1943, p. 185, Fig.
136; Simonsen, 1974, p. 38, Plate 24, Fig. 5, Plate 25, Figs. 1–3;
Hallegraeff, 1986, pp. 70 and 72.

Generic characters:
    Cells solitary or united into colonies by valve surface of foot pole.
    Cells more or less twisted.
    Cells heteropolar, head pole with spines, and foot pole wedge shaped
        in valve and girdle views.
    Sternum narrow near valve apices and usually wider in the rest of the
        valve.
    Irregularly located marginal structures, each evident as an empty space
        in the place of an areola.[20]
    Foramina of about the same size as reticulate vela (**Plate 54**, SEM).
    The two lips of the labiate process dissimilar in size (**Plate 54, Fig. 1a**;
        parrot beak shaped process in Hallegraeff, 1986) (SEM).

KEY TO SPECIES

1a. Head pole bluntly rounded or almost square, with two spines, the other
    end (foot pole) smoothly tapering, areolae in two or more rows along
    each margin 2
1b. Head pole in valve view narrowly club shaped, with one spine, the other
    end suddenly constricting before a fairly long tapering apex, large marginal
    areolae in one row along each valve margin, sternum wide except near
    the tapering apex. . . . . . . *L. elongatum*[21] (Grunow) Hasle comb. nov.

---

[20] These structures may be difficult to observe with LM or may be lacking. With EM they are
    seen as flattened bubbles of about the same size as an areola, with a central opening and
    positioned on the internal valve surface (Plate 54, Figs. 1a–1c, 2b, and 3b, SEM).
[21] Basionyms: *Thalassiothrix elongata* Grunow in Van Hureck, *Thalassiothrix delicatula* Cupp and
    Thalassiothrix mediterranea var. *pacifica* Cupp, respectively. Synonym of *Lioloma elongatum*:
    Thalassiothrix vanhoeffenii Heiden.

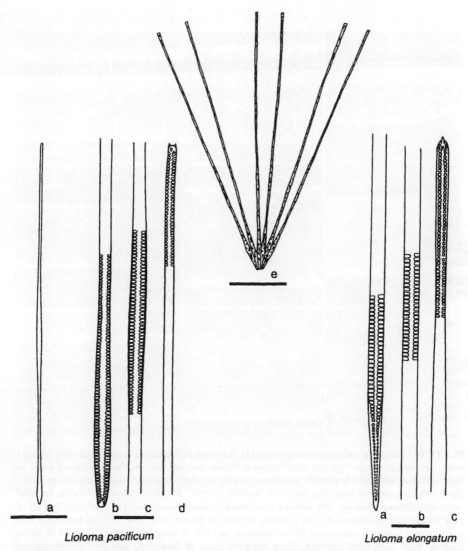

PLATE 53 *Lioloma pacificum* (a) valve outline. Scale bar = 100 μm; (b) foot pole, with valve structure; (c) median part of valve; (d) head pole with two apical spines. Scale bar = 10 μm; (e) colony. Scale bar = 100 μm. After Cupp (1943). *Lioloma elongatum* : (a) foot pole; (b) median part of valve; (c) head pole with one apical spine. Scale bar = 10 μm.

2a. Valve widened at head pole (valve and girdle views), gradually narrower until about one-third cell length from head pole, then wider and about the same width until tapering foot pole. Cell distinctly twisted. Structure very delicate. Solitary . . . . . *L. delicatulum*[21] (Cupp) Hasle comb. nov.

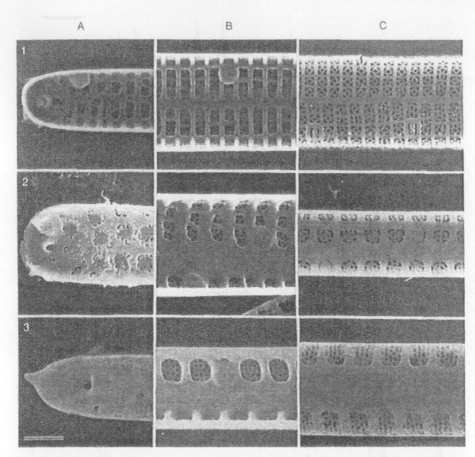

PLATE 54 Scanning electron micrographs. (1) *Lioloma delicatulum:* (a) internal view of head pole, two apical spines, "parrot beak"-shaped labiate process, two or three marginal rows of foramina, and one "bubble shaped" structure; (b) part of valve, internal view with narrow sternum, three marginal rows of foramina, and one "bubble-shaped" structure; (c) part of valve, external view with narrow sternum, vela divided into several compartments, and external openings (?) of two bubble-shaped structures. (2) *Lioloma pacificum:* (a) external view of head pole with two apical spines, external opening of labiate process, and vela in irregular pattern; (b) part of valve, internal view with wide sternum, three marginal rows of foramina, and two bubble-shaped structures; (c) part of valve, external view with wide sternum, and two marginal rows of vela. (3) *Lioloma elongatum:* (a) external view of head pole with one apical spine and external opening of labiate process; (b) part of valve, internal view with wide sternum, one marginal row of foramina, and one bubble-shaped structure; (c) part of valve, external view with wide sternum, and one marginal row of vela. Scale bar = 1 μm.

TABLE 65   Morphometric Data of *Lioloma* spp.

Species	Pervalvar axis (μm)	Apical axis (μm)	Transapical axis (μm)	Areolae in 10 μm
L. delicatulum	—	1120–1920	1–4	19–24
L. elongatum	4–6.6	990–2040	3–4	8–14
L. pacificum	1.8–7	525–1076	1.5–5	14–19

Note. —, No data. Size of pervalvar and transapical axes varies along the cell length in girdle as well as in valve views (Plate 53); the ranges given in Table 65 indicate the variation along the apical axis of the particular species.

2b. Valve width almost the same from head pole until somewhat enlarged for a short distance about one-third from wedge-shaped, blunt-pointed foot pole. Cells in star or fan-shaped colonies . . . .
. . . . *L. pacificum*[21] (Cupp) Hasle comb. nov.

Distribution:
   L. delicatulum—warm water region to temperate—off Portugal, NW Africa, Mediterranean, Gulf of Mexico, Indian Ocean, off California, South Pacific.
   L. elongatum—warm water region—Java, Indian Ocean, Gulf of Thailand.
   L. pacificum—warm water region to temperate—Mediterranean, Indian Ocean, South Atlantic.
Remarks: *Lioloma elongatum* is readily distinguished in LM by the coarser areolae (Table 65), the single apical spine, and the long tapering foot pole (Plate 53); *L. pacificum* is distinguished by the blunt, not especially widened head pole with two small spines and the more or less distinct widening near the foot pole (Plate 53), and *L. delicatulum* by the delicate areolation (Table 65), and in valve view, the wide head-pole.

   The diatom we identify as *Lioloma elongatum* has the morphological features illustrated for *T. elongata* in Van Heurck (1880–1885) and described and illustrated for *T. vanhoeffenii* in Heiden & Kolbe (1928).

Genus *Thalassionema* Grunow ex Mereschkowsky 1902 (Plates 55, 56, and 57, Table 66)
Type: *Thalassionema nitzschioides* (Grunow) *Mereschkowsky*
Basionym: *Synedra nitzschioides* Grunow.
Synonym: *Thalassiothrix nitzschioides* (Grunow) Grunow in Van Heurck.
References: Grunow, 1862, p. 403, Plate 5, Figs. 18a and 18b; Grunow, 1863, p. 140, Plate 5, Figs. 18a–18c; Greville, 1865a, p. 4, Plate 5, Figs. 5 and 6; Cleve & Grunow, 1880, p. 109; Van Heurck, 1880–1885, Plate 37, Figs. 11–13, Plate 43, Figs. 7–10; Mereschkowsky, 1902, p. 78, 1903

Thalassionema

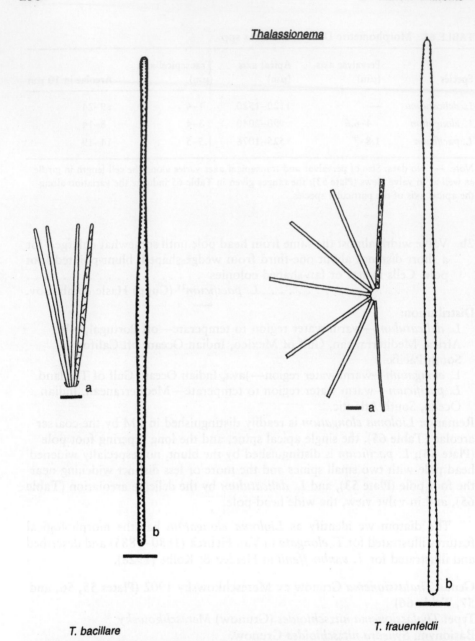

T. bacillare

T. frauenfeldii

PLATE 55 *Thalassionema bacillare*: (a) colony in valve view. Scale bar = 20 μm; (b) valve view with marginal structure. Scale bar = 10 μm. *Thalassionema frauenfeldii*: (a) colony in girdle view. After Cupp (1943); Scale bar = 20 μm. (b) valve view with marginal structure and apical spine. Scale bar = 10 μm.

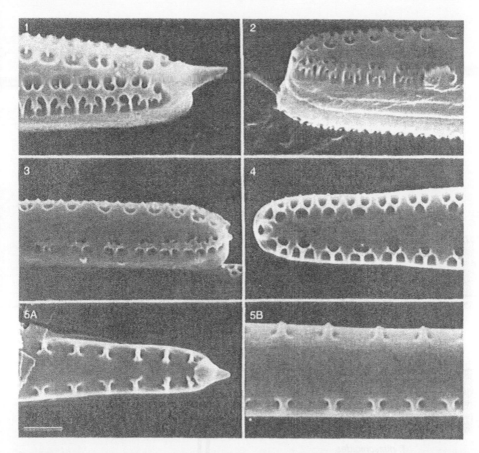

PLATE 56 Scanning electron micrographs. (1) *Thalassionema javanicum:* external view of pointed valve end, apical spine, one marginal row of areolae, and elaborate vela. (2) *Thalassionema nitzschioides:* external view of rounded valve end, wide sternum, one marginal row of areolae, and elaborate vela. (3) *Thalassionema bacillare:* external view of slightly dilated cell end, wide sternum, one marginal row of areolae and elaborate vela. (4) *Thalassionema pseudonitzschioides:* external view of the more narrow valve end, base of apical spine, wide sternum, one marginal row of areolae, and less elaborate vela (reduced ?). (5) *Thalassionema frauenfeldii:* external view with one marginal row of areolae crossed by a simple bar; (a) tapering end with apical spine; (b) part of valve with wide sternum. Scale bar = 1 μm.

pp. 91 and 178; Heiden & Kolbe, 1928, p. 564, Plate 6, Fig. 121; Meister, 1932, p. 25, Fig. 52; Cupp, 1943, p. 182, Fig. 133; Kolbe, 1955, p. 178; Hustedt, 1959, pp. 244 and 247, Figs. 725 and 727; Hasle & Mendiola, 1967; Drebes, 1974, p. 103, Fig. 85c; Schuette & Schrader, 1982; Hallegraeff, 1986, pp. 60–65.

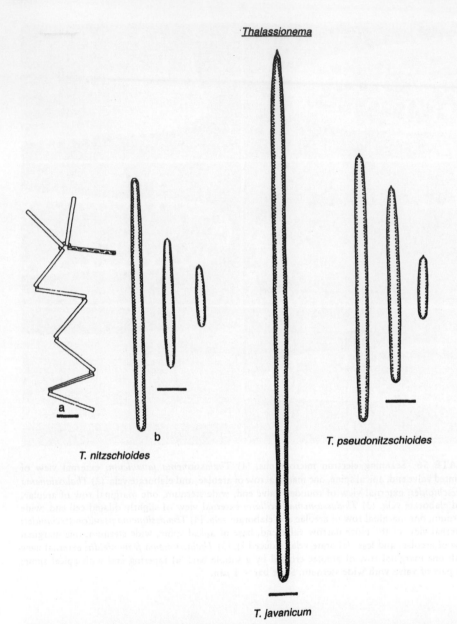

Thalassionema

T. nitzschioides

b

T. pseudonitzschioides

T. javanicum

PLATE 57 *Thalassionema nitzschioides:* (a) chain in girdle view. Scale bar = 20 μm; (b) valves showing size variation and marginal areolae. Scale bar = 10 μm. *Thalassionema javanicum:* valve with apical spine and areolae. Scale bar = 10 μm. *Thalassionema pseudonitzschioides:* valves, showing size variation, apical spine, and marginal areolae. Scale bar = 10 μm.

TABLE 66 Morphometric Data of *Thalassionema* spp.

Species	Pervalvar axis (μm)	Apical axis (μm)	Transapical axis (μm)	Areolae in 10 μm
T. bacillare	2–3	97–230	1.3–4[a]	7–8[b]
T. frauenfeldii	—[d]	54–200	2–4	5–9
T. javanicum	—	142–180	1.2–4	10–12
T. nitzschioides	—	10–110[c]	2–4	10–12
T. pseudonitzschioides	1–2	10–200	2–4	10–12

[a] 3–4 μm in the middle of the valve, 1.7–2.7 μm at the ends, 1.3–2.0 μm in the narrow parts close to the valve ends (Hasle & Mendiola, 1967); Hallegraeff (1986) gives 400 μm as the maximum length and 5.5–10 areolae in 10 μm.
[b] Greater number closer to the apices.
[c] The maximum length is after Hustedt (1959); Hasle & Mendiola (1967) measured 10–70 μm.
[d] —, No data.

Generic characters:

Cells in girdle view rectangular.

Cells isopolar or heteropolar.

Cells in valve view varying from smoothly dilated in the center (acicular, spindle shaped) to linear, or distinctly dilated in the center and at the apices, or one apex rounded and the other slightly tapering.

Sternum wide.

One marginal row of areolae.

Areolae circular.

Internal openings (foramina) of the areolae smaller than the external openings (SEM).

External openings of areolae crossed by a simple silicified bar (strut) or a pattern of crossing bars (struts; **Plate 56**; SEM). The simple bar is usually coarse and may be discerned with LM, whereas the more complicated pattern is delicate and is only occasionally and partly discernible with LM.

KEY TO SPECIES

1a. Valve ends dissimilar in width and/or shape. 2
1b. Valve ends similar in width and shape 3

2a. Valve margins generally straight or slightly convex. Valve width approximately the same along the whole valve length except near the apices 4

2b. Whole valve almost narrowly clavate: one end acutely club shaped, the other end pointed and ending in a spine. Valve narrow near the club-shaped end, wider in the middle and close to the tapering apex. Marginal structure visible with LM as short ribs between less silicified inter-spaces. . . . *T. javanicum*[22] (Grunow in Van Heurck) Hasle comb. nov.

3a. Valves linear to narrowly lanceolate in outline, presence of apical spine variable, marginal structure visible with LM as ribs . . . . *T. nitzschioides* (Grunow) Mereschkowsky

3b. Valves more or less expanded in the middle and less often at the ends, areolae visible with LM as circular or subcircular holes, a structure sometimes visible within the holes . . . . *T. bacillare*[22] (Heiden in Heiden & Kolbe) Kolbe

4a. Valves linear in outline, one end more or less tapering and usually with a spine, the other end broader, rounded, and sometimes slightly expanded, areolae crossed by a simple strongly silicified bar, discernible with LM . . . . *T. frauenfeldii*[22] (Grunow) Hallegraeff

4b. Valves linear in outline except for the one tapering end, usually with a spine, both apices smoothly rounded, marginal structure visible with LM as short ribs. . . . . *T. pseudonitzschioides*[22] (Schuette & Schrader) Hasle comb. nov.

Distribution:

T. bacillare—warm water region.

T. frauenfeldii—warm water region to temperate.

T. javanicum—warm water region to temperate.

T. nitzschioides—cosmopolitan but not in the high Arctic and Antarctic.

T. pseudonitzschioides—warm water region?

Remarks: *Thalassiothrix pseudonitzschioides* was said to differ from *T. nitzschioides* "by having heteropolar apices" (Schuette & Schrader, 1982). Hallegraeff (1986) regarded the two species as conspecific, probably interpreting the heteropolarity as the presence of one apical spine per valve.

---

[22] Basionyms: *Thalassiothrix frauenfeldii* var. *javanica* Grunow in Van Heurck, *Spinigera bacillaris* Heiden in Heiden & Kolbe; *Asterionella frauenfeldii* Grunow and *Thalassiothrix pseudonitzschioides* Schuette & Schrader, respectively. Synonym: *Thalassiothrix (fauenfeldii* [sic!] var.) *javanica* (Grunow) Cleve and *Thalassiothrix frauenfeldii* (Grunow) Grunow in Cleve & Grunow, respectively.

Based on the areola structure *T. pseudonitzschioides* certainly belongs to *Thalassionema*. The valve outline with one apex more pointed and the other more rounded exhibits only inconsiderable variation from the smallest to the largest specimens (Schuette & Schrader, 1982; G. Hasle and E. Syvertsen, personal observations), and distinguishes *T. pseudonitzschioides* as a separate species. Grunow's drawings of *T. frauenfeldii* and *T. javanicum* (Van Heurck, 1880–1885, Plate 37, Figs. 12 and 13) point out the difference in the marginal structure which refers to the density of areolae as well as the velum structure seen with EM. There is a certain possibility that *Asterionella synedraeformis* (Greville, 1865a) is an earlier synonym of *T. frauenfeldii* var. *javanica*. The variety was raised in rank by Cleve (1990c) which antedates *T. javanica* Hustedt in Meister, 1932 (see VanLandingham, 1967–1979, p. 4001). It should be noted that, in agreement with Hallegraeff (1986), weight here has been put on the presence and absence of marginal spines and the areola structure and not on the polarity of cells as the distinction between *Thalassiothrix* and *Thalassionema*.

Genus *Thalassiothrix* Cleve & Grunow 1880 (Plates 58 and 59, Table 67)
Type: *Thalassiothrix longissima* Cleve & Grunow (conservation proposed by Silva & Hasle, 1993).
Synonym: *Synedra thalassiothrix* Cleve ["*Thalassothrix*"].
References: Cleve, 1873b, p. 22, Plate 4, Fig. 24; Cleve & Grunow, 1880, p. 108; Karsten, 1905, p. 124, Plate 17, Fig. 12; Cupp, 1943, p. 184, Fig. 134; Hustedt, 1959, p. 247, Fig. 726; Hasle, 1960, p. 19, Fig. 6, Plate 5, Figs. 46–48; Hallegraeff, 1986, pp. 64 and 66–69; Hasle & Semina, 1987; Silva & Hasle, 1993.

Generic characters:
 Cells solitary or in radiating colonies.
 Cells straight or slightly curved or sigmoid.
 Cells usually strongly twisted.
 Cells isopolar or heteropolar.
 Valves more or less inflated in the middle and near the apices.
 Sternum wide and sometimes narrower near the apices.
 One marginal row of areolae.
 External openings of areolae elongate, with LM appearing as short marginal striae.
 Internal openings (foramina) of areolae smaller than the external openings (Plate 59, Fig. 1a; SEM).

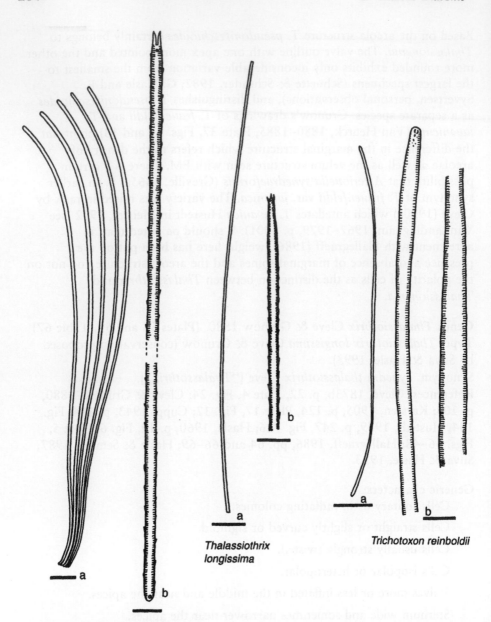

Thalassiothrix antarctica

PLATE 58 *Thalassiothrix antarctica*: (a) colony in girdle view. Scale bar = 100 μm; (b) valve showing foot and head poles, marginal spines, and areolation. After Heiden & Kolbe (1928). Scale bar = 10 μm. *Thalassiothrix longissima*: (a) single cell. Scale bar = 100 μm; (b) valve end. Scale bar = 10 μm. After Hustedt (1959). *Trichotoxon reinboldii*: (a) whole cell. After Van Heurck (1909). Scale bar = 100 μm; (b) parts of valve with marginal structure. Scale bar = 10 μm.

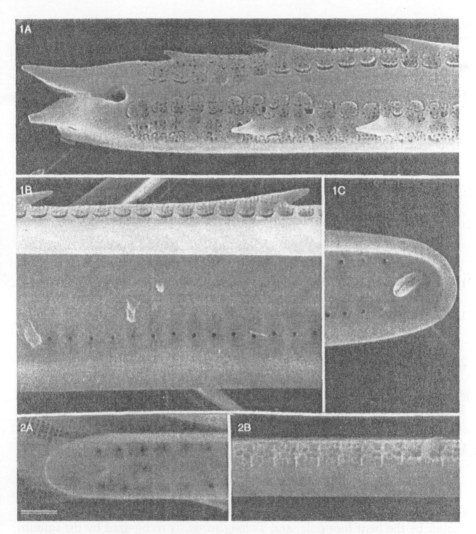

PLATE 59 Scanning electron micrographs. (1) *Thalassiothrix antarctica:* (a) external view, head pole with two heavy apical spines, external opening of labiate process, and marginal spines in the middle of the vela; (b) partial internal view, one marginal row of small foramina, and marginal spines; (c) internal view of rounded, spineless foot pole with labiate process. (2) *Thalassiothrix longissima:* (a) internal view, valve end with serrated protrusions, and labiate process; (b) external side view showing vela and one marginal spine below the vela. Scale bar = 1 μm.

External openings of areolae covered by reticulate vela (SEM) divided
more or less distinctly into two compartments by a longitudinal bar
running parallel to valve margins just discernible with LM.

TABLE 67 Morphometric Data of *Thalassiothrix* spp.

Species	Apical axis (μm)	Transapical axis (μm)	Areolae in 10 μm	Marginal spines in 10 μm
T. antarctica	420–5680	1.5–6	12–17	1–2
T. gibberula	365–1022	2–8	13–16	2–3
T. longissima	530–4000	2.5–6	11–16	1–3

Marginal spines located in the middle of the vela on the longitudinal bars (LM) or on the border between the vela and the unperforate margin of the valve mantle (Plate 59, Figs. 1a and 2b).

KEY TO SPECIES

1a. Cells heteropolar, one end with two winged spines, the other with no spines, smoothly rounded marginal spines in the middle of the vela. . 2
1b. Cells isopolar, either end with serrated protrusions, solitary, slightly curved, marginal spines at the lower edge of the vela. . . . .
. . . . *T. longissima* Cleve & Grunow
2a. Cells in radiating colonies, joined by bent foot poles and a shorter or longer part of the cell length, sigmoid in girdle view . . . .
. . . . *T. antarctica* Schimper ex Karsten
2b. Cells solitary, usually straight, inflated in the middle. . . . .
. . . . *T. gibberula* Hasle

Distribution:
T. antarctica—southern cold water region.
T. gibberula—warm water region.
T. longissima—northern cold water region to temperate.
Remarks: The colony-forming *T. antarctica* is readily recognized in water mount by the bent foot pole, and *T. longissima,* is recognized by the smooth curvature of the cell. As cleaned valves in permanent mounts *T. longissima* and *T. antarctica* are distinguished by the shape of the valve ends and the difference in location of the marginal spines (Hasle & Semina, 1987, Figs. 2–12, *T. longissima*; Figs. 34–43, *T. antarctica*). The heteropolarity and the location of the marginal spines in the middle of the vela make *T. gibberula* morphologically closer to *T. antarctica* than to *T. longissima*. The apparent lack of colony formation, the straight cells, and the inflation in the middle (Hasle, 1960, Fig. 6) characterize *T. gibberula*. The sternum of *T. lanceolata* Hustedt is narrower and the marginal spines are more widely spaced than those in *T. gibberula* (Simonsen, 1987, Plate 667, Figs. 7–11). A warm water species under description (Hasle, manuscript in preparation) differs

from the *Thalassiothrix* species mentioned by the shape of the foot pole. *Thalassiothrix gibberula* was described from water samples of the equatorial Pacific (Hasle, 1960). The description was accompanied by a figure and a Latin diagnosis but no holotype was indicated (see Taxonomic Appendix).

Genus *Trichotoxon* F. M. Reid & F. E. Round 1988 (Plate 58)
Type: *Trichotoxon reinboldii* (Van Heurck) Reid & Round.
Monospecific genus.

Trichotoxon reinboldii (Van Heurck) Reid & Round
Basionym: *Synedra reinboldii* Van Heurck.
Synonym: *Synedra pelagica* Hendey.
References: Van Heurck, 1909, p. 23, Plate 3, Fig. 35; Hendey, 1937, p. 335; Hasle & Semina, 1988, p. 189, Figs. 67–75; Reid & Round, 1988.
Cells solitary or in pointed, ovoid colonies formed by cells attached at either end. Cells bow shaped and not twisted. Valves expanded in central part and less at the ends. Cell ends isopolar with no apical spines. No marginal spines. Internal openings of areolae much smaller than the external openings (SEM). External vela reticulate.
Morphometric data: Apical axis, 800–3500 μm; transapical axis of mid-section, 5–8 μm, and of the ends, 3.5–6.6 μm (Reid & Round, 1987).
Distribution: Southern cold water region.
How to identify: Diatoms of the family Thalassionemataceae in intact colonies, occasionally also as single whole cells, may be identified in water mounts. Permanent mounts of material cleaned of organic matter may be needed in critical cases, e.g., to distinguish between *T. bacillare* and *T. frauenfeldii*.

Suborder Bacillariineae—Raphid pennate diatoms

Terminology specific to raphid pennate diatoms (Mann, 1978; Ross et al., 1979; Round et al., 1990): (**Fig. 18**)

Raphe system—one or two longditudinal slits through the valve wall.

Central raphe ending—central end of the raphe slit when the raphe system consists of two slits.

Central pore—a pore-like expansion of the central raphe ending.

Central nodule—bridge of silica separating the two raphe slits, often thicker than the rest of the valve.

Stauros—a central nodule that is expanded transapically and reaches or almost reaches the margin of the valve.

Terminal nodule—a thickening at the apical end of a raphe.

Helictoglossa—an inwardly projecting lipped structure terminating the raphe on the inner side of the valve.

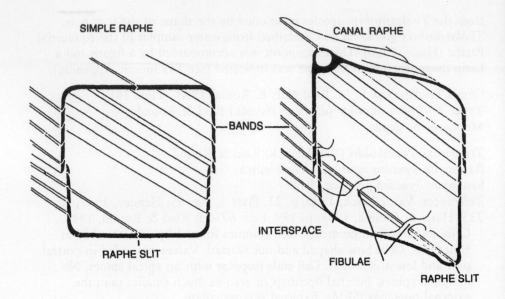

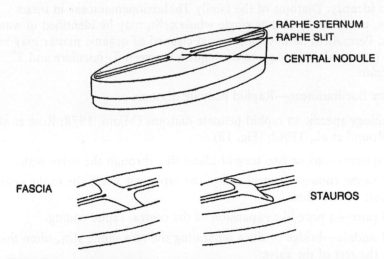

FIGURE 18 Schematic illustration of simple and canal raphes with terminology.

Raphe-sternum—the usually unperforate strip of silica, often thickened pervalvarly, which contains the raphe (Mann, 1978, p. 27).

Family Achnanthaceae Kützing 1844

Diatoms of this family have heterovalvar cells. One of the valves has a raphe with two longitudinal slits. The other valve has no raphe or only short slits, with the raphe being filled out by silica during the formation of the new valves (Round et al., 1990, p. 33). The cells are more or less genuflexed in the transapical axis. *Achnanthes taeniata* belongs to Arctic and Baltic Sea plankton, evidently as the only truly planktonic species of this family.

Achnanthes taeniata Grunow in Cleve & Grunow (Plate 60)
References: Cleve & Grunow, 1880, p. 22, Plate 1, Fig. 5; Hustedt, 1959, p. 382, Fig. 828; Hasle & Syvertsen, 1990b, p. 289, Figs. 12–22.
    Girdle view: Cells only slightly genuflexed, raphe valve on the inside of the curvature. Cells in ribbons; vegetative cells attached along entire valve face; apices of resting spores in chains not attached. One (?) H-shaped chloroplast along the girdle.
    Valve view: Valve linear with rounded apices. Raphe straight, sternum narrow.
Morphometric data: Apical axis, 10–40 μm; transapical axis, 4–6 μm; transapical striae, ca. 25 in 10 μm.
Distribution: Northern cold water region and the Baltic Sea.
How to identify: *Achnanthes taeniata* may easily be confused with *Fragilariopsis* species and with *Navicula* and *Nitzschia* species-forming ribbons. In some cases the shape of the chains is distinctive but in most cases all these diatoms must be examined in valve view, preferably in permanent mounts of cleaned material.

Family Phaeodactylaceae J. Lewin 1958

Genus *Phaeodactylum* Bohlin 1897
Type: *Phaeodactylum tricornutum* Bohlin.
Monospecific genus.

Phaeodactylum tricornutum Bohlin (Plate 60)
Synonym: *Nitzschia closterium* W. Smith f. *minutissima* Allen & Nelson.
References: Bohlin, 1897, pp. 519 and 520, Fig. 9; Allen & Nelson, 1910, p. 426; Wilson, 1946; Hendey, 1954; Lewin, 1958; Round et al., 1990, p. 560; Gutenbrunner et al., 1994, p. 129, Figs. 2–5.
    Solitary. Three types of cells: ovate (naviculoid), fusiform, and, more rarely, triradiate. Ovate cells motile with one siliceous valve per cell.
    Fusiform cells nonmotile and lack a siliceous valve. One chloroplast.
Morphometric data: Ovate cells 8 μm in apical axis, 3 μm in transapical axis; striae not resolved with LM; fusiform cells up to 25–35 μm long.

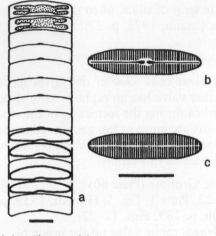

Achnanthes taeniata

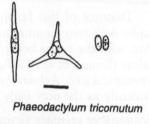

Phaeodactylum tricornutum

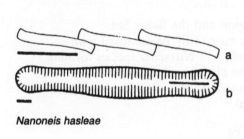

Nanoneis hasleae

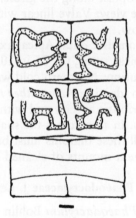

Meuniera membranacea

PLATE 60 *Achnanthes taeniata:* (a) chain with resting spores and chloroplasts; (b) valve with raphe; (c) rapheless valve. Scale bars = 10 μm. *Phaeodactylum tricornutum:* Three cell types. After Wilson (1946). Scale bar = 10 μm. *Nanoneis hasleae:* (a) stepped chain in girdle view. Scale bar = 10 μm; (b) valve view. Scale bar = 1 μm. *Meuniera membranacea:* short chain in girdle view with chloroplasts. After Sournia (1968) and Gran (1908). Scale bar = 10 μm.

Distribution: Intertidal rock pools—probably cosmopolitan.
How to identify: Siliceous valves, recognized only by the raphe, are found in cleaned material mounted in a medium of a high refractive index.
Phase or interference contrast will most likely be needed. The presence of nonsiliceous cells can be verified by "elimination" methods since they will

disappear in acid-cleaned material or, sometimes be seen with LM as an unstructured cell wall.

Remarks: The fusiform cell type was, in the past, frequently confused with the pennate diatom *N. closterium* (= *Cylindrotheca closterium*), e.g., the Plymouth strain "*Nitzschia closterium f. minutissima.*" This strain was used for decades in diatom physiology studies until a microscopical examination showed that it was the fusiform *Phaeodactylum tricornutum*.

Incertae sedis (Raphid diatoms)

Genus *Nanoneis* R. E. Norris 1973
Type: *Nanoneis hasleae* R. E. Norris.
Monospecific genus.

Nanoneis hasleae R. E. Norris (Plate 60)
Reference: Norris, 1973.
   Girdle view: Valve surfaces slightly convex to flat or concave, with the concavity present between the apices and the middle part of the valve. Cells occurring in irregular chains with a short overlap of cell ends.
   Valve view: Broadly elliptical to linear. Valve structure not resolved with LM. Raphe central and extending from one pole to near center; raphe of opposite valve extending from the opposite pole to near center.
Morphometric data: Apical axis, 5–12 μm; transapical axis, 1–1.5 μm; transapical interstriae, ca. 40 in 10 μm (TEM).
Distribution: Warm water region—open ocean.
How to identify: Electron microscopy may be needed.

Family Naviculaceae Kützing 1844

The genera treated here under the family name Naviculaceae have been placed in a variety of families in the classification systems by Glezer et al. (1988) and Round et al. (1990). In a manual to be used for species identification, like this chapter, we prefer Simonsen's (1979) broader delineation of Naviculaceae. Naviculaceae differs from Achnanthaceae and Phaeodactylaceae by being isovalvar; both valves of a cell have a "naviculoid" raphe not subtended by the fibulae present in Bacillariaceae.

Navicula is the largest of all diatom genera with 1860 "acceptable" and 2000 "unacceptable species," mainly bottom living forms (Mann, 1986, p. 216). Many of the few marine planktonic *Navicula* species were transferred to other genera, especially after Cox (1979) typified and emended the description of *Navicula sensu stricto*.

Tropidoneis is another genus under revision. Patrick & Reimer (1975) found that the name *Tropidoneis* had to be rejected in favor of *Plagiotropis*. *Plagiotropis* has about 30 species (Paddock, 1990) and is the largest of the genera to which the former *Tropidoneis* species have been transferred. It comprises brackish water and marine, mostly bottom living species.

Pachyneis with one planktonic tropical and subtropical species was suggested as a possible transition form between *Haslea,* the former *Navicula fusiformes sensu* Hustedt, 1961, and some planktonic *Tropidoneis* species (Simonsen, 1974). *Pleurosigma* includes marine species, some of which are planktonic.

For practical reasons the genera here referred to as Naviculaceae are divided into four groups.

A. Former and present *Navicula* species.
   1. Former *Navicula* sp. in ribbons: *Meuniera membranacea.*
   2. Present *Navicula* spp. in ribbons: *N. granii, N. pelagica, N. septentrionalis,* and *N. vanhoeffenii* (Table 68).
   3. Former solitary *Navicula* spp.: *Haslea* spp. (Table 69).
   4. Present solitary *Navicula* spp.: *N. directa, N. distans, N. transitans* var. *derasa,* and *N. transitans* var. *derasa* f. *delicatula* (Table 70).
B. *Pleurosigma* spp.: *P. directum, P. normanii,* and *P. simonsenii* (Table 71).
C. The *Tropidoneis* group.
   1. Former *Navicula* sp., usually solitary: *Ephemera planamembranacea.*
   2. Former *Tropidoneis* spp., usually in ribbons.
      a. Valves lying in girdle view, vaulted to a high ridge: *Banquisia* and *Membraneis* (Table 72).
      b. Valves lying in girdle or valve view; valve with only low ridge: *Manguinea* and *Plagiotropis* (Table 72).

D. *Incertae sedis* (Naviculaceae):*Pachyneis.*

A. Former and present *Navicula* species.

Common characters:

   Valves linear, lanceolate, or elliptical.

   Raphe generally straight.

   Raphe not raised on a ridge.

   Stauros or stauros-like structure present in some species.

Characters showing differences between species:

   The presence or absence of chains.

   Number and shape of chloroplasts.

   Valve striation pattern.

   Extension of stauros or stauros like structure.

   The presence or absence of raphe fins.

1. Former *Navicula* species in ribbons.

Genus *Meuniera* P. C. Silva nom. nov.
Type: *Meuniera membranacea* (Cleve) P. C. Silva comb. nov.
Monospecific genus.

Meuniera membranacea (Cleve) P. C. Silva comb. nov. (**Plate 60**)
Basionym: *Navicula (Stauroneis) membranacea* Cleve.
Synonyms: *Stauropsis membranacea* (Cleve) Meunier; *Stauroneis membranacea* (Cleve) Hustedt.
References: Cleve, 1897a, p. 24, Plate 2, Figs. 25–28; Meunier, 1910, p. 319, Plate 33, Figs. 37–40; Cupp, 1943, p. 193, Fig. 142; Hustedt, 1959, p. 833, Fig. 1176; Hendey, 1964, p. 221, Plate 21, Fig. 3; Paddock, 1986, p. 89, Figs. 1–8.

    Girdle view: Rectangular, valves flat or slightly concave in the center. Cell wall weakly silicified. Stauros narrow and distinct in girdle view. Raphe fins at corners of cells in girdle view (LM). Four ribbon-like and folded chloroplasts per cell, two along each side of the girdle.

    Valve view: Valves narrow and elliptical with pointed ends. Structure barely visible with LM. Stauros extending from the central nodule to valve margin.

Morphometric data: Pervalvar axis, 30–40 μm; apical axis, 50–90 μm.
Distribution: Temperate.
How to identify: *Meuniera membranacea* is readily recognized with LM in girdle view by the distinct stauros and the typical chloroplasts.
Remarks: *Stauropsis* Reichenbach 1860 had been used for an orchid genus; therefore, the diatom genus had to be given a new name (see Taxonomic Appendix).

2. Present *Navicula* spp. in ribbons (**Plate 61, Table 68**).

    Meunier (1910) regarded the following four marine, mainly planktonic species to belong to the genus he described as *Stauropsis*. Paddock (1986) disagreed after having reviewed them on the basis of LM and EM data. Characters showing differences between the species are evident from the key and the illustrations.
References: Grunow, 1884, p. 105, Plate 1, Fig. 48; Cleve, 1896a, p. 11, Plate 1, Fig. 9; Gran, 1897a, p. 21, Plate 1, Figs. 1–3; Jørgensen, 1905, p. 107, Plate 7, Fig. 25; Gran, 1908, p. 123, Figs. 167–170; Meunier, 1910, p. 321, Plate 33, Figs. 26, 27, and 33–36; Cleve-Euler, 1952, p. 25, Fig. 1381; Heimdal, 1970, Figs. 1–11; Syvertsen, 1984; Hasle & Syvertsen, 1990b, p. 288, Figs. 1–4.

Navicula

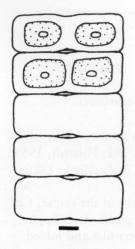

N. granii

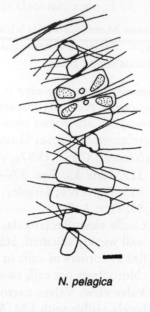

N. pelagica

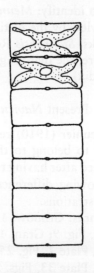

N. septentrionalis

N. vanhoeffenii

TABLE 68 Morphometric Data of *Navicula* spp. in Ribbons

Species	Pervalvar axis (μm)	Apical axis (μm)	Transapical axis (μm)	Striae in 10 μm
N. granii	—[a]	50–57	—	—
N. pelagica	5–9	10–50	6–7	50–56[b]
N. septentrionalis	—	20–30	—	—
N. vanhoeffenii	—	13–65	4–5	54–63[b]

[a] —, No data.
[b] Measured in TEM.

KEY TO SPECIES

1a. Cells in ribbons partly or not in touch .2
1b. Cells in tight ribbons 3
2a. Cells united by a mucilage pad in the central nodule area with threads ("setae") attached, two comma- or pear-shaped chloroplasts per cell. . .
. . . . *N. pelagica* Cleve
2b. Cells in colonies widely separated, two plate-like chloroplasts per cell . .
. . . . *N. vanhoeffenii* Gran
3a. Cells united by the entire valve faces, one chloroplast per cell, with four arms and large central pyrenoid . . . .*N. septentrionalis* (Grunow) Gran
3b. Cells not touching at apices and in the center, two rectangular chloroplasts per cell, each with large central pyrenoid . . *N. granii* (Jørgensen) Gran

Basionyms: *Stauroneis septentrionalis* Grunow; *Stauroneis granii* Jørgensen.
Synonyms: *Stauropsis pelagica* (Cleve) Meunier; *Stauropsis vanhoeffenii* (Gran) Meunier; *Stauropsis granii* (Jørgensen) Meunier; and *Navicula quadripedis* Cleve-Euler [regarded by VanLandingham (1975, p. 2796) as the valid name for *Navicula septentrionalis* Gran, 1908 (Fig. 167)].
Distribution: Northern cold water region
How to identify: These species may be distinguished in chains in water mounts.
Remarks: The setae-like structures in *N. pelagica* are dislocated girdle bands (Syvertsen, 1984). *Navicula pelagica* is also distinguished by the

PLATE 61 *Navicula granii*: ribbon in girdle view with characteristic chloroplasts. *Navicula pelagica*: twisted chain in girdle view with dislocated bands. *Navicula vanhoeffenii*: ribbon in girdle view, cells not touching, square chloroplasts. *Navicula septentrionalis*: ribbon in girdle view, lens-shaped structure between cells, chloroplasts lobed. Scale bars = 10 μm.

arrangement of cells in chains, which is rotated approximately 50° on the chain axis in relation to its neighboring cell.

3. Former solitary *Navicula* spp.

Genus *Haslea* Simonsen 1974
Type: *Haslea ostrearia* (Gaillon) Simonsen.
Basionym: *Vibrio ostrearius* Gaillon.
Synonym: *Navicula ostrearia* Turpin in Bory de St.-Vincent.
References: Hustedt, 1961, p. 34 as *Navicula fusiformes,* Simonsen, 1974, p. 46; Cox, 1979.

Haslea comprises 13 species (Simonsen, 1974), all of which are marine; some of them are planktonic and some are associated with a substratum, e.g., ice (Poulin, 1990). Three, apparently more commonly recorded planktonic species, distinguished by size, fineness of structure, and distribution, are included here.

Generic characters:
Cells solitary (or in mucilage tubes = benthic forms).

Cell wall weakly silicified.

Cells generally fusiform in girdle and valve views.

Valves narrow and linear to lanceolate, valve ends pointed.

Transverse and longitudinal striae crossed at right angles.

Raphe central pores small and approximate.

Chloroplasts, two plate like (*H. ostrearia*).

Von Stosch (1986) observed many small bacilliform or roundish chloroplasts in *H. gigantea* and two in *H. wawrikae* which in old cultures seemed to divide into numerous small platelets.

Haslea gigantea (Hustedt) Simonsen (Table 69)
Basionym: *Navicula gigantea* Hustedt.

TABLE 69 Morphometric Data of *Haslea* spp.

Species	Apical axis (μm)	Transapical axis (μm)	Striae in 10 μm	
			Transverse	Longitudinal
H. gigantea	300–419	32–54	16–17	20–24
H. trompii	70–160	10–14	28–30	17–20
H. wawrikae	286–560	4–6.5	18–21	ca. 40

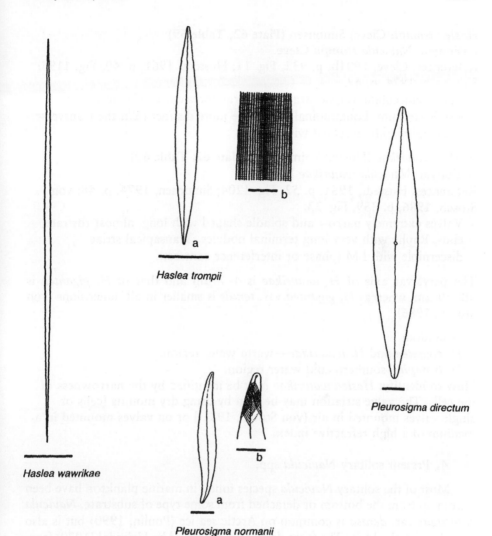

PLATE 62 *Haslea wawrikae:* valve outline. After Sournia (1968). Scale bar = 50 μm. *Haslea trompii:* (a) valve view. Scale bar = 50 μm; (b) valve structure. After Hustedt (1961). Scale bar = 10 μm. *Pleurosigma normanii:* (a) valve view. Scale bar = 20 μm; (b) valve end, structure. After Cupp (1943). Scale bar = 10 μm. *Pleurosigma directum:* valve view. After Peragallo (1891). Scale bar = 20 μm.

References: Hustedt, 1961, p. 40, Fig. 1194; Simonsen, 1974, p. 47, Plate 31, Fig. 1; von Stosch, 1986, p. 333, Fig. 20.

Valves lanceolate. Raphe straight, central pores extremely close. No visible sternum. Striation discernible with LM (interference contrast).

Haslea trompii(Cleve) Simonsen (Plate 62, Table 69)
Basionym: *Navicula trompii* Cleve.
References: Cleve, 1901b, p. 932, Fig. 11; Hustedt, 1961, p. 40, Fig. 1195;
Simonsen, 1974, p. 47.

Valves lanceolate. Raphe straight; central pores extremely close. No
visible sternum. Longitudinal interstriae more distinct than the transverse
ones and readily observed with LM.

Haslea wawrikae (Hustedt) Simonsen (Plate 62, Table 69)
Basionym: *Navicula wawrikae* Hustedt.
References: Hustedt, 1961, p. 52, Fig. 1204; Simonsen, 1974, p. 48; von
Stosch, 1986, p. 339, Fig. 23.

Valves extremely narrow and spindle shaped with long, almost rostrate,
ends. Raphe with very long terminal nodules. Transapical striae
discernible with LM (phase or interference contrast).

The pervalvar axis of *H. wawrikae* is 4–7 μm and that of *H. gigantea* is
30–70 μm, whereas *H. gigantea* var. *tenuis* is smaller in all dimensions (von
Stosch, 1986).

Distribution:
  H. gigantea and *H. wawrikae*—warm water region.
  H. trompii—southern cold water region.
How to identify: *Haslea wawrikae* may be identified by the narrowness of
the cells. The valve striation may be seen by using dry mounts [cells or
single valves mounted in air (von Stosch, 1986)] or on valves mounted in a
medium of a high refractive index.

4. Present solitary *Navicula* spp.

Most of the solitary *Navicula* species found in marine plankton have been
stirred up from the bottom or detached from some type of substrate. *Navicula
transitans* var. *derasa* is common on Arctic sea ice (Poulin, 1990) but is also
found in the plankton. The form *delicatula,* described by Heimdal (1970) from
a fjord in northern Norway, is common in Norwegian coastal waters in general.
Navicula directa and *N. distans* are primarily bottom dwelling species but are
often found in the plankton (Gran & Angst, 1931; Hendey, 1964).

PLATE 63   *Navicula directa:* (a) valve with chloroplasts; (b) valve structure. After Gran & Angst
(1931). *Navicula distans:* (a) valve with chloroplasts; (b) valve structure. After Gran & Angst
(1931). *Navicula transitrans:* (a) var. *derasa,* valve with chloroplasts. After Heimdal (1970);
(b) var. *derasa,* valve structure. After Cleve & Grunow (1880); (c) f. *delicatula,* valve with typical
chloroplasts. After Heimdal (1970). Scale bars = 10 μm.

Navicula - valve view

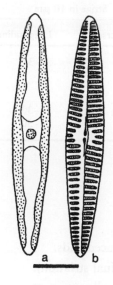

N. directa

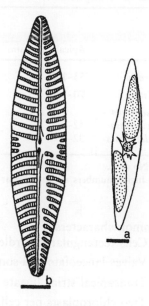

N. distans

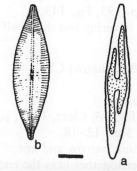

N. transitrans var. *derasa*

f. *delicatula*

TABLE 70   Morphometric Data of Solitary *Navicula* spp.

| | | Transapical axis | Striae in 10 μm | |
Species	Apical axis (μm)	(μm)	Transverse	Longitudinal
N. directa	53–120	7–12	7–11	30–33
N. distans	70–130	14–20	5–6	—[a]
N. transitans				
var. *derasa*	42–96	9–21	8.5–12; 10–16[b]	24–30
f. *delicatula*	32–49	7–9	15–16	24–30

[a] —, No data.
[b] The higher numbers are found near the valve ends.

Common characters:
    Cells rectangular in girdle view.

    Valves lanceolate and sometimes with slightly produced ends.

    Transapical striae lineate (crossed by finer longitudinal striation).

    Two chloroplasts per cell, one at each side of the girdle.

Navicula directa (W. Smith) Ralfs in Pritchard (**Plate 63, Table 70**)
Basionym: *Pinnularia directa* W. Smith.
References: Smith, 1853, p. 56, Plate 18, Fig. 172; Pritchard, 1861, p. 906;
Gran & Angst, 1931, p. 499, Fig. 87; Heimdal, 1970, Figs. 25–28.
    Valves narrow and lanceolate with subacute ends. Raphe–sternum
    indistinct. Striae parallel and uniformly spaced throughout the whole
    valve. Each chloroplast covering the girdle from end to end.

Navicula distans (W. Smith) Ralfs in Pritchard (**Plate 63, Table 70**)
Basionym: *Pinnularia distans* W. Smith.
References: Smith, 1853, p. 56, Plate 18, Fig. 169; Pritchard, 1861, p. 907;
Gran & Angst, 1931, p. 499, Fig. 86; Cupp, 1943, p. 193, Fig. 143.
    Valves lanceolate. Striae radiate. Each chloroplast covering less than half
    of each side of the girdle.

Navicula transitans var. **derasa** (Grunow, in Cleve & Grunow) Cleve
(**Plate 63, Table 70**)
Basionym: *Navicula derasa* Grunow in Cleve & Grunow.
References: Cleve & Grunow, 1880, p. 39, Plate 2, Fig. 46; Cleve, 1883, p.
467, Plate 36, Figs. 31, 33, and 37; Heimdal, 1970, Figs. 12–18.
    Valves lanceolate with slightly rostrate ends. Sternum narrow and not
    centrally expanded. Striae parallel to radiate, closer together near the ends
    than in the middle of the valves. Chloroplasts often asymmetrical and not
    covering the whole length of each side of the girdle.

Navicula transitans var. *derasa* f. *delicatula* Heimdal (Plate 63, Table 70)
Reference: Heimdal, 1970, p. 72, Figs. 30–37.
Valves lanceolate to elliptical, mostly with slightly rostrate ends. Striae parallel and uniformly spaced throughout the valve. Each of the two chloroplasts covering most of each side of the girdle.
Distribution: Uncertain.
How to identify: Size and position of the chloroplasts are useful distinctive characters. Positive identification requires examination of cleaned valves mounted in a medium of a high refractive index. Due to the coarse silicification *N. directa* and *N. distans* are best examined in brightfield illumination.

B. Genus *Pleurosigma* W. Smith 1852

Type: *Pleurosigma angulatum sensu* W. Smith emend. Sterrenburg.
Basionym: *Navicula angulata* Quekett *pro parte quoad typum.*
Synonyms: See Sterrenburg, 1991a.

The genus was last monographed 100 years ago (Peragallo, 1891), and many more taxa have been introduced since then without a revision of the genus. Recently Cardinal et al. (1989) introduced criteria for species characterization, revealed by SEM, and Sterrenberg (1991b) pointed out LM criteria suitable for taxonomic purposes.

There are some 250 major taxonomic entries for *Pleurosigma* in VanLandingham (1978), 90 of them are listed as valid. According to the information given by Cupp (1943), Hendey (1964), and Simonsen (1974), less than one-tenth appear more or less regularly in the plankton. Two species are treated here, viz. *Pleurosigma normanii,* the most common and most widely spread of all *Pleurosigma* species, found from the tropics to the polar seas, often in the plankton (Hendey, 1964), and *P. directum,* which has been characterized as an "almost cosmopolitan plankton species" (Simonsen, 1974, p. 45). A third species, *P. simonsenii,* was described as a planktonic species from the Indian Ocean (Simonsen, 1974) and later recorded as abundant in the phytoplankton of the western English Channel (Boalch & Harbour, 1977).

Distinctive characters:

Valves more or less flattened, gently sigmoid, or almost straight.

Valve outline lanceolate.

Raphe straight or more or less sigmoid and central.

Three striae systems: one transverse and two oblique.

Two or four elongated chloroplasts, often extremely convoluted and lying under valve face rather than along girdle, many pyrenoids per chloroplast (Cox, 1981).

Characters showing differences between subgroups or species:
Shape of valve.

Curvature of raphe.

Crossing angle of striae.

Areolation at valve apices.

Color in standardized darkfield (Sterrenburg, 1991b).

Pleurosigma directum Grunow in Cleve & Grunow (**Plate 62, Table 71**)
References: Cleve & Grunow, 1880, p. 53; Peragallo, 1891, p. 14, Plate 5,
Fig. 29; Simonsen, 1974, p. 45, Plate 29, Fig. 2.
Valves rhombo–lanceolate to elliptic–lanceolate.
Raphe almost straight.

Pleurosigma normanii Ralfs in Pritchard (**Plate 62, Table 71**)
References: Pritchard, 1861, p. 919; Cupp, 1943, p. 196, Fig. 148;
Sterrenburg, 1991b, Fig. 2.
Valves broadly lanceolate, slightly sigmoid, with subacute ends. Raphe
nearly central, sigmoid with single curvature. Raphe–sternum or central
nodule dilated transversely. Crossing angle of striae greater at center than
toward valve apices. Color in standardized darkfield deep blue with
silverish center (Sterrenburg, 1991b).

Pleurosigma simonsenii Hasle nom. nov. (**Table 71**)
Synonym: *Pleurosigma planctonicum* Simonsen.
References: Simonsen 1974, p. 46, Plate 30; Boalch & Harbour, 1977.
Valves slightly sigmoid near the ends and flat; ends acute and not
protracted. Raphe straight, sigmoid before the ends, central in the middle,
eccentric before the ends, central pores close together. Raphe–sternum
narrow and not centrally expanded. Valve thin and striation hardly visible
with LM. No color in standardized darkfield (Sterrenburg, personal
communication).

TABLE 71 Morphometric Data of *Pleurosigma* spp.

Species	Apical axis (µm)	Transapical axis (µm)	Striae in 10 µm	
			Transverse	Oblique
P. directum	180–270	44	—	18.5
P. normanii	90–220	28–36	19–22	16–19
P. simonsenii	300–600	40–75	28–30	ca. 30

Note. —, No data.

Distribution: All three species are probably cosmopolitan.
How to identify: If possible to see at all with LM, the striation can only be seen on cleaned valves mounted in a medium of a high refractive index or also occasionally as a dry mount. The large size and the extremely delicate valve striation characterize *P. simonsenii*.
Remarks: *Pleurosigma planctonicum* Simonsen is a later homonym of *P. planctonicum* Cleve-Euler (see Taxonomic Appendix). *Gyrosigma* Hassall, separated from *Pleurosigma* by having longditudinal and transverse (no oblique) striae, is another large genus with a few species occasionally occurring in marine and brackish water plankton, e.g., *G. macrum* (W. Smith) Griffith & Henfrey (Sterrenburg, personal communication).

C. The "Tropidoneis" group

Termimology mainly used for the *Tropidoneis* group (Paddock & Sims, 1981; Paddock, 1988).

Raphe ridge—a simple angular elevation of the valve which bears the raphe and raises it, but which lacks specialized supporting structures.

Valve face—the whole valve surface.

Greater and lesser parts of valve face—the two parts of valve face divided by the raphe, termed "greater face" and "lesser face," respectively, by Paddock (1988).

Raphe fins—paired external small vane-like (blade, plate-like) siliceous structures, each shaped as a shark's dorsal fin (SEM) and arising from the sternum (also present in *Stauropsis*).

Notes: Paddock & Sims (1981, p. 178): "In practice it proved extremely difficult to decide whether some diatoms have a raphe 'raised upon a keel' or whether the valve of the diatom which is 'highly vaulted' is merely laterally compressed to an extreme degree." Paddock (1988, p. 14): "It would seem preferable to distinguish the raphe ridge [defined above] from the raphe keel in which specialized supporting structures i.e., fibulae are present." The shape of the raphe fins will scarcely be seen with LM although their presence should be discernible in girdle view, especially when phase or interference contrast optics are used.

Common characters:
   Intact frustules as well as single valves usually lying in broad girdle view.

   Valves linear to lanceolate.

   Valves and raphe straight and not sigmoid.

   Raphe more or less ridged and raised above the general level of the valve.

Raphe without fibulae.

Large helictoglossae.

Characters showing differences between genera:

Shape of frustule in girdle view.

Valves vaulted or ridged.

Shape of valve ridge in girdle view.

Size of the two parts of the valve face on each side of the raphe (equal, subequal, or unequal in area).

The presence or absence of raphe fins.

1. Former *Navicula*—usually solitary

Genus *Ephemera* Paddock 1988
Type: *Ephemera planamembranacea* (Hendey) Paddock.
Monospecific genus.

Ephemera planamembranacea (Hendey) Paddock (**Plate 64**)
Basionym: *Navicula planamembranacea* Hendey.
References: Hendey, 1964, p. 188, Text Fig. 8; Paddock, 1988, p. 86, Plate 31.
  Cells usually solitary. Valves highly vaulted and flattened in transapical
  plane. Raphe separating the valve face into two unequal parts. Central
  nodule slightly depressed, usually with four stronger and more widely
  spaced interstriae, reaching valve margin and producing "the impression
  of a pseudo-stauros" (Hendey, 1964, p. 188). Small raphe fins near valve
  ends and one on each side of the central nodule. Numerous small and
  rounded chloroplasts.
Morphometric data: Pervalvar axis, 15–30 μm; apical axis, 66–90 μm;
transapical axis, 8–10 μm; transapical striae, 28–30 in 10 μm.
Distribution: North Atlantic Ocean.
Remarks: Due to the shape of the cell it is usually seen in girdle view where
the central and terminal nodules of the raphe are more distinct than the
delicate striae. *Ephemera planamembranacea* is more similar to
"*Tropidoneis*" than to *Navicula sensu stricto* and is therefore dealt with
under this group, although, in the past, it has never been referred to
Tropidoneis. It is also similar to *Stauropsis* but differs from this genus in
chloroplast number and shape and by usually being solitary.

PLATE 64 *Ephemera planamembranacea*: cell in girdle view with four stronger, central interstriae, small paired raphe fins, and large helictoglossae. After Paddock (1988). *Banquisia belgicae*:
cell in girdle view with raphe fins. *Membraneis challengerii*: cell in girdle view with reinforced
striae, raphe fins, and enlarged helictoglossae. Scale bars = 10 μm.

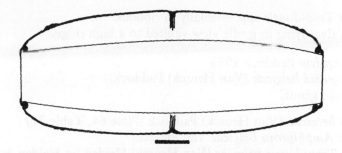

Ephemera planamembranacea

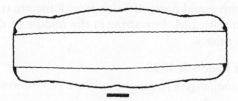

Banquisia belgicae

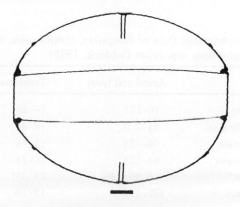

Membraneis challengeri:

2. Former *Tropidoneis* spp.—usually in ribbons.
   2a. Valves lying in girdle view vaulted to a high ridge.

Genus *Banquisia* Paddock 1988
Type: *Banquisia belgicae* (Van Heurck) Paddock.
Monospecific genus.

***Banquisia belgicae* (Van Heurck) Paddock (Plate 64, Table 72)**
Basionym: *Amphiprora belgicae* Van Heurck.
Synonym: *Tropidoneis belgicae* (Van Heurck) Heiden (in Heiden & Kolbe, 1928).
References: Van Heurck, 1909, p. 14, Plate 1, Figs. 11 and 15; Heiden & Kolbe, 1928, p. 655, Plate 4, Figs. 98 and 99; Paddock, 1988, p. 79, Plate 28; Paddock, 1990, p. 153, Fig. 10.
   Raphe biarcuate in girdle view: a depression at central nodule and another at about one-third of the raphe's length from the valve apices. Poles steep. A pair of raphe fins at or near lowest point in raphe outline. Valves narrow with equal parts of valve face. Punctate transverse valve striae clearly seen with LM. Interstriae in the middle of the valve slightly broader than the others.

Genus *Membraneis* Paddock 1988
Type: *Membraneis challengeri* (Grunow, in Cleve & Gronow) Paddock.

Generic characters:
   Cells more or less lens shaped in girdle view.
   Valves with unequal or subequal parts of valve face.

Characters showing differences between species:
   Valve outline in girdle view.

TABLE 72   Morphometric Data of *Banquisia, Membraneis, Manguinea,* and *Plagiotropis* spp. (after Paddock, 1988)

Species	Apical axis (μm)	Transverse striae in 10 μm
Banguisia belgicae	70–125	16–20
Membraneis challengeri	85–270	18–24
Membraneis imposter	80–125	15–16
Manguinea fusiformis	48–135	26–28
Manguinea rigida	60–80 (120)	24 (18)
Plagiotropis gaussii	102–159	15–18

Note. Numbers in parentheses occasionally found.

The presence or absence of raphe fins.

The presence or absence of reinforced striae at the central area.

Coarseness of valve striation.

Membraneis challengeri Grunow, in Cleve & Grunow Paddock (Plate 64, Table 72)
Basionym: *Navicula challengeri* Grunow in Cleve & Grunow.
Synonym: *Tropidoneis antarctica* (Grunow in Cleve & Möller) Cleve (for other synonyms see Paddock, 1988, p. 81).
References: Cleve & Grunow, 1880, p. 64; Cleve, 1894b, p. 24; Paddock, 1988, p. 81, Plate 29; Paddock, 1990, p. 153, Fig. 11.
Valve outline in girdle view a smooth curve from end to end. Raphe fins present about one-eighth of the valve length from the ends of the valve. Enlarged helictoglossae. Reinforced striae at central nodule varying in number and development. Flat unequal parts of valve face. Punctate valve striae just visible with LM.

Membraneis imposter Paddock (Table 72)
Reference: Paddock, 1988, p. 84, Plate 30; Paddock, 1990, p. 153, Figs. 12 and 13.
Raphe ridge in girdle view convex and slightly depressed at central nodule, poles sloping, slightly concave shoulder at about one-eighth of valve's length from each pole. No raphe fins and no reinforced striae at the central area. The two parts of valve face markedly unequal. Valve striae coarsely punctate.

2b. Valves lying in girdle or valve view; valves with only a low ridge.

Genus *Manguinea* Paddock 1988
Type: *Manguinea fusiformis* (Manguin) Paddock.

Generic characters:
Valves long and narrow with unequal faces.

Ridge not lobed.

Raphe fins or similar structures present.

Valve striae with separate puncta delicate and often difficult to see with LM.

Characters showing differences between species:
Valve outline in girdle view.

Size of raphe fins.

Size of helictoglossae.

Manguinea fusiformis

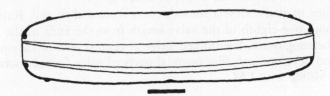

Manguinea rigida

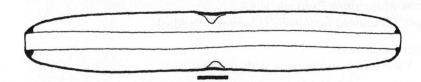

Plagiotropis gaussii

PLATE 65 *Manguinea fusiformis*: cell in girdle view and enlarged helictoglossae and fin like structures. After Paddock (1988). *Manguinea rigida*: cell in girdle view with small raphe fins. After Paddock (1988) *Plagiotropis gaussii*: Cell in girdle view. After Paddock (1988). Scale bars = 10 μm

Manguinea fusiformis (Manguin) Paddock (**Plate 65, Table 72**)
Basionym: *Tropidoneis fusiformis* Manguin.
References: Manguin, 1957, p. 130, Plate 6, Fig. 39; Paddock, 1988, p. 88, Plate 32; Paddock, 1990, p. 153, Fig. 14.
 Frustule fusiform in girdle view. Valve outline slightly narrowed before poles, poles appearing slightly attenuated. Raphe fin-like structures positioned in the narrow parts. Enlarged helictoglossae.

Manguinea rigida (M. Peragallo) Paddock (**Plate 65, Table 72**)
Basionym: *Amphiprora rigida* M. Peragallo.
Synonym: *Tropidoneis glacialis* Heiden in Heiden & Kolbe.
References: Van Heurck, 1909, Plate 1, Fig. 19; Peragallo, 1924, p. 21; Heiden & Kolbe, 1928, p. 656, Plate 5, Fig. 100; Paddock, 1988, p. 90, Plate 33; Paddock, 1990, p. 153, Fig. 15.
 Frustules in girdle view with parallel sides and bluntly rounded ends. Raphe fins small. Helictoglossae moderately sized.

Genus *Plagiotropis* Pfitzer 1871 emend. Paddock 1988
Type: *Plagiotropis baltica* Pfitzer.

The following species may be found in plankton:

Plagiotropis gaussii (Heiden, in Heiden & Kolbe) Paddock (**Plate 65, Table 72**)
Basionym: *Tropidoneis gaussii* Heiden in Heiden & Kolbe.
References: Heiden & Kolbe, 1928, p. 656, Plate 5, Fig. 102; Paddock, 1988, p. 63, Plate 24; Paddock, 1990, p. 152, Figs. 5 and 6.
 Valves delicate with narrow transapical axis and lying in girdle view. Whole frustules slightly waisted. Valve ridge very low and gently curving. Valve outline narrowed to a slight shoulder some distance before the valve poles. Raphe fins absent. Helictoglossae unusually large. Punctate transverse valve striae. Puncta visible with LM.
Distribution: Southern cold water region.
How to identify: Most of these species are easier to identify in girdle than in valve view. Acid-cleaned material studied with phase or interference contrast is recommended.
Remarks: Some of these species are similar to *Amphiprora* spp.
(= *Entomoneis* Ehrenberg in Patrick & Reimer, 1975) in gross morphology. *Amphiprora* occurs on brackish and marine sediments, occasionally in freshwater (Round et al., 1990); some species referred to *Amphiprora* in the past are extremely abundant in connection with ice in the Arctic and the Antarctic. *Amphiprora* and *Tropidoneis* have usually been regarded as closely related; EM observations disproved this, the main distinction being the internal structure of the keel of the valve (Paddock & Sims, 1981).

D. Genus *Pachyneis* Simonsen 1974

Type: *Pachyneis gerlachii* Simonsen.
Monospecific genus.

Pachyneis gerlachii Simonsen
References: Simonsen, 1974, p. 49, Plates 33 and 34; Paddock, 1986, p. 94,
Figs. 19 and 20.
   Girdle view: Cells almost elliptical.
   Valve view: Fusiform and highly vaulted. Valve membrane extremely
   delicate with a varying number of longitudinal folds, two to four on each
   side of the raphe. Raphe straight except for arching sideways near the
   central nodule. Vane-like structures (raphe fins?) near the poles. Striae
   parallel (interference contrast).
Morphometric data: Apical axis, 100–370 µm; transapical axis,
ca. 25–80 µm; 30–32 transapical striae in 10 µm.
Distribution: Warm water region.
How to identify: The most characteristic feature is probably the longitudinal
folds; high contrast and high resolution are needed to reveal the striation.

Family Bacillariaceae Ehrenberg 1831

   The genera dealt with here were all placed in Nitzschiaceae Grunow 1860
by Simonsen (1979) and, with the exception of one genus, also by Glezer et
al. (1988), whereas Round et al. (1990) used the older name Bacillariaceae.
Only a few of the 15 genera included in Bacillariaceae *sensu* Round et al. are
represented in marine plankton. *Bacillaria paxillifera* appears occasionally in
plankton of shallow waters being swept up from the bottom. *Cylindrotheca*
has its main distribution in and on mud; however, *C. closterium* (= *Nitzschia
closterium*) has been recorded from a variety of habitats, including marine
plankton. *Neodenticula* has several fossil but only one living marine planktonic
species. *Nitzschia* had around 900 nomenclaturally valid species at the time
when Mann (1986) wrote the paper "*Nitzschia* subgenus *Nitzschia* (Notes for
a monograph of the Bacillariaceae, 2)." Of special interest to marine
planktologists is the fact that his EM observations of the subgenus, including
the generitype *N. sigmoidea* (Nitzsch) W. Smith, indicate that most of the
marine planktonic so-called *Nitzschia* species are probably too remote
morphologically from the generitype to fit into the genus. This is especially
true for *Fragilariopsis* and *Pseudo-nitzschia* which in this chapter will be treated
as separate genera.

Terminology specific to Bacillariaceae (Anonymous, 1975; Ross et al., 1979;
Mann, 1978): (Fig. 18)

Canal raphe system consisting of

Raphe canal—a space on the inner side of the raphe cut off to a greater or lesser extent from the rest of the interior of the frustule.

Fibula—a bridge of silica between portions of the valve on either side of the raphe (= keel punctum).

Interspace—the space between two fibulae.

Central interspace—the space between the two central fibulae.

Keel—the summit of the ridge bearing the raphe.

A central interspace larger than the others usually indicates the presence of two raphe slits, i.e., the presence of central raphe endings and a central nodule. Since a larger central interspace is observed with LM, but not always the corresponding central raphe endings and a central nodule, "central larger interspace" is a repeatedly used term in this chapter. In the past the wording was "die beiden mittleren Kielpunkte weiter voneinander entfernt" (e.g., Hustedt, 1958a) and "the two keel puncta in the middle more widely spaced than the others" (e.g., Hasle, 1965a,b). It should also be noted that the same authors used the term "pseudonodulus" instead of "central nodule" for the Bacillariaceae.

Characters of the genera dealt with:

Cells in chains of various types or more seldom solitary.

Cells rectangular or spindle shaped in girdle view.

Valves elongate although variable in outline.

Raphe with bridges of silica cross-linking the valve beneath the raphe ("canal raphe").

Raphe usually strongly eccentric running along one valve margin.

Chloroplasts usually two plates, one toward each pole of the cell.

Resting spores uncommon.

KEY TO GENERA

1a. Raphe system eccentric 2
1b. Raphe system central . . . . *Bacillaria*, p. 293
2a. Cells usually in chains 3
2b. Cells usually solitary 4
3a. Cells united by valve surfaces into ribbons 5
3b. Cells united by overlap of valve ends into stepped chains . . . .
. . . . *Pseudo-nitzschia*, p. 307
4a. Frustules usually spirally twisted, valves lightly silicified . . . .
. . . . *Cylindrotheca*, p. 293
4b. Frustules not spirally twisted . . . . *Nitzschia, pro parte*, p. 324

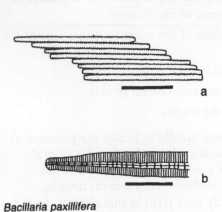

Bacillaria paxillifera

Cylindrotheca closterium

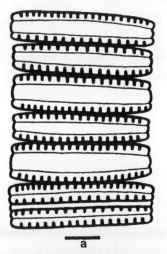

Fragilariopsis kergulensis

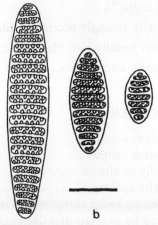

PLATE 66 *Bacillaria paxillifera*: (a) part of chain in girdle view. Scale bar = 50 μm; (b) valve end with raphe and striae. After Cupp (1943). Scale bar = 10 μm. *Cylindrotheca closterium*: (a) cell in girdle view; (b) valve with fibulae visible. Scale bar = 10 μm. *Fragilariopsis kerguelensis*: (a) ribbon in girdle view, curved valve faces, strong interstriae. From Hasle (1968b). Scale bar = 10 μm. (b) larger specimen heteropolar, medium sized and smaller specimens isopolar, and transverse striae with two distinct rows of poroids. Scale bar = 10 μm.

5a. Valve interior with transapical pseudosepta . . . . *Neodenticula*, p. 305
5b. Valve interior without pseudosepta . . . . *Fragilariopsis*, p. 295

Genus *Bacillaria* J. F. Gmelin 1791
Type: *Bacillaria paradoxa* J. F. Gmelin.
Correct name: *Bacillaria paxillifera* (O. F. Müller) Hendey.

Bacillaria paxillifera (O. F. Müller) Hendey (**Plate 66**)
Basionym: *Vibrio paxillifer* O. F. Müller.
Synonyms: *Bacillaria paradoxa* J. F. Gmelin; *Nitzschia paradoxa* (J. F. Gmelin) Grunow in Cleve & Grunow.
References: Müller, 1783, pp. 277 and 286; Gmelin, 1791, p. 3903; Cleve & Grunow, 1880, p. 85; Cupp, 1943, p. 206, Fig. 159; Hendey, 1951, p. 74; Hendey, 1964, p. 274, Plate 21, Fig. 5; Drum & Pankratz, 1966; Round et al., 1990, p. 608.

Girdle view: Cells rectangular; cells in colonies sliding along one another to form a linear array to retract into a tabular array.
Valve view: Valves linear lanceolate with produced ends. Raphe system slightly keeled. Raphe continuous from pole to pole. Fibulae strong. Valve surface with transverse parallel striae.
Morphometric data: Apical axis, 70–115 μm; transapical axis, 5–6 μm; 7–9 fibulae and 20–21 striae in 10 μm.
Distribution: Probably cosmopolitan.
How to identify: *Bacillaria* may consist of more than one species (Round et al., 1990). *Bacillaria paxillifera* seems to be the only one recorded from plankton, however. Due to the unique type of motile colonies the species may be identified in water mounts. Permanent mounts may be needed for identification of single valves.

Genus *Cylindrotheca* Rabenhorst 1859
Type: *Cylindrotheca gerstenbergeri* Rabenhorst.
Correct name: *Cylindrotheca gracilis* (Brébisson in Kützing) Grunow (*vide* Van Heurck, 1880–1885, p. 186).
The usual characterization of the genus *Cylindrotheca* is
Frustules cylindrical, fusiform, twisted about the apical axis, and rotating when in motion. Very weakly silicified, girdle bands narrow and numerous, and the valves appear to be hyaline in LM.

Reimann & Lewin (1964) redefined the genus and transferred *N. closterium* (Ehrenberg) W. Smith to *Cylindrotheca* probably based on a similarity with *Cylindrotheca* in raphe structure (TEM) and the weakly silicified valves.

Cylindrotheca closterium (Ehrenberg) Lewin & Reimann (Plate 66)
Basionym: *Ceratoneis closterium* Ehrenberg.
Synonym: *Nitzschia closterium* (Ehrenberg) W. Smith.
References: Ehrenberg, 1841a, p. 144, Plate 4, Fig. 7; Smith, 1853, p. 42,
Plate 15, Fig. 120; Cupp, 1943, p. 200, Fig. 153; Hendey, 1964, p. 283,
Plate 21, Fig. 8; Hasle, 1964, p. 16, Text Figs. 1–10, Plate 7, Figs. 1–12,
Plate 9, Figs. 1–9, Plate 10, Figs. 1–4; Reimann & Lewin, 1964, p. 289,
Plate 124, Figs. 1–4, Plate 125, Figs. 1–4, Plate 126, Figs. 1–3; Takano,
1990, pp. 320–321; Hasle & Medlin, 1990a, p. 177, Plate 23.1,
Figs. 1–4.
Emended description (Reimann & Lewin, 1964, p. 289)
    Frustules in the fusiform part not twisted, in the rostra not or only
    slightly twisted about the apical axis. Valve face weakly silicified, almost
    imperforate, traversed by more or less transapical silicified thickenings.
    Raphe traversed by a series of fibulæ, joined directly to the valve face.
    One of the edges of the valve bordering the fissure minutely serrate.
    Fissure interrupted at the center. Two chromatophores.

    It should be kept in mind that the valves of the other *Cylindrotheca*
species examined by the same authors consisted of a canal raphe solely, and
the frustules were twisted two to three times about the apical axis. The lack
of, or the insignificant twisting of the *C. closterium* cell, together with the
presence of a striated valve face, indicates a rather isolated position of this
diatom within *Cylindrotheca*. There are, therefore, still reasons for not using
Reimann & Lewin's combination but instead to retain the name *Nitzschia
closterium* until more thorough studies of marine planktonic species of the
"*N. closterium*" shape from various types of habitats have been performed.
It is true, however, that the species in question does not belong to *Nitzschia
sensu stricto* (see also Medlin, 1990 for rRNA molecule studies of *C.
closterium* and *Nitzschia* spp.).
Morphometric data: Apical axis, 30–400 μm; transapical axis, 2.5–8 μm;
fibulae, 10–12 in 10 μm; interstriae, 70–100 in 10 μm (TEM; Hasle, 1964).
Distribution: Cosmopolitan?—planktonic and common on seaweeds and
polar ice.
How to identify: The identification of *C. closterium* may have caused more
problems and more confusion than the identification of any other diatom
enountered in marine plankton (see *Lennoxia faveolata* and *Phaeodactylum
tricornutum*). When working with coastal material the most common
problem is distinguishing between *C. closterium* and the coarser *N.
longissima*; examination of water mounts is usually not sufficient. Not too
thoroughly acid-cleaned material is better since this often shows the two
delicate valves and the many bands of *C. closterium* lying together,
seemingly twisted around each other in the rostrate ends.

Genus *Fragilariopsis* Hustedt in Schmidt emend. Hasle 1993
Type: *Fragilariopsis antarctica* (Castracane) Hustedt in A. Schmidt.
Basionym: *Fragilaria antarctica* Castracane.
Correct name: *Fragilariopsis kerguelensis* (O'Meara) Hustedt *(vide* Hustedt, 1952, p. 294).

Glezer et al. (1988) and Round et al. (1990) regarded *Fragilariopsis* as a separate genus. Since this is also done in this chapter, the new names and combinations introduced by Hasle (1972c, 1974, see the synonymy list) are inappropriate.

Fragilariopsis has a canal raphe which justifies the placement in the family Bacillariaceae. The raphe system as seen with EM is simple in construction compared to that of *N. sigmoidea* and several other *Nitzschia* species (Mann, 1986). The raphe is not raised above the valve surface, the external canal wall is not poroid, and the fibulae are small and not extending across the valve (Hasle, 1965a, 1968c, 1972c; Mann, 1978). In further contrast to the subgenus *Nitzschia*, *Fragilariopsis* has no conopea (flaps of silica extending out from near the raphe), most species are narrow, not sigmoid in girdle view, and the valve striae mostly have two rows of poroids.

The little information available from EM observations on the *Fragilariopsis* girdle demonstrates a certain unconformity that may to some extent be related to the silicification of the cell wall. The lightly silicified *F. oceanica* has several distinctly striated intercalary bands, whereas those of the more heavily silicified *F. cylindrus* and *F. curta,* and especially *F. kerguelensis,* have one or perhaps two bands with one row of perforations (Hasle, 1965a, 1972c; Medlin & Sims, 1993).

All *Fragilariopsis* species examined by Hasle (1965a) are present in polar waters and are dealt with in the handbook "Polar Marine Diatoms" (Medlin & Priddle, 1990) in which a key to species is constructed (Hasle & Medlin, 1990b), as was also done by Hasle (1965a). In this chapter we include only the species more commonly recorded from plankton. For the examination of samples collected from or near sea ice Hasle & Medlin (1990b) should be consulted.

Generic characters:

Cells rectangular in girdle view.

Cells in ribbons united by the entire or the greater part of the valve surface.

Raphe strongly eccentric.

Raphe not raised above the general level of the valve.

Approximately equal numbers of interstriae and fibulae.

Fibulae often more distinct than interstriae (LM).

Central larger interspace lacking in most species.

Valve face more or less flattened and not undulated.

Valves narrowly elliptical to lanceolate to broadly elliptical or subcircular, or linear to sublinear.

Apical axis often heteropolar.

Valve poles usually bluntly rounded.

Striae parallel except near poles.

Striae with two rows of poroids, seldom one or more than two.

Chloroplasts—two plates lying along the girdle, one on either side of the median transapical plane.

Resting spores rare (one species).

Characters showing differences between species:

Valve outline.

Polarity of apical axis.

The presence or absence of central larger interspace.

Structure of striae, discernible or not discernible with LM.

A. Valves elliptical to lanceolate.
  1. Apical axis often heteropolar; apices generally rounded: *F. kerguelensis, F. ritscheri,* and *F. atlantica* (Table 73).
  2. Apical axis isopolar; apices more or less rounded: *F. oceanica* and *F. pseudonana* (Table 73).
  3. Apical axis isopolar; apices generally pointed: *F. rhombica* and *F. separanda* (Table 73).
B. Valves linear to sublinear.
  1. Apical axis isopolar: *F. cylindrus* and *F. cylindriformis* (Table 73).
  2. Apical axis heteropolar: *F. curta* (Table 73).
C. Valves semilanceolate: *F. doliolus* (Table 73).

A. Valves elliptical to lanceolate.

1. Apical axis often heteropolar, apices generally rounded.

Fragilariopsis kerguelensis (O'Meara) Hustedt (**Plate 66, Table 73**)
Basionym: *Terebraria kerguelensis* O'Meara.
Synonyms: *Fragilariopsis antarctica* (Castracane) Hustedt in A. Schmidt; *Nitzschia kerguelensis* (O'Meara) Hasle.
References: O'Meara, 1877, p. 56, Plate 1, Fig. 4; Castracane, 1886, p. 56, Plate 25, Fig. 1; Schmidt, 1913, Plate 299, Figs. 9–14; Hustedt, 1952, p. 294; Hustedt, 1958a, p. 162, Figs 121–127; Hasle, 1965a, p. 14, Plate 4,

TABLE 73 Morphometric Data of *Fragilariopsis* spp.

Species	Apical axis (μm)	Transapical axis (μm)	Striae and fibulae in 10 μm
Group A			
F. kerguelensis	10–76	5–11	4–7
F. ritscheri	22–57	8–9	6–11
F. atlantica	20–43	7–8	18–21
F. oceanica	10–41	ca. 6	12–15
F. pseudonana	4–20	3.5–5	18–22
F. rhombica	8–53	7–13	8–16
F. separanda	10–33	8–13	10–14
Group B			
F. cylindrus	3–48	2–4	13–17
F. cylindriformis	3–13	ca. 2	16–20
F. curta	10–42	3.5–6	9–12
Group C			
F. doliolus	30–70	5–8	9–14

Figs. 11–18, Plate 7, Fig. 9; Hasle, 1968c, Figs. 1, 2, and 7–9; Hasle, 1972c, p. 115; Sournia et al., 1979, Fig. 33; Hasle & Medlin, 1990b, p. 181, Plate 24.2, Figs. 11–18.

Girdle view: Valve face slightly curved; valve faces of sibling cells united for the greater part of the cell length. Interstriae coarse, penetrating deeply into the cell interior, readily seen with LM.

Valve view: Larger (ca. 40–70 μm) specimens narrowly elliptical with heteropolar apical axis, smaller specimens broadly lanceolate to broadly elliptical with isopolar apical axis. Transverse striae parallel, slightly curved near apices, two rows of large poroids, readily seen with LM. No central larger interspace (raphe uninterrupted from pole to pole, SEM).

Remarks: The coarse interstriae shaped as anvils in cross section and thus similar to pseudosepta indicate a close relationship to *N. seminae*.

Fragilariopsis ritscheri Hustedt (Plate 67, Table 73)
Synonym: *Nitzschia ritscheri* (Hustedt) Hasle.
References: Hustedt, 1958a, p. 164, Figs. 133, 136, and 153; Hustedt, 1958b, p. 205, Figs. 16 and 17; Hasle, 1965a, p. 20, Plate 4, Figs. 1–7, Plate 7, Fig. 8; Hasle, 1968c, Fig. 10; Hasle, 1972c, p. 115; Hasle & Medlin, 1990b, p. 181, Plate 24.2, Figs. 1–10.

Valve view: Larger (ca. 40–70 μm) specimens narrowly elliptical, apical axis slightly heteropolar, one pole broadly rounded, the other more pointed; smaller specimens more broadly elliptical with inconsiderable heteropolarity. Valves broad in relation to their length; middle part of

Fragilariopsis

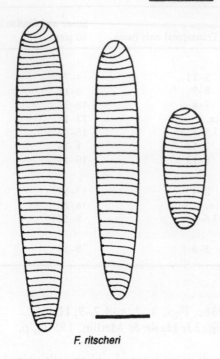

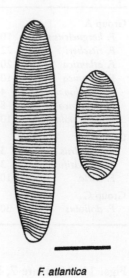

F. atlantica

F. ritscheri

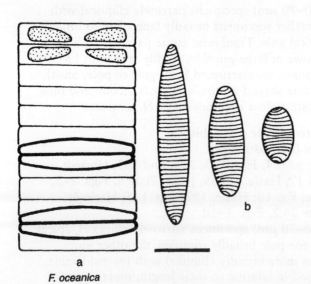

a

F. oceanica

b

F. pseudonana

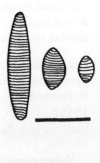

valve with slightly curved margins. Transverse striae with two rows of small poroids discernible in cleaned valves mounted in a medium of a high refractive index. No central larger interspace.

Fragilariopsis atlantica Paasche (Plate 67, Table 73)

Synonym: *Nitzschia paaschei* Hasle.

References: Paasche, 1961, p. 199, Fig. 1b, Plate 1, b–f; Hasle, 1965a, p. 9, Plate 1, Figs. 1–5, Plate 2, Figs. 1–3; Hasle, 1974, p. 426; Hasle & Medlin, 1990b, p. 181, Plate 24.1, Figs. 1–5.

Girdle view: Valves flat, cells fairly low; valves of sibling cells in ribbons united along their whole length with no interspace. Interstriae weak, discernible with LM.

Valve view: Largest (ca. 40 μm) specimens broadly linear with more or less obtuse apices, slightly heteropolar apical axis; medium-sized valves lanceolate with slightly pointed apices and isopolar apical axis; smallest specimens broadly elliptical with broadly rounded apices. Valves broad in relation to their length. Central larger interspace present; central nodule visible with LM (cleaned valves mounted in a medium of a high refractive index). Transverse interstriae more curved in smaller than in larger specimens. Stria structure not resolved with LM.

2. Apical axis isopolar; apices more or less rounded.

Fragilariopsis oceanica (Cleve) Hasle (Plate 67, Table 73)

Basionym: *Fragilaria oceanica* Cleve.

Synonyms: *Fragilaria arctica* Grunow in Cleve & Grunow; *Nitzschia grunowii* Hasle.

References: Cleve, 1873b, p. 22, Plate 4, Fig. 25; Cleve & Grunow, 1880, p. 110, Plate 7, Fig. 124; Hustedt, 1959, p. 148, Fig. 662; Hasle, 1965a, p. 11, Plate 1, Figs. 15–19, Plate 2, Figs. 6 and 7; Hasle, 1972c, p. 115; Hasle & Medlin, 1990b, p. 181, Plate 24.1, Figs. 15–19.

Girdle view: Cells in straight or sometimes curved ribbons; no interspace between valves of sibling cells. Pervalvar axis often high compared to that of the other *Fragilariopsis* spp. Mantle fairly deep, better silicified than valve face.

Valve view: Largest specimens narrowly elliptical with slightly elongated ends, medium-sized specimens more lanceolate with rounded ends, and

---

PLATE 67 *Fragilariopsis ritscheri*: valves showing size variation, striation, and heteropolarity. *Fragilariopsis atlantica*: valves showing size variation, slight heteropolarity, striation, and central interspace. *Fragilariopsis oceanica*: (a) ribbon with resting spores; (b) valves showing size variation and isopolarity. *Fragilariopsis pseudonana*: Valves showing size variation, striation, and isopolarity. Scale bars = 10 μm.

smallest specimens broadly elliptical. Interstriae on valve face weakly silicified, more strongly silicified on valve mantle. Raphe along the bend between valve face and mantle or slightly displaced to the mantle; central larger interspace present. Structure of striae not resolved with LM.
Resting spore valves structured as vegetative valves but heavily silicified.

Fragilariopsis pseudonana (Hasle) Hasle (**Plate 67, Table 73**)
Basionym: *Nitzschia pseudonana* Hasle.
Synonyms: *Fragilaria nana* Steemann Nielsen in Holmes, 1956, *pro parte*;
Fragilariopsis nana (Steemann Nielsen) Paasche.
References: Holmes, 1956, p. 47, Fig. 17 *pro parte*; Paasche 1961, p. 201, Fig. 1a, Plate 1, Fig. a; Hasle, 1965a, p. 22, Plate 1, Figs. 7–14, Plate 4, Figs. 20 and 21; Hasle, 1974, p. 427; Hasle & Medlin, 1990b, p. 181, Plate 24.1, Figs 7–14; Hasle, 1993, p. 317.
   Valve view: Largest (ca. 15–20 µm) specimens narrowly elliptical and smaller specimens more lanceolate with more or less rounded ends. Interstriae straight in middle part of the valves and curved closer to the apices. Fibulae and interstriae, but not striae structure resolved with LM. No central larger interspace.
Remarks: Since *Fragilaria nana* Steemann Nielsen, basionym of *Fragilariopsis nana* (Steemann Nielsen) Paasche, is synonomous with *Fragilariopsis cylindrus*, the name *Fragilariopsis nana* cannot be used. No type was indicated for either *Fragilariopsis pseudonana* (Hasle, 1965a) or for *Nitzschia pseudonana* (Hasle, 1972c, p. 115); the first valid name is therefore *Nitzschia pseudonana* Hasle in Hasle (1974).

3. Apical axis isopolar; apices generally pointed.

Fragilariopsis rhombica (O'Meara) Hustedt (**Plate 68, Table 73**)
Basionym: *Diatoma rhombica* O'Meara.
Synonym: *Nitzschia angulata* Hasle.
References: O'Meara, 1877, p. 55, Plate 1, Fig. 2; Hustedt, 1952, p. 296, Figs. 6 and 7; Hustedt, 1958a, p. 163, Figs. 113–120; Hustedt, 1958b, p. 205, Figs. 8–11; Hasle, 1965a, p. 24, Plate 9, Figs. 1–6; Hasle, 1972c, p. 115; Hasle & Medlin, 1990b, p. 181, Plate 24.1, Fig. 6, Plate 24.2, Fig. 19, Plate 24.4, Figs. 1–6.
   Valve view: Largest and medium-sized specimens either lanceolate or linear with parallel margins tapering toward pointed ends and smallest specimens almost circular. Valves broad in relation to their length. Interstriae straight in the middle linear or convex part of the valve and curved in the tapering parts. Striae with two alternating rows of poroids discernible with LM. No central larger interspace.

Fragilariopsis

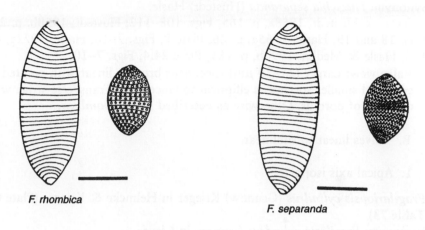

F. rhombica

F. separanda

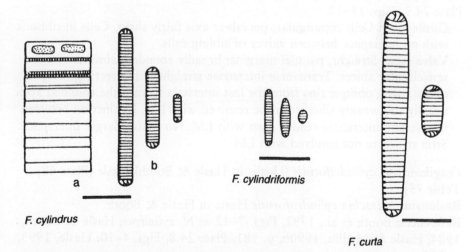

a

F. cylindrus

b

F. cylindriformis

F. curta

PLATE 68 *Fragilariopsis rhombica*: two valves of different size, biseriate striae indicated in smaller valve. *Fragilariopsis separanda*: two valves, uniseriate striae indicated in smaller valve. *Fragilariopsis cylindrus*: (a) ribbon in girdle view, chloroplasts in top cell; (b) valves showing size variation. *Fragilariopsis cylindriformis*: valves showing variation in size and outline. *Fragilariopsis curta*: valves of different size and heteropolarity. Scale bars = 10 μm.

Fragilariopsis separanda Hustedt (Plate 68, Table 73)
Synonym: *Nitzschia separanda* (Hustedt) Hasle.
References: Hustedt, 1958a, p. 165, Figs. 108–112; Hustedt, 1958b, p. 207,
Figs. 18 and 19; Hasle, 1965a, p. 26, Plate 9, Figs. 7–10; Hasle, 1972c, p.
115; Hasle & Medlin, 1990b, p. 181, Plate 24.4, Figs. 7–10.
  Valve view: Larger (25–30 μm) specimens broadly linear with pointed
  ends and smaller specimens elliptical to lanceolate. Transverse striae with
  one row of poroids. Otherwise as described for *F. rhombica.*

  B. Valves linear to sublinear.

  1. Apical axis isopolar.

Fragilariopsis cylindrus (Grunow) Krieger in Helmcke & Krieger (Plate 68,
Table 73)
Basionym: *Fragilaria cylindrus* Grunow in Cleve.
Synonym: *Nitzschia cylindrus* (Grunow) Hasle.
References: Grunow in Cleve & Möller, 1882, No. 314; Cleve, 1883,
p. 484, Plate 37, Figs. 64a–64c; Helmcke & Krieger, 1954, p. 17, Plate
187; Hustedt, 1958a, p. 162, Figs. 145 and 146; Hustedt, 1959, p. 152, Fig.
665; Hasle, 1965a, p. 34, Plate 12, Figs 6–12; Hasle, 1968c, Fig. 6; Hasle,
1972c, p. 115; Hasle & Medlin, 1990b, p. 181, Plate 24.6, Figs. 6–12,
Plate 24.8, Figs. 11–13.
  Girdle view: Cells rectangular; pervalvar axis fairly short. Cells in ribbons
  with no interspace between valves of sibling cells.
  Valve view: Straight, parallel margins; broadly rounded, almost
  semicircular apices. Transverse interstriae straight in the rectangular part
  of the valve; oblique ribs from the last interstria toward the rounded apex
  usually too weakly silicified to be resolved with LM. Distinction between
  fibulae and interstriae readily seen with LM. No central larger interspace.
  Stria structure not resolved with LM.

Fragilariopsis cylindriformis (Hasle, in Hasle & Booth) Hasle (Plate 68,
Table 73)
Basionym: *Nitzschia cylindroformis* Hasle in Hasle & Booth.
References: Booth et al., 1982, Figs. 7–12 as *N. cylindrus*; Hasle & Booth,
1984; Hasle & Medlin, 1990b, p. 181, Plate 24.8, Figs. 4–10; Hasle, 1993,
p. 316.
  Girdle view: Cells in ribbons with no interspace between valves of sibling
  cells.
  Valve view: Largest specimens linearly oblong (similar to *F. cylindrus*),
  medium-sized specimens narrowly elliptical to lanceolate (like *F.*

pseudonana), and the smallest specimens broadly elliptical to subcircular. Fibulae and interstriae resolved with LM with fibulae more distinct than interstriae. Interstriae parallel and straight, except close to the poles. No central larger interspace.

2. Apical axis heteropolar.

Fragilariopsis curta (Van Heurck) Hustedt (Plate 68, Table 73)
Basionym: *Fragilaria curta* Van Heurck.
Synonym: *Nitzschia curta* (Van Heurck) Hasle.
References: Van Heurck, 1909, p. 24, Plate 3, Fig. 37; Hustedt, 1958a, p. 160, Figs. 140–144, 159; Hustedt, 1958b, p. 201, Figs. 2–4; Hasle, 1965a, p. 32, Plate 12, Figs. 2–5; Hasle, 1972c, p. 115; Hasle & Medlin, 1990b, p. 181, Plate 24.6, Figs. 2–5.
Valve view: Apical axis heteropolar with one pole slightly narrower than the other. Valve margins more or less parallel, tapering toward the narrower of the two broadly rounded poles. Interstriae and fibulae readily observed with LM; fibulae coarser than interstriae. Interstriae straight in middle part of valve and curved near the poles with some additional apical ribs, these ribs as well as stria structure just visible with LM. No central larger interspace.

C. Valves semilanceolate.

Fragilariopsis doliolus (Wallich) Medlin & Sims (Plate 69)
Basionym: *Synedra doliolus* Wallich.
Synonym: *Pseudoeunotia doliolus* (Wallich) Grunow in Van Heurck.
References: Wallich, 1860, p. 48, Plate 2, Fig. 19; Van Heurck, 1880–1885, Plate 35, Fig. 22; Cupp, 1943, p. 190, Fig. 140; Hustedt, 1958b, p. 199, Fig. 1; Hustedt, 1959, p. 259, Fig. 737; Hasle, 1972c, Fig. 9; Hasle 1976a, Figs. 29–31; Medlin & Sims, 1993.
Girdle view: Cells rectangular in broad girdle view, united into curved ribbons by the valve surfaces; no interspace between sibling cells. Open bands, three unperforated and one with one row of perforations (Medlin & Sims, 1993). Chloroplasts—two plates, one on either side of the median transapical plane.
Valve view: Valves semilanceolate with bluntly rounded ends. One side of valve (ventral) straight and seldom slightly convex; the other side (dorsal) more strongly convex, gradually decreasing toward the ends, and near the ends often slightly constricted. Transverse striae with two alternating rows of poroids (LM). Interstriae thickened vertically on external and internal surfaces of valve face and mantle (SEM). Canal raphe (SEM)

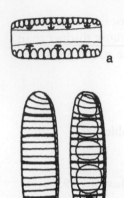

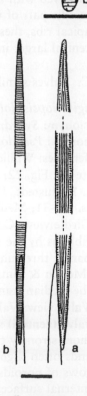

Neodenticula seminae

Fragilariopsis doliolus

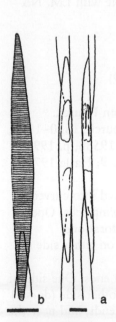

Pseudo-nitzschia seriata f. seriata

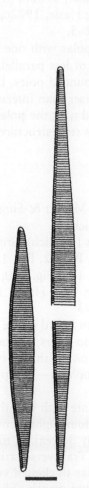

P. australis

P. pungens

along bend between valve face and mantle, along either dorsal or ventral side, indicating a diagonal location of the raphe systems on the frustule. Some irregularities between the middle fibulae observed with LM and SEM.

Remarks: SEM investigations (Hasle, 1976a) confirmed Hustedt's (1958b) LM observation of a canal raphe in *F. doliolus*; central raphe endings, reported by Hustedt, were not seen (see also Medlin & Sims, 1993). The expressed curvature giving the ribbons the appearance of a section of a barrel may be due to a more shallow valve mantle along the straight rather than the curved valve margin.

Distribution:

F. kerguelensis, F. ritscheri, F. rhombica, and *F. separanda*—southern cold water region, plankton.

F. atlantica and *F. cylindriformis*—northern cold water region, plankton.

F. doliolus—warm water region.

F. pseudonana—cosmopolitan, plankton.

F. curta—southern cold water region, plankton and ice.

F. oceanica—northern cold water region, plankton and ice(?).

F. cylindrus—southern and northern cold water regions, plankton and ice.

Fragilariopsis kerguelensis is particularly abundant in the southern ocean but has been recorded as far north as the Cape Verde Islands (Heiden & Kolbe, 1928).

How to identify: *Fragilariopsis doliolus* is readily identified in colonies and as single cells in water mounts due to the unique shape of the ribbons and the valve outline. *Fragilariopsis kerquelensis* may be discriminated from other planktonic species by its coarse silification, also when observed in girdle view. The other planktonic species have to be examined in valve view to secure a positive identification. Material cleaned of organic matter and mounted in a medium of a high refractive index may be required.

Genus *Neodenticula* Akiba & Yanagisawa 1986
Type: *Neodenticula kamtschatica* (Zabelina) Akiba & Yanagisawa.
Basionym: *Denticula kamtschatica* Zabelina.

PLATE 69 *Neodenticula seminae:* (a) single cell in girdle view with pseudosepta; (b) valves with striation and pseudosepta and valve with attached band. After Simonsen & Kanaya (1961). *Fragilariopsis doliolus:* (a) ribbon in broad girdle view; (b) valve with striation and areolae indicated. After Cupp (1943). *Pseudo-nitzschia seriata* f. *seriata:* (a) stepped chains in valve and girdle views; (b) valve view. From Hasle (1972d). *Pseudo-nitzschia australis:* valves of different size. From Hasle (1972d). *Pseudo-nitzschia pungens:* partial chains in (a) girdle view and (b) valve view. From Hasle (1972d). Scale bars – 10 μm.

Terminology specific to *Neodenticula:*

Pseudoseptum—diaphragm-like ingrowth of valve, in this genus shaped as a vertical wall penetrating into the cell interior and separating the lumen of the valve into chambers.

Neodenticula seminae (Simonsen & Kanaya) Akiba & Yanagisawa (Plate 69)
Basionym: *Denticula seminae* Simonsen & Kanaya.
Synonyms: *Denticula marina* Semina; *Denticulopsis seminae* (Simonsen & Kanaya) Simonsen.
References: Semina, 1956, p. 82, Figs. 1 and 2; Simonsen & Kanaya, 1961, p. 503, Plate 1, Figs. 26–30; Hasle, 1972c, Figs. 3, 4, and 8; Simonsen, 1979, p. 65; Semina, 1981b; Akiba & Yanagisawa, 1986, p. 491, Plate 24, Figs. 1–11, Plate 26, Figs. 1–10; Yanagisawa & Akiba, 1990, p. 263, Plate 7, Figs. 45–49; Medlin & Sims, 1993, Figs. 13–21.
   Girdle view: Cells rectangular with rounded corners, interstriae, and strong pseudosepta, the latter widening toward cell interior. Bands septate.
   Valve view: Valves linear to elliptical, usually with broadly rounded ends. Valve surface with transverse striae, the areolation of the striae usually too fine to be resolved with LM. Pseudosepta seen as sharp, distinct lines with focus on external valve surface and as wide crossbars with focus on their distal parts; both ends of a crossbar connected with the thick valve wall by a suture. Pseudosepta sometimes short in pervalvar direction and without crossbars. A varying number of striae between two pseudosepta. One interstria often branched into two fibulae (SEM). Raphe continuous along the edge of valve face scarcely discernible with LM. Raphe systems of the two valves on opposite sides of the frustule.
Morphometric data: Apical axis, 10–60 μm, transapical axis, 4–9 μm; pseudosepta, 2–4 in 10 μm; 1–4 striae and interstriae between two pseudosepta and 8–12(?) fibulae in 10 μm.
Distribution: Common in North Pacific Ocean, also reported from tropical parts of the Indian and Pacific Oceans (Semina, 1981b).
How to identify: The species may be identified as whole cells in water mounts.
Remarks: Controversial opinions exist concerning the crossbars being parts of the pseudosepta and thus of the valves or of the valvocopula septum. Based on our own published and unpublished observations we agree with Simonsen (1979) and Akiba & Yanagisawa (1986) that although the bands are septate, the crossbars do belong to the pseudosepta. Another peculiarity is the presence of open and closed bands in *N. seminae* (Hasle, 1972c, Fig. 4; Akiba & Yanagisawa, 1986, Plate 24, Figs. 6 and 7, Plate 26, Figs. 7 and 8; Yanagisawa & Akiba, 1990, Plate 7, Fig. 49; Medlin & Sims, 1993, Figs. 15, 16, 20, and 21). Medlin & Sims (1993) suggested that the closed bands might be more commonly associated with valves with deeper pseudosepta.

Genus *Pseudo-nitzschia* H. Peragallo in H. & M. Peragallo
**Lectotype: *Pseudo-nitzschia seriata* (Cleve) H. Peragallo (in H. & M.
Peragallo, 1897–1908) (*vide* Fryxell et al., 1991, p. 243).**

The genus was erected for *Nitzschia sicula* with the varieties *bicuneata*
and *migrans*, and *Nitzschia seriata* with the variety *fraudulenta*—all pelagic
forms with obscure keels ("carènes très obscures"), if present at all (H. &
M. Peragallo, 1897–1908, p. 263). *Pseudo-nitzschia seriata* was furthermore
characterized by a fine striation and cells in long filaments, and *N. sicula* by
coarser striation and cells probably free living. The raphe system of *Pseudo-
nitzschia* as revealed with EM is like that of *Fragilariopsis*, not being
elevated above the general level of the valve, and lacking conopea and
lacking poroids in the external canal wall. Some of the species have striae
with two rows of poroids like *Fragilariopsis*, whereas others have more than
two rows and still others have only one row. Some *Pseudo-nitzschia* species
are rectangular in girdle view like *Fragilariopsis*, but many are fusiform, and
in valve view in general longer and narrower than *Fragilariopsis*. The
distinction between the two genera is especially manifest in the stepped
colonies and in the tendency of heteropolarity of the transapical axis of
Pseudo-nitzschia. The girdle structure may be a third distinctive character.
The intercalary bands of *Pseudo-nitzschia* are open, distinctly pointed,
narrow, and mostly striated. The striae are similar in structure to those on
the valve with one or more rows of poroids (Hasle, 1965b; Hasle et al.,
1995). This type of band structure together with other morphological
characters such as the structure of the raphe system, characterize *Pseudo-
nitzschia* as a natural group separate from *Nitzschia* (Hasle, 1994).

Pseudo-nitzschia is a geographically widely distributed genus (Hasle,
1965b, 1972d) restricted to marine plankton. More detailed studies of the
distribution of taxa referred to *Pseudo-nitzschia* spp. are of special interest,
however, since *P. multiseries* is well established as being capable of
producing the neurotoxin, domoic acid, in growth-limiting conditions in
stationary stage. This has been reported from at least two areas with
different climatic and hydrographic conditions (Bates et al., 1989; Fryxell et
al., 1990). Domoic acid was also detected in clams and mussels in the Bay
of Fundy, eastern Canada, at a time when *P. pseudodelicatissima* was the
predominant phytoplankton species (Martin et al., 1990), and *P. australis*
has been documented as a producer of domoic acid in California waters
(Buck et al., 1992; Garrison et al., 1992). *Pseudo-nitzschia delicatissima* and
P. seriata have been shown to produce domoic acid in cultures (Smith et al.,
1991; Lindholm et al., 1994). Domoic acid may be a worldwide threat, on
temperate coasts at least, but action can be taken to alleviate the potential
danger by monitoring the phytoplankton and utilizing temporary closings of
selected fisheries in target areas when necessary for the duration of a bloom.

Generic characters:
   Cells strongly elongate, rectangular, or fusiform in girdle view.

   Cells in stepped chains united by shorter or longer overlap of valve
      ends.

   Chains motile.

   Raphe strongly eccentric.

   Raphe not raised above the general level of the valve.

   Valve face interstriae often more than one to each fibula.

   Central larger interspace in most species.

   Valve face slightly curved or flattened, not undulated.

   Valve narrowly lanceolate to fusiform and linear with rounded or
      pointed ends.

   Transapical axis heteropolar in some species.

   Stria structure usually too delicate to resolve with LM.

   Chloroplasts—two plates, lying along the girdle, one on either side of
      the median transapical plane.

   Resting spores unknown.

Characters showing differences between species:
   Valve outline.

   Width of valve (= length of transapical axis).

   Polarity of transapical axis.

   Linear density of interstriae versus fibulae.

   Size of central interspace.

   Shape of valve ends in girdle and valve views.

   Length of overlap of cell ends.

KEY TO SPECIES (based on light microscope observations)

1a. Transapical axis wider than 3 μm 2
1b. Transapical axis narrower than 3 μm. 9
2a. Central larger interspace present 3
2b. Central larger interspace absent. 6
3a. Outline of valve asymmetrical in the apical axis; one margin almost
      straight, the other curved. 4
3b. Outline of valve symmetrical in the apical axis 5
4a. Fibulae and interstriae discernible with the light microscope . *P. heimii*
4b. Fibulae but not interstriae discernible with the light microscope . . . . .
      . . . .*P. subpacifica*

5a. Valve outline lanceolate; margins not parallel; fibulae and interstriae equal in number .... *P. fraudulenta*

5b. Valve outline with parallel margins for the greater part of the valve length; fibulae and interstriae unequal in number .. *P. subfraudulenta*

6a. Outline of valve asymmetrical in the apical axis; one margin almost straight, the other curved. .... *P. seriata*

6b. Outline of valve symmetrical in the apical axis 7

7a. The presence of poroids in the striae discernible with the light microscope .... *P. pungens*

7b. The presence of poroids in the striae not discernible with the light microscope 8

8a. Transapical axis wider than 5 μm; valve apices rostrate; outline of smaller valves tends to be asymmetrical in the apical axis ...... *P. australis*

8b. Transapical axis narrower than 5 μm; valve outline lanceolate with pointed (not rostrate) apices................... *P. multiseries*

9a. Central larger interspace present 10

9b. Central larger interspace absent, valve with prolonged projections (rostrate). 17

10a. Valve inflated in the middle 11

10b. Valve otherwise. 14

11a. Valve with prolonged projections. .... *P. prolongatoides*

11b. Valves otherwise 12

12a. Valves slightly inflated before tapering ends, interstriae not discernible with the light microscope. .... *P. inflatula*

12b. Valve ends obtuse, interstriae distinct 13

13a. Valves comparatively wide (length ca. 12–30 times width), valve ends broadly obtuse .... *P. turgidula*

13b. Valves narrow (length ca. 27–60 times width) ..... *P. turgiduloides*

14a. Valves comparatively wide (ca. 3 μm) and tapering parts of valves long .... *P. cuspidata*

14b. Valves otherwise 15

15a. Valves slightly lanceolate, fibulae and central larger interspace barely visible with the light microscope, interstriae not visible........... .... *P. delicatissima*

15b. Valves linear, tapering parts of valves short 16

16a. Fibulae and central larger interspace clearly visible with the light microscope, interstriae also occasionally visible .... *P. pseudodelicatissima*

16b. Fibulae and interstriae clearly visible with the light microscope. .... *P. lineola*

17a. One valve margin straight or slightly concave, the other convex, especially in the middle part of the valve .... *P. subcurvata*

17b. Valve narrow, lanceolate to needle shaped. .... *P. granii*

The species may also be grouped according to numerical data, shape of valves and stria structure, the latter mainly revealed with TEM.

A. Valves fairly wide compared to length (transapical axis ca. 3 μm or more).
   1. Approximately equal numbers of fibulae and interstriae; fibulae therefore often indistinct; no central larger interspace: *P. seriata, P. australis, P. pungens,* and *P. multiseries* (Table 74).
   2. Slightly more interstriae than fibulae; central larger interspace present: *P. pungiformis, P. fraudulenta,* and *P. subfraudulenta* (Table 74).
   3. More than one interstria per fibula; central larger interspace present; transapical axis heteropolar: *P. heimii* and *P. subpacifica* (Table 74).
B. Valves narrow compared to cell length (transapical axis ca. 3 μm or less); approximately two interstriae for each fibula.
   1. Circular poroids.
      Central larger interspace present: *P. delicatissima, P. lineola, P. prolongatoides, P. turgidula,* and *P. turgiduloides* (Table 75).
   2. Square poroids.
      a. Central larger interspace present: *P. cuspidata, P. pseudodelicatissima,* and *P. inflatula* (Table 75).
      b. No central larger interspace: *P. granii* and *P. subcurvata* (Table 75).

A. Transapical axis ca. 3 μm or more.

This group of larger (= wider in valve view) *Pseudo-nitzschia* species has been called the "*Nitzschia seriata* complex" (Hasle, 1965b). They all have a tendency to appear in girdle view in water mounts, and with *P. pungens* as the only possible exception, the distinction between species can only be seen in valve view. *Pseudo-nitzschia seriata* seems for some obscure reasons to have been regarded by many planktonologists as the only species of this size group appearing in stepped chains. The whole group may have been identified as "*Nitzschia seriata*" on many occasions, and the numerous records of *N. seriata* from all oceans should be regarded with sceptisism.

   1. Approximately equal numbers of interstriae and fibulae; no central larger interspace.

Pseudo-nitzschia seriata (Cleve) H. Peragallo in H. & M. Peragallo f. *seriata* (Plate 69, Table 74)
Basionym: *Nitzschia seriata* Cleve.
References: Cleve, 1883, p. 478, Fig. 75; H. & M. Peragallo, 1897–1908, p. 300, Plate 72, Fig. 28; Cupp, 1943, p. 201, Fig. 155; Hasle, 1965b, p. 8, Plate 1, Fig. 1, Plate 3, Figs. 1–7 and 10; Hasle & Medlin, 1990c,

TABLE 74 Morphometric Data of *Pseudo-nitzschia* spp. of the "*Nitzschia seriata* Complex" (Hasle, 1965b, 1971)

Species	Apical axis (μm)	Transapical axis (μm)	Striae in 10 μm	Fibulae in 10 μm
Group A1				
P. seriata f. seriata	91–160	5.5–8	14–18	14–18
P. seriata f. obtusa	61–100	4.5–5.3	15–20	15–20
P. australis	75–144	6.5–8	12–18	12–18
P. pungens	74–142	3–4.5	9–15	9–15
P. multiseries	68–140	4–5	10–13	10–13
Group A2				
P. pungiformis	96–145	4–5	14–20	12–18
P. fraudulenta	64–111	4.5–6.5	18–24	12–24
P. subfraudulenta[a]	65–106	5–7	23–26	14–17
Group A3				
P. heimii	67–120	4–6	19–26	11–16
P. subpacifica	33–70	5–7	28–32	15–20

[a] Mentioned in the text for comparison.

p. 169, Plate 22.1, Figs 1–7 and 10; Hasle, 1994, Fig. 3; Hasle et al., 1996, Figs. 10–14 and 45–49.

Girdle view: Linear to fusiform with distinctly pointed ends; overlap of cells in chains one-third to one-fourth of cell length.

Valve view: One margin of middle part almost straight and the other convex; both margins tapering toward more or less rounded apices. Interstriae of well-silicified specimens discernible in water mounts and readily seen on cleaned valves on permanent mounts. Canal raphe along either curved or straight margin.

Pseudo-nitzschia seriata f. *obtusa* (Hasle) Hasle
Basionym: *Nitzschia seriata* f. *obtusa* Hasle.
References: Hasle, 1965b, p. 10, Plate 3, Figs. 8, 9, and 11; Hasle, 1974, p. 426; Hasle & Medlin, 1990c, p. 169, Plate 22.1, Figs. 8, 9, and 11; Hasle 1993, p. 319; Hasle et al., 1996, Figs. 15 and 50.

Valve view: Same as the nominate form but smaller and with distinctly rounded apices. Specimens in which the valve structure is not resolved may be confused with *Thalassionema nitzschioides*.

Pseudo-nitzschia australis Frenguelli (Plate 69, Table 74)
Synonym: *Nitzschia pseudoseriata* Hasle.
References: Frenguelli, 1939, p. 217, Plate 2, Fig. 13; Hasle, 1965b, p. 11, Plate 5, Figs. 1–6, Plate 6, Fig. 1; Rivera, 1985, p. 13, Figs. 14–18; Hasle et al., 1995, Figs. 16 and 51–54.

Girdle view: Fusiform; overlap of cells in chains ca. one fourth of cell length (Rivera, 1985, Figs. 15 and 16). Bands strongly silicified with transverse ribs somewhat more closely spaced than those on the valves (Hasle, 1965b, Plate 5, Fig. 6).

Valve view: Larger (apical axis more than ca. 100 μm) specimens with slightly rostrate ends. Middle, about one third part of valve with more or less parallel margins, sometimes less distinct in smaller specimens. Fibulae more distinct than interstriae, both readily seen with LM.

Remarks: Rivera (1985) rejected *Nitzschia pseudoseriata* (= *P. australis*) as a taxon of its own, separated from *Nitzschia seriata,* his argument being the great variation in his Chilean material. *Pseudo-nitzschia australis* is larger, generally coarser, and more symmetrical with respect to the apical plane than *P. seriata.* Also, the valve striae of *P. australis* bear two rows of poroids and those of *P. seriata* three or four, and the poroids of *P. australis* are larger than those of *P. seriata* (Hasle, 1965b; Hasle et al., 1995).

Pseudo-nitzschia pungens (Grunow ex Cleve) Hasle (**Plate 69, Table 74**)
Basionym: *Nitzschia pungens* Grunow ex Cleve.
References: Grunow (in Cleve & Möller, 1882), No.307; Cleve, 1897a, p. 24, Plate 2, Fig. 23; Cupp, 1943, p. 202, Fig. 156; Hasle, 1965b, p. 12, Plate 1, Figs. 4 and 5, Plate 5, Figs. 7–9; Takano & Kuroki, 1977, p. 42, Figs. 1 and 4–8; Rivera, 1985, p. 12, Figs. 1–3; Takano & Kikuchi, 1985; Takano, 1990, pp. 328–329; Hasle, 1993, p. 319; Hasle et al., 1996, Figs. 3–6 and 30–37; Hasle, 1995, Fig. 1.

Girdle view: Fusiform, pervalvar axis up to 8 μm. Fibulae and/or ends of interstriae distinct. Overlap of cells in chains considerable, close to one-third or more of cell length.

Valve view: Larger specimens linear with distinctly pointed ends, and smaller specimens more fusiform. Strongly silicified. Interstriae visible in water mounts. Fibulae distinct on cleaned valves on permanent mounts; striae biseriate; the two rows of poroids discernible with LM under optimum optical conditions (phase contrast, oil immersion).

Pseudo-nitzschia multiseries (Hasle) Hasle (**Table 74**)
Basionym: *Nitzschia pungens* f. *multiseries* Hasle.
Synonym: *Pseudo-nitzschia pungens* f. *multiseries* (Hasle)
References: Hasle, 1965b, p. 14, Plate 2, Figs. 1 and 2, Plate 5, Figs. 10–12; Hasle, 1974, p. 426; Takano & Kuroki, 1977, p. 43, Figs. 2 and 9–13; Subba Rao & Wohlgeschaffen, 1990; Fryxell et al., 1990, p. 171; Takano, 1990, pp. 328–329; Hasle, 1993, p. 319; Hasle et al., 1996, Figs. 7–9 and 38–44; Hasle, 1995, Fig. 2; Manhart et al., 1995.

Valve view: Multiseriate striae, not resolved with LM. Smaller and medium-sized specimens on an average broader compared to valve length than the same sized specimens of *P. pungens.*

Remarks: Since the poroids do not resolve with LM, EM is necessary to ascertain the identification of this species. Takano & Kikuchi (1985) reported on *P. pungens* cells with swellings from the girdle and undulated valves from eutrophic marine Japanese waters. Subba Rao & Wohlgeschaffen (1990) found similar cells of *P. multiseries* in cultures and in naturally very dense populations. Beaked and lobed valve margins, such as those shown in these investigations, are common in dense, unhealthy cultures of *Pseudo-nitzschia* in general (Hasle, 1965b, Plate 2, Fig. 9) and are also comparable to the forms of *Synedropsis hyperborea* found in nature (Grunow, 1884, Plate 2, Figs. 4–6) and in cultures (Hasle et al., 1994). The occurrence of aberrant cells of *P. multiseries* should be no obstacle for a correct identification since the cells retain most of their specific morphological characters.

2. Slightly more interstriae than fibulae; central larger interspace present.

Pseudo-nitzschia pungiformis (Hasle) Hasle (Table 74)
Basionym: *Nitzschia pungiformis* Hasle.
References: Hasle, 1971, p. 143, Figs. 1–8; Simonsen, 1974, p. 54, Plate 40, Figs. 1 and 2; Hasle, 1993, p. 319.
Valve view: Larger specimens with almost parallel margins for their greater length; smaller specimens fusiform, valve ends slightly pointed. Fibulae, interstriae, and the central larger interspace readily recognized on cleaned valves on permanent mounts (LM). Number of fibulae and interstriae generally the same in the middle part of the valve with a slight deviation toward the valve ends. The biseriate structure of the striae not resolved with LM.

Pseudo-nitzschia fraudulenta (Cleve) Hasle (Plate 70, Table 74)
Basionym: *Nitzschia fraudulenta* Cleve.
Synonym: *Pseudo-nitzschia seriata* var. *fraudulenta*
(Cleve) H. Peragallo in H. & M. Peragallo.
References: Cleve, 1897b, p. 300, Fig. 11; H. & M. Peragallo, 1897–1908, p. 300, Plate 72, Fig. 29; Hasle, 1965b, p. 15, Plate 1, Figs. 2 and 3, Plate 6, Figs. 5–10; Takano & Kuroki, 1977, p. 43, Figs. 14–18; Rivera, 1985, p. 15, Figs. 43–46; Hasle, 1993, p. 318; Hasle, 1994, Fig. 4; Hasle et al., 1996, Figs. 17–19 and 57–61.
Girdle view: Linear to fusiform; overlap of cells in chains fairly short.
Valve view: Fusiform. Weakly silicified; valve structure not discernible in water mounts. Fibulae and central nodule distinct on cleaned valves on permanent mounts; interstriae barely discernible (LM).
Remarks: *Pseudo-nitzschia subfraudulenta* (Hasle) Hasle (basionym: *Nitzschia subfraudulenta*) is a less frequently recorded species which is very

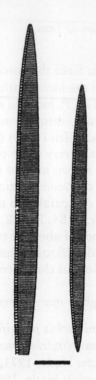

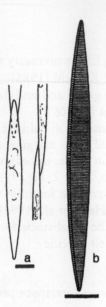

P. fraudulenta

P. subfraudulenta

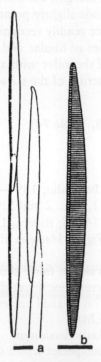

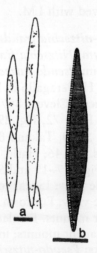

P. heimii

P. subpacifica

similar to *P. fraudulenta* but differs in morphometric data and by having more linear valves, especially in the middle part (Hasle et al., 1996, Fig. 62).

3. More than one interstria per fibula; central larger interspace present; transapical axis heteropolar; canal raphe along either convex or straight margin.

Pseudo-nitzschia heimii Manguin (Plate 70, Table 74)
Synonym: *Nitzschia heimii* (Manguin) Hasle.
References: Manguin, 1957, p. 131, Plate 6, Fig. 42; Manguin, 1960, p. 332, Plate 19, Figs. 223 and 224; Hasle, 1965b, p. 21, Plate 1, Figs. 6 and 7, Plate 10, Figs. 9–13; Simonsen, 1974, p. 51, Plate 36, Figs. 4 and 5; Hasle & Medlin, 1990c, p. 169, Plate 22.2, Figs. 9–13; Hasle et al., 1996, Figs. 20 and 65–69.

Girdle view: Linear, tapering toward somewhat sigmoid and obliquely truncated poles; overlap of cells in chains one-fourth to one-fifth of cell length.
Valve view: Largest specimens almost fusiform; smaller specimens with one convex and one straight margin attenuated toward broadly rounded, obtuse ends, sometimes with a constriction near apices. Fibulae, central larger interspace, central nodule and interstriae readily seen on cleaned valves on permanent mounts (LM).

Pseudo-nitzschia subpacifica (Hasle) Hasle (Plate 70, table 74)
Basionym: *Nitzschia subpacifica* Hasle.
References: Hasle, 1965b, p. 20, Plate 1, Figs. 9 and 10, Plate 10, Figs. 1–8; Hasle, 1974, p. 427; Simonsen, 1974, p. 55, Plate 41, Figs. 1–3; Hasle, 1993, p. 320.

Girdle view: Slightly convex margins; sigmoid, obliquely truncated cell ends; overlap of cells in chains one-fifth to one-sixth of cell length.
Valve view: One margin convex, the other more or less straight especially in the middle two-thirds of valve length; valve ends more or less pointed. Canal raphe somewhat indented in the middle; central larger interspace and central nodule observable on cleaned valves on permanent mounts, interstriae barely discernible (LM).

PLATE 70   *Pseudo-nitzschia fraudulenta:* (a) partial chains in valve and girdle views; (b) valve with striation and central larger interspace. *Pseudo-nitzschia subfraudulenta:* valves with striation and central larger interspace. *Pseudo-nitzschia heimii:* (a) partial chains in valve and girdle views; (b) valve with striation and central larger interspace. *Pseudo-nitzschia subpacifica:* (a) partial chains in valve and girdle views; (b) valve with striation and central larger interspace. From Hasle (1972d). Scale bars = 10 μm.

Distribution (Hasle 1972d):

P. seriata f. *seriata*—northern cold water region to temperate.

P. seriata f. *obtusa*—northern cold water region.

P. australis, P. subfraudulenta, and *P. subpacifica*—warm water region to temperate.

P. pungiformis—warm water region.

P. pungens, P. multiseries, P. fraudulenta, and *P. heimii*—cosmopolitan. *Pseudo-nitzschia heimii* is particularly abundant in the subantarctic. The presence of *P. australis* in the north Pacific should be noted (Hasle, 1972d, Fig. 2, confirmed by personal observations of net samples from Monterey Bay, August 7, 1967 and July 14, 1982, and in September 1991 by Work et al., 1991).

How to identify: In general, the species of the *"Nitzschia seriata* complex" can only be identified to genus and not to species when examined in girdle view. A few of the larger species are coarsely structured and are identifiable in valve view in water mounts. Most of them have to be examined as cleaned valves mounted in a medium of a high refractive index.

B. Transapical axis ca. 3 μm or less.

With a few exceptions the diatoms of this so-called *"Nitzschia delicatissima* complex" (Hasle, 1965b) are delicate forms with a very fine structure. Electron microscopy reveals two types of stria structure, either well-defined circular

TABLE 75 Morphometric Data of *Pseudo-nitzschia* spp. of the *"Nitzschia delicatissima* Complex" (Hasle, 1964, 1965b)

Species	Apical axis (μm)	Transapical axis (μm)	Striae in 10 μm	Fibulae in 10 μm
Group B1				
P. delicatissima	40–76	ca. 2	36–40	19–25
P. lineola	56–112	1.8–2.7	22–28	11–16
P. prolongatoides	20–70	0.5–2.5	30–35	15–18
P. turgidula	30–80	2.5–3.5	23–28	13–18
P. turgiduloides[a]	63–126	1.2–2.7	17–21	10–13
Group B2				
P. cuspidata	30–80	ca. 3	29?–37[b]	16?–22
P. inflatula[a]	60–100	1.5–2.5	32–35	18–21
P. pseudodelicatissima	59–140	1.5–2.5	30–46?	16–26
P. granii	25–79	1.5–2.5	44–49	12–18
P. subcurvata	47–113	1.5–2.5	44–49	12–18

[a] Mentioned in the text for comparison.
[b] ?, Data uncertain.

poroids in a varying number of rows or one row of more square poroids. These narrower *Pseudo-nitzschia* species are, like the larger ones, almost indistinguishable in girdle view and have most likely often been confused with *N. delicatissima* as the best known of the complex.

1. Circular poroids; central larger interspace present.

Pseudo-nitzschia delicatissima (Cleve) Heiden in Heiden & Kolbe (Plate 72, Table 75)
Basionym: *Nitzschia delicatissima* Cleve.
Synonym: *Nitzschia actydrophila* Hasle.
References: Cleve, 1897a, p. 24, Plate 2, Fig. 22; Heiden & Kolbe, 1928, p. 672; Hasle, 1965b, p. 35, Plate 2, Fig. 10, Plate 15, Figs. 19–23; Hasle, 1976b, p. 102, Figs 1–5; Hasle & Medlin, 1990c, p. 169, Plate 22.5, Figs. 19–23; Hasle et al., 1996, Figs. 23–25 and 70–75.
   Girdle view: Linear; cells slightly sigmoid, narrow, and truncated; overlap of cells in chains short (about one-ninth of cell length).
   Valve view: Spindle shaped with slightly rounded apices. Fibulae and central larger interspaces barely visible with LM. Striae with two rows of poroids (TEM).
Remarks: *Pseudo-nitzschia pseudodelicatissima* and *P. delicatissima* are readily distinguished by their stria structure resolved with EM. The distinction is harder to recognize with LM. The more pointed cell ends in girdle view, the more tapering valve ends, and the more linear valve outline of *P. pseudodelicatissima* compared to *P. delicatissima* do help to ascertain the identification of the two species.

Pseudo-nitzschia lineola (Cleve) Hasle (Plate 71, Table 75)
Basionym: *Nitzschia lineola* Cleve.
Synonym: *Nitzschia barkleyi* Hustedt.
References: Cleve, 1897b, p. 300, Fig. 10; Hustedt, 1952, p. 293, Figs. 13 and 14; Hasle, 1965b, p. 29, Plate 12, Figs. 15–19; Simonsen, 1974, p. 53, Plate 39, Figs. 4–6; Hasle & Medlin, 1990c, p. 169, Plate 22.3, Figs. 15–21; Hasle, 1993, p. 319.
   Girdle view: Linear to slightly lanceolate with pointed, somewhat sigmoid ends; overlap of cell ends in chains one-fifth to one-sixth of cell length.
   Valve view: Linear with slightly pointed apices. Fibulae coarse; striae with one, and in parts of the valve two, rows of porioids (TEM), punctate when examined with LM.

Pseudo-nitzschia prolongatoides (Hasle) Hasle (Plate 73, Table 75)
Basionym: *Nitzschia prolongatoides* Hasle.
Synonym: *Nitzschia prolongata* Manguin non *Nitzschia prolongata* Hustedt (see Hasle, 1965b, p. 27).

Pseudo-nitzschia

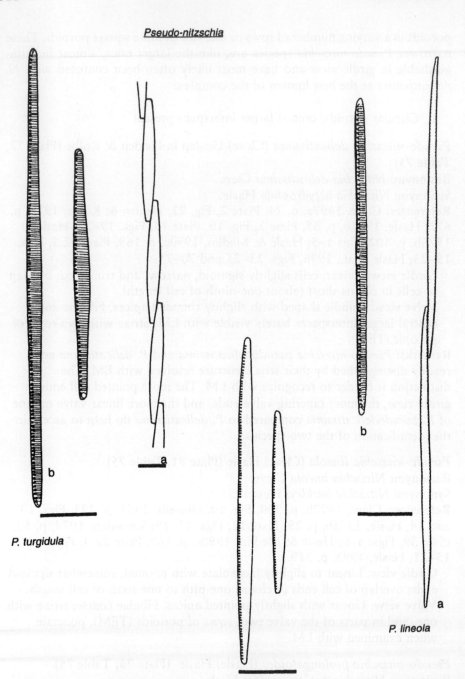

P. turgidula

P. lineola

P. cuspidata

References: Manguin, 1957, p. 132, Plate 7, Fig. 46; Hasle, 1965b, p. 25, Plate 2, Fig. 11, Plate 12, Figs. 7 and 8; Hasle & Medlin, 1990c, p. 169, Plate 22.3, Figs. 7 and 8; Hasle, 1993, p. 319.

Girdle view: Almost linear, possibly expanded in the middle; overlap of cell ends extremely short (about 1/15 of cell length).

Valve view: Expanded middle part occupying about one-third of cell length; rostrate parts with slightly wider apices. Fibulae, central larger interspace with central nodule, and delicate transverse interstriae discernible with LM. Striae with two to three rows of circular poroids resolved with TEM.

Pseudo-nitzschia turgidula (Hustedt) Hasle (Plate 71, Table 75)
Basionym: *Nitzschia turgidula* Hustedt.
References: Hustedt, 1958a, p. 182, Figs. 172 and 173; Hasle, 1965b, p. 24, Plate 1, Fig. 11, Plate 2, Fig. 3, Plate 12, Figs. 1–6; Hasle & Medlin, 1990c, p. 169, Plate 22.3, Figs. 1–6; Hasle 1993, pp. 320.

Girdle view: Margins slightly convex tapering toward truncated cell ends; overlap of cells in chains short, about one-sixth of cell length.

Valve view: Larger specimens linear except for a middle expansion; smaller specimens rhomboid to lanceolate with rounded ends. Fibulae and interstriae distinct. Striae with two rows of poroids (TEM).

Remarks: Due to the truncated cell ends and the short overlap of cells in chains *P. turgidula* may be identified in girdle view. *Pseudo-nitzschia turgiduloides* (Hasle) Hasle (basionym: *Nitzschia turgiduloides,* synonym: *Pseudo-nitzschia barkleyi* var. *obtusa* Manguin nom. nud.) is a larger and more coarsely structured species similar to *P. turgidula* in shape of cell ends.

2a. Square poroids, central larger interspace present.

Pseudo-nitzschia cuspidata (Hasle) Hasle (Plate 71, Table 75)
Basionym: *Nitzschia cuspidata* Hasle.
References: Hasle, 1965b, p. 34, Plate 2, Fig. 6, Plate 15, Figs. 9–18; Hasle, 1974, p. 427; Simonsen, 1974, p. 51, Plate 39, Figs. 1–3; Hasle, 1993, p. 318.

Girdle view: Linear to slightly lanceolate tapering to narrow, truncated ends; overlap of cells in chains one-fifth to one-sixth of cell length.

Valve view: Narrowly lanceolate, tapering parts long; apices narrow with truncated ends. Transapical axis wide compared with other species of this

---

PLATE 71 *Pseudo-nitzschia turgidula:* (a) chain in girdle view; (b) two valves of different size with striation and central larger interspace. *Pseudo-nitzschia cuspidata:* two valves with fibulae and central larger interspace. *Pseudo-nitzschia lineola:* (a) chain in girdle view; (b) valve with striation and central larger interspace. Scale bars = 10 μm.

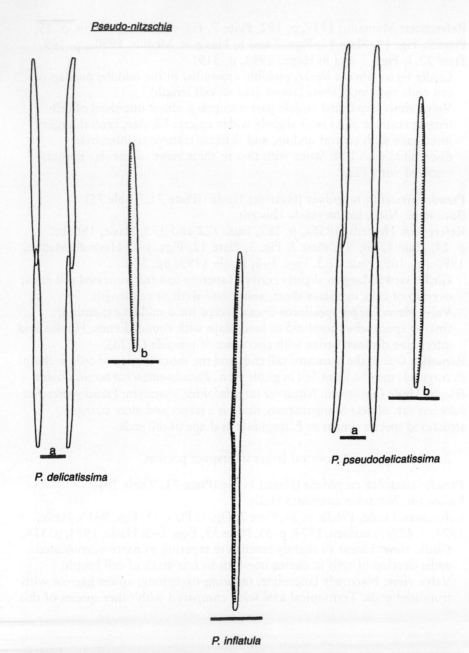

PLATE 72 *Pseudo-nitzschia delicatissima:* (a) chains in valve and girdle views, truncated cell ends in girdle view; (b) valve with fibulae and central larger interspace. *Pseudo-nitzschia inflatula:* valve with fibulae and central larger interspace. *Pseudo-nitzschia pseudodelicatissima:* (a) chains in girdle and valve views; (b) valve view with fibulae and central larger interspace. Scale bars = 10 μm.

complex. Fibulae readily seen with LM; interstriae occasionally discernible with LM.

Remarks: *Pseudo-nitzschia cuspidata* is recognized by its comparatively wide, mainly lanceolate valves tapering along most of its length.

Pseudo-nitzschia pseudodelicatissima (Hasle) Hasle (**Plate 72, Table 75**)
Basionym: *Nitzschia pseudodelicatissima* Hasle.
Synonym: *Nitzschia delicatula* Hasle non *Nitzschia delicatula* Skvortzow (*vide* Hasle & Mendiola, 1967, p. 115).
References: Hasle, 1965b, p. 37, Plate 17, Figs. 1–16; Hasle, 1976b, p. 103; Takano & Kuroki, 1977, p. 44, Figs. 21–25; Hasle & Medlin, 1990c, p. 169, Plate 22.4, Figs. 1–16, Plate 22.6, Fig. 1; Hasle, 1993, p. 319; Hasle et al., 1996, Figs. 26–29 and 76–81.
  Girdle view: Linear with pointed ends.
  Valve view: Linear to almost linear, tapering for a longer or shorter distance toward narrow, rounded apices. Fibulae readily seen with LM; interstriae occasionally discernible with LM.
Remarks: *Pseudo-nitzschia pseudodelicatissima* is distinguished from *P. lineola* by a more delicate valve structure and from the rest of the *Nitzschia delicatissima* complex by the narrow linear valves. *Pseudo-nitzschia inflatula* (Hasle) Hasle (basionym: *Nitzschia inflatula*) is morphologically closely related to *P. pseudodelicatissima* but differs by being distinctly inflated in a short middle part and more or less distinctly inflated a short distance from the pointed apices (**Plate 72**; Hasle & Medlin, 1990c, Plate 22.5, Figs. 1–8).

2b. Square poroids; no central larger interspace.

Pseudo-nitzschia granii (Hasle) Hasle var. *granii* (**Plate 73, Table 75**)
Basionym: *Nitzschia granii* Hasle.
References: Hasle, 1964, p. 31, Plate 13, Fig. 10, Plate 14, Figs. 1–4; Hasle, 1974, p. 426; Hasle, 1993, p. 318.
  Girdle view: Approximately linear.
  Valve view: Spindle shaped, more or less gradually tapering into prolonged projections; apices rounded. Fibulae of irregular distances. Interstriae not resolved with LM.

Pseudo-nitzschia granii var. *curvata* Hasle.
Basionym: *Nitzschia granii* var. *curvata* Hasle.
References: Hasle, 1964, p. 32, Plate 13, Figs. 11 and 12, Plate 14, Figs. 5–7; Hasle, 1974, p. 426.
  Var. *curvata* differs from the nominate variety by having curved cells (probably in girdle view) and broader, somewhat rhomboid valves.
Remarks: *Pseudo-nitzschia granii* occurs as single cells associated with colonies of the prymnesiophyte *Phaeocystis* and in the plankton.

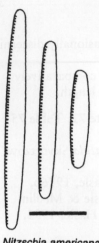

Nitzschia americana

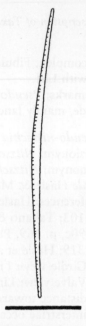

Pseudo-nitzschia subcurvata

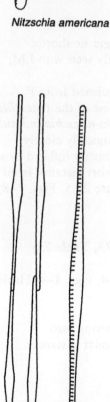

a   b

P. prolongatoides

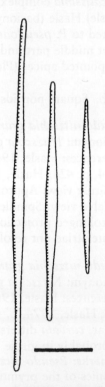

P. granii var. granii

Pseudo-nitzschia subcurvata (Hasle) G. Fryxell in Fryxell et al. (Plate 73, Table 75)

Basionym: *Nitzschia subcurvata* Hasle.

References: Hasle, 1964, p. 28, Plate 12, Figs. 14–16, Plate 13, Figs. 1 and 3; Hasle, 1974, p. 426; Hasle & Medlin, 1990a, p. 177, Plate 23.2, Figs. 6–8, Fryxell et al., 1991.

Girdle view: Overlap of cell ends very short.

Valve view: Middle part (one-fourth to one-half of cell length) dilated, abruptly attenuated toward long, slender projections; one side of valve straight or slightly concave and the other convex, especially in middle part of the valve. Fibulae of irregular distances, interstriae not resolved with LM.

Remarks: The species was transferred to the section *Pseudo-nitzschia* by Fryxell et al. (1991) based on type of colony formation, valve shape, and the nature of the raphe and fibulae.

Distribution:

P. granii—northern cold water region to temperate?

P. cuspidata—warm water region to temperate.

P. prolongatoides, P. subcurvata, and *P. turgiduloides*—southern cold water region.

P. turgidula, P. lineola, and *P. inflatula*—cosmopolitan.

The information on distribution is mainly from Hasle (1964, 1965b) and represents only fragmentary observations. *Pseudo-nitzschia delicatissima* and *P. pseudodelicatissima* are common, occasionally as the predominant diatom species, in North Atlantic plankton, and may be cosmopolitan although without reliable identifications from the Antarctic. *Pseudo-nitzschia granii* was described from the Norwegian Sea and has most probably a wide distribution in the North Atlantic region. Small needle-shaped *Nitzschia* species in or on *Phaeocystis* colonies have been reported from the Norwegian Sea and the North Sea since the beginning of this century (see Hasle, 1964, p. 30). In July, 1988 it was found together with *Phaeocystis* as far north as the King's Bay, Spitzbergen, ca. 80°N (G. Hasle and B. Heimdal, personal observations).

How to identify: In general, single valves mounted in a medium of a high refractive index examined with an oil immersion lens under darkfield or phase contrast illumination are required to identify the *Nitzschia*

PLATE 73 *Nitzschia americana:* valves showing size variation and fibulae. *Pseudo-nitzschia subcurvata:* valve with fibulae. *Pseudo-nitzschia prolongatoides:* (a) partial chains in valve and girdle views; (b) valve with fibulae and central larger interspace. *Pseudo-nitzschia granii:* var. *granii:* Valves, size variation, fibulae. Scale bars = 10μm.

delicatissima complex to species. In these circumstances the fibulae are resolved, and in some species the interstriae are also resolved. The structure of the striae is discernible in coarsely silicified specimens of a few species. **Remarks:** Takano (1993) described *Nitzschia multistriata* from inlets of southern Japan, forming a bloom in summer 1991. It forms stepped chains, has no central nodule, more interstriae than fibulae, valve striae with two rows of circular poroids, and is thus similar to *P. delicatissima* except for the lack of a central nodule. Takano (1995, p. 74) made the new combination *Pseudo-nitzschia multistriata*.

Incertae sedis (*Nitzschia*)

With the exception of *Nitzschia longissima*, records of the following species are in recent literature. Their morphology is poorly known and, as a consequence, so is their taxonomy. Our EM observations are fragmentary but are used here as a starting point for further investigations. As far as we have observed these species all have an eccentric raphe; they are marine and occur in plankton although that may not be their primary habitat.

A. Valve outline linear to lanceolate to slightly rostrate.

Nitzschia americana Hasle (Plate 73)
References: Hasle, 1964, p. 41, Plate 14, Figs. 13–19; Hasle, 1993, p. 318.
   Girdle view: Linear.
   Valve view: Linear with slightly convex margins; valve ends obtusely rounded. Fibulae more conspicuous than interstriae. Striae with two rows of small poroids (EM).
Remarks: *Nitzschia americana* is probably primarily epiphytic on *Chaetoceros* setae (G. Hasle and E. Syvertsen, personal observations) and possibly epiphytic on other diatoms. Being confused with a mophologically similar, probably undescribed species occurring in stepped colonies, *N. americana* was incorrectly transferred to *Pseudo-nitzschia* (Hasle, 1993).
Mophomeric data: Apical axis, 16–40 μm; transapical axis, ca. 3 μm; 18–20 fibula and 27–30 striae in 10 μm.
Distribution: Cosmopolitan (?).

PLATE 74 *Nitzschia sicula:* valves, var. *bicuneata* to the left, var. *rostrata* to the right. *Nitzschia longissima:* (a) whole cell; (b) valve view. *Nitzschia bicapitata:* valve with striae and central larger interspace. *Nitzschia braarudii:* valve with striae and slightly central larger interspace. Scale bars = 10 μm.

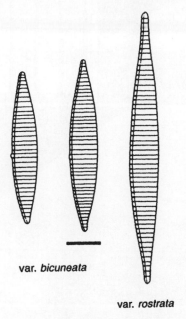

var. *bicuneata*

var. *rostrata*

N. sicula

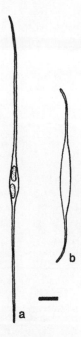

N. longissima

N. bicapitata

N. braarudil

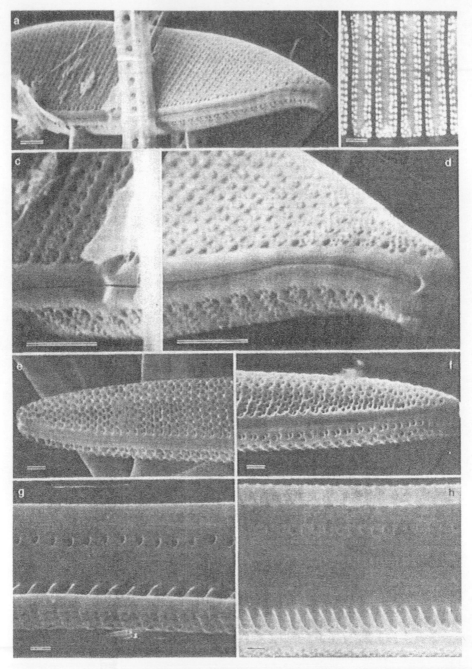

PLATE 75 *Nitzschia sicula* (a–d): (a, c, and d) SEM of the same valve. (a) Showing stria structure on valve face and mantle and location of the raphe; (c) central depression with raphe endings; (d) valve apex with terminal raphe ending; (b) TEM of striae. *Nitzschia marina* (e–h) SEM; (e and f) external views of raphe and striae; (g and h) internal views of fibulae (lower part), valve surface, and perforations along the other margin. Scale bars = 1 μm.

TABLE 76   Morphometric Data of *Nitzschia sicula* Varieties

Variety	Apical axis (µm)	Transapical axis (µm)	Striae and fibulae in 10 µm
sicula	23–121	5–8	7.5–11
bicuneata	32–39	6	8–12
rostrata	50–90	5–8	7–10
migrans	21–68	10–12	9–11

Nitzschia sicula (Castracane) Hustedt var. *sicula, N. sicula* var. *rostrata* Hustedt; *N. sicula* var. *bicuneata* Grunow in Cleve & Möller Hasle; *N. sicula* var. *migrans* (Cleve) Hasle (**Plates 74 and 75, Table 76**)
Basionyms: *Synedra sicula* Castracane; *Rhaphoneis? (Raphoneis) Diatoma? bicuneata* Grunow in Cleve & Möller; *Nitzschia migrans* Cleve.
Synonyms: *Pseudo-nitzschia sicula* (Castracane) H. Peragallo (in H. & M. Peragallo, 1897–1908); *Pseudo-nitzschia sicula* var. *bicuneata* (Grunow in Cleve & Möller) H. Peragallo in H. & M. Peragallo; *Pseudo-nitzschia sicula* var. *migrans* (Cleve) H. Peragallo in H. & M. Peragallo.
References: Castracane, 1875, p. 34, Plate 6, Fig. 7; Cleve & Möller, 1879, Nos. 208–210; Cleve, 1897b, p. 300, Fig. 9; H. & M. Peragallo, 1897–1908, p. 299, Plate 72, Figs. 25–27, Plate 82, Fig. 28; Heiden & Kolbe, 1928, p. 671; Hustedt, 1958a, p. 180, Figs. 128–132; Hasle, 1960, p. 26, Fig. 16; Hasle, 1964, p. 38, Figs. 11–13, Plate 5, Fig. 8, Plate 13, Figs. 14a and 14b, Plate 14, Fig. 22, Plate 16.
The varieties differ in valve outline, particularly in the shape of the apices. The eccentric canal raphe, the flattened, not undulated valves with a fairly coarse striation, and the equal numbers of fibulae and interstriae are common features. These diatoms have been recorded from widespread localities but evidently in small cell numbers which may be the reason why their fine structure, taxonomy and distribution are poorly known.
Nitzschia sicula var. *sicula,* var. *bicuneata,* and var. *rostrata* are lanceolate in valve outline with slightly rounded apices; the nominate variety has more obtuse ends, var. *bicuneata* has slightly protracted apices, and var. *rostrata* has rostrate valve ends. *Nitzschia sicula* var. *migrans* differs from the other varieties by being linear in the greater part of the valve length. Central raphe endings in a circular depression, a complex stria structure differing along the valve length, consisting of rows of small poroids close to the interstriae, and the absence of poroids in the external canal wall were documented with TEM and SEM (**Plate 75, a–d**; Hasle, 1964).
Distribution: Probably warm water region to temperate.

How to identify: These diatoms may be identified in water mounts in valve view but not in girdle view.

Remarks: Hustedt (1958a) made the combination *Nitzschia sicula* (Castracane) Hustedt. If, however, *Pseudo-nitzschia* was regarded as a separate genus, *P. sicula* was, in his opinion, the correct name. Mann (1978), on the other hand, found no problems in accomodating *N. sicula* in *Fragilariopsis*. In light of the present information on the stria structures of *Pseudo-nitzschia* and *Fragilariopsis*, in addition to no observations of a possible chain formation in *N. sicula*, we prefer to retain the species with its varieties in *Nitzschia*. We have no educated opinion, however, as to which section it belongs.

Nitzschia marina Grunow in Cleve & Grunow (**Plate 75**)
Basionym: *Nitzschia angustata* var. *marina* Grunow in Cleve & Möller, (*N. angusta* var. *marina* Cleve & Grunow, 1880, p. 70).
Synonyms: *Synedra gaussi* Heiden in Heiden & Kolbe; *Pseudo-nitzschia hustedtii* Meister.
References: Cleve & Möller, 1878, Nos. 154 and 155; Cleve & Grunow, 1880, p. 70; Van Heurck, 1880–1885, Plate 57, Figs. 26 and 27; H. & M. Peragallo, 1897–1908, p. 272, Plate 72, Fig. 24; Heiden & Kolbe, 1928, p. 561, Plate 5, Fig. 112; Meister, 1937, p. 272, Plate 11, Fig. 2; Kolbe, 1954, p. 40, Plate 3, Figs. 38–40; Kolbe, 1957, p. 39, Plate 1, Fig. 7; Simonsen, 1992, p. 22, Plate 18, Figs. 1–7.

   Valve view: Linear to lanceolate with rounded apices. Valve wall extremely coarsely silicified. Valve mantle high and rounded. Valve face and mantle similar in structure: externally with biseriate striae and slightly raised interstriae. Striation just discernible with LM. Raphe not raised above valve surface; external canal wall unperforated except for the raphe slit. Fibulae strong and curved as seen on valves in internal views. Valve wall apparently double layered, the internal layer unperforated except for one row of holes along the non-raphe bearing margin.
Morphometric data: Apical axis, 53–353 μm; transapical axis, 8–12 μm, fibulae and striae, 10–12 in 10 μm.
Distribution: Warm water region.

Nitzschia kolaczeckii Grunow
References: Grunow, 1867, p. 18; Grunow, 1877, p. 173, Plate 94, Fig. 10; Hustedt in A. Schmidt, 1924, Plate 349, Figs. 38 and 39; Kolbe, 1955, p. 174, Plate 2, Figs. 22 and 23; Hasle, 1960, p. 24, Plate 5, Figs. 50a–50c.

   Valve view: lanceolate with slightly outstretched apices. The two raphes of a frustule diagonally located. Central interspace slightly larger than the rest. Valve with three stria systems, one transverse and two oblique ones as in *Pleurosigma*.

Morphometric data: Apical axis, 67–120 μm; transapical axis, 7.5–11 μm; eight or nine fibulae in 10 μm; 13–16 oblique, and 17–18 transverse striae in 10 μm.

Distribution: Warm water region.

How to identify: *Nitzschia marina* and *N. kolaczeckii* may be identified in water mounts in valve view but most likely not in girdle view.

Remarks: Cleve & Grunow (1880) placed *N. marina* in *Tryblionella* and *N. kolaczeckii* in *Lanceolatae*. *Nitzschia sicula* and *N. marina* have similar valve and raphe structures as seen with LM. There is also a certain similarity between the two species in external valve structure as revealed with SEM. Information on the internal valve structure of *N. sicula* is lacking, however. The only available information on the valve structure of *N. kolaczeckii* is from LM investigations. Based on light micrographs (Hasle, 1960, Plate 5, Fig. 50c) it may be assumed that the central circular depression with raphe endings is also present in *N. kolaczeckii*.

B. Valves with prolonged projections.

Nitzschia longissima (Brébisson, in Kützing) Ralfs in Pritchard (Plate 74)
Basionym: Ceratoneis longissima Brébisson in Kützing.
References: Kützing, 1849, p. 891; Pritchard, 1861, p. 783, Plate 4, Fig. 23; Cupp, 1943, p. 200, Fig. 154; Hendey, 1964, p. 283; Hasle, 1964, p. 20, Plate 10, Figs. 5–7; Hasle & Medlin, 1990a, p. 177, Plate 23.1, Figs. 5–7.

Valve view: Linear to lanceolate, tapering to very long projections. Raphe with fibulae and central larger interspace distinct in mounted cleaned material. External canal wall poroid (TEM). Transverse striae and interstriae hardly visible with LM.

Morphometric data: Apical axis, 125–450 μm; transapical axis, 6–7 μm; fibulae 6–14? in 10 μm; striae, ca. 16 in 10 μm (Cupp, 1943), 52–60 in 10 μm (Hasle, 1964).

Distribution: Cosmopolitan (?), inshore waters.

How to identify: As mentioned previously *C. closterium* and *N. longissima* may easily be confused when examined as entire cells in water mounts. Acid-cleaned material mounted in a medium of a high refractive index, or often EM observations, may be needed to distinguish between the extremely weakly silicified *C. closterium* with the numerous narrow bands and the simple canal raphe and the more coarsely silicified *N. longissima* with fibulae connected with silicified strips running parallel to the raphe slit (Hasle, 1964, Plate 10).

Remarks: Detailed morphological studies are needed before a good definition of this and similar freshwater and brackish water species can be given. Cleve & Grunow (1880, p. 100) placed *N. longissima* in *Nitzschiella*, which they characterized as a group with "Schalen mit excentrischem Kiele und lang vorgezogenen Spitzen."

C. Valves with capitate ends.

Nitzschia bicapitata Cleve (Plate 74, Table 77)
References: Cleve, 1901b, p. 933, Fig. 12; Hustedt, 1958a, p. 169, Figs. 176–190; Hasle, 1960, p. 21, Fig. 10; Hasle, 1964, p. 37, Plate 5, Fig. 7, Plate 14, Figs. 8–12, Plate 15, Figs. 3–6; Simonsen, 1974, p. 50, Plate 35, Figs. 3–15; Kaczmarska & Fryxell, 1986, p. 242, Fig. 4; Kaczmarska et al., 1986, p. 1859, Fig. 10.

> **Valve view:** Valves lanceolate with capitate ends less pronounced in smaller specimens. Fibulae and central larger interspace readily observed on cleaned mounted valves; striae with one row of areolae discernible on well silicified specimens. Central nodule and external canal wall without poroids observed with EM.

Remarks: Kaczmarska & Fryxell (1986, p. 237) carried out a detailed morphological study of "the diatom known in the literature as Nitzschia bicapitata Cl." SEM investigations showed three morphotypes of N. bicapitata Cleve sensu lato and a new species, N. bifurcata Kaczmarska & Licea. The epithet of the new species refers to the stria structure revealed with SEM. Whereas the rest of the N. bifurcata valve has striae with one row of areolae, like in N. bicapitata, the margin opposite to the raphe has two rows of alternating small pores.

Nitzschia braarudii Hasle (Plate 74, Table 77)
Synonym: Nitzschia capitata Heiden in Heiden & Kolbe.
References: Heiden & Kolbe, 1928, p. 666, Plate 7, Fig. 151; Hasle, 1960, p. 22, Fig. 11, Plate 7, Figs. 58–63; Hasle, 1964, p. 35, Plate 2, Fig. 1, Plate 14, Fig. 20, Plate 15, Figs. 1 and 2; Taylor, 1967, p. 450, Fig. 30;

TABLE 77   Morphometric Data of Nitzschia spp.

Species/author	Apical axis (μm)	Transapical axis (μm)	Striae in 10 μm	Fibulae in 10 μm
N. bicapitata				
Cleve, 1901b	12–16	3–5	26	13
Hustedt, 1958a	6–30	3–3.5	26–28	12–14
Hasle, 1964	6–30	2.5–5.5	18–28	12–16
N. bifurcata[a]	7–31	3–6	22–28	10–16
N. braarudii	35–63	3–5	22–30	10–15

[a] Mentioned in the text for comparison; morphometric data from Kaczmarska & Fryxell (1986).

Simonsen, 1974, p. 50, Plate 35, Fig. 2; Kaczmarska et al., 1986, p. 1860, Fig. 15; Simonsen, 1992, p. 83, Plate 83, Figs. 5–7.

Valve view: Central part with straight or slightly rounded margins varying in length with length of valve. Valve narrowing toward capitate ends. Fibulae, a slightly larger central interspace, and interstriae resolved with LM of acid cleaned mounted valves. Stria structure consisting of one row of areolae, a central nodule, and external canal wall without poroids were observed with EM.

Remarks: *Nitzschia braarudii* and *N. bicapitata* belong to a large group of species characterized by capitate ends and mainly found in tropical and subtropical open ocean plankton (Hasle, 1960; Simonsen, 1974; Kaczmarska & Fryxell, 1986, 1994; Kaczmarska et al., 1986). Simonsen (1992) examined the type material of *N. capitata* and isotype material of *N. braarudii* and concluded that they were conspecific. He also found that the name *N. capitata* was preoccupied and that therefore the name *N. braarudii* could be used. Hasle (1960) gave a Latin diagnosis and illustrations of *N. braarudii* and Equatorial and subantarctic Pacific Ocean as type locality. The description is validated in the Taxonomic Appendix (this chapter) giving the holotype, isotypes and the type locality.

How to identify: Kaczmarska & Fryxell's (1986) investigation demonstrates the problems involved in identifying marine, planktonic *Nitzschia* species of this group, and also when the fibulae and interstriae of cleaned mounted specimens are visible with LM.

Distribution: Warm water region to temperate. *Nitzschia bicapitata* has been found between 66°N and 62°S (Hasle, 1976a). *Nitzschia braarudii* seems to have a more restricted longitudinal distribution.

METHODOLOGY

Detailed procedures for collection, preservation, and examination of phytoplankton have been described in the UNESCO Phytoplankton manual (Sournia, 1978). Because of the wide range of habitats, cell sizes, degree of silicification, and fineness, a variety of methods are required for collection, preparation, and microscopy of diatoms. The current diatom literature contains a multitude of methods, particularly for cleaning and mounting (e.g., Ricard, 1987; Round et al., 1990). The selection mentioned below is intended as a guide and represents simple, general methods to obtain material adequate for identification at the species level.

COLLECTION AND CONCENTRATION (Table 78)

Sampling by water bottle is the recommended method (Sournia, 1978, p. 33) to obtain a correct picture of the quantitative composition of the phyto-

TABLE 78   Collection

Method	Use	Disadvantage
Water bottle	Enumeration of cells; all species	Few rare species
Pump	Quantitative and qualitative; large amount of material for enumeration and identification	Sometimes biased species composition; damage to some species
Membrane filtering	In combination with one of the methods above; samples for direct cell enumeration and microscopy (LM, SEM)	May be selective; damage to some specimens
Net	Large quantities of material for identification	Selective, depending on mesh size, net proportions and composition of the plankton

plankton. Theoretically, a water bottle sample contains all but the rarest organisms in the water mass sampled and includes the whole size spectrum from the largest entities, like diatom colonies, to the smallest single cells. Similar results can be obtained by pump sampling (Sournia, 1978, p. 41), which samples much larger quantities of water allowing the collection of the rarer species. The technique has its disadvantages, however, e.g., breaking up colonies, breaking off large *Chaetoceros* setae, and breaking into pieces long pennate cells like *Thalassiothrix* spp. In contrast to these quantitative methods, sampling by plankton nets (Sournia, 1978, p. 50) is highly selective, depending on the mesh size of the gauze, net towing speed, and the species present in the water. *Chaetoceros* setae, for instance, may form a fine network inside the gauze, and very small single cells, which in other cases pass through the meshes, are retained. On the other hand, nets with very fine meshes (e.g., 5 or 10 μm) often filter too little water to provide an adequate diatom sample. As a compromise, the most useful mesh size for collecting diatoms is 25 μm.

Net hauls have the advantage of a simultaneous collection and concentration of the plankton providing sufficient quantities for species identification. Water bottle and pump samples in most cases have to be concentrated. The smaller the subsample, the fewer number of rare species will be obtained. On the other hand, there is no point in concentrating large quantities of a sample rich in one or a few species. Concentration by settling (Sournia, 1978, p. 88), centrifugation (Sournia, 1978, p. 98), and filtration (Sournia, 1978, p. 108) are the most used methods.

The rich plankton in the marginal ice zone and the ice-covered waters of the polar seas has attracted particular attention during the past decades. Scuba diving has shown that at least parts of the plankton flora begin their spring development on the undersurface of the ice. In addition, there is a particular rich and highly specialized subice flora in polar waters which may become part of the plankton when the ice melts (Syvertsen, 1991). An electric suction pump or "vacuum cleaner" used in the study of ice zoo benthos (Lønne, 1988) has been modified to sample the subice flora and algae found in cracks and crevices. A plankton net is placed in front of or behind the impeller, and the water is pushed at low speed through the net. The action is gentle and the algal cells are not damaged.

UNIALGAL CULTURES AS A MEANS FOR SPECIES IDENTIFICATION

Natural samples often yield cells at approximately the same stage in the life cycle of the species in question. This is presumably due to a coordination of processes, such as sexual reproduction (auxospore formation in diatoms), caused by chemical and physical factors in the environment. For similar reasons, diatom resting spores are often formed only at certain times. For studies of morphology and taxonomy, it is desirable to know the whole range of morphological variation including the effect of size variation and resting spore formation. This information is normally not obtained by studying a few natural samples. Cultures in which auxospore and resting spore formation can be induced may increase the information, although a single clone does not necessarily reflect the whole range of variation of a species. It is important to note, however, that a single clonal culture does not necessarily reflect the whole possible range in variation of a population or species.

Dilution cultures (Sournia, 1978, p. 218) or crude cultures may serve as a base for unialgal cultures. The best universal method to establish unialgal diatom cultures is to isolate single cells by a micropipette (Stein, 1973, p. 53) into an appropriate synthetic or enriched seawater growth medium (Stein, 1973, p. 25).

It is important to keep in mind that dense unialgal cultures as well as dense natural populations may contain aberrant forms. This seems to be especially the case for *Pseudo-nitzschia* spp. and *Synedropsis hyperborea* and related species (Grunow, 1884; Takano & Kikuchi, 1985; Subba Rao & Wohlgeschaffen, 1990; Hasle et al., 1994).

PRESERVATION AND STORAGE (Table 79)

If possible, diatom samples should be studied immediately after sampling for information on colony formation and chloroplasts. However, in most cases, due to practical reasons they have to be preserved for later studies. A pH lower than 7 is preferable to hinder dissolution of the siliceous structures. The most

TABLE 79   Preservation

Agent	Solution	Comments
Formaldehyde/acetic acid	Equal volumes of p.a. grade 40% HCHO and 100% acetic acid	20 ml of the solution to 70 ml net sample; 2 ml of the solution to 100 ml water sample (= 0.4% HCHO)
Formaldehyde, alkaline	Equal volumes of 40% HCHO and distilled water; to 1 liter solution 100 g hexamin	As described above
Lugol's solution, acidic	Dissolve in 1 liter distilled water 100 g KI, then 50 g I₂, and finally add 100 ml glacial acetic acid	For water samples, 0.2–0.4 ml to 100 ml sample; for net samples, add to a weakly brown color

Note. Samples should be preserved immediately after collection and stored in glass bottles, preferably in darkness and at low temperature. Iodine will oxidize with time, and samples preserved with Lugol's solution need regular attention. Diatom frustules stored in alkaline solution may dissolve with time.

commonly used preservatives are formaldehyde neutralized with hexamethy-lenetetramine (hexamin) or acidified with acetic acid, and the Lugol's solution (Sournia, 1978, p. 69).

For storage of diatom samples over a longer period of time containers (bottles, jars) of moderate glass quality should be used. Silica dissolved from the containers apparently helps to keep the diatom siliceous wall intact. Metal caps and lids should be avoided. The samples should not be exposed to temperatures much higher than ca. 15°C and should be kept away from bright light, especially when preserved with Lugol's or other iodine solutions.

PREPARATION FOR LIGHT MICROSCOPY (Tables 80 and 81)

Examination of raw (not cleaned) material in a water mount or embedded in a resin may give sufficient information to identify a number of common planktonic diatoms, e.g., *Chaetoceros* spp. and *Rhizosolenia* spp. (Sournia, 1978, p. 137). These diatoms are identified by their gross morphology and/or special structures like the *Chaetoceros* setae and the shape of the *Rhizosolenia* valves and processes. This procedure is often ineffective for revealing the essential morphological structures of other genera, e.g., the areolation and processes of *Coscinodiscus* and *Thalassiosira* and the striation and raphe structure of *Navicula* and *Pseudo-nitzschia*. Cell content and the organic part of the cell wall obscure the image of the valve structures and have to be removed. The

TABLE 80 Cleaning of Diatom Material

Method	Procedure	Comments
von Stosch's method	Concentrate sample to near dryness, add equal amount of HNO_3 and 3× sample amount of H_2SO_4, boil for ca. 3 min, cool, and rinse with distilled water until free of acid	Fast method, usually with good result
Simonsen's method	Rinse sample with distilled water, add an equal amount of $KMnO_3$, agitate, leave for 24 hr, add an equal amount of HCl, heat until the sample becomes clear or only slightly colored, and rinse until free of acid	More time consuming, reliable
UV, H_2O_2 enzymes	Useful to obtain intact frustules	Difficult and time consuming

Note. Rinse and concentrate samples by settling or centrifugation at approximately 4000 rpm: small and lightly silicified valves and bands need long settling/centrifugation time; for settling, usually overnight; for centrifugation, up to half an hour.

structures are best seen on single valves. This is a further reason to remove the organic material which keeps the various frustular elements together.

By acid cleaning (Table 80) the diatom frustule separates into single valves and bands free from organic material. Ultraviolet and enzyme techniques (Table 80) usually give intact frustules cleaned of organic material. The refractive index of silica and water is about the same, and to increase the contrast the cleaned material is embedded in a resin of a higher refractive index than that of silica (Table 81). The normal procedure thus includes (1) removal of the preservative by repeated centrifugation and decanting with distilled water, (2) cleaning, followed by (3) embedding in a mounting medium.

PREPARATION FOR ELECTRON MICROSCOPY

The method of preparation for EM depends on whether the objective of the investigation is to study the cell interior (cell organelles, etc.) or the morphology of the siliceous cell wall components. Only the latter is relevant to this chapter.

For a few species (usually the more heavily silicified ones) intact frustules may be obtained by simply air drying a drop of raw material after thorough rinsing with distilled water. Although delicate specimens will be crushed by

TABLE 81 Embedding for Light Microscopy

Media	Refractive index	Solvent	Manufacturer
Clearax	1.67	Xylene, acetone	G. T. Gurr
Naphrax	1.72	Xylene, toluene, acetone	Northern Biological Supplies
Pleurax	1.74	Alcohol	(See von Stosch, 1974)

Note. A drop of the cleaned sample in distilled water is left to dry on a coverslip (0.17 ± 0.02 mm thick) cleaned with alcohol. The medium is applied to the dry sample and left to dry. The cover slip is "picked up" by pressing a cleaned microscope slide gently toward it. Thereafter, the slide with the cover slip is heated to ca. 50–70°C for a few minutes to remove air bubbles and harden the medium. Rinsed but not cleaned material to be embedded in Pleurax has to first be brought into 100% alcohol. The sample is then concentrated to near dryness, mixed with Pleurax, and put onto a coverslip. The coverslip is placed face down on a microscope slide and left to dry at room temperature or gently heated on a hot plate till the solvent has evaporated. Addresses of manufacturers cited: G. T. Gurr Ltd/Baird and Tatlock Ltd, Freshwater Road, Chadwell Heath, Romford, Essex RM1 1HA, UK; Northern Biological Supplies, 31 Cheltenham Avenue, Ipswich, Suffolk IP1 4LN, UK.

this treatment, useful information, especially on the girdle, can be obtained. Freeze-drying (Williams, 1953) is an alternative, but the most reliable results are obtained with critical point drying (Cohen, 1974). The equipment and methods are developed for general use and not especially for diatoms. For diatoms particular vessels to contain the sample during the drying have to be constructed. The end product is usually a dry powder that subsequently may be further prepared for SEM or TEM.

On a routine basis acid-cleaned material prepared for LM is used for both TEM and SEM. For TEM, a drop of an aqueous suspension of cleaned material is put onto a formvar-coated copper grid (Sournia, 1978, p. 138). After air drying, it can be studied with TEM without further treatment. The same procedure can be followed for SEM, except that the sample suspension is put onto a small glass coverslip or other smooth material. After air drying, the material has to be coated with a metal (usually Au/Pd, Pt) film in an evaporating or sputter device.

MICROSCOPY

LM

Identification of diatoms in water samples is usually best done by using phase contrast optics, which reveal especially well lightly silicified structures, like delicate *Chaetoceros* setae, and also the organic chitan threads in Thalassiosiraceae. Brightfield or differential interference contrast (DIC) may be preferable for the study of cell organelles. Ten or 25× objectives should be sufficient to recognize common species. If the goal is to pick out single cells, an objective with a long working distance is preferable. Normally, a compound microscope with a 10× objective is the best choice, allowing enough space for the use of a micropipette and sufficient magnification to control the isolation of the single cell. A dissecting (binocular) microscope will provide even more space for operating the pipette.

The study of the finer structures of the silica wall requires cleaned, embedded material and a 40× dry or, preferably, a 100× oil immersion objective. Phase contrast and/or DIC are recommended for the examination of lightly silicified diatoms to increase the contrast. For extremely lightly silicified diatoms this is best done with a so-called negative phase contrast objective. For the examination of heavily silicified diatoms there is no need to increase the contrast and brightfield is a better choice. For a superficial scanning of a water or permanent mount a darkfield illumination, obtained by a 10× objective and the phase contrast condensor in a position corresponding to a 100× objective, is useful for showing the siliceous elements distinctly against a black background.

TEM

This type of electron microscopy of cleaned diatom material allows the highest resolution and thus reveals most details. The contrast between heavily silicified parts of the valves and less silicified structures calls for caution during the printing of TEM pictures, however. Finer details like vela, are easily lost at the cost of coarser details that need a different printing. It also must be kept in mind that three dimensional structures are "flattened" into a two-dimensional picture, and the two surfaces of the diatom wall are seen as one image, except when stereo pictures are used. This makes interpretation of the image demanding and considerable experience is often needed to arrive at the right conclusions.

SEM

This type of electron microscopy gives an apparent three-dimensional image with only one of the surfaces revealed. This provides the possibility to differentiate between the various valve processes, provided the internal surface of cleaned single valves are examined. SEM also shows the respective positions of areola vela and foramina. On the other hand, the construction of the cell wall cannot be studied unless broken parts are available. The resolution is often poor. However, the use of high-resolution low-voltage SEM may be an improvement since this instrument allows high resolution without beam damage to the specimens (Navarro, 1993).

WHAT TO LOOK FOR—GENERAL HINTS FOR
IDENTIFICATION AND PREPARATION

To identify intact (whole) diatoms we first of all need to know which side the cell is viewed: valve, broad or narrow girdle view, or from an angle. Intact single cells with a short pervalvar axis tend to lie valve side up under the coverslip (*Coscinodiscus radiatus* and *Pleurosigma*). Diatoms with a pervalvar axis longer than the cell diameter or the apical axis turn girdle side upwards (*Corethron* and *Rhizosolenia*). Some colony types are normally seen in girdle view in a water mount (*Chaetoceros, Fragilariopsis,* and *Thalassiosira*); others may show either valve or girdle side (*Pseudo-nitzschia, Asterionellopsis,* and *Thalassionema*).

Cylindrical and discoid diatoms are readily recognized by the general circular outlines in valve view and rectangular outlines in girdle view. The distinction between the two views of the more or less spindle-shaped cells (e.g. *Pseudo-nitzschia, Thalassionema,* and *Thalassiothrix*) is more problematic. The shape of cell ends will usually help, being straight cut off or pointed in girdle view and more obtuse in valve view.

Once the orientation of the cell is ascertained, the next step is to look for outstanding features like setae (Chaetocerotaceae), shape of linking processes (*Skeletonema*), and in unpreserved material, organic threads from the valve (Thalassiosiraceae) or gliding movements indicating the presence of a raphe.

Frustular elements cleaned of organic material may also be oriented in various ways in a permanent mount. Flattened valves with a low mantle will usually be seen in valve view (some *Coscinodiscus* spp., most *Navicula* spp.), while valves with a high mantle and/or protuberances may appear in girdle view (*Eucampia,* and *Rhizosolenia*). Lightly silicified bands shaped as those in *Rhizosolenia* and *Stephanopyxis* often lie with girdle side up. The more heavily silicified bands, especially those with a septum, tend to be seen as ellipses or circles (*Rhabdonema* and *Thalassiosira*).

Valves and bands are often seen in strewn permanent mounts in a view that does not provide the necessary information for identification. The single frustular element may be brought into the desirable position under the microscope by using a micropipette before drying. Alternatively, they may be turned by means of a needle when dried to the coverslip. The single element may also be turned after the resin is added if utilizing one that does not need to be heated (e.g., Pleurax).

TAXONOMIC APPENDIX*

NEW GENUS

Lioloma Hasle gen. nov.
   Cells long, narrow, twisted. Valve ends dissimilar, both ends with a labiate process, one end with apical spine(s). Marginal spines absent. Internal foramina and external vela about the same size. Sternum variable in width along the apical axis, no raphe.

---

* All nomenclatural novelties given in this Appendix were first reported and validated in "Identifying Marine Diatoms and Dinoflagellates" C. Tomas (ed.) Academic Press, 1996.

Cellulae longae, angustae, tortae. Poli valvae dissimiles, uterque rimoportula una munitus, unum extremum cum spina (spinis). Nullae spinae marginales. Foramina interna et vela externa eiusdem prope magnitudinis. Sternum variae latitudinis axem apicalem sequens, sine raphe. Typus: *Lioloma elongatum* (Grunow) Hasle comb. nov.

NEW NAMES

Meuniera P. C. Silva, nom. nov. pro *Stauropsis* Meunier, 1910 (*Microplankton Barents et Kara*, p. 318) non *Stauropsis* Reichenbach, 1860 (*Hamburger Garten- und Blumenzeitung* 16: 117. Orchidaceae).
Type: M. *membranacea* (Cleve) P. C. Silva comb. nov.

Pleurosigma simonsenii Hasle, nom. nov. pro *Pleurosigma planctonicum* Simonsen, 1974 ("*Meteor*" *Forschungsergebnisse, Reihe D*. 19:46, Plate 30); non *P. planctonicum* Cleve-Euler, 1952 (*Kungliga Svenska Vetenskapsakademiens Handlingar* ser. 4, 3(3):23, Fig. 1374).

VALIDATION OF NAMES

New Taxa

Hemidiscaceae Hendey, 1937 emend. Simonsen, 1975 ex Hasle
Cells cylindrical to discoid, usually solitary. Valve outline circular, slightly elliptical, ovate or semi-circular. Valve face and mantle often different in structure. One marginal ring of large labiate processes (rimoportulae), similar in shape and generally also similar in size. A pseudonodulus usually present. Cellulae discoiideae vel cylindricae, plerumque solitariae. Valvae circulares, parum ellipticae, ovatae vel semicirculares. Frons valvae e limbo saepe structura differens, annulum marginalem rimoportularum forma amplitudineque similarium habens. Pseudonodulus plerumque adest.

Nitzschia braarudii Hasle
Description: Hasle, 1960 (*Skrifter utgitt av Det Norske Videnskaps-Akademi i Oslo I. Matematisk-Naturvidenskapelige Klasse* 2:22, Fig. 11, Plate 7, Figs. 58–63),
Type locality: 52°50'S, 90°03'W.
Holotype: IMBB slide No. 78 conserved in Section of Marine Botany, Department of Biology, University of Oslo.
Isotypes: IMBB slides Nos. 79, 80, BRM Zv 2/95, BM 78510.

Thalassiothrix gibberula Hasle
Description: Hasle, 1960 (*Skrifter utgitt av Det Norske Videnskaps-Akademi i Oslo I. Matematisk-Naturvidenskapelige Klasse* 2:19, Fig. 6, Plate 5, Figs. 46–48.
Type locality: 00°–02°N, 145°W.

Holotype: IMBB slide No. 102 conserved in Section of Marine Botany,
Department of Biology, University of Oslo.

NEW NOMENCLATURAL COMBINATIONS

Dactyliosolen fragilissimus (Bergon) Hasle comb. nov.
Basionym: *Rhizosolenia fragilissima* Bergon, 1903 (*Bulletin de la Société scientifique d'Arachon* 6: 49, Plate 1, Figs. 9 and 10).

Dactyliosolen phuketensis (Sundström) Hasle comb. nov.
Basionym: *Rhizosolenia phuketensis* Sundström, 1980 (*Botaniska Notiser* 133: 579, Figs. 1 and 3).

Guinardia cylindrus (Cleve) Hasle comb. nov.
Basionym: *Rhizosolenia cylindrus* Cleve, 1897 (*A Treatise on the Phytoplankton of the Atlantic and Its Tributaries*, p. 24, Plate 2, Fig. 12).

Guinardia delicatula (Cleve) Hasle comb. nov.
Basionym: *Rhizosolenia delicatula* Cleve, 1900 (*Kongliga Svenska Vetenskaps-Akademiens Handlingar* 32(8): 28, Fig. 11).

Guinardia striata (Stolterfoth) Hasle comb. nov.
Basionym: *Eucampia striata* Stolterfoth, 1879 (*Journal of Royal Microscopical Society* 2:835).
Synonym: *Rhizosolenia stolterfothii* H. Peragallo, 1888 (*Bulletin de la Société Histoire Naturelle Toulouse* 22: 82, Plate 6, Fig. 44).

Lioloma elongatum (Grunow) Hasle comb. nov.
Basionym: *Thalassiothrix elongata* Grunow in Van Heurck, 1881 (*Synopsis des Diatomées de Belgique,* Plate 37, Fig. 9).

Lioloma delicatulum (Cupp) Hasle comb. nov.
Basionym: *Thalassiothrix delicatula* Cupp, 1943 (*Bulletin of the Scripps Institution of Oceanography of the University of California* 5:188, Fig. 137).

Lioloma pacificum (Cupp) Hasle comb. nov.
Basionym: *Thalassiothrix mediterranea* var. *pacifica* Cupp, 1943 (*Bulletin of the Scripps Institution of Oceanography of the University of California* 5:185, Fig. 136).

Meuniera membranacea (Cleve) P. C. Silva comb. nov.
Basionym: *Navicula membranacea* Cleve, 1897 (*A treatise on the phytoplankton of the Atlantic and its tributaries*, p. 24, Plate 2, Figs. 25–28).

Thalassionema javanicum (Grunow) Hasle comb. nov.
Basionym: *Thalassiothrix frauènfeldii* var. *javanica* Grunow in Van Heurck, 1881 (*Synopsis des Diatomées de Belgique,* Plate 37, Fig. 13).
Synonyms: *Thalassiothrix (fauenfeldii* [sic!] var.) *javanica* (Grunow) Cleve, 1900 (*Göteborgs Kungliga Vetenskaps-och Vitterhets-Samhälles Handlingar Fjärde följden* 3(3):357); *Thalassiothrix javanica* (Grunow) Hustedt ex Meister, 1932 (*Kieselalgen aus Asien,* p. 25).

Thalassionema pseudonitzschoiides (Schuette & Schrader) Hasle comb. nov.
Basionym: *Thalassiothrix pseudonitzschioides* Schuette & Schrader, 1982 (*Bacillaria* 5:214, Figs. 1–30).

COMMON DIATOM SYNONYMS

Actinocyclus actinochilus (Ehrenberg) Simonsen 1982
= *Charcotia actinochila* (Ehrenberg) Hustedt 1958a

Actinocyclus curvatulus Janisch (in A. Schmidt, 1878)
= *Coscinodiscus curvatulus* var. *subocellatus* Grunow 1884
= *Actinocyclus subocellatus* (Grunow) Rattray 1890b

Actinocyclus kützingii (A. Schmidt) Simonsen 1975
= *Coscinodiscus kützingii* A. Schmidt 1878

Actinocyclus normanii (Gregory) Hustedt 1957
= *Coscinodiscus normanii* Gregory (in Greville, 1859a)

Actinocyclus normanii f. *subsalsus* (Juhlin-Dannfelt) Hustedt 1957
= *Coscinodiscus subsalsus* Juhlin-Dannfelt 1882

Actinocyclus octonarius Ehrenberg 1838
= *Actinocyclus ehrenbergii* Ralfs (in Pritchard, 1861)

Actinoptychus senarius (Ehrenberg) Ehrenberg 1843
= *Actinocyclus senarius* Ehrenberg 1838
= *Actinoptychus undulatus* (Bailey) Ralfs (in Pritchard, 1861)

Asterionellopsis glacialis (Castracane) F. E. Round (in Round et al., 1990)
= *Asterionella glacialis* Castracane 1886
= *Asterionella japonica* Cleve (in Cleve & Möller, 1882)

Asterionellopsis kariana Grunow (in Cleve & Grunow) F. E. Round (in Round et al., 1990)
= *Asterionella kariana* Grunow (in Cleve & Grunow, 1880)

Asterolampra marylandica Ehrenberg 1844a
= *Asterolampra vanheurckii* Brun 1891

Asteromphalus arachne (Brébisson) Ralfs (in Pritchard, 1861)
= *Spatangidium arachne* Brébisson 1857

Attheya septentrionalis (Østrup) Crawford (in Crawford et al., 1994)
= *Chaetoceros septentrionalis* Østrup 1895
= *Gonioceros septentrionalis* (Østrup) Round, Crawford, & Mann 1990

Azpeitia africana (Janisch ex A. Schmidt) G. Fryxell & T. P. Watkins (in Fryxell et al., 1986b)
= *Coscinodiscus africanus* Janisch ex A. Schmidt 1878

Azpeitia neocrenulata (VanLandingham) G. Fryxell & T. P. Watkins (in Fryxell et al., 1986b)
= *Coscinodiscus crenulatus* Grunow 1884
= *Coscinodiscus neocrenulatus* VanLandingham 1968

Azpeitia nodulifera (A. Schmidt) G. Fryxell & P. A. Sims (in Fryxell et al., 1986b)
= *Coscinodiscus nodulifer* A. Schmidt 1878

Azpeitia tabularis (Grunow) G. Fryxell & P. A. Sims (in Fryxell et al., 1986b)
= *Coscinodiscus tabularis* Grunow 1884

Bacteriastrum furcatum Shadbolt 1854
= *Bacteriastrum varians* Lauder 1864a

Bacterosira bathyomphala (Cleve) Syvertsen & Hasle (in Hasle & Syvertsen, 1993)
= *Coscinodiscus bathyomphalus* Cleve 1883
= *Bacterosira fragilis* Gran 1900

Bacillaria paxillifera (O. F. Müller) Hendey 1964
= *Bacillaria paradoxa* Gmelin 1791
= *Nitzschia paradoxa* (Gmelin) Grunow in Cleve & Grunow 1880

Banquisia belgicae (Van Heurck) Paddock 1988
= *Amphiprora belgicae* Van Heurck 1909
= *Tropidoneis belgicae* (Van Heurck) Heiden (in Heiden & Kolbe, 1928)

Bleakeleya notata (Grunow) F. E. Round (in Round et al., 1990)
= *Asterionella bleakeleyi* var. *notata* Grunow 1867

Brockmanniella brockmannii (Hustedt) Hasle, von Stosch, & Syvertsen 1983
= *Plagiogramma brockmannii* Hustedt 1939

Cerataulina pelagica (Cleve) Hendey 1937
= *Cerataulina bergonii* (H. Peragallo) Schütt 1896

Cerataulina bicornis (Ehrenberg) Hasle (in Hasle & Sims, 1985)
= *Cerataulina compacta* Ostenfeld (in Ostenfeld & Schmidt, 1901)
= *Cerataulina daemon* (Greville) Hasle (in Hasle & Syvertsen, 1980)

Chaetoceros peruvianus Brightwell 1856
= *Chaetoceros chilensis* Krasske 1941

Chaetoceros diadema (Ehrenberg) Gran 1897b
= *Chaetoceros subsecundus* (Grunow) Hustedt 1930

Chaetoceros minimus (Levander) Marino, Giuffre, Montresor, &
   Zingone 1991
= *Rhizosolenia minima* Levander 1904
= *Monoceros isthmiiformis* Van Goor 1924

Chaetoceros tenuissimus Meunier 1913
= *Chaetoceros simplex* var. *calcitrans* Paulsen 1905
= *Chaetoceros galvestonensis* Collier & Murphy 1962
= *Chaetoceros calcitrans* f. *pumilus* Takano 1968

Chaetoceros throndsenii var. *throndsenii* (Marino, Montresor & Zingone)
   Marino, Giuffre, Montresor, & Zingone 1991
= *Miraltia throndsenii* Marino, Montresor, & Zingone 1987

Cylindrotheca closterium (Ehrenberg) Reimann & Lewin 1964
= *Nitzschia closterium* (Ehrenberg) W. Smith 1853

Dactyliosolen blavyanus (H. Peragallo) Hasle 1975
= *Guinardia blavyana* H. Peragallo 1892

Dactyliosolen fragilissimus (Bergon) Hasle comb. nov.
= *Rhizosolenia fragilissima* Bergon 1903

Dactyliosolen phuketensis (Sundström) Hasle comb. nov.
= *Rhizosolenia phuketensis* Sundström 1980

Dactyliosolen tenuijunctus (Manguin) Hasle 1975
= *Rhizosolenia tenuijuncta* Manguin 1957

Delphineis karstenii (Boden) G. Fryxell (in Fryxell & Miller, 1978)
= *Fragilaria karstenii* Boden 1950

Delphineis surirella (Ehrenberg) G. W. Andrews 1981
= *Rhaphoneis surirella* (Ehrenberg) Grunow (in Van Heurck, 1881)

Delphineis surirelloides (Simonsen) G. W. Andrews 1981
= *Rhaphoneis surirelloides* Simonsen 1974

Detonula pumila (Castracane) Gran 1900
= *Schroederella delicatula* (H. Peragallo) Pavillard 1913
= *Thalassiosira condensata* Cleve 1900a

Ephemera planamembranacea (Hendey) Paddock 1988
= *Navicula planamembranacea* Hendey 1964

Eucampia antarctica (Castracane) Mangin 1915
= *Eucampia balaustium* Castracane 1886

Fragilariopsis atlantica Paasche 1961
= *Nitzschia paaschei* Hasle 1974

Fragilariopsis curta (Van Heurck) Hustedt 1958a
= *Fragilaria curta* Van Heurck 1909
= *Nitzschia curta* (Van Heurck) Hasle 1972c

Fragilariopsis cylindrus (Grunow) Krieger (in Helmcke & Krieger, 1954)
= *Fragilaria cylindrus* Grunow (in Cleve & Möller, 1882)
= *Nitzschia cylindrus* (Grunow) Hasle 1972c

Fragilariopsis cylindriformis (Hasle in Hasle & Booth) Hasle 1993
= *Nitzschia cylindroformis* Hasle (in Hasle & Booth, 1984)

Fragilariopsis doliolus (Wallich) Medlin & Sims 1993
= *Synedra doliolus* Wallich 1860
= *Pseudoeunotia doliolus* (Wallich) Grunow (in Van Heurck, 1881)

Fragilariopsis kerguelensis (O'Meara) Hustedt 1952
= *Fragilariopsis antarctica* (Castracane) Hustedt (in A. Schmidt, 1913)
= *Nitzschia kerguelensis* (O'Meara) Hasle 1972c

Fragilariopsis oceanica (Cleve) Hasle 1965a
= *Fragilaria oceanica* Cleve 1873b
= *Nitzschia grunowii* Hasle 1972c

Fragilariopsis pseudonana (Hasle) Hasle 1993
= *Fragilariopsis nana* (Steemann Nielsen) Paasche 1961 *pro parte*
= *Nitzschia pseudonana* Hasle 1974

Fragilariopsis rhombica (O'Meara) Hustedt 1952
= *Nitzschia angulata* Hasle 1972c

Fragilariopsis ritscheri Hustedt 1958a
= *Nitzschia ritscheri* (Hustedt) Hasle 1972c

Fragilariopsis separanda Hustedt 1958a
= *Nitzschia separanda* (Hustedt) Hasle 1972c

Guinardia cylindrus (Cleve) Hasle comb. nov.
= *Rhizosolenia cylindrus* Cleve 1897a

Guinardia delicatula (Cleve) Hasle comb. nov.
= *Rhizosolenia delicatula* Cleve 1900b

Guinardia striata (Stolterfoth) Hasle comb. nov.
= *Eucampia striata* Stolterfoth 1879
= *Rhizosolenia stolterfothii* H. Peragallo 1888

Guinardia tubiformis (Hasle) Hasle comb. nov.
= *Rhizosolenia tubiformis* Hasle 1975

Haslea gigantea (Hustedt) Simonsen 1974
= *Navicula gigantea* Hustedt 1961

Haslea trompii (Cleve) Simonsen 1974
= *Navicula trompii* Cleve 1901b

Haslea wawrikae (Hustedt) Simonsen 1974
= *Navicula wawrikae* Hustedt 1961

Helicotheca tamesis (Shrubsole) Ricard 1987
= *Streptotheca tamesis* Shrubsole 1890

Lauderia annulata Cleve 1873a
= *Lauderia borealis* Gran 1900

Leptocylindrus mediterraneus (H. Peragallo) Hasle 1975
= *Dactyliosolen mediterraneus* H. Peragallo 1888

Lioloma elongatum (Grunow) Hasle comb. nov.
= *Thalassiothrix elongata* Grunow (in Van Heurck, 1881)

Lioloma delicatulum (Cupp) Hasle comb. nov.
= *Thalassiothrix delicatula* Cupp 1943

Lioloma pacificum (Cupp) Hasle comb. nov.
= *Thalassiothrix mediterranea* var. *pacifica* Cupp 1943

Manguinea fusiformis (Manguin) Paddock 1988
= *Tropidoneis fusiformis* Manguin 1957

Manguinea rigida (M. Peragallo) Paddock 1988
= *Amphiprora rigida* M. Peragallo 1924
= *Tropidoneis glacialis* Heiden (in Heiden & Kolbe, 1928)

Membraneis challengeri (Grunow, in Cleve & Grunow)
    Paddock 1988
= *Navicula challengeri* Grunow (in Cleve & Grunow, 1880)
= *Tropidoneis antarctica* (Grunow, in Cleve & Möller) Cleve 1894b

Meuniera membranacea (Cleve) P. C. Silva comb. nov.
= *Stauropsis membranacea* (Cleve) Meunier 1910
= *Stauroneis membranacea* (Cleve) Hustedt 1959

Minutocellus polymorphus (Hargraves & Guillard) Hasle, von Stosch, &
    Syvertsen 1983
= *Bellerochea polymorpha* Hargraves & Guillard 1974

Navicula granii (Jørgensen) Gran 1908
= *Stauroneis granii* Jørgensen 1905

Navicula septentrionalis (Grunow) Gran 1908
= *Stauroneis septentrionalis* Grunow 1884
= *Navicula quadripedis* Cleve-Euler 1952

Neodelphineis indica (F. J. R. Taylor) Tanimura (1992)
= *Synedra indica* F. J. R. Taylor 1967

Neodenticula seminae (Simonsen & Kanaya) Akiba & Yanagisawa 1986
= *Denticula seminae* Simonsen & Kanaya 1961
= *Denticulopsis seminae* (Simonsen & Kanaya) Simonsen 1979

Nitzschia braarudii Hasle sp. nov.
= *Nitzschia capitata* Heiden (in Heiden & Kolbe, 1928)

Nitzschia marina Grunow (in Cleve & Grunow, 1880)
= *Synedra gaussii* Heiden (in Heiden & Kolbe, 1928)
= *Pseudo-nitzschia hustedtii* Meister 1937

Odontella aurita (Lyngbye) C. A. Agardh 1832
= *Biddulphia aurita* (Lyngbye) Brébisson (in Brébisson & Godey, 1838)

Odontella litigiosa (Van Heurck) Hoban (in Hoban et al., 1980)
= *Biddulphia litigiosa* Van Heurck 1909

Odontella longicruris (Greville) Hoban 1983
= *Biddulphia longicruris* Greville 1859b

Odontella mobiliensis (Bailey) Grunow 1884
= *Biddulphia mobiliensis* Grunow (in Van Heurck, 1882)

Odontella sinensis (Greville) Grunow 1884
= *Biddulphia sinensis* Greville 1866

Odontella weissflogii (Janisch) Grunow 1884
= *Biddulphia weissflogii* Grunow (in Van Heurck, 1882)
= *Biddulphia striata* Karsten 1905

Palmeria hardmaniana Greville 1865b
= *Hemidiscus hardmanianus* (Greville) Mann 1907

Paralia sulcata (Ehrenberg) Cleve 1873b
= *Melosira sulcata* (Ehrenberg) Kützing 1844

Phaeodactylum tricornutum Bohlin 1897
= *Nitzschia closterium* (Ehrenberg) W. Smith f. *minutissima* Allen &
　　Nelson 1910

Plagiogrammopsis vanheurckii (Grunow) Hasle, von Stosch, &
　　Syvertsen 1983
= *Plagiogramma vanheurckii* Grunow (in Van Heurck, 1881)

Plagiotropis gaussii (Heiden in Heiden & Kolbe) Paddock 1988
= *Tropidoneis gaussii* Heiden (in Heiden & Kolbe, 1928)

Planktoniella blanda (A. Schmidt) Syvertsen & Hasle (in Hasle & Syvertsen,
　　1993)
= *Coscinodiscus blandus* A. Schmidt 1878
= *Coscinodiscus latimarginatus* Guo 1981
= *Thalassiosira blanda* Desikachary & Gowthaman (in Desikachary, 1989)
= *Thalassiosira bipartita* (Rattray) Hallegraeff 1992

Planktoniella muriformis (Loeblich, III, Wight, & Darley) Round 1972
= *Coenobiodiscus muriformis* Loeblich, III, Wight, & Darley 1968

Pleurosigma simonsenii Hasle nom. nov.
= *Pleurosigma planctonicum* Simonsen 1974

Porosira pseudodenticulata (Hustedt) Jousé (in Jousé et al., 1962)
= *Coscinodiscus pseudodenticulatus* Hustedt 1958a

Proboscia inermis (Castracane) Jordan & Ligowski 1991
= *Rhizosolenia inermis* Castracane 1886
= *Rhizosolenia alata* f. *inermis* (Castracane) Hustedt *sensu* Hendy 1937

Proboscia truncata (Karsten) Nöthig & Ligowski 1991
= *Rhizosolenia truncata* Karsten 1905
= *Rhizosolenia alata* f. *curvirostris* Gran 1900

Proboscia alata (Brightwell) Sundström 1986
= *Rhizosolenia alata* Brightwell 1858a

Pseudo-nitzschia australis Frenguelli 1939
= *Nitzschia pseudoseriata* Hasle 1965b

Pseudo-nitzschia cuspidata (Hasle) Hasle 1993
= *Nitzschia cuspidata* Hasle 1974

Pseudo-nitzschia delicatissima (Cleve) Heiden (in Heiden & Kolbe, 1928)
= *Nitzschia delicatissima* Cleve 1897a
= *Nitzschia actydrophila* Hasle 1965b

Pseudo-nitzschia fraudulenta (Cleve) Hasle 1993
= *Nitzschia fraudulenta* Cleve 1897b
= *Pseudo-Nitzschia seriata* var. *fraudulenta* H. Peragallo (in H. & M. Peragallo, 1900)

Pseudo-nitzschia heimii Manguin 1957
= *Nitzschia heimii* (Manguin) Hasle 1965b

Pseudo-nitzschia inflatula (Hasle) Hasle 1993
= *Nitzschia inflatula* Hasle 1974

Pseudo-nitzschia lineola (Cleve) Hasle 1993
= *Nitzschia lineola* Cleve 1897b
= *Nitzschia barkleyi* Hustedt 1952

Pseudo-nitzschia multiseries (Hasle) Hasle 1995
= *Pseudo-nitzschia pungens* f. *multiseries* (Hasle) Hasle 1993
= *Nitzschia pungens* f. *multiseries* Hasle 1974

Pseudo-nitzschia prolongatoides (Hasle) Hasle 1993
= *Nitzschia prolongatoides* Hasle 1965b

Pseudo-nitzschia pseudodelicatissima (Hasle) Hasle 1993
= *Nitzschia pseudodelicatissima* Hasle 1976b
= *Nitzschia delicatula* Hasle 1965b

Pseudo-nitzschia pungens (Grunow ex Cleve) Hasle 1993
= *Nitzschia pungens* Grunow ex Cleve 1897a

Pseudo-nitzschia pungiformis (Hasle) Hasle 1993
= *Nitzschia pungiformis* Hasle 1971

Pseudo-nitzschia seriata (Cleve) H. Peragallo (in H. & M. Peragello,
  1900) f. *seriata*
= *Nitzschia seriata* Cleve 1883

Pseudo-nitzschia seriata f. *obtusa* (Hasle) Hasle 1993
= *Nitzschia seriata* f. *obtusa* Hasle 1974

Pseudo-nitzschia subcurvata (Hasle) G. Fryxell (in Fryxell et al., 1991)
= *Nitzschia subcurvata* Hasle 1964

Pseudo-nitzschia subfraudulenta (Hasle) Hasle 1993
= *Nitzschia subfraudulenta* Hasle 1974

Pseudo-nitzschia subpacifica (Hasle) Hasle 1993
= *Nitzschia subpacifica* Hasle 1974

Pseudo-nitzschia turgidula (Hustedt) Hasle 1993
.= *Nitzschia turgidula* Hustedt 1958a

Pseudo-nitzschia turgiduloides (Hasle) Hasle 1993
= *Nitzschia turgiduloides* Hasle 1965b
= *Pseudo-nitzschia barkleyi* var. *obtusa* Manguin 1960 (no Latin diagnosis)

Pseudosolenia calcar-avis (Schultze) Sundström 1986
= *Rhizosolenia calcar-avis* Schultze 1859

Rhizosolenia antennata (Ehrenberg) Brown 1920
= *Rhizosolenia hebetata* f. *bidens* Heiden (in Heiden & Kolbe, 1928)

Rhizosolenia borealis Sundström 1986
= *Rhizosolenia styliformis* var. *oceanica* Wimpenny 1946 *pro parte*

Rhizosolenia hebetata f. *hebetata* Bailey 1856
= *Rhizosolenia hebetata* f. *hiemalis* Gran 1904

Rhizosolenia hebetata f. *semispina* (Hensen) Gran 1904
= *Rhizosolenia semispina* Hensen 1887

Rhizosolenia imbricata Brightwell 1858a
= *Rhizosolenia shrubsolei* Cleve 1881
= *Rhizosolenia imbricata* var. *shrubsolei* (Cleve) Schröder 1906

Rhizosolenia polydactyla Castracane 1886
= *Rhizosolenia styliformis* var. *oceanica* Wimpenny 1946 *pro parte*

Rhizosolenia styliformis Brightwell 1858a
= *Rhizosolenia styliformis* var. *longispina* Hustedt (in A. Schmidt, 1914)

Stellarima microtrias (Ehrenberg) Hasle & Sims 1986b
= *Symbolophora microtrias* Ehrenberg 1844b
= *Coscinodiscus symbolophorus* Grunow 1884
= *Coscinodiscus furcatus* Karsten 1905

Stellarima stellaris (Roper) Hasle & Sims 1986b
= *Coscinodiscus stellaris* Roper 1858

Synedropsis hyperborea (Grunow) Hasle, Medlin, & Syvertsen 1994
= *Synedra hyperborea* Grunow 1884
= *Synedra hyperborea* var. *flexosa* Grunow 1884
= *Synedra hyperborea* var. *rostellata* Grunow 1884

Thalassionema bacillare (Heiden, in Heiden & Kolbe) Kolbe 1955
= *Spinigera bacillaris* Heiden (in Heiden & Kolbe, 1928)

Thalassionema frauenfeldii (Grunow) Hallegraeff 1986
= *Thalassiothrix frauenfeldii* (Grunow) Grunow (in Cleve & Grunow, 1880)

Thalassionema javanicum (Grunow, in Van Heurck) Hasle comb. nov.
= *Thalassiothrix frauenfeldii* var. *javanica* Grunow (in Van Heurck, 1881)

Thalassionema nitzschioides (Grunow) Mereschkowsky 1902
= *Thalassiothrix nitzschioides* Grunow (in Van Heurck, 1881)

Thalassionema pseudonitzschioides (Schuette & Schrader) Hasle comb. nov.
= *Thalassiothrix pseudonitzschioides* Schuette & Schrader 1982

Thalassiosira angulata (Gregory) Hasle 1978a
= *Thalassiosira decipiens* (Grunow) Jørgensen 1905 (in Hustedt, 1930; Cupp, 1943)

Thalassiosira anguste-lineata (A. Schmidt) G. Fryxell & Hasle 1977
= *Coscinodiscus anguste-lineatus* A. Schmidt 1878
= *Thalassiosira polychorda* (Gran) Jørgensen 1899
= *Coscinosira polychorda* (Gran) Gran 1900

Thalassiosira antarctica var. *borealis* G. Fryxell, Ducette, & Hubbard 1981
= *Thalassiosira fallax* Meunier 1910

Thalassiosira delicatula Ostenfeld (in Borgert, 1908)
= *Thalassiosira coronata* Gaarder 1951
non *Thalassiosira delicatula* Hustedt 1958a

Thalassiosira dichotomica (Kozlova) G. Fryxell & Hasle 1983
= *Porosira dichotomica* Kozlova 1967

Thalassiosira eccentrica (Ehrenberg) Cleve 1904
= *Coscinodiscus eccentricus* Ehrenberg 1841a

Thalassiosira gracilis var. *expecta* (VanLandingham) G. Fryxell & Hasle 1979a
= *Thalassiosira delicatula* Hustedt 1958a
= *Thalassiosira expecta* VanLandingham 1978

Thalussiosira guillardii Hasle 1978b
= *Cyclotella nana* Guillard clone 7-15 (in Guillard & Ryther, 1962)

Thalassiosira hendeyi Hasle & G. Fryxell 1977b
= *Coscinodiscus hustedtii* Müller-Melchers 1953

Thalassiosira hyperborea (Grunow) Hasle (in Hasle & Lange, 1989)
= *Coscinodiscus* (*lacustris* var?) *hyperboreus* Grunow 1884

Thalassiosira lentiginosa (Janisch, in A. Schmidt) G. Fryxell 1977
= *Coscinodiscus lentiginosus* Janisch (in A. Schmidt, 1878)

Thalassiosira leptopus (Grunow, in Van Heurck) Hasle & G. Fryxell 1977b
= *Coscinodiscus lineatus* Ehrenberg 1839
= *Coscinodiscus* (lineatus var.?) *leptopus* Grunow (in Van Heurck, 1883)

Thalassiosira mediterranea (Schröder) Hasle 1972b
= *Coscinosira mediterranea* Schröder 1911
= *Thalassiosira stellaris* Hasle & Guillard (in Fryxell & Hasle, 1977)

Thalassiosira minima Gaarder 1951
= *Coscinosira floridana* Cooper 1958
= *Thalassiosira floridana* (Cooper) Hasle 1972b

Thalassiosira minuscula Krasske 1941
= *Thalassiosira monoporocyclus* Hasle 1972a

Thalassiosira oceanica Hasle 1983a
= *Cyclotella nana* Guillard clone 13-1 (in Guillard & Ryther, 1962)

Thalassiosira oestrupii (Ostenfeld) Hasle 1972b
= *Coscinosira oestrupii* Ostenfeld 1900

Thalassiosira poroseriata (Ramsfjell) Hasle 1972b
= *Coscinosira poroseriata* Ramsfjell 1959

Thalassiosira pseudonana Hasle & Heimdal 1970
= *Cyclotella nana* Guillard clone 3H (in Guillard & Ryther, 1962)

Thalassiosira punctigera (Castracane) Hasle 1983b
= *Thalassiosira angstii* (Gran) Makarova 1970

Thalassiosira ritscheri (Hustedt) Hasle (in Hasle & Heimdal, 1970)
= *Coscinodiscus ritscheri* Hustedt 1958a

Thalassiosira tumida (Janisch) Hasle (in Hasle et al., 1971)
= *Coscinodiscus tumidus* Janisch (in A. Schmidt, 1878)

Thalassiosira weissflogii (Grunow) G. Fryxell & Hasle 1977
= *Thalassiosira fluviatilis* Hustedt 1926

Toxarium hennedyanum (Gregory) Pelletan 1889
= *Synedra hennedyana* Gregory 1857

Toxarium undulatum Bailey 1854
= *Synedra undulata* (Bailey) Gregory 1857

Trichotoxon reinboldii (Van Heurck) Reid & Round 1988
= *Synedra reinboldii* Van Heurck 1909
= *Synedra pelagica* Hendey 1937

INDEX OF DIATOM TAXA

Synonyms in boldface

Achnanthaceae, 10, 26, 269, 271
Achnanthales,* 26
Achnanthes, 26
A. taeniata, 11, 269
Actinocyclus, 17, 25, 99, 116, 119
A. actinochilus, 118–119, 121–122
A. circellus,118–119, 121
A. curvatulus,118 121–122
A. ehrenbergii, 120
A. exiguus, 118, 122
A. kützingii,118–119, 121
A. normanii, 118, 121
A. normanii f. normanii, 119
A. normanii f. subsalsus, 119
A. octonarius, 117–118, 120–121
A. octonarius var. tenellus,* 120
A. sagittulus, 118, 120–121
A. senarius, 141
A. spiritus, 118, 122–123
A. subocellatus, 121
A. subtilis, 118, 120–121
A. vestigulus, 118, 123
Actinoptychus, 25, 140
A. senarius, 140
A. undulatus, 140
Adoneis, 26, 248
A. pacifica, 248
Amphiprora,* 289
A. belgicae, 28
A. rigida, 287
Arcocellulus, 25, 177–178, 183
A. cornucervis, 179
A. mammifer, 178–179
Asterionella
A. bleakeleyi var. notata, 243
A. frauenfeldii, 262
A. glacialis, 241
A. japonica, 241
A. kariana, 241, 243
A. socialis,* 243
A. synedraeformis,* 263
Asterionellopsis, 26, 241, 243, 338
A. glacialis, 243
A. kariana, 243

Asterolampra, 25, 133
A. grevillei, 133
A. marylandica, 133
A. roperiana, 139
A. vanheurckii, 133
Asterolampraceae, 25, 131
Asterolamprales,* 131
Asteromphalus, 25, 131
A. arachne, 137, 140
A. darwinii,* 133, 140
A. elegans, 135, 137, 140
A. flabellatus, 133, 135
A. heptactis, 135, 137, 140
A. hookeri, 139, 140
A. hyalinus, 135, 139–140
A. parvulus, 135, 139–140
A. roperianus, 135, 139–140
A. rossii,* 140
A. sarcophagus, 133, 135, 140
Attheya, 184, 221, 225
A. decora,* 225
A. longicornis*, 225
A. septentrionalis, 225
Attheyaceae,* 225
Aulacodiscaceae,* 140
Aulacodiscus, 140
A. argus, 141
Azpeitia, 25, 98, 116–117, 123
A. africana, 124, 126
A. antiqua, 123
A. barronii, 124, 126, 128
A. neocrenulata, 124, 126, 128
A. nodulifera, 126, 128
A. tabularis, 126, 128
A. temperi,* 123
Azpeitiaceae,* 116

Bacillaria, 5, 293
B. paradoxa, 291, 293
B. paxillifera, 290–291, 293
Bacillariaceae, 27, 290, 295
Bacillariales, 26
Bacillariineae, 26

* Mentioned in the text but not discussed as a separate item

Bacillariophyceae, 6

Bacillariophycidae,* 26

Bacterastroidea,* 190

Bacteriastrum, 10–11, 13, 25, 184, 186, 225

  B. comosum, 188–189

  B. delicatulum, 186, 189

  B. elongatum, 188–189

  B. furcatum, 186, 189

  B. hyalinum, 186, 188–189

  B. varians, 186

Bacterosira, 23, 31

  B. bathyomphala, 31, 36–37

  B. fragilis, 31

Banquisia, 27, 272, 284

  B. belgicae, 284

Bellerochea, 26, 225, 227

  B. horologicalis, 229

  B. horologicalis var. _recta,_ 229

  B. malleus, 227, 229

  B. polymorpha, 182

  B. yucatanensis, 229

Biddulphia, 168

  B. aurita, 236

  B. litigiosa, 239

  B. longicruris, 239

  B. mobiliensis, 239

  B. sinensis, 239

  B. striata, 239

  B. weissflogii, 239

Biddulphiaceae,* 97, 184

Biddulphiales, 23, 27, 177

Biddulphiineae, 25, 168

Biddulphiophycidae,* 236

Bleakeleya, 26, 243

  B. notata, 243

Brockmanniella, 25, 177–179, 183

  B. brockmannii, 179

Cerataulina, 25, 169

  C. bergonii, 169

  C. bicornis, 171

  C. compacta, 171

  C. daemon, 171

  C. dentata, 171

  C. pelagica, 168–169, 171

Cerataulus

  C. bergonii, 169

Ceratoneis

  C. closterium, 293

  C. longissima, 329

Chaetoceraceae,* 25, 184, 338

Chaetoceros, 6, 10–11, 13, 25, 27, 184, 186, 225, 332, 334, 336, 338

  C. aequatorialis, 191, 193, 196

  C. affinis, 190, 193, 218, 220

  C. affinis var. _willei,*_ 218

  C. anastomosans, 192, 210

  C. armatus,* 184

  C. atlanticus, 190–191, 196, 198

  C. bactereastroides, 190

  C. borealis, 193, 198, 203

  C. borealis var. _densus,_ 201

  C. brevis,* 210

  C. calcitrans,* 221

  C. calcitrans f. _pumilus,_ 223

  C. castracanei, 191, 198, 203

  C. ceratosporus, 193, 221

  C. ceratosporus var. _brachysetus,_ 223

  C. chilensis, 195

  C. cinctus, 193, 211, 215

  C. coarctatus, 191, 198, 203

  C. compressus, 190, 192, 206–207

  C. compressus var. _hirtisetus,*_ 206

  C. concavicornis, 190–191, 198, 201, 203

  C. convolutus, 190, 198, 203

  C. constrictus, 190, 192, 209–210

  C. costatus, 193, 218

  C. criophilus, 191, 193, 196

  C. curvisetus, 190, 192, 210–211, 216

  C. dadayi, 191, 201, 203

  C. danicus, 191, 194–195

  C. debilis, 190–192, 211

  C. decipiens, 190, 192, 204, 206–207

  C. densus, 191, 201, 203

  C. diadema, 190, 193, 220

  C. dichaeta, 189, 191

  C. didymus, 192, 207, 209–210

  C. distans var. _subsecundus,_ 220

  C. diversus, 190, 193, 216

  C. eibenii, 11, 191, 201, 203

  C. externus, 210

  C. flexuosus, 192, 203

  C. furcellatus, 203, 213, 215

  C. galvestonensis, 223

  C. gracilis,* 221

  C. hendeyi, 203

  C. holsaticus, 193

  C. karianus, 193, 221

  C. laevis, 216

  C. lorenzianus, 192, 204, 206, 210

C. *laciniosus,* 191–192, 209–210
C. *lauderi,* 192, 207
C. *messanensis,* 193, 216
C. *minimus,* 193, 224
C. *mitra,* 192, 204, 206–207
C. *muelleri,* * 221
C. *neglectus,* 207
C. *pelagicus,* 210
C. *pendulus,* 193
C. *peruvianus,* 191, 195–196
C. *protuberans,* * 209
C. *pseudocurvisetus,* 192, 211
C. *radians,* 193, 211, 215, 221
C. *radicans,* 193
C. *rostratus,* 191, 196
C. *seiracanthus,* 193, 220
C. *septentrionalis,* 184, 225
C. *similis,* 192, 210
C. *simplex,* 193, 223
C. *simplex* var. *calcitrans,* 223
C. *socialis,* 190, 193, 221
C. *subsecundus,* 220
C. *subtilis,* 221, 223
C. *tenuissimus,* 191, 193, 223
C. *teres,* 192, 206
C. *tetrachaeta,* * 189, 201
C. *tetrastichon,* 192, 203
C. *throndsenii,* 193, 224
C. *throndsenii* var. *throndsenii,* 224
C. *throndsenii* var. *trisetosus,* 224
C. *tortissimus,* 215
C. *vanheurckii,* 210
C. *wighamii,* 193, 221
C. *willei,* 218

Chaetocerotaceae, 184

Chaetocerotales,* 184

Chaetocerotophycidae,* 93
Charcotia
   C. *actinochila,* 119
Chromophyta,* 6

Chrysophyta,* 6
Climacodium, 25, 169, 171
   C. *biconcavum,* 172
   C. *frauenfeldianum,* 172
Coenobiodiscus
   C. *muriformis,* 40, 346
Cocconeis
   C. *amphiceros,* 252
Corethraceae,* 96

Corethrales,* 96

Corethron, 23, 96, 338
   C. *criophilum,* 96–97
   C. *inerme,* 97
Corethrophycidae,* 96

Coscinodiscaceae, 97, 116

Coscinodiscineae, 23, 28
Coscinodiscus, 23, 27, 98, 334
   C. *actinochilus,* 119
   C. *africanus,* 124
   C. *alboranii,* 102, 106, 110
   C. *anguste-lineatus,* 71
   C. *argus,* 99, 103–105, 109
   C. *asteromphalus,* 102, 104–105
   C. *bathyomphalus,* 31
   C. *bipartita,* 41
   C. *bioculatus,* 80
   C. *blandus,* 40–41
   C. *bouvet,* 97, 103, 109–110
   C. *centralis,* 102, 104–105
   C. *concinniformis,* 102, 105–106, 110
   C. *concinnus,* 98, 102, 105–106
   C. *crenulatus,* 124
   C. *curvatulus,* *121–122
   C. *curvatulus* var. *inermis,* * 122
   C. *curvatulus* var. *genuina,* * 122
   C. *curvatulus* var. *karianus,* * 122
   C. *curvatulus* var. *minor,* * 122
   C. *curvatulus* var. *subocellatus,*
        121–122
   C. *eccentricus,* 62
   C. *furcatus,* 111, 113, 115
   C. *gazellae,* 111
   C. *gracilis,* 82
   C. *granii,* 103, 109–110
   C. *hustedtii,* 62
   C. *hyalinus,* 69
   C. *jonesianus,* 102, 107, 110
   C. *kützingii,* 119
   C. *lacustris,* * 74
   C. *(lacustris* var.?*) hyperboreus,* 74
   C. *lacustris* var. *pelagicus,* * 74
   C. *latimarginatus,* 346
   C. *lentiginosus,* 79
   C. *lineatus,* 73, 75
   C. *(lineatus* var.?*) leptopus,* 75
   C. *marginatus,* 102, 107
   C. *neocrenulatus,* 124
   C. *nodulifer,* 126
   C. *normanii,* 119

C. *polyacanthus* var. *balticus*, 68
C. *polychordus*, 71
C. *praelineatus*, 75
C. *pseudodenticulatus*, 41
C. *pseudolineatus*, 75
C. *radiatus*, 103, 107, 338
C. *rex*, 111
C. *ritscheri*, 86
C. *sol*, 40
C. *stellaris*, 113
C. *striatus*, 34
C. *subglobosus*, 68–69
C. (*subglobosus* var.?) *antarcticus*, 68
C. *sublineatus*,* 73
C. *subsalsus*, 119
C. *symbolophorus*, 113, 115
C. *tabularis*, 126
C. *trioculatus*, 37
C. *tumidus*, 70
C. *wailesii*, 98, 102, 105–106, 110

Coscinosira
  C. *floridana*, 65
  C. *mediterranea*, 74
  C. *oestrupii*, 83
  C. *polychorda*, 71
  C. *poroseriata*, 87

Creswellia
  C. *palmeriana*, 102
  C. *turris*,* 102

Cyclotella, 23, 30, 33
  C. *caspia*, 33, 34
  C. *cryptica*,* 31
  C. *distinguenda*,* 31
  C. *litoralis*, 31, 34
  C. *meneghiniana**
  C. *nana*, 77, 79
  C. *striata*, 33–34
  C. *stylorum*, 33–34
  C. *tecta*,* 33

Cylindrotheca, 27, 291, 293
  C. *closterium*, 184, 270, 290, 294, 329
  C. *gerstenbergeri*,* 293
  C. *gracilis*, 293

Cymatosiraceae, 7, 10, 25, 177

Cymatosirales,* 177

Cymatosirophycidae,* 177
Cymatosira, 25, 177–178, 181, 183
  C. *belgica*, 182
  C. *lorenziana*, 8, 10, 181–182

Dactyliosolen, 25, 141–142, 144, 165
  D. *antarcticus*, 165–166, 168
  D. *blavyanus*, 166–167
  D. *fragilissimus*, 166–168, 341
  D. *mediterraneus*, 95
  D. *phuketensis*, 166–168, 341
  D. *tenuijunctus*, 166–167
Delphineis, 26, 247–248, 251
  D. *angustata*,* 248
  D. *karstenii*, 249
  D. *surirella*, 249
  D. *surirelloides*, 249
Denticella
  D. *regia*, 239
Denticula
  D. *kamtschatica*, 305
  D. *marina*, 305
  D. *seminae*, 305
Denticulopsis
  D. *seminae*, 305
Detonula, 23, 31, 34
  D. *confervacea*, 11, 35–36
  D. *moseleyana*, 35–36
  D. *pumila*, 34–37
Diatoma, 5
  D. *aurita*, 236
  D. *rhombica*, 300
Dicladia
  D. *antennata*, 147
  D. *groenlandica*, 204
  D. *mitra*, 204
Ditylum, 26, 225, 227, 230
  D. *brightwellii*, 230–231
  D. *buchananii*, 231
  D. *pernodii*, 231
  D. *sol*, 231
  D. *trigonum*, 230

Entomoneis,* 289
Ephemera, 27, 284
  E. *planamembranacea*, 272, 284
Ethmodiscaceae,* 97
Ethmodiscales,* 97
Ethmodiscus, 23, 97–98, 109
  E. *gazellae*, 111
  E. *gigas*, 109
  E. *punctiger*, 59
  E. *rex*, 111
Eucampia, 25
  E. *antarctica*, 175–176
  E. *antarctica* var. *recta*,* 176
  E. *balaustium*, 175

E. cornuta, 175–176
E. groenlandica, 176
E. striata, 163, 340
E. zodiacus, 172, 175
E. zodiacus f. *cylindricornis,** 175
Eunotia
   *E. soleirolii,** 11
Eupodiscaceae, 26, 236
Eupodiscus
   E. jonesianus, 107
   E. subtilis, 120
   E. tesselatus, 130

*Fasciculigera,** 45
*Fragilaria,** 7, 246
   F. antarctica
   F. arctica, 299
   F. curta, 302
   F. cylindrus, 302
   F. karstenii, 249
   F. nana, 300
   F. oceanica, 299
   *F. pectinalis,** 246
   F. striatula, 26, 246
   F. unipunctata, 244
Fragilariaceae, 241
Fragilariineae, 241
Fragilariophyceae,** 26
Fragilariopsis, 27, 246, 269, 290–291,
   295–296, 305, 307, 338
   F. atlantica, 298–299, 303
   F. antarctica, 294, 296
   F. curta, 295–296, 303, 305
   F. cylindriformis, 296, 302–303
   F. cylindrus, 295–296, 300, 302, 305
   F. doliolus, 296, 303, 305
   F. kerguelensis, 294–296, 303, 305
   F. nana, 300
   F. oceanica, 11, 295–296, 299, 305
   F. pseudonana, 296, 300, 305
   F. rhombica, 296, 300, 303
   F. ritscheri, 296, 297, 303
   F. separanda, 296, 301–302
Gaillonella
   G. sulcata, 91
Gonioceros
   G. septentrionalis, 225
Gossleriella, 25, 115
   G. tropica, 116
Gossleriellaceae,** 116
Guinardia, 25, 141–142, 144, 160
   G. blavyana, 167

G. cylindrus, 161, 165, 340
G. delicatula, 161, 165, 168, 340
G. flaccida, 131, 161, 163, 165
G. striata, 161, 163, 165, 168, 340
G. tubiformis, 161, 165
Gyrosigma, 283
   G. macrum, 283

Haslea, 26, 271, 275–276
   H. gigantea, 276–277
   H. gigantea var. *tenuis,** 278
   *H. ostrearia,** 275
   H. trompii, 278
   H. wawrikae, 276, 278
Helicotheca, 26, 226–227, 234, 236
   H. tamesis, 234
Heliopelthaceae, 25, 140
Hemiaulaceae, 168, 184
Hemiaulales,** 168
Hemiaulus, 25, 169, 176
   H. hauckii, 177
   H. indicus, 177
   H. membranaceus, 177
   *H. proteus,** 176
   H. sinensis, 177
Hemidiscaceae, 22, 116–117, 130,
   339
Hemidiscus, 25, 116, 128, 131
   H. cuneiformis, 128
   H. hardmanianus, 107
   H. kanayanus, 128
Hyalochaete, 192, 204
Hyalosira
   *H. delicatula,** 244

*Inconspicuae,** 45

Lanceolatae, 329
Lauderia, 23, 31, 36
   L. annulata, 36–37
   L. borealis, 36
   L. confervacea, 35
   L. fragilis, 31
   L. moseleyana, 35
   L. pumila, 34
Lauderiaceae,** 30
Lauderiopsis, 131
   *L. costata,** 131
Lennoxia, 25, 184
   L. faveolata, 184, 294
Leptocylindraceae, 93
Leptocylindrales,** 93

Leptocylindrus, 23, 27, 93, 131
  L. *curvatulus,**96
  L. *danicus,* 13, 93, 94–96, 168
  L. *danicus* var. *adriaticus,** 96
  L. *mediterraneus,* 94–96
  L. *minimus,* 13, 94–95
Lioloma, 26, 253, 254, 338
  L. *delicatulum,* 257, 340
  L. *elongatum,* 254, 257, 338, 340
  L. *pacificum,* 257, 340
Liriogramma,* 140
Lithodesmiaceae, 26, 225–226
Lithodesmiales,* 225
Lithodesmiophycidae,* 225
Lithodesmioides, 26, 225–227, 231, 234
  L. *minuta,* 332
  L. *polymorpha,* 231–232
Lithodesmium, 26, 225–227, 232
  L. *duckerae,* 233, 234
  L. *intricatum,* 234
  L. *undulatum,* 232–233, 234
  L. *variabile,* 234
Manguinea, 27, 272, 287
  M. *fusiformis,* 289
  M. *rigida,* 289
Melosiraceae, 23, 88, 93
Melosira, 23, 88–89
  M. *arctica,* 89, 241, 246
  M. *costata,* 44
  M. *hyperborea,* 89
  M. *nummuloides,* 89
  M. *subsalsa,* 44
  M. *sulcata,* 91
Membraneis, 27, 272, 286
  M. *challengeri,* 286–287
  M. *imposter,* 287
Meuniera, 27
  M. *membranacea,* 272–273, 339–340
Micropodiscus
  M. *weissflogii,* 75
Microtabella,* 244
Minidiscus, 23, 30, 37, 39
  M. *comicus,* 37
  M. *chilensis,* 37, 39
  M. *trioculatus,* 37, 39
Minutocellus, 25, 177–178, 182–183
  M. *polymorphus,* 182–183
  M. *scriptus,* 182–183
Miraltia
  M. *throndsenii,* 224

Mölleria
  M. *cornuta,* 175
Monoceros
  M. *isthmiiformis,* 224

Nanoneis, 27, 271
  N. *hasleae,* 271
Navicula, 27, 269, 272, 275, 278, 334
  N. *angulata*
  N. *challengeri,* 287
  N. *derasa,* 280
  N. *directa,* 272, 278, 280
  N. *distans,* 272, 280–281
  N. *gigantea,* 274
  N. *granii,* 272, 275
  N. *(Stauroneis) membranacea,* 273, 340
  N. *ostrearia,** 275
  N. *pelagica,* 272–273, 275
  N: *planamembranacea,* 284
  N. *quadripedis,* 275
  N. *septentrionalis,* 272, 275
  N. *transitans* var. *derasa,* 272, 278, 280
  N. *transitans* var. *derasa* f. *delicatula,*
    272, 278, 281
  N. *trompii,* 272
  N. *vanhoeffenii,* 271–272
  N. *wawrikae,* 278
Naviculaceae, 27, 271–272
Naviculae
  N. *fusiformes,** 271
Naviculales,* 26
Neodelphineis, 26, 247, 249, 251
  N. *indica,* 249
  N. *pelagica,* 247, 249
Neodenticula, 27, 290–291, 305
  N. *kamtschatica,* 305
  N. *seminae,* 306
Neostreptotheca, 26, 225, 227, 235
  N. *subindica,* 235–236
  N. *torta,* 236
  N. *torta* f. *torta,** 236
  N. *torta* f. *triangularis,** 236
Nitzschiaceae,* 290
Nitzschia, 27, 269, 290–291, 295, 307,
    323–324, 328
  N. *actydrophila,* 316
  N. *americana,* 324
  N. *angulata,* 300
  N. *angusta* var. *marina,* 330
  N. *angustata* var. *marina,* 330
  N. *barkleyi,* 317

N. *bicapitata*, 330–331
N. *bifurcata*,* 330
N. *braarudii*, 330–331, 339
N. *capitata*, 319, 330–331
N. *closterium*, 184, 230, 290, 293, 294
N. *closterium* f. *minutissima*, 269, 271
N. *curta*, 302
N. *cuspidata*, 319
N. *cylindriformis*, 302
N. *cylindrus*, 302
N. *delicatissima*, 316–317, 323
N. *delicatula*, 321
N. *fraudulenta*, 313
N. *granii*, 321
N. *granii* var. *curvata*, 321
N. *grunowii*, 299
N. *heimii*, 314
N. *inflatula*, 321
N. *kerguelensis*, 296
N. *kolaczeckii*, 328–329
N. *lineola*, 317
N. *longissima*, 294, 324, 329–330
N. *marina*, 328–329
N. *migrans*, 327
N. *multistriata*,* 323
N. *paaschei*, 297
N. *paradoxa*, 293
N. *prolongata*, 317
N. *prolongatoides*, 317
N. *pseudodelicatissima*, 321
N. *pseudonana*, 300
N. *pseudoseriata*, 311
N. *pungens*, 312
N. *pungens* f. *multiseries*, 312
N. *pungiformis*, 313
N. *ritscheri*, 247
N. *separanda*, 300
N. *seriata*, 305, 310–311, 316
N. *seriata* f. *obtusa*, 311
N. *sicula*, 305, 329
N. *sicula* var. *bicuneata*, 325, 327
N. *sicula* var. *migrans*, 325, 327
N. *sicula* var. *rostrata*, 327
N. *sicula* var. *sicula*, 325, 327
N. *sigmoidea*,* 300
N. *subcurvata*, 321
N. *subfraudulenta*, 313
N. *subpacifica*, 314
N. *turgidula*, 319
N. *turgiduloides*, 319
Nitzschiella, 330

Odontella, 26, 168, 236
O. *aurita*, 236, 239
O. *litigiosa*, 239
O. *longicruris*, 239
O. *mobiliensis*, 239
O. *regia*, 239
O. *sinensis*, 239
O. *weissflogii*, 239
Orthosira
O. *angulata*, 51
O. *marina*, 89
Pachyneis, 27, 271–272, 290
P. *gerlachii*, 289–290
Palmeria, 23, 97, 111–112
P. *hardmaniana*, 111–112
P. *ostenfeldii*, 112–113
Paralia, 10, 86–87
P. *marina*, 89
P. *sulcata*, 91
Paraliaceae,* 88
Paraliales,* 88
Phaeoceros, 190–191, 193, 204
Phaeocystis,* 321, 323
Phaeodactylaceae, 26, 269, 271
Phaeodactylum, 26, 269
P. *tricornutum*, 269, 271, 294
Pinnularia
P. *directa*, 278
P. *distans*, 280
Plagiogramma
P. *brockmannii*, 179
P. *vanheurckii*, 183
Plagiogrammopsis, 25, 177–178, 183
P. *vanheurckii*, 183
Plagiotropis, 27, 271–272, 289
P. *baltica*, 289
P. *gaussii*, 289
Planktoniella, 23, 30, 39
P. *blanda*, 40
P. *muriformis*, 40
P. *sol*, 40
Planktoniellaceae, 30
Pleurosigma, 27, 272, 281, 338
P. *angulatum*,* 281
P. *directum*, 272, 281–282
P. *normanii*, 272, 281–282
P. *planctonicum*, 282, 339
P. *simonsenii*, 272, 281–282, 339
Podosira
P. *hormoides* var. *glacialis*, 41
P. (?) *subtilis*, 58

Porosira, 17, 23, 31, 41
P. denticulata, 41, 43
P. dichotomica, 65
P. glacialis, 41, 43
P. pentaportula, 41, 43
P. pseudodenticulata, 41, 43
Proboscia, 25, 141–142, 159
P. alata, 159
P. eumorpha, 159
P. inermis,* 159
P. subarctica, 159
P. truncata,* 159
Pseudoeunotia
P. doliolus, 303
Pseudoguinardia, 25, 130
P. recta, 130–131
Pseudo-nitzschia, 27, 290, 306–307, 310–312, 323, 328, 333, 334, 338
P. australis, 309, 311–312, 315–316
P. barkleyi var. obtusa, 319
P. cuspidata, 309, 319, 323
P. delicatissima, 307, 309–310, 317, 323, 324
P. fraudulenta, 308–309, 313, 315
P. granii, 309–310, 321, 323
P. granii var. curvata, 321
P. granii var. granii, 321
P. heimii, 308–309, 314–316
P. hustedtii, 328
P. inflatula,* 309–310, 321, 323
P. lineola, 309–310, 317, 323
P. multistriata, 324
P. multiseries, 307, 309, 312, 315
P. prolongatoides, 309–310, 317, 323
P. pseudodelicatissima, 307, 309–310, 317, 321, 323
P. pungens, 308–309, 312, 315
P. pungens f. multiseries, 312
P. pungiformis, 309, 313, 315
P. seriata, 306, 308–310, 312
P. seriata f. obtusa, 311, 315
P. seriata f. seriata, 315
P. seriata var. fraudulenta, 305, 313
P. sicula,* 328
P. sicula var. bicuneata, 305, 327
P. sicula var. migrans, 305, 327
P. subcurvata, 309–310, 323
P. subfraudulenta,* 308–309, 315
P. subpacifica, 308–309, 315
P. turgidula, 308, 310, 319, 323
P. turgiduloides,* 309–310, 323

Pseudopodosiraceae,* 88
Pseudosolenia, 25, 141–142, 160
P. calcar-avis, 160
Pyxidicula,* 92
P. aculeata,* 91
P. mediterranea,* 92
Pyxidiculaceae,* 88
Pyxillaceae,* 93

Rhaphoneidaceae, 26, 247, 252
Rhaphoneis, 26, 247, 251
R. amphiceros, 251
R. angustata, 248
R. surirella, 343
R. surirelloides, 249
R. ?(Raphoneis) Diatoma? bicuneata, 327
Rhizomonas
R. setigera, 94
Rhizosolenia, 6, 25, 27, 141, 144, 338
R. acicularis, 144, 147
R. acuminata, 146, 153, 155
R. alata, 159
R. antennata f. antennata, 144, 147, 150
R. antennata f. semispina, 149–150
R. arafurensis,* 159
R. bergonii, 146, 155
R. bidens, 147
R. borealis, 144, 147
R. calcar-avis, 160
R. castracanei, 146, 155
R. castracanei var. castracanei, 151
R. castracanei var. neglecta, 151
R. clevei, 146, 150, 155
R. clevei var. clevei, 151
R. clevei var. communis, 151
R. crassa, 146, 150–151, 155
R. curva, 147
R. curvata, 144, 147
R. cylindrus, 161, 340
R. debyana, 146, 151, 155
R. decipiens, 146, 156
R. delicatula, 161, 340
R. fallax, 146, 156
R. (?) flaccida, 161
R. formosa, 144, 146
R. fragilissima, 167, 341
R. hebetata, 141
R. hebetata f. bidens, 147
R. hebetata f. hebetata, 149
R. hebetata f. hiemalis, 149

R. *hebetata* f. *semispina*, 149, 150
R. *hyalina*, 146, 151, 155
R. *imbricata*, 146, 155–156
R. *imbricata* var. *shrubsolei*, 155
R. *indica*,* 159
R. *inermis*, 159
R. *minima*, 224
R. *ostenfeldii*, 146, 155
R. *pellucida*, 151
R. *polydactyla*, 144
R. *polydactyla* f. *polydactyla*, 150
R. *polydactyla* f. *squamosa*, 150
R. *phuketensis*, 167, 341
R. *pungens*, 146, 157
R. *robusta*, 146, 158–159
R. *semispina*, 149
R. *setigera*, 146, 157
R. *shrubsolei*, 155
R. *sima* f. *sima*, 144, 150
R. *sima* f. *silicea*, 150
R. *simplex*, 146, 155
R. *stolterfothii*, 163, 340
R. *striata*, 146, 156
R. *styliformis*, 144, 146–147
R. *styliformis* f. *latissima*, 146
R. *styliformis* var. *longispina*, 146
R. *styliformis* var. *oceanica*, 147, 150
R. *temperei*, 146, 151, 155
R. *temperei* var. *acuminata*, 153
R. *tenuijuncta*, 167
R. *truncata*, 159
R. *tubiformis*, 163
Rhizosoleniaceae, 25, 116, 141, 142

Rhizosoleniales,* 93, 141

Rhizosoleniineae, 25, 141

Rhizosoleniophycidae,* 141
Roperia, 25, 116–117, 130
R. *tesselata*, 130

Sargassum,*252
Schroederella
S. *delicatula*, 34
Skeletonema, 10, 23, 31, 43, 338
S. *barbadense*,* 43–44
S. *costatum*, 44–45, 186
S. *menzelii*, 44–45
S. *pseudocostatum*, 45
S. *subsalsum*, 44–45
S. *tropicum*, 44
Skeletonemataceae,* 30

Skeletonemopsis, 44
Solenicola setigera, 44
Spatangidium,* 140
S. *arachne*, 137, 140
S. *flabellatum*, 135
S. *heptactis*, 137
Spinigera
S. *bacillaris*, 262
Stauroneis
S. *granii*, 275
S. *membranacea*, 273
S. *septentrionalis*, 275
Stauropsis, 273, 284
S. *granii*, 275
S. *membranacea*, 273
S. *pelagica*, 275
S. *vanhoeffenii*
Stellarima, 11, 25, 113
S. *microtrias*, 113, 115
S. *stellaris*, 113, 115
Stellarimaceae, 25, 113, 115–116

Stephanodiscaceae,* 30
Stephanopyxis, 20
S. *nipponica*, 92–93
S. *palmeriana*, 92–93
S. *turris*, 45, 92–93
Stephanopyxidaceae,* 88

Stictodiscoideae,* 97
Streptotheca, 225–226
S. *tamesis*, 234
Streptothecaceae,* 226
Striatella, 26, 241, 244
S. *unipunctata*, 244
Symbolophora
S.? *microtrias*, 113
S. *furcata*, 113
Symbolophoraceae,* 116
Syndendrium
S. *diadema*, 220
Synedra
S. *doliolus*, 303
S. *gaussii*, 328
S. *hennedyana*, 252
S. *hyperborea*, 246, 333
S. *hyperborea* var. *flexosa*, 246
S. *hyperborea* var. *rostellata*, 246
S. *indica*, 251
S. *nitzschioides*, 257
S. *pelagica*, 267
S. *reinboldii*, 267

S. *sicula*, 327
S. *thalassiothrix*, 263
S. *undulata*, 252
Synedropsis, 26, 241, 246
S. *hyperborea*, 241, 246–247
S. *hyperboreoides*,* 247
S. *recta*,* 247
Syringidium
S. *bicorne*, 171

Tangentales,* 45
Terebraria
T. *kerguelensis*, 296
Tessella
T. *interrupta*,* 244
Thalassionema, 253, 256, 338
T. *bacillare*, 262, 267
T. *frauenfeldii*, 262, 267
T. *javanicum*, 262, 340
T. *nitzschioides*, 257, 262, 311
T. *pseudonitzschioides*, 262, 340
Thalassionemataceae, 26, 253
Thalassiosira, 6–7, 11, 23, 26–27, 31, 45, 98, 334, 338
T. *aestivalis*, 48, 56, 59
T. *allenii*, 48, 51, 59
T. *angstii*, 58
T. *angulata*, 48, 51, 57, 59, 61, 70, 73
T. *anguste-lineata*, 50, 71, 75
T. *antarctica*, 11, 50, 69
T. *antarctica* var. *antarctica*, 66, 71
T. *antarctica* var. *borealis*, 66, 68, 71
T. *australiensis*,* 74
T. *australis*, 11, 13, 50, 71, 75
T. *baltica*, 68, 71
T. *binata*,* 48, 50, 56, 59, 65
T. *bioculata*, 50, 80, 82, 84, 86
T. *bipartita*, 41
T. *blanda*, 40
T. *bramaputrae*,* 74
T. *bulbosa*, 48, 52–53, 59
T. *chilensis*,* 61
T. *condensata*, 34, 36
T. *conferta*, 48, 54, 56, 59, 65
T. *confusa*,* 50, 87–88
T. *constricta*,* 50, 69–71
T. *coronata*, 59
T. *curviseriata*, 50, 63, 65
T. *decipiens*,* 48
T. *decipiens*, 8, 48, 51
T. *delicatula*, 50, 59, 63, 83
T. *dichotomica*, 50, 63

T. *diporocyclus*, 50, 56, 61, 63
T. *eccentrica*, 50, 52, 62–63
T. *endoseriata*, 50, 86–87
T. *expecta*, 83
T. *fallax*, 66
T. *ferelineata*,* 48, 59
T. *floridana*, 65
T. *fluviatilis*, 75
T. *fragilis*,* 50, 56, 63
T. *frenguellii*,* 50, 87
T. *gerloffii*,* 50, 70–71, 73
T. *gessneri*,* 74
T. *gracilis*, 50, 82, 86
T. *gracilis* var. *gracilis*, 83
T. *gracilis* var. *expecta*, 83
T. *gravida*, 50, 68–70
T. *guillardii*, 50, 77
T. *hendeyi*, 50, 61–62
T. *hibernalis*,* 59
T. *hispida*,* 48, 56, 59
T. *hyalina*, 50, 69–71
T. *hyperborea*, 50, 74–75
T. *hyperborea* var. *hyperborea*,* 74
T. *hyperborea* var. *pelagica*,* 74
T. *karenae*,* 50, 69, 71
T. *kushirensis*,* 50, 73, 75
T. *lacustris*,* 74
T. *latimarginata*,* 87
T. *lentiginosa*, 50, 79–80
T. *leptopus*, 50, 75
T. *licea*,* 48, 58–59
T. *lineata*, 50, 80
T. *lineoides*,* 50, 80
T. *lundiana*,* 58
T. *mala*, 48, 54, 59
T. *mediterranea*, 50, 74–75
T. *mendiolana*, 50, 62, 63
T. *minima*, 50, 65
T. *minuscula*, 48, 54, 56, 58–59
T. *monoporocyclus*, 54
T. *nordenskioeldii*, 11, 45, 48, 51, 56, 59, 65
T. *oceanica*, 48, 57, 59, 79
T. *oestrupii*, 50, 83, 86
T. *oestrupii* var. *oestrupii*,* 83
T. *oestrupii* var. *venrickae*,* 83
T. *pacifica*, 48, 57, 59, 61, 70
T. *partheneia*, 48, 57, 59
T. *perpusilla*, 50, 83–84, 86

T. polychorda, 71
T. poro-irregulata, 50, 88
T. poroseriata, 50, 86–87
T. proschkinae, 50, 84, 86
T. proschkinae var. *spinulata,** 84
T. pseudonana, 50, 79
T. punctigera, 48, 58–59
T. ritscheri, 50, 86, 88
T. rosulata,* 50, 84, 86
T. rotula, 10, 50, 70–71
T. simonsenii, 41, 50, 63
T. spinulata,* 84
T. stellaris, 74
T. subtilis, 48, 56, 58–59
T. tealata,* 50, 65
T. tenera, 48, 59
T. trifulta, 50, 87–88
T. tumida, 50, 70–71
T. weissflogii, 50, 74–75
Thalassiosiraceae, 23, 74–75
Thalassiosirales,* 30
Thalassiosirophycidae,* 30
Thalassiothrix, 26, 253, 263, 267, 332, 338
T. acuta,* 263
T. antarctica, 264
T. delicatula, 340
T. elongata, 254, 257, 340
T. (fauenfeldii var.) *javanica,* 340
T. gibberula, 266–267, 339
T. heteromorpha,* 263
T. heteromorpha var. *mediterranea,** 263
T. lanceolata,* 266
T. longissima, 263, 266
T. mediterranea,* 263

T. mediterranea var. *pacifica,* 254, 340
T. nitzschioides, 257
T. pseudonitzschioides, 262, 340
T. vanhoeffenii,* 254, 257
Thalassotrix
Toxariaceae, 252
Toxarium, 26
T. hennedyanum, 252–253
T. undulatum, 252–253
Triceratiaceae,* 236
Triceratiales,* 236
Triceratium, 231
T. brightwellii, 231
T. intricatum, 234
T. malleus, 227
T. sol, 231
Trichotoxon, 26, 253, 267
T. reinboldii, 267
Tripodiscus
T. argus, 141
Tropidoneis, 271–272, 283, 284
T. antarctica, 287
T. belgicae, 286
T. fusiformis, 287
T. glacialis, 289
T. gaussii, 289
Tryblionella,* 329
Urosolenia, 141
Vibrio ostrearius, 275
Vibrio paxillifer, 293
Xanthophyceae,* 6
Zygoceros
Z. (*Denticella?*) *mobiliensis,* 239
Z. *surirella,* 249

REFERENCES

Agardh, C. A. 1832. "Conspectus criticus diatomacearum," Vol. 4, pp. 49–66. Berlingianiis, Lund.

Akiba, F., & Yanagisawa, Y. 1986. 7. Taxonomy, morphology and phylogeny of the Neogene diatom zonal marker species in the middle-to-high latitudes of the North Pacific. *In* "Initial Reports of the Deep Sea Drilling Project" (H. Kagami, D. E. Karig, W. T. Coulborn, *et al.,* eds.), Vol. 87, pp. 483–554.

Allen, E. J., & Nelson, E. W. 1910. On the artificial culture of marine plankton. *Journal of the Marine Biological Association of the United Kingdom* 8:421–474.

Allen, W. E., & Cupp, E. E. 1935. Plankton diatoms of the Java Sea. *Annales du Jardin Botanique de Buitenzorg* 44(2):101–174.

Andersen, R. A., Medlin, L. K., & Crawford, R. M. 1986. An investigation of the cell wall components of *Actinocyclus subtilis* (Bacillariophyceae). *Journal of Phycology* 22:466–479.

Andrews, G. W. 1977. Morphology and stratigraphic significance of *Delphineis*, a new marine diatom genus. *Beiheft zur Nova Hedwigia* 54:243–260.

Andrews, G. W. 1981. Revision of the diatom genus *Delphineis* and morphology of *Delphineis surirella* (Ehrenberg) G. W. Andrews, n. comb. *In* "Proceedings of the 6th Symposium on Recent and Fossil Diatoms" (R. Ross, ed.), pp. 81–90. Koeltz, Koenigstein.

Andrews, G. W., & Rivera, P. 1987. Morphology and evolutionary significance of *Adoneis pacifica* gen. et sp. nov. *Diatom Research* 2:1–14.

Anonymous, 1975. Proposals for a standardization of diatom terminology and diagnoses. *Beiheft zur Nova Hedwigia* 53:323–354.

Bailey, J. W. 1851. Microscopical observations made in South Carolina, Georgia and Florida. *Smithsonian Contributions to Knowledge* 2(8):1–48.

Bailey, J. W. 1854. Notes of new American species and localities of microscopical organisms. *Smithsonian Contributions to Knowledge* 7(3):1–16.

Bailey, J. W. 1856. Notice of microscopic forms found in the soundings of the Sea of Kamtschatka. *The American Journal of Science and Arts Second Series* 22:1–6.

Bailey, L. W. 1861. Notes on new species of microscopical organisms, chiefly from the Para River, South America. *Boston Journal of Natural History* 7(3):329–352.

Barber, H. G., & Haworth, E. Y. 1981. A guide to the morphology of the diatom frustule. *Freshwater Biological Association Scientific Publication* 44:1–112.

Bates, S. S., Bird, C. J., Freitas, A. S. W. De, Foxall, R., Gilgan, M., Hanic, L. A., Johnson, G. R., McCulloch, A. W., Odense, P., Pocklington, R., Quilliam, M. A., Sim, P. G., Smith, J. C., Subba Rao, D. V., Todd, E. C. D., Walter, J. A., & Wright, J. L. C. 1989. Pennate diatom *Nitzschia pungens* as the primary source of domoic acid, a toxin in shellfish from Eastern Prince Edward Island, Canada. *Canadian Journal of Fisheries and Aquatic Sciences* 46:1203–1215.

Bergon, P. 1903. Études sur la flora diatomique du bassin d'Arcachon et des parages de l'Atlantique voisins des cette station. *Bulletin de la Société Scientifique d'Arcachon* 6:39–112.

Bethge, H. 1928. Über die Kieselalge *Sceletonema subsalsum*. *Berichte der deutschen botanischen Gesellschaft* 46:340–347.

Boalch, G. T. 1971. The typification of the diatom species *Coscinodiscus concinnus* Wm. Smith and *Coscinodiscus granii* Gough. *Journal of the Marine Biological Association of the United Kingdom* 51:685–695.

Boalch, G. T. 1975. The Lauder species of the diatom genus *Bacteriastrum*. *Beiheft zur Nova Hedwigia* 53:185–189.

Boalch, G. T., & Harbour, D. S. 1977. Observations on the structure of a planktonic *Pleurosigma*. *Beiheft zur Nova Hedwigia* 54:275–280.

Boden, B. P. 1950. Some marine diatoms from the west coast of South Africa. *Transactions of the Royal Society of South Africa* 32:321–434.

Bohlin, K. 1897. Zur Morphologie und Biologie einzelliger Algen. *Öfversigt af Kongliga Vetenskaps-Akademiens Förhandlingar 1897* 54(9):507–529.

Booth, B. C., Lewin, J., & Norris, R. E. 1982. Nanoplankton species predominant in the subarctic Pacific in May and June 1978. *Deep-Sea Research* 29:185–200.

Borgert, A. 1908. Bericht über eine Reise nach Ostafrika und dem Victoria Nyansa nebst Bemerkungen über einen kurzen Aufenthalt auf Ceylon. *Sitzungsberichte der naturwissenschaftlichen Abteilung der Niederrheinischen Gesellschaft für Natur- und Heilkunde zu Bonn 1907*:12–33.

Boyer, C. S. 1927. Synopsis of North American Diatomaceae. *Proceedings of the Academy of Natural Sciences of Philadelphia* 78 (1926 Supplement):1–228.

Braarud, T. 1935. The "Øst" Expedition to the Denmark Strait 1929. II. The phytoplankton and its conditions of growth. *Hvalrådets Skrifter* 10:1–173.

Brightwell, T. 1856. On the filamentous, long-horned diatomacæ, with a description of two new species. *Quarterly Journal of Microscopical Science* 4:105–109.

Brightwell, T. 1858a. Remarks on the genus "Rhizosolenia" of Ehrenberg. *Quarterly Journal of Microscopical Science* 6:93–95.

Brightwell, T. 1858b. Further observations on the genera *Triceratium* and *Chaetoceros*. *Quarterly Journal of Microscopical Science* 6:153–155.

Brightwell, T. 1860. On some of the rarer or undescribed species of diatomacæ. Part II. *Quarterly Journal of Microscopical Science* 8:93–96.

Brooks, M. 1975. Studies on the genus *Coscinodiscus* I–III. *Botanica Marina* 18:1–13, 15–27, 29–39.

Brown, N. E. 1920. New and old diatoms from the Antarctic regions. *English Mechanic and World of Science* 111:210–211, 219–220, 232–233.

Brun, J. 1891. Diatomées espèces nouvelles marines, fossiles ou pélagiques. *Mémoires de la Société de physique et d'histoire naturelle de Genève* 31, *Seconde partie* 1:1–47.

Buck, K. R., Uttal-Cooke, L., Pilskaln, C. H., Roelke, D. L., Villac, M. C., Fryxell, G. A., Cifuentes, L., & Chaves, F. P. 1992. Autecology of the diatom *Pseudo-nitzschia australis*, a domoic acid producer, from Monterey Bay, California. *Marine Ecology Progress Series* 84: 293–302.

Cardinal, A., Poulin, M., & Bérard-Therriault, T. 1989. New criteria for species characterization in the genera *Donkinia*, *Gyrosigma* and *Pleurosigma* (Naviculaceae, Bacillariophyceae). *Phycologia* 28:15–27.

Castracane, F. 1875. Contribuzione alla florula delle diatomee del Mediterraneo ossia esame del contenuto nello stomaco di una Salpa pinnata, pescata a Messina. *Atti dell'Accademia Pontificia de'Nuovi Lincei* 28:377–396.

Castracane, F. 1886. Report on the diatomaceæ collected by H.M.S. Challenger during the years 1873–1876. Report on the scientific Results of the voyage of H.M.S. Challenger during the Years 1873–76. *Botany* 2(4):I–III, 1–178.

Chang, Tsang-Pi, & Steinberg, C. 1989. Identifizierung von nanoplanktischen Kieselalgen (Centrales, Bacillariophyceae) in der Rott und im Rott-Stausee (Bayern, Bundesrepublik Deutschland). *Archiv für Protistenkunde* 137:111–129.

Cleve, P. T. 1873a. Examination of diatoms found on the surface of the Sea of Java. *Bihang till Kongliga Svenska Vetenskaps-Akademiens Handlingar* 1(11):1–13.

Cleve, P. T. 1873b. On diatoms from the arctic Sea. *Bihang till Kongliga Svenska Vetenskaps-Akademiens Handlingar* 1(13):1–28.

Cleve, P. T. 1881. On some new and little known diatoms. *Kongliga Svenska Vetenskaps-Akademiens Handlingar* 18(5):1–28.

Cleve, P. T. 1883. Diatoms collected during the expedition of the Vega. *Vega-Expeditionens Vetenskapliga Iakttagelser* 3:455–517.

Cleve, P. T. 1889. Pelagiske diatomeer från Kattegat. *Det évidenskabelige Udbytte af Kanonbaaden "Hauchs" Togter i de danske Have indenfor Skagen i Aarene 1883–86*:53–56.

Cleve, P. T. 1894a. Planktonundersökningar. *Bihang till Kongliga Svenska Vetenskaps-Akademiens Handlingar* 20(3,2):1–16.

Cleve, P. T. 1894b. Synopsis of the naviculoid diatoms. *Kongliga Svenska Vetenskaps-Akademiens Handlingar* 26(2):1–194.

Cleve, P. T. 1896a. Diatoms from Baffins Bay and Davis Strait. *Bihang till Kongliga Svenska Vetenskaps-Akademiens Handlingar* 22(3,4):1–22.

Cleve, P. T. 1896b. Redogörelse för de svenska hydrografiske undersökningarne februari 1896. V. Planktonundersökningar: Vegetabiliskt plankton. *Bihang till Kongliga Svenska Vetenskaps-Akademiens Handlingar* 22(5):1–33.

Cleve, P. T. 1897a. "A treatise on the phytoplankton of the *Atlantic and its tributaries*," pp. 28. Upsala, Sweden.

Cleve, P. T. 1897b. Report on the phytoplankton collected on the expedition of H.M. S. "Research" 1896. *Fifteenth Annual Report of the Fishery Board for Scotland* 3:296–304.

Cleve, P. T. 1900a. Notes on some Atlantic plankton-organisms. *Kongliga Svenska Vetenskaps-Akademiens Handlingar*. 34(1):1–22.

Cleve, P. T. 1900b. The plankton of the North Sea, the English Channel, and the Skagerak in 1898. *Kongliga Svenska Vetenskaps-Akademiens Handlingar* 32(8):1–53.

Cleve, P. T. 1900c. The seasonal distribution of Atlantic plankton organisms. *Göteborgs Kungliga Vetenskaps- och Vitterhets-Samhälles Handlingar Fjärde följden* 3:1–369.

Cleve, P. T. 1901a. Plankton from the Indian Ocean and the Malay Archipelago. *Kongliga Svenska Vetenskaps-Akademiens Handlingar* 35(5):11–58.

Cleve, P. T. 1901b. Plankton from the southern Atlantic and the southern Indian ocean. *Öfversigt af Kongliga Vetenskaps-Akademiens Förhandlingar 1900* 57:919–938.

Cleve, P. T. 1904. Plankton table for the North Sea. Bulletin, Part D, May 1904. *Conseil Permanent pour l'Exploration de la Mer, Bulletin des Résultats aquit pendant les courses périodiques* 1903–1904:216.

Cleve, P. T., & Grunow, A. 1880. Beiträge zur Kenntniss der arctischen Diatomeen. *Kongliga Svenska Vetenskaps-Akademiens Handlingar* 17(2):1–121.

Cleve, P. T., & Möller, J. D. 1878. "Diatoms," Part 3, Nos. 109–168, pp. 1–9. Uppsala, Sweden.

Cleve, P. T., & Möller, J. D. 1879. "Diatoms," Part 4, Nos. 169–216, pp. 1–7. Uppsala, Sweden.

Cleve, P. T., & Möller, J. D. 1882. "Diatoms," Part 6, Nos. 277–324, pp. 1–6. Uppsala, Sweden.

Cleve-Euler, A. 1912. Das Bacillariaceen-Plankton in Gewässern bei Stockholm III. Über Gemeinden des schwach salzigen Wassers und eine neue Charakterart derselben. *Archiv für Hydrobiologie und Planktonkunde* 7:500–514.

Cleve-Euler, A. 1937. Undersökningar över Öresund. 24. Sundets plankton. 1. Sammansättning och fördelning. *Lunds Universitets Årsskrift N. F. Avdelning 2* 33(9):1–50.

Cleve-Euler, A. 1952. Die Diatomeen von Schweden und Finnland. *Kungliga Svenska Vetenskapsakademiens Handlingar Fjärde Serien* 3(3):1–153.

Cohen, A. L. 1974. Critical point drying. *In* "Principles and Techniques of Scanning Electron Microscopy 1" (A. M. Hyat, ed.), pp. 44–112. Van Nostrand–Reinhold, New York.

Collier, A., & Murphy, A. 1962. Very small diatoms: Preliminary notes and description of Chaetoçeros galvestonensis. *Science* 136:780–781.

Comber, T. 1896. On the occurrence of endocysts in the genus *Thalassiosira*. *Journal of the Royal Microscopical Society* 9:489–491.

Cooper, I. C. G. 1958. A new diatom from Fort Myers, Florida, U.S.A. *Revue Algologique* 2:125–128.

Cox, E. J. 1979. Taxonomic studies on the diatom genus *Navicula* Bory: The typification of the genus. *Bacillaria* 2:137–153.

Cox, E. J. 1981. The use of chloroplasts and other features of the living cell in the taxonomy of naviculoid diatoms. *In* "Proceedings of the 6th Symposium on Recent and Fossil Diatoms" (R. Ross, ed.), pp. 115–133. Koeltz, Koenigstein.

Cox, E. J., & Ross, R. 1981. The striae of pennate diatoms. *In* "Proceedings of the 6th Symposium on Recent and Fossil Diatoms" (R. Ross, ed.), pp. 267–278. Koeltz, Koenigstein.

Crawford, R. M. 1975. The taxonomy and classification of the diatom genus *Melosira* C. Ag. 1. The type species *M. nummuloides* C. Ag. *British Phycological Journal* 10:323–338.

Crawford, R. M. 1979. Taxonomy and frustular structure of the marine centric diatom *Paralia sulcata. Journal of Phycology* 15:200–210.

Crawford, R. M. 1988. A reconsideration of *Melosira arenaria* and *M. teres;* Resulting in a proposed new genus *Ellerbeckia. In* "Algae and the Aquatic Environment" (F. E. Round, ed.), pp. 413–433. Biopress, Bristol.

Crawford, R. M., Gardner, C., & Medlin, L. K. 1994. The genus *Attheya.* I. A description of four taxa, and the transfer of *Gonioceros septentrionalis* and *G. armatus. Diatom Research* 9:27–51.

Crawford, R. M., Sims, P. A., & Hajos, M. 1990. The morphology and taxonomy of the centric diatom genus *Paralia. Paralia siberica* comb. nov. *Diatom Research* 5:241–251.

Cupp, E. E. 1943. Marine plankton diatoms of the west coast of North America. *Bulletin of the Scripps Institution of Oceanography of the University of California* 5:1–238.

de Brébisson, A. 1857. Refermant la description de quelques nouvelles Diatomées dans le guano de Pérou, et formant le genre *Spatangidium. Bulletin de la Société Linnéenne de Normandie Caen Série 1* 2:292–298.

de Brébisson, A., & Godey, A. 1838. "Consideration sur les diatomées et essai d'une classification des genre et des espéces appartenant á cette famille," pp. 20, Brée L'Ainé, Falaise.

Delgado, M., & Fortuño, J.-M. 1991. Atlas de fitoplancton del Mar Mediterráneo. *Scienta Marina* 55 (Suppl. 1):1–133.

Desikachary, T. V. 1986–1989. "Atlas *of diatoms 1–6*," 809 plates. Madras Science Foundation, Madras.

Dickie, G. 1852. Notes on the algae. *In* "Journal of a Voyage in Baffin's Bay and Barrow Straits in the Years 1850–1851 2." (P. C. Sutherland, ed.) pp. cxci–cc. Longman, Brown, Green, and Longmans, London.

Drebes, G. 1972. The life history of the centric diatom *Bacteriastrum hyalinum* Lauder. *Beiheft zur Nova Hedwigia* 39:95–110.

Drebes, G. 1974. "Marines Phytoplankton. Eine Auswahl der Helgoländer Planktonalgen (Diatomeen, Peridineen)," pp. 186. Georg Thieme, Stuttgart.

Drebes, G. 1977. Sexuality. *Blackwell Scientific Publications, Botanical Monographs* 13:250–283.

Drum, R. W., & Pankratz, H. S. 1966. Locomotion and raphe structure of the diatom *Bacillaria. Nova Hedwigia* 10:315–317.

Duke, E. L., Lewin, J., & Reimann, B. E. F. 1973. Light and electron microscope studies of diatom species belonging to the genus *Chaetoceros* Ehrenberg. I. *Chaetoceros septentrionale* Oestrup. *Phycologia* 12:1–9.

Ehrenberg, C. G. 1838. "Die Infusionsthierchen als vollkommene Organismen." pp. i–xvii, 548. Leopold Voss, Leipzig.

Ehrenberg, C. G. 1839. Über die Bildung der Kreidefelsen und des Kreidemergels durch unsichtbare Organismen. *Abhandlungen der königlichen Akademie der Wissenschaften zu Berlin* 1838:59–147.

Ehrenberg, C. G. 1841a. Über noch jetzt zahlreich lebende Thierarten der Kreidebildung und den Organismus der Polythalamien. *Abhandlungen der königlichen Akademie der Wissenschaften zu Berlin* 1839:81–174.

Ehrenberg, C. G. 1841b. "Characteristic von 274 neuen Arten von Infusorien" (the article has no title.) *Deutsche Akademie der Wissenschaften zu Berlin* 1840:197–219.

Ehrenberg, C. G. 1843. Verbreitung und Einfluss des mikroskopischen Lebens in Süd- und Nord-Amerika. *Abhandlungen der königlichen Akademie der Wissenschaften zu Berlin* 1841:291–445.

Ehrenberg, C. G. 1844a. 2 neue Lager von Gebirgsmassen aus Infusorien als Meeres-Absatz in Nord-Amerika und eine Vergleichung derselben mit den organischen Kreide-Gebilden in

Europa und Afrika. *Deutsche Akademie der Wissenschaften zu Berlin, Berichte* Februar 1844:57–97.

Ehrenberg, C. G. 1844b. Einige vorläufige Resultate seiner Untersuchungen der ihm von der Südpolreise des Capitain Ross. *Deutsche Akademie der Wissenschaften zu Berlin, Berichte* Mai 1844:182–207.

Ehrenberg, C. G. 1844c. Über das kleinste Leben in den Hochgebirgen von Armenien und Kurdistan. *Deutsche Akademie der Wissenschaften zu Berlin, Berichte* Juni 1844:252–275.

Ehrenberg, C. G. 1845a. Novorum generum et specierum brevis definitio. *Deutsche Akademie der Wissenschaften zu Berlin, Berichte* November 1845:357–377.

Ehrenberg, C. G. 1845b. Neue Untersuchungen über das kleinste Leben als geologisches Moment. Mit kurzer Charakteristik von 10 neuen Genera und 66 neuen Arten. *Deutsche Akademie der Wissenschaften zu Berlin, Berichte* Februar 1845:53–88.

Ehrenberg, C. G. 1854. "Mikrogeologie," pp. 374 (atlas, pp. 88; 40 plates). Leopold Voss, Leipzig.

Estep, K. W., Rey, F., Bjørklund, K., Dale, T., Heimdal, B. R., Van Hertum, A. J. W., Hill, D., Hodell, D., Syvertsen, E. E., Tangen, K., & Throndsen, J. 1992. *Deus creavit; linneaus disposuit:* An international effort to create a catalogue and expert system for the identification of protistan species. *Sarsia* 77:275–285.

Evensen, D. L., & Hasle, G. R. 1975. The morphology of some *Chaetoceros* (Bacillariophyceae) species as seen in the electron microscope. *Beiheft zur Nova Hedwigia* 53:153–174.

Farr, E. R., Leussink, J. A., & Stafleu, F. A. 1979. Index Nominum Genericorum (Plantarum) 1–3 *Regnum vegetabile* 102:I–XXVI, 1–1896.

Farr, E. R., Leussink, J. A., & Zijlstra, G. 1986. Index Nominum Genericorum (Plantarum) Supplementum I. *Regnum vegetabile* 113:VII–XIII, 1–126.

Feibicke, M., Wendker, S., & Geissler, U. 1990. *Thalassiosira proschkinae* Makarova—A contribution to its morphology and autecology. *Beiheft zur Nova Hedwigia* 100:155–169.

Ferrario, M. E. 1988. Ultrastructure de deux taxa de la famille Thalassiosiraceae. *Cryptogamie, Algologie* 9:311–318.

Florin, M.-B. 1970. The fine structure of some pelagic fresh water diatom species under the scanning electron microscope. I. *Svensk Botanisk Tidsskrift* 64:51–64.

French, F. W., III, & Hargraves, P. E. 1985. Spore formation in the life cycles of the diatoms *Chaetoceros diadema* and *Leptocylindrus danicus*. *Journal of Phycology* 21:477–483.

French, F. W., III, & Hargraves, P. E. 1986. Population dynamics of the spore-forming diatom *Leptocylindrus danicus* in Narragansett Bay, Rhode Island. *Journal of Phycology* 22:411–420.

Frenguelli, J. 1939. Diatomeas del Golfo de San Matias. *Revista del Museo de La Plata* 2:201–226.

Fryxell, G. A. 1975. Morphology, taxonomy, and distribution of selected diatom species of *Thalassiosira* Cleve in the Gulf of Mexico and Antarctic waters, pp. 189. Dissertation for the degree of Doctor of Philosophy, Texas A&M University. College Station, TX.

Fryxell, G. A. 1976. The position of the labiate process in the diatom genus *Skeletonema*. *British Phycolgical Journal* 11:93–99.

Fryxell, G. A. 1977. *Thalassiosira australis* Peragallo and *T. lentiginosa* (Janisch) G. Fryxell, comb.nov.: Two antarctic diatoms (Bacillariophyceae). *Phycologia* 16:95–104.

Fryxell, G. A. 1978a. The diatom genus *Thalassiosira*: *T. licea* sp. nov. and *T. angstii* (Gran) Makarova, species with occluded processes. *Botanica Marina* 21:131–141.

Fryxell, G. A. 1978b. Proposal for the conservation of the diatom *Coscinodiscus argus* Ehrenberg as the type of the genus. *Taxon* 27:122–125.

Fryxell, G. A. 1978c. Chain-forming diatoms: Three species of Chaetoceraceae. *Journal of Phycology* 14:62–71.

Fryxell, G. A. 1983. New evolutionary patterns in diatoms. *BioScience* 33:92–98.

Fryxell, G. A. 1988. Polymorphism in relation to environmental conditions as exemplified by clonal cultures of *Thalassiosira tumida* (Janisch) Hasle. *In* "Proceedings of the 9th International Diatom Symposium" (F. E. Round, ed.) pp. 61–73. Biopress, Bristol, and Koeltz Scientific Books, Koenigstein.

Fryxell, G. A. 1989. Marine phytoplankton at the Weddell Sea ice edge: Seasonal changes at the specific level. *Polar Biology* 10:1–18.

Fryxell, G. A. 1990. Family Hemidiscaceae: The genera *Actinocyclus* and *Azpeitia*. *In* "Polar Marine Diatoms" (L. K. Medlin & J. Priddle, eds.), pp. 111–113. British Antarctic Survey, Cambridge.

Fryxell, G. A., & Ashworth, T. K. 1988. The diatom genus *Coscinodiscus* Ehrenberg: Characters having taxonomic value. *Botanica Marina* 31:359–374.

Fryxell, G. A., & Hasle, G. R. 1971. *Corethron criophilum* Castracane: Its distribution and structure. *Antarctic Research Series 17, Biology of the Antarctic Seas* 4:335–346.

Fryxell, G. A., & Hasle, G. R. 1972. *Thalassiosira eccentrica* (Ehrenb.) Cleve, *T. symmetrica* sp. nov., and some related centric diatoms. *Journal of Phycology* 8:297–317.

Fryxell, G. A., & Hasle, G. R. 1977. The genus *Thalassiosira*: Some species with a modified ring of central strutted processes. *Beiheft zur Nova Hedwigia* 54:67–89.

Fryxell, G. A., & Hasle, G. R. 1979a. The genus *Thalassiosira*: Species with internal extensions of the strutted processes. *Phycologia* 18:378–393.

Fryxell, G. A., & Hasle, G. R. 1979b. The genus *Thalassiosira*: *T. trifulta* sp. nova and other species with tricolumnar supports on strutted processes. *Beiheft zur Nova Hedwigia* 64:13–32.

Fryxell, G. A., & Hasle, G. R. 1980. The marine diatom *Thalassiosira oestrupii*: Structure, taxonomy and distribution. *American Journal of Botany* 67:804–814.

Fryxell, G. A., & Hasle, G. R. 1983. The antarctic diatoms *Thalassiosira dichotomica* (Kozlova) comb.nov. and *T. ambigua* Kozlova. *Polar Biology* 2:53–62.

Fryxell, G. A., & Johansen, J. R. 1990. Family Thalassiosiraceae. Section 2: The genus *Thalassiosira* from the Antarctic. *In* "Polar Marine Diatoms" (L. K. Medlin & J. Priddle, eds.) pp. 98–103. British Antarctic Survey, Cambridge.

Fryxell, G. A., & Medlin, L. K. 1981. Chain forming diatoms: Evidence of parallel evolution in *Chaetoceros*. *Cryptogamie, Algologie* 2:3–29.

Fryxell, G. A., & Miller, W. I., III 1978. Chain-forming diatoms: Three araphid species. *Bacillaria* 1:113–136.

Fryxell, G. A., & Prasad, A. K. S. K. 1990. *Eucampia antarctica* var. *recta* (Mangin) stat.nov. (Biddulphiaceae, Bacillariophyceae): Life stages at the Weddell Sea ice edge. *Phycologia* 29:27–38.

Fryxell, G. A., & Semina, H. J. 1981. *Actinocyclus exiguus* sp. nov. from the southern parts of the Indian and Atlantic Oceans. *British Phycological Journal* 16:441–448.

Fryxell, G. A., Doucette, G. J., & Hubbard, G. F. 1981. The genus *Thalassiosira*: The bipolar diatom *T. antarctica* Comber. *Botanica Marina* 24:321–335.

Fryxell, G. A., Gould, R. W., & Watkins, T. P. 1984. Gelatinous colonies of the diatom *Thalassiosira* in Gulf Stream warm core rings including *T. fragilis*, sp. nov. *British Phycological Journal* 19:141–156.

Fryxell, G. A., Hasle, G. R., & Carty, S. V. 1986a. *Thalassiosira tumida* (Janisch) Hasle: Observations from field and clonal cultures. *In* "Proceedings of the 8th International Diatom Symposium" (M. Ricard, ed.), pp. 11–21. Koeltz, Koenigstein.

Fryxell, G. A., Sims, P. A., & Watkins, T. P. 1986b. *Azpeitia* (Bacillariophyceae): Related genera and promorphology. *Systematic Botany Monographs* 13:1–74.

Fryxell, G. A., Prasad, A. K. S. K., & Fryxell, P. A. 1989. *Eucampia antarctica* (Castracane) Mangin (Bacillariophyta): Complex nomenclature and taxonomic history. *Taxon* 38: 638–640.

Fryxell, G. A., Reap, M. E., & Valencic, D. L. 1990. *Nitzschia pungens* Grunow f. *multiseries* Hasle: Observations of a known neurotoxic diatom. *Beiheft zur Nova Hedwigia* 100:171–188.

Fryxell, G. A., Garza, S. A., & Roelke, D. L. 1991. Auxospore formation in an Antarctic clone of *Nitzschia subcurvata* Hasle. *Diatom Research* 6:235–245.

Gaarder, K. R. 1951. Bacillariophyceae from the "Michael Sars" North Atlantic Deep-Sea Expedition 1910. *Report on the Scientific Results of the "Michael Sars" North Atlantic Deep-Sea Expedition 1910* 2(2):1–36.

Gardner, C., & Crawford, R. M. 1994. A description of *Plagiogrammopsis mediaequatus* Gardner & Crawford, sp. nov. (Cymatosiraceae, Bacillariophyta) using light and electron microscopy. *Diatom Research* 9:53–63.

Garrison, D. L., Conrad, S. M., Eilers, P. P., & Waldron, M. 1992. Confirmation of domoic acid production by *Pseudo-nitzchia australis* (Bacillariophyceae) cultures. *Journal of Phycology* 28:604–607.

Gaul, U., Geissler, U., Henderson, M., Mahoney, R., & Reimer, C. W. 1993. Bibliography on the fine-structure of diatom frustules (Bacillariophyceae). *Proceedings of The Academy of Natural Sciences of Philadelphia* 144:69–238.

Gersonde, R., & Harwood, D. M. 1990. Lower Cetaceous diatoms from ODP Leg 113 site 693 (Weddell Sea). 1. Vegetative cells. *In* "Proceedings of the Ocean Drilling Program, Scientific Results" (P. F. Baker, J. P. Kennett *et al.*, eds.), vol. 113, pp. 365–402. Ocean Drilling Program, College Station, TX.

Glezer, Z. I. 1983. Taxonomical significance of characters in diatoms in the light of elaboration of a new Bacillariophyta classification. *Botanicheskii Zhurnal* 68:993–1002. [In Russian]

Glezer, S. I., Makarova, I. V. (editor-in-chief), Moisseeva, A. I., & Nikolaev, V. A. 1988. "The Diatoms of the USSR, Fossil and Recent 2(1):Pyxidiculaceae, Thalassiosiropsidaceae, Triceratiaceae, Thalassiosiraceae." *In* "NAUKA," pp. 116, 61 plates. Leningrad. [In Russian]

Gmelin, J. F. 1791. "Systema Naturae," 13th Ed., 1(6):3021–3910. Georg Emanuel Beer, Leipzig.

Gombos, A. M., Jr. 1980. The early history of the diatom family Asterolampraceae. *Bacillaria* 3:227–272.

Gough, L. H. 1905. Report on the plankton of the English Channel in 1903. *North Sea Fisheries Investigations, Committee Report* 2:325–377.

Gran, H. H. 1897a. Bacillariaceen vom kleinen Karajakfjord. *Bibliotheca Botanica* 8(42):13–24.

Gran, H. H. 1897b. Protophyta: Diatomaceæ, Silicoflagellata and Cilioflagellata. *The Norwegian North-Atlantic Expedition 1876–1878, Botany*, 1–36.

Gran, H. H. 1900. Bemerkungen über einige Planktondiatomeen. *Nyt Magazin for Naturvidenskaberne* 38:103–128.

Gran, H. H. 1904. Die Diatomeen der arktischen Meere. I. Die Diatomeen des Planktons. *Fauna Arctica* 3:509–554.

Gran, H. H. 1908. Diatomeen. *Nordisches Plankton* 19:1–146.

Gran, H. H. 1915. The plankton production of the North European waters in the spring of 1912. *Bulletin Planktonique pour l'année 1912*, 1–142.

Gran, H. H., & Angst, E. C. 1931. Plankton diatoms of Puget Sound. *Publications Puget Sound Biological Station* 7:417–519.

Gran, H. H., & Yendo, K. 1914. Japanese diatoms. I. On *Chaetoceras* II. On *Stephanopyxis*. *Videnskapsselskapets Skrifter I. Matematisk-Naturvidenskapelig Klasse* 8:3–29.

Gregory, W. 1857. On new forms of marine Diatomaceæ, found in the Firth of Clyde and in Loch Fine. *Transactions of the Royal Society of Edinburgh* 21:473–542.

Greuter, W., Brummitt, R. K., Farr, E., Kilian, N., Kirk, P. M., & Silva, P. C. 1993. Names in current use for extant plant genera 3. *Regnum Vegetabile* 129:i–xxvii, 1–1464.

Greville, W. 1859a. Descriptions of new species of British Diatomaceæ, chiefly observed by the late Professor Gregory. *Quarterly Journal of Microscopical Science* 7:79–86.

Greville, R. K. 1859b. Descriptions of diatomaceæ observed in Californian guano. *Quarterly Journal of Microscopical Science* 7:155–166.

Greville, R. K. 1860. A monograph of the genus *Asterolampra,* including *Asteromphalus* and *Spatangidium. Transactions of the Microscopical Society of London New Series* 8:102–125.

Greville, R. K. 1862. Descriptions of new and rare diatoms. Series 5. *Transactions of the Microscopical Society of London New Series* 10:18–29.

Greville, R. K. 1864. Descriptions of new species of diatoms from the South Pacific. *Transactions of the Botanical Society of Edinburgh* 8:233–238.

Greville, R. K. 1865a. Descriptions of new and rare diatoms. Series 14. *Transactions of the Microscopical Society of London New Series* 13:1–10.

Greville, R. K. 1865b. Descriptions of new genera and species from Hong Kong. *The Annals and Magazine of Natural History Series 3* 16(91):1–7.

Greville, R. K. 1866. Descriptions of new and rare diatoms. Series 19. *Transactions of the Microscopical Society of London New Series* 14:77–86.

Grunow, A. 1862. Die österreichischen Diatomaceen. *Verhandlungen der kaiserlich-königlichen zoologisch-botanischen Gesellschaft in Wien* 12:315–588.

Grunow, A. 1863. Ueber einige neue und ungenügend bekannte Arten und Gattungen von Diatomaceen. *Verhandlungen der kaiserlich-königlichen zoologisch–botanischen Gesellschaft in Wien* 13:137–162.

Grunow, A. 1867. Diatomeen auf *Sargassum* von Honduras, gesammelt von Lindig. *Hedwigia* 6:1–8, 17–37.

Grunow, A. 1868. Algen. *Reise seiner Majestät Fregatte Novaria um die Erde Botanischer Theil* 1:1–104.

Grunow, A. 1877. I. New diatoms from Honduras. *Transactions of the Royal Microscopical Society* 18:165–186.

Grunow, A. 1878. Algen und Diatomaceen aus dem Kaspischen Meere. *In* "O. Schneider: Naturwissenschaftliche Beiträge zur Kenntniss der Kaukasusländer," pp. 98–160.

Grunow, A. 1879. New species and varieties of Diatomaceæ from the Caspian Sea. Translated with additional notes by F. Kitton. *Journal of the Royal Microscopical Society* 2:677–691.

Grunow, A. 1884. Die Diatomeen von Franz Josefs-Land. *Denkschriften der kaiserlichen Akademie der Wissenschaften. Mathematisch-naturwissenschaftliche Classe. Wien* 28:53–112.

Guiffré, G., & Ragusa, S. 1988. The morphology of *Chaetoceros rostratum* Lauder (Bacillariophyceae) using light and electron microscopy. *Botanica Marina* 31:503–510.

Guillard, R. R. L., & Ryther, J. H. 1962. Studies of marine planktonic diatoms I. *Cyclotella nana* Hustedt, and *Detonula confervacea* (Cleve) Gran. *Canadian Journal of Microbiology* 8:229–239.

Guillard, R. R. L., Carpenter, E. J., & Reimann, E. F. 1974. *Skeletonema menzelii* sp. nov., a new diatom from the western Atlantic Ocean. *Phycologia* 13:131–138.

Guo, Yujie (Kuo Yuchieh) 1981. Studies on the planktonic *Coscinodiscus* (diatoms) of the South China Sea. *Studia Marina Sinica* 18:149–180.

Gutenbrunner, S. A., Thalhamer, J., & Schmid, A-M. M. 1994. Proteinaceous and immunochemical distinctions between the *oval* and *fusiforn* morphotypes of *Phaeodactylum tricornutum* (Bacillariophyceae). *Journal of Phycology* 30:129–136.

Håkansson, H. 1989. A light and electron microscopical investigation of the type species of *Cyclotella* (Bacillariophyceae) and related forms, using original material. *Diatom Research* 4:255–267.

Håkansson, H., & Ross, R. 1984. Proposals to designate conserved types for *Cymbella* C. Agardh and *Cyclotella* (Kützing) Brébisson, and to conserve *Rhopalodia* O. Müller against *Pyxidicula* Ehrenberg (all Bacillariophyceae). *Taxon* 33:525–531.

Hallegraeff, G. M. 1984. Species of the diatom genus *Thalassiosira* in Australian waters. *Botanica Marina* 27:495–513.

Hallegraeff, G. M. 1986. Taxonomy and morphology of the marine plankton diatoms *Thalassionema* and *Thalassiothrix*. *Diatom Research* 1:57–80.

Hallegraeff, G. M. 1992. Observations on the mucilaginous diatom *Thalassiosira bipartita* (Rattray) comb.nov. from the tropical Indo-west Pacific. *Diatom Research* 7:15–23.

Hargraves, P. E. 1976. Studies on marine plankton diatoms. III. Structure and classification of *Gossleriella tropica*. *Journal of Phycology* 12:285–291.

Hargraves, P. E. 1979. Studies on marine plankton diatoms IV. Morphology of *Chaetoceros* resting spores. *Beiheft zur Nova Hedwigia* 64:99–120.

Hargraves, P. E. 1982. Resting spore formation in the marine diatom *Ditylum brightwellii* (West) Grun. ex V. H. *In* "Proceedings of the Seventh International Diatom Symposium" (D. G. Mann, ed.), pp. 33–46.

Hargraves, P. E. 1990. Studies on marine planktonic diatoms. V. Morphology and distribution of *Leptocylindrus minimus* Gran. *Beiheft zur Nova Hedwigia* 100:47–60.

Hargraves, P. E., & French, F. W. 1983. Diatom resting spores: Significance and strategies. *In* "Survival Strategies of the Algae" (G. A. Fryxell, ed.), pp. 49–68. Cambridge University Press, Cambridge.

Hargraves, P. E., & Guillard, R. R. L. 1974. Structural and physiological observations on some small marine diatoms. *Phycologia* 13:163–172.

Hasle, G. R. 1960. Phytoplankton and ciliate species from the tropical Pacific. *Skrifter utgitt av Det Norske Videnskaps-Akademi i Oslo I. Matematisk-Naturvidenskapelig Klasse* 2:1–50.

Hasle, G. R. 1964. *Nitzschia* and *Fragilariopsis* species studied in the light and electron microscopes. I. Some marine species of the groups *Nitzschiella* and *Lanceolatae*. *Skrifter utgitt av Det Norske Videnskaps-Akademi i Oslo. I. Matematisk-Naturvidenskapelig Klasse. Ny Serie* 16:1–48.

Hasle, G. R. 1965a. *Nitzschia* and *Fragilariopsis* species studied in the light and electron microscopes. III. The genus *Fragilariopsis*. *Skrifter utgitt av Det Norske Videnskaps-Akademi i Oslo. I. Matematisk-Naturvidenskapelig Klasse. Ny Serie* 21:1–49.

Hasle, G. R. 1965b. *Nitzschia* and *Fragilariopsis* species studied in the light and electron microscopes. II. The group *Pseudo-nitzschia*. *Skrifter utgitt av Det Norske Videnskaps-Akademi i Oslo. I. Matematisk-Naturvidenskapelig Klasse. Ny Serie* 18:1–45.

Hasle, G. R. 1968a. The valve processes of the centric diatom genus *Thalassiosira*. *Nytt Magasin for Botanikk* 15:193–201.

Hasle, G. R. 1968b. Marine diatoms. *Antarctic Map Folio Series* 10:6–8.

Hasle, G. R. 1968c. Observations on the marine diatom *Fragilariopsis kerguelensis* (O'Meara) Hust. in the scanning electron microscope. *Nytt Magasin for Botanikk* 15:205–208.

Hasle, G. R. 1969. An analysis of the phytoplankton of the Pacific Southern Ocean: Abundance, composition, and distribution during the Brategg Expedition, 1947–1948. *Hvalrådets Skrifter* 52:1–168.

Hasle, G. R. 1971. *Nitzschia pungiformis* (Bacillariophyceae), a new species of the *Nitzschia seriata* group. *Norwegian Journal of Botany* 18:139–144.

Hasle, G. R. 1972a. *Thalassiosira subtilis* (Bacillariophyceae) and two allied species. *Norwegian Journal of Botany* 19:111–137.

Hasle, G. R. 1972b. The inclusion of *Coscinosira* Gran (Bacilariophyceae) in *Thalassiosira* Cleve. *Taxon* 21:543–544.

Hasle, G. R. 1972c. *Fragilariopsis* as a section of the genus *Nitzschia* Hassall. *Beiheft zur Nova Hedwigia* 39:111–119.

Hasle, G. R. 1972d. The distribution of *Nitzschia seriata* Cleve and allied species. *Beiheft zur Nova Hedwigia* 39:171–190.

Hasle, G. R. 1973a. Some marine plankton genera of the diatom family Thalassiosiraceae. *Beiheft zur Nova Hedwigia* 45:1–49.

Hasle, G. R. 1973b. Morphology and taxonomy of *Skeletonema costatum* (Bacillariophyceae). *Norwegian Journal of Botany* 20:109–137.

Hasle, G. R. 1973c. The "mucilage pore" of pennate diatoms. *Beiheft zur Nova Hedwigia* 45:167–186.

Hasle, G. R. 1974. Validation of the names of some marine planktonic species of *Nitzschia* (Bacillariophyceae). *Taxon* 23:425–428.

Hasle, G. R. 1975. Some living marine species of the diatom family Rhizosoleniaceae. *Beiheft zur Nova Hedwigia* 53:99–140.

Hasle, G. R. 1976a. The biogeography of some marine planktonic diatoms. *Deep-Sea Research* 23:319–338.

Hasle, G. R. 1976b. Examination of diatom type material: *Nitzschia delicatissima* Cleve, *Thalassiosira minuscula* Krasske, and *Cyclotella nana* Hustedt. *British Phycological Journal* 11:101–110.

Hasle, G. R. 1977. Morphology and taxonomy of *Actinocyclus normanii* f. *subsalsa* (Bacillariophyceae). *Phycologia* 16:321–328.

Hasle, G. R. 1978a. Some *Thalassiosira* species with one central process. *Norwegian Journal of Botany* 25:77–110.

Hasle, G. R. 1978b. Some freshwater and brackish water species of the diatom genus *Thalassiosira* Cleve. *Phycologia* 17:263–292.

Hasle, G. R. 1979. *Thalassiosira decipiens* (Grun.) Jørg. (Bacillariophyceae). *Bacillaria* 2:85–108.

Hasle, G. R. 1980. Examination of *Thalassiosira* type material: *T. minima* and *T. delicatula* (Bacillariophyceae). *Norwegian Journal of Botany* 27:167–173.

Hasle, G. R. 1983a. The marine planktonic diatoms *Thalassiosira oceanica* sp. nov. and *T. partheneia*. *Journal of Phycology* 19:220–229.

Hasle, G. R. 1983b. *Thalassiosira punctigera* (Castr.) comb. nov., a widely distributed marine planktonic diatom. *Nordic Journal of Botany* 3:593–608.

Hasle, G. R. 1983c. Proposal to conserve the generic name *Cerataulina* H. Perag. ex Schütt over *Syringidium* Ehrenb. *Taxon* 32:474–475.

Hasle, G. R. 1990. The planktonic marine diatom *Thalassiosira mediterranea* (synonym *Thalassiosira stellaris*). *Diatom Research* 5:415–418.

Hasle, G. R. 1993. Nomenclatural notes on marine planktonic diatoms. The family Bacillariaceae. *Beiheft zur Nova Hedwigia* 106:315–321.

Hasle, G. R. 1994. *Pseudo-nitzschia* as a genus distinct from *Nitzschia* (Bacillariophyceae). *Journal of Phycology* 30:1036–1039.

Hasle, G. R. 1995. *Pseudo-nitzschia pungens* and *P. multiseries* (Bacillariophyceae): Nomenclatural history, morphology and distribution *Journal of Phycology* 31(3):428–435.

Hasle, G. R., & Booth, B. C. 1984. *Nitzschia cylindroformis* sp. nov., a common and abundant nanoplankton diatom of the eastern subarctic Pacific. *Journal of Plankton Research* 6:493–503.

Hasle, G. R., & Evensen, D. L. 1975. Brackish-water and fresh-water species of the diatom genus *Skeletonema* Grev. I. *Skeletonema subsalsum* (A. Cleve) Bethge. *Phycologia* 14:283–297.

Hasle G. R., & Fryxell, G. A. 1977a. *Thalassiosira conferta* and *T. binata*, two new diatom species. *Norwegian Journal of Botany* 24:239–248.

Hasle, G. R., & Fryxell, G. A. 1977b. The genus *Thalassiosira*: Some species with a linear areola array. *Beiheft zur Nova Hedwigia* 54:15–66.

Hasle, G. R., & Fryxell, G. A. 1995. Taxonomy of Diatoms. In "Manual on Harmful Marine Microalgae" (G. M. Hallegraeff, Danderson, D. M., Cembella, A. D., eds.) pp. 339–364. IOC Manuals and Guides No. 33, UNESCO, Paris.

Hasle, G. R., & Heimdal, B. R. 1968. Morphology and distribution of the marine centric diatom *Thalassiosira antarctica* Comber. *Journal of the Royal Microscopical Society* 88:357–369.

Hasle, G. R., & Heimdal, B. R. 1970. Some species of the centric diatom genus *Thalassiosira* studied in the light and electron microscopes. *Beiheft zur Nova Hedwigia* 31:543–581.

Hasle, G. R., & Lange, C. B. 1989. Freshwater and brackish water *Thalassiosira* (Bacillariophyceae): Taxa with tangentially undulated valves. *Phycologia* 28:120–135.

Hasle, G. R., & Lange, C. B. 1992. Morphology and distribution of *Coscinodiscus* species from the Oslofjord, Norway, and the Skagerrak, North Atlantic. *Diatom Research* 7:37–68.

Hasle, G. R., & Medlin, L. K. 1990a. Family Bacillariaceae: The genus *Nitzschia* section *Nitzschiella*. In "Polar Diatoms" (L. K. Medlin & J. Priddle, eds.), pp. 177–180. British Antarctic Survey, Cambridge.

Hasle, G. R., & Medlin, L. K. 1990b. Family Bacillariaceae: The genus *Nitzschia* section *Fragilariopsis*. In "Polar Diatoms" (L. K. Medlin & J. Priddle, eds.), pp. 181–196. British Antarctic Survey, Cambridge.

Hasle, G. R., & Medlin, L. K. 1990c. Family Bacillariaceae: The genus *Nitzschia* section *Pseudonitzschia*. *In* "Polar Diatoms" (L. K. Medlin & J. Priddle, eds.), pp. 169–176. British Antarctic Survey, Cambridge.

Hasle, G. R., & de Mendiola, B. R. E. 1967. The fine structure of some *Thalassionema* and *Thalassiothrix* species. *Phycologia* 6:107–125.

Hasle, G. R., & Semina, H. J. 1988. The marine planktonic diatoms *Thalassiothrix longissima* and *Thalassiothrix antarctica* with comments on *Thalassionema* spp. and *Synedra reinboldii*. *Diatom Research* 2:175–192.

Hasle, G. R., & Sims, P. A. 1985. The morphology of the diatom resting spores *Syringidium bicorne* and *Syringidium simplex*. *British Phycological Journal* 20:219–225.

Hasle, G. R., & Sims, P. A. 1986a. The diatom genus *Coscinodiscus* Ehrenb.: *C. argus* Ehrenb. and *C. radiatus* Ehrenb. *Botanica Marina* 29:305–318.

Hasle, G. R., & Sims, P. A. 1986b. The diatom genera *Stellarima* and *Symbolophora* with comments on the genus *Actinoptychus*. *British Phycological Journal* 21:97–114.

Hasle, G. R., & Syvertsen, E. E. 1980. The diatom genus *Cerataulina*: Morphology and taxonomy. *Bacillaria* 3:79–113.

Hasle, G. R., & Syvertsen, E. E. 1981. The marine diatoms *Fragilaria striatula* and *F. hyalina*. *Striae* 14:110–118.

Hasle, G. R., & Syvertsen, E. E. 1984. *Coscinodiscus pseudolineatus* Pant. and *Coscinodiscus praelineatus* Jousé as synonyms of *Thalassiosira leptopus* (Grun.) Hasle & G. Fryxell. *In* "Proceedings of the 7th International Diatom Symposium" (D. G. Mann, ed.), pp. 145–155. Koeltz, Koenigstein.

Hasle, G. R., & Syvertsen, E. E. 1990a. Family Thalassiosiraceae. Section 1: The genera *Skeletonema*, *Porosira*, *Bacterosira*, and *Detonula* and the genus *Thalassiosira* from the Arctic. *In* "Polar Marine Diatoms" (L. K. Medlin & J. Priddle, eds.), pp. 83–89. British Antarctic Survey, Cambridge.

Hasle, G. R., & Syvertsen, E. E. 1990b. Arctic diatoms in the Oslofjord and the Baltic Sea, a bio- and palaeographic problem? *In* "Proceedings of the 10th International Diatom Symposium" (H. Simola, ed.), pp. 285–300. Koeltz, Koenigstein.

Hasle, G. R., & Syvertsen, E. E. 1990c. Family Coscinodiscaceae: The genus *Stellarima*. *In* "Polar Marine Diatoms" (L. K. Medlin & J. Priddle, eds.), p. 109. British Antarctic Survey, Cambridge.

Hasle, G. R., & Syvertsen, E. E. 1993. New nomenclatural combinations of marine diatoms. The families Thalassiosiraceae and Rhaphoneidaceae. *Beiheft zur Nova Hedwigia* 106:297–314.

Hasle, G. R., Heimdal, B. R., & Fryxell, G. A. 1971. Morphologic variability in fasciculated diatoms as exemplified by *Thalassiosira tumida* (Janisch) Hasle, comb. nov. *Biology of the Antarctic Seas 4, Antarctic Research Series* 17:313–333.

Hasle, G. R., von Stosch, H. A., & Syvertsen, E. E. 1983. Cymatosiraceae, a new diatom family. *Bacillaria* 6:9–156.

Hasle, G. R., Sims, P. A., & Syvertsen, E. E. 1988. Two Recent *Stellarima* species: *S. microtrias* and *S. stellaris* (Bacillariophyceae). *Botanica Marina* 31:195–206.

Hasle, G. R., Medlin, L. K., & Syvertsen, E. E. 1994. *Synedropsis* gen. nov., a genus of araphid diatoms associated with sea ice. *Phycologia* 33:248–270.

Hasle, G. R., Lange, C. B., & Syvertsen, E. E. 1995. A review of *Pseudo-nitzschia,* with special reference to the Skagerrak, North Atlantic, and adjacent waters. *Helgoländer Meeresuntersuchungen*, 50, (in press).

Heiden, H., & Kolbe, R. W. 1928. Die marinen Diatomeen der Deutschen Südpolar-Expedition 1901–3. *Deutsche Südpolar-Expedition 1901–1903* 8:447–715.

Heimdal, B. R. 1970. Morphology and distribution of two *Navicula* species in Norwegian coastal waters. *Nytt Magasin for Botanikk* 17:65–75.

Heimdal, B. R. 1971. Vegetative cells and resting spores of *Thalassiosira constricta* Gaarder (Bacillariophyceae). *Norwegian Journal of Botany* 18:153–159.

Heimdal, B. R. 1983. Phytoplankton and nutrients in the waters north-west of Spitsbergen in the autumn of 1979. *Journal of Plankton Research* 5:901–918.

Helmcke, J.-G., & Krieger, W. 1953. "Diatomeenschalen im elektronenmikroskopischen Bild 1," pp. 19, plates 1–102. Cramer, Weinheim.

Helmcke, J.-G., & Krieger, W. 1954. "Diatomeenschalen im elektronenmikroskopischen Bild 2," pp. 24, plates 103–200. Cramer, Weinheim.

Hendey, N. I. 1937. The plankton diatoms of the Southern Seas. *Discovery Reports* 16:151–364.

Hendey, N. I. 1951. Littoral diatoms of Chichester harbour with special reference to fouling. *Journal of the Royal Microscopical Society* 71:1–86.

Hendey, N. I. 1954. Note on the Plymouth "*Nitzschia*" culture. *Journal of the Marine Biological Association of the United Kingdom* 33:335–339.

Hendey, N. I. 1964. An introductory account of the smaller algae of British coastal waters. Part 5: Bacillariophyceae (Diatoms). *In* "Ministry of Agriculture, Fisheries and Food. Fishery Investigations Series IV," pp. 317. HMSO, London.

Hensen, V. 1887. Ueber die Bestimmung der Plankton's oder des im Meere treibenden Materials an Pflanzen und Thieren. *Bericht der Commission zur wissenschaftlichen Untersuchung der deutschen Meere, in Kiel* 5:1–107.

Hernández-Becerril, D. U. 1990. Observations on the morphology and distribution of the planktonic diatom *Neodelphineis pelagica*. *British Phycological Journal* 25:315–319.

Hernández-Becerril, D. U. 1991a. The morphology and taxonomy of species of the diatom genus *Asteromphalus* Ehr. *Bibliotheca Diatomologica* 23:1–55, 33 plates.

Hernández-Becerril, D. U. 1991b. The morphology and taxonomy of the planktonic diatom *Chaetoceros coarctatus* Lauder (Bacillariophyceae). *Diatom Research* 6:281–287.

Hernández-Becerril, D. U. 1991c. Note on the morphology of *Chaetoceros didymus* and *C. protuberans*, with some considerations on their taxonomy. *Diatom Research* 6:289–297.

Hernández-Becerril, D. U. 1992a. Reinstatement of the diatom genus *Spatangidium* (Bacillariophyta): The type species *S. arachne*. *Phycologia* 31:278–284.

Hernández-Becerril, D. U. 1992b. Two new species of the diatom genus *Chaetoceros* (Bacillariophyceae). *Plant Systematics and Evolution* 181:217–226.

Hernández-Becerril, D. U. 1992c. Observations on two closely related species, *Chaetoceros tetrastichon* and *C. dadayi* (Bacillariophyceae). *Nordic Journal of Botany* 12:365–371.

Hernández-Becerril, D. U. 1993a. Note on the morphology of two planktonic diatoms: *Chaetoceros bacteriastroides* and *C. seychellarus*, with comments on their taxonomy and distribution. *Botanical Journal of the Linnean Society* 111:117–128.

Hernández-Becerril, D. U. 1993b. Study of the morphology and distribution of two planktonic diatoms: *Chaetoceros paradoxus* and *Ch. filiferus* (Bacillariophyceae). *Cryptogamic Botany* 3:169–175.

Hernández-Becerril, D. U., Castillo, M. E. del, & Villa, M. A. L. 1993. Observations on *Chaetoceros buceros* (Bacillariophyeae), a rare tropical planktonic species collected from the Mexican Pacific. *Journal of Phycology* 29:811–818.

Herzig, W. N., & Fryxell, G. A. 1986. The diatom genus *Thalassiosira* Cleve in the Gulf Stream core rings: Taxonomy, with *T. intrannula* and *T. lineoides*, spp. nov. *Botanica Marina* 29:11–25.

Hoban, M. A. 1983. Biddulphoid diatoms. II: The morphology and systematics of the pseudocellate species, *Biddulphia biddulphiana* (Smith) Boyer, *B. alternans* (Bailey) Van Heurck, and *Trigonium arcticum* (Brightwell) Cleve. *Botanica Marina* 26:271–284.

Hoban, M. A., Fryxell, G. A., & Buck, K. R. 1980. Biddulphioid diatoms: Resting spores in Antarctic *Eucampia* and *Odontella*. *Journal of Phycology* 16:591–602.

Holmes, R. W. 1956. The annual cycle of phytoplankton in the Labrador Sea, 1950–51. *Bulletin of the Bingham Oceanographic Collection* 16(1):1–74.

Hustedt, F. 1926. *Thalassiosira fluviatilis* nov. spec., eine Wasserblüte im Wesergebiet. *Berichte der deutschen botanischen Gesellschaft* 43:565–567.

Hustedt, F. 1930. Die Kieselalgen Deutschlands, Österreichs und der Schweiz. *Dr. L. Rabenhorsts Kryptogamen-Flora von Deutschland, Österreich und der Schweiz* 7(1):1–920.

Hustedt, F. 1939. Die Diatomeenflora des Küstengebietes der Nordsee vom Dollart bis zur Elbenmündung. *Abhandlungen herausgegeben vom Naturwissenschaftlichen Verein zu Bremen* 31:571–677.

Hustedt, F. 1952. Diatomeen aus der Lebensgemeinschaft des Buckelwals (Megaptera nodosa Bonn.). *Archiv für Hydrobiologie* 46:286–298.

Hustedt, F. 1955. Marine littoral diatoms of Beaufort, North Carolina. *Bulletin Duke University Marine Station* 6:1–67.

Hustedt, F. 1956. Diatomeen aus dem Lago de Maracaibo in Venezuela. *Ergebnisse der deutschen limnologischen Venezuela-Expedition 1952* 1:93–140.

Hustedt, F. 1957. Die Diatomeenflora des Flusssystems der Weser im Gebiet der Hansestadt Bremen. *Abhandlungen herausgegeben vom Naturwissenschaftlichen Verein zu Bremen* 34:181–440.

Hustedt, F. 1958a. Diatomeen aus der Antarktis und dem Südatlantik. *Deutsche antarktische Expedition 1938/39* 2:103–191.

Hustedt, F. 1958b. Phylogenetische Untersuchungen an Diatomeen. *Österreichische botanische Zeitschrift* 105:193–211.

Hustedt, F. 1959. Die Kieselalgen Deutschlands, Österreichs und der Schweiz. *Dr. L. Rabenhorsts Kryptogamen-Flora von Deutschland, Österreich und der Schweiz* 7(2):1–845.

Hustedt, F. 1961. Die Kieselalgen Deutschlands, Österreichs und der Schweiz. *Dr. L. Rabenhorsts Kryptogamen-Flora von Deutschland, Österreich und der Schweiz* 7(3):1–816.

Ikari, J. 1927. On *Bacteriastrum* of Japan. *The Botanical Magazine, Tokyo* 41:421–431.

Johansen, J. R., & Fryxell, G. A. 1985. The genus *Thalassiosira* (Bacillariophyceae): Studies on species occurring south of the Antarctic Convergence Zone. *Phycologia* 24:155–179.

Jørgensen, E. 1899. Protophyten und Protozoën im Plankton aus der norwegischen Westküste. *Bergens Museums Aarbog for 1899* 6:1–112, LXXXIII.

Jørgensen, E. 1905. Protist plankton of Northern Norwegian fiords. *Bergens Museums Skrifter* 1905:49–151.

Jordan, R. W., Ligowski, R., Nöthig, E.-M., & Priddle, J. 1991. The diatom genus *Proboscia* in Antarctic waters. *Diatom Research* 6:63–78.

Jousé, A. P. 1968. Species novae Bacillariophytorum in sedimentis fundi Oceani Pacifici et Maris Ochotensis inventae. *Novitates Systematicae Plantarum non Vascularium* 1968:12–21.

Jousé, A. P., Koroleva, G. S., & Nagaeva, G. A. 1962. Diatoms in the surface layer of sediment in the Indian sector of the Antarctic. *Trudy IOAN* 61:19–92 [In Russian, title and abstract in English]

Juhlin-Dannfelt, H. 1882. On the diatoms of the Baltic Sea. *Bihang till Kongliga Svenska Vetenskaps-Akademiens Handlingar* 6(21):1–52.

Kaczmarska, I., & Fryxell, G. A. 1986. The diatom genus *Nitzschia*: Morphologic variation of some small bicapitate species in two Gulf Stream warm core rings. *In* "Proceedings of the 8th International Diatom Symposium" (M. Ricard, ed.), pp. 237–255. Koeltz, Koenigstein.

Kaczmarska, I., & Fryxell, G. A. 1994. The genus *Nitzschia*: Three new species from the Equatorial Pacific Ocean. *Diatom Research* 9:87–98.

Kaczmarska, I., Fryxell, G. A., & Watkins, T. P. 1986. Effect of two Gulf Stream warm-core rings on distributional patterns of the diatom genus *Nitzschia*. *Deep-Sea Research* 33:1843–1868.

Karayeva, N. I., & Dzhafarova, S. K. 1984. On the morphology of *Coscinodiscus granii* (Bacillariophyta). *Botanicheskii Zhurnal* 70:1078–1082. [In Russian]

Karsten, G. 1905. Das Phytoplankton des Antarktischen Meeres nach dem Material der deutschen Tiefsee-Expedition 1898–1899. *Deutsche Tiefsee-Expedition 1898–1899* 2(2):1–136.

Karsten, G. 1907. Das indische Phytoplankton. Nach dem Material der Deutschen Tiefsee-Expedition 1889–1899. *Deutsche Tiefsee-Expedition 1889–1899* 2(2):221–538.

Koch, P., & Rivera, P. 1984. Contribucion al conocimiento de las diatomeas Chilenas III el genero *Chaetoceros* Ehr (subgenero *Phaeoceros* Gran). *Gayana, Botanica* 41:61–84.

Kolbe, R. W. 1954. Diatoms from equatorial Pacific cores. *Reports of the Swedish Deep-Sea Expedition* 6(1).1–49.

Kolbe, R. W. 1955. Diatoms from equatorial Atlantic cores. *Reports of the Swedish Deep-Sea Expedition* 7(3):150–184.

Kolbe, R. W. 1957. Diatoms from equatorial Indian Ocean cores. *Reports of the Swedish Deep-Sea Expedition* 9(1):1–50.

Körner, H. 1970. Morphologie und Taxonomie der Diatomeengattung *Asterionella*. *Nova Hedwigia* 20:557–724.

Kozlova, O. G. 1967. De speciebus Bacillariophytorum novis e partibus Antarcticis Oceanorum Indici et Pacifici. *Novitates Systematicae Plantarum non Vascularium* 1967:54–62.

Krasske, G. 1941. Die Kieselalgen des chilenischen Küstenplanktons. *Archiv für Hydrobiologie* 38:260–287.

Kützing, F. T. 1844. "Die kieselschaligen Bacillarien oder Diatomeen," pp. 152. Ferd. Förstmann, Nordhausen, 1865, Zweiter Abdruck.

Kützing, F. T. 1849. "Species Algarum," pp. 922. Leipzig.

Lange, C. B. 1985. Spatial and seasonal variations of diatom assemblages off the Argentinian coast (South Western Atlantic). *Oceanologica Acta* 8:361–369.

Lange, C. B., & Mostajo, E. L. 1985. Phytoplankton (diatoms and silicoflagellates) from the southwestern Atlantic Ocean. *Botanica Marina* 28:469–476.

Lange, C. B., & Syvertsen, E. E. 1989. *Cyclotella litoralis* sp. nov. (Bacillariophyceae), and its relationship to *C. striata* and *C. stylorum*. *Nova Hedwigia* 48:341–356.

Lange, C. B., Hasle, G. R., & Syvertsen, E. E. 1992. Seasonal cycle of diatoms in the Skagerrak, North Atlantic, with emphasis on the period 1980–1990. *Sarsia* 77:173–187.

Lauder, H. S. 1864a. On new diatoms. *Transactions of the Microscopical Society of London New Series* 12:6–8.

Lauder, H. S. 1864b. Remarks on the marine diatomaceæ found at Hong Kong, with descriptions of new species. *Transactions of the Microscopical Society of London New Series* 12:75–79.

Lebour, M. V. 1930. "The Planktonic Diatoms of Northern Seas," pp. 244. Ray Society, London.

Lee, J. H., & Lee, J. Y. 1990. A light and electron microscopy study on the marine diatom *Roperia tesselata* (Roper) Grunow. *Diatom Research* 5:325–335.

Levander, K. M. 1904. Zur Kenntnis der Rhizosolenien Finlands. *Meddelanden af Societas pro Fauna et Flora Fennica* 30:112–117.

Lewin, J. C. 1958. The taxonomic position of *Phaeodactylum tricornutum*. *The Journal of General Microbiology* 18:427–432.

Lewin, J. C., & Norris, R. E. 1970. Surf-zone diatoms of the coasts of Washington and New Zealand (*Chaetoceros armatum* T. West and *Asterionella* spp.). *Phycologia* 9:143–149.

Loeblich, A. R., III, Wight, W. W., & Darley, W. M. 1968. A unique colonial marine centric diatom *Coenobiodiscus muriformis* gen. et sp. nov. *Journal of Phycology* 4:23–29.

Lønne, O. J. 1988. A diver operated electric suction sampler for sympagic (= under ice) invertebrates. *Polar Research* 6:135–136.

Lundholm, N., Skov, J., Pocklington, R., & Moestrup, Ø. 1994. Domoic acid, the toxic amino acid responsible for amnesic shellfish poisoning, now in *Pseudo-nitzschia seriata* (Bacillariophyceae) in Europe (Research Note). *Phycologia* 33:475–478.

Lyngbye, H. C. 1819. "Tentamen Hydrophytologiae Danicae," pp. 248. Gyldendal, Copenhagen.

Makarova, I. V. 1970. Ad taxonomiam specierum nonnullarum generis *Thalassiosira* Cl. *Novitates Systematicae Plantarum non Vascularium* 7:13–20.

Makarova, I. V. 1977. Proposals for a standardization of diatom terminology and diagnosis. *Botanicheskii Zhurnal* 62:192–213. [in Russian]

Makarova, I. V. 1985. On the morphology and taxonomy of two species of the genus *Coscinodiscus* (Bacillariophyta). *Botanicheskii Zhurnal* 70:51–54. [in Russian]

Makarova, I. V. 1988. "Diatomovye vodorosli morei SSSR: Rod Thalassiosira Cl," pp. 115. Akademiya NAUK SSSR, Leningrad.

Makarova, I. V., Genkal, S. I., & Kuzmin, G. V. 1979. Species of the genus *Thalassiosira* Cl. (Bacillariophyta), found in continental waterbodies of the U.S.S. R. *Botanicheskii Zhurnal* 64:921–927. [In Russian, title and abstract in English]

Mangin, L. 1910. Sur quelques algues nouvelles ou peu connues du phytoplancton de l'Atlantique. *Bulletin de la Société Botanique de France* 57:344–383.

Mangin, L. 1915. Phytoplankton de l'Antarctique. Expédition du "Pourquoi-Pas?" 1908–1910. *Deuxième Expédition Antarctique Francaise 1908–1910, Sciences naturelles: Documents scientifiques*, 1–96.

Mangin, L. 1917. Sur le *Chaetoceros criophilus* Castr., espèce caractéristique des mers antarctique. *Comptes rendus des séances de l'Académie des Sciences* 164:1–9.

Mangin, L. 1922. Phytoplancton antarctique. Expédition antarctique de la "Scotia" 1902–1904. *Memoires de l'Academie Sciences. Paris, Série 2* 57(2):1–134.

Manguin, E. 1957. Premier inventaire des diatomées de la Terre Adélie antarctique. Espèces nouvelles. *Revue Algologique, Serie 2* 3:111–134.

Manguin, E. 1960. Les diatomées de la Terre Adélie. Campagne du "Commandant Charcot" 1949–1950. *Annales des Sciences Naturelles Botanique* 12:223–363.

Manhart, J. R., Fryxell, G. A., Villac, M. C. & Segura, L. Y. 1995. *Pseudo-nitzschia pungens* and *P. multiseries* (Bacillariophyceae): Nuclear rDNAs and species boundaries. *Journal of Phycology*, 31:421–427.

Mann, A. 1907. Report on the diatoms of the Albatross Voyages in the Pacific Ocean, 1888–1904. *Contributions from the United States National Herbarium* 10(5):221–419.

Mann, D. G. 1978. "Studies in the family Nitzschiaceae (Bacillariophyta)," pp. XXXIII, 386, 146 plates, Ph.D. Dissertation, University of Bristol.

Mann, D. G. 1981. Sieves and flaps: Siliceous minutiae in the pores of raphid diatoms. *In* "Proceedings of the Sixth Symposium on Recent and Fossil Diatoms" (R. Ross, ed.), pp. 279–300. Koeltz, Koenigstein.

Mann, D. G. 1986. *Nitzschia* subgenus *Nitzschia* (notes for a monograph of the Bacillariaceae (2). *In* "Proceedings of the 8th International Diatom Symposium" (M. Ricard, ed.), pp. 215–226. Koeltz, Koenigstein.

Marino, D., Montresor, M., & Zingone, A. 1987. *Miraltia throndsenii* gen. nov., sp. nov., a planktonic diatom from the Gulf of Naples. *Diatom Research* 2:205–211.

Marino, D., Giuffre, G., Montresor, M., & Zingone, A. 1991. An electron microscope investigation on *Chaetoceros minimus* (Levander) comb. nov. and new observations on *Chaetoceros throndsenii* (Marino, Montresor and Zingone) comb. nov. *Diatom Research* 6:317–326.

Martin, J. L., Haya, K., Burridge, L. E., & Wildish, D. J. 1990. *Nitzchia pseudodelicatissima*— A source of domoic acid in the Bay of Fundy, eastern Canada. *Marine Ecology Progress Series* 67:177–182.

Medlin, L. K. 1990. Comparison of restriction endonuclease sites in the small subunit 16S-like rRNA gene from the major genera of the family Bacillariaceae. *Diatom Research* 5:63–71.

Medlin, L. K., & Priddle, J. 1990. "Polar Marine Diatoms," pp. 214. British Antarctic Survey, Cambridge.

Medlin, L. K., & Sims, P. A. 1993. The transfer of *Pseudoeunotia doliolus* to *Fragilariopsis*. *Beiheft zur Nova Hedwigia* 106:323–334.

Medlin, L. K., Elwood, H. J., Stickel, S., & Sogin, M. L. 1991. Morphological and genetic variation within the diatom *Skeletonema costatum* (Bacillariophyta): Evidence for a new species, *Skeletonema pseudocostatum. Journal of Phycology* 27:514–524.

Medlin, L. K., Williams, D. M., & Sims, P. A. 1993. The evolution of the diatoms (Bacillariophyta). I. Origin of the group and assessment of the monophyly of its major divisions. *European Journal of Phycology* 28:261–275.

Meister, F. 1932. "Kieselalgen aus Asien," pp. 56. Verlag von Gebrüder Borntraeger, Berlin.

Meister, F. 1937. Seltene und neue Kieselalgen II. *Berichte der Schweizerischen Botanischen Gesellschaft* 47:258–276.

Mereschkowsky, C. 1902. Liste des Diatomées de la Mer Noire. *Scripta Botanica Horti Universitatis Petropolitanae* 19:1–42.

Mereschkowsky, C. 1903. Les types de l'endochrome chez les Diatomées. *Scripta Botanica Horti Universitatis Petropolitanae* 21:107–193.

Meunier, A. 1910. "Microplankton des Mers de Barents et de Kara," Duc d'Orleans Campagne Arctique de 1907, pp. 355. Bulens, Brussels.

Meunier, A. 1913. Microplancton de la Mer Flamande. 1ére partie: Le genre *"Chaetoceros"* Ehr. *Mémoires du Musée Royal d'Histoire Naturelle de Belgique* 7(2):1–49.

Müller, O. F. 1783. Om et besynderligt Væsen i Strandvandet. *Nye Samling af det Kongelige Danske Videnskabernes Selskabs Skrifter* 2:277–286.

Müller-Melchers, F. C. 1953. New and little known diatoms from Uruguay and the South Atlantic coast. *Comunicaciones Botanicas del Museo de Historia Natural de Montevideo* 3(30):1–11.

Nagumo, T., & Ando, K. 1985. Alluvium diatoms from Arakawa Lowland, Saitama Perfecture (3). The frustular morphology and taxonomic studies of *Cyclotella* sp. and *C. stylorum* Brightwell. *Bulletin of Nippon University, General Education* 14:205–215.

Nagumo, T., & Kobayasi, H. 1985. Fine structure of three freshwater and brackish water species of the genus *Cyclotella* (Bacillariophyceae) *C. atomus, C. caspia* and *C. meduanae. Bulletin of Plankton Society of Japan* 32:101–109.

Navarro, J. N. 1993. Three-dimensional imaging of diatom ultrastructure with high resolution low-voltage SEM. *Phycologia* 32:151–156.

Nikolaev, V. A. 1983. On the genus *Symbolophora* (Bacillariophyta). *Botanicheskii Zhurnal* 68:1123–1128.

Nikolaev, V. A. 1984. To the construction of the system of centric diatoms (Bacillariophyta). *Botanicheskii Zhurnal* 69:1468–1474. [In Russian]

Norris, R. E. 1973. A new planktonic diatom, Nanoneis hasleae gen. et sp. nov. *Norwegian Journal of Botany* 20:321–325.

O'Meara, E. 1877. On the diatomaceous gatherings made at Kerguelen's Land. *Journal of the Linnean Society (Botany)* 15:55–59.

Ostenfeld, C. 1899. Plankton. *In* "Iagttagelser over overfladevandets temperatur, saltholdighed og plankton paa islandske og grönlandske skibsrouter i 1899" (M. Knudsen & C. Ostenfeld, eds.), pp. 1–93. Bianco Luno, Copenhagen.

Ostenfeld, C. H. 1900. Plankton in 1899. *In* "Iagttagelser over overfladevandets temperatur, saltholdighed og plankton paa islandske og grönlandske skibsrouter i 1899" (M. Knudsen & C. Ostenfeld, eds.), pp. 43–93. Bianco Luno, Copehagen.

Ostenfeld, C. H. 1901a. Iagttagelser over plankton diatomeer. *Nyt Magazin for Naturvidenskaberne* 39:287–302.

Ostenfeld, C. H. 1901b. Phytoplankton fra det Kaspiske Hav (Phytoplankton from the Caspian Sea). *Videnskablige Meddelelser fra den naturhistoriske Forening i København* 1901: 129–139.

Ostenfeld, C. H. 1902. Marine plankton diatoms. *Botanisk Tidsskrift* 25(1):219–245.

Ostenfeld, C. H. 1903. Phytoplankton from the sea around the Færöes. *In* "Botany of the Færöes Part 2," pp. 559–611. Det Nordiske Forlag, Ernst Bojesen, Copenhagen.

Ostenfeld, C. H. 1910. List of diatoms and flagellates. *Meddelelser om Grønland* 43:259–285.

Ostenfeld, C. H. 1915. A list of phytoplankton from the Boeton Strait, Celebes. *Dansk Botanisk Arkiv* 2(4):1–18.

Ostenfeld, C. H., & Schmidt, J. 1901. Plankton fra det Røde Hav og Adenbugten. (Plankton from the Red Sea and the Gulf of Aden.) *Videnskablige Meddelelser fra den naturhistoriske Forening i København* 1901:142–182.

Østrup, E. 1895. Marine Diatoméer fra Østgrønland. *Meddelelser om Grønland* 18:395–476.

Paasche, E. 1961. Notes on phytoplankton from the Norwegian Sea. *Botanica Marina* 2:197–210.

Paasche, E. 1973. Silicon and the ecology of marine plankton diatoms. I. *Thalassiosira pseudonana* (*Cyclotella nana*) grown in a chemostat with silicate as limiting nutrient. *Marine Biology* 19:117–126.

Paasche, E., Johansson, S., & Evensen, D. 1975. An effect of osmotic pressure on the valve morphology of the diatom *Skeletonema subsalsum*. *Phycologia* 14:205–211.

Paddock, T. B. B. 1986. Observations on the genus *Stauropsis* Meunier and related species. *Diatom Research* 1:89–98.

Paddock, T. B. B. 1988. *Plagiotropis* Pfitzer and *Tropidoneis* Cleve, a summary account. *Bibliotheca Diatomologica* 16:1–152.

Paddock, T. B. B. 1990. Family Naviculaceae: The genera *Plagiotropis*, *Banquisia*, *Membraneis*, and *Manguinea*. *In* "Polar Marine Diatoms" (L. K. Medlin & J. Priddle, eds.), pp. 151–155. British Antarctic Survey, Cambridge.

Paddock, T. B. B., & Sims, P. A. 1981. A morphological study of keels of various raphe-bearing diatoms. *Bacillaria* 4:177–222.

Patrick, R., & Reimer, C. W. 1975. The diatoms of the United States 2(1). *Monographs of the Academy of Natural Sciences of Philadelphia* 13:vii–ix, 1–213.

Paulsen, O. 1905. On some Peridineæ and Plankton-Diatoms. *Meddelelser fra Kommisionen for Havundersøgelser Serie Plankton* 1(3):1–7.

Pavillard, J. 1911. Observations sur les Diatomées; (2e série). *Bulletin de la Société Botanique de France* 58:21–29.

Pavillard, J. 1913. Observations sur les Diatomées. (2e série). *Bulletin de la Société Botanique de France* 60:126–133.

Pavillard, J. 1916. Recherches sur les diatomées pélagiques du Golfe du Lion. *Travail de l'Institut de Botanique de l'Université de Montpellier, Série mixte—Mémoire* 5:7–62.

Pavillard, J. 1921. Sur la reproduction du *Chætoceros eibenii* Meunier. *Comptes Rendus des Séances de l'Académie des Sciences* 172:469–471.

Pavillard, J. 1924. Observations sur les diatomées; (4e série). Le genre *Bacteriastrum*. *Bulletin de la Société Botanique de France* 71:1084–1090.

Pavillard, J. 1925. Bacillariales. *Report on the Danish Oceanographical Expeditions 1908–1910 to the Mediterranean and adjacent seas* 2:1–72.

Pelletan, J. 1889. Les Diatomées Histoire Naturelle, Préparation, Classification et description des principales espèces. *Journal de Micrographie* 2:364.

Peragallo, H. 1888. Diatomées de la Baie de Villefranche. *Bulletin de la Société Histoire Naturelle Toulouse* 22:1–100.

Peragallo, H. 1891. Monographie du genre *Pleurosigma*. *In* "Extrait du Diatomiste 1890–91" (J. Tempère, ed.), pp. 1–35.

Peragallo, H. 1892. Monographie du genre *Rhizosolenia* et de quelques genres voisins. *In* "Le Diatomiste" (J. Tempère), Vol. 1, pp. 79–82, 99–117.

Peragallo, H., & Peragallo, M. 1897–1908. "Diatomées marines de France" (M. J. Tempère, ed.), pp. 492, I–XII, pp. 48, 135 plates. Grez-sur-Loing.

Peragallo, M. 1924. "Les diatomées marines de la première expédition Antarctique Francaise (1903–05)," pp. 27. Masson et Cie, Paris.

Poulin, M. 1990. Family Naviculaceae: Arctic species. *In* "Polar Marine Diatoms" (L. K. Medlin & J. Priddle, eds.), pp. 137–148. British Antarctic Survey, Cambridge.

Priddle, J., & Fryxell, G. 1985. "Handbook of the common plankton diatoms of the Southern Ocean," pp. 159. British Antarctic Survey, Cambridge.

Priddle, J., & Thomas, D. P. 1989. *Coscinodiscus bouvet* Karsten—A distinctive diatom which may be an indicator of changes in the Southern Ocean. *Polar Biology* 9:161–167.

Priddle, J., Jordan, R. W., & Medlin, L. K. 1990. Family Rhizosoleniaceae. *In* "Polar Marine Diatoms" (L. K. Medlin & J. Priddle, eds.), pp. 115–127. British Antarctic Survey, Cambridge.

Pritchard, R. 1861. "A History of Infusoria, Including the *Desmidiaceae and Diatomaceae, British and Foreign,*" pp. 968, XII. Whittaker, London.

Proschkina-Lavrenko, A. I. 1953. Species *Chaetoceros* novae et curiosae Maris Nigri I. *Botanicheskie materialy, Otdela sporovykhrastenii. Akademiya NAUK SSSR* 9:46–56.

Proschkina-Lavrenko, A. I. 1955. "Diatomovye vodorosli planktona Chernogo Morya," pp. 222. Akademiya NAUK SSSR Botanicheskii Institut, Moskva–Leningrad. [in Russian]

Proschkina-Lavrenko, A. I. 1961. Diatomeae novae e Mari Nigro (Ponto Euxino) et Azoviano (Maeotico). *Notulae systematicae e sectione cryptogamica URSS* 14:33–39.

Ramsfjell, E. 1959. Two new phytoplankton species from the Norwegian Sea, the diatom *Coscinosira poroseriata,* and the dinoflagellate *Goniaulax parva. Nytt Magasin for Botanikk* 7:175–177.

Rattray, J. 1890a. A revision of the genus *Coscinodiscus* and some allied genera. *Proceedings of the Royal Society of Edinburgh* 16:449–692.

Rattray, J. 1890b. A revision of the genus *Actinocyclus* Ehrb. *Journal of the Quekett Microscopical Club, Series 2* 4:137–212.

Reid, F. M. H., & Round, F. E. 1988. The Antarctic diatom *Synedra reinboldii:* Taxonomy, ecology and transfer to a new genus, *Trichotoxon. Diatom Research* 2:219–227.

Reimann, B. E. F., & Lewin, J. C. 1964. The diatom genus *Cylindrotheca* Rabenhorst. *Journal of the Royal Microscopical Society* 83:283–296.

Ricard, M. 1987. "Atlas du phytoplancton marin 2: Diatomophycees," pp. 297. Centre National de la Recherche Scientifique, Paris.

Rines, J. E. B., & Hargraves, P. E. 1986. Considerations of the taxonomy and biogeography of *Chaetoceros ceratosporus* Ostf. and *Chaetoceros rigidus* Ostf. *In* "Proceedings of the 8th International Diatom Symposium" (M. Ricard, ed.), pp. 97–112. Koeltz, Koenigstein.

Rines, J. E. B., & Hargraves, P. E. 1988. The *Chaetoceros* Ehrenberg (Bacillariophyceae) flora of Narragansett Bay, Rhode Island, U.S.A. *Bibliotheca Phycologica* 79:1–196.

Rines, J. E. B., & Hargraves, P. E. 1990. Morphology and taxonomy of *Chaetoceros compressus* Lauder var. *hirtisetus* var. nova, with preliminary consideration of closely related taxa. *Diatom Research* 5:113–127.

Rivera, P. 1981. Beiträge zur Taxonomie und Verbreitung der Gattung *Thalassiosira* Cleve. *Bibliotheca Phycologica* 56:1–220, 71 plates.

Rivera, P. 1983. A guide for references and distribution for the class Bacillariophyceae in Chile between 18°28'S and 58°S. *Bibliotheca Diatomologica* 3:1–386.

Rivera, P. 1985. Las especies del genero *Nitzschia* Hassall, seccion *Pseudo-nitzschia* (Bacillariophyceae), en las aguas marinas chilenas. *Gayana Botanica* 42:9–38.

Rivera, P., & Koch, P. 1984. Contributions to the diatom flora of Chile II. *In* "Proceedings of the 7th International Diatom Symposium" (D. G. Mann, ed.), pp. 279–298. Koeltz, Koenigstein.

Rivera, P. S., Avaria, S., & Barrales, H. L. 1989. *Ethmodiscus rex* collected by net sampling off the coast of northern Chile. *Diatom Research* 4:131–142.

Roper, F. C. S. 1858. Notes on some new species and varieties of British marine diatomaceæ. *Quarterly Journal of Microscopical Science* 6:17–25.

Ross, R., & Sims, P. A. 1971. Generic limits in the Biddulphiaceae as indicated by the scanning electron microscope. *In* "Scanning Electron Microscopy" (V. H. Heywood, ed.), pp. 155–176. Academic Press, New York.

Ross, R., & Sims, P. A. 1973. Observations on family and generic limits in the Centrales. *Beiheft zur Nova Hedwigia* 45:97–128.

Ross, R., Sims, P. A., & Hasle, G. R. 1977. Observations on some species of the Hemiauloideae. *Beiheft zur Nova Hedwigia* 54:179–213.

Ross, R., Cox, E. J., Karayeva, N. I., Mann, D. G., Paddock, T. B. B., Simonsen, R., & Sims, P. A. 1979. An amended terminology for the siliceous components of the diatom cell. *Beiheft zur Nova Hedwigia* 64:513–533.

Round, F. E. 1972. Some observations on colonies and ultra structure of the frustule of *Coenobiodiscus muriformis* and its transfer to *Planktoniella*. *Journal of Phycology* 8:221–231.

Round, F. E. 1973. On the diatom genera *Stephanopyxis* Ehr. and *Skeletonema* Grev. and their classification in a revised system of the Centrales. *Botanica Marina* 16:148–154.

Round, F. E. 1980. Forms of the giant diatom *Ethmodiscus* from the Pacific and Indian Oceans. *Phycologia* 19:307–316.

Round, F. E., & Crawford, R. M. 1981. The lines of evolution of the Bacillariophyta. 1. Origin. *Proceedings of the Royal Society London B* 211:237–260.

Round, F. E., & Crawford, R. M. 1984. The lines of evolution of the Bacillariophyta. 2. The centric series. *Proceedings of the Royal Society London B* 221:169–188.

Round, F. E., & Crawford, R. M. 1989. Phylum Bacillariophyta. *In* "Handbook of Protoctista" (L. Margulis, J. O. Corliss, M. Melkonian, & D. J. Chapman, eds.), pp. 574–596. Jones and Barlett, Boston.

Round, F. E., Crawford, R. M., & Mann, D. G. 1990. "The Diatoms, Biology & Morphology of the Genera," pp. 747. Cambridge University Press, Cambridge.

Sancetta, C. 1987. Three species of *Coscinodiscus* Ehrenberg from North Pacific sediments examined in the light and scanning electron microscopes. *Micropaleontology* 33:230–241.

Sancetta, C. 1990. Occurrence of Thalassiosiraceae (Bacillariophyceae) in two fjords of British Columbia. *Beiheft zur Nova Hedwigia* 100:199–215.

Schmid, A.-M. 1990. Interclonal variation in the valve structure of *Coscinodiscus wailesii* Gran et Angst. *Beiheft zur Nova Hedwigia* 100:101–119.

Schmid, A.-M., & Volcani, B. E. 1983. Wall morphogenesis in *Coscinodiscus wailesii* Gran and Angst. I. Valve morphology and development of its architecture. *Journal of Phycology* 19:387–402.

Schmidt, A., continued by Schmidt, M., Fricke, F., Heiden, H., Müller, O., & Hustedt, F. 1874–1959. "Atlas der Diatomaceenkunde." Reisland, Leipzig.

Schrader, H.-J. 1972. *Thalassiosira partheneia*, eine neue Gallertlager bildende zentrale Diatomee. *"Meteor" Forschungs-Ergebnisse Reihe D* 10:58–64.

Schröder, B. 1906. Beiträge zur Kenntnis des Phytoplanktons warmer Meere. *Vierteljahrsschrift der Naturforschenden Gesellschaft in Zürich* 51:319–377.

Schröder, B. 1911. Adriatisches Phytoplankton. *Aus den Sitzungsberichten der kaiserlichen Akademie der Wissenschaften in Wien, Mathematische-naturwissenschaftliche Klasse* 70(1):1–57.

Schuette, G., & Schrader, H. 1982. *Thalassiothrix pseudonitzschioides* sp. nov.: A common pennate diatom from the Gulf of California. *Bacillaria* 5:213–223.

Schultze, M. 1858. Innere Bewegungserscheinungen bei Diatomeen der Nordsee aus den Gattungen *Coscinodiscus, Denticella, Rhizosolenia. Archiv für Anatomie, Physiologie und wissenschaftliche Medicin* 1858:330–342.

Schultze, M. 1859. Innere Bewegungserscheinungen bei Diatomeen der Nordsee aus den Gattungen *Coscinodiscus, Denticella, Rhizosolenia. Archiv für Anatomie, Physiologie und wissenschaftliche Medicin* 1858:330–342. Translated as Phenomena of internal movements in diatomaceæ of the North Sea, belonging to the genera *Coscinodiscus, Denticella*, and *Rhizosolenia*. *Quarterly Journal of Microscopical Science* 7:13–21.

Schütt, F. 1892. Das Pflanzenleben der Hochsee. *Ergebnisse der Plankton-Expedition der Humboldt-Stiftung* IA:243–314.

Schütt, F. 1895. Arten von *Chaetoceros* und *Peragallia*. Ein Beitrag zur Hochseeflora. *Berichte der deutschen botanischen Gesellschaft* 13:35–48.

Schütt, F. 1896. Bacillariales. In "Die natürlichen Pflanzenfamilien" (A. Engler & K. Prantl, eds.), Vol. I, 1b, pp. 31–150.

Semina, H. J. 1956. De specie nova generis Denticula Ktz. notula. Notulae systematicae e sectione cryptogamica URSS 11:82–84.

Semina, H. J. 1981a. De specie nova Thalassiosirae Cl. (Bacillariophyta) ex Antarctica notula. Novitates systematicae plantarum non vascularium 18:31–33.

Semina, H. J. 1981b. Morphology and distribution of a tropical Denticulopsis. In "Proceedings of the 6th Symposium on Recent and Fossil Diatoms" (R. Ross, ed.), pp. 179–187.

Semina, H. J., & Sergeeva, O. M. 1980. Morphology of certain diatom species from the White Sea. Biologiya Morya 1:34–37. [In Russian]

Shadbolt, G. 1854. A short description of some new forms of diatomaceæ from Port Natal. Transactions of the Microscopical Society of London 2:13–18.

Shrubsole, W. H. 1890. On a new diatom in the estuary of the Thames. Journal of the Quekett Microscopical Club, Serie 2 4:259–262.

Silva, P. C., & Hasle, G. R. 1993. Proposal to conserve Thalassiothrix Cleve et Grunow (Bacillariophyceae) with a conserved type. Taxon 42:125–127.

Silva, P. C., & Hasle, G. R. 1994. Proposal to conserve Thalassiosiraceae against Lauderiaceae and Planktoniellaceae (Algae). Taxon 43:287–289.

Simonsen, R. 1972. Über die Diatomeengattung Hemidiscus Wallich und andere Angehörige der sogenannten "Hemidiscaceae." Veröffentlichungen des Instituts für Meeresforschung in Bremerhaven 13:265–273.

Simonsen, R. 1974. The diatom plankton of the Indian Ocean Expedition of RV "Meteor" 1964–1965. "Meteor" Forschungsergebnisse Reihe D 19:1–107.

Simonsen, R. 1975. On the pseudonodulus of the centric diatoms, or Hemidiscaceae reconsidered. Beiheft zur Nova Hedwigia 53:83–94.

Simonsen, R. 1979. The diatom system: Ideas on phylogeny. Bacillaria 2:9–71.

Simonsen, R. 1982. Note on the diatom genus Charcotia M. Peragallo. Bacillaria 5:101–116.

Simonsen, R. 1987. "Atlas and catalogue of the diatom types of Friedrich Hustedt. 1, Catalogue, pp. 525; 2 & 3, atlas, 772 plates. Cramer, Berlin–Stuttgart.

Simonsen R. 1992. The diatom types of Heinrich Heiden in Heiden & Kolbe 1928. Bibliotheca Diatomologica 24:1–100.

Simonsen, R., & Kanaya, T. 1961. Notes on the marine species of the diatom genus Denticula Kütz. Internationale Revue gesamten Hydrobiologie 46:498–513.

Sims, P. A. 1990. The fossil diatom genus Fenestrella, its morphology, systematics and palaeogeography. Beiheft zur Nova Hedwigia 100:277–288.

Sims, P. A. 1994. Skeletonemopsis, a new genus based on the fossil species of the genus Skeletonema Grev. Diatom Research 9:387–410.

Sims, P. A., & Hasle G. R. 1990. The formal establishment of the family Stellarimaceae Nikolaev ex Sims & Hasle. Diatom Research 5:207–208.

Sims, P. A., Fryxell, G. A., & Baldauf, J. G. 1989. Critical examination of the diatom genus Azpeitia: Species useful as stratigraphic markers for the Oligocene and Miocene Epochs. Micropaleontology 35:293–307.

Smith, W. 1853–1856. "A Synopsis of the British Diatomaceae," pp. 89, 107. Van Voorst, London.

Smith, J. C., Pauley, K., Cormier, P., Angus, R., Odense, P., O'Neil, D., Quilliam, M. A., & Worms, J. 1991. Population dynamics and toxicity of various species of Dinophysis and Nitzschia from the southern Gulf of St. Lawrence. In "Proceedings of the Second Canadian Workshop on Harmful Marine Algae. Bedford Institute of Oceanography, Dartmouth, Nova Scotia. October 2–4 1990" (D. C. Gordon, Jr, ed.) p. 24. (Canadian Technical Report on Fisheries and Aquatic Sciences 1799).

Sournia, A. 1968. Diatomées planctoniques du Canal de Mozambique et de l'ile Maurice. Mémoire O.R.S.T.O.M. 31:1–120.

Sournia, A. (ed.) 1978. Phytoplankton manual. *In* "Monographs on Oceanographic Methodology 6," pp. 337.UNESCO, Paris.

Sournia, A., Grall, J.-R., & Jacques, G. 1979. Diatomées et dinoflagellés d'une coupe méridienne dans le sud de l'océan Indien. *Botanica Marina* 22:183–198.

Sournia, A., Chrêtiennot-Dinet, M.-J., & Ricard, M. 1991. Marine Phytoplankton: How many species in the world? *Journal of Plankton Research* 13:1093–1099.

Stein, J. R. (ed.) 1973. "Handbook of Phycological Methods," pp. 448. Cambridge University Press, London.

Sterrenburg, F. A. S. 1991a. Studies on the genera *Gyrosigma* and *Pleurosigma* (Bacillariophyceae). The Typus Generis of *Pleurosigma,* some presumed varieties and imitative species. *Botanica Marina* 34:561–573.

Sterrenburg, F. A. S. 1991b. Studies on the genera *Gyrosigma* and *Pleurosigma* (Bacillariophyceae). Light microscopical features for taxonomy. *Diatom Research* 6:367–389.

Stolterfoth, H. 1879. On a new species of the genus *Eucampia. Journal of Royal Microscopical Society* 2:835–836.

Strelnikova, N. I., & Nikolaev, V. A. 1986. A contribution to the revision of the genera *Stephanopyxis* and *Pyxidicula. Botanicheskii Zhurnal, Akademiya NAUK SSSR* 71:950–953. [In Russian]

Subba Rao, D. V., & Wohlgeschaffen, G. 1990. Morphological variants of *Nitzschia pungens* Grunow f. *multiseries* Hasle. *Botanica Marina* 33:545–550.

Sundström, B. G. 1980. *Rhizosolenia phuketensis* sp. nov. and *Rhizosolenia stolterfothii* H. Peragallo (Bacillariophyceae). *Botaniska Notiser* 133:579–583.

Sundström, B. G. 1984. Observations on *Rhizosolenia clevei* Ostenfeld (Bacillariophyceae) and *Richelia intracellularis* Schmidt (Cyanophyceae). *Botanica Marina* 27:345–355.

Sundström, B. G. 1986. The marine diatom genus Rhizosolenia, pp. 117, 39 plates. Doctoral Dissertation. Lund University, Lund, Sweden.

Syvertsen, E. E. 1977. *Thalassiosira rotula* and *T. gravida*: Ecology and morphology. *Beiheft zur Nova Hedwigia* 54:99–112.

Syvertsen, E. E. 1979. Resting spore formation in clonal cultures of *Thalassiosira antarctica* Comber, *T. nordenskioeldii* Cleve and *Detonula confervacea* (Cleve) Gran. *Beiheft zur Nova Hedwigia* 64:41–63.

Syvertsen, E. E. 1984. Morphology of the arctic, planktonic *Navicula pelagica* (Bacillariophyceae), with special reference to the "setae." *Nordic Journal of Botany* 4:725–728.

Syvertsen, E. E. 1985. Resting spore formation in the Antarctic diatoms *Coscinodiscus furcatus* Karsten and *Thalassiosira australis* Peragallo. *Polar Biology* 4:113–119.

Syvertsen, E. E. 1986. *Thalassiosira hispida* sp. nov., a marine planktonic diatom. *In* "Proceedings of the 8th International Diatom Symposium" (M. Ricard, ed.), pp. 33–42. Koeltz, Koenigstein.

Syvertsen, E. E. 1991. Ice algae in the Barents Sea: Types of assemblages, origin, fate and role in the ice-edge phytoplankton bloom. *Polar Research* 10:277–287.

Syvertsen, E. E., & Hasle, G. R. 1982. The marine planktonic diatom *Lauderia annulata* Cleve, with particular reference to the processes. *Bacillaria* 5:243–256.

Syvertsen, E. E., & Hasle, G. R. 1983. The diatom genus *Eucampia*: Morphology and taxonomy. *Bacillaria* 6:169–210.

Syvertsen, E. E., & Hasle, G. R. 1984. *Thalassiosira bulbosa* Syvertsen, sp. nov., an Antarctic marine diatom. *Polar Biology* 3:167–172.

Syvertsen, E. E., & Hasle, G. R. 1988. *Melosira arctica* in the Baltic Sea and in the Oslofjord (a preliminary note). *USDQR Report* 12:79–84.

Syvertsen, E. E., & Lange, C. B. 1990. *Porosira pentaportula* Syvertsen & Lange, sp. nov. (Bacillariophyceae), a marine planktonic diatom. *Beiheft zur Nova Hedwigia* 100:143–151.

Takahashi, K., Jordan, R. & Priddle, J. 1994. The diatom genus *Proboscia* in subarctic waters. *Diatom Research* 9:411–428.

Takano, H. 1956. Harmful blooming of minute cells of *Thalassiosira decipiens* in coastal water in Tokyo Bay. *Journal of Oceanographical Society of Japan* 12:63–67.

Takano, H. 1965. New and rare diatoms from Japanese marine waters—I. *Bulletin of Tokai Regional Fisheries Research Laboratory* 42:1–10.

Takano, H. 1968. On the diatom *Chaetoceros calcitrans* (Paulsen) emend. and its dwarf form *pumilus* forma nov. *Bulletin of Tokai Regional Fisheries Research Laboratory* 55:1–7.

Takano, H. 1976. Scanning electron microscopy of diatoms—II. *Thalassiosira mala* Takano. *Bulletin of Tokai Fisheries Research Laboratory* 87:57–65.

Takano, H. 1979. New and rare diatoms from Japanese marine waters—III. *Lithodesmium variabile* sp. nov. *Bulletin of Tokai Regional Fisheries Research Laboratory* 100:35–43.

Takano, H. 1980. New and rare diatoms from Japanese marine waters—V. *Thalassiosira tealata* sp. nov. *Bulletin of Tokai Regional Fisheries Research Laboratory* 103:55–63.

Takano, H. 1981. New and rare diatoms from Japanese marine waters—VI. Three new species in Thalassiosiraceae. *Bulletin of Tokai Regional Fisheries Research Laboratory* 105:31–43.

Takano, H. 1982. New and rare diatoms from Japanese marine waters—VIII. *Neodelphineis pelagica* gen. et sp. nov. *Bulletin of Tokai Regional Fisheries Research Laboratory* 106:45–51.

Takano, H. 1983. New and rare diatoms from Japanese marine waters—XI. Three new species epizoic on copepods. *Bulletin Tokai Regional Fisheries Research Laboratory* 111:23–35.

Takano, H. 1984. Scanning electron microscopy of diatoms—VII. *Odontella aurita* and *O. longicruris*. *Bulletin of Tokai Regional Fisheries Research Laboratory* 113:79–85.

Takano, H. 1985. Two new diatoms in the genus *Thalassiosira* from Japanese marine waters. *Bulletin of Tokai Regional Fisheries Research Laboratory* 116:1–11.

Takano, H. 1990. Diatoms. *In* "Red tide organisms in Japan—An illustrated taxonomic guide" (Y. Fukuyo, H. Takano, M. Chihara, & K. Matsuoka, eds.), pp. 162–331. Uchida Rokakuho, Tokyo.

Takano, H. 1993. Marine Diatom *Nitzschia multiseriata* sp. nov. Common at Inlets of Southern Japan. *Diatom, The Japanese Journal of Diatomology* 8:39–41.

Takano, H., & Kikuchi, K. 1985. Anomalous cells of *Nitzschia pungens* Grunow found in eutrophic marine waters. *Diatom, The Japanese Journal of Diatomology* 1:18–20.

Takano, H., & Kuroki, K. 1977. Some diatoms in the section *Pseudo-nitzschia* found in coastal waters of Japan. *Bulletin of Tokai Regional Fisheries Rearch Laboratory* 91:41–51.

Taylor, F. J. R. 1967. Phytoplankton of the south western Indian Ocean. *Nova Hedwigia* 12:433–476.

Thomas, D. P., & Bonham, P. I. 1990. Family Leptocylindraceae: the genus *Corethron*. *In* "Polar Marine Diatoms" (L. K. Medlin & J. Priddle, eds.), pp. 105–107. British Antarctic Survey, Cambridge.

Thomsen, H. A., Buck, K. R., Marino, D., Sarno, D., Hansen, L. E. Østergaard, J. B., & Krupp, J. 1993. *Lennoxia faveolata* gen. et sp. nov. (Diatomophyceae) from South America, California, West Greenland and Denmark. *Phycologia* 32:278–283.

Van Goor, A. C. J. 1924. Über einige neue und bemerkenswerte Schwebealgen. *Recueil des Travaux Botaniques Néerlandais* 21:297–328.

Van Heurck, H. 1880–1885. "Synopsis des Diatomées de Belgique," pp. 235, 132 Suppl. A, B, and C plates. Antwerp.

Van Heurck, H. 1909. Diatomées. *In* "Expédition Antarctique Belge. Résultates du Voyage du S. Y. Belgica 1897–1899," pp. 126. Buschmann. Antwerp.

Vanlandingham, S. L. 1967–1979. "Catalogue of the Fossil and Recent Genera and Species of Diatoms and Their Synonyms. 1–8," pp. 4653. Cramer, Lehre.

Villareal, T. A., & Fryxell, G. A. 1983. The genus *Actinocyclus* (Bacillariophyceae): Frustule morphology of *A. sagittulus* sp. nov. and two related species. *Journal of Phycology* 19:452–466.

Von Stosch, H. A. 1974. Pleurax, seine Synthese und seine Verwendung zur Einbettung und Darstellung der Zellwände von Diatomeen, Peridineen und anderen Algen, sowie für eine neue Methode zur Elektivfärbung von Dinoflagellaten-Panzern. *Archiv für Protistenkunde* 116(1,2):132–141.

Von Stosch, H. A. 1975. An amended terminology of the diatom girdle. *Beiheft zur Nova Hedwigia* 53:1–28.

Von Stosch, H. A. 1977. Observations on *Bellerochea* and *Streptotheca*, including descriptions of three new planktonic diatom species. *Beiheft zur Nova Hedwigia* 54:113–166.

Von Stosch, H. A. 1980. The two *Lithodesmium* species (Centrales) of European waters. *Bacillaria* 3:7–20.

Von Stosch, H. A. 1982. On auxospore envelopes in diatoms. *Bacillaria* 5:127–156.

Von Stosch, H. A. 1986. Some marine diatoms from the Australian region, especially from Port Phillip Bay and tropical north-eastern Australia. *Brunonia* 8:293–348.

Von Stosch, H. A. 1987. Some marine diatoms from the Australian region, especially from Port Phillip Bay and tropical north-eastern Australia. II. Survey of the genus *Palmeria* and of the family Lithodesmiaceae including the new genus *Lithodesmioides*. *Brunonia* 9:29–87.

Von Stosch, H. A., & Fecher, K. 1979. "Internal thecae" of *Eunotia soleirolii* (Bacillariophyceae): Development, structure and function as resting spores. *Journal of Phycology* 15:233–243.

Von Stosch, H. A., Theil, G., & Kowallik, K. V. 1973. Entwicklungsgeschichtliche Untersuchungen an zentrischen Diatomeen. 5. Bau und Lebenszyklus von *Chaetoceros didymum*, mit Beobachtungen über einige andere Arten der Gattung. *Helgoländer wissenschaftlichen Meeresuntersuchungen* 25:384–445.

Wallich, G. C. 1860. On the siliceous organisms found in the digestive cavities of the salpæ and their relation to the flint nodules of the chalk formation. *Transactions of the Microscopical Society New Series* 8:36–55.

Watkins, T. P., & Fryxell, G. A. 1986. Generic characterization of *Actinocyclus*: Consideration in light of three new species. *Diatom Research* 1:291–312.

West, T. 1860. Remarks on some diatomaceæ, new or imperfectly described, and a new desmid. *Transactions of the Microscopical Society of London New Series* 8:147–153.

Williams, D. M. 1985. Morphology, taxonomy and interrelationships of the ribbed araphid diatoms from the genera *Diatoma* and *Meridion*. *Bibliotheca Diatomologica* 8:1–228, 27 plates.

Williams, D. M. 1986. Comparative morphology of some species of *Synedra* Ehrenb. with a new definition of the genus. *Diatom Research* 1:131–152.

Williams, D. M., & Round, F. E. 1986. Revision of the genus *Synedra* Ehrenb. *Diatom Research* 1:313–339.

Williams, D. M., & Round, F. E. 1987. Revision of the genus *Fragilaria*. *Diatom Research* 2:267–288.

Williams, R. C. 1953. A method for freeze drying for electron microscopy. *Experimental Cell Research* 4:188–201.

Wilson, D. P. 1946. The triradiate and other forms of *Nitzschia closterium* (Ehrenberg) Wm. Smith, forma *minutissima* of Allen and Nelson. *Journal of the Marine Biological Association of the United Kingdom* 26:235–270.

Wimpenny, R. S. 1946. The size of diatoms II. Further observations on *Rhizosolenia styliformis* (Brightwell). *Journal of the Marine Biological Association of the United Kingdom* 26:271–284.

Wiseman, J. D. H., & Hendey, N. I. 1953. The significance and diatom content of a deep-sea floor sample from the neighbourhood of the greatest oceanic depth. *Deep-Sea Research* 1:47–59.

Work, T. M., Beal, A. M., Fritz, L., Quilliam, M. A., Silver, M., Buck, K., & Wright, J. L. C. 1993. Domoic acid intoxication of brown pelicans and cormorants in Santa Cruz, California. *In* "Toxic Phytoplankton Blooms in the Sea" (T. J. Smayda & Y. Shimizu, eds.), pp. 643–664. Elsevier.

Yanagisawa, Y., & Akiba, F. 1990. Taxonomy and phylogeny of the three marine diatom genera, *Crucidenticula, Denticulopsis,* and *Neodenticula. Bulletin of the Geological Survey of Japan* 41:197–301.

Zacharias, O. 1905. *Rhizosolenia curvata mihi,* eine neue Planktondiatomee. *Archiv für Hydrobiologie und Planktonkunde* 1:120–121.

Zeitschel, B. 1982. Zoogeography of pelagic marine protozoa. *Annals Institute Oceanographicque.* Paris 58:91–116.

REFERENCE ADDED IN PROOF:

Tanimura, Y. 1992. Seasonal changes in flux and species composition of diatoms: Sediment trap results from Northwest Pacific, August 1986-November 1988. *Bulletin National Science Museum, Tokyo.* Series C. 18:121–154.

Chapter 3

Dinoflagellates

Karen A. Steidinger in Collaboration with Karl Tangen

INTRODUCTION

GENERAL CHARACTERISTICS

The first evidence of dinoflagellates in the fossil record dates back to the Silurian. Free-living marine dinoflagellates are thus a very old and successful haplontic group of eukaryotic microorganisms adapted to a variety of pelagic and benthic habitats from arctic to tropical seas and estuaries as well as fresh to hypersaline waters. Many species are cosmopolitan and probably represent a complex of ecological strains. Based on morphology and cytology, their motile phase is distinctive at the light microscope level by the placement of two dimorphic flagella and the presence of a permanent fingerprint-like nucleus having continually condensed chromosomes. Biochemically, in addition to chlorophylls *a* and *c*, photosynthetic dinoflagellates have unique accessory pigments. Some species also produce potent neurotoxins.

At the ultrastructural level, dinoflagellates have a common thecal or cell covering structure that, along with their flagellar and nuclear characters, differentiates them from other algal groups. The theca can be smooth and relatively unornamented, as in some *Gymnodinium,* or it can constitute a cell wall of

Identifying Marine Phytoplankton

polysaccharide plates with spines and flanges, as in *Pyrodinium*. Its basic structure is a series of membranes, sometimes with a pellicle layer and microtubules. Thecal vesicles usually constitute the second and third membranes (outer to inner), and can be empty, contain additional membranes, or, in the case of most armored forms, contain polysaccharides such as cellulose, mannose, or galactose. The term amphiesma is synonymous with theca, cortex, or cell covering; however, since much of the historical literature uses epitheca and hypotheca in taxonomic accounts, the term "theca" is the preferred term in this chapter in place of amphiesma or cell covering. Armored forms, either under stress or during reproduction (e.g., gamete formation and gamete fusion), are often capable of shedding their "armor" or a portion of the thecal membrane components leaving a viable spheroplast (Adamich & Sweeney, 1976) or a reproductive cell such as a gamete.

Dodge (1966) termed the dinoflagellate nucleus "mesokaryotic" because it possessed characters intermediate between the coiled DNA areas (2.5 nm) of prokaryotic bacteria and the well-defined eukaryotic nucleus. The 1N nucleus in nonparasitic species has condensed, banded chromosomes during interphase, a permanent nuclear envelope, a permanent nucleolus, and chromosomes attached to the nuclear envelope. The chromosomes are mostly DNA as demonstrated in electron micrographs in which the application of DNAase left little residue (Leadbeater, 1967). Biochemical analyses reveal the absence of typical histones and nucleosomes in most dinoflagellates analyzed although some basic nuclear proteins are present (Rizzo, 1987). Mitotic division, in most free-living species, is typically endonuclear starting with microtubules invading cytoplasmic channels and ending with typical anaphase and telophase.

The number of chromosomes per species or species complex varies from <10 to >500, with many species having upward of 100 for a 1N cell. Additionally, chromosome counts are often expressed as ranges due to fragmentation, aneuploidy, and technique problems. Recently developed fluorescence techniques to quantify DNA using DAPI can also be used to differentiate 1N from 2N cells and perhaps sibling species.

Although dinoflagellate nuclei are not characteristically eukaryotic because they lack histones, nucleosomes, and maintain continually condensed chromosomes during mitosis, this group of microalgae does have typical eukaryotic organelles such as chloroplasts, mitochondria, and golgi bodies.

Nutritional modes of dinoflagellates can vary from autotrophic to mixotrophic whereby photosynthetic cells may be phagotrophic—even on other dinoflagellates (Kimor, 1981; Gaines & Elbrächter, 1987). Heterotrophic species can have specialized structures, such as peduncles, used in phagocytizing other organisms. Food reserves in dinoflagellates are typically unsaturated fatty acids, starch, or both in the same individual cell (Dodge, 1973). Not all dinoflagellates are photosynthetic, particularly large open water pelagic species belonging to the genera *Protoperidinium* and *Gymnodinium* which can be holozoic. Nearly half of the known extant dinoflagellate species are heterotrophic. Very few pelagic heterotrophic species reach high abundance; notable exceptions are *Protoperidinium depressum* and *Oxyrrhis marina*, which seasonally can attain large concentrations in estuarine or nearshore waters. Benthic heterotrophic species can also attain high concentrations, but their distribution, abundance, basic biology, and ecology remains relatively unstudied. Dinoflagellate bloom species that recur in specific areas are obviously adapted or acclimatized to their environmental regimes, e.g., water temperature, salinity, light, water circulation patterns, and nutrients. Light, for example, can influence vertical distribution, photosynthetic rates and efficiencies, chromatic adaptation and pigments, cellular nutrient pools, and metabolic pathways. Vertical migration behavior is thought to be related to positive geotaxis and secondarily influenced by light (Levandowsky & Kaneta, 1987). Temperature can influence photosynthetic and division rates, uptake and respiration rates, cell size, and successional patterns through interspecific competitive adaptations. Physiologically, dinoflagellates have varied strategies relative to the uptake and storage of resources, organic N and P utilization (e.g., alkaline phosphatase and hydroxamate siderophores) and utilization of vitamins, trace metals, and other growth factors. They also produce ectocrines that can inhibit other species, thus imparting a competitive advantage. Many of the environmental variables act synergistically on phytoplankton populations, not only on the dinoflagellates themselves but on associated bacterial populations.

Loeblich (1976) and Taylor (1980), among others advanced the theory that the Prorocentrales constitute ancestral dinoflagellates which evolved into the advanced Peridiniales and Gymnodinales through division of and then loss of plates. This theory may be supported by two lines of evidence, namely, the position of flagella compared to other algal groups and lower chromosome numbers as in *Prorocentrum*. A contrary position that unarmored forms preceded armored forms also exists. This speculation is supported by the theory of multiserial endosymbioses wherein photosynthetic dinoflagellates originated through heterotrophic forms phagocytizing chloroplast-bearing cells, perhaps prasinophytes or chrysophytes (Steidinger & Cox, 1980; Bujak & Williams, 1981). Support for this theory rests in the fossil record and the presence of thecal vesicles in unarmored forms. Another theory (Loeblich III in Spector, 1984) places *Oxyrrhis* as the ancestral dinoflagellate because of its cytological, biochemical, and reproductive features. However, without a better understanding of *Arpylorus* and fossil acritarchs from the Silurian as well as life cycles of extant dinoflagellates, any evolutionary projections are speculative.

Known toxic marine dinoflagellates, consisting of less than 60 of nearly 2000 extant species, vary little from nontoxic free-living dinoflagellates except (1) the majority are photosynthetic estuarine or neritic forms; (2) most probably produce benthic, sexual resting stages; (3) most are capable of producing monospecific or near monospecific populations above background levels, suggesting competitive advantages through exclusion; and (4) all produce bioactive water-soluble and/or lipid-soluble substances that are cytolytic, hemolytic, hepatotoxic, or neurotoxic in activity depending on chemical structure and conversion state.

The question of what is a biological species still remains, particularly when dinoflagellate morphotypes and genotypes are discussed. Today, biochemical genetic markers are being pursued to assess species complexes and the distinctness of geographical isolates that have evolved over time/space. Morphology and cytology are still valid criteria, and in most cases the only available criteria, for separating dinoflagellate species, but these species should be termed morphospecies.

For more detailed reviews on dinoflagellate morphology, cytology, biochemistry, physiology, behavior, reproduction, and ecology see Dodge (1973), Steidinger & Cox (1980), Spector (1984), Pfiester & Anderson (1987), Taylor (1987), Steidinger & Vargo (1988), Larsen & Sournia (1991), Fensome et al. (1993), and Steidinger (1993).

DINOFLAGELLATES: EUKARYOTIC UNICELLS

Flagella: Two dimorphic; nonautofluorescent.

Pigments: Chl a, c_2, betacarotene, peridinin, fucoxanthin, and its derivatives, and other xanthophylls. Some species with chl c_1, chl b, phycobiliproteins, and/or other pigments, e.g., 19' hexanoyloxy–fucoxanthin associated with old or new symbioses.

Nutrition: Autotrophic, auxotrophic, and heterotrophic (saprophytic, phagocytic, and parasitic).

Food Reserves: Starch (polyglucan-like) and oil (long-chained unsaturated fatty acids—C14, C16, C18, and C22).

Sterols: >35 sterols, dinosterol, amphisterol, cholesterol, and 4,24-dimethylcholestanol dominate.

Theca: Multiple membrane complex w/vesicles, some species with scales.

Nucleus: Permanent nuclear envelope and nucleolus; continually condensed, banded chromosomes; phased DNA replication; most species haplont; mitosis closed; 4 to 500 chromosomes. No basic histones nor nucleosomes.

Reproduction: Asexual by binary fission; fission along predetermined plane; sexual—meiosis mostly postzygotic, isogamous to anisogamous, homothallic and heterothallic species, one heterothallic species w/ uniparental inheritance.

Habitat: Marine → freshwater, pelagic, benthic, symbionts, including parasites.

TERMINOLOGY AND MORPHOLOGY

The following diagrams, text, and charts on dinoflagellate structure and interpretation are organized into convenient topics and layers of resolution in order to build upon existing knowledge. In the text, "aka" is equivalent to "also known as."

MORPHOLOGICAL TYPES

Desmokont: A dinoflagellate cell type in which two dissimilar flagella emerge from the anterior part of the cell. See dinokont.

Dinokont: A dinoflagellate cell type in which two flagella are inserted ventrally; one flagellum is transverse and housed in a cingulum (= girdle) and the other is longitudinal and housed in a sulcus. The transverse flagellum provides propulsion and the longitudinal flagellum provides direction. **(Figs. 1 and 2).**

GENERAL CELL TERMS

Antapex: The posterior-most part of the cell body excluding spines, lists, and similar structures. This specific point of the cell may be difficult to determine. See Fig. 1B and **Plate 50** schematic.

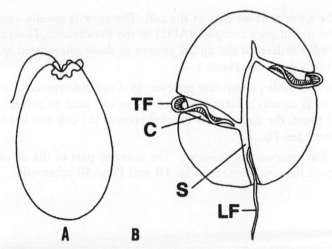

FIGURE 1 (A) Lateral view of a desmokont cell illustrating anterior location of two dissimilar flagella. (B) Ventral view of dinokont cell type illustrating location of two dissimilar flagella, both housed in furrows. LF, longitudinal flagellum; TF, transverse flagellum; C, cingulum; S, sulcus.

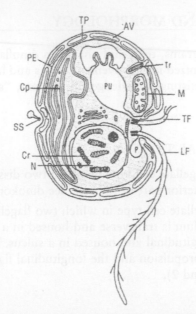

FIGURE 2 Lateral view of dinokont cell illustrating insertion of flagella and typical dinophycean organelles. Thecal vesicle (AV) with thecal plate material (TP), chloroplast (Cp), chromosome (Cr), Golgi apparatus (G), longitudinal flagellum (LF), mitochondrion (M), nucleus (N), pellicle layer (PE), pusule (PU), striated strand (SS) of transverse flagellum (TF), trichocyst (Tr). Redrawn from Taylor (1980).

Apex: The anterior-most part of the cell. The apex is usually associated with the apical pore complex (APC) in the Peridiniales, Gonyaulacales, and similar orders, or the apical groove in those unarmored species that have this feature. See Plates 1–7.

Cingulum aka girdle, transverse groove: In dinokont-type cells, this structure is usually a furrow encircling the cell once or several times; if several times, the cingulum is twisted around the cell and is considered "torsion." See Fig. 7.

Epitheca aka epicone aka episome: The anterior part of the dinokont-type cell above the cingulum. See Fig. 1B and Plate 50 schematic.

PLATE 1 (A) *Katodinium glaucum;* (B) *Gymnodinium pulchellum;* (C) *Gymnodinium mikimotoi;* (D) *Gyrodinium* sp.; (E) *Gyrodinium instriatum;* (F) *Gyrodinium pepo.* Scale = 10 μm. All micrographs courtesy of Dr. Haruyoshi Takayama.

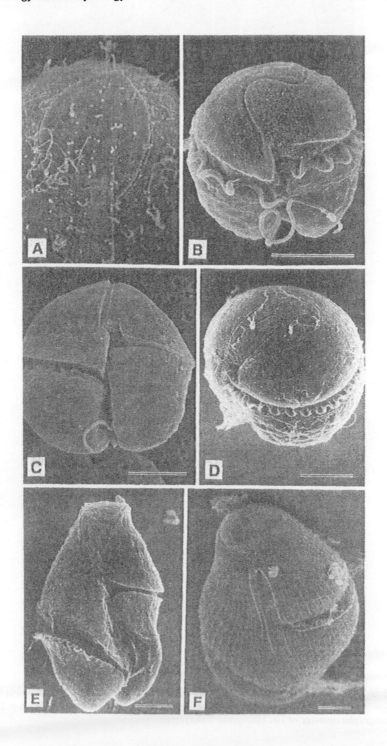

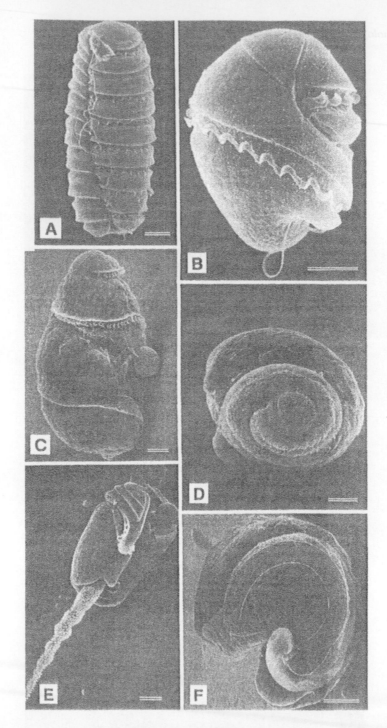

PLATE 2 (A) *Polykrikos schwartzii;* (B) *Nematodinium armatum;* (C) *Warnowia* sp. 1;
(D) *Warnowia* sp. 1.; (E) *Erythropsidinium agile;* (F) *Erythropsidinium agile.* Scale = 10 μm.
All micrographs courtesy of Dr. Haruyoshi Takayama.

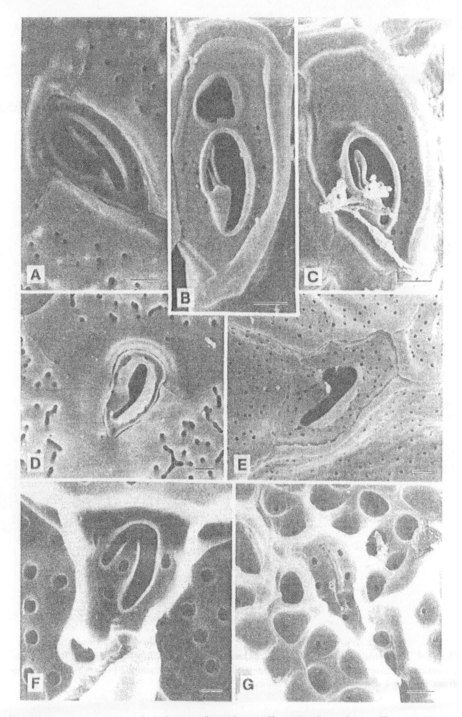

PLATE 3 (A) *Alexandrium balechii*; (B) *Alexandrium affine*; (C) *Alexandrium affine*; (D) *Alexandrium foedum*; (E) *Alexandrium monilatum*; F) *Goniodoma polyedricum*; (G) *Ceratocorys armata* (scale = 2 μm). Scale = 1 μm unless otherwise indicated.

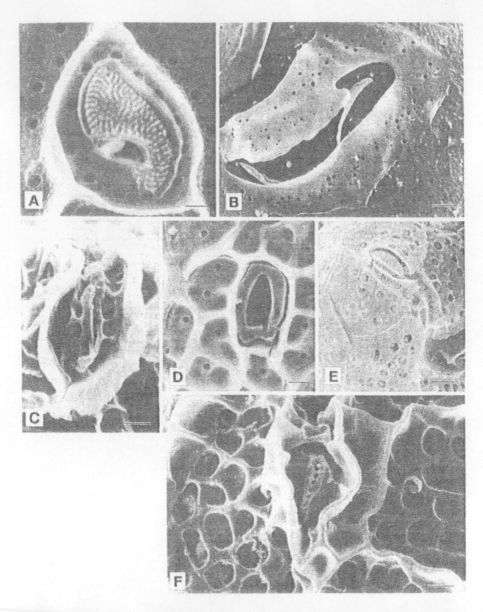

PLATE 4   (A) *Pyrodinium bahamense* var. *bahamense;* (B) *Fragilidium heterolobum;* (C) *Gonyaulax verior;* (D) *Gonyaulax grindleyi;* (E) *Gonyaulax* sp. (scale = 2 μm); (F) *Lingulodinium polyedrum.* Scale = 1 μm unless otherwise indicated.

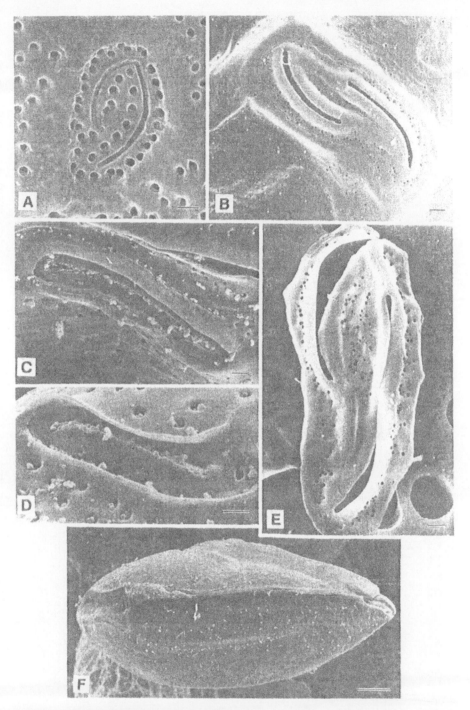

PLATE 5 (A) *Gambierdiscus toxicus;* (B) *Pyrophacus horologium;* (C) *Ostreopsis heptagona;* (D) *Coolia monotis;* (E) *Pyrophacus steinii;* F) *Ostreopsis heptagona* (scale = 10 μm). Scale = 1 μm unless otherwise indicated.

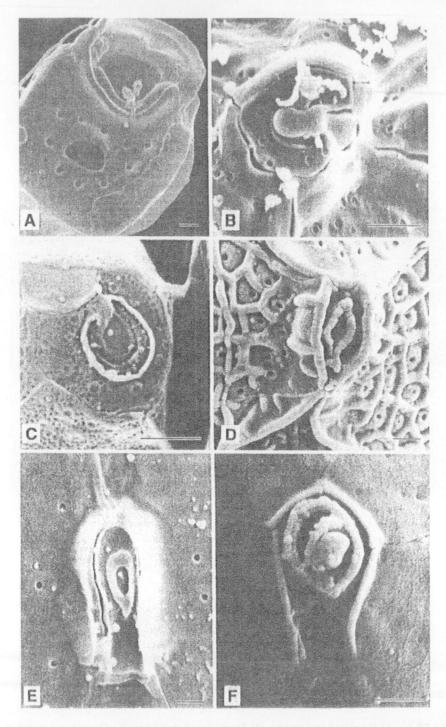

PLATE 6   (A) *Ceratium vultur;* (B) *Heterocapsa triquetra;* (C) *Heterocapsa niei;* (D) *Protoperidinium divergens;* (E) *Diplopelta bomba;* F) *Scrippsiella trochoidea.* Scale = 1 μm.

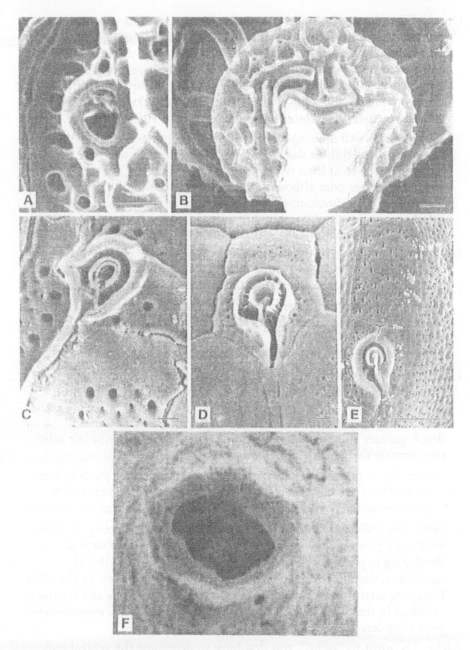

PLATE 7 (A) *Oxytoxum scolopax;* (B) *Corythodinium constrictum;* (C) *Paleophalacroma unicinctum;* (D) *Blepharocysta* sp.; (E) *Lissodinium orcadense;* (F) *Podolampas bipes.* Scale = 1 μm. D, E, and F courtesy of Consuelo Carbonell-Moore.

Dorsal: The dorsal side (back) of the dinokont-type cell is opposite the ventral side (front) where the sulcus is located. When using the light microscope to observe dinoflagellates, it is important to know whether you are focused on the ventral surface or whether you have focused through the dorsal side to the inside of the ventral surface. In the latter case, you will be observing a reversed image. Also see ventral. See **Plate 18,** *Gymnodinium mikimotoi,* **Plates 24, 41,** and **50** schematic.

Flagellar pore(s): All dinoflagellates at some time in their life cycle have two dissimilar flagella; these flagella either emerge through one pore or two separate pores. In at least two species of *Prorocentrum,* the two flagella emerge from one pore although there are two pores in the periflagellar plate area. In the dinokont-type cell there are typically two pores in the sulcal area although often only one pore is visible because the other pore is hidden by the ventral edge, the peduncle, or a sulcal plate.

Hypotheca aka hypocone aka hyposome: The posterior part of a dinokont-type cell below the cingulum.

Lateral: The lateral view of a cell is the left or right side view which usually represents the depth contour of a cell. In dinokonts, to orient to left and right, focus on the ventral surface with the sulcus and think of the left as the left side of the cell, not the left side of the microscope field. See **Plate 18,** *Gymnodinium breve,* and **Plate 50** schematic.

Left: Plate series in armored dinokonts are tabulated or counted starting from the left side of the cell so that if a reference article mentions the shape of the sixth precingular as a diagnostic character, you have to be able to identify the 6″ from other precingular plates. Unarmored dinoflagellates also have left sides; for example, the posterior left side may contain the nucleus. See right and lateral terms.

Peduncle: A cytoplasmic appendage located near the flagellar pores in some photosynthetic as well as nonphotosynthetic species. It is an extensible organelle associated with phagotrophy and may also have other functions.

Pellicle: The pellicle of some dinoflagellates is a chemically resistant layer which can give rigidity to an ecdysial cell or to other cells such as developing hypnozygotes. If present, it is below the outer thecal membrane, the top vesicle membrane, and the thecal plates. In the older literature, unarmored dinoflagellates with apparently thick cell coverings or ridges in the covering (e.g., in some *Gyrodinium* and *Gymnodinium*) were considered pelliculate and rigid. See **Fig. 2.**

Right: To orient to "right," you first have to determine the ventral and dorsal surfaces of the cell. If you can equate a dinoflagellate cell with the human body, the right side is equivalent to the right arm, right leg, etc. If you were looking directly at the ventral surface of a dinokont cell it would be on your left. However, if you have a reversed image either by focusing through a cell or because of the optics of the microscope, then the cell's right side appears as the left side. This is a critical determination and presents problems to those first working with dinoflagellates. See left term.

Sulcus: The longitudinal area on the ventral surface of the cell that forms a more or less pronounced furrow or depression that houses the longitudinal flagellum. In some armored species the sulcus, which is mainly in the hypotheca, invades the epitheca as a plate (the anterior sulcal or sa). In unarmored species it can be associated with an apical groove that has its origin anteriorly.

Ventral: The ventral side of the dinokont-type cell is identified by the presence of the sulcus and the juncture of the cingulum–sulcus. It is also the place of flagellar insertion in dinokonts.

MICROANATOMY

Nematocyst: The ejectile organelles of *Nematodinium, Warnowia,* and some *Polykrikos.* Structurally, they have component parts called the posterior body and anterior operculum that are further divided into identifiable parts. They are often arranged radially or subradially in the cell. The nematocysts of *Nematodinium* and *Polykrikos* are structurally different. These organelles should not be equated to the stinging cells called cnidoblasts of cnidarians (**Fig. 3**).

Ocellus aka ocelloid: This organelle occurs in the Warnowiaceae and consists of a hyalosome, a dorsally pigmented melanosome, and an ocelloid chamber. This is a complex photoreceptor that can focus images and is sensitive to light. This organelle is typically located in the left part of the cell (**Fig. 4**).

Pusule: Organelle of variable complexity which may be regarded as a specialized vacuole opening through the cell surface usually in the flagellar area (**Fig. 5**).

Theca aka amphiesma, cell covering, cell wall: All dinoflagellates have a membrane system encompassing the whole cell consisting of a complex of three to six membranes. Different authors use different terms for this complex based on their own preference. The use of the term theca makes the use of epitheca and hypotheca more practical. See **Fig. 2.**

Thecal plates: Plates of armored dinoflagellates are composed of cellulose or some other polysaccharide microfibrils. These plates are formed in thecal vesicles which may or may not contain other membrane or osmophilic-staining structures.

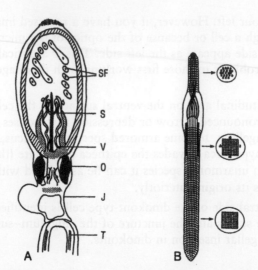

FIGURE 3   (A) Nematocyst of *Polykrikos* with joining piece (J), the operculum (O), the striker (S), the spiral filament (SF), a toeniocyst (T), and a valve (V). Redrawn from Taylor (1980). (B). Another ejectile organelle called a trichocyst (redrawn from Bouck & Sweeney (1966).

Thecal vesicles: These polygonal vesicles are membrane bound and usually compressed against one another in the cell covering or theca. The vesicles represent the second and third membrane profiles seen in cross section with an electron microscope.

CHARACTERS USED IN IDENTIFYING PROROCENTROID DESMOKONT CELLS

Metacytic growth zone: The cell growth that occurs at the suture between the two valves of the Prorocentrales or the fissure halves of the Dinophysiales. When this usually horizontally striated zone is at its maximum extent, the cell will be at its greatest depth or width, respectively. See **Fig. 6.**

Periflagellar plates: Consist of the anterior plates or platelets around two pores, one of which is a flagellar pore. These plates typically indent the right valve.

Pores aka trichocyst pores: Pores are openings or channels in the theca of desmokonts and dinokonts that can be involved in pinocytosis, extrusion of trichocysts or mucocysts, and other active processes. Pore number and location are variable within a species, but in many groups, the pattern is genetically determined and consequently a reliable, but variable character for identification of species, e.g., within *Prorocentrum*.

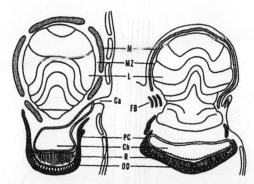

FIGURE 4   Ocelli of *Nematodinium* and *Erythropsidinium* with the canal (Ca), fibrillar bands (FB), lens (L), mitochrondrion (M), microtubular zone (MZ), pigment granules (OD), pigment cup (PC), chamber (Ch), and the retinoid (R). Redrawn from Taylor (1980).

Valves: In the Prorocentrales or thick-walled desmokonts, two opposing halves of the theca are called valves. The right valve is the one most indented anteriorly by the periflagellar plates.

CHARACTERS USED IN IDENTIFYING DINOKONT CELLS

Unarmored

Unarmored dinokont cells can be identified to family, genus, and species by a combination of characters using (1) size, shape, and proportions of living or well-fixed whole cells; (2) cingular position, displacement, and overhang; (3) sulcal placement and intrusion; (4) the presence/absence of thecal ridges; (5) the presence/absence of an apical groove, its shape, and relationship to the sulcus; (6) the presence/absence of a peduncle and position; and (7) the presence and location of organelles, e.g. nucleus, nematocysts, and ocelloid. Unarmored species can be biconical, round, ovoid, oblong, posteriorly bilobed, and various other descriptive shapes (**Fig. 7**).

Apical groove aka acrobase: A groove composed of many thecal vesicles resolved by silver staining; originally described as an acrobase with subcomponent parts. The groove itself, without the substructure, is called the apical groove and is located at the anterior part of many unarmored dinoflagellate cells. It can be straight, curved, or looped and extends posteriorly both on the ventral and on the dorsal sides but on the ventral surface it does not exceed the juncture with the sulcus. See **Plates 1** and **2**.

Circular cingulum: A cingulum which is not displaced and in which the proximal end meets the distal end. See **Fig. 7C**.

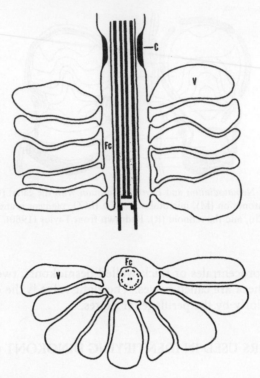

FIGURE 5 Pusule with flagellar pore constriction (C), flagellar canal (Fc), and pusule vesicles (V). Redrawn from Dodge & Crawford (1968).

Displaced cingulum: A cingulum in which the distal end is either above (ascending) or below (descending) the proximal end which is always on the left side of the cell. See Figs. 7A and 7E (descending) and **Plate 54,** *P. pellucidum.*

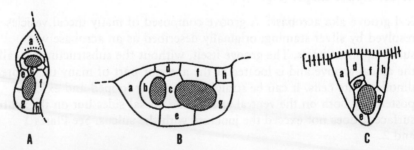

FIGURE 6 (A) *Prorocentrum micans,* (B) *Prorocentrum marinum,* and (C) *Prorocentrum lima.* Redrawn from Taylor (1980).

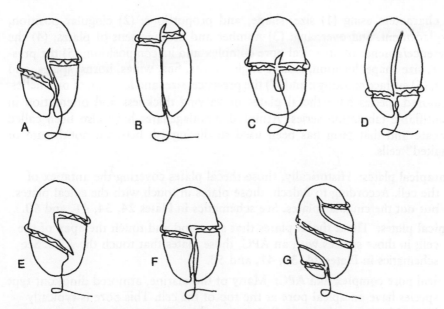

FIGURE 7 (A) Displaced with sulcal intrusion onto the epitheca; (B) premedian; (C) median; (D) postmedian; (E) overhang; (F) displaced without cingular intrusion onto the epitheca; (G) torsion with the cingulum and sulcus going around the cell more than once.

Median cingulum aka equatorial girdle: The cingulum which is located approximately at the midpoint of the cell. See Fig. 7C.

Postmedian cingulum: A situation in dinokont-type cells when the cingulum is below the midpoint of the cell. See Fig. 7D.

Premedian cingulum: In dinokont-type cells when the cingulum is above the midpoint of the cell. See Fig. 7B.

Unarmored aka naked, athecate: Dinokont-type cells that do not have identifiable plate series and do not have apical pore complexes but may have an apical groove. More recently, scanning electron microscopy has revealed identifiable plate series in "gymnodinioid" dinokonts and those with an apical pore complex are actually armored. *Symbiodinium* species, although they have thin plates, also have apical grooves and are transitional between unarmored and armored.

Ventral ridge aka ventral flange: An identifiable ridge on the right side of the sulcal intrusion onto the epitheca that structurally has a microtubular complex.

Armored

Armored dinokont cells, e.g., Dinophysiales, Gonyaulacales, and Peridiniales, can be identified to order, family, genus, and species by a combination of characters using (1) size, shape, and proportions; (2) cingular position, displacement, and overhang; (3) number and arrangement of plates; (4) the presence/absence of an apical pore complex and its composition; (5) the presence, size shape, location, and/or angle of lists, fins, wings, horns, spines, and ribs; (6) surface markings; and (7) the presence, size, and location of organelles. Armored species have thecal plates of varying thickness and orientation in identifiable tabulation series. Armored dinoflagellates have also been called thecate and that term has been used to distinguish between nonthecate or "naked" cells.

Antapical plates: Historically, those thecal plates covering the antapex of the cell. According to Balech, those plates in touch with the sulcal plates but not the cingular plates. See schematics in **Plates 24, 34, 41,** and **50.**

Apical plates: Those thecal plates that surround and touch the apex of the cell; in those species with an APC, those plates that touch the APC. See schematics in **Plates 24, 34, 41,** and **50.**

Apical pore complex aka APC: Many of the marine, armored dinokont-type species have an apical pore at the top of the cell. This pore is typically located in a special plate called a pore plate. The pore is not always a round or oval hole, but can be one or two slits. If the pore is a hole, then it may have a closing or cover plate or what Balech calls a canopy. The canopy appears to be the outer membrane of the thecal membrane complex and covers the pore, whereas the closing plate is a separable plate. Often the closing plate is detachable and may be missing in prepared specimens. In addition, there can be a ventral apical plate or what has been called a canal plate or an X plate. This plate is always posterior and ventral to the pore plate. See schematics in **Plates 24, 34, 41,** and **50** and **Plates 3–7.**

Areolate: In armored dinoflagellates, ornamentation on thecal plates that approximates deep or shallow depressions with or without raised sides. The sides may be polygonal or round and are closely appressed. Areolae can contain pores, even double pores. See **Plate 11,** *D. acuta,* and **Plate 14,** *P. rotundatum.*

Armored aka thecate: Armored species have thecal plates of varying thickness and orientation in identifiable tabulation series. At one time, armored dinoflagellates were also called thecate and that term was used to distinguish between armored and naked cells.

Attachment pore: Unarmored and armored dinoflagellates that form chains or longitudinally connected cells that swim as a chain, often have cytoplasmic contact between cells. In such cases with the armored species, an anterior attachment pore is typically located in the pore plate of the APC in addition to the apical pore. In *Alexandrium,* the attachment pore in relation to the apical pore can be a diagnostic character. The posterior attachment pore is typically located in the posterior sulcal plate and can also be a diagnostic character. See schematic of **Plate 34.**

Denticulate: Thecal markings or surface ornamentations consisting of spines having a broad base which taper to a narrow point like a tooth.

Depression aka pit: Surface ornamentation on thecal plates. A depression typically with a rim and not appressed to another depression. Depressions can have pores. See **Fig. 11.**

First apical plate: The plate in armored dinokonts that is typically situated directly above the sa plate. In addition, it is typically directly below the APC in those species that have this feature. Some species have what is called a "displaced" first apical plate not in direct contact with the APC but connected to it by a vertical suture. The displaced 1' is directly above the sa plate. The 1' plate can be four to six sided and is used as a diagnostic character in *Protoperidinium.* See **Fig. 8** and **Plate 34** for schematic.

Homologous plates: The tabulation of plates occurs in defined series or groups. To look at relationships between species and higher ranks in the armored types, a plate in the precingular series may equate to a 1' that is displaced from the apex or a sulcal plate may equate to a postcingular. Plate overlap patterns and other patterns are being used to infer homologous structures based on form.

Horns: In armored species only, these prominent apical or antapical extensions of the cytoplasm are covered by thecal plates. An apical horn forms the apex of the cell and is formed by apical plates. Antapical horns are typically formed by antapical plates; however, in *Ceratium,* they are formed by both antapical and posterior intercalary plates. Therefore, in *Ceratium* they are called hypothecal horns. Horns can be hollow or portions can be solid. See **Plates 24 and 50** schematics.

Intercalary plates: In armored species, these plates are those located between the precingular and the apical series (anterior intercalaries) or those plates between the postcingular and the antapical series (posterior intercalaries). Anterior intercalary plates are by definition not in contact with the apex or the cingulum, and posterior intercalary plates are not in

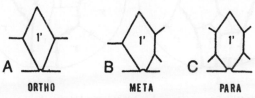

A ORTHO   B META   C PARA

FIGURE 8 (A) Ortho; (B) meta; (C) para. Redrawn from Taylor (1980).

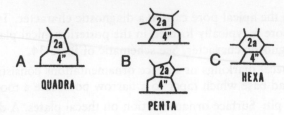

FIGURE 9   (A) Quadra; (B) penta; (C) hexa. Redrawn from Taylor (1980).

touch with the cingulum but may touch the antapical plates. The second anterior intercalary (2a) may be four to six sided, i.e., quadra, penta, or hexa, and is used as a diagnostic character in *Protoperidinium*. See Fig. 9 and **Plates 41 and 50** schematics.

List aka wing, flange: Membranous extensions of armored dinoflagellates, e.g., cingular lists and sulcal lists that extend beyond the cell boundary. Sometimes these extensions are curved and the space created harbors pigmented symbiotic cyanobacteria. See **Plate 10** schematic and **Plate 13, and Plate 48,** *D. lenticula*.

Plate overlap aka imbrication: Typically, in the Peridiniales, thecal plates adjoin neighbor plates with bevelled peripheral plate edges; the angle of the bevel depends on whether the plate is the overlying or underlying plate. Plate overlap or imbrication patterns are conservative characters and are being used to infer plate homologies (**Fig. 10**).

Reticulae: Surface ornamentation on thecal plates where raised straight or irregular lines cross one another forming a network or mesh of varying shape and size. The reticulae may even be incomplete (**Fig. 11**).

Ribs: Supports for sulcal lists, e.g., in the Dinophysiales, and cingulum lists of many *Protoperidinium* species. See **Plates 11–14.**

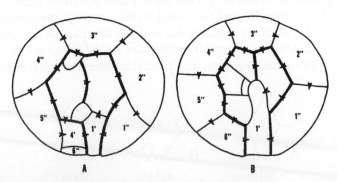

FIGURE 10   Plate overlap of gonyaulacoids with arrows showing direction of overlap. (A) Fossil genus; (B) *Lingulodinium polyedra*. Redrawn from Gocht (1981).

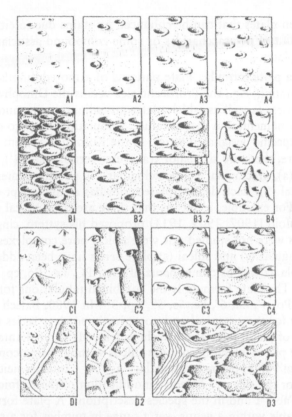

FIGURE 11 (A) A1–A4 show irregularly arranged pores of the same or varying size; (B) B1–B4 show depressions and pores and depressions and spines; (C) C1–C4 show pores associated with variously shaped reliefs; (D) D1–D3 show different combinations of reticular reliefs and simple pores. Redrawn from Andreis et al. (1982).

Spines: Solid protuberances on the thecal plates that usually taper to a point; they can be short or long or broad or narrow.

Striae: Surface ornamentation on unarmored or armored dinoflagellates that appear as longitudinal lines or ridges; on armored species the striae can be interrupted by pores and may be associated with other markings such as reticulations. See **Plate 20,** *Gyrodinium spirale,* and **Plate 42,** *Gonyaulax fragilis.*

Sutures: In armored species, visible linear boundaries between plates that usually indicate the end of one plate and the beginning of another. In many species, the substance that holds plates together is susceptible to chemical agents, such as Chlorox, and plates can be physically separated by using such chemicals. In light microscopy, sutures typically appear as lines between plates that can be enhanced by stains due to either the higher amount of polysaccharides or some compound associated with the binding of plates.

Tabulation aka plate formula: Plate patterns of armored dinoflagellates reflect orientation, e.g., the Prorocentrales versus the Dinophysiales versus the Peridiniales, as well as which plates are adjoining one another or what type ornamentation they produce. Tabulation refers to counting the plates in a specific designated series. In the Kofoidian System of plate nomenclature, there are six major transverse series: apical ('), anterior intercalary (a), precingular ("), postcingular (‴), posterior intercalary (p), and antapical (‴′). Each defined plate series has a superscript designation or a letter. For example, 3' means that there are three apical (') plates. At the time Kofoid (1907, 1909, 1911) proposed his system, cingular and sulcal plates were not studied nor were apical pore complexes. More recently, cingular (c) and sulcal (s) plate series have been added to the plate formula as well as components of the APC, i.e., Po, cp, and X, if they occur. The plate formula is the combined tabulations for the plate series, e.g., Po, 4', 0a, 6", 8c, 5s, 5‴, 1p, 3‴′. In 1980, Balech suggested a clarification for the antapical and posterior intercalary series that involves positioning of plates in relation to the sulcus. He defined antapical as ". . . those plates which border the sulcus without being connected with the cingulum" and posterior intercalary as ". . . touches neither the cingulum nor the sulcus." Balech's modification of the Kofoidian scheme is used in **Table 1** and in the species descriptions. A plate formula is generally stable within a genus, yet a range in number for a specific series can occur due to splitting of plates or other factors. Older species descriptions typically do not include the sulcal and cingular plates. There are other tabulation schemes, e.g., Eaton (1980) and Taylor-Evitt in Evitt (1985). Eaton's is the simplest with three series in the epitheca and three series in the hypotheca: precingular (plates touch cingulum), apical (plates touch precingulars), apical closing (plates anterior to apicals), postcingular (plates touch cingulum), antapical (plates touch antapex), and antapical closing (plates posterior to antapicals). Intercalaries do not exist in Eaton's scheme, but cingulars and sulcals are counted.

Transitional plate—t: A small plate located in the cingulum–sulcus juncture. If the t plate is on the left side of the cell in the cingulum, it is really the first cingular plate and is counted as such. Occasionally, you will see cingular tabulations that read 3 + 1t, the tabulation should read 4c. If the t plate is on the right side of the cell at the cingulum–sulcus juncture, it is a sulcal plate and is so tabulated **(Fig. 12)**.

Ventral pore: In some gonyaulacoids, a ventral pore may be present at the juncture of the first apical plate (1') and an anterior intercalary or another apical plate. Sometimes the pore is in one of the apical plates or an intercalary. The presence of a ventral pore or its placement may be diagnostic for certain species, also in the epitheca of dinophysoids. See **Plate 34** schematic.

Vermiculate: Plate ornamentation patterns in armored species that are raised worm-like markings. Markings such as reticulae, areolae, vermiculae, etc., may not be visible on newly formed thecae because they are among the last thecal ornamentation characters to be formed.

"x" Plate: In *Crypthecodinium* and *Hemidinium* the distal end of the cingulum does not meet the proximal end, it ends on the right ventral surface. In *Crypthecodinium,* the plate that occupies what would be a cingular plate (and a postcingular) in other armored dinokonts is called an x plate. See **Plate 34,** *C. cohnii.*

Life Cycles

Life cycle: A continuum of phases and cell types in the reproduction and growth of a species. The life cycle usually contains at least an asexual phase in which a cell can divide by binary fission and produce two similar cells. It may also contain a sexual phase in which gametes fuse to form zygotes (2N) and these zygotes produce 1N cells (**Fig. 13**).

Archeopyle: This is the area of a cyst where the cell will emerge during excystment. If the cyst does not split or separate at the cingulum, it usually will have an archeopyle and an operculum associated with the escapement hole; the operculum may even be hinged. See **Plate 51,** cysts.

Asexual: Reproduction where a 1N cell produces two to four cells with the same chromosome number. This can be by binary fission of a motile stage or a nonmotile stage. In many armored dinokonts, the original cell splits along predetermined sutures and then each half produces a new half with new thin plates (**Figs. 14 and 15**).

Coccoid: Nonmotile stages of the life cycle, usually referring to vegetative phase cells that dominate the life cycle and can be thick walled.

Diploid: A cell (zygote) that has a nucleus with two sets of chromosomes (2N).

Ecdysis: The process of shedding or casting off the thecal plates or armor whereby one of the internal membranes, the innermost membrane, becomes the new outer membrane. When the induction of ecdysis is due to stress and the stress is removed, the spheroplast can regenerate a new theca. Not all ecdysis is due to stress, it can be involved in fusion of armored gametes or other reproductive processes like cell division.

Encystment: There are several types of encystment. Stressed cells can "round up" and settle out of the water column and yet be viable if the

TABLE 1 Kofoidian Plate Formulae for Selected Armored Dinoflagellate Genera

Genus	Po	cp	X	'	a	"	c	s	'''	P	''''	Text location
Centrodinium Kofoid	+	0	0	2	3	7	5	?	5	0	2	p. 516
Corythodinium Loeblich Jr. & Loeblich III	+	0	0	3	2	7	5	4?	5	0	1	pp. 516–517; Plates 7a and 45
Oxytoxum Stein	+	0	0	5	0	5	5	?	5	0	1	pp. 517–519; Plates 7a and 45
Amphidiniopsis Woloszynska	+	0	0	4	1–3	5–6	3?	?	5	0	2	p. 546
Roscoffia Balech	+	0	0	4	0	5	5	3	5	0	2	pp. 546–547
Adenoides Balech	+	0	0	3	0	5	5	6	5	0	4	p. 550
Thecadinium Kofoid & Skogsberg	+	0	0	3	1	5?	5?	6	3	0	2	p. 547; Plate 55
Cladopyxis Stein	+	0	0	3	3	6	6	7	6	0	2	p. 487
Paleophalacroma Schiller	+	0	0	4	3	6	6	6	6	0	2	p. 487; Plate 7c
Coolia Schmidt	+	0	0	3(4)	0	7(6)	6	6?	5	1	2	pp. 513 and 515; Plates 5d and 44
Ostreopsis Munier	+	0	0	3(4)	0	7(6)	6	6?	5	1	2	pp. 515; Plates 5c, 5f, and 44
Amphidoma Stein	+	0	0?	6	0?	6	6	4?	6	0	2	p. 504; Plate 41
Gambierdiscus Adachi & Fukuyo	+	0	0	4	0	6	6	8	6	0	2	p. 501; Plate 5a
Amylax Munier	+	0?	0	3	3	6	6	7–8	6	0	2	p. 504; Plate 41
Ceratocorys Stein	+	0?	0	3	1	5	6	10	5	0	1	p. 482; Plates 3g and 30
Ceratium Schrank	+	0?	0	4	0	5–6	5–6	2+	6	0	2	pp. 470–482; Plates 6 and 24–33
Gonyaulax Diesing	+	0?	0	3	2	6	6	7	6	0	2	pp. 506–507; Plates 4e, 41 and 42
Lingulodinium (Stein) Dodge	+	0?	0	3	3	6	6	7	6	0	2	pp. 509–510; Plates 4f and 43
Schuettiella Balech	+	0?	0	2	1	6	6	9	6	0	2	p. 512; Plate 43
Spiraulax Kofoid	+	0?	0	3	2	6	6	7	6	0	2	p. 512
Fragilidium Balech ex Loeblich III	+	+	0	4–5	0	7–9	9–11	6–8	7–8	1	2	pp. 520–521; Plate 4b

Taxon												Reference
Pyrophacus Stein	+	+	0	5–9	0	7–15	9–16	8	8–17	0–15	3	p. 523; Plates 5b, 5e, and 46
Alexandrium Halim	+	+	0	4	0	6	6	9–10	5	0	2	pp. 488–500; Plates 3 and 34–39
Goniodoma Stein	+	+	0	4	0	6	6	6	6	0	2	p. 501; Plates 3f and 40
Heterodinium Kofoid	+	+	0	3	2	6	6	?	6	0	3	p. 513; Plate 44
Protoceratium Bergh	+	+	0	3	0	6	6	6	6	0	2	pp. 419–426; Plates 8 and 9
Pyrocystis Murray ex Haeckel	+	+	0	4	0	6	6	5–7	5	0	2	pp. 519–520; Plate 45
Pyrodinium Plate	+	+	0	4–5	0	6	6	6	6	0	2	p. 503; Plates 4 and 40
"Phantom"	+	+	+	4	1	5	6	4	5	0	2	p. 550; Plate 55
Peridiniella Kofoid & Michener ex Balech	+	0	+	4	3–4	7	6	6–7	6	0	2	p. 510; Plate 43
Ensiculifera Balech	+	0	+	4	3	7	5	5	5	0	2	pp. 523–524; Plate 47
Pentapharsodinium Indelicato & Loeblich III	+	0	+	4	3?	7	5	4	5	0	2	pp. 524–526; Plate 47
Scrippsiella Balech ex Loeblich III	+	0	+	4	3	7	6	4–5	5	0	2	pp. 525–527; Plates 6 and 47
Boreadinium Dodge & Hermes	+	0	+	4	1	7	4	5	5	0	1	p. 527
Diplopelta Stein ex Jørgensen	+	0	+	4	1	6	4	6	5	0	2	pp. 527–529; Plates 6 and 48e
Diplopsalis Bergh	+	0	+	3	1	6	4	5	5	0	1	p. 529; Plate 48a
Diplopsalopsis Meunier	+	0	+	4	1	7	4	6	5	0	2	p. 529; Plate 48f
Oblea Balech ex Loeblich, Jr., & Loeblich III	+	0	+	3	1	6	4	6	5	0	2	pp. 529–530; Plate 48g
Preperidinium Mangin	+	0	+	4	1	7	4	5	5	0	1	p. 530; Plate 48b
Protoperidinium Bergh	+	0	+	4	2–3	7	4	6	5	0	2	pp. 534–546; Plates 6d and 50–54
Heterocapsa Stein	+	+	+	6	3	7	6	5	5	0–1	2	pp. 530–531; Plates 6b, 6c, and 49
Crypthecodinium Biecheler	0	0	0	4	3	5 + x	6	5	5	0	3	pp. 516–518
Gotius Abé	0	0	0	4	1	6	4	5	5	0	2	p. 529; Plate 48

(continues)

TABLE 1 (*Continued*)

Genus	Po	cp	X	′	a	″	c	s	‴	p	⁗	Text location
Peridinium Ehrenberg	0/+	0	0/+	4	3	7	5–6	5–6	5	0	2	p. 539
Blepharocysta Ehrenberg (no cingulum)	+	+	+	3	1	5	3	4	4–5	0	1	p. 533; Plates 7d and 49
Lissodinium Matzenguer emend. Carbonell-Moore (no cingulum)	+	+	+	3	1	5	3	5	5	0	1	p. 533; Plates 7e and 49
Podolampas Stein (no cingulum)	+	+	+	3	1	5	3	5	5	0	1	p. 534

Selected Armored Dinoflagellate Genera without Kofoidian Plate Formulae

	Plate descriptions	Text location
Prorocentrum Ehrenberg	Left and right valves; 5–14 periflagellar pore plates; no kofoidian series	pp. 419–426; Plates 8 and 9
Dinophysis Ehrenberg	All Dinophysiales have a similar plate formula of 18 or 19 Plates; 4E 2A 4C 4-SS 4H	pp. 428–434; Plates 11–17
Phalacroma Jørgensen		pp. 437–439; Plates 14–16
Ornithocercus Stein		pp. 436–437; Plates 13 and 16
Histioneis Stein		p. 434; Plate 13
Amphisolenia Stein		pp. 426–428; Plate 10

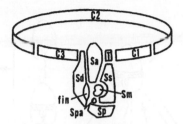

FIGURE 12 The "t" plate is on the left side of the cell and is therefore a cingular plate, c1. Redrawn from Taylor (1980).

conditions of stress are removed. Others involve temporary cysts for asexual reproduction, flotation, and other functional aspects of individual survival. Yet another type of encystment involves sexual reproduction and the production of thick-walled hypnozygotes which can remain encysted for months, even years. These resting cysts or hypnozygotes have their own classification system, distinguishing morphological characters, and applied terminology (see Fig. 13).

Excystment: When the hypnozygote matures and is ready to produce a motile cell from the resting cell, a naked meiocyte will emerge from the archeopyle or opening in the wall. This emerging cell will either be flagellated or amoeboid. Typically, this cell will undergo meiosis and produce four vegetative 1N cells that are motile in the water column (see Fig. 13).

Gametes: In armored and unarmored species a 1N cell that fuses with another 1N cell to produce a zygote (2N). Gametes can be like-sized cells and morphologically similar cells (isogametes) or unlike-sized cells and morphologically similar or dissimilar (anisogametes) (Fig. 16).

Haploid: Vegetative or gametic cells that have one set of chromosomes. Dinoflagellates are haplonts meaning that their dominant stage is haploid (1N) and their zygotic sexual stages are diploid (2N), with the exception of *Noctiluca*.

Hypnozygote aka dinocyst: This is a diploid zygotic stage in the sexual life cycle of dinoflagellates. All but one dinoflagellate species is a haplont with the planozygote and hypnozygote being the only 2N stage. Dinoflagellates with a hypnozygote are typically dimorphic in that the 1N and 2N zygote stages are morphologically dissimilar. The hypnozygote is a resting, nonmotile stage that settles to the sea bottom and may be dormant for some time depending on entrainment of internal biological clocks or other triggering mechanisms. It is not known whether all dinocysts are zygotes (see Fig. 13).

Planozygote: A motile zygote (2N) cell produced by the fusion of (1N) gametes. Typically, the planozygote is morphologically similar to the vegetative (1N) cell (see Fig. 13).

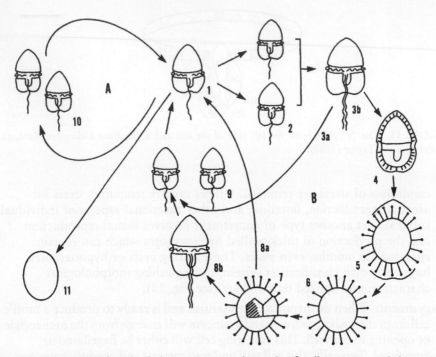

FIGURE 13   General life cycle of a hypnozygote-producing dinoflagellate.(A) Asexual phase with
motile, planktonic vegetative cell (1) which divides by binary fission (10) and sometimes forms
nonmotile temporary cysts (11); (B) Sexual phase with motile, planktonic vegetative cell (1) which
produces gamete pairs (2) that fuse to form a planozygote (3). The planozygote can produce
motile, planktonic vegetative cells (9) by meiosis (3a) or cyst formation (4) may occur and proceed
(5) by expansion of the cyst to form a resting cyst (hypnozygote) (6). Excystment (7) can produce
a planozygote (8b) which continues to divide to produce a motile vegetative cell (1) or division
can take place completely during excystment (8a) to produce a motile vegetative cell (1) directly.
Redrawn from Dale (1986).

Sexual: Sexual reproduction involves the production of gametes that fuse to
   produce a zygote. Sexual reproduction has been documented in the
   laboratory for >30 species where most produce hypnozygotes. Some only
   produce planozygotes and it is not known whether the planozygote stage
   is a "resting stage" or if the conditions for hypnozygote formation were
   not simulated.

IDENTIFICATION OF SPECIES

   The following dinoflagellate species descriptions are presented by order
then alphabetically by family. Planktonic and benthic species are detailed to

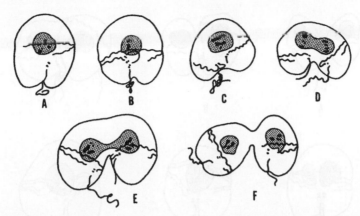

FIGURE 14 Asexual division of *Gyrodinium uncatenum* with typical mitotic phases. Redrawn from Coats (1984).

provide the reader with the diversity of morphological characters that are needed to differentiate between genera or species. Many benthic species are tychoplanktonic and found at some time in the water column. Also, many

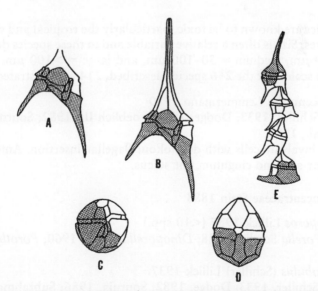

FIGURE 15 In asexual division of armored species, division can occur along predetermined lines creating daughter cells with specific plate series; shaded areas indicate the portion inherited from the parent cell. (A) Indistinct tabulation of apical horn of young freshwater *Ceratium furcoides*; (B) apical horn tabulation evident on older cell of the same *Ceratium*; (C) *Goniodoma sphericum*; (D) freshwater *Peridinium lubiensiforme*; (E) *Ceratium* gametes. Redrawn from Evitt (1985).

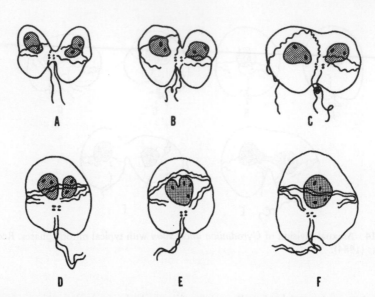

FIGURE 16 Gamete fusion in *Gyrodinium uncatenum*. (A) Like gametes (isogametes) fuse until in the nuclei fuse (E) and produce a planozygote (F). The planozygote will form a hypnozygote that is round and thick walled. Redrawn from Coats et al. (1984).

benthic species are known to be toxic, particularly the tropical and subtropical representatives. Size is often a relative variable and in these species descriptions small = <50 μm, medium = 50–100 μm, and large = >100 μm. These are approximate scales. Of the 246 species described, 214 are illustrated in plates.

Order Prorocentrales Lemmermann 1910
References: Schiller, 1933; Dodge, 1982; Loeblich III, 1982; Sournia, 1986; Fensome et al., 1993.
  Armored, bivalvate cells with desmokont flagellar insertion. Anterior periflagellar area; no cingulum nor sulcus.

Family Prorocentraceae Stein 1883

Genus *Mesoporos* Lillick 1937 (<10 spp.)
Synonyms: *Porella* Schiller 1928; *Dinoporella* Halim 1960; *Porotheca* Silva 1960.
Type: *M. globulus* (Schiller) Lillick 1937.
References: Schiller, 1933; Dodge, 1982; Sournia, 1986; Subrahmanyan, 1966.
  Armored, small round to ovoid cells somewhat compressed laterally. Bivalvate; similar to *Prorocentrum* but with a large central cone-shaped pore in each valve. Surface markings present. Chloroplast(s) present.

Remarks: Resolution of the periflagellar area may help further separate this genus from *Prorocentrum*.

Mesoporos perforatus (Gran) Lillick 1937 (not illustrated)
Surface covered with small pyramid-shaped papillae; area around the central pore in both valves is clear of papillae.
Distribution: Neritic to oceanic; cold temperate to tropical waters.

Genus *Prorocentrum* Ehrenberg 1833 (<50 spp.)
Synonym: *Exuviaella* Cienkowski 1881.
Type: *P. micans* Ehrenberg 1833.
References: Abé, 1967a; Bursa, 1959; Dodge, 1975; Faust, 1974, 1990a,b, 1991, 1993a,c, 1994; Honsell & Talarico, 1985; Hulburt, 1965; Loeblich et al., 1979; Norris and Berner, 1970; Schiller, 1933; Steidinger, 1983; Steidinger & Williams, 1970; Tafall, 1942; von Stosch, 1980:
Armored. Small to medium-sized cells that vary from spheroid to pyriform in valve view. Cells with chloroplasts. Valves can be convex to concave in lateral view. Cell with two anterior dissimilar flagella. In some, both emerge out of one flagellar pore. The second pore between periflagellar plates may be associated with mucoid production and attachment. In others, flagella emerge from two separate pores. Some species with anterior winged spines, termed "tooth," or anterior short projections. Surface markings vary from pores to areolae to spines. Internally, many *Prorocentrum* have a central pyrenoid, a posterior nucleus, and anterior vacuoles. Two opposing valves, one designated left and the other right. Small periflagellar plates (5–14) indent the anterior margin of the right valve; 8 most common.
Remarks: Surface markings in conjunction with shape and size and indentation of the right valve are conservative and diagnostic characters that can be used to separate species when cells are not too aged. If the outer membrane is still attached to scanning electron microscopy (SEM) prepared specimens, however, it can obscure surface features such as pores. Older cells often have megacytic zones, altered outlines, and muted surface features due to additional polysaccharide deposition. As with the gonyaulacoids, the first thecal markings to appear are sutures separating plates and valves, then pores, then reticulations or areolations. Von Stosch (1980), like Dodge (1983), considered the Prorocentraceae evolutionarily advanced and originating from plate reduction. This hypothesis was advanced based on the periflagellar plates and similarities with the Dinophysiales which were thought to be ancestral. Other dinoflagellate workers, e.g., Loeblich III (1976) and F. J. R. Taylor (1980), present the Prorocentraceae as primitive and ancestral. Species of this genus are differentiated based on the following characters: size and shape, the presence of apical processes, shape of periflagellar area, number and pattern

of periflagellar plates, number and size of aerolae or poroids, pore pattern, and markings on intercalary bands.
Distribution: Species can be planktonic or benthic/epiphytic (tychoplanktonic); if the latter they can attach to a substratum via a mucoid holdfast or thread. Many plankton species are bloom species and many benthic species are toxic and can reach high cell densities/cm².

Prorocentrum arcuatum Issel 1928 (Plate 8)
  Medium-sized, lanceolate cell that is broadest above the median point. Posterior portion attenuated and sometimes twisted. Cell with a long anterior spine that is broad at the base. Surface reported with shallow depressions.
Distribution: Planktonic. Warm temperate and tropical waters.

Prorocentrum balticum (Lohmann) Loeblich III 1970 (Plate 9)
  Small, <20 μm round to ovoid cell in valve view. In side view, round and not flattened. Periflagellar area bordered by two apical projections. Valves covered with many interconnected spines; pores appear rimmed and scattered.
Remarks: Surface pattern as illustrated by Dodge (1985) is different than any form found in the *minimum* complex. Because of its small size, records of *P. balticum* may actually include closely related, but undescribed species, as could *P. minimum*. Toxic?
Distribution: Planktonic. Neritic; worldwide distribution.

Prorocentrum belizeanum Faust (Plate 9)
  Medium-sized, round to broadly oval cell with about an average of 950 areolae on each valve. Areolae are <1 μm in diameter. Valves with pores but not in center. Raised anterior ridge on left valve as in *P. hoffmannianum*. Toxic.
Remarks: This species can easily be confused with *P. concavum* and *P. hoffmannianum*. It differs from *P. concavum* by having prominent areolae in the center of both valves. These areolae can fill in with age. It differs from *P. hoffmannianum* by having a periflagellar area similar to *P. lima* and smaller but more thecal areolae.
Distribution: Benthic. Tropical coastal waters; recently described.

Prorocentrum compressum (Bailey) Abé ex Dodge 1975 (Plate 9)
  Small to medium-sized, broadly ovate cell in valve view; compressed in side view. Characteristic anterior feature of what appear to be two opposed short spines bordering the periflagellar area. Valves covered with pores and shallow depressions. Ornamentation more developed centrally.
Remarks: The anterior projections are extensions of the periflagellar plates, like collars.

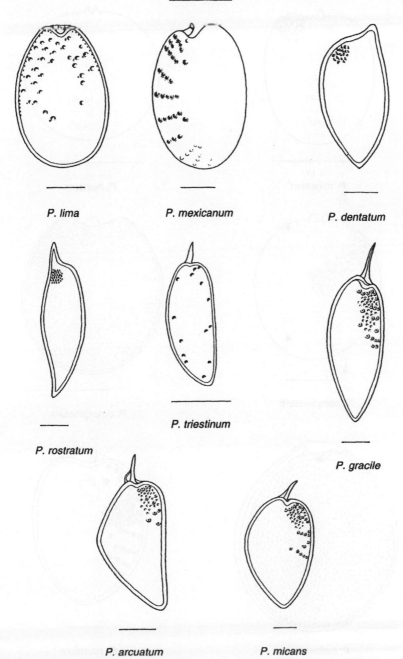

P. lima

P. mexicanum

P. dentatum

P. rostratum

P. triestinum

P. gracile

P. arcuatum

P. micans

PLATE 8 *Prorocentrum lima, P. mexicanum, P. dentatum, P. rostratum, P. triestinum, P. gracile, P. arcuatum,* and *P. micans.* Scale = 10 μm.

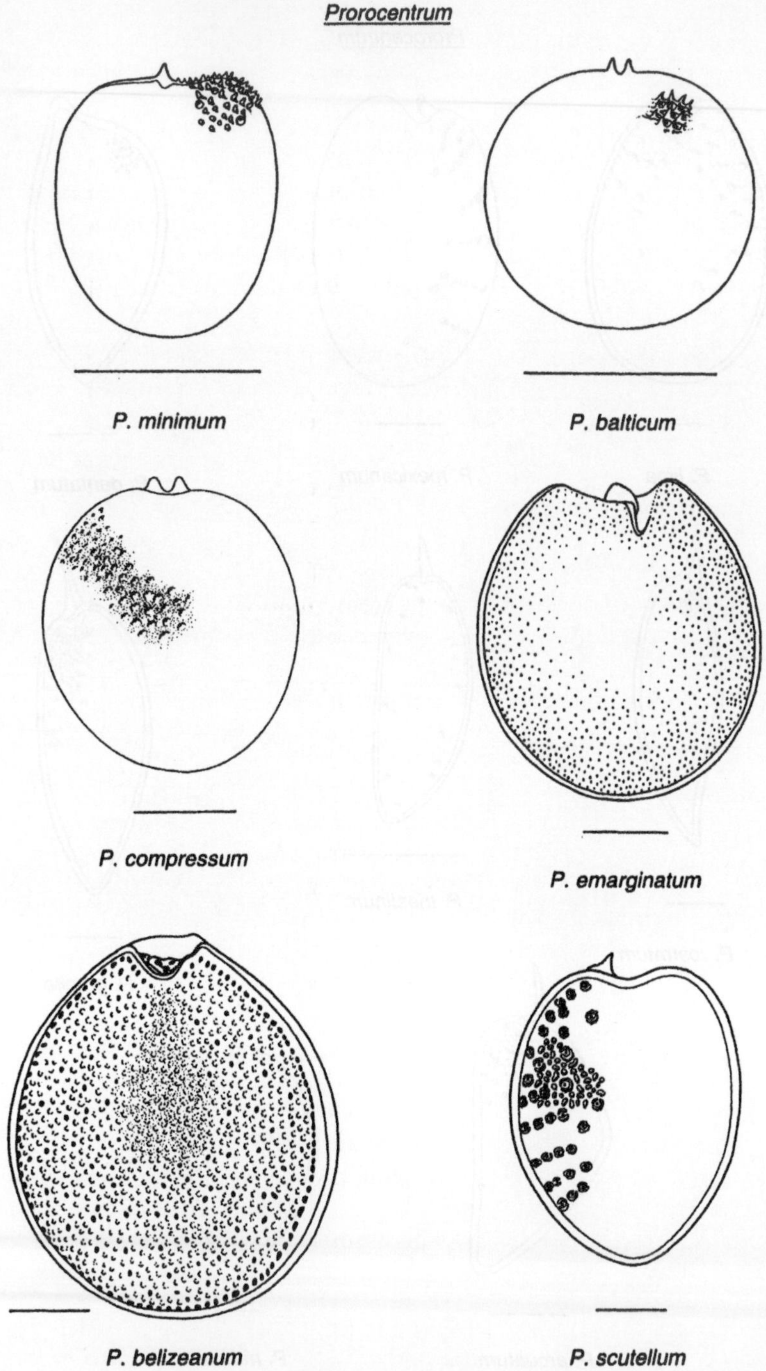

P. minimum

P. balticum

P. compressum

P. emarginatum

P. belizeanum

P. scutellum

PLATE 9 *Prorocentrum minimum*, P. *balticum*, P. *compressum*, P. *emarginatum*, P. *belizeanum*, and P. *scutellum*. Scale = 10 μm.

Distribution: Mostly planktonic. Neritic, oceanic; cosmopolitan in cold temperate to tropical waters.

Prorocentrum concavum Fukuyo 1981 (not illustrated)
Medium-sized, broadly ovoid cell with prominent surface areolae. Valves with scattered pores in areolae. The anterior periflagellar area is a narrow V-shaped indentation of the right valve composed of eight platelets and two pores. The left valve is slightly indented as well. Both valves are centrally devoid of areolae. Toxic.
Remarks: This species is very closely related to *P. hoffmannianum* Faust, which has less areolae (about 700 per valve), and *P. belizeanum* (about 950 areolae per valve). The shallow areolae in the central portion can be obscured in older cells.
Distribution: Benthic; can be tychoplanktonic. Tropical and neritic waters.

Prorocentrum dentatum Stein 1883 (Plate 8)
Small to medium-sized, broadly oblong cell that is attenuated posteriorly. In valve view, sides mostly straight from anterior to median point. Anterior with shoulder. Valves covered with almost evenly spaced, broad-based spines.
Distribution: Planktonic. Oceanic; cold temperate to warm temperate waters; worldwide.

Prorocentrum emarginatum Fukuyo 1981 (Plate 9)
Small oval cell with anterior margin broadly excavated in valve view. The right valve contains the periflagellar plates which fill a cuneiform indentation that curves at the distal end. A short, angled flange at one side of the periflagellar area appears as a thick, winged spine. In young cells there are small pores and the pores that are postmedian are arranged radially. In older cells, the pore pattern can be obscured by depressions that contain pores but the radial pattern is not obvious. Young valves smooth, but older valves can have poroids and pustules. No prominent marginal pore series.
Remarks: This is another case in which field specimens can be misidentified because of differences between young and old cells and the effects of aging on surface markings. In culture the aging process can be followed within a population.
Distribution: Benthic; can be tychoplanktonic. Warm temperate to tropical coastal waters.

Prorocentrum gracile Schütt 1895 (Plate 8)
Synonym: *P. hentschelii* Schiller 1933.
Small to medium-sized, elongate *P. micans*-like cell that is more than twice as long as broad. Pyriform rather than heart shaped with pointed

posterior end in valve view. Valves with shallow poroids and postmedian radial pore fields as in *P. micans*. Long, winged anterior spine adjacent to periflagellar area.

Remarks: This species has been misidentied frequently as *P. micans*, *P. hentschelii*, *P. redfeldii*, or *P. rostratum*.

Distribution: Principally neritic and estuarine; cosmopolitan in cold temperate to tropical waters.

Prorocentrum lima (Ehrenberg) Dodge 1975 (Plate 8)

Small to medium-sized obovate cell that is broadest postmedian. Cell with a central pyrenoid and a posterior nucleus. Periflagellar area with eight plates; anterior flange attached to plate. Each valve with about 50–80 marginal pores and about 60–100 evenly spaced pores on the valve surface. In older cells, surface can become vermiculate. Both valves are anteriorly indented, but the right valve has a shallow V-shaped or triangular excavation with the protruding flange. Toxic.

Remarks: This species can be confused with a variety of similar *Prorocentrum* with a triangular periflagellar area and oval or ovoid shape. For example, it can be confused with *P. concavum* at the light microscope level unless the edge effect or marginal pore effect is studied. It also can be confused with *P. maculosum* which has marginal pores and scattered depressions.

Distribution: Neritic and estuarine. Benthic/epiphytic; can be tychoplanktonic. Worldwide distribution.

Prorocentrum mexicanum Tafall 1942 (Plate 8)

Synonyms: *P. maximum* Schiller 1937; *P. rhathymum* Loeblich, Sherley, & Schmidt 1979.

Small oval cell with delicate appearance including very lightly pigmented, stranded cytoplasm. Nucleus typically postmedian, but not posterior. Periflagellar area with well developed winged spine. Postmedian trichocyst pores are radially arranged in deep, round to oval depressions that create diagonal furrows. These radiating rows of pores can appear as spines at the light microscope level. The pores can appear hooded or collared. Toxic.

Remarks: As with other *Prorocentrum*, surface features can change with age.

Distribution: Neritic and estuarine. Benthic; can be tychoplanktonic. Tropical and subtropical waters.

Prorocentrum micans Ehrenberg 1833 (Plate 8)

Medium-sized, pyriform to heart-shaped cell. Typically, in valve view, cell will have one convex side and one arched side. The convex arch profile is typically in the middle of the cell. In lateral view, the cell is flattened.

Valves with shallow depressions and postmedian radial pore fields as in some other *Prorocentrum* species.
Remarks: One of the most common and variant species in the genus tolerating salinities above 90‰ in salt lagoons in Caribbean islands.
Distribution: Planktonic; neritic and estuarine, but found in oceanic environments. Cosmopolitan in cold temperate to tropical waters.

Prorocentrum minimum (Pavillard) Schiller 1933 **(Plate 9)**
Synonyms: *P. mariae-lebourae* (Parke & Ballantine, 1957) Loeblich III 1970; *P. triangulatum* Martin 1929.
Small oval to triangular-shaped cell in valve view; flattened in side view. Short apical spine sometimes observable. Valves with short, evenly shaped broad-based spines that can appear as rounded papillae depending on angle of view. Two sized pores present; smaller pores scattered. Larger pores at bases of some peripheral spines; pores appear hooded. Toxic.
Remarks: This species has been confused with *P. balticum* but it is larger with a different shape and surface markings. Both species bloom.
Distribution: Planktonic. Mostly estuarine but also neritic. Cosmopolitan in cold temperate to tropical waters. Toxic; has caused shellfish poisoning and fish kills.

Prorocentrum rostratum Stein 1883 **(Plate 8)**
Medium-sized cell, five or six times as long as broad. Elongate, sometimes curved profile in valve view with an anterior broad rostrum that is part of the valve and not the periflagellar area. Posterior end pointed. Valve surface with shallow depressions and probably pores.
Remarks: Sometimes confused with a species that has an apical spine, e.g., *P. gracile.*
Distribution: Planktonic. Warm water neritic species; worldwide distribution.

Prorocentrum sabulosum Faust 1994 (not illustrated)
Medium-sized oval cell with about an average of 390 areolae (>1 μm) on each valve. Trichocyst pores oblong and located in some of the areolae. Triangular periflagellar area with flared apical collar.
Remarks: Can easily be confused with *P. hoffmannianum,* but differs in number of areolae and in configuration and structure of periflagellar plates.
Distribution: Benthic, tychoplanktonic. Known from type locality, Belize.

Prorocentrum scutellum Schröder 1900 **(Plate 9)**
Small to medium-sized cell. Rounded, heart-shaped species with unique valve markings composed of radial pore fields where each pore is partially circled by raised markings. In between the pores are irregular markings that look like wax droppings. Broadly curved, winged anterior spine part of periflagellar area.
Distribution: Neritic or estuarine; Arctic to tropical waters.

Prorocentrum triestinum Schiller 1918 (Plate 8)
Synonym: *P. redfeldii* Bursa 1959.
   Small posteriorly pointed cell resembling a thin, narrow *P. micans;*
   usually <30 μm in length. Depressions few and mainly peripherally
   located. Cell with a thin anterior spine.
Distribution: Oceanic and neritic; worldwide distribution.

Order Dinophysiales Lindemann 1928
References: Schiller, 1933; Tai & Skogsberg, 1934; Loeblich III, 1982;
Sournia, 1986; Fensome et al., 1993.
   Laterally flattened cells with a dinokont flagellar orientation and a
   premedian cingulum. Cingulum and sulcus often with wide lists supported
   by ribs. A total of 18 or 19 plates in apical (A), epithecal (E), cingular (C),
   sulcal (S), and hypothecal (H) series. Artificially includes Citharistaceae.
Remarks: The three families of this order discussed here are
Amphisoleniaceae Lindemann, Dinophysiaceae Stein, and Oxyphysiaceae
Sournia.

Family Amphisoleniaceae Lindemann 1928

Genus *Amphisolenia* Stein 1883 (<50 spp.)
Type: *A. globifera* Stein 1883.
References: Abé, 1967c; Balech, 1980, 1988a; Kofoid & Skogsberg, 1928;
Schiller, 1933; Sournia, 1986; F. J. R. Taylor, 1976.
   Armored. Large, fusiform cells up to >1 mm in length. Cells have often
   been interpreted as stick figures with a head, neck, shoulder, narrow body
   sometimes with an inflated midregion, and a foot. Portion of plate
   surfaces covered with pores. Chloroplasts present; sometimes external
   cyanobacteria symbionts in the cingular area. The "head" is composed of
   the epitheca and cingulum and species of the genus have the
   representative 18 plates: 4E, 2A, 4C, 4S, and 4H. The neck and shoulder
   have the sulcal plates and lists; Sa and Sd are elongate plates. Cingulum
   circular.
Remarks: This genus is distinctive and not easily confused with any other
dinoflagellate form. As in the photosynthetic *Dinophysis,* the chloroplasts of
Amphisolenia probably originated by serial endosymbiosis of other
microalgae.

Amphisolenia bidentata Schröder 1900 (Plate 10)
   Between 500 and 1000 μm with a prominent spined heel and two front
   spines on the foot. Posterior part of cell with foot slightly curved.
Distribution: Oceanic, sometimes associated with upwelling; cosmopolitan
in warm temperate to tropical waters.

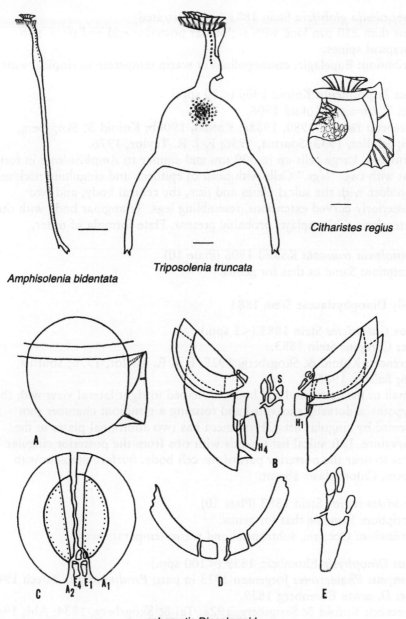

Citharistes regius

Triposolenia truncata

Amphisolenia bidentata

schematic Dinophysoid

PLATE 10 *Amphisolenia bidentata* (scale = 100 μm). Triposolenia truncata and *Citharistes regius* (scale = 10 μm) redrawn from Kofoid and Skogsberg (1928); schematic Dinophysoid (after Balech (1980). (A) Lateral view; (B) dissected ventral area; (C) apical view; (D) singular plates; (E) sulcal plates. (A–E, see Dinophysiales, pg. 426).

Amphisolenia globifera Stein 1883 (not illustrated)
Less than 250 μm long with a globular posterior end and up to two
antapical spines.
Distribution: Eupelagic; cosmopolitan in warm temperate to tropical waters.

Genus *Triposolenia* Kofoid 1906 (<10 spp.)
Type: *T. truncata* Kofoid 1906.
References: Balech, 1980, 1988a; Kofoid, 1906b; Kofoid & Skogsberg,
1928; Schiller, 1933; Sournia, 1986; F. J. R. Taylor, 1976.
Armored. Large cells up to 300 μm and similar to *Amphisolenia* in form
but with two "legs." Cells with head of epitheca and cingulum; neck and
shoulder with the sulcal plates and lists, the central body, and two
posteriorly curved extensions, resembling legs. Triangular body with three
extensions. Chloroplasts probably present. Plate formula of order.

Triposolenia truncata Kofoid 1906 (Plate 10)
Description: Same as that for genus.

Family Dinophysiaceae Stein 1883

Genus *Citharistes* Stein 1883 (<5 spp.)
Type: *C. regius* Stein 1883.
References: Kofoid & Skogsberg, 1928; F. J. R. Taylor, 1976; Sournia,
1986; Balech, 1988a.
Small to medium-sized cell body, C-shaped in right lateral view with the
hypotheca dorsally excavated and forming a symbiont chamber (not
formed by cingular lists). Hypotheca has two additional plates in the
curvature. Left sulcal list extends with ribs from the posterior cingular
area to near the posterior part of the cell body. Surface areolate with
pores. Chloroplasts absent.

Citharistes regius Stein 1883 (Plate 10)
Description: Same as that for genus.
Distribution: Oceanic, subtropical and warm temperate seas.

Genus *Dinophysis* Ehrenberg 1839 (<100 spp.)
Synonyms: *Phalacroma* Jörgensen 1923 in part; *Prodinophysis* Balech 1944.
Type: *D. acuta* Ehrenberg 1839.
References: Kofoid & Skogsberg, 1928; Tai & Skogsberg, 1934; Abé, 1967;
Steidinger & Williams, 1970; Norris & Berner, 1970; Balech, 1976b,c,
1980, 1988a; F. J. R. Taylor, 1976; Rampi & Bernhard, 1980; Dodge,
1982; Hallegraeff & Lucas 1988.
Dinophysis and *Phalacroma* species reportedly overlap morphologically,
but they can be separated by the development and direction of the

cingular lists in combination with the height and shape of the epitheca. *Dinophysis* species have a distinctive funnel-shaped anterior cingular list. Many of the species can also be separated by the presence or absence of chloroplasts and relative distribution (Hallegraeff & Lucas, 1988), but there are several exceptions. Species in this genus have an apical pore formed by two apical plates. No doubt genetic and immunoassay studies will identify the level of relatedness among these two genera.

Remarks: To differentiate species use the following suite of conservative characters: dorsal and ventral (lateral) cell curvature, relative length of cell, length and shape of left and right sulcal lists, positioning of the three ribs that support the left sulcal list, ventral view, and dorsal–ventral depth of epitheca. Use the same characters for *Phalacroma* after differentiating the two genera.

Dinophysis acuminata Claparède & Lachmann 1859 (Plate 11)

Small to medium species, almost oval or elliptical in shape. Posterior profile is rounded. Left sulcal list well developed, extends beyond the midpoint of the cell, and is of equal depth. Surface with areolae, each with a pore. Type E. Toxic.

Remarks: This species can be confused with *D. sacculus*, *D. norvegica*, *D. ovum*, and *D. punctata*. Also, *D. skagii* and *D. lachmannii* are probably variants of *D. acuminata*. Cell shape, sulcal list development, and possibly surface markings may help to differentiate species, but there still can be confusion. Within this group or complex, surface markings can range from pores to depressions with scattered pores to depressions each with a pore to areolae each with a pore and appearance may depend on age of the cell as in *Prorocentrum*.

Distribution: Neritic; typically cold and warm temperate waters, worldwide.

Dinophysis acuta Ehrenberg 1839 (Plate 11)

Large, robust cell with a rounded dorsal curvature and a posterior broad V-shaped lateral profile. The left sulcal list extends about two-thirds of the body length and ends at or above the deepest portion of the cell below the midpoint. The R3 is at or above this point. Surface with areolations; type E. Toxic.*

Remarks: This species can be easily confused with *D. norvegica*. The distinction between the two species can be made by determining whether the deepest portion of the cell is two-thirds the cell length or one-half and determining the length of the left sulcal list in relation to the cell length.

Distribution: Oceanic and neritic; cold temperate, worldwide.

* R designation refers to ribs in sulcal lists.

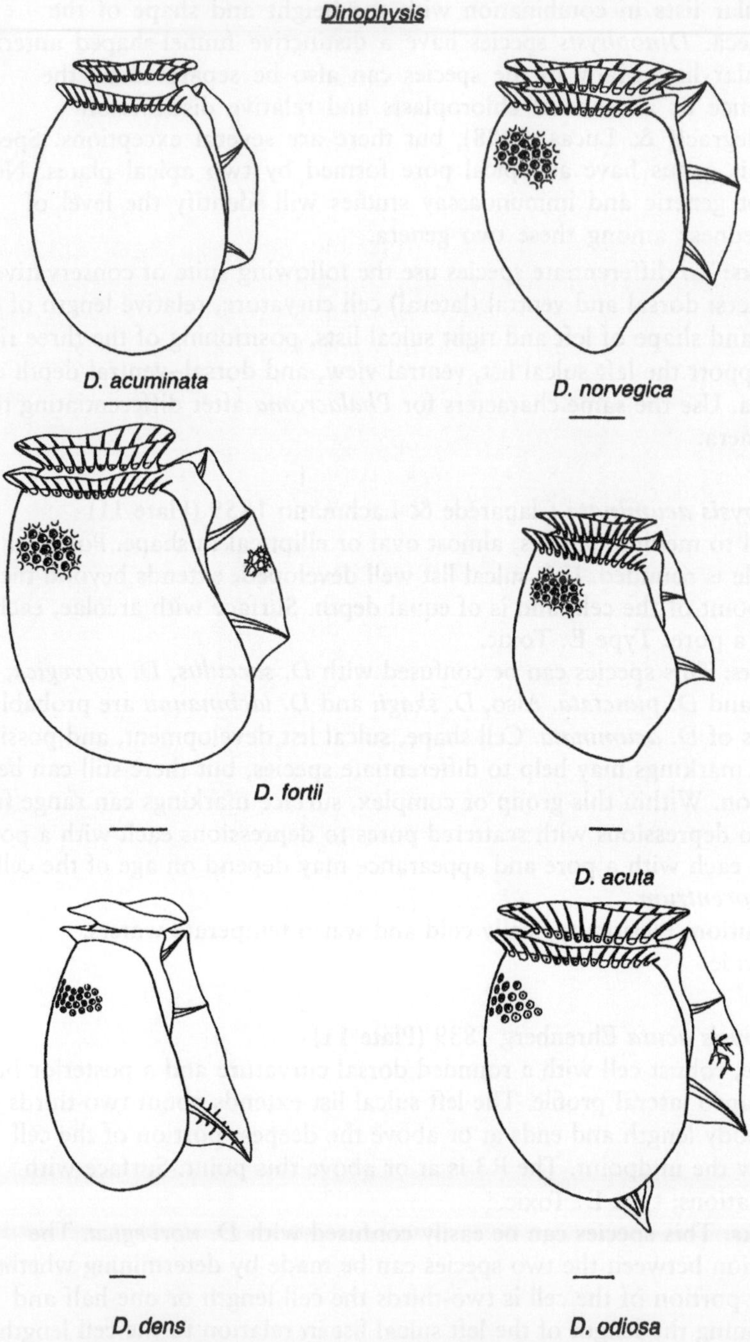

Dinophysis

D. acuminata

D. norvegica

D. fortii

D. acuta

D. dens

D. odiosa

PLATE 11 _Dinophysis acuminata, D. norvegica, D. fortii, D. acuta, D. dens,_ and _D. odiosa._ Scale = 10 μm.

Dinophysis caudata Saville-Kent 1881 (Plate 12)
  Medium-sized species that has a characteristic posterior finger-like
  process; cells often occur in pairs, dorsally attached. Dorsal contour is
  gradually curved, whereas the ventral margin in lateral profile is generally
  straight along the main body. The posterior process varies in length and
  shape and the left sulcal list extends the length of the main body. Surface
  with areolations; type E. Toxic.
Remarks: This species superficially resembles *D. tripos* and *D. diegensis. D.
diegensis* has been called a variety of *D. caudata.*
Distribution: Neritic and estuarine in warm temperate to tropical waters,
worldwide; rarely found in cold water, possibly an intruder in warm water
masses.

Dinophysis dens Pavillard 1915 (Plate 11)
  Small to medium-sized species with curved dorsal margin and angled
  posterior ventral margin below the R3. Cell can look like a small *D.
  acuta,* but it does not have the prominent postmedian depth that
  characterizes *D. acuta.* The left sulcal list typically has surface markings
  like reticulations or vermiculae, particularly between R1 and R2; depth of
  list is unequal and the distal margin is partially curved.
Remarks: This species is thought to be a variant or a gamete of *D. acuta*
(Bardouil et al., 1991; MacKenzie, 1992). *Dinophysis dens* and *D. acuta*
have been observed attached together as couplets in wild samples.
Distribution: Warm and cold water species; worldwide distribution.

Dinophysis diegensis Kofoid 1907 (not illustrated)
  Medium-sized cell with a slightly curved dorsal margin and an elongated,
  attenuated hypotheca. Left sulcal list narrow and of equal depth which
  extends to the point of attenuation. Main body of equal depth. Surface
  markings of type E.
Remarks: Similar to *D. caudata* but can be distinguished from that species
by the width and shape of the main body and the left sulcal list.
Distribution: Estuarine sometimes neritic; warm temperate waters.

Dinophysis fortii Pavillard 1923 (Plate 11)
  Medium-sized cell, broadly subovoid, widest posteriorly. Dorsal margin
  curved and ventral margin almost straight. Left sulcal list long and can be
  up to four-fifths of the cell length. Right sulcal list also well developed
  and can extend beyond the R2. Surface with deep poroids, each with a
  pore. Surface markings of type E. Toxic.
Distribution: Oceanic and neritic; cold temperate to tropical waters,
worldwide distribution.

Dinophysis

D. caudata

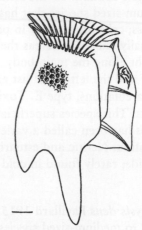

D. tripos

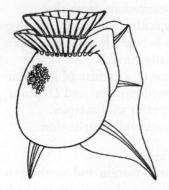

D. uracantha

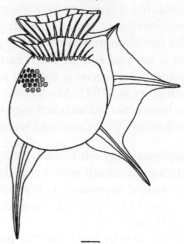

D. schuettii

D. hastata

Dinophysis hastata Stein 1883 (Plate 12)

Medium-sized cell with characteristic left sulcal list and R3 and curved posterior spine, with list, directed ventrally. The posterior spine is composed of fused ribs and its development may depend on the age of the cell. The R3 is typically curved distally. The shape of the main body is ovoid. Surface markings of depressions with scattered pores, type B.

Remarks: There are several species that can be confused with *D. hastata.* This is a nonphotosynthetic *Dinophysis,* but it may be phagotrophic and contain pigment.

Distribution: Neritic; warm temperate to tropical waters and rarely found in cold water; worldwide distribution.

Dinophysis norvegica Claparède & Lachmann 1859 (Plate 11)

Large, robust cell with fully rounded dorsal curvature and a straight-angled lateral profile to the lower half of the ventral margin. The deepest part of the cell is about midway between the lower cingular list and the antapex, and the R3 of the left sulcal list occurs at this point or just above it. Plate surfaces with large areolae with pores, type E.
Posterior and dorsal margins sometimes with protuberances or thick extensions called bosses. Toxic.

Remarks: See *D. acuta* for a comparison.

Distribution: Neritic; cold water species.

Dinophysis odiosa (Pavillard) Tai & Skogsberg 1934 (Plate 11)

Medium-sized species, subovoid in shape and with a ribless posterior sail. Left sulcal list of unequal depth and reticulations and vermiculae often between R2 and R3. Posterior sail rarely double. Chloroplasts absent.

Distribution: Cold temperate to warm temperate waters.

Dinophysis schuettii Murray & Whitting 1899 (Plate 12)

Small-sized species, round to subovate with a curved, reinforced posterior sail that has a median rib joined to the marginal ribs. Left sulcal list between R2 and R3 concave and shorter than ribs. Posterior sail dorsal and to the right. Chloroplasts absent. Surface markings of type B of Hallegraeff & Lucas (1988).

Remarks: Can be confused with *D. swezyii* and *D. uracantha.* This is a nonphotosynthetic *Dinophysis,* but it may be phagotrophic and contain some pigment.

PLATE 12 *Dinophysis caudata, D. tripos, D. uracantha, D. schuettii,* and *D. hastata.* Scale = 10 μm.

Distribution: Oceanic; warm temperate to tropical waters; worldwide
distribution. Considered a shade species.

Dinophysis tripos Gourret 1883 (Plate 12)
   Large cell with two posterior V-shaped processes, one extends almost the
   length of the main body and is midway between the ventral and dorsal
   margins, while the other is short and dorsal. The left sulcal list is broader
   posteriorly and has a straight margin. Type E surface markings.
   Toxic.
Remarks: The species can be distinquished from *D. caudata* by the shape of
the left sulcal list and the presence of the two posterior processes.
Distribution: Neritic, estuarine, and oceanic; warm temperate to tropical
species, rarely found in cold waters.

Dinophysis uracantha Stein 1883 (Plate 12)
   Medium-sized species similar to *D. scheuttii* but without median rib in
   posterior sail, and left sulcal list between R2 and R3 not concave, but
   with a straight edge. Left sulcal list posterior margin with a proximal
   lobe. Chloroplasts absent.
Distribution: Oceanic, tropical to warm temperate waters; worldwide
distribution. Considered a shade species.

Genus *Histioneis* Stein 1883 (<100 spp.)
Synonym: *Parahistioneis* Kofoid & Skogsberg 1928
Type: *H. remora* Stein 1883.
References: Kofoid & Michener, 1911; Kofoid & Skogsberg, 1928;
Rampi & Bernhard, 1980; F. J. R. Taylor, 1976, 1987; Balech, 1988.
   Armored, small to large subcircular, reniform or subreniform,
   dinophysoid cell bodies with ornate list and rib systems and large cingular
   chamber; poorly characterized species but readily recognizable group.
   Posterior cingular list often cup-shaped with largest portion of cingulum
   being posterior and with vertical support ribs. Anterior cingular area
   reduced, sometimes to a funnel with a small anterior cingular list.
   Reduced membraneous right sulcal list but extensive left sulcal list.
   Chloroplasts absent. Surface markings of pores to areolae.

Histioneis depressa Schiller 1933 (Plate 13)
   Cell small and reniform in lateral view with anterior cingular funnel;
   circular cingular chamber appears fused to the prominent left sulcal list.
   Left sulcal list sac like and directed dorsally; subovoid U-shaped space
   midpoint to ventral on posterior margin of cell body without surface
   reticulations and rib spurs.
Remarks: F. J. R. Taylor (1976) suggested that this species may be a variant
of *H. mitchellana*.
Distribution: Oceanic; tropical and warm temperate waters; cosmopolitan.

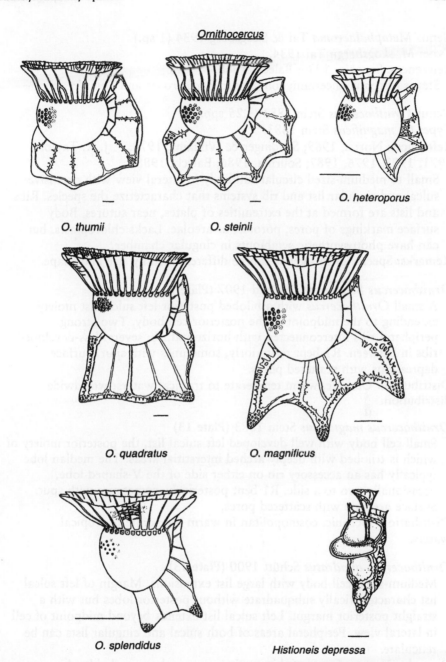

PLATE 13 *Ornithocercus thumii, O. steinii, O. heteroporus, O. quadratus, O. magnificus, O. splendidus. Histioneis depressa* (redrawn from Taylor, 1976). Scale = 10 μm.

Genus *Metaphalacroma* Tai & Skogsberg 1934 (1 sp.)
Type: *M. skogsbergii* Tai 1934.
References: Sournia, 1986; Balech, 1988.
  Status of genus uncertain; compare to *Phalacroma* and *Sinophysis*.

Genus *Ornithocercus* Stein 1883 (<25 spp.)
Type: *O. magnificus* Stein 1883.
References: Norris, 1969; Steidinger & Williams, 1970; F. J. R. Taylor,
1971, 1973, 1976, 1987; Sournia, 1986; Balech, 1988a.
  Small to medium-sized circular full body in lateral view with extensive
  sulcal and cingular list and rib systems that characterize the species. Ribs
  and lists are formed at the extremities of plates, near sutures. Body
  surface markings of pores, poroids, or areolae. Lacks chloroplasts, but
  can have photosynthetic symbionts in cingular chamber.
Remarks: Species of this genus can be differentiated by size and shape.

Ornithocercus heteroporus Kofoid 1907 (Plate 13)
  A small *Ornithocercus* with a bilobed posterior left sulcal list moiety
  extending to the midpoint of the posterior cell body. Two strong
  peripheral ribs interconnected with horizontal rib; several less-developed
  ribs in between. R1 bent posteriorly, sometimes with spur. Surface
  depressions with scattered pores.
Distribution: Oceanic, warm temperate to tropical waters; worldwide
distribution.

Ornithocercus magnificus Stein 1883 (Plate 13)
  Small cell body with well developed left sulcal list, the posterior moiety of
  which is trilobed with deeply arched interstitial areas. The median lobe
  typically has an accessory rib on either side of the V-shaped lobe,
  occasionally two to a side. R1 bent posteriorly, sometimes with spur.
  Surface areolate with scattered pores.
Distribution: Oceanic; cosmopolitan in warm temperate to tropical
waters.

Ornithocercus quadratus Schütt 1900 (Plate 13)
  Medium-sized cell body with large list extensions. Margin of left sulcal
  list characteristically subquadrate without posterior lobes but with a
  straight posterior margin. Left sulcal list extends beyond midpoint of cell
  in lateral view. Peripheral areas of both sulcal and cingular lists can be
  reticulate.
Remarks: Various forms of this species occur and are probably phenotypic
variants.
Distribution: Oceanic; warm temperate to tropical waters; worldwide
distribution.

Ornithocercus splendidus Schütt 1893 (Plate 13)
  The most ornate *Ornithocercus* described with very extensive, posterior reticulate cingular list and a posteriorly bilobed left sulcal list that reaches midpoint of the posterior cell margin in lateral view and is directed ventrally. Each prominent lobe with mass of sculptured thecal material and a well-developed rib. Other less-developed ribs present. Surface with pores.
Remarks: Very distinctive species.
Distribution: Oceanic; warm temperate to tropical waters; worldwide distribution.

Ornithocercus steinii Schütt 1900 (Plate 13)
  Medium-sized body with four weakly developed posterior lobes to the left sulcal list; ribs not interconnected. Right sulcal list does not extend beyond cell body posterior. Surface areolate.
Remarks: Often confused with *O. thumii* which has three posterior lobes and a rib to the median lobe.
Distribution: Oceanic; cosmopolitan in warm temperate to tropical waters.

Ornithocercus thumii (Schmidt) Kofoid & Skogsberg 1928 (Plate 13)
  Small cell body. Right sulcal list extends beyond posterior cell margin. Left sulcal list with three lobes to posterior moiety and a rib to the median lobe. Other ribs present and ribs interconnected. Cell surface areolate with pores.
Remarks: See *O. steinii*.
Distribution: Neritic, oceanic; warm temperate to tropical waters; worldwide distribution.

Genus *Phalacroma* Stein 1883 (<100 spp.)
Type: *P. prodictyum* Stein 1883
References: Kofoid & Skogsberg, 1928; Tai & Skogsberg, 1934; Abé, 1967b; Balech, 1976b; Steidinger & Williams, 1970; Rampi & Bernhard, 1980; Hallegraeff & Lucas, 1988.
  Medium to large cells. Epitheca detectable above the anterior cingular list in lateral view. Cingular lists typically narrow and directed horizontally rather than the anterior cingular list forming a funnel-shaped fan. Most species nonphotosynthetic and of oceanic distribution.
Remarks: See *Dinophysis*. Many authors consider *Phalacroma* to be synonymous with *Dinophysis*.

Phalacroma argus Stein 1883 (Plate 14)
  Medium-sized, egg-shaped cell with the widest part at the cingulum. R3 not well developed or may be absent.

Phalacroma

P. rotundatum

P. argus

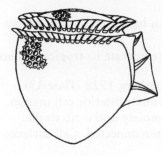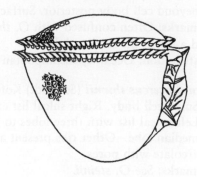

P. mitra

P. cuneus

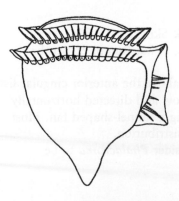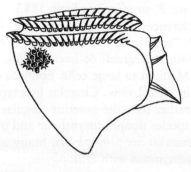

P. favus

P. rapa

Distribution: Oceanic; warm temperate to tropical waters; worldwide distribution.

Phalacroma circumsutum Karsten 1907 **(Plate 17)**
Characteristic triangular ribbed posterior sail not connected to the left sulcal list. Ventral edge of left sulcal list to R3 and angled ventrally. Surface markings of shallow depressions with scattered pores.

Distribution: Neritic and pelagic; warm temperate to tropical waters; worldwide distribution.

Phalacroma cuneus Schütt 1895 **(Plate 14)**
Medium-sized cell that is broadest anteriorly and is narrowest posteriorly. Left sulcal list is curved distally. Surface reticulate with a pore in almost every depression.
Distribution: Oceanic; warm temperate to tropical waters; worldwide distribution.

Phalacroma favus Kofoid & Michener 1911 **(Plate 14)**
Distinctive medium-sized cell with posterior finger-like projection.
Distribution: Warm water species; subtropical, tropical, and occasionally warm temperate waters; worldwide distribution.

Phalacroma mitra Schütt 1895 **(Plate 14)**
Posterior portion of hypothecal plates concave from R3 to anatapex. Produces okadaic acid. Toxic.
Remarks: Similar to *P. rapa.*
Distribution: Warm temperate to tropical oceanic and neritic waters; worldwide distribution.

Phalacroma rapa Jorgensen 1923 **(Plate 14)**
In lateral view, the left ventral magin from R1 to R3 is angled making the left sulcal list extend out at about a 45–60° angle perpendicular to the depth axis.
Remarks: Similar to *P. mitra.*
Distribution: Oceanic and neritic; temperate to tropical waters; worldwide distribution.

Phalacroma rotundatum (Claparède & Lachmann) Kofoid & Michener 1911 **(Plate 14)**

PLATE 14 *Phalacroma rotundatum, P. argus, P. mitra, P. cuneus, P. favus,* and *P. rapa.* Scale = 10 μm.

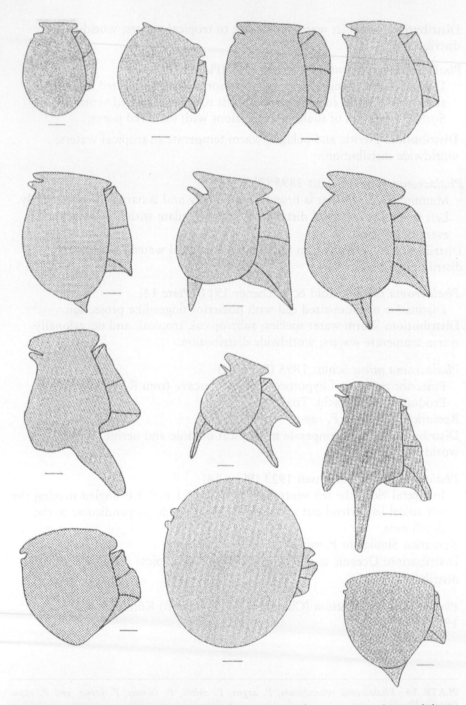

PLATE 15 Silhouettes of *Dinophysis* and *Phalacroma* species for comparison of size and shape. The scale = 10 μ.

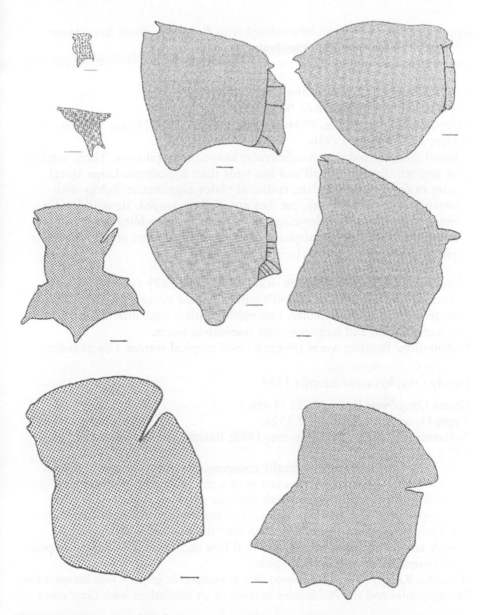

PLATE 16 Silhouettes of *Phalacroma* and *Ornithocercus* species for comparison of size and shape. The scale = 10 μ.

Small to medium-sized cell. In lateral view, body almost round with concave ventral and dorsal edges. The left sulcal list extends from greater than one-half to almost three-fourths the length of the hypotheca. Produces okadaic acid. Toxic.

Remarks: This species can be confused with *P. rudgei* which has a more prominent epitheca and a larger body size.
Distribution: Cold and warm waters; cosmopolitan.

Genus *Sinophysis* Nie & Wang 1944 (<5 spp.)
Type: *S. microcephala* Nie & Wang 1944.
References: Nie & Wang, 1944; Balech, 1967a, 1980, 1988a; Saunders & Dodge, 1984; Faust, 1993b.
Small dinophysoid with a subcircular shape in lateral view. Theca areolate or smooth. Epitheca small and less wide than hypotheca. Large apical pore in right epithecal plate; epithecal plates asymmetric. Sulcus with narrow left and right lists, but lists not well developed; sigmoid left list more prominent. Posterior cingular list vertical resembling a collar; anterior list thinner and slightly upturned. Chloroplasts absent, but feeds phagotrophically. Nucleus located posteriorly.

Sinophysis microcephala Nie & Wang 1944 (Plate 17)
Theca areolate (mean of ca. 460) with scattered pores in areolae. Several larger pores between areolae. Left epitheca plate with different surface ornamentation and large openings resembling pores.
Distribution: Benthic; warm temperate and tropical waters. Cosmopolitan.

Family Oxyphysaceae Sournia 1984

Genus *Oxyphysis* Kofoid 1926 (1 spp.)
Type: *O. oxytoxoides* Kofoid 1926.
References: Kofoid, 1926; Sournia, 1986; Balech, 1988a; Fensome et al., 1993.
Armored, medium-sized, laterally compressed, subfusiform cell. Length <4× width. Most of cell areolate with scattered pores. Epitheca well developed and asymmetrical with anterior spine on right; not as wide as hypotheca. Ventral pore near apex. Cingulum barely displaced and slightly premedian. Hypotheca with laterally convex sides; attenuated with antapex directed ventrally. Sulcal lists short and not well developed. Chloroplasts absent; heterotrophic.
Remarks: Kofoid (1926) considered this monospecific genus a link between the Dinophysiales and the Peridiniales because of its similarities with *Oxytoxum*.

Oxyphysis oxytoxoides Kofoid 1926 (Plate 17)
Description: Same as that for genus.
Remarks: This species is common in many waters but often overlooked; probably misidentified as a *Oxytoxum*. It may be that the coastal cold water form and the warm temperate estuarine/coastal form are two different species.

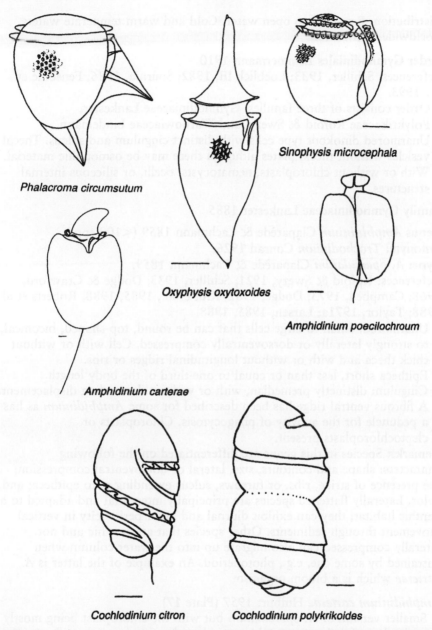

Phalacroma circumsutum

Sinophysis microcephala

Oxyphysis oxytoxoides

Amphidinium poecilochroum

Amphidinium carterae

Cochlodinium citron

Cochlodinium polykrikoides

PLATE 17 Scale = 10 µm unless otherwise indicated. *Phalacroma doryphorum, Oxyphysis oxytoxoides, Sinophysis microcephala. Amphidinium carterae* redrawn from Dodge & Crawford (1968) (scale = 1 µm); *Amphidinium poecilochroum* (scale= 2 µm); *Cochlodinium citron* redrawn from Kofoid & Swezy (1921); *C. polykrikoides.*

Distribution: Estuarine to open water. Cold and warm temperate waters, worldwide distribution.

Order Gymnodiniales Lemmermann 1910

References: Schiller, 1933; Loeblich III, 1982; Sournia, 1986; Fensome et al., 1993.

Order consists of three families: Gymnodiniaceae Lankester, Polykrikaceae Kofoid & Swezy, and Warnowiaceae Lindemann.

Unarmored dinokont type cells with distinct cingulum and sulcus. Thecal vesicles without thecal plates although there may be osmophilic material. With or without chloroplasts, nematocysts, ocelli, or siliceous internal structures.

Family Gymnodiniaceae Lankester 1885

Genus *Amphidinium* Claparède & Lachmann 1859 (<100 spp.)
Synonym: *Trochodinium* Conrad 1926.
Type: *A. operculatum* Claparède & Lachmann 1859.
References: Kofoid & Swezy, 1921; Schiller, 1933; Dodge & Crawford, 1968; Campbell, 1973; Dodge, 1982; Klut et al., 1985, 1988; Roberts et al., 1988; Taylor, 1971a; Larsen, 1985, 1988.

Unarmored. Small to large cells that can be round, top-shaped, biconical, to strongly laterally or dorsoventrally compressed. Cell with or without thick theca and with or without longitudinal ridges or ribs.

Epitheca short, less than or equal to one-third of the body length.
Cingulum distinctly premedian, with or without descending displacement.
A fibrous ventral ridge has been described for some *Amphidinium* as has a peduncle for the activity of phagocytosis. Chloroplasts or cleptochloroplasts present.

Remarks: Species in this genus are differentiated on the following characters: shape and contours, size, lateral or dorsoventral compression; the presence of striae, ribs, or furrows, sulcus extending onto epitheca; and color. Laterally flattened species are principally interstitial and adapted to a benthic habitat; they can exhibit diurnal and lunar periodicity in vertical movement through sediments. Other species that are benthic and not laterally compressed can also migrate up into the water column when entrained by some cue, e.g., photoperiod. An example of the latter is *A. carterae* which is a bloom organism.

Amphidinium carterae Hulburt 1957 (Plate 17)

Smaller version of *A. operculatum* but with the chloroplast being mostly parietal, supporting a central pyrenoid, and appearing to have less pigment. Cell surface with glycocalyx. Toxic.

Remarks: One of the most common *Amphidinium* that undergoes daily vertical migrations into the water column from the benthos.
Distribution: Cosmopolitan in temperate and tropical waters.

Amphidinium operculatum Claparede & Lachmann 1859 (not illustrated)
Synonym: *A. klebsii* Kofoid & Swezy 1921
Small, elliptical to round cell with premedian cingulum in a Y
configuration with sulcus, not a T configuration. Cingulum not complete.
Epitheca asymmetric and directed to the left, tongue or beak shaped in
dorsal view. Cell dorsoventrally compressed. Chloroplast(s) multilobed
with lamellae running through a central pyrenoid; chloroplast pigment
often obscures other cell organelles. Nucleus posterior in hypotheca.
Toxic.
Remarks: Schiller (1933) and Campbell (1973) among others, consider *A.
klebsii* a separate species based on shape of the epitheca. Dodge (1982) and
Larsen (1985) consider the two species to be synonymous.
Distribution: Cosmopolitan in temperate to tropical estuarine and coastal
waters.

Amphidinium poecilochroum Larsen 1985 (Plate 17)
Small, oblong to round cell with premedian cingulum in a T configuration
with sulcus. Cingulum slightly displaced and descending. Epitheca small
and not as wide as hypotheca. Sulcus reaches the rounded antapex.
Heterotrophic/mixotrophic nutrition; cleptochloroplasts present.
Distribution: Recently described benthic species from the Danish
Wadden Sea.

Genus *Cochlodinium* Schütt 1896 (<50 spp.)
Type: *C. strangulatum* (Schütt) Schütt 1896.
References: Kofoid & Swezy, 1921; Schiller, 1933; Steidinger & Williams,
1970; Campbell, 1973; Takayama, 1985; Sournia, 1986.
Unarmored. Gyrodinioid cells, small to medium sized, with greater than
1.5× cingular rotation. Usually with apical groove. Single cell or chain-
forming species. Body sometimes appears twisted due to cingular torsion.
Nematocysts and ocelli absent. Chloroplasts present or absent.
Encystment common.
Remarks: Species in this genus are differentiated based on the following
characters: shape and contours, size, amount of cingular rotation or number
of turns, the presence of striae or ribs, and color.

Cochlodinium citron Kofoid & Swezy 1921 (Plate 17)
Small to medium-sized subellipsoidal cell with broadly rounded epitheca
and hypotheca. Cingulum with more than two turns or rotations. Sulcus
slightly invades epitheca. Apical groove circular, not complete. Periphery
of cell with blue–green rodlets. Chloroplasts absent. Nucleus spherical
and in anterior half of cell.
Distribution: Temperate and subtropical neritic waters.

Cochlodinium polykrikoides Margelef 1961 (Plate 17)
Synonym: *C. heterolobatum* Silva 1967
   Chain forming gyrodinioid; in chain, cell compressed but individual small-
   sized free cells are ellipsoidal. Epitheca conical; hypotheca bilobed.
   Cingulum excavated and displaced; 1.8–1.9 body turns. Associated with
   fish kills. Toxic.
Distribution: Warm temperate and tropical waters; cosmopolitan.

Genus *Gymnodinium* Stein 1878 (>200 spp.)
Synonyms: *Aureodinium* Dodge 1982; *Ceratodinium* Conrad 1926;
Ptychodiscus Stein 1883 in part.
Type: *G. fuscum* (Ehrenberg) Stein 1878.
References: Kofoid & Swezy, 1921; Lackey, 1956; Kimball & Wood, 1965;
Steidinger & Williams, 1970; Campbell, 1973; Dodge, 1982; Larsen, 1985,
1994; Takayama, 1985; Sournia, 1986; Steidinger et al., 1989; Steidinger,
1990, 1993, Elbrachter, 1994.
   Unarmored. Small to large cells with varied morphology from spherical to
   biconical and lobed to pyriform. Single cell or chain-forming species.
   Theca with or without longitudinal ridges or ribs. Cingulum usually
   equatorial or premedian, with or without descending displacement (left
   handed). If displaced, less than one-fifth body length. Sulcus often
   invading epitheca. Apical groove present or absent. Chloroplasts present
   or absent. Color can be green, yellow, brown, blue, or pink.
Remarks: There are *Gymnodinium* that have a cingulum displacement that
can be more or less than one-fifth the cingulum width in both natural and
cultured specimens (see Kimball & Wood 1965; Steidinger & Williams,
1970; and Larsen, 1985 for a discussion on the separation of *Gymnodinium*
and *Gyrodinium*). In addition there are small *Woloszynskia, Heterocapsa,
Scrippsiella* and other estuarine and brackish water species of <15 μm that
are often mistaken as *Gymnodinium* species. The species in this genus are
differentiated based on the following characters: shape and contours, size,
chain formation, the presence and shape of apical groove; cingulum
premedian, median, or postmedian; displacement of cingulum; sulcus
extending onto epitheca; sulcal–apical groove juncture; shape of ventral
ridge; the presence of striae, ribs, or furrows; the presence of chloroplasts;
placement of nucleus; and color.

Gymnodinium abbreviatum Kofoid & Swezy 1921 (Plate 19)
   Medium to large cell, circular in cross section, half as wide as long.
   Epitheca subconical, hypotheca asymmetrical and notched by sulcus at
   antapex. Cingulum premedian and descending 2×. Theca thick and
   striated. Chloroplasts absent but with pink or blue cytoplasm.
Distribution: Oceanic.

Gymnodinium breve Davis 1948 (Plate 18)
Synonym: *Ptychodiscus brevis* (Davis) Steidinger 1979.
Small dorsoventrally flattened cell with rounded epitheca and apical
process or carina directed ventrally. Hypotheca notched and slightly
bilobed. Cell ventrally concave, dorsally convex. Larger cells more
concave but with the right side wider and flatter than the left. Left and
right sides can contract ventrally, presumably from horizontally arranged
microtubules. Cingulum equatorial and descending up to 2x. Cingulum
with longitudinal thecal ridges. Sulcus invades epitheca about one-third
the height. Apical groove starts to the right of the distal epithecal end of
the sulcus and extends onto the dorsal surface; right edge of apical groove
is thicker. The apical groove is not an extension of the sulcus. The ventral
edge is characteristic and somewhat broadly undulating. Nucleus round
and located in posterior left quadrant. Chloroplasts present and
peripheral. Chain formation noted in very dense concentrations, e.g., 10^8
cells liter^{-1}. Toxic.
Remarks: Several species in this complex, e.g., *G. breve* and *G. mikimotoi*,
have similar morphologies and photosynthetic pigments. This complex is
distinct and should be separated out as a new genus based on a suite of
characters, e.g., dorsoventral flattening, dorsal left pore field, dorsal
cingulum pore, cingulum–sulcus juncture, cingulum displacement, apical
groove, and unique pigments.
Distribution: Oceanic to estuarine; warm temperate to tropical. Gulf of
Mexico, southeast coast of the United States entrained in the Gulf Stream.
Also recorded from the West Indies. *G. breve*-like cells recorded from
Japanese, European, Australian, and New Zealand waters.

Gymnodinium catenatum Graham 1943 (Plate 18)
Small cell with varied morphology depending on whether seen as chain
former or single cell. Chains can be made of up to 64 slightly
anterioposteriorly compressed cells. Hypotheca exceeds epitheca; epitheca
truncate, rounded, or abruptly conical. Apical groove runs
counterclockwise around apex; proximal end starts at end of sulcal
intrusion. Theca with reticulate pattern; quilted. Chloroplasts present.
Nucleus large and central. Produces characteristic cyst. Toxic.
Remarks: There is a similar smaller species isolated from the NW rias of
Spain that has been described as *Gyrodinium impudicum* (Fraga et al., 1995)
which is not toxic.
Distribution: Temperate waters; North America, Europe, Australia, and
Japan.

Gymnodinium galatheanum Braarud 1957 (Plate 18)
Small oval, gyrodinioid cell (<20 μm) with a displaced cingulum $3\times$
cingular width. Sulcus slightly invades epitheca. Straight apical groove

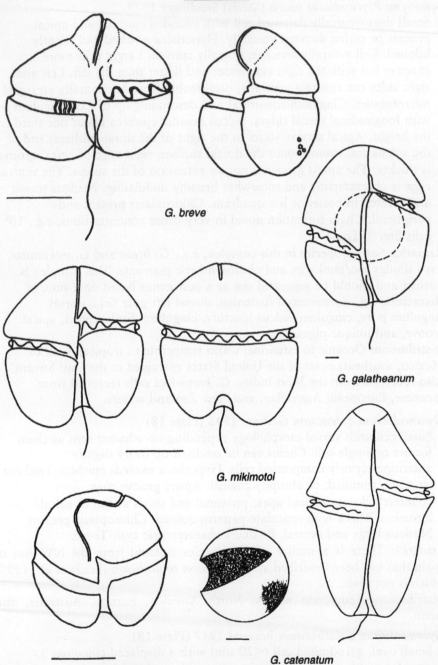

G. breve

G. galatheanum

G. mikimotoi

G. catenatum

G. pulchellum

originates to the left of the sulcal intrusion. Chloroplasts present. Nucleus round and centrally located. Toxic.

Distribution: North Sea, British Isles.

Gymnodinium heterostriatum Kofoid & Swezy 1921 (**Plate 19**)
Medium-sized cell with conical epitheca and pointed hypotheca, not equal. Appears circular in cross section. Cingulum circular to slightly descending and premedian. Sulcus invades epitheca to apex. Theca with longitudinal striae; hypotheca has more striae than epitheca. Chloroplasts absent; heterotrophic nutrition. Position of the nucleus variable.

Distribution: Temperate; neritic waters.

Gymnodinium mikimotoi Miyake & Kominami ex Oda 1935 (**Plates 1 and 18**)
Synonyms: *G. aureolum* Hulburt 1957 in part; *G. nagasakiense* Takayama & Adachi 1985.
Small, broadly oval cell that is dorsoventrally compressed. Hypotheca exceeds epitheca; epitheca is broadly rounded and hypotheca is notched and slightly bilobed. Cingulum slightly premedian and displaced 2× cingular width. Apical groove–sulcus juncture characteristic. Sulcus slightly invades epitheca; immediately to right is the proximal end of the apical groove which extends onto the dorsal epitheca and is straight. Ventral ridge inverted hook shape. Clustered pore field on left dorsal hypotheca. Chloroplasts present. Nucleus ellipsoidal and on left side near periphery. Toxic.

Remarks: This species is easily confused with others in this complex but can be differentiated by the apical groove–sulcus juncture and ventral ridge.

Distribution: Temperate to tropical neritic waters; cosmopolitan.

Gymnodinium pulchellum Larsen 1994 (**Plates 1 and 18**)
Synonym: *Gymnodinium* type '84-K, Onoue et al. 1985
Small, broadly oval cell with slight dorsoventral compression. Cingulum displaced 1–1.5× cingulum widths. Apical groove characteristically reversed S shaped that terminates on the dorsal surface and originates to the right of the ventral ridge. Chloroplasts present. Nucleus ellipsoidal and on the left side. Toxic.

Distribution: Recorded from Japanese, Australian, Tasmanian, and Mediterranean waters.

PLATE 18 Scale = 10 μm unless otherwise indicated. *Gymnodinium breve; G. galathaneum* redrawn from Larsen & Moestrup (1989) (scale = 1 μm); *G. mikimotoi* redrawn from Takayama & Adachi (1984); *G. pulchellum* and *Gymnodinium catenatum.*

Gymnodinium

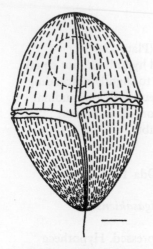

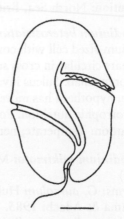

Gymnodinium heterostriatum   Gymnodinium abbreviatum   Gyrodinium uncatenum

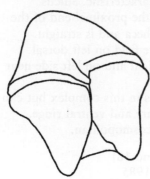

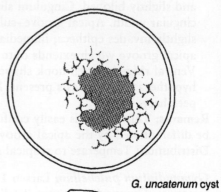

G. uncatenum cyst

Gymnodinium sanguineum

Gymnodinium estuariale

Gyrodinium instriatum

Gymnodinium sanguineum Hirasaka 1922 (Plate 19)
Synonyms: *G. splendens* Lebour 1925; *G. nelsonii* Martin 1929
　　Small to medium-sized pleomorphic cell; typically pentagonal in shape
　　with a broadly conical epitheca and a bilobed hypotheca. Epitheca and
　　hypotheca almost equal. Cingulum displaced 1× cingulum width. Sulcus
　　does not extend onto the epitheca but deeply notches hypotheca. Apical
　　groove present. Chloroplasts present; radiate from center of cell. Nucleus
　　large and central. Associated with fish kills. Toxic(?)
Remarks: This species can vary from about 40 to 75 μm in length and be
dorsoventrally compressed to circular in cross section, from heavily
pigmented to almost colorless or pale yellow. The hypotheca can be
rounded to truncate to prominently bilobed. These observations apply to
natural populations as well as cultures. Often, cultures indicate the
phenotypic variability that a species can express. In nature, such
pleomorphism is rarely seen except in bloom conditions. This species may
form resting stages surrounded by a mucoid halo.
Distribution: Temperate to tropical estuarine and costal waters;
cosmopolitan.

Genus *Gyrodinium* Kofoid & Swezy 1921 (> 100 spp.)
Synonyms: *Gymnodinium* Stein 1878 in part; *Sclerodinium* Dodge 1982;
Spirodinium Schütt 1896.
Type: *G. spirale* (Bergh) Kofoid & Swezy 1921.
References: Kofoid & Swezy, 1921; Schiller, 1933; Hulburt, 1957;
Campbell, 1973; Dodge, 1982; Takayama, 1985; Sournia, 1986; Gardiner
et al., 1989; Larsen, 1994.
　　Unarmored. Small to large-sized cells. Cells fusiform, biconical, ovoid,
　　tear shaped; sometimes compressed dorsoventrally or laterally. Cingulum
　　displaced more than one-fifth body length in a descending left spiral, with
　　or without overhang. Many species with cingulum displaced more than
　　one-third body length. Sulcus often invading epitheca. Some species with
　　apical groove. Chloroplasts present or absent. Pigmentation and nutrition
　　vary. In phagotrophic species, food vacuoles often present.
Remarks: *Gyrodinium* and *Gymnodinium* are distinguished based on
whether the cingulum displacement is more or less than one-fifth the body
length. Steidinger & Williams (1970) and others have discussed why this is
not a conservative character for generic separation. There are other
characters such as fusiform, ovoid, or conical, in conjunction with more

---

PLATE 19　*Gymnodinium heterostriatum* and *G. abbreviatum* redrawn from Kofoid & Swezy
(1921); *G. sanguineum, Gyrodinium uncatenum,* and *G. estuariale* redrawn from Hulburt (1957);
G. instriatum. Scale = 10 μm.

than one-third the body length that may separate out the *Gyrodinium* based on the type species. Currently, the species in this genus are differentiated by the following characters: shape and contours; size; the presence and shape of apical groove; the presence of striae, ribs, or furrows; total displacement of cingulum; sulcus extending onto epitheca; torsion of sulcus; the presence of chloroplasts; lateral compression of body; and color.

Gyrodinium aureolum Hulburt 1957 (not illustrated, see **Plate 18**)
Remarks: Same description as that for *Gymnodinium mikimotoi* in which the cingulum displacement can be less than one-fifth the body length or equal to, or exceed, one-fifth. The original species as described by Hulburt had a centrally located round nucleus, whereas *G. mikimotoi* has an ellipsoidal nucleus located on the left side of the cell at the periphery. The question remains whether or not there are two distinct species. Toxic.
Distribution: North Atlantic Ocean.

Gyrodinium estuariale Hulburt 1957 (**Plate 19**)
    Small (<20 μm) oval cell with conical epitheca and slightly lobed or subtrapezoidal hypotheca; right side of hypotheca longer than left. Cingulum excavated and wide; displaced about 1× cingulum width and less than one-third body length.
Remarks: This species has an APC and probably is armored.
Distribution: Common species in temperate and tropical estuaries; cosmopolitan.

Gyrodinium fissum (Levander) Kofoid & Swezy 1921 (not illustrated)
    Similar to *G. instriatum*. Small to medium-sized ellipsoidal cell. Theca with longitudinal striations or ridges; twice as many on the hypotheca as on the epitheca. Cell circular in cross section. Kofoid & Swezy (1921) describe the cytoplasm as granular and vacuolate with radially arranged pale green rods at the cell periphery. These may or may not be chloroplasts.
Distribution: Neritic species in temperate waters.

Gyrodinium instriatum Freudenthal & Lee 1963 (**Plates 1 and 19**)
    Small to medium-sized, broadly oval cell; broadest at the middle. Slightly dorsoventrally compressed. Epitheca convex or slightly concave or truncate. Hypotheca bilobed with sulcus extending to antapex. Cingulum excavated but of narrow width; displaced more than one-third body length. Sulcus invades epitheca. Anterior portion of epitheca often without pigment; contains nucleus. Chloroplasts present; pigmentation yellow–brown. Forms cysts.

Remarks: Has been confused with *G. fissum* which has thecal ridges or striations.

Distribution: Common species in temperate and tropical estuarine and neritic waters; cosmopolitan.

Gyrodinium lachryma (Meunier) Kofoid & Swezy 1921 (Plate 20)
   Large, tear-shaped cell with attenuated, pointed epitheca directed to the right. Cingulum descending and displaced >10 cingular widths. Sulcus extends almost to antapex. Surface with light striations. Chloroplasts absent.
Distribution: Oceanic and coastal, boreal and cold temperate waters.

Gyrodinium pingue (Schütt) Kofoid & Swezy 1921 (not illustrated)
   Small to medium-sized oval-shaped species. Epitheca and hypotheca conical; hypotheca wider. Cell circular in cross section. Cingulum displaced more than one-third body length. Sulcus invades epitheca and extends to antapex. Theca with fine striations or ribs. Chloroplasts absent. Food vacuoles present. Nucleus spherical and near-median. Cytoplasm diffuse pale green; short blue–green rodlets at the cell periphery.
Distribution: Neritic, temperate waters of the Atlantic and Pacific Oceans.

Gyrodinium spirale (Bergh) Kofoid & Swezy 1921 (Plate 20)
   Medium to large-sized spindle-shaped asymmetric cell with slight longitudinal twist. Cingulum excavated and narrow; displaced more than one-third body length. Epitheca curved to right with pointed apex, anatapex slightly bilobed with right side longer than left. Chloroplasts absent. Nucleus elongate and median. Food vacuoles present.
Remarks: This may represent a species complex with several similar species.
Distribution: Temperate to subtropical waters; cosmopolitan.

Gyrodinium uncatenum Hulburt 1957 (Plate 19)
   Medium-sized cell, slight lateral compression. Epitheca rounded, hypotheca bilobed. Cingulum displaced more than one-third body length. Sulcus sigmoid and invades epitheca, slightly curved to left. Chloroplasts present. Nucleus spherical and in epitheca. Forms cysts.
Distribution: Estuarine in temperate waters.

Genus *Katodinium* Fott 1957 (<30 spp.)
Synonym: *Massartia* Conrad 1926.
Type: *K. nieuportense* (Conrad) Loeblich Jr. & Loeblich III 1966.
References: Kofoid & Swezy, 1921; Fott, 1957; Loeblich III, 1965; Loeblich & Loeblich, 1966; Dodge & Crawford, 1970; Campbell, 1973;

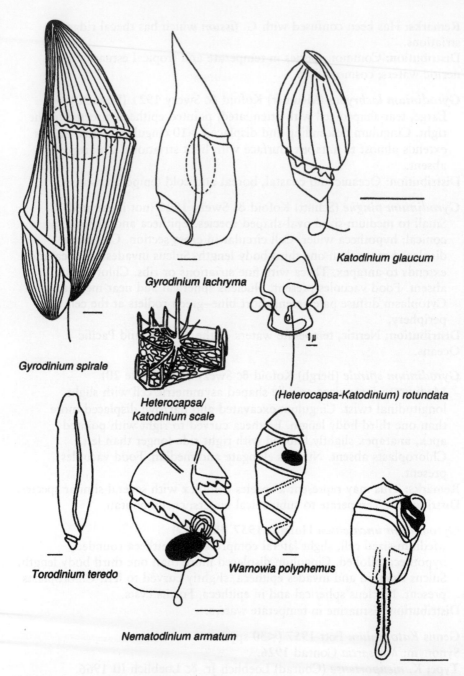

Gyrodinium lachryma

Katodinium glaucum

Gyrodinium spirale

Heterocapsa/
Katodinium scale

(Heterocapsa–Katodinium) rotundata

1 μ

Torodinium teredo

Warnowia polyphemus

Nematodinium armatum

Erythropsidinium agile

Spero & Moree, 1981; Spero, 1982; Takayama, 1985; Sournia, 1986; Hansen, 1989; Popovsky & Pfiester, 1990.

Unarmored/armored. Small gymnodinioid cells with a postmedian cingulum which is often difficult to delineate. Cells pear shaped or inverted top shaped, pendulate, club, or "mushroom" shaped. Theca with or without ridges or ribs. Epitheca exceeds hypotheca by 2x. Thecal vesicles with thin plates. Outer membrane covered with characteristic scales similar to *Heterocapsa*. Triangular scales with six radiating ribs and nine peripheral spines. Chloroplasts present or absent.

Remarks: This genus contains many small freshwater and estuarine species. TEM by Dodge & Crawford (1970) and Hansen (1989) showed thin thecal plates in thecal vesicles of *K. rotundatum* which has recently been transferred to *Heterocapsa* (Hansen, 1995). Also, recent SEM of several other estuarine dinoflagellates (5 to 15 µm), which would be characterized as *Gymnodinium* or *Katodinium* by light microscopy, reveals that they are armored and have plate patterns and tabulations fitting the Peridiniales, but they lack scales. At least some *Katodinium* are probably armored. Until the type species is studied in more detail, the placement of *Katodinium* is in question.

Katodinium fungiforme (Anissimova) Loeblich III 1965 (not illustrated)
Small cell with descending cingulum, 1–1.25×. Epitheca rounded, hypotheca asymmetrical. Epitheca broader than hypotheca. No apparent plates in thecal vesicles, although there is osmophilic material.

Remarks: Common estuarine species.
Distribution: Benthic and planktonic. Widely distributed in estuarine waters; cosmpolitan.

Katodinium glaucum (Lebour) Loeblich III 1965 (Plates 1 and 20)
Synonym: *Spirodinium glaucum* Lebour 1917.
Small fusiform cell theca thick, with about 20 longitudinal ribs on epitheca and 2 or 3 on hypotheca; epitheca has circular apical groove. cingulum displaced 4 or 5× cingulum width and about one-fifth of body. Epitheca exceeds hypotheca in both length and width. Chloroplasts absent; food vacuoles present. Nucleus round and in epitheca.

Remarks: This is an easily recognizable species.
Distribution: Common estuarine species; temperate to tropical waters. Cosmopolitan.

PLATE 20 Scale = 10 µm unless otherwise indicated. *Gyrodinium spirale* and *G. lachryma* redrawn from Kofoid & Swezy (1921); *Katodinium glaucum; K. rotundatum* (scale = 1 µm); and *Katodinium* scale redrawn from Hansen (1989); *Torodinium teredo, Nematodinium armatum, Warnowia polyphemus,* and *Erythropsidinium agile* (scale = 50 µm) redrawn from Kofoid & Swezy (1921).

Katodinium rotundatum (Lohmann) Loeblich III 1965 (Plate 20)
  Small "mushroom"-shaped cell in which epitheca exceeds hypotheca in
  both length and width. Reported to have apical pore or opening in an
  apical groove structure.
Distribution: Common estuarine species in temperate to tropical waters;
cosmopolitan.

Genus *Torodinium* Kofoid & Swezy 1921 (<5 spp.)
Type: *T..teredo* (Pouchet) Kofoid & Swezy 1921.
References: Kofoid & Swezy, 1921; Schiller, 1933; Steidinger & Williams,
1970; Campbell, 1973; Dodge, 1982; Sournia, 1986.
  Unarmored. Distinctive small to medium-sized cell. Body cigar shaped.
  Epitheca exceeds hypotheca (>5×) and sulcus extends almost entire
  length of cell. Sulcus and cingulum join posteriorly to form antapical
  loop. Chloroplasts present.

Torodinium teredo (Pouchet) Kofoid & Swezy 1921 (Plate 20)
  Cell length >3× width. Cell with brown pigmentation. Slow swimmer.
Remarks: Can be distinquished from *T. robustum* by the latter's size (<3×
width), green pigmentation.
Distribution: Wide distribution in temperate and tropical waters.

Genus *Lepidodinium* Watanabe, Suda, Inouye, Sawaguchi, & Chihara 1990
Type: *L. viride* Watanabe et al. 1990.
Reference: Watanabe et al., 1990.
  Unarmored. Small gymnodinioid cell with an apical groove. Surface
  covered with organic scales. Theca without plate material in vesicles.
  Cingulum descending and displaced about 1×.
Remarks: The closely related genus *Gymnodinium* was not described with
body scales and the type species lacks scales. The theca of the type species,
G. fuscum, contains vesicles with a similar membrane profile just above the
inner vesicle membrane as in *Lepidodinium.*

Lepidodinium viride Watanabe et al. 1990 (not illustrated)
  Small, subglobular cell which is compressed dorsoventrally. Apical groove
  extends from the sulcus to the apex and curves counterclockwise around
  the apex. Short club-shaped or finger-shaped structure in anterior sulcal
  region. Green color due to vestigial symbiont that has chlorophylls a and
  b. Cell does not have two nuclei, just one dinokaryon.
Distribution: Pacific Ocean; temperate and neritic.

Family Polykrikaceae Kofoid & Swezy 1921

Genus *Pheopolykrikos* Chatton emend. Matsuoka & Fukuyo 1986 (<5 spp.)

Type: *P. beauchampii* Chatton 1933.

References: Chatton, 1933; Schiller, 1933; Takayama, 1985; Matsuoka & Fukuyo, 1986; Sournia, 1986.

Unarmored. Small to medium-sized two-celled pseudocolony or as single cell. Gyrodinioid cell with broadly rounded epitheca, bilobed to rounded hypotheca. Cingulum premedian and displaced, left-handed. Epitheca with apical groove. Psuedocolonial cell with one nucleus per zooid centrally located. Cell not dorsoventrally compressed and lacks nematocysts. Chloroplasts present. Produces cysts.

Remarks: Cysts of this species differ from cysts of *Polykrikos* species, e.g., *P. kofoidii*.

Pheopolykrikos hartmannii (Zimmerman) Matsuoka & Fukuyo 1986 (Plate 21)

Cell surface without ridges. Cingulum displaced 1.5–2× width and about one-third length of body.

Remarks: This species may be synonymous with *Polykrikos barnegatensis* Martin 1929.

Distribution: Temperate and tropical waters; cosmopolitan.

Genus *Polykrikos* Bütschli 1873 (<5 spp.)

Type: *P. schwartzii* Bütschli 1873.

References: Kofoid & Swezy, 1921; Schiller, 1933; Steidinger & Williams, 1970; Morey-Gaines & Ruse, 1980; Harland, 1981; Dodge, 1982; Takayama, 1985; Sournia, 1986; Fensome et al., 1993.

Unarmored. Medium to large-sized pseudocolonial cell of 4 to 16 zooids or single cells. Gyrodinioid or gymnodinioid zooids closely appressed longitudinally without cellular septations, each component with its own cingulum but all sharing a common, continuous sulcus. Cingulum equatorial or slightly displaced. Apex with apical groove. Species with >2 zooids usually have one nucleus per 2 zooids. Chloroplasts absent. Nematocysts present or absent. Food vacuoles often present.

Remarks: Active phagocytic nutrition, feeding on other dinoflagellates such as *Ceratium, Protoperidinium, Scrippsiella, Gonyaulax,* and others, as well as diatoms.

Polykrikos kofoidii Chatton 1914 (Plate 21)

Single cells oval, longer than wide. Theca with striae or ridges. Resemble *Gyrodinium pellucidum*. Pseudocolony composed of 4, 8, or 16 zooids.

Polykrikos

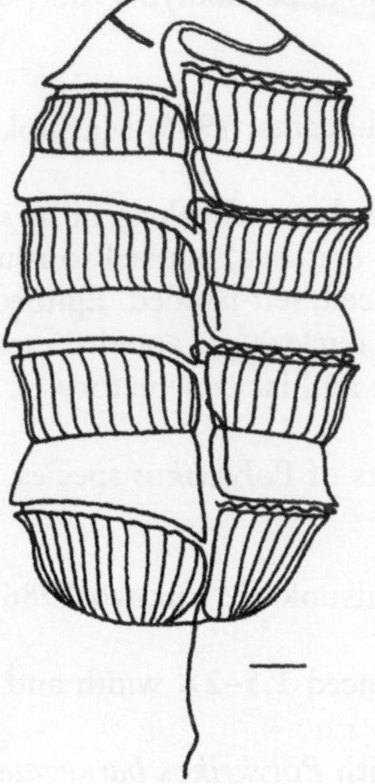

P. schwartzii

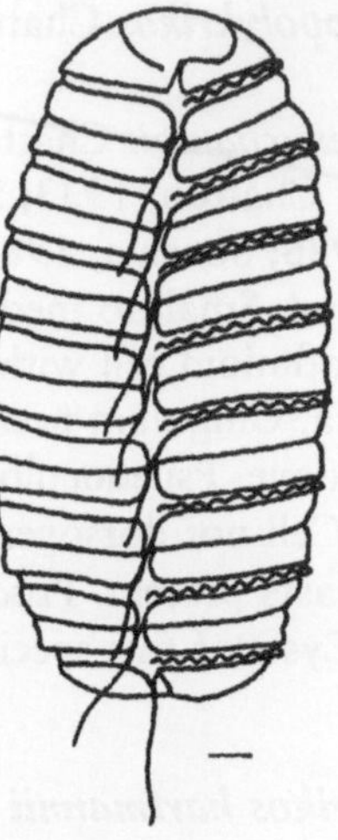

P. kofoidii

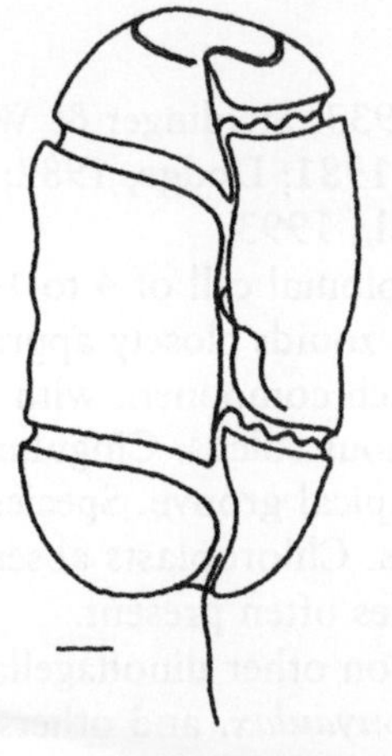

Pheopolykrikos hartmannii

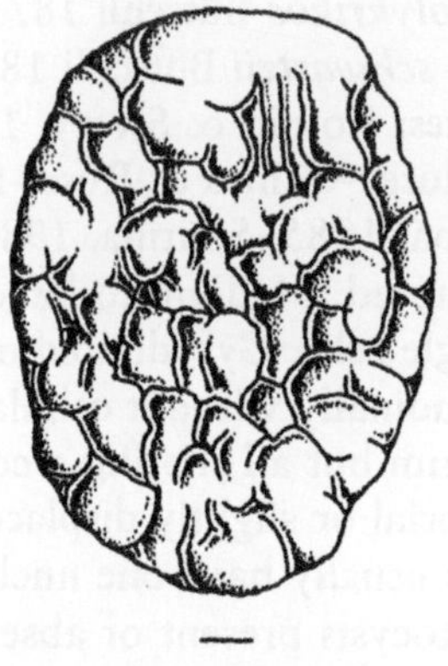

P. schwartzii cyst

PLATE 21   *Polykrikos kofoidii* and *P. schwartzii* redrawn from Kofoid and Swezy (1921); *Pheopolykrikos hartmannii* redrawn from Hulburt (1957); *Polykrikos* cyst. Scale = 10 μm.

One nucleus per 2 zooids. Hypotheca of zooids with ridges or furrows. Forms characteristic cysts.
Remarks: Easily distinguishable from *P. schwartzii* (**Plate 21**) which lacks ridges in the hypotheca and has an almost equatorial cingulum. In addition, the cysts of these two species differ.
Distribution: Temperate to tropical waters; cosmpolitan.

Family Warnowiaceae Lindemann 1928

Genus *Erythropsidinium* Silva 1960 (10 spp.)
Synonym: *Erythropsis* Hertwig 1884.
Type: *E. agile* (Hertwig) Silva 1960.
References: Kofoid & Swezy, 1921; Schiller, 1933; Silva, 1960; Elbrächter, 1979; Takayama, 1985; Sournia, 1986.
   Unarmored. Small to large amphidinioid cells with short epitheca; with or without apical carina; may have apical groove. Cingulum can be displaced up to 20× cingular width. Sulcus invades epitheca; no torsion. Possesses contractile tentacle, simple or compound pigment masses (red, brown, black), and simple or compound lenses. Pigment mass and ocellus median or anterior. Chloroplasts absent. Tentacle with or without stylet.
Distribution: Warm temperate to tropical seas.

Erythropsidinium agile (Hertwig) Silva 1960 (**Plates 2 and 20**)
   Large cell with long tentacle posteriorly directed. Ocellus simple and pigment mass red. Nucleus ellipsoidal and central.
Remarks: There are probably several different species described as *E. agile*.

Genus *Nematodinium* Kofoid & Swezy 1921 (<5 spp.)
Synonyms: *Nematopsides* Greuet 1978; *Pouchetia* Schütt 1895.
Type: *N. partitum* Kofoid & Swezy 1921.
References: Kofoid & Swezy, 1921; Lebour, 1925; Martin, 1929; Schiller, 1933; Hulburt, 1957; Francis, 1967; Mornin & Francis, 1967; Sournia, 1986.
   Small to large gyrodinioid cells with 1.5 cingular rotations. Sulcus with torsion creating posterior loop, may invade epitheca. Like *Warnowia* in that cells have melanosome and ocellus. Ocellus can be dispersed in small spherical lenses or in one. Genus characterized by the presence of nematocysts. Chloroplasts absent.

Nematodinium armatum (Dogiel) Kofoid & Swezy 1921 (**Plate 20**)
   Small subellipsoidal cell with cingulum displaced more than one-third body length. Sulcal torsion creates posterior loop; sulcus invades epitheca. Black pigment mass near posterior loop. Single lens near melanosome and with concentric rings. Nucleus round or oval and in anterior half of cell.

Remarks: This species has been described with three different size ranges for three different geographic areas and it is suspected that there is homonomy.
Distribution: Estuarine and coastal; temperate to warm waters.

Genus *Warnowia* Lindemann 1928 (25 spp.)
Synonyms: *Pouchetia* Schütt 1895; *Protopsis* Kofoid & Swezy 1921.
Type: *W. fusus* (Schütt) Lindemann 1928.
References: Kofoid & Swezy, 1921; Lebour, 1925; Schiller, 1933; Hulburt, 1957; Takayama, 1985; Sournia, 1986.
   Unarmored. Small to large-sized gyrodinioid cell. Cingulum with one or two turns; body may appear twisted. Some species with sulcal torsion creating posterior loop. Apical groove present (**Plate 2**). Ocellus and red or black pigment masses present, usually median or posterior. Nematocysts and chloroplasts absent.
Remarks: Species of this genus are differentiated by the following characters: shape and contours, size, fragmented, simple or compound lens, scattered or concentrated pigment mass, shape of pigment mass, position of ocellus, cingular displacement, the presence of striae or ribs, and color.

Warnowia polyphemus (Pouchet) Schiller 1933 (**Plate 20**)
   Medium-sized ellipsoidal cell. Cingulum with two turns. Ocellus in anterior part of cell with laminated lens and pigment mass.
Distribution: Oceanic and coastal; cold and warm temperate waters.

Order Suessiales Fensome et al. 1993
Reference: Fensome et al., 1993.
   One fossil and one extant family. Extant species principally coccoid cells living as symbionts in marine invertebrates. Biflagellated motile cell dinokont type and armored with thin plates in theca vesicles. Plates arranged in seven horizontal series. The cingulum consists of two series. *Symbiodinium* species appear to be transitional between unarmored and armored.

Family Symbiodiniaceae Fensome et al. 1993

Genus *Symbiodinium* Freudenthal 1962 (<25 spp.)
Synonym: *Zooxanthella* Brandt 1881.
Type: *S. microadriaticum* Freudenthal 1962.
References: Freudenthal, 1962; Loeblich & Sherley, 1979; Schoenberg & Trench, 1980; Blank & Trench, 1986; Trench & Blank, 1987; Palincsar et al., 1988; Rowan & Powers, 1991; Fensome et al., 1993.
   Species differentiated using size and cytology of coccoid stages and size and morphometrics of the dinospores. Chloroplasts present in both coccoid and dinospore stages.

Remarks: This genus was originally described as unarmored or "naked"; however, with SEM resolution of the theca of small dinospores, species are actually thinly armored with >50 plates arranged in Kofoidian series. The sutures of the plates may be delimited by nodules or papillae over the sutures. As in some *Woloszynskia*, the cingulum has a tiered series of cingular plates, and the epitheca has an apical groove or acrobase, not an apical pore complex. These characters put *Symbiodinium* in between unarmored and armored genera.

Symbiodinium microadriaticum Freudenthal 1962 (**Plate 22**)
  Very small (<10 μm), ovoid cell with rounded apex and antapex.
  Epitheca and hypotheca almost of equal length; epitheca exceeds
  hypotheca in width. Cingulum wide.
Remarks: Many species of this endosymbiotic genus appear to be host specific, although more than one dinoflagellate species can be endosymbiotic per host. Referred to as "zooxanthellae" because of their pigmentation and symbiotic nature with marine invertebrates.
Distribution: Coccoid cells endosymbiotic in coelenterates and flatworms; particularly in coral reef and tropical areas. Worldwide.

Order Ptychodiscales
References: Sournia, 1986; Elbrächter, 1993; Fensome et al., 1993.
  Gymnodinioid cell with well-developed pellicle.
Remarks: Fensome et al. (1993) place the following genera and *Berghiella* Kofoid & Michener 1911 and *Sclerodinium* Dodge 1981 in the order Ptychodiscales which they characterize as having a strongly developed pellicle. Sournia (1986) considered *Sclerodinium* a synonym of *Gyrodinium* and *Balechina* a synonym of *Gymnodinium*. Unfortunately, there is confusion with the family Kolkitziellaceae which is in the Peridiniales.

Family Ptychodiscaceae

Genus *Balechina* Loeblich Jr. & Loeblich III 1968 (<5 spp.)
Type: *B. pachydermata* (Kofoid & Swezy) Loeblich & Loeblich 1968.
References: Kofoid & Swezy, 1921; Steidinger & Williams, 1970; F. J. R. Taylor, 1976; Sournia, 1986; Fensome et al., 1993.
  Unarmored. Large subellipsoidal cell with thick, rigid theca; linear thecal ridges and aerolations discernible. Circular in cross section. Cingulum median and descending 0.5–2×. Sulcus invades epitheca. Apex hyaline. Chloroplasts absent.

Balechina coerulea (Dogiel) F. J. R. Taylor 1976 (**Plate 22**)
  Large gymnodinioid cell with blue, blue–green, green or pink pigmentation. More thecal ridges on hypotheca than epitheca.

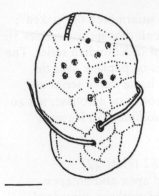

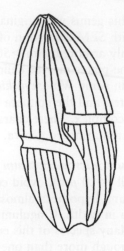

Symbiodinium microadriaticum

Balechina coerulea

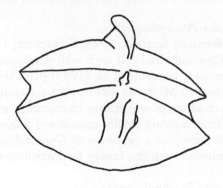

Herdmania litoralis   Ptychodiscus noctiluca

PLATE 22   Scale = 20 μm unless otherwise indicated. *Symbiodinium microadriaticum* (scale = 2 μm); *Balechina coerulea* modified from Balech (1988a); *Herdmania litoralis* (scale = 10 μm) redrawn from Dodge (1981a); *Ptychodiscus noctiluca* redrawn from Balech (1988a).

Remarks: Distinctive species.
Distribution: Coastal and open water. Warm water species.

Genus *Herdmania* Dodge 1981 (1 sp.)
Type: *H. litoralis* Dodge 1981.

References: Herdmann, 1922; Dodge, 1981a.
  Unarmored; maybe armored. Dodge characterizes the genus as having a thin theca with a few large plates. Small round cell, dorsoventrally compressed. Apex with apical notch or perhaps apical groove. Cingulum incomplete; similar to *Crypthecodinium*; ends on ventral right side.

Herdmania litoralis Dodge 1981 (Plate 22)
Description: Same as that for genus.
Remarks: If this species is armored, it needs to be removed from the above family.
Distribution: Sand dweller; British Isles.

Genus *Ptychodiscus* Stein 1883 (1 sp.)
Type: *P. noctiluca* Stein 1883.
References: Schiller, 1933; Boalch, 1969; F. J. R. Taylor, 1976; Sournia, 1986; Steidinger, 1990; Fensome et al., 1993.
  Unarmored. Medium-sized gymnodinioid cell flattened anterioposteriorly with a prominent apical carina that extends from the ventral surface onto the dorsal. Hypotheca conical; exceeds epitheca in width and length. Well-developed cingulum with ridges; slightly descending. Carina and anterior cingular edge hyaline. Theca pelliculate and resistant. Reported with and without chloroplasts and with two flagella emerging from one flagellar pore.

Ptychodiscus noctiluca Stein 1883 (Plate 22)
Description: Same as that for genus.
Remarks: This species is quite variable in anterioposterior compression and form. Boalch (1969) speculated that *P. carinatus* and *P. inflatus* were synonymous with *P. noctiluca* and that confusion arose with preserved, swollen, and distorted specimens.
Distribution: Coastal and oceanic; cold temperate to tropical waters.

Order Noctilucales Haeckel 1894
References: Loeblich III, 1982; Sournia, 1986; Fensome et al., 1993.
  Large, free-living unarmored cells that are morphologically modified for floatation and highly vacuolate. Cell with a cytosome and phagotrophic nutrition. Flagella reduced or absent.
Remarks: Three families: Kofoidiniaceae F. J. R. Taylor, Leptodiscaceae F. J. R. Taylor, and Noctilucaceae Kent.

Family Kofoidiniaceae F. J. R. Taylor 1976

Genus *Kofoidinium* Pavillard 1928 (<5 spp.)
Type: *K. velleloides* Pavillard 1928.

References: Cachon & Cachon, 1967a; Steidinger & Williams, 1970;
F. J. R. Taylor, 1976; Sournia, 1986; Fensome et al., 1993.
   Unarmored. Large inflated noctilucoid cell with epitheca, cingulum,
   sulcus, and two flagella. Sequential immature stages resemble
   Gymnodinium and then *Amphidinium*. There are six stages including the
   mature stage which is asymmetric and amphidinium-like with a laterally
   compressed hypotheca and an anterior, extracellular dome capable of
   rotation. The hypotheca with the sulcus extending along the posterior
   margin is sometimes referred to as a velum. The hemispherical dome is
   transparent. Chloroplasts absent.

Kofoidinium velleloides Pavillard 1928 (Plate 23)
Remarks: Distinctive species.
Distribution: Coastal and open water; temperate to tropical. Cosmopolitan.

Genus *Pomatodinium* Cachon & Cachon-Enjumet 1966 (1 sp.)
Type: *P. impatiens* Cachon & Cachon-Enjumet 1966.
References: Cachon & Cachon-Enjumet, 1966; Sournia, 1986; Fensome et
al., 1993.
   Unarmored. Large noctilucoid cell with cingulum, sulcus, two flagella,
   and an extracellular anterior dome. In ventral view the cell is bilobed
   with the sulcus associated with the right lobe; the left lobe is J shaped
   extending below and under the right lobe. The hemispherical dome is
   transparent and rotates. The hypotheca is contractile and changes form.
   Chloroplasts absent.

Pomatodinium impatiens J. Cachon & Cachon-Enjumet 1966 (Plate 23)
Remarks: Distinctive species.
Distribution: British Isles and Mediterranean.

Genus *Spatulodinium* Cachon & Cachon 1976 (1 sp.)
Type: *S. pseudonotiluca* (Pouchet) Cachon & Cachon ex Loeblich &
Loeblich 1969
References: Cachon & Cachon, 1969; Loeblich & Loeblich, 1969.
   Similar to *Kofoidinium* but with a ventral nonretractile tentacle.
   Immature stages gymnodinioid.

Spatulodinium pseudonoctiluca (Pouchet) Cachon & Cachon ex
Loeblich & Loeblich 1969 (Plate 23)
Remarks: Distinctive species.
Distribution: Mediterranean.

Family Leptodiscaceae F. J. R. Taylor 1987
Remarks: This family contains *Leptodiscus* (type genus), *Abedinium*,
Cachoninium, *Craspedotella*, *Cymbodinium*, *Petalodinium*, and

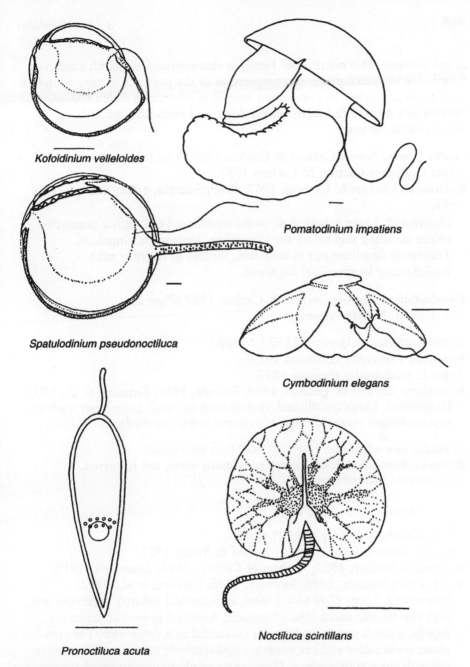

Kofoidinium velleloides

Pomatodinium impatiens

Spatulodinium pseudonoctiluca

Cymbodinium elegans

Pronoctiluca acuta

Noctiluca scintillans

PLATE 23 *Kofoidinium velleloides* (scale = 10 µm) redrawn from Cachon & Cachon (1968); *Pomatodinium impatiens* (scale = 10 µm) redrawn from Cachon & Cachon-Enjumet (1966); *Spatulodinium pseudonoctiluca* (scale = 20 µm) redrawn from Cachon & Cachon (1968); *Cymbodinium elegans* (scale = 100 µm) redrawn from Cachon & Cachon (1969); *Pronoctiluca acuta* (scale = 10 µm) redrawn from Schiller (1933); *Noctiluca scintillans* (scale = 500 µm) redrawn from Balech (1988a).

Scaphodinium. This noctilucoid family is characterized by width exceeding length due to anterioposterior compression or the presence of two wing-like extensions just below the cingulum; extension sometimes called vellum. Species lack a rotating hemispherical dome and tentacle. Species have encapsulated nucleus.

Genus *Cymbodinium* Cachon & Cachon 1967 (1 sp.)
Type: *C. elegans* Cachon & Cachon 1967.
References: Cachon & Cachon, 1967, 1969; Sournia, 1986; Fensome et al., 1993.
   Unarmored. Large spherical to ovoid noctilucoid cell with a contractile velum initiating just below the premedian, incomplete cingulum.
   Transverse flagellum not in cingulum; located in flagellar tube.
   Rudimentary longitudinal flagellum.

Cymbodinium elegans Cachon & Cachon 1967 (Plate 23)
Distribution: Mediterranean.

Genus *Leptodiscus* Hertwig 1877 (<5 spp.)
Synonym: *Pratjetella* Lohmann 1919.
Type: *L. medusoides* Hertwig 1877.
References: Cachon & Cachon, 1969; Sournia, 1986; Fensome et al., 1993.
   Unarmored. Large noctilucoid, disk-shaped cell with prominent nucleus.
   Gymnodinium stages known. Movement similar to medusa.

Leptodiscus medusoides Hertwig 1877 (not illustrated)
Remarks: Distinctive fragile species. In lateral view, cell is curved.
Distribution: Mediterranean.

Family Noctilucaceae Kent 1881

Genus *Noctiluca* Suriray 1836 (1 sp.)
Type: *N. scintillans* (Macartney) Kofoid & Swezy 1921.
References: Schiller, 1933; Cachon & Cachon, 1969; Zingmark, 1970;
Soyer, 1970; Sournia, 1986; Sweeney, 1978; Fensome et al., 1993.
   Unarmored. Large (200 to >1 mm), subspherical inflated vegetative cell with two flagella and a striated tentacle. A ventral groove contains the flagella, a tooth and tentacle and is connected to a cytostome. The cytoplasm is vacuolate and can contain photosynthetic symbionts. Vegetative cell with eukaryotic nucleus. Gametes gymnodinioid; with dinokaryotic nucleus. Chloroplasts absent; phagotrophic. Forms blooms.

Noctiluca scintillans (Macartney) Kofoid & Swezy 1921 (Plate 23)
Remarks: Distinctive species.
Distribution: Neritic; cosmopolitan in cold and warm water.

Genus *Pronoctiluca* Fabre-Domergue 1889 (<5 spp.)
Synonym: *Protodinifer* Kofoid & Swezy 1921
Type: *P. pelagica* Fabre-Domergue 1889.
References: Kofoid & Swezy, 1921; Steidinger & Williams, 1970; F. J. R. Taylor, 1976; Sournia, 1986; Fensome et al., 1993.
    Small to medium-sized fusiform cell with premedian indistinct cingulum and anterior tentacle. Tentacle flexible and without cytoplasm. Sulcus exists on epitheca only. Anterior nucleus reported with condensed chromosomes. Chloroplasts absent.
Remarks: The taxonomic placement of this genus in the Pyrrhophyta has been questioned by Kofoid & Swezy (1921) and Fensome et al. (1993).

Pronoctiluca acuta (Lohmann) Schiller 1933 (Plate 23)
Distribution: Coastal and oceanic. Temperate and warm water; worldwide distribution.

Order Lophodiniales Dodge 1984
References: Loeblich III, 1982; Fensome et al., 1993.
    Unarmored freshwater species with a theca of many hexagonal vesicles not arranged in series except for the cingulum where there can be one or two rows of vesicles.
Remarks: Loeblich III (1982) considers this a family in the Peridiniales and Fensome et al. (1993) place it in the Ptychodiscales.

Family Lophodiniaceae Osorio-Tafall 1941

Genus *Woloszynskia* Thompson 1951 (>5 spp.)
Synonym: *Aureodinium* Dodge, 1967.
Type: *W. reticulata* Thompson 1951.
References: Thompson, 1950; Crawford & Dodge, 1971; Crawford et al., 1970; Roberts & Timpano, 1989; Popovsky & Pfiester, 1990.
    Unarmored. Small to medium-sized cells of a gymnodinioid shape. Up to hundreds of detectable hexagonal vesicles; some species with fibrillar plate material in vesicles; plates not arranged in Kofoidian series. With or without apical groove. Contains chloroplasts, a nonmembrane bound stigma or eyespot, and other dinoflagellate organelles such as a pusule.
Remarks: A ventral ridge lies in the cingular and sulcal areas near the flagellar pores and contains fibrous material and may be homologous to the right sulcal plate in the Peridiniaceae. Species in this genus are found in brackish estuarine waters and can be identified as *Gymnodinium, Katodinium,* and *Amphidinium,* particularly those that are <15 μm.

Order Brachydiniales Loeblich III ex Sournia 1984
References: Loeblich III, 1982; Sournia, 1972, 1986; Fensome et al., 1993.

Unarmored pelliculate cell with four or five radiating elongate, flexible extensions; the anterior extension may be reduced. No sulcus observed.
Remarks: Fensome et al. (1993) place this group in the Ptychodiscales.

Family Brachydiniaceae Sournia 1972

Genus *Brachydinium* F. J. R. Taylor 1963 (<5 spp.)
Type: *B. capitatum* F. J. R. Taylor 1963.
References: F. J. R. Taylor, 1963; Sournia, 1972, 1986; Fensome et al., 1993.

Unarmored. Large, pelliculate cell with four radiating arms from hypotheca that are moveable; dorsoventrally flattened. Epitheca with apical process or reduced arm. Two flagella and an incomplete cingulum present. Ovoid nucleus occupies most of central body. Chloroplasts present.

Brachydinium capitatum F. J. R. Taylor 1963 (Plate 24)
Remarks: Distinctive species.
Distribution: Mediterranean Sea.

Order Actiniscales Sournia 1984

References: Loeblich III, 1982; Sournia, 1986; Fensome et al., 1993.
Unarmored gymnodinioid cell type with internal skeletal elements. Where known, elements are siliceous.
Remarks: Loeblich III (1982) and Fensome et al. (1993) place this group in the Gymnodiniales.

Family Actiniscaceae Kützing 1844

Genus *Actiniscus* Ehrenberg 1841 (<5 spp.)
Synonyms: *Diaster* Meunier 1919; *Gymnaster* Schütt 1891.
Type: *A. pentasterias* (Ehrenberg) Ehrenberg 1854.
References: Bursa, 1969; Steidinger & Williams, 1970; Sournia, 1986; Larsen & Sournia, 1991; Hansen, 1993.

Unarmored. Small gymnodinioid cell with siliceous internal skeleton of two star-shaped pentasters surrounding the nucleus. Cingulum median and descending about one cingulum width. Sulcus slightly indents the epitheca. Chloroplasts absent, but with other typical dinophycean organelles.
Remarks: It is not known whether representatives of this genus have an apical groove.

Actiniscus pentasterias (Ehrenberg) Ehrenberg 1854 (Plate 24)
Distribution: Subtropical to cold temperate in oceans even as marine relics in freshwater arctic lakes.

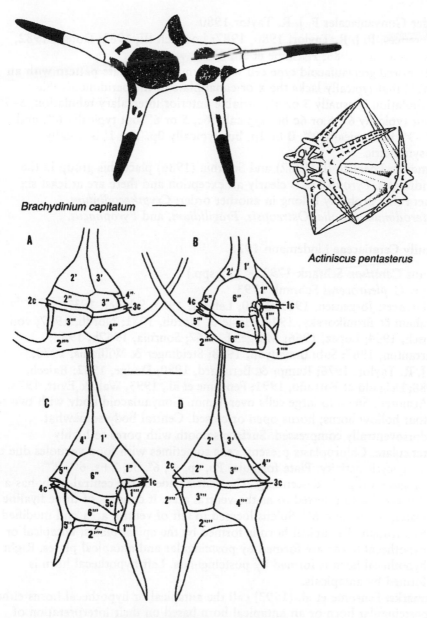

PLATE 24 *Brachydinium capitatum* (scale = 10 μm) redrawn from Sournia (1972); *Actiniscus* pentasters; schematic *Ceratium* redrawn from Evitt and Wall (1975) (B and C are ventral views, A and D are dorsal views).

Order Gonyaulacales F. J. R. Taylor 1980
References: F. J. R. Taylor, 1980, 1987; Loeblich III, 1982; Dodge, 1982,
1989; Sournia, 1986; Fensome et al., 1993.
  Armored gonyaulacoid type cell with asymmetrical plate pattern with an
  APC that typically lacks the x or canal plate of the peridinioids. Plate
  tabulation is usually 3 or 4′, variable anterior intercalary tabulation, 5–7″
  but typically 6″, 5 or 6c but typically 6c, 5 or 6‴ but typically 6‴, and
  1–3″″ but typically 2″″, 0 or 1p, but typically 0p. The 1′ is usually
  asymmetric.
Remarks: Loeblich III (1982) and Sournia (1986) place this group in the
Peridiniales. *Pyrophacus* is clearly an exception and there are at least six
genera that probably belong in another order: *Crypthecodinium,*
Heterodinium, Coolia, Ostreopsis, Fragilidium, and *Pyrophacus.*

Family Ceratiaceae Lindemann 1928

Genus *Ceratium* Schrank 1793 (<125 spp.)
Type: *C. pleuroceras* Schrank 1793.
References: Jörgensen, 1911, 1920; Lebour, 1925; Schiller, 1937;
Graham & Bronikovsky, 1944; Hasle & Nordli, 1951; Nordli, 1957; von
Stosch, 1964; Lopez, 1966; Margelef, 1967; Sournia, 1967b, 1986;
Yarranton, 1967; Subrahmanyan, 1968; Steidinger & Williams, 1970;
F. J. R. Taylor, 1976; Rampi & Bernhard, 1980; Dodge, 1982; Balech,
1988; Delgado & Fortuño, 1991; Fensome et al., 1993, Wall & Evitt, 1975.
  Armored. Small to large cells over 1 mm. Gonyaulacoid body with two to
  four hollow horns; horns open or closed. Central body somewhat
  dorsoventrally compressed. Surface smooth with pores to highly
  reticulate. Chloroplasts present; cells sometimes with food vacuoles due to
  phagocytic activity. Plate formula: Po, cp, 4′, 6″, 5c, 2+s, 6‴, 2″″.
  Cingulum slightly descending. The ventral side of the central body has a
  depressed area referred to as the ventral area. It consists of three hyaline
  plates, 6″, 5c, and 6‴. Sulcus located to left of ventral area and modified
  to a trough. The apical horn is formed by the apical plates; antapical or
  hypothecal horns are formed by postcingular and antapical plates. Right
  hypothecal horn is formed by postcingulars. Left hypothecal horn is
  formed by antapicals.
Remarks: Fensome et al. (1993) call the antapical or hypothecal horns either
a postcingular horn or an antapical horn based on their interpretation of
plates that form the horns. Species in this genus are differentiated on the
following characters: the presence or absence of an apical horn; the presence
of inflated epitheca; development, direction, and curvature of horns; relation
of left and right hypothecal horns; hypothecal horns with open or closed
ends; total length of cell; shape and width of body; epitheca in relation to

hypotheca; cell contour; and surface markings. The best view for cell outline and horn curvature and direction is a dorsal view. Plates 31–33 have silhouettes of *Ceratium* species in dorsal view. Compare them with their ventral view counterparts in Plates 25–29 and visualize how the preceeding characters can be used to differentiate species. Some species have been documented to produce anisogametes. The smaller or male gamete often does not resemble the vegetative stage and can be misidentified as another species, such as *C. lineatum,* or similar species with posteriorly directed hypothecal horns.

Ceratium arcticum (Ehrenberg) Cleve 1901 (Plate 26)

Large *C. horridum*-like cell with apical horn directed to the right. Hypothecal horns open distally, slightly curved and directed outward from the body; not substantially curved below the straight posterior body margin.

Remarks: Similar to *Ceratium longipes.*

Distribution: Cold water species; North Atlantic and North Pacific Oceans, and Arctic Ocean.

Ceratium arietinum Cleve 1900 (Plate 27)

Synonym: *C. bucephalum* Cleve, 1897.

Large *C. tripos*-like cell with rounded epithecal shoulders and slightly offset apical horn. Hypothecal horns curved and attenuated; ends closed. Curvature gentle and almost equal distant from apical horn; right horn longer. Surface with pores.

Distribution: Neritic, oceanic; warm temperate to tropical waters; worldwide distribution.

Ceratium candelabrum (Ehrenberg) Stein 1883 (Plate 26)

Large *C. furca*-like cell with body wider than long. Epithecal shoulders angularly tapering into straight, offset apical horn. Hypothecal horns divergent and directed posteriorly beyond posterior margin; left horn longer and often serrated. Surface with pores; sometimes with thecal flanges. Can form chains. Several varieties.

Distribution: Oceanic; warm temperate to tropical waters; worldwide distribution.

Ceratium carriense Gourret 1883 (Plate 29)

Largest *C. massiliense*-like cell with slightly offset but straight apical horn formed from rounded epitheca. Left hypothecal horn gently or broadly curves from body and distal end (open) is directed anteriorly. Right horn curves proximally and then is directed to right. Several varieties.

Distribution: Oceanic; warm temperate to tropical waters; worldwide distribution.

Ceratium contortum (Gourret) Cleve 1900 (**Plate 27**)
Synonym: *C. arcuatum* (Gourret) Cleve.
  Large *C. tripos*-like cell with offset apical horn directed to right. Right
  hypothecal horn curved and bent or recurved past apical horn. Horns
  attenuated, closed ends. Posterior body margin straight or slightly
  rounded. Several varieties.
Remarks: Highly variable species; see Sournia (1967).
Distribution: Oceanic and coastal. Cosmopolitan in warm temperate to
tropical waters; worldwide distribution.

Ceratium declinatum (Karsten) Jörgensen 1911 (**Plate 26**)
  Large *C. tripos*-like cell with epitheca that abruptly forms an apical horn.
  The left epithecal shoulder is almost straight; right shoulder is rounded.
  Posterior body margin is rounded. Curved hypothecal horns (closed)
  reach anteriorly just beyond epithecal body; right horn longer. Several
  varieties.
Distribution: Oceanic; temperate to tropical waters; worldwide distribution.

Ceratium furca (Ehrenberg) Claparède & Lachmann 1859 (**Plate 25**)
  Large species with two unequal, parallel or slightly divergent hypothecal
  horns; right horn shorter than left. Hypothecal horns serrated. Epitheca
  gradually tapers into apical horn.
Remarks: Sometimes confused with *C. hircus* even though *C. hircus* is
smaller, restricted to estuarine/coastal warm waters, and has almost equal
hypothecal horns and a more robust body. The right hypothecal horn of *C.
hircus* is directed ventrally. *C. hircus* is dark brown to brown-yellow in the
living state and *C. furca* is typically more yellow than brown.
Distribution: Principally coastal, but found in estuarine and oceanic
environments; cosmopolitan in cold temperate to tropical waters.

Ceratium fusus (Ehrenberg) Dujardin 1841 (**Plate 25**)
  Large fusiform cell with a fully developed apical horn, a fully developed
  left hypotheca horn, and a rudimentary right hypothecal horn. Left horn
  slightly curved to straight. Epitheca tapers gently into slightly curved
  apical horn. Surface with linear markings.
Remarks: Similar to *C. extensum* which is considerably larger (>1 mm)
with a long, straight left hypothecal horn.
Distribution: Oceanic to estuarine; principally coastal. Cosmopolitan in cold
temperate to tropical waters.

Ceratium gibberum Gourret 1883 (**Plate 27**)
  Large *C. tripos*-like cell with offset straight apical horn and angular right
  epithecal shoulder. Hypothecal horns (closed) curved and directed
  anteriorly or right horn recurved past apical horn. Posterior body margin

Ceratium

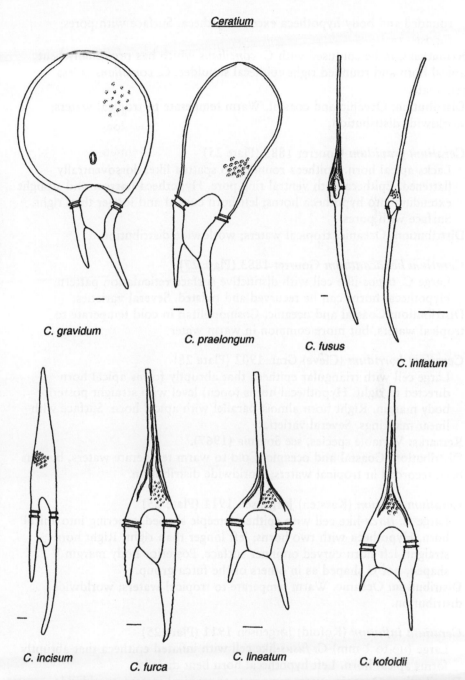

PLATE 25 Scale = 10 μm unless otherwise indicated. *Ceratium gravidum, C. praelongum, C. fusus, C. inflatum* (scale = 100 μm), *C. incisum, C. furca, C. lineatum,* and *C. kofoidii.*

rounded and body hypotheca exceeds epitheca. Surface with pores; rugose.
Remarks: Can be confused with *C. concilians* which has proximally bent apical horn and rounded right epithecal shoulder. *C. concilians* is less reticulate.
Distribution: Oceanic and coastal. Warm temperate to tropical waters; worldwide distribution.

Ceratium gravidum Gourret 1883 (Plate 25)

Lacks apical horn. Epitheca rounded to spatula like; dorsoventrally flattened. Epitheca with ventral ring pore. Hypotheca narrow and straight extending into hypotheca horns; left horn curved and longer than right. Surface with pores.
Distribution: Oceanic; tropical waters; worldwide distribution.

Ceratium hexacanthum Gourret 1883 (Plate 27)

Large *C. tripos*-like cell with distinctive surface reticulation pattern. Hypothecal horns can be recurved and twisted. Several varieties.
Distribution: Coastal and oceanic. Cosmopolitan in cold temperate to tropical waters, but more common in warm water.

Ceratium horridum (Cleve) Gran 1902 (Plate 28)

Large cell with triangular epitheca that abruptly forms apical horn directed to right. Hypothecal horns (open) level with straight posterior body margin. Right horn almost parallel with apical horn. Surface with linear markings. Several varieties.
Remarks: Variable species, see Sournia (1967).
Distribution: Coastal and oceanic. Cold to warm temperate waters, but has been recorded in tropical waters; worldwide distribution.

Ceratium incisum (Karsten) Jørgensen 1911 (Plate 25)

Large *C. furca*-like cell with epitheca steeple shaped, tapering into apical horn. Hypotheca with two horns; left longer than right. Right horn straight, left horn curved on inner surface. Posterior body margin V shaped, not U shaped as in others of the furca group.
Distribution: Oceanic. Warm temperate to tropical waters; worldwide distribution.

Ceratium inflatum (Kofoid) Jørgensen 1911 (Plate 25)

Large (up to 1 mm) *C. fusus*-like cell with inflated epitheca that abruptly forms apical horn. Left hypothecal horn bent distally.
Distribution: Oceanic; warm temperate to tropical waters; worldwide distribution.

Ceratium kofoidii Jørgensen 1911 (Plate 25)
Large, delicate *C. furca*-like cell with triangular epitheca abruptly forming apical horn. Hypothecal horns needle shaped, serrated, and parallel to slightly divergent. Surface with linear markings.
Distribution: Oceanic; warm temperate to tropical waters.

Ceratium limulus Gourret 1883 (Plate 28)
Large *C. tripos*-like cell of condensed form. Short straight apical horn sitting on angled and humped epithecal shoulders. Hypothecal horns (closed) nearly symmetrical and left closely appressed to body. Posterior body margin rounded. Surface rugose.
Remarks: Can be confused with *C. paradoxides* which is smaller and has an apical horn directed to the right. Also, the hypothecal horns of *C. paradoxides* are away from the body.
Distribution: Oceanic. Warm water species; worldwide distribution.

Ceratium lineatum (Ehrenberg) Cleve 1899 (Plate 25)
Medium-sized *C. furca*-like cell with apical horn formed from triangular epitheca and directed to the right. Hypothecal horns (closed ends) divergent and directed posteriorly beyond posterior body margin. Left horn longer. Body shape narrowly pentagonal. Surface with pores and linear markings.
Remarks: This species can be confused with *C. pentagonum* and similar species.
Distribution: Neritic, oceanic; cold temperate to tropical waters. There may be distinct warm water forms.

Ceratium longipes (Bailey) Gran 1902 (Plate 26)
Large *C. horridum*-like cell with open hypothecal horns. Epitheca tapers into an apical horn that is bent to the right. Hypothecal horns (open) are level with the straight posterior body margin. Several forms.
Remarks: This is a variable species and is very similar to *C. horridum*.
Distribution: Coastal. Arctic to cold temperate waters.

Ceratium lunula (Schimper) Jörgensen 1911 (Plate 29)
Large robust cell sometimes with symmetrical hypothecal horns (closed). Epitheca triangular abruptly forming straight apical horn. Hypothecal horns parallel with straight posterior body margin. Right horn curved but directed to right.
Distribution: Coastal and oceanic. Warm temperate to tropical waters; worldwide distribution.

Ceratium macroceros (Ehrenberg) Vanhöffen 1897 (Plate 29)
Large *C. massiliense*-like cell with angular box-like body that abruptly forms an offset apical horn directed to the right. Left and right

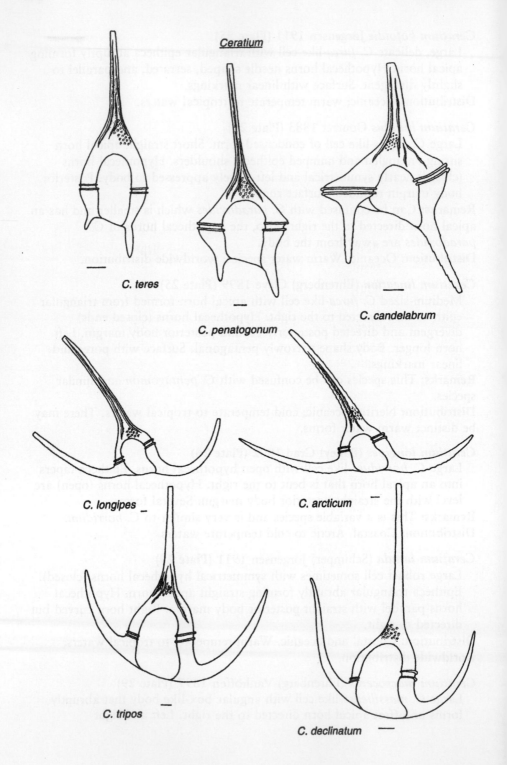

Ceratium

C. teres

C. candelabrum

C. penatogonum

C. longipes

C. arcticum

C. tripos

C. declinatum

hypothecal horns (open ended) proximally of same width and formed abruptly from body; extend beyond straight posterior body margin almost equally before curving or bending. Left horn curves anteriorly; right horn bends to the right and almost parallels the apical horn.
Remarks: Can easily be confused with *C. massiliense*, but the hypothecal horn development and the right-directed apical horn clearly distinguish it.
Distribution: Oceanic and coastal. Cold temperate to tropical waters; worldwide distribution.

Ceratium massiliense (Gourret) Jørgensen 1911 (Plate 33)
Large cell with rounded epitheca and apical horn slightly directed to the right. Typical diagonal straight posterior body margin that makes proximal end of left horn wider than that of the right horn. Left horn curves and is almost immediately directed anteriorly. Right horn bends and is directed to the right. Hypothecal horns with open ends.
Distribution: Coastal and oceanic. Cosmopolitan in warm temperate to tropical waters.

Ceratium pentagonum Gourret 1883 (Plate 26)
Large *C. furca*-like cell with broad pentagonal central body. Epithecal shoulders angular and abruptly forming straight apical horn. Hypothecal horns almost parallel; extend beyond posterior margin. Apical horn can be short or long. Surface with raised ornamentation. Several varieties.
Distribution: Oceanic, warm temperate to tropical waters; worldwide distribution.

Ceratium platycorne Daday 1888 (Plate 30)
Large *C. horridum*-like cell with characteristic inflated hypothecal horns (open); horns almost symmetrical. Surface with pores and lists; horns serrated at margins.
Distribution: Oceanic. Warm water species.

Ceratium praelongum (Lemmermann) Kofoid 1907 (Plate 25)
Large cell without apical horn but with expanded epitheca that is spatula like. Epitheca dorsoventrally flattened with ventral ring pore. Proximal hypotheca same width as epitheca and extending into two parallel hypothecal horns (closed ends). Surface with pores.
Distribution: Oceanic; tropical waters; worldwide distribution.

PLATE 26 *Ceratium teres, C. pentagonum, C. candelabrum, C. longipes, C. arcticum, C. tripos,* and *C. declinatum*. Scale = 10 μm.

Ceratium pulchellum Schröder 1906 (Plate 27)
Large delicate *C. tripos*-like cell with long straight apical horn and short hypothecal horns (closed). Posterior body margin rounded. Right hypothecal horn extends about two or three girdle widths and can be close to the body. Left horn is slightly curved and reaches as high as the cingulum.
Distribution: Oceanic. Subtropical to tropical waters; worldwide distribution.

Ceratium ranipes Cleve 1900 (Plate 30)
Large *C. horridum*-like cell with characteristic finger-like extensions from hypothecal horns (open). Apical horn directed to right. Horns partially serrated along margins.
Distribution: Oceanic. Warm temperate to tropical waters; worldwide distribution.

Ceratium symmetricum Pavillard 1905 (Plate 28)
Large *C. tripos*-like cell with rounded epithecal shoulders and straight apical horn. Hypothecal horns almost equal distance from apical horn and symmetrical; horns attenuated with closed ends. Hypothecal horns can approach length of apical horn. Posterior body margin rounded. Surface with pores and fine markings.
Distribution: Warm temperate to tropical waters; worldwide distribution.

Ceratium teres Kofoid 1907 (Plate 26)
Large, delicate *C. furca*-like cell with characteristic hypothecal horns. Triangular epitheca with slightly rounded shoulders abruptly forming straight apical horn. Hypothecal horns short and slightly swollen in middle; left longer than right. Surface with pores.
Distribution: Oceanic; warm temperate to tropical waters, but rare; worldwide distribution.

Ceratium trichoceros (Ehrenberg) Kofoid 1908 (Plate 29)
Large, delicate *C. massiliense*-like cell with characteristic horn development. Apical and hypothecal horns (open ends) in parallel plane.
Distribution: Coastal and oceanic. Cosmopolitan in warm temperate to tropical waters; worldwide distribution.

Ceratium tripos (O. F. Müller) Nitzsch 1817 (Plate 26)
Large cell with triangular epitheca that abruptly forms straight apical horn. When hypothecal horns (closed) well developed, almost parallel with apical horn. In other forms, right horn can be directed to the right. Surface with linear markings; rugose.
Remarks: Variable species, see Sournia (1967).
Distribution: Coastal and oceanic. Cosmopolitan in cold temperate to tropical waters; worldwide distribution.

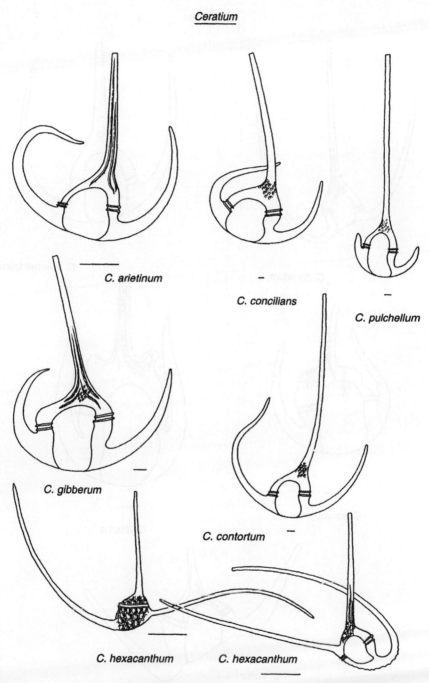

Ceratium

C. arietinum

C. concilians

C. pulchellum

C. gibberum

C. contortum

C. hexacanthum   C. hexacanthum

PLATE 27 Scale = 10 μm unless otherwise indicated. *Ceratium arietinum*, C. *concilians*, C. *pulchellum*, C. *gibberum*, C. *contortum*, and C. *hexacanthum* (scale = 100 μm).

Ceratium

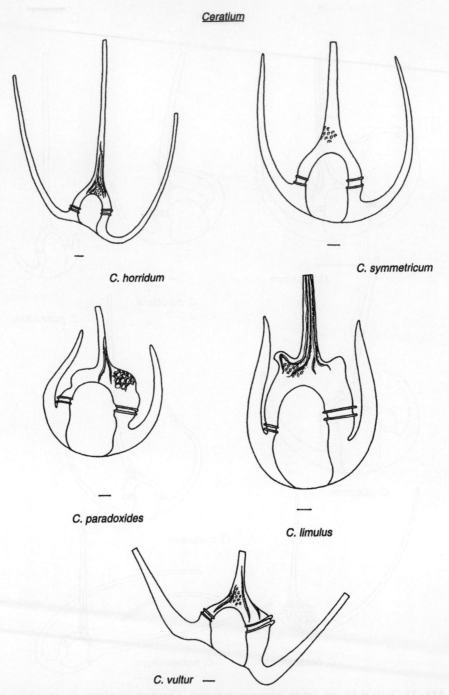

C. horridum

C. symmetricum

C. paradoxides

C. limulus

C. vultur —

PLATE 28 *Ceratium horridum, C. symmetricum, C. paradoxides, C. limulus,* and *C. vultur.*
Scale = 10 μm.

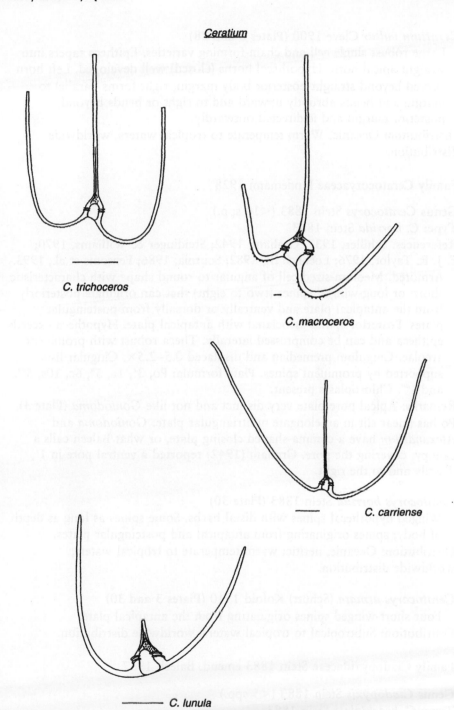

Ceratium

C. trichoceros

C. macroceros

C. carriense

C. lunula

PLATE 29 Scale = 100 μm unless otherwise indicated. *Ceratium trichoceros, C. macroceros* (scale = 10 μm), *C. carriense,* and *C. lunula.*

Ceratium vultur Cleve 1900 (Plates 6 and 28)
   Large robust single cell and chain-forming varieties. Epitheca tapers into
   straight apical horn. Hypothecal horns (closed) well developed. Left horn
   curved beyond straight posterior body margin; right forms parallel to
   margin and bends abruptly upward and to right or bends beyond
   posterior margin and is directed outwardly.
Distribution: Oceanic. Warm temperate to tropical waters; worldwide
distribution.

Family Ceratocoryaceae Lindemann 1928

Genus *Ceratocorys* Stein 1883 (<10 spp.)
Type: *C. horrida* Stein 1883.
References: Schiller, 1937; Graham, 1942; Steidinger & Williams, 1970;
F. J. R. Taylor, 1976; Loeblich III, 1982; Sournia, 1986; Fensome et al., 1993.
   Armored. Medium-sized cell of angular to round shape with characteristic
   short- or long-winged spines (two to eight) that can originate posteriorly
   from the antapical plate and ventrally or dorsally from postcingular
   plates. Posterior spines associated with antapical plate. Hypotheca exceeds
   epitheca and can be compressed laterally. Theca robust with prominent
   areolae. Cingulum premedian and displaced 0.5–2.5×. Cingular lists
   supported by prominent spines. Plate formula: Po, 3′, 1a, 5″, 6c, 10s, 5‴,
   and 1⁗. Chloroplasts present.
Remarks: Apical pore plate very distinct and not like *Goniodoma* (Plate 3).
Po has linear slit in an elongate nontriangular plate. *Goniodoma* and
Alexandrium have a comma-shaped closing plate, or what Balech calls a
canopy, covering the pore. Graham (1942) reported a ventral pore in 1′,
distally and to the right.

Ceratocorys horrida Stein 1883 (Plate 30)
   Winged hypothecal spines with distal barbs. Some spines as long as depth
   of body; spines originating from antapical and postcingular plates.
Distribution: Oceanic, neritic; warm temperate to tropical waters;
worldwide distribution.

Ceratocorys armata (Schütt) Kofoid 1910 (Plates 3 and 30)
   Four short-winged spines originating from the antapical plate.
Distribution: Subtropical to tropical waters; worldwide distribution.

Family Cladopyxidaceae Stein 1883 emend. Balech 1967

Genus *Cladopyxis* Stein 1883 (<5 spp.)
Type: *C. brachiolata* Stein 1883.

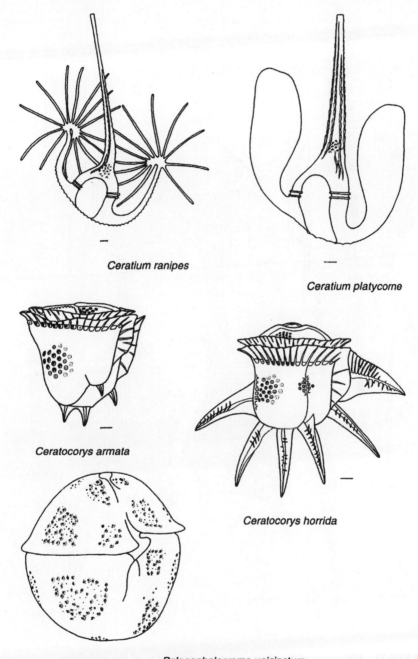

Ceratium ranipes

Ceratium platycorne

Ceratocorys armata

Ceratocorys horrida

Palaeophalacroma unicinctum

PLATE 30 *Ceratium ranipes, C. platycorne, Ceratocorys armata, C. horrida,* and *Palaeophala-croma unicinctum.* Scale = 10 μm.

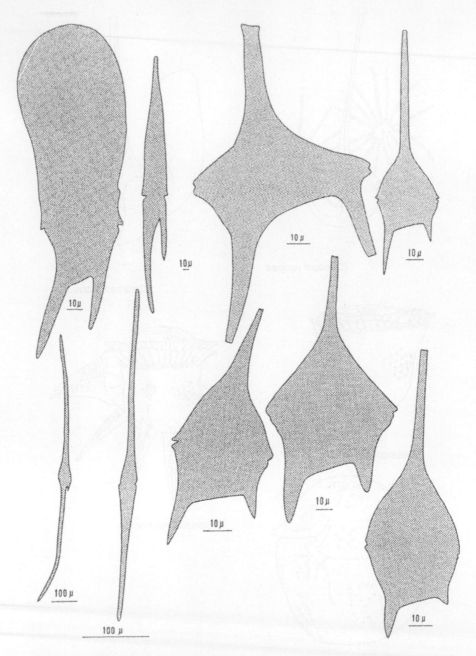

PLATE 31   Silhouettes of *Ceratium* species for comparison of size and shape.

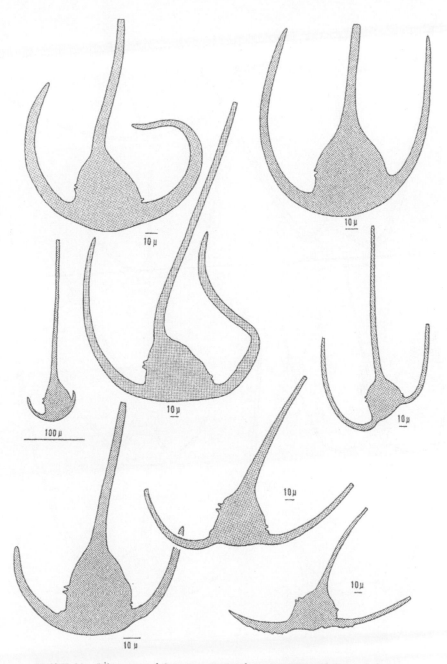

PLATE 32 Silhouettes of *Ceratium* species for comparison of size and shape.

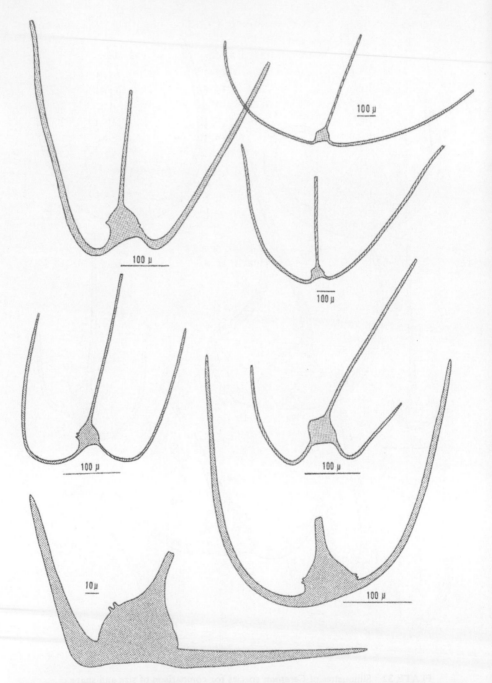

PLATE 33   Silhouettes of *Ceratium* species for comparison of size and shape. Top left drawing represents a dorsal view of *C. massiliense,* its ventral view is not illustrated in **Plate 29.**

References: Balech, 1964a, 1967; Schiller, 1937; Loeblich III, 1982; Sournia, 1986; Below, 1987b.
Armored. Cell small to medium sized and round to oval with characteristic thick extensions (arms) that can be pointed or furcated at the tip. Typically, between 3 and 12 arms are present. Extensions are similar to the hystrichospherid fossil dinoflagellate extensions. Theca with pores. Cingulum is premedian and only slightly displaced. Plate formula: Po, 3′, 3a, 7″, 6c, 7s, 6′″, and 2″″. The sa has been interpreted as the 1″ on some occasions. Chloroplasts present.
Remarks: This genus is closely related to *Paleophalacroma*.

Cladopyxis brachiolata Stein 1883 (not illustrated)
Cell small to medium sized with 10 strongly developed furcated arms on plates 3–6″, 2′″, 3–6′″, and 1–2″″.
Remarks: Compare to *Palaeophalacroma*.
Distribution: Oceanic; tropical and warm temperate waters of oceans and seas.

Genus *Palaeophalacroma* Schiller 1928 (>5 spp.)
Synonyms: *Epiperidinium* Gaarder 1954; *Sinodinium* Nie 1945.
Type: *P. verrucosum* Schiller 1928.
References: Balech, 1967; Loeblich III, 1982; Dodge, 1982; Sournia, 1986; Below, 1987.
Armored. Cell small, subspherical. Cingulum premedian, descending, about 1X. Anterior cingular list present, posterior list absent. Sulcus not excavated. Theca with large and small pores. Plate formula: Po, 4′, 3a, 7″, 6c, 6s, 6′″, and 2″″. 1′ narrow as in *Gonyaulax*. Chloroplasts usually present.

Paleophalacroma unicinctum Schiller 1928 (Plates 7 and 30)
Small ovoid cell with numerous chloroplasts and a centrally located nucleus.
Distribution: Tropical and warm temperate oceans.

Family Crypthecodiniaceae Biecheler 1952

Genus *Crypthecodinium* Biecheler 1938 (<5 spp.)
Type: *C. setense* Biecheler 1938.
References: Biecheler, 1952; Javornicky, 1962; Kubai & Ris, 1969; Loeblich III, 1982; Sournia, 1986; Beam & Himes, 1987; Fensome et al., 1993.
Armored. Cell small and spherical to ovoid with thin thecal plates in a modified Kofoidian series. The cingulum is descending and incomplete; the distal end does not meet the sulcus. Species are phagocytic and have a peduncle; chloroplasts absent. Forms immobile reproductive cysts. Plate

formula: 4′, 3a, 5″, 6c, 5s, 5′″, 3 ″″, and an x plate in the right hypotheca adjacent to the end of the cingulum.

Remarks: Genus aptly named; however, there are several genera with thin thecae. The plate pattern with the X plate is unique.

Crypthecodinium cohnii (Seligo) (Plate 34)
Remarks: Reported to have 52 sibling species, 7 major and 45 that were only found once.
Distribution: Brackish to marine, littoral; found associated with decaying macroalgae from the North Sea to the Caribbean.

Family Goniodomataceae Lindemann 1928

Genus *Alexandrium* Halim 1960 (>30 spp.)
Synonyms: *Gonyaulax* Diesing 1866; *Pyrodinium* Plate 1906; *Gessnerium* Halim 1967 ex Halim 1969; *Protogonyaulax* Taylor 1979.
Type: *A. minutum* Halim 1960. The type species for the *Gessnerium* subgenus is *A.* (*Gessnerium*) *monilatum* (Howell) Loeblich III 1970.
References: Lebour, 1925; Whedon & Kofoid, 1936; Halim, 1960a; Steidinger, 1971, 1990, 1993; F. J. R. Taylor, 1975, 1979, 1984; Schmidt & Loeblich, 1979; Dale, 1983; Balech, 1985a,1989, 1990a,b, 1992a, 1993, 1994, 1995; Balech & Tangen, 1985; Kita & Fukuyo, 1988; Larsen & Moestrup, 1989; Steidinger & Moestrup, 1990; Hallegraeff, 1991; Fensome et al., 1993.

This genus has extensive synonymy due to continual scrutiny of toxic species causing public health, economic, and ecological problems. The genus is composed of two related subgenera: *Alexandrium* and *Gessnerium*. *Alexandrium*: armored. Cells typically spherical to hemispherical to oval to slightly biconical, but without horns or spines. Plate formula: Po, cp, 4′, 0a, 6″, 6c, 9 or 10s, 5′″, and 2″″. Descending median cingulum without overhang or contortion, displaced 1–1.5 girdle widths. Surface markings include pores, reticulae, and vermiculae. Thecae can be thin and delicate to rugose. Cytoplasm includes elongate to C-shaped nucleus and all species contain chloroplasts. In this subgenus, the Po touches the 1′, but the connection is sometimes obscured in older cells with plate overlap growth of 2′ and 4′. Also, in some species of the

---

PLATE 34 *Crypthecodinium cohnii* (scale = 5 μm) redrawn from Biecheler (1952); schematic of *Alexandrium* with ventral, apical, and antapical views, apical pore complex and 1′, and sulcal plates. **sa,** anterior sulcal; **sma,** median anterior sulcal; **ssa,** acessory left sulcal; **smp,** posterior median sulcal; **ssp,** posterior left sulcal plate; **sp,** posterior sulcal; **sdp,** posterior right sulcal; **sacp,** posterior accessory sulcal; **sda,** anterior right sulcal; **saca,** anterior accessory sulcal. Redrawn from Balech (personal communication).

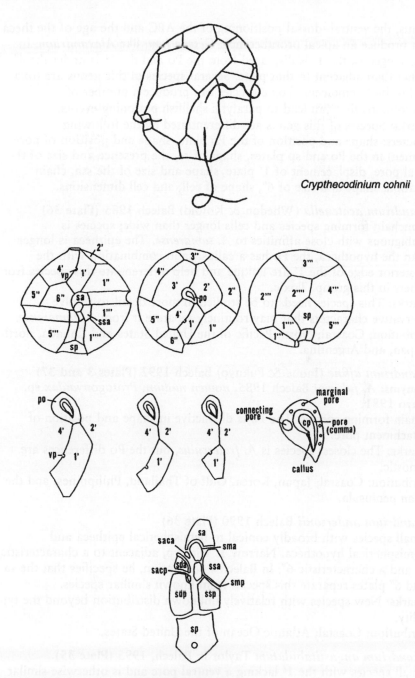

Crypthecodinium cohnii

schematic *Alexandrium*

genus, the ventral–dorsal positioning of the APC and the age of the theca can produce an apical protuberance. *Gessnerium:* like *Alexandrium,* in this subgenus, the 1' is displaced from the Po and the 1' is under the 4' rather than adjacent to that plate. Several species of this genus are toxic and/or bioluminescent. Toxic species can produce a number of neurotoxins that can lead to paralytic shellfish poisoning events.
Remarks: Species of this genus are differentiated by the following characters: shape and position of the Po plate, shape and position of pores (foramen) in the Po and sp plates, shape of sa, the presence and size of the ventral pore, displacement of 1' plate, shape and size of the ssa, chain formation, shape and size of 6", shape of cell, and cell dimensions.

Alexandrium acatenella (Whedon & Kofoid) Balech 1985 (Plate 36)
   Nonchain forming species and cells longer than wide; species is ambiguous with close affinities to *A. tamarense.* The epitheca is longer than the hypotheca, the Po has a callus, sa in combination with the posterior edge of the 1' are unique and help differentiate this species from others in this group. Toxic.
Remarks: This species needs to be studied in more detail to resolve conservative characters that may routinely separate it from *A. tamarense.*
Distribution: Coastal; North Pacific of the United States and Canada, north of Japan, and Argentina.

Alexandrium affine (Inoue & Fukuyo) Balech 1992 (Plates 3 and 37)
Synonyms: *A. fukuyoi* Balech 1985, *nomen nudum*; *Protogonyaulax* sp. Fukuyo 1981.
   Chain-forming species. The Po is distinctive in shape and position of attachment pore.
Remarks: The closest species is *A. fraterculus,* but the Po differences are diagnostic.
Distribution: Coastal; Japan, Korea, Gulf of Thailand, Philippines, and the Iberian peninsula.

Alexandrium andersonii Balech 1990 (Plate 36)
   Small species with broadly conical to hemispherical epitheca and hemispherical hypotheca. Narrow 1' with vp, adjacent to a characteristic sa and a characteristic 6". In Balech's description, he specifies that the sa and 6" plates separate this species from known similiar species.
Remarks: New species with relatively unknown distribution beyond the type locality.
Distribution: Coastal; Atlantic Ocean of the United States.

Alexandrium angustitabulatum Taylor in Balech, 1995 (Plate 35)
   Small species with the 1' lacking a ventral pore and is otherwise similar to *A. minutum.* The shape of the 1' and the narrow 6" are diagnostic when used in conjunction with APC and sulcal plates.

Alexandrium

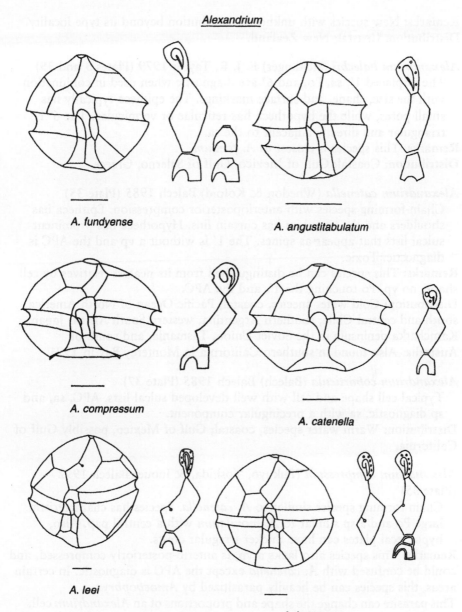

A. fundyense

A. angustitabulatum

A. compressum

A. catenella

A. leei

A. kutnerae

PLATE 35 *Alexandrium fundyense, A. angustitabulatum, A. compressum, A. catenella, A. leei,* and *A. kutnerae.* Scale = 10 μm. Redrawn from Balech (1995).

Remarks: New species with unknown distribution beyond its type locality.
Distribution: Coastal; New Zealand.

Alexandrium balechii (Steidinger) F. J. R. Taylor 1979 (Plates 3 and 39)
   The displaced 1', sa, Po, and 6" are diagnostic when used in conjunction
   with the size, shape, and surface markings. The epitheca typically has
   small pores, while the hypotheca has reticulae or vermiculae. The 6" is
   triangular and directly adjacent to the sa.
Remarks: This species is closest to *A. foedum*.
Distribution: Coastal; Gulf of Mexico, Gulf of Salerno, Greece.

Alexandrium catenella (Whedon & Kofoid) Balech 1985 (Plate 35)
   Chain-forming species with anterioposterior compression. Epitheca has
   shoulders and the cingulum has curtain fins. Hypotheca has prominent
   sulcal lists that appear as spines. The 1' is without a vp and the APC is
   diagnostic. Toxic.
Remarks: This species can be distinguished from its nearest relatives by cell
shape, no vp, Po touching the 1', and the APC.
Distribution: Cold water species, coastal, Pacific Ocean of North America,
south and central Chile, southern Argentina, western South Africa, Japan,
Kamchatka peninsula in the Soviet Union, Tasmania, and south of
Australia. Also found in southern California in Monterey Bay in 1991.

Alexandrium cohorticula (Balech) Balech 1985 (Plate 37)
   Typical cell shape and cell with well-developed sulcal lists. APC, sa, and
   sp diagnostic. sa with a precingular component.
Distribution: Warm water species, coastal; Gulf of Mexico, possibly Gulf of
California.

Alexandrium compressum (Fukuyo, Yoshida, & Inoue) Balech 1992
(Plate 35)
   Chain-forming species similar to *A. catenella*. Species has characteristic
   large Po and a sp similar to *A. monilatum* with a central pap. Also,
   hypothecal plates can have thicker irregular crests.
Remarks: This species also lacks a vp, is anterio-posteriorly compressed, and
could be confused with *A. catenella* except the APC is diagnostic. In certain
areas, this species can be heavily parasitized by *Amoebophrya*.
This parasite can change the shape and proportions of an *Alexandrium* cell.
Distribution: Coastal; Pacific Ocean off Japan and southern California.

Alexandrium concavum (Gaarder) Balech 1985 (Plate 36)
   Medium- to large-sized pentagonal cell with conical epitheca and concave
   sides in ventral view.
Distribution: Oceanic; warm temperate to tropical.

Alexandrium

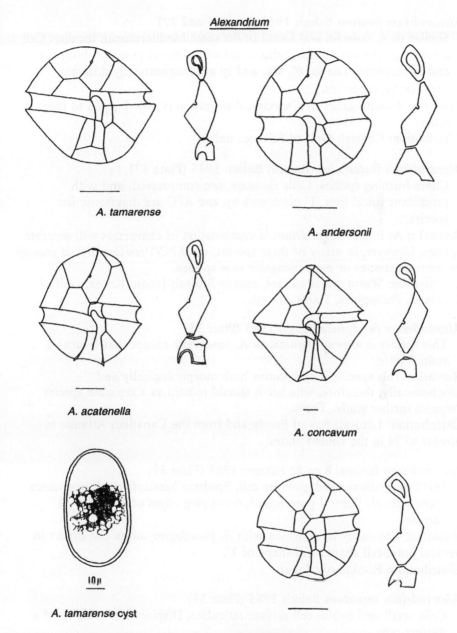

A. tamarense

A. andersonii

A. acatenella

A. concavum

10 μ

A. tamarense cyst

A. lusitanicum

PLATE 36 *Alexandrium tamarense, A. andersonii, A. acatenella, A. concavum, A. tamarense cyst,* and *A. lusitanicum.* Scale = 10 μm. Redrawn from Balech (1995).

Alexandrium foedum Balech 1990 (Plates 3 and 39)
   Similar to *A. balechii* and found in the same Mediterranean locality. Cell
   more angular with a thicker theca and with distinct pores on both epi-
   and hypothecae. The sa, 6″, ssa, and sp are diagnostically different
   between these species.
Remarks: As with other new species, distribution is now limited to type
locality.
Distribution: Coastal; Gulf of Salerno, Italy.

Alexandrium fraterculus (Balech) Balech 1985 (Plate 37)
   Chain-forming species. Cells elongate, not compressed, and with
   prominent sulcal lists. 1′ plate with vp and APC are diagnostic for
   species.
Remarks: As in all *Alexandrium*, a combination of characters will separate
species; however, in many of these species, the APC/1′/sa complex is enough
to separate species or even recognize new species.
Distribution: Warm water species, mainly littoral; Japan, Korea, Gulf of
Thailand, Philippines, South Atlantic.

Alexandrium fundyense Balech 1985 (Plate 35)
   This species is almost identical to *A. tamarense* except that it lacks a
   ventral pore.
Remarks: This species is ambiguous both morphologically and
biochemically; therefore, whether it should remain as a separate species
requires further study. Toxic.
Distribution: Littoral; Bay of Fundy and from the Canadian Atlantic to
almost 41°N in the United States.

Alexandrium hiranoi Kita & Fukuyo 1988 (Plate 39)
   Small to medium-sized globular cell. Epitheca hemispherical; hypotheca
   asymmetrical. Ventral pore round, occupying edges of 1′ and 4′. 1′
   narrow.
Remarks: Can easily be confused with *A. pseudogonyaulax* but differs in
ventral pore, cell shape, and shape of 1′.
Distribution: Rookpools; Japan.

Alexandrium insuetum Balech 1985 (Plate 38)
   Cells small and ovoid; cell surface reticulate. Displaced 1′ with vp of a
   distinct type.
Remarks: As more specimens are studied, variations will be confirmed.
Although Yuki and Yoshimatsu (1990) reported eight sulcal plates, Balech
(1995) reports 9s.
Distribution: Coastal; Korea and Japan.

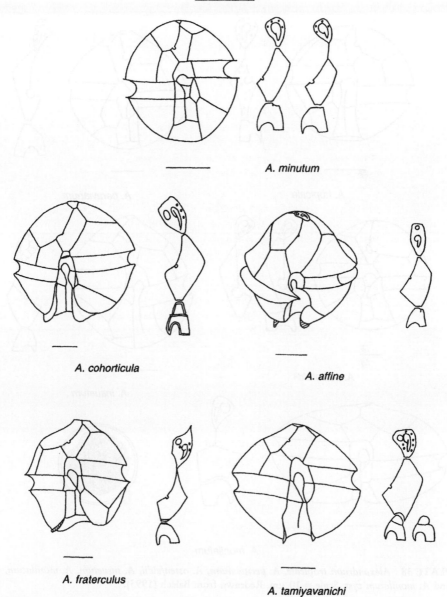

PLATE 37 *Alexandrium minutum, A. cohorticula, A. affine, A. fraterculus,* and A. *tamiyavanichi.*
Scale = 10 μm. Redrawn from Balech (1995).

Alexandrium

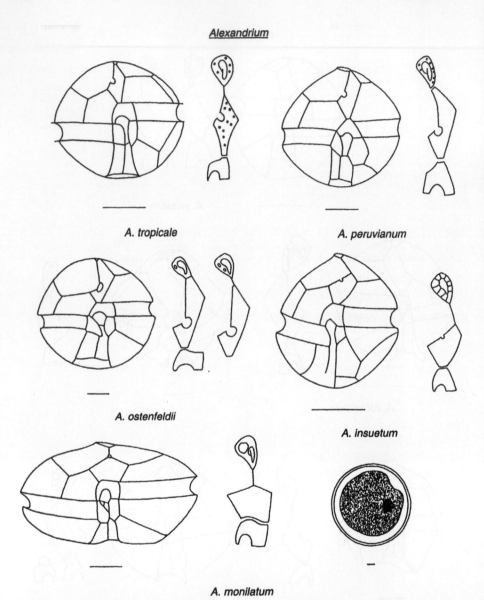

A. tropicale

A. peruvianum

A. ostenfeldii

A. insuetum

A. monilatum

PLATE 38 *Alexandrium tropicale, A. peruvianum, A. ostenfeldii, A. insuetum, A. monilatum, and A. monilatum cyst. Scale = 10 μm. Redrawn from Balech (1995).*

Alexandrium kutnerae (Balech) Balech 1985 (Plate 35)

Large spherical cell with an interesting sa that has a precingular component. Ventral pore in 1' can be at the edge of the anterior right

side or it can be a short distance into the 1' connected by a line; it is a variable character.
Distribution: Coastal; Atlantic Ocean off South America from approximately 24°S to 38°S.

Alexandrium leei Balech 1985 (Plate 35)

Large, conical cell with a longer left hypothecal lobe. Lacks sulcal lists but does have a ventral pore that is variable in placement. The ssa is different than in other similar species.
Remarks: Most similar species is *A. kutnerae.*
Distribution: Coastal; Korea, Japan, Gulf of Thailand, and the Philippines.

Alexandrium lusitanicum Balech 1985 (Plate 36)

Similar to type species and other small related species like *A. andersonii* and *A. angustitabulatum*; however, the sa of this species is of a different shape. Toxic.
Distribution: Coastal; Iberian peninsula.

Alexandrium margalefii Balech 1992 (Plate 39)

One of the most distinct species described. 1' displaced and only slightly higher than 6". The vp is at the juncture of the 4' and 1' in the left corner. The Po has a well-developed callus. The sp is unique.
Distribution: Distribution unknown beyond the type locality of Ria de Vigo, Spain.

Alexandrium minutum Halim 1960 (Plate 37)

Synonym: *A. ibericum* Balech 1985.
Alexandrium with a typical 1' and ventral pore and connection of the 1' to the Po. This connection may be obscured by growth overlap of 2' and 4' plates or may appear as a thin line connecting the 1' and Po. Characteristic sa and ssa. Epithecal profile goes from hemispherical to conical. Different isolates have different surface markings. Toxic.
Remarks: This species can be reticulated, either light or heavy markings.
Distribution: Coastal; Alexandria Harbor, Egypt; Bay of Naples, Italy; Bay of Izmir, Turkey; Greece; Spain; Portugal; France; south of England; New York state, United States; and south of Australia.

Alexandrium monilatum (Howell) F. J. R. Taylor 1979 (Plates 3 and 38)

Synonyms: *Gonyaulax monilata* Howell 1953; *Gessnerium mochimaensis* Halim 1967; *Gessnerium monilata* (Howell) Loeblich III 1970; *Pyrodinium monilatum* (Howell) F. J. R. Taylor 1976.

Chain former in nature; in culture short chains or single cells. The displaced 1' plate is broadly pentagonal and in culture varies in shape and placement. In clonal culture, the species isolates from Florida coastal

Alexandrium

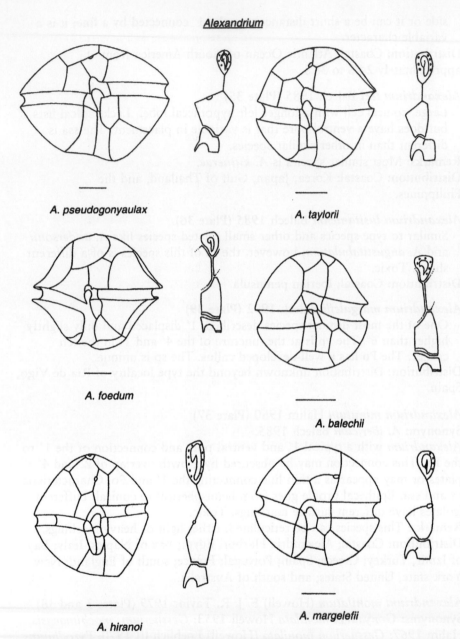

A. pseudogonyaulax

A. taylorii

A. foedum

A. balechii

A. hiranoi

A. margelefii

PLATE 39 *Alexandrium pseudogonyaulax, A. taylorii, A. foedum, A. balechii, A. hiranoi,* and *A. margalefii.* Scale = 10 μm. Redrawn from Balech (1995).

waters can have a small ventral pore at the apex of the 1' where the 1', 2', and 4' plates meet. The sp plate is distinctive in shape, size, and location of the pap. The sp plate is concave and recessed; it also has radial raised markings. Toxic.

Remarks: Distinct from other *Gessnerium* by the APC and sp. Distinct from other chain formers by displaced 1' and anterioposterior compression of cells in chains. Plates are thin and difficult to discern; therefore, it is best to use DIC or fluorescent microscopy if available. This species is the type for the subgenus.

Distribution: Coastal and estuarine; Gulf of Mexico, Caribbean Sea, Atlantic Ocean off Florida and the Chesapeake Bay, and Ecuador.

Alexandrium ostenfeldii (Paulsen) Balech & Tangen 1985 **(Plate 38)**
Synonyms: *Goniodoma ostenfeldii* Paulsen 1904; *Gonyaulax tamarensis* Lebour var. *globosa* Braarud 1945; *Gonyaulax ostenfeldii* (Paulsen) Paulsen 1949; *Heteraulacus ostenfeldii* (Paulsen) Loeblich III 1970; *Gonyaulax globosa* (Braarud) Balech 1971; *Gonyaulax trygvei* Parke 1976; *Protogonyaulax globosa* (Braarud) Taylor 1979; *Gessnerium ostenfeldii* (Paulsen) L. Loeblich and Loeblich III 1979; *Pyrodinium phoneus* Woloszynska and Conrad 1939; *Triadinium ostenfeldii* (Paulsen) Dodge 1981.

Large species with characteristic large ventral pore and thin theca. Toxic.
Remarks: Resembles *A. peruvianum,* but is slightly larger with distinctions in the 1', sa, and APC.
Distribution: Coastal; European Atlantic from Norway to Spain; Iceland; Baring Sea; Kara Sea; Chukchi Sea; Kamchatka peninsula of the Soviet Union; Washington state, United States; and Alexandria Harbor, Egypt.

Alexandrium peruvianum (Balech & Mendiola) Balech & Tangen 1985 **(Plate 38)**
Large cell with prominent ventral pore as in *A. ostenfeldii.* In some cells, the APC is raised and forms an apical protuberance; character is variable. Right margin of the 1' is noticeably curved, while in *A. ostenfeldii,* the right margin is more angular.
Distribution: Coastal; Peru and New York state, United States.

Alexandrium pseudogonyaulax (Biecheler) Horiguchi ex Kita & Fukuyo 1992 **(Plate 39)**
This is a medium-sized cell with a characteristically displaced 1', sa, and sp. The sp is unique with an anterior truncated hook. The sa has a precingular component that is not demarcated and fits in a notch of the 6" and 1'. Toxic.

Remarks: Similar to *A. satoanum* Yuki & Fukuyo 1992.
Distribution: Brackish lagoons; southern France, Portugal, Norway, Japan, and Italy.

Alexandrium tamarense (Lebour) Balech 1992 **(Plate 36)**
Synonyms: *Gonyaulax tamarensis* Lebour 1925; *Gonyaulax tamarensis* var. *excavata* Braarud 1945; *Gonyaulax excavata* (Braarud) Balech 1971; *Gessnerium tamarensis* (Lebour) Loeblich III and L. Loeblich 1979; *Protogonyaulax tamarensis* (Lebour) F. J. R. Taylor 1979; and *A. excavatum* (Braarud) Balech and Tangen 1985.
 Small to medium cells with size and shape variability. Most typically, epitheca is broadly conical with slight shoulders. The hypotheca is somewhat trapezoidal and posteriorly concave. Often the left side is longer than the right and the hypothecal profile is skewed. The Po has a strongly developed callus and the 1' has a small ventral pore. The sp is short and not attenuated. Toxic and nontoxic strains.
Remarks: This species may still have homonyms that sort out morphologically, biochemically, or geographically with new studies.
Distribution: Coastal; western Europe from Norway to the Iberian peninsula including the British Isles, the Atlantic Ocean of the United States from Maine to 40°N, Argentina, Japan, Korea, Gulf of Thailand, Canadian Pacific, Venezuela, Barents Sea, Kamchatka peninsula in the Soviet Union, and south of Taiwan.

Alexandrium tamiyavanichi Balech 1994 **(Plate 37)**
 This species has curtain fins, and distinctive APC, 1', and sa. Toxic.
Remarks: New species of unknown distribution beyond its type locality.
Distribution: Coastal; Gulf of Thailand and the Philippines.

Alexandrium taylorii Balech 1994 **(Plate 39)**
 Subspherical cell with angled hypothecal profile. Cingulum with curtain fins. Characteristic 1' with anterior vp at juncture of 1', 2' and 4'. Unusual sacp and characteristic sp.
Remarks: New species with unknown distribution beyond its type locality.
Distribution: Bay of Arcachon, France.

Alexandrium tropicale Balech 1985 **(Plate 38)**
Synonyms: *Gonyaulax excavata* (Braarud) Balech 1971; *G. tamarensis* var. *excavata* Braarud (not *G. excavata* (Braarud) Balech 1971).
 Small cell with unique type of 1', otherwise character combinations have to be used to distinquish this rounded cell from a species such as *A. tamarense*.
Remarks: Found in waters >27°C.
Distribution: Oceanic, warm water species; west equatorial Atlantic.

Genus *Gambierdiscus* Adachi & Fukuyo 1979 (1 sp.)
Type: *G. toxicus* Adachi & Fukuyo 1979.
References: Adachi & Fukuyo, 1979; F. J. R. Taylor, 1979; Besada et al., 1982; Steidinger, 1983, 1993; Sournia, 1986; Fukuyo et al., 1990; Fensome et al., 1993.

Medium to large species in monophyletic genus in which the cells are anterioposteriorly compressed and the profile is sublenticular. Plate formula: Po, 4', 6", 6c, 8s, 6'", 2"". Similar to *Goniodoma* but with unique APC. The displaced 1' and the 6" are small, adjacent, and right above the sulcal plates. The cingulum is displaced; however, the distal side recurves and confuses displacement. Chloroplasts present.

Gambierdiscus toxicus Adachi & Fukuyo 1979 (**Plate 40**)
Species with wide cingular lists and thick thecae with areolae. Toxic.
Distribution: Benthic and epiphytic; can be tychoplanktonic. Found on coral reefs to shallow lagoons and bays of tropical and subtropical areas of the oceans. Can be abundant even when attached to macroalgae; attaches by mucoid strands from sulcal area.

Genus *Goniodoma* Stein 1883 (>5 spp.)
Synonyms: *Heteraulacus* Diesing 1850 in part; *Triadinium* Dodge 1981 in part.
Type: *G. acuminatum* (Ehrenberg) Stein 1883.
References: Schiller, 1937; Balech, 1980, 1988a; Dodge, 1981a, 1982; F. J. R. Taylor, 1976; Steidinger, 1990, 1993.

Armored. Cells similar to *Alexandrium*, but different in size, thickness of theca, surface markings, size and shape of plates, position of plates, and the presence of strong cingular lists. Plate formula: Po, cp, 4', 6", 6c, 6s, 6'", and 2"". The 1' is a characteristic shape and is displaced. The connection between the Po and 1' in *Alexandrium* is either direct or in a straight line. In *Goniodoma*, the connection is not straight and appears as two lines at almost right angles. Also, in *Goniodoma* the Po is facing horizontally to the left side of the cell. Large vp in 1' toward the anterior right margin. Chloroplasts present. About half of the current *Alexandrium* species were transferred from *Gonyaulax* and *Goniodoma*.

Goniodoma polyedricum (Pouchet) Jørgensen 1899 (**Plates 3 and 40**)
Species with very characteristic polyhedral shape, deep set areolae, and cingular lists. Cingulum descending less than two widths. Thecal sutures with crests.
Distribution: Oceanic; cosmopolitan in subtropical to tropical waters; worldwide distribution.

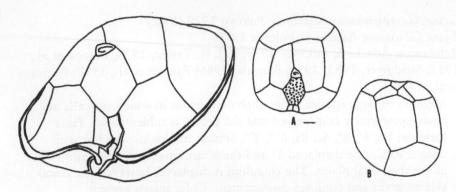

Gambierdiscus toxicus

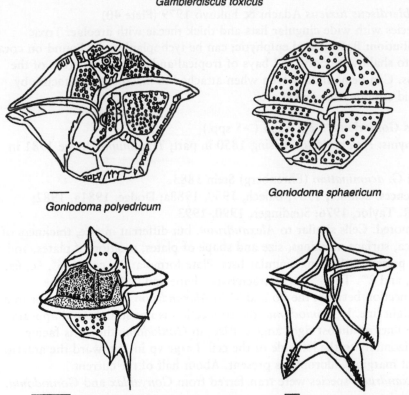

Goniodoma polyedricum

Goniodoma sphaericum

Pyrodinium bahamense

Acanthogonyaulax spinifera

PLATE 40   *Gambierdiscus toxicus* (A and B redrawn from Fukuyo, 1981b); *Goniodoma polyedricum, Goniodoma sphaericum, Pyrodinium bahamense* var. *bahamense,* and *Acanthogonyaulax spinifera* redrawn from Graham (1942). Scale = 10 μm.

Goniodoma sphaericum Murray & Whitting 1899 (**Plate 40**)
   Cell rugose and spherical with prominent pores.
Remarks: The division line for this species is probably similar to
Alexandrium.
Distribution: Worldwide distribution.

Genus *Pyrodinium* Plate 1906 (1 sp.)
Type: *P. bahamense* Plate 1906.
References: Schiller, 1937; Tafall, 1942; Buchanan, 1968; Wall & Dale,
1969; Steidinger & Williams, 1970; Steidinger et al., 1980; Sournia, 1986;
Balech, 1985b; Fensome et al. 1993.
   Armored. Monotypic genus with two varieties. Cells medium sized with
   distinct apical and antapical spines and lists even when cells are in chains.
   One variety can form long chains in nature. Plate formula: Po, cp, 4-5',
   6", 6c, 6s, 6' ", and 2"". Plate sutures with crests. Chloroplasts present.

Pyrodinium bahamense Plate 1906 var. *bahamense* (**Plate 40**)
   Variety with prominent apical protuberance or horn and an apical spine
   with a list. Thecal surface markings differ between the two varieties.
Remarks: This variety is nontoxic but is bioluminescent. It has a distinct
cyst type with an archeopyle. The variety is responsible for massive
bioluminescence in certain tropical and subtropical inshore waters. Forms
cysts.
Distribution: Subtropical to tropical waters; Atlantic.

Pyrodinium bahamense var. *compressum* (Böhm) Steidinger, Tester, &
   Taylor 1980 (not illustrated)
   Variety similar to *P. bahamense* var. *bahamense* but cells typically in
   chains and slightly anterioposteriorly compressed with associated
   reductions. Pores larger than in other variety. Toxic. Forms cysts.
Remarks: This variety, so far, has only been identified from the Pacific and
Indian Oceans, not the Atlantic. Balech (1985b) does not separate out this
toxic variety and believes that the species is highly variable.
Distribution: Warm water species; Pacific.

Family Gonyaulacaceae Lindemann 1928

Genus *Acanthogonyaulax* (Kofoid) Graham 1942 (1 sp.)
Synonyms: *Ceratocorys* Stein 1883 in part; *Gonyaulax* Diesing 1866 in part.
Type: *A. spinifera* (Murray & Whitting) Graham 1942.
References: Graham, 1942; Sournia, 1986.

Armored. Medium- to large-sized cell with prominent hypothecal spines and lists. Cingulum premedian and descending 3×. Plate formula: Po, 3', 7(6)"?, 6c, 9 (10)s, 6' ", and 2"". Interpretation of small precingulars and anterior sulcal plates is confused. 1' with a ventral pore.

Remarks: Species is characteristic but could be confused with *Ceratocorys*.

Acanthogonyaulax spinifera (Murray & Whitting) Graham 1942 (**Plate 40**)
Distribution: Oceanic; tropical.

Genus *Amphidoma* Stein 1883 (<10 spp.)
Synonyms: *Pavillardinium* De-Toni 1936 in part; *Murrayella* Kofoid 1907.
Type: *A. nucula* Stein 1883.
References: Kofoid, 1907b; Dodge, 1982; Dodge & Saunders, 1985b; Sournia, 1986.

Armored. Cells biconical or attenuated in shape; Po, X(?), 6', 6", 6c, 4s(?), 6' ", and 2"". No anterior intercalaries. Small to medium-sized cells; typically with a descending cingulum (<1×) and an antapical process. Chloroplasts present. APC distinctive. Most species in this genus are not well documented except for the type.

Amphidoma nucula Stein 1883 (not illustrated)
Theca heavily marked with reticulae. 1' " small.
Distribution: Subtropical to tropical waters; Atlantic Ocean.

Genus *Amylax* Meunier 1910 (<5 spp.)
Synonym: *Gonyaulax* Diesing 1866 in part.
Type: *A. lata* Meunier 1910.
References: Kofoid, 1911; Balech, 1977b; Dodge, 1982, 1989; Sournia, 1986; Fensome et al., 1993.

Armored. Small to medium cells with a tapered apical horn and at least one antapical spine. Cell compressed dorsoventrally. Plate formula: Po, 3', 3a, 6", 6c, 7 or 8s, 6' ", and 2"". Narrow 1' with ventral pore in posterior right side. APC with round Po, not oval as in *Gonyaulax*. Chloroplasts present.

Amylax triacantha (Jörgensen) Sournia 1984 (**Plate 41**)
Synonym: *Gonyaulax triacantha* Jörgensen 1899.
Cell small, subpyriform with a rather straight antapical profile.
Remarks: Close to *A. buxus* (Balech) Dodge 1989, but epithecal and hypothecal profiles differ. In *A. buxus*, there are more rounded profiles. Division products as in *Gonyaulax* with the anterior daughter cell retaining 1–3', 1 or 2a, 1 or 2", 1–3c, 1–3' ", and the sa.
Distribution: Coastal; cold water species; Pacific and Atlantic Oceans.

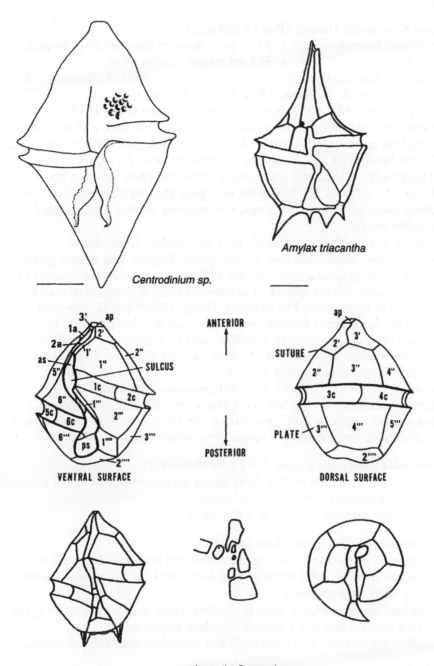

PLATE 41 *Centrodinium sp.; Amylax triacantha* (after Dodge, 1989); schematic of *Gonyaulax* showing ventral and dorsal surfaces (modified from Eaton 1980); schematic of *Gonyaulax* with sulcal plates (redrawn from Balech, 1980). Scale = 10 μm.

Genus *Gonyaulax* Diesing 1866 (>100 spp.)
Synonyms: *Steiniella* Schütt 1895 in part; *Amylax* Meunier 1910 in part.
Type: *G. spinifera* (Claparède & Lachmann) Diesing 1866.
References: Kofoid, 1911; Schiller, 1937; Graham, 1942; Steidinger, 1968, 1971; Steidinger & Williams, 1970; F. J. R. Taylor, 1976; Balech, 1977b, 1988a; Dodge, 1982, 1989; Sournia, 1986; Fensome et al., 1993.

Armored, cells subspherical to biconical to fusiform with a descending cingulum up to six cingulum widths, with or without cingulum overhang. 1' can be sigmoid and narrow to rhomboidal as in *Protoperidinium*. Theca varies in thickness and surface markings. Plate formula: Po, 3', 2a, 6", 6c, 7s, 6'", and 2''''. APC with oval pore plate or elongate oval pore plate, often with linear markings. Chloroplasts present. Documented benthic cysts.

Remarks: Even with the confusion over *Gonyaulax, Goniodoma, Protoceratium, Lingulodinium, Alexandrium, Amylax,* and related genera, the APC of these genera and even the APC within a genus can be used to further separate related species or groups. The genus *Gonyaulax* may still be too broad in application. For example, Dodge (1989) put *G. polyedra* with 3a into the fossil genus *Lingulodinium*. Asexual division products for *G. spinifera* (the type species), *G. grindleyi,* and *G. digitale* are as follows. Anterior daughter cell retains 1–3', 1 or 2a, 1 or 2", 1–3c, 1–3'", sa; the posterior daughter cell retains 3–6', 4–6c, 4 or 5'", 1 or 2'''', and the rest of the sulcal plates (Evitt, 1985). Currently, species of this genus are differentiated based on the following characters: size and shape, plate tabulation and pattern, APC (more than one type), cingulum displacement and overhang, development of apical process/horn, and ornamentation.

Gonyaulax alaskensis Kofoid 1911 (not illustrated)
Cells broad and deep with a short apical horn or modified apical process. Narrow cingulum with ≥3× displacement.
Distribution: Oceanic to coastal. Cold water species.

Gonyaulax fragilis (Schütt) Kofoid 1911
Small oval-shaped cell. Cingulum median and displaced 1×; cingulum ends can overlap. APC extends onto dorsal surface. Thin theca striated; more striae on hypotheca.
Remarks: Can be confused with *G. hyalina* which is more robust in shape but very delicate and has a greater cingulum displacement.
Distribution: Oceanic and coastal. Warm temperate and tropical waters.

Gonyaulax grindleyi Reinecke 1967 (**Plates 4 and 42**)
Synonym: *Protoceratium reticulatum* (Claparède & Lachmann) Bütschli 1885.

Species with 1a and ventral pore in right margin of 1'. Polyhedral-shaped cell with prominent reticulations that obscure plates unless disassociated. Forms cysts.

Remarks: The question of whether this is a *Protoceratium* or a *Gonyaulax* may await interpretation of the "lost" small 3' and whether the existing 3' is the 2a plate. However, the APC is similar to that of an *Alexandrium* and not a *Protoceratium*, even though both Po plates have cp components. Asexual division products as in *Gonyaulax*.

Distribution: Neritic, estuarine; cold temperate to subtropical waters.

Gonyaulax polygramma Stein 1883 (Plate 42)

Elongate cell with tapered epitheca; epitheca angular with short to moderate apical horn. Hypotheca symmetrically rounded or truncate. 1' with vp 1' narrow and recessed. Po elliptical and when initially dissected remains attached to the 1'; Po does not extend onto dorsal side. Mature cells with thecal reticulae and striae. Cells typically have two antapical spines. Cingulum descending about 1.5× without an overhang.

Remarks: The characteristic striae and shape distinguish this species from all other *Gonyaulax* in the polygramma group. In newly developing cells, thecal maturation progresses from formation of sulcal and cingular grooves to plates and sutures, to thecal pores, to linear striae and antapical spines, to poroids, and finally to reticulae.

Distribution: Neritic, oceanic; cosmopolitan in cold temperate to tropical waters; worldwide distribution.

Gonyaulax scrippsae Kofoid 1911 (Plate 42)

Small cell with short to moderate apical horn. Cell and hypotheca rounded; without antapical spines. Descending cingulum with slight overhang. Raised thecal markings present. Forms cysts.

Distribution: Neritic, oceanic (rare); worldwide distribution.

Gonyaulax spinifera (Claparède & Lachmann) Diesing 1866 (Plate 42)

Synonym: *Peridinium spiniferum* Claparède & Lachmann 1859.

Small *Gonyaulax* with prominent, excavated cingulum that is descending by at least two cingulum widths and has an overhang. The angled, conical epitheca has a short to moderate apical horn. Ventral pore between 2a and 3', not in right margin of 1'. Two antapical spines typically present, but not diagnostic. Forms cysts.

Remarks: Often, *G. spinifera*, *G. digitale*, and *G. diegensis* are confused. These three species can be distinguished by stoutness of apical horn, width of cingulum and amount of displacement, and general shape. *Gonyaulax diegensis* is round with a widely displaced narrower cingulum. *Gonyaulax digitale* typically has epithecal shoulders and a moderate apical horn.

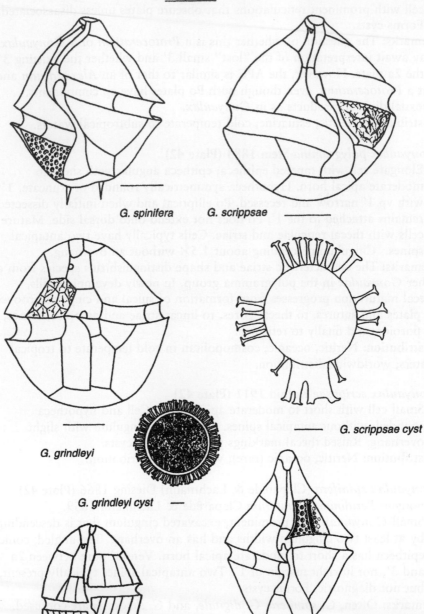

Gonyaulax

G. spinifera   G. scrippsae

G. grindleyi

G. scrippsae cyst

G. grindleyi cyst

G. fragilis

G. polygramma

Gonyaulax spinifera produces polymorphic cysts. In all likelihood, the taxon *G. spinifera* represents several species or sibling species.
Distribution: Neritic, estuarine, oceanic; cosmopolitan; Pacific Ocean, Mediterranean Sea, Atlantic Ocean, and the Gulf of Mexico.

Gonyaulax verior Sournia 1973 (not illustrated)
Synonyms: *Amylax diacantha* Meunier 1919; *Gonyaulax diacantha sensu* Schiller 1937.

Gradually tapered epitheca with a short apical horn that is flanged, apical pore complex area with a tear-shaped Po. The 1' has a ventral pore at the juncture of the 2a and 3'. Cell dorsoventrally compressed with two prominent antapical spines. At low magnification appears as thin, delicate cell.
Remarks: This species is truly synonymous with *G. diacantha* and is rarely confused with any other species.
Distribution: Brackish water species. Temperate to tropical waters; cosmopolitan.

Genus *Lingulodinium*
Synonym: *Gonyaulax polyedra* Stein 1883.
Cyst synonym: *L. machaerophorum* (Deflandre & Cookson) Wall 1967.
Type: *Lingulodinium polyedrum* (Stein) Dodge 1989
References: Kofoid, 1911; Dodge, 1989; Fensome et al. 1993.

Cell polyedral shaped, without antapical spines and apical horn. Descending cingulum without overhang. Theca is reticulate to areolate with pores in the depressions. Plate formula: Po, 3', 3a, 6", 6c, 7s, 6' ", and 2" ". The APC is characteristic of *Gonyaulax*, e.g., *G. verior* and *G. digitale*. The raised inner elliptical rim of the Po can be closely appressed and appears as a lattice or protuberances.
Remarks: If plate reduction is the evolutionary path of armored dinoflagellates, then this species may be an ancestral form with 3a that was later reduced to 2a in more recent species. Clearly, the 1a of the typical *Gonyaulax* cell is large enough to split in two making three anterior intercalaries of approximately the same size. It is not uncommon to find cells in artificial culture where one plate has split into two. Perhaps a species retains the genetic code for the old pattern and reversion is possible.

PLATE 42 Scale = 10 μm unless otherwise indicated. *Gonyaulax spinifera, G. scrippsae, G. scrippsae* cyst, *G. grindleyi, G. grindleyi* cyst, *G. fragilis* (scale = 20 μm), and *G. polygramma.*

Lingulodinium polyedrum (Stein) Dodge 1989 (Plate 43)
Description: Same as that for genus. Forms distinctive cysts. Toxic.
Distribution: Neritic; warm temperate to tropical waters.

Genus *Peridiniella* Kofoid & Michener 1911 emend. Balech 1977 (<5 spp.)
Synonyms: *Gonyaulax* Diesing 1866 in part; *Amylax* Meunier 1910 in part.
Type: *P. sphaeroidea* Kofoid & Michener 1911.
References: Balech, 1977b, 1980; Dodge, 1987.

· Armored. Epitheca with a *Protoperidinium* plate tabulation and pattern,
hypotheca with a *Gonyaulax* plate tabulation; Po, x, 4', 3 or 4a, 7", 6c, 6
or 7s, 6'", 2"". Descending cingulum, displaced about one width. Theca
with surface markings, e.g., reticulations. sa similar to *Alexandrium* type.
Chloroplasts present.

Peridiniella catenata (Levander) Balech 1977 (Plate 43)
Synonyms: *Gonyaulax catenata* (Levander) Kofoid 1911; *Amylax catenata*
(Levander) Meunier 1910.
Reference: Kofoid, 1911b.

Single or catenate cells; shape varies with habit. Small species with 4a.
Remarks: The 2a and 3a plates are small as if one plate split into two or
they are ancestral plates in the lineage of *Peridiniella*. For example, in the
type species there are 3a plates, the 2a plate could be a reduction and
reorientation of two smaller plates.
Distribution: Brackish cold water species that can form blooms.

Genus *Protoceratium* Bergh 1881 (>10 spp.)
Synonym: *Gonyaulax* Diesing 1866 in part.
Type: *P. aceros* Bergh 1881.
References: Reinecke, 1967; Wall & Dale, 1968; von Stosch, 1969a;
Steidinger & Williams, 1970; F. J. R. Taylor, 1976; Dodge, 1982.

Small, oval to broadly biconical cell. Heavy reticulations or areolations
obscure the plate pattern which can only be determined upon treatment
or dissection. Plate formula: Po, 3', 0a, 6", 6c, 6s, 6'", and 2"", following
Balech (1988a). Cingulum descending <l cingulum width. The 1' lacks a
vp and the Po is round with a crescent-shaped pore. The APC
differentiates this group from *Gonyaulax* and *Alexandrium*, although
Protoceratium species have an sa that approaches an *Alexandrium*.
Chloroplasts present.

Protoceratium spinulosum (Murray & Whitting) Schiller 1937 (Plate 43)
Crested reticulae appear as spines at crest junctures. APC with round to
oval Po with crescent-shaped pore. Closing plate possible, but not
observed. Pore not comma shape of *Alexandrium*.

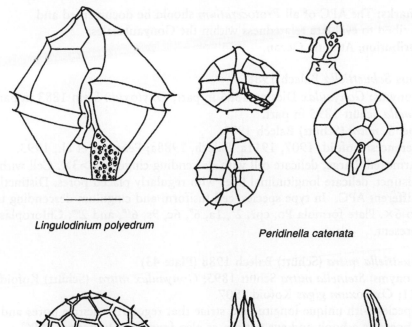

Lingulodinium polyedrum

Peridinella catenata

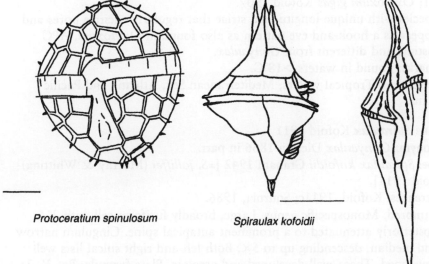

Protoceratium spinulosum   *Spiraulax kofoidii*

Schuetiella mitra

PLATE 43 Scale = 10 μm unless otherwise indicated. *Lingulodinium polyedrum*; *Peridinella catenata* (redrawn from Balech, 1980), *Protoceratium spinulosum*, *Spiraulax kofoidii* (redrawn from Graham, 1942) (scale = 50 μm), *Schuetiella mitra* (scale = 50 μm) redrawn from Graham (1942).

Remarks: The APC of all *Protoceratium* should be documented and described to evaluate relatedness within the Gonyaulacales.
Distribution: Atlantic Ocean.

Genus *Schuettiella* Balech 1988
Synonyms: *Gonyaulax* Diesing 1866 in part; *Oxytoxum* Stein 1883 in part; *Steinella* Schütt 1895 in part.
Type: *S. mitra* (Schütt) Balech 1988.
References: Kofoid, 1907, 1911a; Balech, 1988a; Fensome et al., 1993.
  Armored. Large, delicate cell with descending cingulum >3×. Cell with distinct, delicate longitudinal lines with regularly placed pores. Distinctly different APC. In type species, cell fusiform and cingulum descending up to 6×. Plate formula Po, cp?, 2', 1a, 6", 6c, 9s, 6''', and 2''''. Chloroplasts present.

Schuettiella mitra (Schütt) Balech 1988 (Plate 43)
Synonyms: *Steinella mitra* Schütt 1895; *Gonyaulax mitra* (Schütt) Kofoid 1911; *Oxytoxum gigas* Kofoid 1907.
  Species with unique longitudinal striae that regularly encircle pores and appear as a hook-and-eye pattern as also found in *Oxytoxum*. APC distinct and different from *Gonyaulax*.
Remarks: Found in waters >18°C.
Distribution: Tropical waters; Mediterranean Sea, Atlantic and Pacific Oceans.

Genus *Spiraulax* Kofoid 1911
Synonym: *Gonyaulax* Diesing 1866 in part.
Type: *Spiraulax kofoidii* Graham 1942 [=*S. jolliffei* (Murray & Whitting) Kofoid 1911].
References: Kofoid, 1911c; Sournia, 1986.
  Armored. Monospecific genus. Large, broadly fusiform cell with 2'''' posteriorly attenuated to a prominent antapical spine. Cingulum narrow and median, descending up to 5×. Both left and right sulcal lists well developed. Theca well developed and areolate. Plate formula: Po, 3', 2a, 6", 6c, 7s, 6''', and 2''''. Po not yet described. Chloroplasts present.
Remarks: Although *Spiraulax* is close to *Gonyaulax*, it differs in plate pattern. The 2a and 1" are situated above the sa plate and the 1' is not the narrow long characteristic plate it is in *Gonyaulax*.

Spiraulax kofoidii Graham 1942 (Plate 43)
Description: Same as that for genus.
Distribution: Oceanic, widely distributed in tropical and subtropical waters.

Family Heterodiniaceae Lindemann 1928

Genus *Heterodinium* Kofoid 1906 (>50 spp.)
Synonym: *Peridinium* sec. Murray & Whitting 1899 in part.
Type: *H. scrippsae* Kofoid 1906.
References: Kofoid, 1906a; Kofoid & Adamson, 1933; F. J. R. Taylor, 1976; Rampi & Bernhard, 1980; Sournia, 1986; Balech, 1988a.
   Armored. Small to large cells, some species compressed dorsoventrally; epitheca conical, hypotheca conical with spines or two prominent horns. Individual reticulae with usually one, maybe two, poroids, poroids with multiple pores, or reticulae with multiple pores. Metacytic areas can be very wide and relatively free of markings. Plate formula: Po, cp, 3', 2a, 6", 6c, 6''', and 3''''. 1a is ventral with a pore and the 2a is dorsal. Cingulum descending with or without overhang; however, posterior cingular list absent or greatly reduced as that in *Palaeophalacroma*. Chloroplasts present or absent.

Heterodinium milneri (Murray & Whitting) Kofoid 1906 (not illustrated)
   Cell with descending cingulum, about 2.5× displacement, premedian with prominent overhang; epitheca shorter than hypotheca and conical with an apical process. Hypotheca with winged spines. Cell not dorsoventrally flattened as in some other *Heterodinium*.
Remarks: Due to ventral overlap of apical plates, clarification of the apical series and the anterior intercalaries needs to be made, at least in this species, e.g., in what plate is the vp located?
Distribution: Oceanic; cold temperate to tropical waters.

Heterodinium rigdenae Kofoid 1906 (Plate 44)
   Cell dorsoventrally flattened with two antapical horns. Epitheca with lateral crests. Cingulum equatorial and descending, <1.5×, no overhang. *Protoperidinium* outline.
Distribution: Warm temperate to tropical waters; Pacific Ocean, Indian Ocean, and Mediterranean Sea.

Family Ostreopsidaceae Lindemann 1928

Genus *Coolia* Meunier 1919 (1 sp.)
Synonyms: *Ostreopsis* Schmidt 1901 in part; *Glenodinium* Ehrenberg 1837 in part.
Type: *C. monotis* Meunier 1919.
References: Schiller, 1937; Biecheler, 1952; Balech, 1956; Fukuyo, 1981b; Besada et al., 1982; Dodge, 1982; Norris et al., 1985; Faust, 1992; Fensome et al., 1993; Steidinger, 1993.
   Armored. Cell small, slightly compressed anterioposteriorly and oval in ventral view. Axes off center, nearly oblique; cingulum descending and in

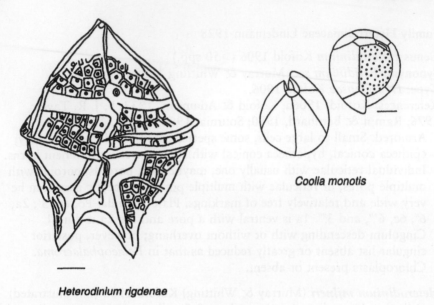

Coolia monotis

Heterodinium rigdenae

Ostreopsis

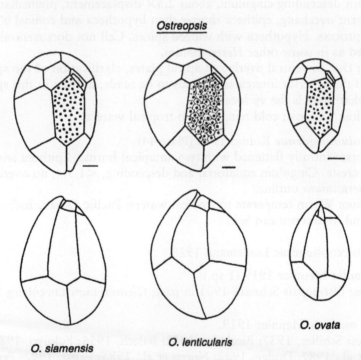

O. ovata

O. lenticularis

O. siamensis

PLATE 44 *Heterodinium rigdenae, Coolia monotis, Ostreopsis lenticularis, Ostreopsis ovata,* and *Ostreopsis siamensis* redrawn from Fukuyo (1981b). Scale = 10 μm.

lateral view appears straight. Plate formula: Po, 3'(4'), 7"(6"), 6c, ≧6s, 5''', 1p, and 2''''. Vp between 1" and 7". Apex displaced dorsally and to the left. In apical view, epitheca smaller than hypotheca. Chloroplasts present.

Coolia monotis Meunier 1919 (Plates 5 and 44)

Remarks: The number and configuration of sulcal plates between *Ostreopsis* and *Coolia* needs study as does the 1" plate because this plate may be homologous to a 1' as Besada et al. (1982) have suggested for *O. ovata* based on plate overlap patterns and the presence of a vp at the right margin of the plate. The hypothecal plate pattern is entirely different than that in *Ostreopsis* and the cell shape is different. Two distinct size classes occur in culture. Toxic.

Distribution: Planktonic, benthic, and epiphytic; brackish water species; temperate to tropical waters; worldwide distribution.

Genus *Ostreopsis* Schmidt 1901 (<5 spp.)

Type: *O. siamensis* Schmidt 1901.

Armored. Medium to large-sized cells, anterioposteriorly compressed, tear shaped in apical view and tapering ventrally. Cingulum not descending and epitheca not noticeably smaller than hypotheca in apical view. Same plate formula as in *Coolia* and with a vp between 1" and 7" or 1' and 6", depending on interpretation of plates. Cingular area in lateral view undulating or straight. Apex displaced to the dorsal surface, left side. Characteristic Po with slit. Chloroplasts present.

Ostreopsis heptagona Norris, Bomber, & Balech, 1985 (Plate 5)

Medium-sized species with characteristic 1' that is seven sided, not six sided as in other species.

Distribution: Known from Florida coastal waters.

Ostreopsis lenticularis Fukuyo 1981 (Plate 44)

Medium-sized lenticular cell without undulating cingulum in lateral view; broadly ovate.

Remarks: Similar to type species, but lacks undulation of cingulum in lateral view and the theca has numerous fine pores. Po plate of *Ostreopsis* species longer than that of *Coolia*. Toxic.

Distribution: Tycoplanktonic, benthic, or epiphytic; tropical shallow waters to offshore reefs.

Ostreopsis ovata Fukuyo 1981 (Plate 44)

Small to medium-sized ovate cell in dorsoventral view but narrow. Theca with scattered pores. Eight sulcal plates.

Remarks: Documentation in this species presents a good argument for the 1" being a displaced 1' with a ventral pore.

Distribution: Tycoplanktonic, benthic, or epiphytic; tropical shallow waters to offshore reefs.

Ostreopsis siamensis Schmidt 1901 (Plate 44)
    Medium-sized ovoid cell. Cingulum undulates in lateral view. Theca with scattered pores. Toxic.
Distribution: Benthic, epiphytic; can be tychoplanktonic. Warm waters.

Family Oxytoxaceae Lindemann 1928

References: Kofoid, 1907; Schiller, 1937; Gaarder, 1954; Steidinger & Williams, 1970; F. J. R. Taylor, 1976; Dodge & Saunders, 1985a; Balech, 1988; Sournia, 1986; Fensome et al., 1993.

Genus *Centrodinium* Kofoid 1907 (<10 spp.)
Synonyms: *Pavillardinium* De-Toni 1936 in part; *Murrayella* Kofoid 1907 in part.
Type: *C. elongatum* Kofoid 1907.
    Armored. Large biconical to attenuated cell that is laterally compressed, particularly in the midbody. Hypotheca attenuated into antapical horn where plate sutures or overlap appear as torsion in the horn. Epitheca sometimes attenuated into apical horn. Cingulum premedian to median and descending about 1×. The sa plate invades the epitheca. Theca with little ornamentation to areolate, sometimes hyaline and delicate. Tentative plate formula: Po, 2', 3a, 7'', 5c, ?s, 5''', and 2''''. Chloroplasts present.
Remarks: This genus has been confused with *Murrayella*-like cells and *Murrayella* was synonymized with *Pavillardinium* because it was previously occupied botanically and *Pavillardinium* was synonymized with *Corythodinium*. The total number of plates (no matter the series) differs between *Centrodinium* and *Corythodinium*. Also, it would appear that *Centrodinium* has a Po at the apex and that *Corythodinium* does not have an apical pore plate, but does have an identifiable dorsal epithecal pore.

Centrodinium elongatum Kofoid 1907 (not illustrated)
    Cingulum median and descending. Epitheca truncate with short horn; hypotheca longer and attenuated into almost cylindrical antapical horn that is gradually curved to the left.
Distribution: Tropical waters; Eastern Pacific Ocean.

Genus *Corythodinium* Loeblich Jr. & Loeblich III 1966 (<25 spp.)
Synonyms: *Pyrgidium* Stein 1883; *Oxytoxum* Stein 1883 in part; *Pavillardinium* De-Toni in part.
Type: *C. tesselatum* (Stein) Loeblich Jr. & Loeblich III 1966.
    Armored. Small to large biconical to elongate cell with sculptured theca. Cingulum anterior to median and descending about l–1.5×; cingulum

prominently excavated and with narrow lists. Although the epitheca is
shorter than the hypotheca and the cell can be biconical or tapered, the
anterior cingular list is almost the same diameter as the posterior cingular
list. The sa plate is obovate or angled and prominently invades the
epitheca. Plate formula is tentative: Po, 3′, 2a, 6″, 5c, 4(?)s, 5‴, and
1‴″(or 1p). Chloroplasts present.

Remarks: This genus is recognized by Balech (1988a) and Sournia, 1986;
but Dodge and Saunders (1985a) synonymized it with *Oxytoxum*. Until
plate analyses are completed with dissected specimens, these two genera
should be kept separate. The apical pore referred to by Dodge and Saunders
that always occurs in their 4′ may not be a separate apical pore plate but
rather a large pore in the 2a. Plate designations vary among authors.

Corythodinium constrictum (Stein) F. J. R. Taylor 1976 (Plate 7)
    Small to medium cell. Cingulum premedian; hypotheca exceeds epitheca
    and ends in an antapical spine. The cell is horizontally constricted about
    midbody and the postcingular plates have longitudinal striae. Areolae
    with pores above the constriction differ from the linear pore fields below
    the constriction.
Remarks: No other species to date resembles this species.
Distribution: Open water species; warm temperate to tropical waters;
worldwide distribution.

Corythodinium tesselatum (Stein) Loeblich Jr. & Loeblich III 1966
(Plate 45)
    Medium-sized cell. Cingulum premedian; hypotheca exceeds epitheca and
    ends in a thick antapical spine. Epitheca with reticulate markings and
    broad hypotheca with characteristic longitudinal striae connecting evenly
    spaced, offset horizontal striae. Horizontal striae with linear field of pores
    on inside margin.
Distribution: Warm temperate to tropical waters; most records from the
Atlantic Ocean.

Genus *Oxytoxum* Stein 1883 (>50 spp.)
Type: *O. scolopax* Stein 1883.
    Armored. Small to large, needle-shaped to top-shaped cell with a
    decidedly anterior epitheca that is drastically narrower and shallower than
    the hypotheca. Cingulum is anterior with little displacement (0–0.5×)
    and the sa plate barely invades the epitheca. Plate formula is thought to
    be the same as that for *Corythodinium*, except some authors consider the
    two anterior intercalaries to be apicals. Certain species less than 15 μm in
    length, probably less than 10 μm. Chloroplasts present.
Remarks: The plate formula and pattern needs to be resolved by separating
plates in both the epitheca and the hypotheca. The presence of a possible

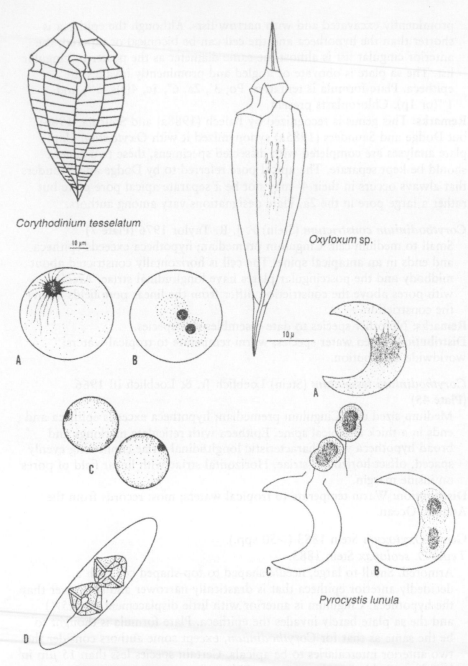

Corythodinium tesselatum

Oxytoxum sp.

A

B

A

C

C

D

Pyrocystis lunula

Pyrocystis noctiluca

PLATE 45 *Corythodinium tesselatum*, redrawn from Dodge & Saunders (1985a); *Oxytoxum* sp., *Pyrocystis noctiluca* life cycle, and *Pyrocystis lunula* life cycle. Both *Pyrocystis* are redrawn from Elbrächter & Drebes (1978).

apical pore plate needs to be demonstrated in order to assign plate series. The sole antapical plate could be interpreted as a posterior intercalary following Balech's modification of the Kofoidian system.

Oxytoxum scolopax Stein 1883 (Plate 7)

Attenuated, spindle-shaped cell with distinct winged lanceolate apical and antapical spines. Hypotheca with linear rows of cylindrical ridges, each with an anterior pore. Medium-sized species. **Plate 45** shows an *Oxytoxum* sp. illustrated that has similar surface markings to *O. scolopax*.
Remarks: One of the most common representatives encountered.
Distribution: Warm temperate to tropical waters; worldwide distribution, but more typical of the Atlantic Ocean.

Oxytoxum variabile Schiller 1937 (not illustrated)

Small cell with reduced, almost flat epitheca. The conical hypotheca has a characteristic hook-and-eye linear pattern of curved ridges and anterior pores.
Distribution: Oceanic; warm temperate to tropical waters; Pacific and Atlantic Oceans.

Family Pyrocystaceae (Schütt) Lemmerman 1899

Genus *Pyrocystis* Murray ex Haeckel 1890 (<25 spp.)
Synonyms: *Gymnodinium* Stein 1878 in part; *Murraycystis* Haeckel, 1890; *Diplodinium* Klebs 1912 in part; *Dissodinium* Klebs in Pascher 1916 in part.
Type: *P. noctiluca* Murray ex Haeckel 1890.
References: Swift & Durbin, 1971; Swift & Wall, 1972; F. J. R. Taylor, 1972; Elbrächter & Drebes, 1978; Pinceman & Gaylor, 1978; Drebes, 1981; Pinceman et al., 1981, 1982; Sournia, 1986; Elbrächter et al., 1987; Fensome et al., 1993.

Unarmored and/or armored depending on life history stage. Dominant vegetative stage is a large, planktonic bladder-shaped cell. The nonflagellated bladder cell which has been called a coccoid stage can be spherical, fusiform, lanceolate, or crescent shaped. Smaller reproductive stages can be armored, biflagellated planospores resembling *Alexandrium* or unarmored nonflagellated, uniflagellated, or biflagellated cells, some of which resemble *Gymnodinium*. Plate formula of armored stages is the same as that for *Alexandrium*. Typically, one or two spore stages are produced within the mother cell and start with contraction of the cytoplasm. Shape of the nucleus of all stages is sausage to horseshoe shaped. Chloroplasts present in all stages. Bioluminescent.
Remarks: Genus has been confused with *Dissodinium*, a genus of planktonic, parasitic dinoflagellates with primary and secondary cyst stages. However, the two genera can easily be distinguished because *Dissodinium*

cysts have large vacuoles and parietal cytoplasm and a spherical nucleus.
The primary cyst, which is spherical, produces 2 to more than 16 smaller
secondary cysts (usually lunate) within the mother cell. Each secondary cyst
produces five or more biflagellated dinospores (gymnodinialean).
Dissodinium stages can be pigmented. Also, Elbrächter et al. (1987)
documented extreme intraspecific morphological variation in the lunate
bladder cell stage and suggested that only four species could be identified
with reasonable certainty: *P. minima, P. gerbaultii, P. obtusa,* and *P.
robusta* if certain characters were present. These authors further suggested
that the remaining lunate specimens should be listed as belonging to the *P.
lunula* species complex or the *P. acuta/lanceolata* species complex.

Pyrocystis noctiluca Murray ex Haeckel 1890 (Plate 45)
Synonyms: *P. pseudonoctiluca* Wyville-Thomson in J. Murray, 1876;
Dissodinium pseudolunula Swift ex Elbrächter & Drebes 1978.
   Dominant vegetative stage is large spherical or subspherical cell, greater
   than 350 μm in diameter when fully developed. However, after release
   from armored planospore, nonmotile cell can be less than 200 μm in
   diameter. Produces one or two biflagellated armored cells that resemble
   Alexandrium. More commonly produces unarmored aplanospores that are
   smaller spheroidal cells which are released and grow to a larger size.
   Spherical cell without large central food vacuole, but with radiating
   cytoplasm.
Remarks: Spherical primary cysts of *Dissodinium* are typically less than
200 μm and typically have food vacuoles.
Distribution: Oceanic species; cosmopolitan in warm temperate to tropical
waters.

Pyrocystis lunula (Schütt) Schütt 1896 (Plate 45)
   Large-sized lunate cell that produces "gymnodinioid" swarmers or
   planospores.
Remarks: This species represents a species complex and field specimens are
difficult to identify unless they are cultured. Elbrächter et al. (1987) suggest
that only the lunate *P. minima* and *P. gerbaultii* can be identified based on
size and shape.
Distribution: Oceanic and coastal. Warm temperate to tropical?

Family Pyrophacaceae Lindemann 1928

Genus *Fragilidium* Balech ex Loeblich III 1965 (<10 spp.)
Synonyms: *Helgolandinium* von Stosch 1969; *Goniodoma* Stein 1883 in
part.
Type: *F. heterolobum* Balech ex Loeblich III 1965.

References: Balech, 1959b, 1988b, 1990a; von Stosch, 1969b; Steidinger &
Williams, 1970; Dodge, 1982; Sournia, 1986; Fensome et al., 1993.
   Armored. As the genus suggests, the cells are fragile and thinly thecate.
   Under stress, ecdysis is common and the thecal plates can appear as a
   fragmented halo surrounding the protoplast or the theca can separate at
   the cingulum. Cell globular, subglobular or spherical in shape; small to
   medium-sized cell. Cingulum equatorial and descending less than 1×.
   Plate formula: Po, cp, 4 or 5', 7–9″, 9–11c, 6–8s, 7 or 8‴, 1p, and
   2⁗. Chloroplasts present and two species known to be phagotrophic
   as well.
Remarks: Interpretation of plates differs among authors. The 1″ can easily
be considered a displaced 1' as in *Alexandrium* or in *Goniodoma*; even the
APC has affinities with *Alexandrium* because of the comma-shaped pore
with a closing plate. The questions become does this belong in the
Pyrophacaceae or the Goniodomataceae and whether or not
Helgolandinium with the slit and ventral pore in the plate over the sa is
distinct from *Fragilidium*. The slit in *F. subglobosum* represents the
anterior edge of a small precingular plate, while the slit in *F. mexicanum* is
just a slit with a terminal vp as in some of the *Alexandrium*. The 1'(1″) of
this genus can resemble the 1' of *Goniodoma* or *Alexandrium*. The sa is
also of the *Alexandrium* type. Knowing plate overlap patterns would help
resolve the question of which plate is 1'.

Fragilidium heterolobum Balech ex Loeblich III 1965 (Plate 4)
   Medium-sized cell. The 1' is of the *Goniodoma* type and the two
   antapical plates are dissimilar in size. The species is distinguished by its
   antapical profile and a prominent left antapical lobe.
Distribution: Coastal; warm temperate to tropical waters; Pacific and
Atlantic Oceans.

Fragilidium fissile Balech 1990 (not illustrated)
   Medium sized spherical cell with broad Po and no vp in the 1'(1″). 1'
   plate of *Alexandrium* type; two antapical plates of dissimilar size.
Remarks: Although this species is similar to *F. subglobosum,* it differs by
not having a vp and it lacks the small precingular plate above the left
margin of the sa, rather, the 1'(1″) sits directly above the sa
Distribution: Recently described species of unknown distribution beyond the
type locality of the Gulf of Salerno, Tyrrhenian Sea.

Fragilidinium mexicanum Balech 1988 (Plate 46)
   Medium-sized broadly conical cell. Cingulum displaced <1×. Displaced
   1' with slit and ventral pore. Five apical plates and seven precingulars
   depending on interpretation. 1p and 7‴ present.

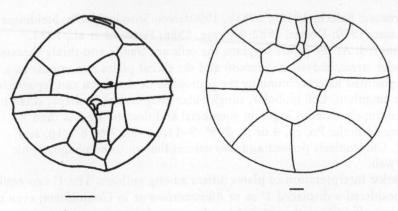

Fragilidium mexicanum   *Pyrophacus horologium*

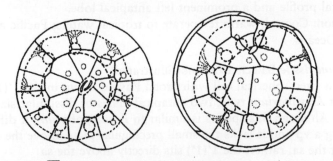

Pyrophacus steinii

PLATE 46 *Fragilidium mexicanum* redrawn from Balech (1988b); *Pyrophacus horologium* and *P. steinii* redrawn from Matsuoka (1985). Scale = 10 μm.

Remarks: Differs from *F. heterolobatum* in shape and plate pattern.
Distribution: Known from the type locality Mexican coastal waters.

Genus *Pyrophacus* Stein 1883 (<5 spp.)
Type: *P. horologium* Stein 1883.
References: Schiller, 1937; Steidinger & Davis, 1967; Steidinger & Williams, 1970; Wall & Dale, 1971; F. J. R. Taylor, 1976; Balech, 1979b, 1980, 1988a.
  Armored. Large biconical to lens-shaped cells, sometimes polar axes oblique. Cingulum narrow, equatorial, and slightly descending. Sulcus short. Theca granulate with pores and growth striae. Plate formula: Po, cp(?), 5–9', 0–8a, 7–15", 9–16c, 8s, 8–17''', 0–15p, and 3''''; follows Balech's 1980 change to Kofoid's designation of antapical plates. Chloroplasts present. Fossil and extant cysts known.
Remarks: Most variable genus for plate tabulation. In the Gonyaulacales and Peridiniales, the hypotheca is typically the more conservative half of the cell in plate pattern and tabulation. However, in *Pyrophacus* the reverse is true, the epitheca is more conservative. Cyst types vary from tuberculate to agglutinous in extant species and Wall and Dale (1971) speculated on fossil representation of agglutinous types.

Pyrophacus horologium Stein 1883 (**Plates 5 and 46**)
  Biconical cell. Most conservative species with little variation in plate formula. Typically the epitheca has 5', 0a, and 9".
Distribution: Oceanic, neritic, estuarine. Cosmopolitan in cold temperate to tropical waters.

Pyrophacus steinii (Schiller) Wall & Dale 1971 (**Plates 5 and 46**)
  Flattened, lenticular cell with attenuated epitheca. Typically, the epitheca has 7', 0a, and 12".
Distribution: Restricted to warm temperate to tropical waters of all oceans.

Order Peridiniales Haeckel 1894
References: Schiller, 1937; Loeblich III, 1982; Sournia, 1986; Fensome et al., 1993.
  Armored dinokonts of varied form. Plate tabulation is usually diagnostic. If the Thecadiniaceae is set aside from the other families in this order, then the APC typically has an X plate, 3–6', typically only 3 or 4', 1–3a, 6 or 7", 4–6c, 4–6s, 5''', and 1 or 2'''' but typically 2''''. The 1' is usually more symmetrical than that in the Gonyaulacales.

Family Calciodinellaceae F. J. R. Taylor 1987

Genus *Ensiculifera* Balech 1967 (<5 spp.)
Type: *E. mexicana* Balech 1967

References: Balech, 1967, 1980, 1988; Cox & Arnott, 1971; Dale, 1978;
Indelicato & Loeblich III, 1986; Sournia, 1986; Matsuoka et al., 1990.
   Armored, small scrippsielloid cells with a conspicuous apical process.
   Photosynthetic. Plate formula: Po, X, 4', 3a, 7", 5c (4c + t), 5s, 5''', and
   2''''. The 1c or t plate with characteristic long, anteriorly directed lance-
   like spine called the ensiculus. sp plate not in contact with cingulum.
   Chloroplasts present. Produces calcareous, spherical cysts.
Remarks: See *Scrippsiella*. The confusion with this genus centers around the
description of the type species which was originally illustrated with 4c
plates. However, newly described species all have had the characteristic 5c
with the 1c or t plate having a long spine internal to the precingular series.

Ensiculifera mexicana Balech 1967 (Plate 47)
Description: Same as that for genus.
Distribution: Gulf of Mexico.

Ensiculifera carinata Matsuoka, Kobayashi, & Gains 1990 (not illustrated)
   Cell body elongate with conical epitheca and concical, laterally
   asymmetric hypotheca. Ensiculus sigmoid, more than half the length of the
   1'. Small antapical spine (carina) at the juncture of the sp, 1'''', and 2''''.
Remarks: This species is differentiated by the presence of an antapical spine
and cell shape.
Distribution: Temperate and tropical coastal waters.

Genus *Pentapharsodinium* Indelicato & Loeblich III 1986 (<5 spp.)
Type: *P. dalei* Indelicato & Loeblich III 1986.
References: Dale, 1977b; Indelicato & Loeblich III, 1986; Horiguchi &
Pienaar, 1991; Lewis, 1991; Montresor et al., 1993.
   Armored, small peridinioid cells similar to other genera in this family and
   producing organic and calcareous cysts. Plate formula: Po, X, 4', 3a(2a?),
   7", 5c (4 + t), 4s, 5''', and 2''''. The sp plate does not contact the
   cingulum.
Remarks: The placement of this genus in this family is tentative. Also, the
species *Protoperidinium quinquecorne* may belong in this genus based on
five cingular and four sulcal plates with the sp not touching the cingulum;
however, it has only two anterior intercalary plates and was recently
transferred to *Peridinium* by Horiguchi & Pienaar (1991). The type species
for *Peridinium* lacks an APC which is a distinctive generic character and
therefore this species should not be assigned to *Peridinium*. It may be that
the generic description for *Pentapharsodinium* should be emended to
accommodate 2a.

Pentapharsodinium dalei Indelicato & Loeblich III 1986 (not illustrated)
Synonym: *Peridinium faeroense* sensu Dale 1977.

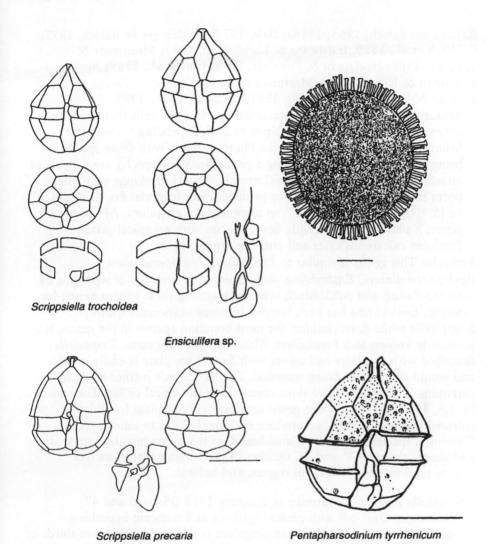

Scrippsiella trochoidea

Ensiculifera sp.

Scrippsiella precaria

Pentapharsodinium tyrrhenicum

PLATE 47 *Scrippsiella trochoidea* and *Ensiculifera* sp. redrawn from Balech (1980); *Scrippsiella precaria* redrawn from Montresor & Zingone (1988); *Pentapharsodinium tyrrhenicum*. Scale = 10 μm.

Cell with distinctive thecal pores surrounded by concentric rings. *P. tyrrhenicum* (see Plate 47) lacks concentric rings around the pores.
Distribution: Temperate coastal areas.

Genus *Scrippsiella* Balech ex Loeblich III 1965 (<25 spp.)
Type: *S. sweeneyae* Balech ex Loeblich III 1965.

References: Balech, 1963, 1988a; Dale, 1977b; Steidinger & Balech, 1977; Loeblich et al., 1979; Indelicato & Loeblich III, 1985; Montresor & Zingone, 1988; Horiguchi & Pinenaar, 1988; Gao et al., 1989a,b; Akselman & Keupp, 1990; Matsuoka et al., 1990; Gao & Dodge, 1991; Honsell & Cabrini, 1991; Lewis, 1991; Banaszak et al., 1993.

Armored, small (<50 μm) characteristic peridinioid cells that occur in either planktonic or benthic habitats or both depending on adaptive behavior. One species is symbiotic. Photosynthetic with some species being mixotrophic and possessing a peduncle. Some species are capable of attaching to substrates by mucoid strands. Thecal markings vary from pores to reticulations to striae to papillae. Plate formula: Po, X, 4′, 3a, 7″, 6c (5 + t), 4 or 5s, 5‴ and 2⁗. sp plate touches cingulum. APC typical for genus; x plate varies in length. Some species with an apical process. Produces calcareous cysts and possibly organic cysts.

Remarks: This genus is similar to *Peridinium, Protoperidinium, Pentapharsodinium, Ensiculifera,* and *Thompsodinium,* but is separable by plate tabulation and APC. Much synonymy among these genera exists; for example, *Ensiculifera* has been lumped by some taxonomists into *Scrippsiella* while *S. trochoidea,* the most common species in the genus, was previously known as a *Peridinium.* Although there are some *Scrippsiella* described with 4s plates and others with 5s, the sm plate is easily missed and would result in 4s being recorded. The APC needs further study to determine if all species have three segments to the apical collar (Toriumi & Dodge, 1993). Species of this genus can have variant plate formulae in cultured specimens; they can produce additonal plates in known series. Currently, species are differentiated based on the following characters: size and shape, shape of 1′ and 2a, number of precingular plates, surface ornamentation, the presence of stigma, and habitat.

Scrippsiella precaria Montresor & Zingone 1988 **(Plates 6 and 47)**
Planktonic. Oval cell with conical epitheca and truncate hypotheca. Epitheca exceeds hypotheca and cingulum is displaced about two-thirds of the width. Theca thin with scattered small pores. Produces spiny calcareous cysts.
Remarks: This species is differentiated by its small 2a plate which is diamond shaped and sits between the 1a, 3a, 4″, and 5″ on the dorsal side.
Distribution: Originally described from the Gulf of Naples, Italy.

Scrippsiella subsalsa (Ostenfeld) Steidinger & Balech 1977 **(not illustrated)**
Epitheca conical and hypotheca trapezoidal; epitheca and hypotheca almost equal. Cell compressed dorsoventrally and longitudinal axis oblique. The 2a and 3a are separated by the 3′. Theca lightly reticulate; reticulae can appear as longitudinal striae.

Remarks: Some of the benthic *Scrippsiella* should be taxonomically separated from the planktonic species based on sulcal plates, noncontact of 2a and 3a, and the APC. One of the characters of *Scrippsiella* is that the sp plate touches the cingulum. In *S. subsalsa*, *S. caponii* (= *P gregarium*), and *S. arenicola* this is not the case.

Distribution: Benthic/epiphytic; tychoplanktonic. Cosmopolitan in warm temperate and tropical estuaries and coastal areas.

Scrippsiella trochoidea (Stein) Loeblich III 1976 **(Plates 6 and 47)**
  Conical epitheca with short, convex apical process and collar; round hypotheca. 1′ is very narrow and slightly asymmetrical. Produces calcareous cysts.

Remarks: This species is differentiated by its pear shape, narrow 1′, and apical process. It is very similar to *S. mimima,* which is smaller, and both have 5s, but they differ in sulcal plate pattern and cingular plate location.

Distribution: Cosmopolitan neritic and estuarine species. This is the most commonly recorded planktonic scrippsielloid dinoflagellate.

Family Kolkwitziellaceae Lindemann 1928
References: Lebour, 1925; Schiller, 1937; Abé, 1941, 1981; Balech, 1964a, 1988; Steidinger & Williams, 1970; F. J. R. Taylor, 1976; Dodge & Hermes, 1981; Sournia, 1986; Matsuoka, 1988; Lewis, 1990; Dale et al., 1993; Dodge & Toriumi, 1993; Elbrächter, 1993; Toriumi & Dodge, 1993.

Remarks: This is a confused group of distinctive species. The confusion arises from interpretation and assignment of plates to a series and the extensive synonymy involved in tracking a single species.

Genus *Boreadinium* Dodge & Hermes 1981 (<5 spp.)
Type: *B. pisiforme* Dodge & Hermes 1981.
  Armored. Small sublenticular to subspherical cell with APC and prominent left sulcal list typical of diplopsalids. Cingulum median and circular. Surface with scattered pores. Chloroplasts absent. Plate formula: Po, X, 4′, 1a, 7″, 4(3+t)c, 5(?)s, 5‴, and 1⁗. Meta. APC of C′ type.

Boreadinium pisiforme Dodge & Hermes 1981 **(Plate 48, schematic)**
Distribution: Coastal; cold temperate.

Genus *Diplopelta* Stein ex Jørgensen 1912
Type: *D. bomba* Stein ex Jørgensen 1912.
Synonym: *Dissodium* Abé 1941 in part.
  Armored. Small to medium-sized spherical cell with APC and prominent left sulcal list as in other diplopsalids. Cingulum median and circular. Surface with scattered pores. Chloroplasts absent. Plate formula: Po, X, 4,

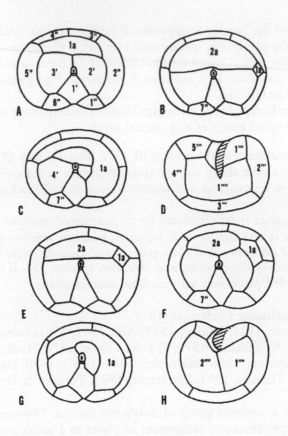

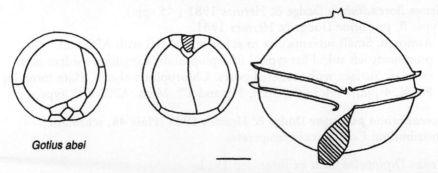

Gotius abei

PLATE 48 (A–H) Schematics redrawn from Dodge & Hermes (1981). (A) *Diplopsalis*;
(B) *Preperidinium*; (C) *Boreadinium*; (D) hypotheca of A–C; (E) *Diplopelta*; (F) *Diplopsalopsis*;
(G) *Oblea*; (H) hypotheca of E–G; *Gotius abei* redrawn from Abé (1981); *Diplopsalis* ventral
view with prominent protruding left sulcal list. (Scale = 10 μm).

1a, 6″, 4(3+t)c, 6(?)s, 5‴, and 2⁗. An alternate formula is 3′and 2a. Ortho. APC of D′ type. Produces cysts.

Diplopelta parva (Abé) Matsuoka 1988 (Plate 48, schematic)
Synonym: *Dissodinium parvum* Abé 1941
Distribution: Coastal; temperate.

Genus *Diplopsalis* Bergh 1881 (<5 spp.)
Synonyms: *Dissodinium* Abé 1941 in part; *Glenodinium* Ehrenberg 1836 in part.
Type: *D. lenticula* Bergh 1881.
Armored. Medium-sized subspherical to lenticular cell with APC and prominent left sulcal list typical of diplopsalids. Surface with scattered pores. Chloroplasts absent. Plate formula: Po, X, 3′, 1a, 6″, 4(3+t)c, 5s, 5‴, and 1⁗. Ortho. APC of D′ type. Produces cysts.

Diplopsalis lenticula Bergh 1881 (Plate 48, schematic)
Distribution: Estuarine to oceanic; cosmopolitan in cold temperate to tropical waters.

Genus *Diplopsalopsis* Meunier 1910 (<5 spp.)
Type: *D. orbicularis* (Paulsen) Meunier 1910.
Armored. Small to medium-sized subspherical cell with APC and prominent left sulcal list of diplopsalids. Surface with scattered pores. Chloroplasts absent. Plate formula: Po, X, 4′, 1a, 7″, 4(3+t)c, 6(?)s, 5‴, and 2⁗. An alternate formula is 3′ and 2a. Ortho. APC of D′ type. Produces cysts.

Diplopsalopsis orbicularis (Paulsen) Meunier 1910 (Plate 48, schematic)
Distribution: Pacific Ocean, North Sea, English Channel, and Danish waters.

Genus *Gotoius* Abé ex Matsuoka 1988 (<5 spp.)
Type: *G. mutsuensis* Abé 1981.
Armored. Medium-sized subspherical to lenticular cell with median, slightly ascending cingulum and small left sulcal list. Surface with scattered pores. Chloroplasts absent. Plate formula: 4′, 1a, 6″, 4(3+t)c, 5s, 5‴, 2⁗; no Po and X. Large 3′ epithecal plate. Ortho. Produces cysts.
Remarks: The large 3′ plate was described as a 2a plate but it appears to be homologous to 3′.

Gotoius abei Matsuoka 1988 (Plate 48)
Subspherical cell. Large apical plates.
Distribution: Recently described from Japanese waters.

Genus *Oblea* Balech ex Loeblich Jr. & Loeblich III 1966 (<5 spp.)
Type: *O. baculifera* Balech ex Loeblich Jr. & Loeblich III 1966.

Armored. Small to medium-sized, globulose or sublenticular cell with
apical pore complex. Prominent left sulcal list as with other diplopsaloids.
Cingulum circular to slightly ascending. Surface with scattered pores.
Typically without chloroplasts. Plate formula: Po, X, 3', 1a, 6", 4(3+t)c,
6(?)s, 5''', and 2''''. Meta 1'. APC of D'. Produces cysts.

Oblea baculifera Balech ex Loeblich Jr. & Loeblich III 1966 (Plate 48,
schematic)
Remarks: Smaller warm water forms exist.
Distribution: Cold water species; South Atlantic Ocean and North Pacific
Ocean.

Genus *Preperidinium* Mangin 1913 (<5 spp.)
Synonyms: *Diplopeltopsis* Pavillard 1913; *Zygabikodinium* Loeblich Jr. &
Loeblich III 1970.
Type: *P. meunieri* (Pavillard) Elbrächter 1993.
  Armored. Small to medium-sized sublenticular to subglobular cell with
  prominent left sulcal list and APC. Cingulum median and circular with
  prominent rib-supported lists. Surface with scattered pores. Chloroplasts
  absent. Plate formula: Po, X, 4', 1a, 7", 4c(3+t), 5s, 5''', and 1''''. An
  alternate formula is 3' and 2a. Ortho. APC of D' type. Produces cysts.

Preperidinium meunieri (Pavillard) Elbrächter 1993 (Plate 48, schematic)
Synonyms: *Peridinium paulsenii* Mangin 1911; *Diplopsalis minor* (Paulsen)
Lindemann 1927; *Zygabikodinium lenticulatum* Loeblich Jr. & Loeblich III
1970.
Distribution: Coastal and estuarine. Temperate to tropical waters;
cosmopolitan.

Family Peridiniaceae Ehrenberg 1828

Genus *Heterocapsa* Stein 1883 (<10 spp.)
Synonym: *Cachonina* Loeblich III 1968.
Type: *H. triquetra* (Ehrenberg) Stein 1883.
References: Loeblich III, 1968; von Stosch, 1969a; Balech, 1977a;
Loeblich et al., 1981; Morrill & Loeblich III, 1981a,b, 1983; Pomroy,
1989.
  Armored. Small (<20 μm) peridinioid that appears unarmored at the
  light microscope level of resolution. Epitheca rounded to conical;
  hypotheca rounded to attenuated. Cingulum slightly displaced and
  descending. Thinly thecate with characteristic body scales. Chloroplasts
  present. Most typical plate formula: Po, cp, X, 6', 3a, 7", 6c, 5s, 5''', 0 or
  1p, 2''''. The 1' is displaced from the Po and lies above the sa. The sa
  plate is small. The X plate is at the anterior left margin of the 6' and in
  contact with the Po. Species can be bloom formers.

Remarks: This genus has been confused with *Gymnodinium, Glenodinium,* and *Katodinium* because plates were not initially detected. The plate formula can differ based on different authors' interpretations of plates; for example, Loeblich et al. (1981) give it as Po, cp, 5', 3a, 7", 6c, 5(7?)s, 5''', and 2'''' for *H. pygmaea.* If you add the work of Morrill and Loeblich III (1981) for *H. triquetra,* the formula would be increased by 1p. The latter authors also demonstrated extreme variation in plate tabulation of cultured specimens. One major interpretation of most authors is that the central plate right above the anterior-most sulcal plates is the sa; however, it is presented here as a displaced 1' above a very small and cryptic sa. See Morrill and Loeblich III (1984) for the application of division lines in supporting their position for the sa plate.

Heterocapsa niei (Loeblich) Morrill & Loeblich III 1981 **(Plates 6 and 49)**
Synonym: *Cachonina niei* Loeblich 1968
   Epitheca and hypotheca equal; slightly compressed dorsoventrally.
Distribution: Estuarine and neritic; cosmopolitan in temperate and tropical waters, often forms spring through fall blooms.

Heterocapsa triquetra (Ehrenberg) Stein 1883 **(Plates 6 and 49)**
   Hypotheca attenuated into horn and with 1p plate.
Distribution: Neritic, estuarine, brackish water, marine, and in some low-salinity water; worldwide distribution.

Genus *Peridinium* Ehrenberg 1832
Type: *P. cinctum* (O. F. Müller) Ehrenberg 1832.
References: Schiller, 1937; Boltovskoy, 1975; Carty & Cox, 1986; Sournia, 1986; Popovsky & Pfiester, 1990; Elbrächter, 1993; Fensome et al., 1993.
   Armored. Small to medium-sized cell of varied shape from spherical to ovoid to lenticular. Type species without an APC. Surface markings varied. Plate formula: 4', 3a, 7", 5c, 5s, 5''', and 2''''.
Remarks: This genus, which is principally freshwater or brackish, is confused as is *Glenodinium, Glenodiniopsis, Durinskia, Peridiniopsis, Kryptoperidinium, Kansodinium* and the diplopsalids. As an example, the species *Kolkwitziella acuta* (Apstein) Elbrächter 1993 has the following synonymy: *Glenodinium acutum, Diplopsalis acuta, Peridinium latum, Entzia acuta, Apsteinia acuta, K. gibbera,* and others. Recently, Fensome et al. (1993) redefined *Glenodinium* (type species *G. cinctum*) and used the family Glenodiniaceae Wiley & Hickson 1909 for *Glenodinium* and *Glenodiniopsis,* which were characterized as having chloroplasts and the plate formula 4', 4a, 7 or 8", 6''', and 2''''.

Peridinium cinctum (O. F. Müller) Ehrenberg 1832 (not illustrated)
Distribution: Freshwater; worldwide.

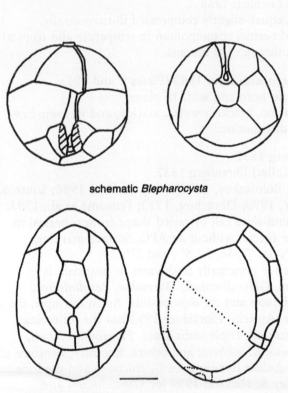

Heterocapsa

Heterocapsa niei

Heterocapsa triquetra

schematic *Blepharocysta*

Lissodinium schilleri

PLATE 49 *Heterocapsa niei* redrawn from von Stosch (1969); *Heterocapsa triquetra;* Schematic *Blepharocysta* redrawn from Carbonell-Moore (1992); *Lissodinium schilleri* redrawn from Carbonell-Moore (1993). Scale = 10 μm.

Family Podolampaceae Lindemann 1928
References: Nie, 1939, 1942; Schiller, 1937; Balech, 1963; Abé, 1966;
Steidinger & Williams, 1970; F. J. R. Taylor, 1976; Dodge, 1982;
Sournia, 1986; Carbonell-Moore, 1991, 1992, 1993a,b; Fensome et al.,
1993.

Genus *Blepharocysta* Ehrenberg 1873 (<10 spp.)
Type: *B. splendor-maris* (Ehrenberg) Ehrenberg 1873.
Armored. Medium-sized cell, spherical or subspherical to oval without
apical attenuation. Homologous cingular area not excavated and cell
without cingular lists but with sulcal lists. Plate formula: Po, cp, X, 3',
1a, 5", 3c, 4 or 5s, 4 or 5''', and 1''''. Postcingular plates without
prominent double pore tract as in *Podolampas*. Plate pore patterns may
help in differentiating species. At least four species can be distinguished
by general shape and shape and position of sulcal lists in lateral view.
Chloroplasts present.

Blepharocysta splendor-maris (Ehrenberg) Ehrenberg 1873 (**Plates 7 and 49**
schematic)
APC diagnostic for genus, maybe slight species differences. Cingular
plates large. Sulcal lists curved, adjacent to antapex. Megacytic areas can
be wide with striae on advancing side of plate overgrowth.
Distribution: Oceanic to coastal, warm water species.

Genus *Lissodinium* Matzenauer emend. Carbonell-Moore 1991
(<25 spp.)
Synonym: *Blepharocysta* Ehrenberg 1873 in part.
Type: *L. schilleri* Matzenauer emend. Carbonell-Moore 1991.
Armored. Medium-sized lentil-shaped cell without a defined cingulum as
in the family but without antapical spines and with a strong lateral
compression. Plate formula: Po, cp, X, 3', 1a, 5", 3c, 5s, 5''', and 1''''. The
apical pore complex is characteristic of the genus with an apical spine
originating from the center of the Po. The collar or rim of the Po is
lower than the outside apical plate flanges, while in *Blepharocysta* the Po
rim can be higher than the outside overlapped flanges. Chloroplasts
present.
Remarks: Species of this genus are differentiated based on the following
characters: cell size and shape, plate size and shape e.g., 1a, 2", 4", the
presence or absence of spines, and type of surface markings. See Carbonell-
Moore (1993) for a taxonomic key.

Lissodinium schilleri Matzenauer 1933 (**Plates 7 and 49 schematic**)
Description: Same as that for genus.
Distribution: Oceanic, subtropical to tropical waters.

Genus *Podolampas* Stein 1883 (<10 spp.)
Synonym: *Parrocelia* Gourret 1883.
Type: *P. bipes* Stein 1883.
    Armored. Large, pear- or top-shaped cell with attenuated epitheca and
    one to three prominent antapical spines. Homologous nonexcavated
    cingular area with three plates and no cingular lists. Plate formula: Po, cp,
    X, 3′, 1a, 5″, 3c, 4–6s, 5‴, and 1⁗. 1a four-sided and small. Antapical
    plate very small and difficult to discern. Postcingular plates with a
    prominent double pore tract. Most species can be separated by shape and
    characters associated with the winged antapical spines. Chloroplasts
    present.
Remarks: The APCs of *Podolampas, Blepharocysta,* and *Lissodinium* are
different as are the sulcal and antapical areas. *Blepharocysta* has a left and
right sulcal list, *Podolampas* has only a left sulcal list, and the type species
of *Lissodinium* has no membraneous sulcal lists although the 1‴ and 5‴
develop a thickened rim that borders the sulcus as would sulcal lists.

Podolampas bipes Stein 1883 (Plate 7)
    Pyriform cell with two prominent, winged antapical spines that recurve.
    The right spine diverges from the cell. Widest point of cell is postmedian.
Distribution: Oceanic; warm temperate to tropical waters; worldwide
distribution.

Podolampas elegans Schütt 1895 (not illustrated)
    Prominent attenuated epitheca and cell with two long, diverging antapical
    spines. Spines with wings and wings connected by median wing. More
    divergent, right antapical spine appears shorter than left.
Distribution: Oceanic; subtropical to tropical waters; worldwide
distribution.

Podolampas palmipes Stein 1883 (Plate 50)
    Narrow podolampoid cell with two unequal antapical spines that are
    winged and not connected. The right spine is slightly divergent and about
    half the length of the left. There can be a tooth-like projection at the
    apex, dorsally situated and perhaps associated with one of the apical
    plates.
Distribution: Oceanic; warm temperate to tropical waters; worldwide
distribution.

Family Protoperidiniaceae F. J. R. Taylor 1987

Genus *Protoperidinium* Bergh 1881 (>250 spp.)
Synonyms: *Peridinium* Ehrenberg 1832 in part; *Archaeperidinium* Jörgensen
1912; *Congruentidium* Abé 1927; *Minuscula* Lebour 1925; *Properidinium*
Meunier 1919; *Glenodinium* in part.

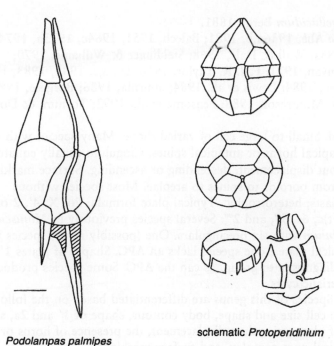

Podolampas palmipes

schematic *Protoperidinium*

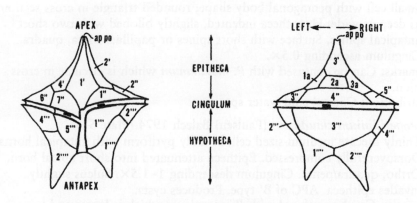

schematic *Protoperidinium*

PLATE 50 *Podolampas palmipes* (scale = 10 µm) and schematic of *Protoperidinium* after Balech (1980); schematic of *Protoperidinium* redrawn from Eaton (1980) showing orientation of cell.

Type: *P. pellucidum* Bergh 1881.
References: Abé, 1936a,b, 1981; Balech, 1951, 1964c, 1973a, 1974, 1976a, 1979c, 1988a; Wall & Dale, 1968; Steidinger & Williams, 1970; Subrahmanyan, 1971; F. J. R. Taylor, 1976; Dodge, 1982, 1983, 1985; Balech et al., 1984; Lewis et al., 1984; Sournia, 1986; Borgese, 1987; Zingone & Montresor, 1988; Fensome et al., 1993; Toriumi & Dodge, 1993.

Armored. Small to large cell of varied shape. Many species with apical and antapical horns or antapical spines. Cingulum usually equatorial with or without displacement; descending or ascending. Surface markings varied from poroids to spines to areolae. Most species without chloroplasts; heterotrophic. Typical plate formula: Po, X, 4', 2 or 3a, 7", (3+t)c, 6s, 5''', and 2''''. Several species previously in *Minuscula* and *Glenodinium* with six precingulars. One (possibly more) species with seven sulcal plates; one species lacks an APC. Shapes of plates 1' and 2a can be diagnostic to group, as can the APC. Some species produce characteristic cysts.

Remarks: Species of this genus are differentiated based on the following characters: cell size and shape, body contour, shape of 1' and 2a, shape and position of plates, cingulum displacement, the presence of horns or spines, type of apical pore complex, and surface markings.

Protoperidinium brevipes (Paulsen) Balech 1974 **(Plate 54)**
Small cell with pentagonal body shape; rounded triangle in cross section; as deep as wide. Hypotheca indented, slightly bilobed with two short antapical spines. Surface with short spines or papillae. Meta, quadra. Cingulum ascending 0.5×.

Remarks: Can be confused with *P. metananum* which is circular in cross section.
Distribution: Coastal cold water species.

Protoperidinium claudicans (Paulsen) Balech 1974 **(Plate 51)**
Thinly thecate medium-sized cell; broadly pyriform with antapical horns. Dorsoventrally compressed. Epitheca attenuated into short apical horn. Ortho, quadra/penta. Cingulum descending 1–1.5×. Sulcus slightly invades epitheca. APC of B' type. Produces cysts.

Remarks: Can be confused with *P. oceanicum* which is larger and has longer apical and antapical horn, greater cingulum inclination dorsally, and the left antapical horn is obviously directed ventrally.
Distribution: Principally coastal and open water, but found in estuarine environments. Temperate to tropical species; cosmopolitan.

Protoperidinium conicoides (Paulsen) Balech 1973 (not illustrated)
Small biconical cell with indented posterior margin that has two winged antapical spines. Cell circular in cross section. Epitheca and hypotheca

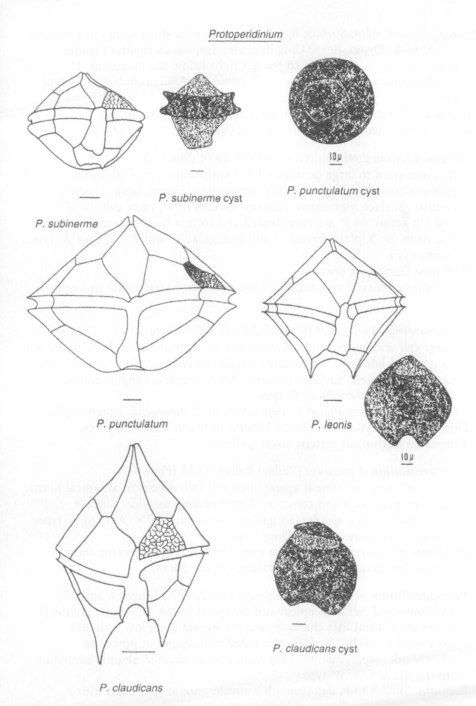

PLATE 51 *Protoperidinium subinerme* and cyst, *P. punctulatum* and cyst, *P. leonis* and cyst, and *P. claudicans* and cyst. Scale = 10 μm.

with convex sides. Surface lightly reticulate with short spines at junctures
of network. Ortho, hexa. Cingulum circular; sulcus slightly invades
epitheca. Sulcus excavated to the left right below the cingulum. 1″
quadrangular; 7″ triangular as in *P. conicum*. APC probably of A′ type.
Forms cysts.
Remarks: Balech (1988a) recognizes a meta form of this species.
Distribution: Coastal; temperate. Worldwide.

Protoperidinium conicum (Gran) Balech 1974 **(Plate 53)**
Medium-sized to large pentagonal cell with short antapical horns. Ventral
epitheca concave. Dorsoventrally compressed. Ortho, hexa. Cingulum
circular. Surface reticulated. Anterior sutures of 1′, right sulture of 1″,
and left suture of 7″ are excentuated and form a thick inverted V suture
line from the X plate. Broad 1′ and triangular 1″ and 7″. APC of A′ type.
Forms cysts.
Remarks: Distinctive species.
Distribution: Coastal and oceanic. Cosmopolitan in temperate to tropical
waters.

Protoperidinium crassipes (Kofoid) Balech 1974 **(Plate 53)**
Large cell with apical and antapical horns; right antapical horn wider and
longer than left horn. Cell almost circular in cross section. Ventral area
slightly excavated. Surface reticulate. Meta, quadra. Cingulum descending
1 or 2x. APC probably of B′ type.
Remarks: Not as angular as *P. depressum* or *P. divergens*. Phagotrophic.
Distribution: Coastal; even forms blooms in warm water estuaries.
Temperate to tropical waters; cosmopolitan.

Protoperidinium depressum (Bailey) Balech 1974 **(Plate 52)**
Large cell with prominent apical horn and two divergent antapical horns.
Epitheca excavated and concave; almost as deep as wide. Surface
reticulated. Ortho, quadra. Cingulum descending >2×. APC of B′ type.
Remarks: Distinctive species. Phagotrophic.
Distribution: Coastal and oceanic; even forms blooms in warm water
estuaries. Temperate to tropical waters; cosmopolitan.

Protoperidinium divergens (Ehrenberg) Balech 1974 **(Plates 6 and 53)**
Medium-sized cell with apical and antapical horns. Divergent antapical
horns with sulcal lists characteristic for species complex. Epitheca
excavated ventrally. Surface reticulated with spines at junctures
of network. Meta, quadra. Cingulum almost circular, slightly ascending
(to 0.5×). APC of B′ type.
Remarks: Similar to *P. depressum* but smaller, not as deep, and with a
meta 1′.

Protoperidinium

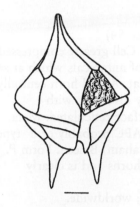

P. oblongum

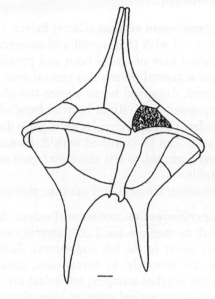

P. oceanicum

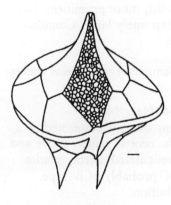

P. depressum

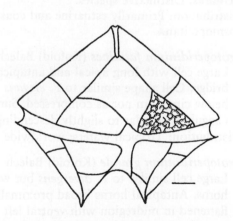

P. conicum

PLATE 52 *Protoperidinium oblongum, P. oceanicum, P. depressum,* and *P. conicum.* Scale = 10 μm.

Distribution: Principally coastal, temperate to tropical waters.
Cosmopolitan.

Protoperidinium elegans (Cleve) Balech 1974 (**Plate 54**)
   Large cell with long apical and antapical horns. Cell greatly compressed
   between base of apical horn and proximal end of antapicals which arise
   from antapical plates; this central area almost flat. Apical horn centrally
   located. Antapical horns almost straight, slightly divergent with
   corrugated ends. Right horn in front of left. Surface reticulated. Meta,
   quadra. Cingular circular or slightly displaced. APC probably of B' type.
Remarks: Can be confused with *P. truncatum* (Graham) Balech. Form *P.
elegans* f. *granulata* has more divergent antapical horns and is clearly
separable.
Distribution: Coastal and oceanic; tropical waters; worldwide.

Protoperidinium excentricum (Paulsen) Balech 1974 (not illustrated)
   Small to medium-sized cell; antedioposteriorly compressed with the apical
   horn offset to the left and ventral. Epitheca concave dorsally, hypotheca
   concave ventrally. In dorsal view, apex-antapex axis is oblique. Sulcus
   deeply notches antapex; left sulcal area, with list, more prominent. Ortho.
   Two asymmetrical anterior intercalaries; 2a extremely large. Cingulum
   circular or slightly ascending. APC of B' type.
Remarks: Distinctive species.
Distribution: Primarily estuarine and coastal. Temperate to tropical waters;
cosmopolitan.

Protoperidinium fatulipes (Kofoid) Balech 1974 (**Plate 54**)
   Large cell with long apical and antapical horns that are connected by a
   bridge. Cell shape similar to *P. elegans* but the central region above and
   below cingulum not as compressed. Surface reticulated. Meta, quadra.
   Cingulum circular to slightly descending. APC probably of B' type.
Distribution: Tropical waters; worldwide distribution.

Protoperidinium grande (Kofoid) Balech 1974 (**Plate 54**)
   Large cell similar to *P. divergens* but with longer apical and antapical
   horns. Antapical horns broad proximally and pointed distally. Cell body
   flattened in midregion with ventral left edge of cingulum straight and not
   curved. Meta, quadra. Cingulum circular to slightly descending. APC of
   B' type.
Distribution: Oceanic; warm temperate to tropical waters; worldwide
distribution.

Protoperidinium leonis (Pavillard) Balech 1974 (**Plate 51**)
   Medium-sized pentagonal or rhombic-shaped species with indented
   posterior margin with two short antapical spines. Cell as long as wide or

slightly longer. Ventral epithecal surface excavated or slightly concave. Epitheca in outline with straight or slightly convex edges. Surface with broad-based spines that can coalesce. Ortho, hexa. Cingulum descending 1 1.5x. 1' characteristic with anterior portion shorter than posterior; 1" and 7" quadrangular. APC of A' type. Forms cysts.
Remarks: Can be easily confused with *P. obtusum.*
Distribution: Coastal and oceanic; temperate to tropical waters; worldwide distribution.

Protoperidinium minutum (Kofoid) Loeblich III 1970 (not illustrated)
Small globular cell with short apical horn. Sulcus expands posteriorly and has prominent, short left sulcal list. Surface with papillae or short spines. Ortho, hexa. Cingulum circular. Only two anterior intercalaries. Produces cysts.
Distribution: Coastal and open water. Cold temperate to warm waters; cosmopolitan.

Protoperidinium nudum (Meunier) Balech 1974 (not illustrated)
Small globular cell, as long as wide. Posterior margin slightly indented with two very short antapical spines formed from sulcal lists. Hypothecal surface markings or faint reticulae with short spines at junctures in network. Ortho, hexa. Cingulum circular; sulcus does not invade epitheca.
Distribution: Coastal and open water; cold temperate waters; Atlantic Ocean.

Protoperidinium oblongum (Aurivillius) Parke & Dodge 1976 (Plate 52)
Medium-sized species. Attenuated pyriform shape with apical and antapical horns. Ventrodorsally compressed with 30–60° inclined cingulum. Ortho, quadra. Cingulum descending 1×. APC of B' type. Produces cysts.
Remarks: This species can easily be confused with *P. oceanicum, P. claudicans,* and *P. steidingerae. Protoperidinium oceanicum* is much larger, while *P. claudicans* is smaller and not as ventrodorsally compressed as *P. oblongum* and *P. oceanicum.* Also, *P. claudicans* is easily distorted with the confinement of a coverslip and usually rounds out even though it is armored. *Protoperidinium steidingerae* lacks the characteristic pore plate complex of all other *Protoperidinium.* This species is often considered a variety or a variant of *P. oceanicum,* and, according to Abé, 1991; the size difference is not applicable.
Distribution: Neritic, oceanic; cosmopolitan in cold temperate to tropical waters.

Protoperidinium obtusum (Karsten) Parke & Dodge 1976 (not illustrated)
Medium-sized species similar to *P. leonis* with descending cingulum, 1×;

shorter anterior margins to 1', and characteristic longitudinal striae on epitheca. Cell wider than long. Striae appear to be formed between rows of pores as in some *Gonyaulax*. APC of A' type.

Remarks: This species is often identified as *P. leonis*.

Distribution: Coastal and oceanic; temperate to tropical. Cosmopolitan.

Protoperidinium oceanicum (VanHöffen) Balech 1974 (Plate 52)

Synonym: *P. murrayii* Kofoid 1907.

Large cell similar in shape to *P. oblongum* but with longer, narrower divergent anatapical horns. Ortho, quadra. Cingulum descending 1–1.5×. Cingulum inclined 30–60°, placed high on dorsal side.

Remarks: See *P. oblongum*.

Distribution: Coastal and oceanic; temperate to tropical waters. Cosmopolitan. Uncommon in cold temperate waters.

Protoperidinium pallidum (Ostenfeld) Balech 1973 (Plate 54)

Medium-sized elongate pyriform cell with two divergent antapical spines and a short apical horn. Dorsoventrally compressed. Surface reticulated. Para, hexa. Cingulum ascending 1×. APC of B' type.

Remarks: Similar to *P. tristylum* and *P. schilleri*, but often confused with *P. pellucidum*. *Protoperidinium pallidum* has been reported to have chloroplasts by F. J. R. Taylor (1976).

Distribution: Coastal and oceanic from cold temperate to warm temperate waters; worldwide distribution.

Protoperidinium parthenopes Zingone & Montresor 1988 (Plate 53)

Small biconical cell; circular in cross section. Surface with raised pores and faint reticulae. Ortho, penta. Cingulum median and slightly ascending, 0.5×. Sulcus extends to antapex. 1' asymmetrical and shifted to right with triangular 7" and quadrangular 1". 1c not transitional as in most other *Protoperidinium*.

Remarks: Similar to *P. americanum* which has 4a, an unusual hexa 2a, and a five-sided 3'. The shape of the 3' probably determines whether there are 4a or 3a, e.g., if it is five-sided there could be 4a, if triangular, then 3a.

Distribution: Recently described species from the Gulf of Naples, Italy.

Protoperidinium pellucidum Bergh 1881 (Plate 54)

Small to medium-sized broadly pyriform cell with short apical horn, two winged antapical spines, and one prominent curved, antapical winged spine that originates from a left sulcal list. Circular in cross section. Surface reticulated. Para, hexa. Cingulum slightly ascending, 0.5×. APC of B' type.

Remarks: Although this species can be confused with *P. pallidum*, it is smaller, circular in cross section, and has a prominent sulcal list that ends in a curved antapical spine.

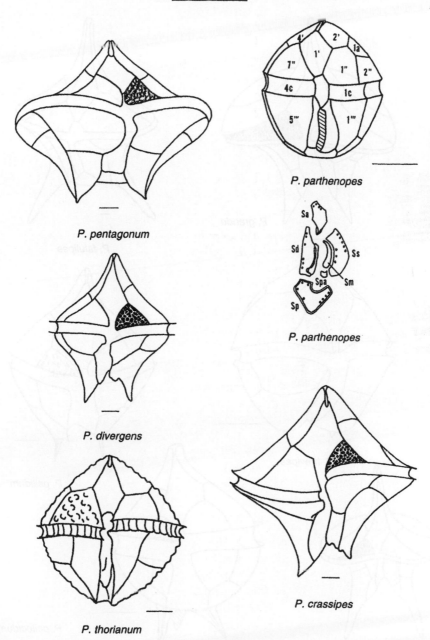

Protoperidinium

P. pentagonum

P. parthenopes

P. parthenopes

P. divergens

P. thorianum

P. crassipes

PLATE 53 *Protoperidinium pentagonum; P. parthenopes* and sulcal plates redrawn from Zingone & Montresor (1988); *P. divergens; P. thorianum* redrawn from Lebour (1925); *P. crassipes*. Scale = 10 μm.

Protoperidinium

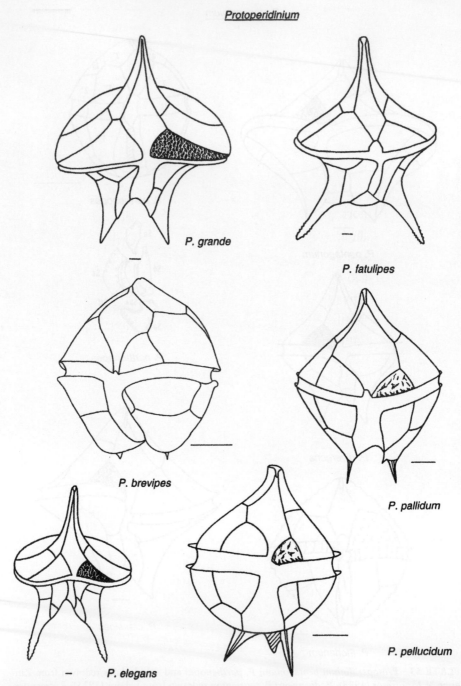

PLATE 54 *Protoperidinium grande, P. fatulipes, P. brevipes, P. pallidum, P. elegans,* and *P. pellucidum.* Scale = 10 μm.

Distribution: Mainly coastal and cosmopolitan in temperate to tropical waters.

Protoperidinium pentagonum (Gran) Balech 1974 (**Plate 52**)
Medium-sized to large, broadly pentagonal cell with truncate posterior margin with short antapical winged spines. Sulcus broad posteriorly, not extending to antapex. Cell in cross section reniform. Ortho, penta/hexa. Cingulum descending 1 or 2×. APC of A' type. Ventral view of 1' and 1" and 7" similar to *P. conicum*. Produces cysts.
Remarks: Distinctive species; can be confused with *P. latissimum* which was originally classified as a variety of this species. *Protoperidinium latissimum* is a broader cell with the epitheca often appearing as an angled roof overhanging the hypotheca; cell in cross section more reniform with the right side wider than the left; sulcus J shaped; and cingulum often convoluted. Wall & Dale (1968) reported a para 1' rather than ortho.
Distribution: Principally coastal, but found in estuarine environments. Cosmopolitan in temperate to tropical waters.

Protoperidinium quarnerense (Schröder) Balech 1974 (not illustrated)
Medium-sized subglobulose cell with button-like APC and two antapical spines. Surface with pores. Meta, quadra/penta. Cingulum ascending, 1–1.5×, with overhang. Sulcus invades epitheca; sa touches 1' as in other armored species.
Remarks: This species was once considered a variety of *P. globulus*.
Distribution: Principally oceanic; cosmopolitan in warm temperate to tropical waters.

Protoperidinium subinerme (Paulsen) Loeblich III 1970 (**Plate 51**)
Medium-sized pentagonal cell with indented posterior margin. Cell almost circular in cross section. Sulcus broader posteriorly, sometimes L shaped because of the wide sp plate. Ortho, hexa. Cingulum circular. Surface reticulate with short spines at junctures of network. Anterior sutures of 1' shorter than posterior and straight or concave in outline. Posterior sutures of 1' can be slightly convex in outline making posterior half of 1' wider. APC of A' type.
Remarks: Can be confused with *Peridinium punctulatum* (Paulsen) Balech (**Plate 51**) which was once considered a variety of this species. *Peridinium punctulatum* is more broadly pentagonal with a rounded, truncate antapex; surface markings are papillae or broad-based short spines; ventral area is more excavated; 2a typically penta but can be hexa; sulcus extends to antapex and is not expanded posteriorly.

Distribution: Coastal and open water. Temperate to tropical waters. Cosmopolitan.

Protoperidinium thorianum (Paulsen) Balech 1974 (Plate 53)
    Medium-sized cell; almost broadly biconical with slightly indented posterior margin. Circular in cross section. Hypotheca exceeds epitheca in length. Sulcus deeply excavated, widening and extending to posterior margin. Surface rugose with pits with raised edges; sometimes reported as papillae. Ortho, hexa. Only two anterior intercalaries; approximately the same size and shape. Cingulum descending about 1×. Produces cysts.
Remarks: Distinctive species but can be confused with *P. avellana* which does not have a rugose cell surface and has a narrower sulcus and a median cingulum.
Distribution: Coastal and open water. Cold temperate to warm water; cosmopolitan.

Family Thecadiniaceae Balech 1956

Genus *Amphidiniopsis* Woloszyńska 1929 (<5 spp.)
Synonym: *Thecadinium* Balech in part.
Type: *A. kofoidii* Woloszyńska 1929.
References: Dodge, 1982; Saunders & Dodge, 1984; Dodge & Lewis, 1986.
    Armored. Small dorsoventrally flattened cell with reduced epitheca. Cingulum–sulcus juncture Y shaped. Sulcus extends to the antapex. Plate formula: Po, 4′, 1–3a, 5 or 6″, 3c?, ?s, 5‴, and 2⁗. Plates typically have ornamentation. Chloroplasts present or absent.

Amphidiniopsis kofoidii Woloszyńska 1929 (not illustrated).
    Not to be confused with *Thecadinium kofoidii*. Epitheca with rounded apex and characteristic curved beak protuberance that creates a central "notch." Rounded or bilobed hypotheca with sulcus broadening toward antapex. Sulcus with central flange/list. Six precingulars and three anterior intercalaries. Thick plates have linear rows of pores or markings. Chloroplasts absent.
Distribution: Benthic, sand dweller, sometimes tychoplanktonic; temperate waters.

Genus *Roscoffia* Balech 1956 (<5 spp.)
Type: *R. capitata* Balech 1956.
References: Balech, 1956; Dodge, 1982.

Armored. Small ovoid cell with premedian, deep cingulum. Epitheca cap shaped. Plate formula: Po?, 4′, 5″, 3c, 3s, 5‴, 1⁗.

Roscoffia capitata Balech 1956
Thecal plates reticulated.
Distribution: Benthic, sand dweller; temperate waters. Sometimes tychoplanktonic.

Genus *Thecadinium* Kofoid & Skogsberg 1928 (>5 spp.)
Synonym: *Phalacroma* Stein 1883 in part.
Type: *T. kofoidii* (Herdman) Schiller 1933.
References: Kofoid & Skogsberg, 1928; Dodge, 1982; Saunders & Dodge, 1984; Faust & Balech, 1993.
Armored. Small, laterally flattened cell with reduced epitheca. Cingulum premedian and sulcus extends to antapex. Can look like *Amphidinium*, but with plates. Plate formula: Po, 3′, 1a, 4″, 5c?, 5s, 3‴, and 1⁗. Reported with or without chloroplasts.

Thecadinium kofoidii (Herdman) Schiller 1933 (Plate 55)
Synonym: *T. petasatum* Dodge 1982
Epitheca triangular in ventral view and a rounded hypotheca in lateral view. Chloroplasts present.
Distribution: Benthic, sand dweller; temperate to tropical waters.

Order Thoracosphaerales Tangen in Tangen et al., 1982
References: Tangen et al., 1982; Inouye & Pienaar, 1983; Sournia, 1986; Fensome et al., 1993.
Dominant stage with a calcareous wall; motile stage unarmored and gymnodinioid. Calcareous stage can undergo nuclear division releasing a binucleate single unarmored cell through an opening in the wall that then can divide to produce biflagellated gymnodinioid cells which round out and become immature coccoid stages.
Remarks: This order has some affinities with the family Calciodinellaceae in the Peridiniales because of a calacareous-walled stage and archeopyle-like opening. Sournia (1986) placed it in the Dinococcales.

Family Thoracosphaeraceae Schiller 1930

Genus *Thoracosphaera* Kamptner 1927 (<5 spp.)
Type: *T. heimii* (Lohmann) Kamptner 1944.
References: Tangen et al., 1982; Inouye & Pineaar, 1983; Sournia, 1986; Fensome et al., 1993.
No armored stages known. Gymnodinioid dinospores produced. Chloroplasts present.
Remarks. This type species was previously thought to be a coccolithophorid.

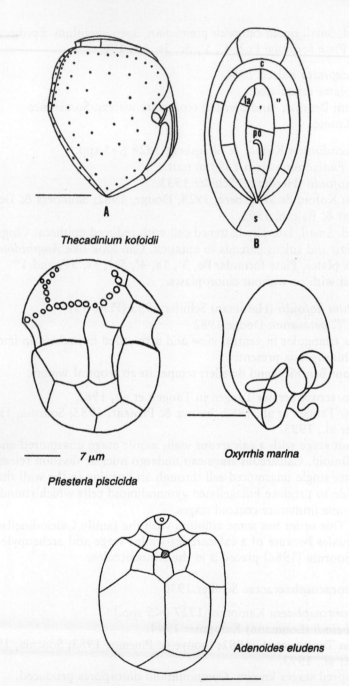

Thecadinium kofoidii

A

B

Pfiesteria piscicida

7 μm

Oxyrrhis marina

Adenoides eludens

PLATE 55 Scale = 10 μm unless otherwise indicated. *Thecadinium kofoidii*; (A) lateral view and (B) apical view redrawn from Saunders & Dodge (1984); phantom (scale = 7 μm), *Oxyrrhis marina* redrawn from Dodge (1982); and *Adenoides eludens* redrawn from Dodge & Lewis (1986).

Thoracosphaera heimii (Lohmann) Kamptner 1944 (not illustrated)
Distribution: Oceanic. Widely distributed in temperate and warm waters.

Order Blastodiniales Schiller 1935
References: Chatton, 1952; Loeblich III, 1982; Cachon & Cachon, 1987; Fensome et al., 1993; Landsberg et al., 1994.
  External parasites on fishes and invertebrates.
  Invertebrate hosts include copepods, appendicularians, siphonophores, polychaetes, pteropods, and coelenterates. Armored and unarmored dinospores known.

Order Syndiniales Loeblich III 1976
References: Loeblich III, 1982; Sournia, 1986; Cachon & Cachon, 1987; Fensome et al., 1993.
  Internal parasite of dinoflagellates, invertebrates, and fish eggs. The trophont stage is multinucleate and the biflagellated infective stage is uninucleate.
Remarks: The genus *Amoebophrya* Koeppen 1894 parasitizes free-living dinoflagellates and the intracellular mature trophont stage appears as a "bee hive" or spiral rings. It can occupy almost the entire cytoplasm.

Order Phytodinales Loeblich III 1970
References: Loeblich III, 1982; Popovsky & Pfiester, 1990; Fensome et al., 1993.
  Mostly freshwater. Dominant stage nonmotile and attached by a stalk, disk, or predetermined area of the cell. Produces unarmored gymnodinioid zoospores or armored zoospores with a Kofoidian plate tabulation. Chloroplasts present.

Family Phytodiniaceae Klebs 1912

Genus *Spiniferodinium* Horiguchi & Chihara 1987
Type: *S. galeiforme* Horiguchi & Chihara 1987
Reference: Horiguchi & Chihara (1987)
  Unicell with nonmotile dominant stage, attached to substrate or immobile, and with rigid, spiny helmet-shaped covering. Contains dinokaryotic nucleus and chloroplasts. Reproduction by gymnodinioid zoospores that are dorsoventrally flattened with a median, circular cingulum.

Spiniferodinium galeiforme Horiguchi & Chihara 1987 (not illustrated)
  Nonmotile phase with cingulum but without flagella, encapsulated by outer shell with spines and mucous. Unarmored small, biflagellated phase without spines and with a circular, median cingulum and a sulcus that notches the hypotheca and invades the epitheca.

Distribution: Recently described benthic and attached species with known distribution limited to the subtropical beach areas Okinawa Prefecture, Japan.

Order Dinotrichales Pascher 1914
References: Loeblich III, 1982; Fensome et al., 1993.
Dinoflagellates with a biphasic life cycle—an immobile, filamentous, benthic, or epiphytic stage and a planktonic gymnodinioid stage.

Order Dinamoebiales Loeblich III 1970
References: Loeblich III, 1982; Burkholder et al., 1992, 1994; Fensome et al., 1993; Landsberg et al., 1995; Steidinger et al., 1996.
Multiphasic life cycle with unicellular stages representing planktonic bi- and triflagellated motile forms of a gymnodinioid or peridinioid type. Benthic or temporarily amoeboid forms with filapodia or lobopodia, and benthic cyst stages of multiple forms. Heterotrophic nutrition.
Remarks: This group is often placed in the Phytodinales because the dominant stage is assumed to be coccoid; the amoeboid stage typically occurs between two other stages and is released through an opening. In this order, the amoeboid stage can be a direct transformation of the flagellated stage.
Distribution: A new species *Pfiesteria piscicida* Steidinger & Burkholder, 1996 (**Plate 55**) of this group was discovered in southeastern U.S. estuarine waters and called the "phantom" dinoflagellate or ambush predator. It produces an exotoxin and causes fish kills; it is nonphotosynthetic and is thinly armored with a Kofoidian plate tabulation. It is often mistaken as a gymnodinioid. Cleptochloroplasts present. Positive identification requires scanning electron microscopy.

UNCERTAIN TAXA

Genus *Adenoides* Balech 1956 (<5 spp.)
Type: *A. eludens* (Herdman) Balech 1956.
References: Balech, 1956; Dodge, 1982; Dodge & Lewis, 1986.
Armored. Small, slightly laterally flattened cell with premedian cingulum and cap-like epitheca. Looks like laterally flattened *Amphidinium* with a short sulcus, but has delicate theca with the plate formula: Po, 3', 5', 5c, 6s, 4''', 5p (4p), and 1'''(2''''). Theca with scattered pores. Pore plate without pore. Chloroplasts present.

Adenoides eludens (Herdman) Balech 1956 (**Plate 55**)
Nearly symmetrical with narrow left sulcal list. Plates 3 and 4p both with one large pore, otherwise no thecal ornamentation.
Distribution: Benthic, sand dweller; temperate waters. Sometimes tychoplanktonic.

Genus *Oxyrrhis* Dujardin 1841 (<5 spp.)
Type: *O. marina* Dujardin 1841.
References: Dodge & Crawford, 1971; Jianfan & Jingyan, 1992; Loeblich III, 1984; Sournia, 1986; Fensome et al., 1993.
> Unarmored. Small ovoid cell; slightly compressed laterally without cingulum and sulcus. Posterior portion of cell with a tentacular lobe and a dorsal flap; two dissimilar flagella (one with fine hairs) arise posteriorly. Scales cover the cell and flagella. Chloroplasts absent; heterotrophic.

Remarks: The question of whether *Oxyrrhis* is a dinoflagellate still remains, particularly with the evidence of a nonconforming morphology, different cortical microtubular cytoskeleton, different flagella, and a different nuclear structure. *Oxyrrhis* appears to be in between dinoflagellates and other eukaryotes.

Oxyrrhis marina Dujardin 1841 (Plate 55)
Remarks: Distinctive species.
Distribution: Marine and brackish inshore waters, coastal rock pools, estuaries, and marshes. Temperate to tropical; cosmopolitan.

Genus *Pleromonas* Pascher 1914 (1 sp.)
Type: *P. erosa* Pascher 1914.
References: Schiller, 1933; Sournia, 1986; Fensome et al., 1993.
> Unarmored but cellulosic thin cell wall. Small, ovoid cells with two apically inserted similar flagella that beat perpendicular to one another. Anterior end truncate with vacuole. Nucleus typically dinokaryotic; chloroplast(s) present with central pyrenoid.

Remarks: Pascher observed cells dividing within the parent wall similar to the longitudinal division of *Prorocentrum* and although free single cells were not observed to have two opposing valves, this monotypic genus may represent very thinly valvate *Prorocentrum* under resting or stress conditions.

Pleromonas erosa Pascher 1914 (not illustrated)
Distribution: Originally, description based on occurrence in an aquarium of marine algae.

TECHNIQUES FOR PREPARATION OF DINOFLAGELLATES FOR IDENTIFICATION

Cell Immobilization and Fixation

There are advantages and disadvantages to using live or dead specimens for observation. Live cells, either field or cultured specimens, provide information on pigmentation, swimming behavior, cytology and location of organelles, reproductive stages, cell orientation, size and shape, and other features. There

are several techniques to immobilize live dinoflagellates to observe their external
and internal characters: (1) glass slides (3 x 1 in.) with a drop of material can
be put in the refrigerator at 4°C without a coverslip for about 3 min; longer
exposure will typically lyse cells. It is important to remember that whole samples
from warm temperate and tropical waters should not be refrigerated or iced,
either in transport or storage, because many cells are fragile and will lyse. It
is preferable to store the samples at ambient temperature or to wrap wet towels
or newspaper around the sample container so that evaporation keeps the sample
cool. These samples can be kept in the dark or low light for 24 to 48 hr;
however, if you are doing enumerations, there will be species composition and
abundance changes with storage. (2) One milliliter of 4% saturated uranyl
acetate (UA) added to 1 liter of sample will typically slow down motile cells,
or the UA can be added to the edge of a coverslip. (3) Methyl cellulose or
MS-222 (a chemical relaxant) can also be added to samples to immobilize
dinoflagellate cells. Another method is to prepare a microscopic slide with
sample and a coverslip and wait for the heat of the microscope lamp to slow
the movement of cells. Cells typically cast off their flagellar first and unarmored
species then round up and lyse; armored cells usually remain intact or shed
their theca (ecdysis). Many unarmored dinoflagellate species do not fix and
preserve well with standard fixatives such as Lugol's solution and formalin;
however, they do fix well with buffered glutaraldehyde (2% GTA with sodium
cacodylate or borate/borax), a combined buffered glutaraldehyde/osmium te-
troxide cold fix (GTA/OsO$_4$) (Steidinger et al., 1989; Landsberg et al., 1994;
1 or 2% OsO$_4$, or even a weak formalin solution (0.5–1%). For scanning
electron microscopy of unarmored and lightly armored cells, the combined,
buffered GTA/OsO$_4$ is the best fixative, followed by critical-point drying, and
coating with gold/palladium (Steidinger et al., 1995).

Armored species are more resistant and most cells fix and preserve well
in 1–4% formalin or Lugol's. Lugol's solution tends to stain the protoplasm
of dinoflagellates to the point that observing surface ornamentation and plates
may be difficult. A few drops of sodium thiosulfate (0.1–1 N) can be used to
destain specimens. In scanning electron microscopy, the outer membrane of
armored cells can obscure plates and plate patterns; therefore, it is sometimes
necessary to remove the outer membrane by using a pretreatment of 40%
ethanol (M. Montresor, personal communication). Also, armored dinoflagellates, such as *Dinophysis, Ceratium, Ornithocercus,* and others with heavy
armor, can be prepared for SEM by fixation in formalin, mounting individual
cells on an SEM stub, rinsing the stub with distilled water 10–20 times under
a dissecting microscope to ensure that specimens are not washed away and
that a salt ring does not develop at the periphery of the stub, air drying, and
coating with gold/palladium.

Temporary and Permanent Slide Mounts

Temporary and permanent slide mounts for preserved armored species can
be made using several simple techniques. Temporary wet mounts can be made

using Vasoline or stopcock grease to seal the edges of the coverslip and this provides a movable mount if you want to reorient the cell under study. Using nail polish as a sealant provides a slightly moveable mount. If you are using an inverted microscope, depression slides with drops of sample can be maintained without a coverslip in a closed petri dish that has a piece of moist sponge or wet toweling. This technique is also suitable for observing and photographing sequential stages in sexual reproduction, e.g., formation of gametes and zygotes, or for observing larger specimens that are often squashed by a coverslip. For larger specimens, raised wet mounts, such as with Vasoline or slivers of cover-slip, can also be used to observe cells. Several permanent mounting media are commercially available, e.g., Permount, and these slide mounts can be sealed with nail polish or a commercial sealant. However, typically with these media, you first have to rinse the sample preparation in deionized or distilled water to remove the salt and then you have to go through several dehydration steps to remove the water before you can use the mounting medium. An alternative mounting medium offers the advantage of not having to do freshwater rinses and dehydration steps; a drop of sample can be mixed with a glycerin jelly preparation. Glycerin jelly is prepared by dissolving 10 g of gelatin in 60 ml of deionized water to which 70 ml of glycerin (glycerol) is added and 1 g of thymol dissolved as a preservative. Stir the mixture carefully to avoid developing bubbles. Specimens are placed on a clean slide with one or two drops of glycerin/jelly mounting medium and covered with a coverslip to dry on a slide warmer at 55°C for 6 hr. Once sealed, these mounts can last >13 years. Permanent mounts in which specimens are fixed in one view (cells cannot be rotated and plates cannot be separated) are not as valuable as preserved specimens kept in sealed vials as wet stack material, providing the material is maintained, e.g., topped up. Photo- and electron micrographs, representing different views, are also valuable reference and voucher material, providing they are cataloged and annotated.

Microscopic Techniques

Brightfield microscopy of armored dinoflagellates often requires chemical staining to observe plate sutures and plate patterns. Trypan blue (0.2–0.3%) or von Stosch's (1969) HI/I_2/chloral hydrate technique are useful chemical stains. Phase contrast and differential contrast microscopy offer optical "staining" that is often superior to any chemical stain for observing organelles, surface ornamentation, and plate patterns. In armored dinoflagellates, plates can be separated using sodium hypochlorite (5%) applied to the edge of the coverslip, but continual observation of individual specimens is required to accurately determine the position of plates that quickly separate from the protoplasm. Graham's (1942) glycerin jelly technique can help reduce the movement of separating plates and maintain them in close proximity to one another. Gentle pressure applied to the coverslip (e.g., with a toothpick or small probe) over the specimen can cause the theca to rupture and the protoplasm to extrude. Freeing the theca from the rest of the cell is the best, and sometimes only, way to identify some armored species because it allows detection of plates without the background interference from pigments obscuring plate sutures. This squash technique is accomplished by first pulling off as much water as possible from under the coverslip using a tissue pressed to one side that will absorb the liquid by capillary action. Then apply physical pressure with the toothpick and gently move the coverslip back and forth to dissociate plates. If using hypochlorite, add it to the edge of the coverslip after draining the excess liquid. Another technique involves fluorescence microscopy for detecting surface markings and plate sutures in armored cells (Fritz & Triemer, 1985). It uses a fluorochrome called Calcofluor White M2R that binds to plate material, such as cellulose, and reveals plate sutures and pores in thinly thecate cells as well as heavily armored cells. This technique is applicable for cells about 20 μm and above. In some species, this technique removes the necessity to separate plates with hypochlorite in order to detect them.

Steidinger (1979) summarized several collection, enumeration, and identification techniques for free-living marine dinoflagellates. Other useful technique papers include Graham (1942; a glycerin jelly technique for studying separated plates and whole armored cells), Burrells (1977; microscopy), Needham (1977; microscopy), Sournia (1978; all techniques), Dale (1979; dinoflagellate cysts), Loper et al. (1980; chromosome squashes), Postek et al. (1980; SEM techniques), Berland (1982; image reversal in microscopy), Pfiester & Holt (1982; chromosome squashes), Guilliard & Keller (1984; culturing), Dodge (1985), Matsuoka et al. (1989; dinoflagellate cyst techniques), and Throndsen (1993; flagellates).

COMMON DINOFLAGELLATE SYNONYMS

Actiniscus pentasterias (Ehrenberg) Ehrenberg
= *Dictyocha pentasterias* Ehrenberg
= *Gymnaster pentasterias* Schütt

Alexandrium acatenella (Whedon & Kofoid) Balech
= *Gonyaulax acatenella* Whedon & Kofoid
= *Gessnerium acatenellum* (Whedon & Kofoid) L. Loeblich & Loeblich III
= *Protogonyaulax acatenella* (Whedon & Kofoid) Taylor

Alexandrium catenella (Whedon & Kofoid) Balech
= *Gonyaulax catenella* Whedon & Kofoid

Alexandrium affine (Inoue & Fukuyo) Balech
= *Protogonyaulax affinis* Inoue & Fukuyo
= *Alexandrium fukuyoi* Balech

Alexandrium balechii (Steidinger) Balech
= *Pyrodinium balechii* (Steidinger) Taylor
= *Gonyaulax balechii* Steidinger

Alexandrium cohorticula (Balech) Balech
= *Protogonyaulax cohorticula* (Balech) Taylor
= *Gessnerium cohorticula* (Balech) L. Loeblich & Loeblich III
= *Gonyaulax cohorticula* Balech

Alexandrium compressum (Fukuyo, Yoshida, & Inoue) Balech
= *Protogonyaulax compressa* Fukuyo, Yoshida, & Inoue
= *Protogonyaulax* sp. Fukuyo

Alexandrium concavum (Gaarder) Balech
= *Gonyaulax concava* (Gaarder) Balech
= *Goniodoma concavum* Gaarder

Alexandrium fraterculus (Balech) Balech
= *Gessnerium fraterculum* (Balech) L. Loeblich & Loeblich III
= *Protogonyaulax fratercula* (Balech) Taylor
= *Gonyaulax fratercula* Balech

Alexandrium kutnerae (Balech)Balech
= *Gonyaulax kutnerae* Balech

Alexandrium monilatum (Howell) Balech
= *Pyrodinium monilatum* (Howell)Taylor
= *Gonyaulax monilatum* (Howell) Loeblich III
= *Gessnerium mochimaensis* Halim
= *Gonyaulax monilata* Howell

Alexandrium minutum Halim
= *Alexandrium ibericum* Balech
= *Pyrodinium minutum* (Halim) Taylor

Alexandrium ostenfeldii (Paulsen) Balech & Tangen
= *Triadinium ostenfeldii* (Paulsen) Dodge
= *Gessnerium ostenfeldii* (Paulsen) L. Loeblich and Loeblich III
= *Protogonyaulax globosa* (Braarud) Taylor
= *Gonyaulax trygvei* Parke
= *Gonyaulax globosa* (Braarud) Balech
= *Heteraulacus ostenfeldii* (Paulsen)Loeblich III
= *Goniaulax ostenfeldii* (Paulsen) Paulsen
= *Goniaulax tamarensis* Lebour var. *globosa* Braarud
= *Pyrodinium phoneus* Woloszynska & Conrad
= *Goniodoma ostenfeldii* Paulsen

Alexandrium peruvianum (Balech & Mendiola) Balech & Tangen
= *Gonyaulax peruviana* Balech & Mendiola

Alexandrium pseudogoniaulax (Biecheler) Horiguchi
= *Goniodoma pseudogoniaulax* Biecheler

Alexandrium tamarense (Lebour) Balech
= ?*Alexandrium excavatum* (Braarud) Balech and Tangen
= *Protogonyaulax tamarensis* (Lebour) Taylor
= *Gessnerium tamarensis* (Lebour) Loeblich III and L. Loeblich
= *Gonyaulax tamarensis* var. *excavata* Braarud
= *Gonyaulax tamarensis* Lebour

Amphidinium operculatum Claperède & Lachmann
= *Amphidinium klebsii* Kofoid & Swezy

Amphidoma nucula Stein
= *Gonyaulax rouchii* Rampi
= *Amphidoma spinosa* (Kofoid) Kofoid & Michener
= *Murrayella spinosa* Kofoid

Amylax triacantha (Jörgensen) Sournia
= *Gonyaulax triacantha* Jörgensen

Balechina coerulea (Dogiel) F. J. R. Taylor
= *Gymnodinium coeruleum* Dogiel

Blepharocysta splendor-maris (Ehrenberg) Stein
= *Peridinium splendor-maris* Ehrenberg

Heterocapsa niei (Loeblich III) Morrill & Loeblich III
= *Cachonina niei* Loeblich III

Ceratium arcticum (Ehrenberg)Cleve
= *Peridinium arcticum* Ehrenberg

Ceratium arietinum Cleve
= *Ceratium bucephalum* Cleve

Ceratium candelabrum (Ehrenberg) Stein
= *Peridinium candelabrum* Ehrenberg

Ceratium declinatum (Karsten)Jörgensen
= *Ceratium tripos declinatum* Karsten

Ceratium furca (Ehrenberg) Claparède & Lachmann
= *Peridinium furca* Ehrenberg

Ceratium fusus (Ehrenberg) Dujardin
= *Peridinium fusus* Ehrenberg

Ceratium hexacanthum Gourret
= *Ceratium reticulatum* (Pouchet)Cleve
= *Ceratium tripos* var. *reticulata* Pouchet

Ceratium hircus Schröder
= *Ceratium furca* var. *hircus* (Schröder) Margalef

Ceratium hirundinella (O. F. Müller) Bergh
= *Bursaria hirundinella* O. F. Müller

Ceratium horridum (Cleve) Gran
= *Ceratium tripos* var. *horridum* Cleve

Ceratium incisum (Karsten) Jörgensen
= *Ceratium furca incisum* Karsten

Ceratium inflatum (Kofoid) Jörgensen
= *Ceratium pennatum* var. *inflata* Kofoid

Ceratium lineatum (Ehrenberg) Cleve
= *Peridinium lineatum* Ehrenberg

Ceratium longipes (Bailey) Gran
= *Ceratium articum* var. *longpipes* (Bailey) Graham & Bronikovsky
= *Peridinium longipes* Bailey

Ceratium lunula (Schimper) Jörgensen
= *Ceratium tripos lunula* Schimper

Ceratium macroceros (Ehrenberg) Cleve
= *Peridinium macroceros* Ehrenberg

Ceratium massiliense (Gourret) Jörgensen
= *Ceratium tripos* var. *massiliense* Gourret

Ceratium praelongum (Lemmermann) Kofoid
= *Ceratium gravidum* var. *praelongum* Lemmermann

Ceratium pulchellum Schröder
= *Ceratium tripos* var. *pulchellum* (Schröder) López
= *Ceratium tripos* subsp. *pulchellum* (Schröder) Peters

Ceratium trichoceros (Ehrenberg) Kofoid
= *Peridinium trichoceros* Ehrenberg

Ceratium tripos (O. F. Müller) Nitzsch
= *Cercaria tripos* O. F. Müller

Ceratocorys armata (Schütt) Kofoid
= *Goniodoma acuminatum* var. *armatum* Schütt

Ceratocorys bipes (Cleve) Kofoid
= *Goniodoma bipes* Cleve

Cochlodinium helix (Pouchet) Lemmermann
= *Gymnodinium helix* Pouchet

Cochlodinium polykrikoides Margalef
= *Cochlodinium heterolobatum* Silva

Coolia monotis Meunier
= *Glenodinium monotis* (Meunier) Biecheler

= *Ostreopsis monotis* (Meunier) Lindemann

Corythodinium compressum (Kofoid) Taylor
= *Oxytoxum compressum* Kofoid

Corythodinium tesselatum (Stein) Loeblich Jr. & Loeblich III
= *Pyrgidium tesselatum* Stein

Crypthecodinium cohnii Seligo in Grasse
= *Crypthecodinium setense* Biecheler
= *Glendodinium cohnii* Seligo in Cohn

Dinophysis hastata Stein
= *Phalacroma hastatum* Pavillard

Dinophysis odiosa (Pavillard) Tai & Skogsberg
= *Prodinophysis odiosa* (Pavillard) Loeblich III
= *Phalacroma odiosum* Pavillard

Dinophysis parvula (Schütt) Balech
= *Prodinophysis parvula* (Schütt) Balech
= *Phalacroma parvulum* (Schütt) Jörgensen
= *Phalacroma porodictyum* var. *parvula* Schütt

Dinophysis porodictyum (Stein) Abé
= *Phalacroma porodictyum* Stein

Dinophysis tripos Gourret
= *Dinophysis caudata* var. *tripos* (Gourret) Gail

Diplopelta parva (Abé) Matsuska
= *Dissodium parvum* Abé

Diplopsalis lenticula Bergh
= *Peridiniopsis lenticula* (Bergh) Starmach
= *Dissodium lenticula* (Bergh) Loeblich III

Diplopsalopsis orbicularis (Paulsen) Meunier
= *Diplopsalis orbicularis* (Paulsen) Paulsen
= *Peridinium orbiculare* Paulsen

Erythropsidinium agile (Hertwig) Silva
= *Erythropsis agile* Hertwig

Fragilidium subglobosum (von Stosch) Loeblich III
= *Helgolandinium subglobosum* von Stosch

Goniodoma polyedricum (Pouchet) Jörgensen
= *Triadinium polyedricum* (Pouchet) Dodge
= *Heteraulacus polyedricus* (Pouchet) Drugg & Loeblich Jr.
= *Peridinium polyedricum* Pouchet

Goniodoma sphaericum Murray & Whitting
= *Triadinium sphaericum* (Murray & Whitting) Dodge

= *Heteraulacus sphaericus* (Murray & Whitting) Loeblich III

Gonyaulax digitale (Pouchet) Kofoid
= *Protoperidinium digitale* Pouchet

Gonyaulax fragilis (Schütt) Kofoid
= *Steiniella fragilis* Schütt

Gonyaulax grindleyi Reinecke
= *Peridiniopsis reticulata* (Claparède & Lachmann) Starmach
= *Protoceratium reticulatum* (Claparède & Lachmann) Bütschli
= *Peridinium reticulatum* Claparède & Lachmann

Gonyaulax spinifera (Claparède & Lachmann) Diesing
= *Peridinium spiniferum* Claparède & Lachmann

Gonyaulax verior Sournia
= *Gonyaulax diacantha* sensu Schiller
= *Amylax diacantha* Meunier

Gymnodinium breve Davis
= *Ptychodiscus brevis* (Davis) Steidinger

Gymnodinium fuscum (Ehrenberg) Stein
= *Peridinium fuscum* Ehrenberg

Gymnodinium mikimotoi Miyaka & Kominami ex Oda
= *Gymnodinium nagasakiense* Takyama & Adachi

Gymnodinium pseudopalustre (Woloszynska) Schiller
= *Gymnodinium palustre* Woloszynska

Gymnodinium sanguineum Hirasaka
= *Gymnodinium nelsonii* Martin
= *Gymnodinium splendens* Lebour

Gyrodinium fissum (Levander) Kofoid & Swezy
= *Gymnodinium fissum* Levander

Gyrodinium spirale (Bergh) Kofoid & Swezy
= *Gymnodinium spirale* Bergh

Herdmania littoralis Dodge
= *Gymnodinium agile* Herdman

Heterocapsa niei (Loeblich) Morrill & Loeblich III
= *Cachonina niei* Loeblich

Heterocapsa rotundata
= *Katodinium rotundatum* (Lohmann) Loeblich III
= *Massartia rotundata* (Lohmann) Schiller
= *Amphidinium rotundatum* Lohmann

Heterocapsa triquetra (Ehrenberg) Balech
= *Peridinium triquetrum* (Ehrenberg) Lebour
= *Properidinium heterocapsa* (Stein) Meunier

= *Glenodinium triquetrum* Ehrenberg

Histioneis diomedeae Kofoid & Michener
= *Parahistioneis diomedeae* (Kofoid & Michener) Kofoid & Swezy

Katodinium nieuportense (Conrad) Loeblich Jr. & Loeblich III
= *Massartia nieuportensis* Conrad

Katodinium glaucum (Lebour) Loeblich III
= *Spirodinium glaucum* Lebour

Lingulodinium polyedrum (Stein) Dodge
= *Gonyaulax polyedra* Stein

Mesoporus globulus (Schiller) Lillick
= *Porella globulus* Schiller

Mesoporus perforatus (Gran) Lillick
= *Dinoporella perforata* (Gran) Halim
= *Porotheca perforata* (Gran) Silva
= *Exuviaella perforata* Gran

Nematodinium armatum (Dogiel) Kofoid & Swezy
= *Pouchetia armata* Dogiel

Noctiluca scintillans (Macartney) Kofoid & Swezy
= *Noctiluca miliaris* Suriray
= *Medusa scintillans* Macartney

Ornithocercus thumii (Schmidt) Kofoid & Skogsberg
= *Parelion thumii* Schmidt

Pentapharsodinium dalei Indelicato & Loeblich III
= *Peridinium faroense* sensu Dale

Peridiniella catenata (Levander) Balech
= *Gonyaulax catenata* (Levander) Kofoid
= *Amylax catenata* (Levander) Meunier
= *Peridinium catenata* Levander

Peridinium cinctum (O. F. Müller) Ehrenberg
= *Vorticella cincta* O. F. Müller

Phalacroma argus Stein
= *Prodinophysis argus* (Stein) Balech
= *Dinophysis argus* (Stein) Abé

Phalacroma cuneus Schütt
= *Prodinophysis cuneus* (Schütt) Balech
= *Dinophysis cuneus* (Schütt) Abé

Phalacroma doryphorum Stein
= *Prodinophysis doryphora* (Stein) Balech
= *Dinophysis doryphora* (Stein) Abé

Phalacroma favus Kofoid & Michener
= *Prodinophysis favus* (Kofoid & Michener) Balech
= *Dinophysis favus* (Kofoid & Michener) Abé

Phalacroma mitra Schütt
= *Prodinophysis mitra* (Schütt) Balech
= *Dinophysis mitra* (Schütt) Abé

Phalacroma rapa Stein
= *Prodinophysis rapa* (Stein) Balech
= *Dinophysis rapa* (Stein) Balech

Phalacroma rotundatum (Claparède & Lachmann) Kofoid & Michener
= *Prodinophysis rotundata* (Claparède & Lachmann) Balech
= *Dinophysis rotundata* Claparède & Lachmann

Pheopolykrikos beauchampii Chatton
= *Polykrikos beauchampii* (Chatton) Loeblich III

Pheopolykrikos hartmannii (Zimmerman) Matsuoka & Fukuyo
= *Polykrikos hartmannii* Zimmermann

Podolampas bipes var. *reticulata* (Kofoid) Taylor
= *Podolampas reticulata* Kofoid

Polykrikos schwartzi Bütschli
= *Polykrikos auricularia* Bergh

Pratjetella medusoides (Hertwig) Loeblich Jr. & Loeblich III
= *Leptodiscus medusoides* Hertwig

Preperidinium meunieri (Pavillard) Elbrächter
= *Diplopsalis minor* (Paulsen) Lindemann
= *Zygabikodinium lenticulatum* Loeblich Jr & Loeblich III
= *Diplopeltopsis minor* (Paulsen)Pavillard
= *Peridinium paulsenii* Mangin
= *Peridinium lenticulatum* Mangin
= *Diplopsalis lenticula* f. *minor* Paulsen

Pronoctiluca acuta (Lohmann) Schiller
= *Rhynchomonas acuta* Lohmann

Prorocentrum balticum (Lohmann) Loeblich III
= *Exuviaella baltica* Lohmann

Prorocentrum compressum (Bailey) Abé ex Dodge
= *Exuviaella compressa* (Bailey) Ostenfeld
= *Pyxidicula compressa* Bailey

Prorocentrum gracile Schütt
= *Prorocentrum hentschelii* Schiller

Prorocentrum lima (Ehrenberg) Dodge

= *Exuviaella marina* var. *lima*(Ehrenberg) Schiller
= *Exuviaella lima* (Ehrenberg) Bütschli
= *Dinopyxis laevis* Stein
= *Cryptomonas lima* Ehrenberg

Prorocentrum marinum (Cienkowski) Loeblich III
= *Exuviaella marina* Cienkowski

Prorocentrum mexicanum Tafall
= *Prorocentrum maximum*
= *Prorocentrum rathymum* Loeblich, Shirley & Schmidt

Prorocentrum minimum (Pavillard) Schiller
= *Exuviaella marie-lebouriae* Parke & Ballantine
= *Prorocentrum marie-lebouriae* (Parke & Ballantine) *Loeblich III*
= *Prorocentrum triangulatum* Martin
= *Exuviaella minima* Pavillard

Prorocentrum rostratum Stein
= *Prorocentrum styliferum* Lohmann
= *Prorocentrum tenue* Lohmann

Prorocentrum triestinum Schiller
= *Prorocentrum redfeldii* Bursa
= *Prorocentrum pyrenoideum* Bursa

Protoceratium spinulosum (Murray & Whitting) Schiller
= *Peridinium spinulosum* Murray & Whitting

Protoperidinium brevipes (Paulsen) Balech
= *Peridinium brevipes* Paulsen

Protoperidinium claudicans (Paulsen) Balech
= *Peridinium claudicans* Paulsen

Protoperidinium compressum (Nie) Balech
= *Peridinium compressum* (Abé) Nie
=   *Congruentidium compressum* Abé

Protoperidinium conicoides (Paulsen) Balech
= *Peridinium conicoides* Paulsen

Protoperidinium conicum (Gran) Balech
= *Peridinium conicum* (Gran) Ostenfeld & Schmidt
= *Peridinium divergens* var. *conica* Gran

Protoperidinium crassipes (Kofoid) Balech
= *Peridinium crassipes* Kofoid

Protoperidinium depressum (Bailey) Balech
= *Peridinium depressum* Bailey

Protoperidinium divaricatum (Meunier) Balech

= *Protoperidinium gainii* (Dangeard) Balech
= *Peridinium gainii* Dangeard
= *Peridinium divaricatum* Meunier

Protoperidinium divergens (Ehrenberg) Balech
= *Glenodinium divergens* (Ehrenberg) Dangeard
= *Peridinium divergens* Ehrenberg

Protoperidinium elegans (Cleve) Balech
= *Peridinium elegans* Cleve

Protoperidinium excentricum (Paulsen) Balech
= *Peridinium excentricum* Paulsen

Protoperidinium fatulipes (Kofoid) Balech
= *Peridinium fatulipes* Kofoid

Protoperidinium grande (Kofoid) Balech
= *Peridinium grande* Kofoid

Protoperidinium latissimum (Kofoid) Balech
= *Peridinium pentagonoides* Balech
= *Peridinium pentagonum* var. *depressum* Abé
= *Peridinium latissimum* Kofoid

Protoperidinium leonis (Pavillard) Balech
= *Peridinium leonis* Pavillard

Protoperidinium mediterraneum (Kofoid) Balech
= *Peridinium steini mediterraneum* Kofoid

Protoperidinium minutum (Kofoid) Loeblich III
= *Peridinium minutum* Kofoid

Protoperidinium nudum (Meunier) Balech
= *Peridinium nudum* Meunier

Protoperidinium oblongum (Aurivillius) Parke & Dodge
= *Peridinium oblongum* (Aurivillius) Cleve
= *Peridinium divergens* var. *oblongum* Aurivillius

Protoperidinium obtusum (Karsten) Balech
= *Peridinium obtusum* Karsten

Protoperidinium oceanicum (VanHöffen) Balech
= *Peridinium oceanicum* VanHöffen

Protoperidinium pallidum (Ostenfeld) Balech
= *Peridinium pallidum* Ostenfeld

Protoperidinium pellucidum Bergh
= *Peridinium pellucidum* (Bergh) Schütt

Protoperidinium pentagonum (Gran) Balech
= *Peridinium pentagonum* Gran

Protoperidinium punctulatum (Paulsen) Balech
= *Peridinium punctalatum* Paulsen

Protoperidinium pyrum (Balech) Balech
= *Peridinium pyrum* Balech

Protoperidinium quarnerense (Schröder) Balech
= *Peridinium quarnerense* Schröder
= *Peridinium globulus* var.Stein

Protoperidinium saltans (Meunier) Balech
= *Peridinium saltans* Meunier

Protoperidinium simulum (Paulsen) Balech
= *Peridinium simulum* Paulsen

Protoperidinium solitarium (Abé) Balech
= *Peridinium solitarium* Abé

Protoperidinium subinerme (Paulsen) Loeblich III
= *Peridinium subinerme* Paulsen

Protoperidinium symmetricum (Halim) Balech
= *Peridinium symmetricum* Halim

Protoperidinium thorianum (Paulsen) Balech
= *Peridinium thorianum* Paulsen

Pyrocystis fusiformis Wyville-Thomson ex Murray
= *Dissodinium fusiforme* (Wyville-Thomson ex Murray) Matzenauer

Pyrocystis lunula (Schütt) Schütt
= *Dissodinium lunula* (Schütt) Pascher
= *Diplodinium lunula* (Schütt) Klebs
= *Gymnodinium lunula* Schütt

Pyrocystis noctiluca Murray ex Schütt
= *Pyrocystis pseudonoctiluca* Wyville-Thomson in J. Murray
= *Dissodinium pseudonoctiluca* Swift ex Elbrächter & Drebes

Pyrodinium bahamense var. *compressum* (Böhm) Steidinger, Tester, &
    Taylor
= *Pyrodinium schilleri* (Matzenauer) Schiller
= *Gonyaulax schilleri* Matzenauer
= *Pyrodinium bahamense* f. *compressa* Böhm

Pyrophacus steinii (Schiller) Wall & Dale
= *Pyrophacus horologicum* var. *steinii* Schiller

Pyrophacus vancampoae (Rossignol) Wall & Dale
= *Tuberculodinium vancampoae* (Rossignol) Wall
= *Pterospermopsis vancampoae* Rossignol

Schuettiella mitra (Schütt) Balech

= *Gonyaulax mitra* (Schütt) Kofoid
= *Oxytoxum gigas* Kofoid
= *Steiniella mitra* Schütt

Sclerodinium calyptoglyphe (Lebour) Dodge
= *Gyrodinium calyptoglyphe* Lebour

Scrippsiella subsalsa (Ostenfeld) Steidinger & Balech
= *Peridinium subsalsum* Ostenfeld

Scrippsiella trochoidea (Stein) Balech
= *Scrippsiella faeroense* (Paulsen) Balech & Soares
= *Peridinium trochoideum* (Stein) Lemmermann
= *Peridinium faeroense* Paulsen
= *Glenodinium trochoideum* Stein

Symbiodinium microadriaticum Freudenthal
= *Zooxanthella microadriatica* (Freudenthal) Loeblich III & Sherley
= *Gymnodinium microadriaticum* (Freudenthal) Taylor

Thecadinium kofoidii (Herdman) Schiller
= *Thecadinium petasatum* Kofoid & Skogsberg
= *Phalocroma kofoidi* Herdman
= *Amphidinium kofoidi* var. *petasatum* Herdman
= *Amphidinium sulcatum* Herdman

Thoracosphaera heimii (Lohmann) Kamptner
= *Thoracosphaera pelagica* Kamptner
= *Syracosphaera heimii* Lohmann

Torodinium teredo (Pouchet) Kofoid & Swezy
= *Gymnodinium teredo* Pouchet

Warnowia fusus (Schütt) Lindemann
= *Pouchetia fusus* Schütt

INDEX OF DINOFLAGELLATE TAXA

Acanthogonyaulax, 503
  A. spinifera, 502–504
Actiniscus, 468
  A. pentasterias, 468–469
Adenoides, 412, 548
  A. eludens, 550
Alexandrium, 406, 413, 482, 488–490,
    501, 506–507, 510, 519, 521
  A. acatenella, 490, 493
  A. affine, 395, 490, 495
  A. andersonii, 490, 493, 497
  A. angustitabulatum, 490–491, 497

A. balechii, 395, 492, 498
A. catenella, 491–492
A. cohorticula, 492, 495
A. compressum, 491–492
A. concavum, 492–493
A. excavatum, 500
A. foedum, 395, 494, 498
A. fraterculus, 490, 494–495
A. fukuyoi, 490
A. fundyense, 491, 494
A. hiranoi, 494, 498
A. ibericum, 497

A. *insuetum*, 494, 496
A. *kutnerae*, 491, 496–497
A. *leei*, 491, 497
A. *lusitanicum*, 493, 497
A. *margalefii*, 498
A. *minutum*, 488, 490, 495, 497
A. *monilatum*, 395, 492, 496–497
A. *ostenfeldii*, 496, 499
A. *peruvianum*, 496, 499
A. *pseudogonyaulax*, 494, 498–499
A. *satoanum*, 500
A. *tamarense*, 490, 493–494, 500
A. *tamiyavanichi*, 495, 500
A. *taylorii*, 498, 500
A. *tropicale*, 496, 500
Amphidinium, 444, 464, 467
   A. *carterae*, 443–444
   A. *globifera*, 426, 428
   A. *klebsii*, 445
   A. *operculatum*, 444–445
   A. *poecilochroum*, 443, 445
Amphidiniopsis, 412, 546
   A. *kofoidii*, 546
Amphidoma, 412, 414, 504
   A. *nucula*, 504–505
Amphisolenia, 426
   A. *bidentata*, 426–427
   A. *globifera*, 426, 428
Amylax, 412, 504, 506
   A. *alata*, 504
   A. *buxus*, 504
   A. *catenata*, 510
   A. *diacantha*, 509
   A. *triacantha*, 504–505
Apsteinia
   A. *acuta*, 531

Balechina, 461
   B. *coerulea*, 461–462
   B. *pachydermata*, 461
Blepharocysta, 399, 414, 532–534
   B. *splendor-maris*, 533
Boreadinium, 413, 527–528
   B. *pisiforme*, 527
Brachydinium, 468
   B. *capitatum*, 468–469

Cachonina
   C. *niei*, 531
Centrodinium, 412, 446, 516
   C. *elongatum*, 516
Ceratium, 407, 412, 457, 469–470, 552

C. *arcticum*, 476
C. *arcuatum*, 471
C. *arietinum*, 471, 479
C. *bucephalum*, 471
C. *candelabrum*, 471, 476
C. *carriense*, 471, 481
C. *concilians*, 479
C. *contortum*, 472, 479
C. *declinatum*, 472, 476
C. *extensum*, 472
C. *furca*, 471–473
C. *furcoides*, 417
C. *fusus*, 472–473
C. *gibberum*, 472, 479
C. *gravidum*, 473, 474
C. *hexacanthum*, 474, 479
C. *hircus*, 472
C. *horridum*, 474, 480
C. *incisum*, 473–474
C. *inflatum*, 473–474
C. *kofoidii*, 473, 475
C. *limulus*, 475, 480
C. *lineatum*, 473, 475
C. *longipes*, 475–476
C. *lunula*, 475, 481
C. *macroceros*, 475, 481
C. *massiliense*, 477
C. *paradoxides*, 475, 480
C. *pentagonum*, 476–477
C. *platycorne*, 477, 480
C. *pleuroceras*, 470
C. *praelongum*, 473, 475, 477
C. *pulchellum*, 478–479
C. *ranipes*, 478, 483
C. *symmetricum*, 478, 480
C. *teres*, 476, 478
C. *trichoceros*, 478, 481
C. *tripos*, 476 478
C. *vultur*, 398, 480, 482
Ceratocorys, 412, 482
   C. *armata*, 395, 482–483
   C. *horrida*, 482–483
Citharistes, 428
   C. *regius*, 427–428
Cladopyxis, 412, 482
   C. *brachiolata*, 487
Cochlodinium, 445
   C. *citron*, 443, 445
   C. *heterolobatum*, 446
   C. *polykrikoides*, 443, 446
   C. *strangulatum*, 445

Coolia, 412, 513, 515
   C. *monotis*, 397, 513–514
Corythodinium, 399, 412, 516
   C. *constrictum*, 516
   C. *tesselatum*, 516–518
Crypthecodinium, 411, 413, 487
   C. *cohnii*, 411, 488–489
   C. *setense*, 487
   C. *cuneus*, 487
Cymbodinium, 466
   C. *elegans*, 465–466

Dinophysis, 415, 426, 528
   D. *acuminata*, 429–430
   D. *acuta*, 406, 428–430
   D. *caudata*, 431–432, 434
   D. *dens*, 430–431
   D. *diegensis*, 431
   D. *fortii*, 430–431
   D. *hastata*, 432–433
   D. *lachmannii*, 429
   D. *lenticula*, 408
   D. *norvegica*, 429–430, 433
   D. *odiosa*, 430, 433
   D. *ovum*, 429
   D. *punctata*, 429
   D. *sacculus*, 429
   D. *schuettii*, 432–433
   D. *skagii*, 429
   D. *swezyii*, 433
   D. *tripos*, 431–432, 434
   D. *uracantha*, 432–434
Diplopelta, 413, 527–528
   D. *bomba*, 398
   D. *parva*, 529
Diplopsalis, 413, 528–529, 552
   D. *acuta*, 531
   D. *lenticula*, 529
   D. *minor*, 528
Diplopsalopsis, 413, 528–529, 530
   D. *orbicularis*, 529
Dissodinium, 519–520, 529
   D. *parvum*, 529
   D. *pseudolunula*, 520

Ensiculifera, 413, 523, 526
   E. *carinata*, 524
   E. *mexicana*, 523–524
Entzia
   E. *acuta*, 531
Erythropsidinium, 403, 459
   E. *agile*, 394, 454, 459

Fragilidium, 412, 520
   F. *mexicanum*, 521–522
   F. *fissile*, 521
   F. *heterolobum*, 396, 521
   F. *subglobosum*, 521

Gambierdiscus, 412, 501–502
   G. *toxicus*, 397, 501
Gessnerium, 488, 490, 499
   G. *mochimaensis*, 497
   G. *monilata*, 497
   G. *ostenfeldii*, 499
   G. *tamarensis*, 500
Glenodinium, 513, 529, 531
   G. *acutum*, 531
   G. *cinctum*, 531
Goniodoma, 413, 482, 501, 506, 521
   G. *acuminata*, 501
   G. *ostenfeldii*, 499
   G. *polyedricum*, 395, 501–502
   G. *sphaericum*, 417, 502–503
Gonyaulax, 396, 412, 457, 504–507
   G. *alaskensis*, 506
   G. *catenata*, 510
   G. *diacantha*, 509
   G. *diegensis*, 507
   G. *digitale*, 506–507, 509
   G. *excavata*, 500
   G. *fragilis*, 409, 506
   G. *globosa*, 499
   G. *grindleyi*, 396, 506, 508
   G. *mitra*, 512
   G. *monilata*, 497
   G. *ostenfeldii*, 499
   G. *polyedra*, 506, 509
   G. *polygramma*, 507–508
   G. *scrippsae*, 507–508
   G. *spinifera*, 506–508
   G. *tamarensis*, 499–500
   G. *tamarensis var. excavata*, 500
   G. *triacantha*, 504
   G. *trygvei*, 499
   G. *verior*, 396, 509
Gotoius, 413, 529
   G. *abei*, 529
   G. *mutsuensis*, 529
Gymnodinium, 387–388, 446, 456, 464, 467, 519, 531
   G. *abbreviatum*, 446, 450
   G. *breve*, 400, 447–448
   G. *catenatum*, 447–448
   G. *fusum*, 446, 456

G. galatheanum, 447–448
G. heterostriatum, 449–450
G. mikimotoi, 392, 400, 447–449,
    452
G. nagasakiense, 449
G. nelsonii, 451
G. pulchellum, 392, 448–449
G. sanguineum, 450–451
G. splendens, 451
Gyrodinium, 392, 400, 451
G. aureolum, 449, 452
G. estuariale, 450, 452
G. fissum, 452
G. instriatum, 450, 452
G. lachryma, 453–454
G. pellucidum, 542
G. pingue, 453
G. pepo, 392
G. spirale, 409, 451, 453–455
G. uncatenum, 417, 450, 453

Herdmania, 463
H. litoralis, 462
Heteraulacus, 501
H. ostenfeldii, 499
Heterocapsa, 413, 446, 530
H. niei, 498, 531–532
H. pigmaea, 531
H. rotundata, 455
H. triquetra, 498, 531–532
Heterodinium, 413, 513
H. milneri, 513
H. rigdenae, 513–514
H. scrippsae, 513
Histioneis, 413, 433
H. depressa, 433–434
H. mitchellana, 433
H. remora, 433

Katodinium, 453, 467, 531
K. fungiforme, 455
K. glaucum, 392, 454–455
K. nieuportense, 453
K. rotundatum, 454, 456
Kofoidinium, 463
K. velelloides, 464–465
Kolkwitziella
K. acuta, 531
K. gibbera, 531

Lepidodinium, 456
L. viride, 456
Leptodiscus, 466
L. medusoides, 466

Lingulodinium, 412, 506, 509
L. machaerophorum, 509
L. polyedrum, 396
Lissodinium, 414, 533–534
L. schilleri, 532–533
L. orcadense, 399

Mesoporos, 418–419
M. globulus, 418
M. perforatus, 419
Metaphalacroma, 436
M. skogsbergii, 436

Nematodinium, 401, 403, 459
N. armatum, 394, 454, 459
N. partitum, 459
Noctiluca, 466
N. scintillans, 465–466

Oblea, 413, 528–529
O. baculifera, 529–530
Ornithocercus, 414, 436, 552
O. heteroporus, 435–436
O. magnificus, 435–436
O. quadratus, 435–436
O. splendidus, 435, 437
O. steinii, 435–437
O. thumii, 435–437
Ostreopsis, 412, 513–515
O. heptagona, 397, 515
O. lenticularis, 514–515
O. ovata, 514–515
O. siamensis, 514–516
Oxyphysis, 442
O. oxytoxoides, 442–443
Oxyrrhis, 389, 551
O. marina, 389, 548, 551
Oxytoxum, 412, 512, 517–518
O. gigas, 512
O. scolopax, 399, 517, 519
O. variabile, 519

Paleophalacroma, 412, 487
P. unicinctum, 399, 483, 487
P. verrucosum, 487
Pentapharsodinium, 413, 524, 526
P. dalei, 524–525
Peridiniella, 413, 510
P. catenata, 510–511
P. sphaeroidea, 510
Peridinium, 414, 457, 513, 524, 526,
    530–531
P. cinctum, 531

P. *faeroense*, 524
P. *latum*, 531
P. *lubiensiforme*, 417
P. *paulsenii*, 530
P. *punctulatum*, 545
P. *spiniferum*, 507
Pheopolykrikos, 457
  P. *beauchampii*, 457
  P. *hartmannii*, 457–458
Phalacroma, 414, 429, 437
  P. *argus*, 437–438
  P. *cuneus*, 438–439
  P. *doryphorum*, 439, 443
  P. *favus*, 438–439
  P. *mitra*, 438–439
  P. *prodictyum*, 436
  P. *rapa*, 438–439
  P. *rotundatum*, 406, 438–439
  P. *verrucosum*, 487
Pleromonas, 551
  P. *erosa*, 451
Podolampas, 414, 533–534
  P. *bipes*, 399, 534
  P. *elegans*, 534
  P. *palmipes*, 534–535
Polykrikos, 400, 402, 457
  P. *barnegatensis*, 457
  P. *kofoidii*, 457–458
  P. *schwartzii*, 394, 457–458
Pomatodinium, 464
  P. *impatiens*, 464–465
Preperidinium, 413, 528
  P. *meunieri*, 530
Pronoctiluca, 467
  P. *acuta*, 465, 467
  P. *pelagica*, 467
Prorocentrum, 389, 400, 402, 413–414,
    418–419, 429
  P. *arcuatum*, 420–421
  P. *balticum*, 420, 422, 425
  P. *belizeanum*, 420, 422–423
  P. *compressum*, 422
  P. *concavum*, 420, 423
  P. *dentatum*, 421, 423
  P. *emarginatum*, 422–423
  P. *gracile*, 421, 424–425
  P. *hentschelii*, 424
  P. *hoffmannianum*, 420, 423, 425
  P. *lima*, 404, 420–421, 424
  P. *maculosum*, 424
  P. *mariae-lebourae*, 425

P. *marinum*, 404
P. *maximum*, 421, 424
P. *mexicanum*, 424
P. *micans*, 404, 421, 424, 426
P. *minimum*, 420, 422, 425
P. *rathymum*, 424
P. *redfeldii*, 416, 424, 426
P. *rostratum*, 421, 424–425
P. *sabulosum*, 425
P. *scutellum*, 422, 425
P. *triestinum*, 421, 426
P. *triangulatum*, 425
Protoceratium, 507, 510
  P. *aceros*, 510
  P. *reticulatum*, 506
  P. *spinulosum*, 510–511
Protogonyaulax, 488
  P. *globosa*, 499
  P. *tamarensis*, 500
Protoperidinium, 388, 407–408, 413,
    506, 510, 513, 526, 534–535
  P. *avellana*, 546
  P. *brevipes*, 536, 544
  P. *claudicans*, 536–537, 541
  P. *conicoides*, 538
  P. *conicum*, 538, 542
  P. *crassipes*, 538, 543
  P. *depressum*, 389, 538–540
  P. *divergens*, 398, 538, 543
  P. *elegans*, 540, 544
  P. *excentricum*, 540
  P. *fatulipes*, 540, 544
  P. *grande*, 540, 544
  P. *latissimum*, 545
  P. *leonis*, 537, 541–542
  P. *metananum*, 536
  P. *minutum*, 541
  P. *murrayii*, 541
  P. *nudum*, 541
  P. *oblongum*, 539, 541–542
  P. *obtusum*, 541–542
  P. *oceanicum*, 536, 539, 541–542
  P. *pallidum*, 536, 542, 544–545
  P. *pellucidum*, 404, 544
  P. *parthenopes*, 542–543
  P. *pellucidum*, 542
  P. *pentagonum*, 539, 545
  P. *punctulatum*, 537
  P. *quarnerense*, 545
  P. *quinquecorne*, 524
  P. *steidingerae*, 541

P. subinerme, 537, 545
P. thorianum, 543, 546
P. truncatum, 540
Ptychodiscus, 446, 463
  P. brevis, 447
  P. carinatus, 463
  P. inflatus, 463
  P. noctiluca, 462–463
Pyrocystis, 413, 519
  P. acuta/lanceolata, 520
  P. gerbaultii, 520
  P. lunula, 518–520
  P. minima, 520
  P. noctiluca, 518, 520
  P. obtusa, 520
  P. pseudonoctiluca, 520
  P. robusta, 520
Pyrodinium, 388, 413, 488, 503
  P. bahamense var. bahamense, 396,
    502–503
  P. bahamense var. compressum, 503
  P. monilatum, 497
  P. phoneus, 499
Pyrophacus, 413, 523
  P. horologium, 397, 522–523
  P. steinii, 397, 522–523

Roscoffia, 412, 546
R. capitata, 546–547

Schuettiella, 412, 512
S. mitra, 511–512
Scrippsiella, 413, 446, 457, 524–525
  S. arenicola, 527
  S. caponii, 527
  S. gregaria, 527
  S. minima, 527
  S. precaria, 525–526
  S. subsalsa, 526

S. sweeneyae, 525
S. trochoidea, 398, 525–527
Sinophysis, 442
  S. microcephala, 442–443
Spatulodinium, 464
  S. pseudonotiluca, 464–465
Spiniferodinium, 549
  S. galeiforme, 549
Spiraulax, 412, 512
  S. jolliffei, 512
  S. kofoidii, 511–512

Spirodinium
  S. glaucum, 455
Steiniella, 506
  S. mitra, 512
Symbiodinium, 405, 460, 462
  S. microadriaticum, 460–461

Thecadinium, 412, 546–547
  T. kofoidii, 546–548
  T. petasatum, 547
Thoracosphaera, 547
  T. heimii, 547, 549
Torodinium, 456
  T. robustum, 456
  T. teredo, 454, 456
Triadinium, 501
  T. ostenfeldii, 499
Triposolenia, 428
  T. truncata, 427–428

Warnowia, 394, 400, 460
  W. polyphemus, 454, 460
  W. fusus, 460
Woloszynskia, 446, 467
  W. reticulata, 467

Zygabikodinium, 530
Z. lenticulatum, 530

REFERENCES

Abé, T. H. 1936a. Report of the biological survey of Mutsu Bay. 29. Notes of the protozoan
  fauna of Mutsu Bay. II. Genus: Peridinium: Subgenus: Archaeperidinium. Science Reports
  of the Tohoku University, 4th Series, Biology 10:639–686.
Abé, T. H. 1936b. Report of the biological survey of Mutsu Bay. 30. Notes on the protozoan
  fauna of Mutsu Bay. III. Subgenus Protoperidinium: Genus Peridinium. Science Reports of
  the Tohoku University, 4th Series, Bioloby 11:19–48.
Abé, T. H. 1941. Studies on the protozoan fauna of Shimoda Bay. I. The Diplopsalis group.
  Records of Oceanographic Works in Japan 12:121–144.

Abé, T. H. 1966. The armoured Dinoflagellata: I. Podolampidae. *Publications of the Seto Marine Biology Laboratory* 14:129-154.

Abé, T. H. 1967a. The armoured Dinoflagellata: II. Prorocentridae and Dinophysidae (A). *Publications of the Seto Marine Biology Laboratory* 14:1369-1389.

Abé, T. H. 1967b. The armoured Dinoflagellata: II. Prorocentridae and Dinophysidae (B). *Dinophysis* and its allied genera. *Publications of the Seto Marine Biology Laboratory* 15:37-78.

Abé, T. H. 1967c. The armoured Dinoflagellata: II. Prorocentridae and Dinophysidae (C). *Ornithocercus, Histioneis, Amphisolenia* and others. *Publications of the Seto Marine Biology Laboratory* 15:79-116.

Abé, T. H. 1981. Studies on the family Peridinidae, an unfinished monograph of the armoured Dinoflagellata. *In* "Publications of the Seto Marine Biological Laboratory Special Publication Series," pp. 409.

Adamich, M., & Sweeney, B. M. 1976. The preparation and characterization of *Gonyaulax* spheroplasts. *Planta* 130:1-6.

Akselman, R., & Keupp, H. 1990. Recent obliquipithonelloid calcareous cysts of *Scrippsiella patagonica* sp. nov. (Peridiniaceae, Dinophyceae) from plankton of the Golfo San Jorge (Patagonia, Argentina). *Marine Micropalaentology* 16:169-179.

Andreis, C., Ciapi, M. D., & Rodondi, G. 1982. The thecal surface of some Dinophyceae: A comparative SEM approach. *Botanica Marina* 25:225-236.

Balech, E. 1951. Deuxieme contribution a la connaissance des Peridinium. *Hydrobiologia* 3:305-330.

Balech, E. 1956. Etude des dinoflagelles du sable de Roscoff. *Revue Algologique* 2:29-52.

Balech, E. 1959. Two new genera of dinoflagellates from California. *Biological Bulletin* 116:195-203.

Balech, E. 1963. Dos dinoflagelados de una laguna salobre de la Argentina. *Notas del Museo de La Plata Zoologia* 20:111-123.

Balech, E. 1964a. El genero *Cladopyxis*. *Hidrobiologia* 1:27-39.

Balech, E. 1964b. El plancton de Mar del Plata durante el periodo 1961-62 (Buenos Aires, Argentina). *Instituto de Biologia Marina, Mar del Plata, Argentine Republic, Boletin* 4:1-49.

Balech, E. 1964c. Tercera contribucion al conocimiento del genero *Peridinium*. *Museo Argentino de ciencias naturales "Bernadino Rivadavia" e Instituto nacional de investigacíon de las ciencias naturales, Revista, Hidrobiología* 1:179-195.

Balech, E. 1967a. Dinoflagelados nuevos o interesantes del Golfo de Mexico y Caribe. *Museo Argentino de ciencias naturales "Bernadino Rivadavia" e Instituto nacional de investigacíon de las ciencias naturales, Revista, Hidrobiología* 2:77-126.

Balech, E. 1967b. *Palaeophalacroma* Schiller, otro miembro de la familia Cladopyxidae (Dinoflagellata). *Neotropica* 13:105-112.

Balech, E. 1973a. Cuarta contribucion al conocimiento del genero *Protoperidinium*. *Museo Argentino de ciencias naturales "Bernadino Rivadavia" e Instituto nacional de investigacíon de las ciencias naturales, Revista, Hidrobiología* 3:347-381.

Balech, E. 1973b. Segunda contribucion al conocimiento del microplancton del Mar de Bellingshausen. *In* "Contrib. Instituto Antartico Argentino," No. 107, pp. 63.

Balech, E. 1974. El genero *Protoperidinium* Bergh, 1881 (*Peridinium* Ehrenberg, 1931, partim). *Museo Argentino de ciencias naturales "Bernadino Rivadavia" e Instituto nacional de investigacíon de las ciencias naturales, Revista, Hidrobiología* 4:1-79.

Balech, E. 1976a. Clave illustrada de dinoflagellados antarticos. *Publications of the Institute Antartic Argentina Buenos Aires* 11:1-99.

Balech, E. 1976b. Notas sobre le genero *Dinophysis* (Dinoflagellata). *Physis A* 35:183-193.

Balech, E. 1976c. Some Norwegian *Dinophysis* species (Dinoflagellata). *Sarsia* 61:75-94.

Balech, E. 1976d. Sur quelques *Protoperidinium* (Dinoflagellata) de Golfe du Lion. *Vie Milieu* 26:27-46.

Balech, E. 1977a. *Cachonina niei* Dinoflagellata and its variations. *Physis A* 36:59–64.

Balech, E. 1977b. Cuatro especies de *Gonyaulax sensu lato*, y consideraciones sobre el genero (Dinoflagellata). *Museo Argentino de ciencias naturales "Bernadino Rivadavia" e Instituto nacional de investigacíon de las ciencias naturales, Revista, Hidrobiología* 5:115–136.

Balech, E. 1979a. El genero *Goniodoma* Stein (Dinoflagellata). *Lilloa* 35:97–109.

Balech, E. 1979b. El genero *Pyrophacus* Stein (Dinoflagellata). *Physis* 38:27–38.

Balech, E. 1979c. Tres dinoflagelados nuevos o interesantes de aguas Brasileñas. *Boletim do Instituto Oceanografico* 28:55–64.

Balech, E. 1980. On thecal morphology of dinoflagellates with special emphasis on cingular and sulcal plates. *Anales del Centro de Ciencias del Mar y Limnologia Universidad Nacional Autonoma de Mexico* 7:57–68.

Balech, E. 1985a. The genus *Alexandrium* or *Gonyaulax* of the tamarensis group. *In* "Toxic Dinoflagellates" (D. M. Anderson, A. W. White, & D. G. Baden, eds.), pp. 33–38. Elsevier, New York.

Balech, E. 1985b. A revision of *Pyrodinium bahamense*, Dinoflagellata. *Review of Palaeobotany and Palynology* 45:17–34.

Balech, E. 1988a. Los Dinoflagelados del Atlantico Sudoccidental. *In "Publicaciones Especiales del Institutto Espanol de Oceanografia,"* No. 1, pp. 310. Madrid, Spain.

Balech, E. 1988b. Una especie nueva del genero *Fragilidium* (Dinoflagellata) de la bahia de Chamela, Jalisco, Mexico. *Anales del Instituto Biologica UNAM, Series Zoologica* 58:479–486.

Balech, E. 1989. Redescription of *Alexandrium minutum* Halim (Dinophyceae) type species of the genus *Alexandrium*. *Phycologia* 28:206–211.

Balech, E. 1990a. Four new dinoflagellates. *Helgoländer Meeresuntersuchungen* 44:387–396.

Balech, E. 1990b. A short diagnostic description of *Alexandrium*. *In* "Toxic Marine Phytoplankton" (E. Graneli, B. Sundstrom, L. Edler, & D. M. Anderson, eds.), pp. 77. Elsevier, New York.

Balech, E. 1994. Three new species of the genus *Alexandrium* (Dinoflagellata). *Transactions of the American Microscopical Society* 113:216–220.

Balech, E. 1995. "The Genus *Alexandrium* Halim (Dinoflagellata)." Sherkin Island Press (in press).

Balech, E., & Tangen, K. 1985. Morphology and taxonomy of toxic species in the tamarensis group (Dinophyceae) *Alexandrium excavatum* (Braarud) comb. nov. and *Alexandrium ostenfeldii* (Paulsen) comb. nov. *Sarsia* 70:333–343.

Banaszak, A. T., Inglesias-Prieto, R., & Trench, R. K. 1993. *Scrippsiella velellae* sp. nov. (Peridiniales) and *Gleodinium viscum* sp. nov. (Phytodiniales), dinoflagellate symbionts of two hydrozoans (Cnidaria). *Journal of Phycology* 29:517–528.

Bardouil, M., Berland, B., Grzebyk, D., & Lassus, P. 1991. L'existence de kystes chez les Dinophysales. *Comptes rendus des séances de L'Academie des sciences Serie III* 312:663–669.

Beam, C. A., & Himes, M. 1987. Electrophoretic characterization of members of the *Crypthecodinium cohnii* (Dinophyceae) species complex. *Journal of Protozoology* 34:204–217.

Below, R. 1987a. Evolution and systematics of dinoflagellate cysts of the order Peridiniales. I. Basic general principles and the subfamily Rhaetogonyaulacoidae, family Peridiniaceae. *Palaeontographica* Abteilung B 205:1–164.

Below, R. 1987b. Evolution and systematics of dinoflagellate cysts of the order Peridiniales. II. Cladophyxiaceae and Valvaeodiniaceae. *Palaeontographica* Abteilung B 206:1–115.

Berland, B. 1982. Image reversal in micorscopy. *Transactions of the American Microscopical Society* 101:174–180.

Besada, E. G., Loeblich, A. R. Jr., & Loeblich, A. R., III. 1982. Observations on tropical, benthic dinoflagellates from ciguatera-endemic areas: *Coolia, Gambierdiscus,* and *Ostreopsis. Bulletin of Marine Science* 32: 723–735.

Biecheler, B. 1952. Recherches sur les peridinens. *Bulletin Biologique de la France et de la Belgique (Suppl.)* 36:1–149.

Blank, R. J., & Trench, R. K. 1986. Nomenclature of endosymbiotic dinoflagellates. *Taxon* 35:286–294.

Boalch, G. T. 1969. The dinoflagellate genus *Ptychodiscus* Stein. *Journal of the Marine Biological Association of the United Kingdom* 49:781–784.

Boltovskoy, A. 1975. Estructura y esteroultrastructura tecal de Dinoflagelados. II. *Peridinium cinctum* (Müller) Ehrenberg. *Physis Buenos Aires* 34:73–84.

Borgese, M. B. 1987. Two armored dinoflagellates from the southwestern Atlantic Ocean: A new species of *Protoperidinium* and a first record and redescription for *Gonyaulax alaskensis* Kofoid. *Journal of Protozoology* 34:332–337.

Bouck, G. B., & Sweeney, B. M. 1966. The fine structure and ontogeny of trichocysts in marine dinoflagellates. *Protoplasma* 61:205–223.

Buchanan, R. J. 1968. Studies at Oyster Bay in Jamaica, West Indies. IV. Observations on the morphology and asexual cycle of *Pyrodinium bahamense* Plate. *Journal of Phycology* 4:272–277.

Bujak, J. P., & Williams, G. L. 1981. The evolution of dinoflagellates. *Canadian Journal of Botany* 59:2077– 2087.

Burkholder, J. M., Noga, E. J., Hobbs, C. W., Glasgow, H. B., Jr., & Smith, S. A. 1992. New "phantom" dinoflagellate is the causative agent of major estuarine fish kills. *Nature* 358:407–410 (see Errata in *Nature* 360:768).

Burkholder, J. M., Glasgow, H. B., Jr., & Steidinger, K. A. 1995. Stage transformations in the complex life cycle of an ichthyotoxic "ambush predator" dinoflagellate. *In* "Harmful Marine Algal Blooms" (P. Lassus, G. Arzul, E. Erard, P. Gentien, & C. Maraillou, eds.), pp. 567–572. Lavoisier, Cachan Cedex, France.

Burrells, W. 1977. "Microscope Technique. A Comprehensive Handbook for General and Applied Microscopy," pp. 574. Halstead Press.

Bursa, A. S. 1959. The genus *Prorocentrum* Ehrenberg. Morphodynamics, protoplasmic structures and taxonomy. *Canadian Journal of Botany* 37:1–30.

Bursa, A. 1969. *Actiniscus canadensis* n. sp., *A. pentasterias* Ehrenberg v. *arcticus* n. var., *Pseudoactiniscus apentasterias* n. gen., n. sp., marine relicts in Canadian arctic lakes. *Journal of Protozoology* 16:411–418.

Cachon, J., & Cachon, M. 1967a. Contribution a l'etude des Noctilucidae Saville-Kent. I. Les Kofoidininae Cachon J. & M. evolution, morphologique et systematique. *Protistologica* 3:427–444.

Cachon, J., & Cachon, M. 1967b. *Cymbodinium elegans* nov. gen. nov. sp. péridinien Noctilucidae Saville-Kent. *Protistologica* 3:313–318.

Cachon, J., & Cachon, M. 1968. *Filodinium hovassei* nov. gen., nov. sp. Péridinien phoretique d'Appendiculaires. *Protistologica* 4:15–18.

Cachon, J., & Cachon, M. 1969. Contribution a l'étude des Noctilucidae Saville-Kent. Évolution, morphologique, cytologie, systematique. *Protistologica* 5:11–33.

Cachon, J., & Cachon, M. 1987. Parasitic dinoflagellates. *In* "The Biology of Dinoflagellates" (F. J. R. Taylor, ed.), pp. 571–610. Blackwell, Oxford.

Cachon, J., & Cachon-Enjumet, M. 1966. *Pomatodinium impatiens* nov. gen. nov. sp. Peridinien Noctilucidae Kent. *Protistologica* 2:23–30.

Campbell, P. H. 1973. Studies on brackish water phytoplankton. University of North Carolina Sea Grant Publication UNC SG-73-07, pp. 406.

Carbonell-Moore, M. C. 1991. *Lissodinium* Matzenauer, emend., based upon the rediscovery of *L. schilleri* Matz., another member of the family Podolampadaceae Lindemann (Dinophyceae). *Botanica Marina* 34:327–340.

Carbonell-Moore, M. C. 1992. *Blepharocysta hermosillai*, sp. nov. a new member of the family Podolampadaceae Lindemann (Dinohyceae). *Botanica Marina* 35:273–281.

Carbonell-Moore, M. C. 1993. Further observations on the genus *Lissodinium* Matz., emend. Carbonell-Moore (Dinophyceae), with descriptions of seventeen new species. *Botanica Marina* 36:561–587.

Carty, S., & Cox, E. R. 1986. *Kansodinium* gen. nov. and *Durinskia* gen. nov.:Two genera of freshwater dinoflagellates (Pyrrhophyta) *Phycologia* 25:197–204.

Chatton, E. 1933. *Pheopolykrikos beauchampii* nov. gen. nov. sp., dinoflagelle polydinide autotrophe, dans l'etang de Thau. *Bulletin de la Societe Zoologique de France* 58:251–254.

Chatton, E. 1952. Classe des Dinoflagelles ou Peridiniens. *In* "Traite de Zoologie, Anatomie, Systematique, Biologie" (P.-P. Grasse, ed.), Vol. 1, fasc. 1, Protozoaires, Generalites, Flagelles, pp. 309–406. Masson et Cie, Paris.

Coats, D. W., Tyler, M. A., & Anderson, D. A. 1984. Sexual processes in the life cycle of *Gyrodinium uncatenum* (Dinophyceae): A morphogenetic overview. *Journal of Phycology* 20:351–361.

Cox, E. R., & Arnott, H. J. 1971. The ultrastructure of the theca of the marine dinoflagellate, *Ensiculifera loeblichii* sp. nov. *In* "Contributions in Phycology" (B. C. Parker & R. M. Brown Jr., eds.), pp. 121–136. Allen Press, Lawrence, KS.

Crawford, R. M., & J. D. Dodge. 1971. The dinoflagellate genus *Woloszynskia*. II. The fine structure of *W. coronata*. *Nova Hedwigia* XXII:699–719.

Crawford, R. M., Dodge, J. D., & Happey, C. M. 1970. The dinoflagellate genus *Woloszynskia*. I. Fine structure and ecology of *W. tenuissimum* from Abbot's pool, Somerset. *Nova Hedwigia* XIX:825–840.

Dale, B. 1977a. Cysts of the toxic red-tide dinoflagellate *Gonyaulax excavata* (Braarud) Balech from Oslofjorden, Norway. *Sarsia* 63:29–34.

Dale, B. 1977b. New observations on *Peridinium faeroense* Paulsen (1905), and classification of small orthoperidinioid dinoflagellates. *British Phycological Journal* 12:241–253.

Dale, B. 1978. Acitarchous cysts of *Peridinium faroense* Paulsen: Implications for dinoflagellate systematics. *Palynology* 2:187–193.

Dale, B. 1979. Collection, preparation and identification of dinoflagellate resting cysts. *In* "Toxic Dinoflagellate Blooms" (D. L. Taylor & H. H. Seliger, eds.), pp. 443–452. Elsevier, New York.

Dale, B. 1983. Dinoflagellate resting cysts: "benthic" plankton. *In* "Survival Strategies of the Algae," (G. A. Fryxell, ed.), pp. 69–136. Cambridge University Press, London.

Dale, B. 1986. Life cycle strategies of oceanic dinoflagellates. *In* "Pelagic Biogeography," (A. C. Pierrot-Bults, S. van der Spoel, B. J. Zahuranec, & R. K. Johnson, eds.), pp. 65–72. UNESCO Technical Papers in Marine Science, Paris.

Dale, B., Montresor, M., Zingone, A., & Zonneveld, K. 1993. The cyst-motile stage relationships of the dinoflagellates *Diplopelta symmetrica* and *Diplopsalopsis latipeltata*. *European Journal of Phycology* 28:129–137.

Delgado, M., & Fortuño, J. M. 1991. Atlas de fitoplancton del Mar Mediterraneo. *Scientia Marina* 55 (Suppl. 1):1–133.

Dodge, J. D. 1966. The dinophyceae. *In* "The Chromosomes of the Algae," (M. B. E. Godward, ed.), pp. 96–115. Edward Arnold, London.

Dodge, J. D. 1973. "The Fine Structure of Algal Cells," pp. 261. Academic Press, London.

Dodge, J. D. 1975. The Prorocentrales (Dinophyceae). II. Revision of the taxonomy within the genus *Prorocentrum*. *Botanical Journal of the Linnean Society* 71:103–125.

Dodge, J. D. 1981. Three new generic names in the Dinophyceae: *Herdmania*, *Sclerodinium*, and *Triadinium* to replace *Heteraulacus* and *Goniodoma*. *British Phycological Journal* 16:273–280.

Dodge, J. D. 1982. "Marine Dinoflagellates of the British Isles," pp. 303. Her Majesty's Stationery Office, London.

Dodge, J. D. 1983. Ornamentation of thecal plates in *Protoperidinium* (Dinophyceae) as seen by scanning electron microscopy. *Journal of Plankton Research* 5:119–127.

Dodge, J. D. 1985. "Atlas of Dinoflagellates: A Scanning Electron Microscope Study," pp. 119. Farrand Press, London.

Dodge, J. D. 1987. A hypothecal pore in some species of *Protoperidinium*, Dinophyceae. *British Phycological Journal* 22:335–338.

Dodge, J. D. 1989. Some revisions of the family Gonyaulacaceae (Dinophyceae) based on a scanning electron microscope study. *Botanica Marina* 32:275–298.

Dodge, J. D., & Crawford R. M. 1968. Fine structure of the dinoflagellate *Amphidinium carteri* Hulburt. *Protistologica* 4:231–242.

Dodge, J. D., & Crawford R. M. 1970. A review of the theca fine structure in the Dinophyceae. *Journal of the Linnean Society of Botany* 63:53–67.

Dodge, J. D., & Crawford, R. M. 1971. Fine structure of the dinoflagellate *Oxyrrhis marina*. I. The general structure of the cell. *Protistologica* 7:295–304.

Dodge, J. D., & Greuet, C. 1987. Dinoflagellate ultrastructure and complex organelles. *In* "The Biology of Dinoflagellates" (F. J. R. Taylor, ed.), Botanical Monographs vol. 21, pp. 92–142. Blackwell, Oxford.

Dodge, J. D., & Hermes, H. 1981. A revision of the *Diplopsalis* group of dinoflagellates (Dinophyceae) based on material from the British Isles. *Botanical Journal of the Linnean Society* 83:15–26.

Dodge, J. D., & Lewis, J. 1986. A further SEM study of armoured sand-dwelling marine dinoflagellates. *Protistologica* 22:221–230.

Dodge, J. D., & Saunders R. D. 1985a. A partial revision of the genus *Oxytoxum* (Dinophyceae) with the aid of scanning electron microscopy. *Botanica Marina* 28:99–122.

Dodge, J. D., & Saunders, R. D. 1985b. An SEM study of *Amphidoma nucula* (Dinophyceae) and description of the thecal plates in *A. caudata*. *Archiv fur Protistenkunde* 129:89–99.

Dodge, J. D., & Toriumi, S. 1993. A taxonomic revision of the *Diplopsalis* group (Dinophyceae). *Botanica Marina* 36:137–147.

Eaton, G. L. 1980. Nomenclature and homology in peridinialean dinoflagellate plate patterns. *Palaeontology* 23:667–688.

Elbrächter, M., & Drebes, G. 1978. Life cycles, phylogeny, and taxonomy of *Dissodinium* and *Pyrocystis* (Dinophyta). *Helgolander wissenschaftliche Meeresuntersuchungen* 31:347–366.

Elbrächter, M. 1979. On the taxonomy of unarmored dinophytes (Dinophyta) from the northwest African upwelling region. *Meteor Forschungsergebnisse Reihe* 30:1–22.

Elbrächter, M. 1993. *Kolkwitziella* Lindemann 1919 and *Preperidinium* Mangin 1913: Correct genera names in the *Diplopsalis*-group (Dinophyceae). *Nova Hedwigia* 56:173– 178.

Elbrächter, M., Hemleben, C., & Spindler, M. 1987. On the taxonomy of the lunate *Pyrocystis* species (Dinophyta). *Botanica Marina* 30:233–241.

Evitt, W. R. 1985. "Sporopollenin Dinoflagellate Cysts: Their Morphology and Interpretation," pp. 333. American Association of Stratigraphic Palynology Association.

Faust, M. A. 1974. Micromorphology of a small dinoflagellate *Prorocentrum mariae-lebouriae* (Parke & Ballantine) comb. nov. *Journal of Phycology* 10:315–332.

Faust, M. A. 1990a. Cysts of *Prorocentrum marinum* (Dinophyceae) in floating detritus at Twin Cays, Belize mangrove habitats. *In* "Toxic Marine Phytoplankton" (E. Graneli, B. Sundstrom, L. Edler, & D. M. Anderson, eds.), pp. 138–143. Elsevier, New York.

Faust, M. A. 1990b. Morphologic details of six benthic species of *Prorocentrum* (Pyrrhophyta) from a mangrove island, Twin Cays, Belize, including two new species. *Journal of Phycology* 26:548–558.

Faust, M. A. 1991. Morphology of ciguatera-causing *Prorocentrum lima* (Pyrrophyta) from widely differing sites. *Journal of Phycology* 27:642–648.

Faust, M. A. 1992. Observations on the morphology and sexual reproduction of *Coolia monotis* (Dinophyceae). *Journal of Phycology* 28:94–104.

Faust, M. A. 1993a. *Prorocentrum belizeanum*, *Prorocentrum elegans* and *Prorocentrum caribbaeum*, three new benthic species (Dinophyceae) from a mangrove island Twin Cays, Belize. *Journal of Phycology* 29:100–107.

Faust, M. A. 1993b. Surface morphology of the marine dinoflagellate *Sinophysis microcephalus* (Dinophyceae) from a mangrove island, Twin Cays, Belize. *Journal of Phycology* 29:355–363.

Faust, M. A. 1993c. Three new benthic species of *Prorocentrum* (Dinophyceae) from Twin Cays, Belize: *P. maculosum* sp. nov., *P. foraminosum* sp. nov. and *P. formosum* sp. nov. *Phycologia* 32:410–418.

Faust, M. A. 1994. Three new benthic species of *Prorocentrum* (Dinophyceae) from Carrie B⎯w Cay, Belize; *P. sabulosum* sp. nov., *P. sculptile* sp. nov., and *P. arenarium* sp. nov. *Journal of Phycology* 30:755–763.

Faust, M. A., & Balech, E. 1993. A further SEM study of marine benthic dinoflagellates from a mangrove island, Twin Cays, Belize including *Plagiodinium belizeanum* gen. et sp. nov. *Journal of Phycology* 29:826–832.

Fensome, R. A., Taylor, F. J. R., Norris, G., Sarjeant, W. A. S., Wharton, D. I., & Williams, G. L. 1993. "A Classification of Living and Fossil Dinoflagellates," pp. 351. Sheridan Press, Hanover, PA.

Fott, B. 1957. Taxonomie drobnohledne flory nasich vod. *Preslia* 29:278–319.

Francis, D. 1967. On the eyespot of the dinoflagellate *Nematodinium*. *Journal of Experimental Biology* 47:495–501.

Fritz, L., & Triemer, R. E. 1985. A rapid simple technique utilizing Calcofluor White M2R for the visualization of dinoflagellate thecal plates. *Journal of Phycology* 21:662–664.

Freudenthal, H. D. 1962. *Symbiodinium* gen. nov. and *Symbiodinium microadriaticum* sp. nov., a zooxanthella: taxonomy, life cycle, and morphology. *Journal of Protozoology* 9:45–52.

Fukuyo, Y. 1981a. *Protogonyaulax* in the coast of Japan. *Akashiwo Kenkyukai Guide Book* 3:1–72.

Fukuyo, Y. 1981b. Taxonomical study on benthic dinoflagellates collected in coral reefs. *Bulletin of the Japanese Society of Scientific Fisheries* 47:967–978.

Fukuyo, Y., Takano, H., Chihara, M., & Matsuoka, K. (eds.) 1990. "Red Tide Organisms in Japan—An Illustrated Taxonomic Guide," pp. 407. Uchida Rokakuho, Tokyo.

Gaarder, K. R. 1954. Dinoflagellatae from the "Michael Sars" North Atlantic deep-sea expedition 1910. *Scientific Research Report of the "Michael Sars" North Atlantic Deep-Sea Expedition 1910* 2:1–62.

Gaines, G., & Elbrächter, M. 1987. Heterotrophic nutrition. *In* "The Biology of Dinoflagellates" (F. J. R. Taylor, ed.), Botanical Monographs No. 21, pp. 224–268. Blackwell, Oxford.

Gao, X., & Dodge, J. D. 1991. The taxonomy and ultrastructure of a marine dinoflagellate, *Scrippsiella minima* sp. nov. *British Phycological Journal* 26:21–31.

Gao, X., Dodge, J. D., & Lewis, J. 1989a. An ultrastructural study of planozygotes and encystment of a marine dinoflagellate *Scrippsiella* sp. *British Phycological Journal* 24:153–165.

Gao, X., Dodge, J. D., & Lewis, J. 1989b. Gamete fusion and mating in the marine dinoflagellate *Scrippsiella* sp. *Phycologia* 28:342–351.

Gardiner, W. E., Rushing, A. E., & Dawes, C. J. 1989. Ultrastructural observations of *Gyrodinium estuariale* (Dinophyceae). *Journal of Phycology* 25:178–183.

Gocht, H. 1981. ≫Direkter≪ nachweis der plattenuberlappung bei *Hystrichogonyaulax clado-phora* (Dinoflagellata, Oberjura). *Neues Jahrbuch fur Geologie und Palaeontologie Monats-hefte* 3:149–156.

Gocht, H., & Netzel, H. 1974. Rasterelecktronenmikroskopische Untersuchungen am Panzer von *Peridinium* (Dinoflagellata). *Archiv fur Protistenkunde* 116:381–410.

Graham, H. W. 1942. Studies on the morphology, taxonomy, and ecology of the Peridiniales. *Carnegie Institution of Washington Publication* 542.

Graham, H. W., & Bronikovsky, N. 1944. The genus *Ceratium* in the Pacific and North Atlantic Oceans. *Carnegie Institution of Washington Publication* 565 5:1–209.

Guillard, R. R. L., & Keller, M. D. 1984. Culturing dinoflagellates. *In* "Dinoflagellates" (D. L. Spector, ed.), pp. 391–442. Academic Press, Orlando.

Halim, Y. 1960a. *Alexandrium minutum,* n. gen. n. sp. dinoflagelle provocant des eaux rouges. *Vie Milieu* 11:102–105.

Halim, Y. 1960b. Étude quantitative et qualitative du cycle ecologique des dinoflagelles dans les eaux de Villefranche-sur-mer. *Annales de l'Institut Oceanographique* 38:1–232.

Hallegraeff, G. M. 1991. "Aquaculturists' Guide to Harmful Australian Microalgae." CSIRO Division of Fisheries, Tasmania, Australia.

Hallegraeff, G. M., & Lucas, I. A. N. 1988. The marine dinoflagellate genus *Dinophysis* (Dinophyceae): Photosynthetic, neritic and non-photosynthetic, oceanic species. *Phycologia* 27:25–42.

Hallegraeff, G. M., Steffensen, D. A., & Wetherbee, R. 1988. Three estuarine Australian dinoflagellates that can produce paralytic shellfish toxins. *Journal of Plankton Research* 10:533–541.

Hansen, G. 1989. Ultrastructure and morphogenesis of scales in *Katodinium rotundatum* (Lohmann) Loeblich (Dinophyceae). *Phycologia* 28:385–394.

Hansen, G. 1993. Light and electron microscopical observations of the dinoflagellate *Actiniscus pentasterias* (Dinophyceae). *Journal of Phycology* 29:486–499.

Harland, R. 1981. Cysts of the colonial dinoflagellate *Polykrikos schwartzii* Bütschli 1873, (Gymnodiniales), from recent sediments, Firth of Forth, Scotland. *Palynology* 5:65–79.

Hasle, G. R., & Nordli, E. 1951. Form variation in *Ceratium fusus* and *tripos* populations in cultures and from the sea. *Avhandlinger Utgitt av det Norske Videnskaps Akademie i Oslo Matematisk - Naturvidenskapelig Klasse* 4:1–25.

Herdman, E. C. 1922. Notes on dinoflagellates and other organisms causing discoloration of the sand at Port Erin. II. *Transactions of the Liverpool Biological Society* 36:15–30.

Honsell, G., & Cabrini, M. 1991. *Scrippsiella spinifera* sp. nov. (Pyrrhophyta): A new dinoflagellate from the northern Adriatic Sea. *Botanica Marina* 34:167–175.

Honsell, G., & Talarico, L. 1985. The importance of flagellar arrangement and insertion in the interpretation of the theca of *Prorocentrum* (Dinophyceae). *Botanica Marina* 28:15–21.

Horiguchi, T., & Chihara, M. 1987. *Spiniferodinium galeiforme,* new genus new species of benthic dinoflagellates, Phytodiniales, Pyrrophyta, from Japan. *Phycologia* 26:478–487.

Horiguchi, T., & Pienaar, R. N. 1988. Ultrastructure of a new sand-dwelling dinoflagellate, *Scrippsiella arenicola,* new species. *Journal of Phycology* 24:426–438.

Horiguchi, T., & Pienaar, R. N. 1991. Ultrastructure of a marine dinoflagellate, *Peridinium quinquecorne* Abé (Peridiniales) from South Africa with particular reference to its chrysophyte endosymbiont. *Botanica Marina* 34:123–131.

Hulburt, E. M. 1957. The taxonomy of unarmored Dinophyceae of shallow embayments on Cape Cod, Massachusetts. *Biological Bulletin* 112:196–219.

Hulburt, E. M. 1965. Three closely allied dinoflagellates. *Journal of Phycology* 1:95–96.

Indelicato, S. R., & Loeblich, A. R., III. 1985. A description of the marine dinoflagellate, *Scrippsiella tinctoria* sp. nov. *Japanese Journal of Phycology* 33:127–134.

Indelicato, S. R., & Loeblich, A. R., III. 1986. A revision of the marine peridinioid genera (Pyyrhophyta) utilizing hypothecal-cingular plate relationships as a taxonomic guide. *Japanese Journal of Phycology* 34:153–162.

Inouye, I., & Pinenaar, R. N. 1983. Observations on the life cycle and microanatomy of *Thoracosphaera heimii* (Dinophyceae) with special reference to its systematic position. *South African Journal of Botany* 2:63–75.

Javornicky, P. 1962. Two scarcely known genera of the class Dinophyceae: *Bernardinium* Chodat and *Crypthecodinium* Biecheler. *Preslia* 34:98–113.

Jianfan, W., & Jingyan, L. 1992. The nucleoskeleton of special dinoflagellate, *Oxyrrhis marina.* *Zoological Research* 13:89–94.

Jørgensen, E. 1911. Die Ceratien. Eine kurze monographie der gattung *Ceratium* Schrank. *Internationale Revue der Gesamten Hydrobiologie und Hydrographie* 4:1–124.

Jørgensen, E. 1920. Mediterranean ceratia. *Report of the Danish Oceanographic Expedition in the Mediterranean,* 2 (Biology) 1:1–110.

Kimball, J. F., Jr., & Wood, E. J. F. 1965. A dinoflagellate with characters of *Gymnodinium* and *Gyrodinium*. *Journal of Protozoology* 12:577–580.

Kimor, B. 1981. The role of phagotrophic dinoflagellates in marine ecosystems. *Kieler Meeresforche Sonderh* 5:164–173.

Kita, T., & Fukuyo, Y. 1988. Description of the gonyaulacoid dinoflagellate *Alexandrium hiranoi* sp. nov. inhabiting tidepools on Japanese Pacific coast. *Bulletin of the Plankton Society of Japan* 35:1–7.

Klut, M. E., Bisalputra, T., & Antia, N. J. 1985. Some cytochemical studies on the cell surface of *Amphidinium carterae* (Dinophyceae). *Protoplasma* 129:93–99.

Klut, M. E., Bisalputra, T., & Antia, N. J. 1988. The use of fluorochromes in the cytochemical characterization of some phytoflagellates. *Histochemical Journal* 20:35–40.

Kofoid, C. A. 1906a. Dinoflagellata of the San Diego region. I: On *Heterodinium* a new genus of the Peridinidae. *University of California Publications in Zoology* 2:341–368.

Kofoid, C. A. 1906b. Dinoflagellata of the San Diego region. II. On *Triposolenia* a new genus of the Dinophysidae. *University of California Publications in Zoology* 3:93–116.

Kofoid, C. A. 1907a. Dinoflagellata of the San Diego region. III. Descriptions of new species. *University of California Publications in Zoology* 3:299–340.

Kofoid, C. A. 1907b. Reports of the scientific results of the expedition to the eastern tropical Pacific in charge of Alexander Agassiz, 9. New species of dinoflagellates. *Bulletin of the Museum of Comparative Zoology at Harvard College* 50:163–207.

Kofoid, C. A. 1909. On *Peridinium steinii* Jørgensen, with a note on the nomenclature of the skeleton of Peridinidae. *Archiv fur Protistenkunde* 16:25–47.

Kofoid, C. A. 1910. A revision of the genus *Ceratocorys* based on skeletal morphology. *University of California Publications in Zoology* 6:177–187.

Kofoid, C. A. 1911a. Dinoflagellata of the San Diego region. IV. The genus *Gonyaulax*, with notes on its skeletal morphology and a discussion of its generic and specific characters. *University of California Publications in Zoology* 8:187–286.

Kofoid, C. A. 1911b. On the skeletal morphology of *Gonyaulax catenata*. *University of California Publications in Zoology* 8:287–294.

Kofoid, C. A. 1911c. Dinoflagellata of the San Diego region, V: *Spiraulax*, a new genus of the Peridinida. *University of California Publications in Zoology* 8:295–300.

Kofoid, C. A. 1926. On *Oxyphysis oxytoxoides* gen. nov., sp. nov. A dinophysoid dinoflagellate convergent toward the peridinioid type. *University of California Publications in Zoology* 28:203–216.

Kofoid, C. A., & Adamson, A. M. 1933. The Dinoflagellata: The family Heterodiniidae of the Peridinoidae. *Memoirs of the Museum of Comparative Zoology, Harvard* 54:1–136.

Kofoid, C. A., & Michener, J. R. 1911. Reports of the scientific results of the expedition to the eastern tropical Pacific in charge of Alexander Agassiz, 22. New genera and species of dinoflagellates. *Bulletin of the Museum of Comparative Zoology at Harvard College* 54:267–302.

Kofoid, C. A., & Skogsberg, T. 1928. The Dinoflagellata: The Dinophysoidae. *Memoirs of the Museum of Comparative Zoology, Harvard* 51:13–760.

Kofoid, C. A., & Swezy, O. 1921. The free-living unarmored dinoflagellates. *Memoirs of the University of California* 5:1–562.

Kubai, D. F., & Ris, H. 1969. Division in the dinoflagellate *Gyrodinium cohnii* (Schiller). A new type of nuclear reproduction. *Journal of Cell Biology* 40:508–528.

Lackey, J. B. 1956. Known geographic range of *Gymnodinium brevis* Davis. *Quarterly Journal of the Florida Academy of Sciences* 19:71.

Landsberg, J. H., Steidinger, K. A., Blakesley, B. A., & Zondervan, R. L. 1994. Scanning electron microscope study of dinospores of *Amyloodinium* cf. *ocellatum*, a pathogenic dinoflagellate parasite of marine fish, and comments on its relationship to the Peridiniales. *Diseases of Aquatic Organisms* 20:23–32.

Landsberg, J. H., Steidinger, K. A., & Blakesley, B. A. 1995. Fish-killing dinoflagellates in a tropical marine aquarium. *In* "Harmful Marine Algal Blooms" (P. Lassus, G. Arzul, E. Erard, P. Gentien, & C. Marcaillou, eds.), pp. 65–70. Lavoisier Cahan Cedex, France.

Larsen, J. 1985. Algal studies of the Danish Wadden Sea. II. A taxonomic study of psammobious dinoflagellates. *Opera Botanica* 79:14–37.

Larsen, J. 1988. An ultrastructural study of *Amphidinium poecilochroum* (Dinophyceae), a phago-trophic dinoflagellate feeding on small species of cryptophytes. *Phycologia* 27:366–377.

Larsen, J. 1994. Unarmored dinoflagellates from Australian waters I. The genus *Gymnodinium* (Gymnodiniales, Dinophyceae). *Phycologia* 33:24–33.

Larsen, J., & Moestrup, Ø. 1989. "Guide to Toxic and Potentially Toxic Marine Algae," pp. 61. The Fish Inspection Service, Ministry of Fisheries, Copenhagen.

Larsen, J., & Sournia, A. 1991. The diversity of heterotrophic dinoflagellates. *In* "The Biology of Free-Living Heterotrophic Flagellates" (D. J. Patterson & J. Larsen, eds.), pp. 313–332. Clarendon Press, Oxford.

Lebour, M. V. 1925. "The Dinoflagellates of the Northern Seas," pp. 250. Marine Biological Association of the United Kingdom, Plymouth, UK.

Levandowsky, M., & Kaneta, P. J. Behaviour in dinoflagellates. *In* "The Biology of Dinoflagellates" (F. J. R. Taylor, ed.), Botanical Monographs Vol. 21, pp. 360–397. Blackwell Oxford.

Lewis, J. 1990. The cyst–theca relationship of *Oblea rotunda* (Diplopsalidaceae, Dinophyceae). *British Phycological Journal* 25:339–351.

Lewis, J. 1991. Cyst–theca relationships in *Scrippsiella* (Dinophyceae) and related orthoperidinioid genera. *Botanica Marina* 34:91–106.

Lewis, J., Dodge, J. D., & Tett, P. 1984. Cyst–theca relationships in some *Protoperidinium* species (Peridiniales) from Scottish sea lochs. *Journal of Micropalaeontology* 3:25–34.

Loeblich, A. R., III. 1965. Dinoflagellate nomenclature. *Taxon* 14:15–18.

Loeblich, A. R., III. 1968. A new marine dinoflagellate genus *Cachonina* in axenic culture from the Salton Sea, California, with remarks on the genus *Peridinium. Proceedings of the Biological Society of Washington* 81:91–96.

Loeblich, A. R., III. 1976. Dinoflagellate evolution: Speculation and evidence. *Journal of Protozool-ogy* 23:111–128.

Loeblich, A. R., III. 1982. Dinophyceae. *In* "Synopsis and Classification of Living Organisms" (S. P. Parker, ed.), Vol. 1, pp. 101–115. McGraw–Hill, New York.

Loeblich, A. R., III, & Loeblich, L. A. 1985. Dinoflagellates: Structure of the amphiesma and re-analysis of thecal plate patterns. *Hydrobiologia* 123:177–180.

Loeblich, A. R., III, & Sherley, J. L. 1979. Observations on the theca of the motile phase of free living and symbiotic isolates of *Zooxanthella microadriatica* new combination. *Journal of the Marine Biological Association of the United Kingdom* 59:195–206.

Loeblich, A. R., III, Sherley, J. L., & Schmidt, R. J. 1979. The correct position of flagellar insertion in *Prorocentrum* and description of *P. rhathymum* sp. nov. (Pyrrhophyta). *Journal of Plankton Research* 1:113–120.

Loeblich, A. R., III, Schmidt, R. J., & Sherley, J. L. 1981. Scanning electron microscopy of *Heterocapsa pygmaea* sp. nov., and evidence for polyploidy as a speciation mechanism in dinoflagellates. *Journal of Plankton Research* 3:67–79.

Loeblich, A. R., Jr., & Loeblich, A. R., III. 1966. Index to the genera, subgenera, and sections of the Pyrrhophyta. *In* "Studies in Tropical Oceaongraphy," Vol. 3, pp. 94. University of Miami, Institute of Marine Science, Miami.

Loeblich, A. R., Jr., & Loeblich, A. R., III. 1969. Index to the genera, subgenera, and sections of the Pyrrhophyta, III. *Journal of Paleontology* 43:193–198.

Loper, C. L., Steidinger, K. A., & Walker, L. M. 1980. A simple chromosome spread technique for unarmored dinoflagellates and implications of polyploidy in algal cultures. *Transactions of the American Microscopical Society* 99:343–346.

López, J. 1966. Variacion y regulacion de la forma en el genero *Ceratium*. *Investigacion Pes-quera* 30:325–427.

MacKenzie, L. 1992. Does *Dinophysis* (Dinophyceae) have a sexual life cycle. *Journal of Phycology* 28:399–406.

Martin, G. W. 1929. Dinoflagellates from marine and brackish waters of New Jersey. *University of Iowa Studies in Natural History* 12:1–32.

Matsuoka, K. 1985. Cyst and thecate forms of *Pyrophacus steinii* (Schiller) Wall et Dale, 1971. *Transactions of the Proceedings of the Palaeontological Society of Japan, N. S.* 140:240–262.

Matsuoka, K. 1988. Cyst–theca relationships in the Diplopsalid group (Peridiniales, Dinophyceae). *Review of Palaeobotany and Palynology* 56:95–122.

Matsuoka, K., & Fukuyo, Y. 1986. Cyst and motile morphology of colonial dinoflagellate *Pheopo-lykrikos hartmannii* (Zimmermann) comb. nov. *Journal of Plankton Research* 8:811–818.

Matsuoka, K., Fukuyo, Y., & Anderson, D. M. 1989. Methods for modern dinoflagellate cyst studies. *In* "Red Tides: Biology, Environmental Science, and Toxicology" (T. Okaichi, D. M. Anderson, & T. Nemato, eds.), pp. 461–479. Elsevier, New York.

Matsuoka, K., Kobayashi, S., & Gaines, G. 1990. A new species of the genus *Ensiculifera* (Dinophyceae); Its cyst and motile forms. *Bulletin of Plankton Society of Japan* 37:127–143.

Montresor, M., & Zingone, A. 1988. *Scrippsiella precaria* sp. nov. (Dinophyceae), a marine dinoflagellate from the Gulf of Naples. *Phycologia* 27:387–394.

Montresor, M., Zingone, A., & Marino, D. 1993. The calcareous resting cyst of *Pentapharsodinium tyrrhenicum* comb. nov. (Dinophyceae). *Journal of Phycology* 29:223– 230.

Morey-Gaines, G., & Ruse R. H. 1980. Encystment and reproduction of the predatory dinoflagellate, *Polykrikos kofoidii* Chatton (Gymnodiniales). *Phycologia* 19:230–232.

Mornin, L., & Francis, D. 1967. The fine structure of *Nematodinium armatum*, a naked dinoflagellate. *Journal of Microscopy* 6:759–772.

Morrill, L. C., & Loeblich, A. R., III. 1981a. A survey for body scales in dinoflagellates and a revision of *Cachonina* and *Heterocapsa* (Pyrrhophyta). *Journal of Plankton Research* 3:53–65.

Morrill, L. C., & Loeblich, A. R., III. 1981b. The dinoflagellate pellicular wall layer and its occurrence in the division Pyrrhophyta. *Journal of Phycology* 17:315–323.

Morrill, L. C., & Loeblich, A. R., III. 1983. Formation and release of body scales in the dinoflagellate genus *Heterocapsa*. *Journal of Marine Biological Association of the United Kingdom* 63:905–913.

Neeham, G. H. 1977. "The Pratical Use of the Microscope Including Photomicrography," pp. 493. C. C. Thomas, Springfield, IL.

Nie, D. 1939. Dinoflagellata of the Hainan region. II. On the thecal morphology of *Blepharocysta*, with a description of a new species. *Contributions of the Biological Laboratory of the Science Society of China, Zoological Series* 13:23–39.

Nie, D. 1942. Dinoflagellata of the Hainan region. IV. On the thecal morphology of *Podolampas* with description of species. *Sinensia* 13:53–60.

Nie, D., & Wang, C. 1944. Dinoflagellata of the Hainan region. VIII. On *Sinophysis microcephalus*, a new genus and species of Dinophysidae. *Sinensia* 15:145–151.

Nordli, E. 1957. Experimental studies in the ecology of Ceratia. *Oikos* 8:200–265.

Norris, D. R. 1969. Thecal morphology of *Ornithocercus magnificus* (Dinoflagellata) with notes on related species. *Bulletin of Marine Science* 19:175–193.

Norris, D. R., & Berner, L. D., Jr. 1970. Thecal morphology of selected species of *Dinophysis* (Dinoflagellata) from the Gulf of Mexico. *Contributions in Marine Science* 15:145–192.

Norris, D. R., Bomber, J. W., & Balech, E. 1985. Benthic dinoflagellates associated with ciguatera from the Florida Keys. I. *Ostreopsis heptagona* sp. nov. *In* "Toxic Dinoflagellates" (D. M. Anderson, A. W. White, & D. G. Baden, eds.) pp. 39–44. Elsevier, New York.

Palincsar, J. S., Jones, W. R., & Palincsar, E. E. 1988. Effects of isolation of the endosymbiont *Symbiodinium microadriaticum* (Dinophyceae) from its host *Aiptasia pallida* (Anthozoa) on cell wall ultrastructure and mitotic rate. *Transactions of the American Microscopical Society* 107:53–66.

Pfiester, L. A., & Anderson, D. M. 1987. Dinoflagellate reproduction. *In* "The Biology of Dinoflagellates" (F. J. R. Taylor, ed.), Botanical Monographs Vol. 21, pp. 611–648. Blackwell, Oxford.

Pfiester, L. A., & Holt, J. R. 1982. A technique for counting chromosomes of armored dinoflagellates, and chromosome numbers of six freshwater dinoflagellate species. *American Journal of Botany* 69:1165–1168.

Pincemin, J. M., & Gayol, P. 1978. Naked and thecate swarmers in clonal *Pyrocystis fusiformis*: Systematic and physiology problems. *Archiv fur Protistenkunde* 120:401– 408.

Pincemin, J. M., Gayol, P., & Salvano, P. 1981. Observations on the thecate stage of the dinoflagellate *Pyrocystis* cf. *fusiformis* (clones NOB2 and 111): Variations in morphology and tabulation. *Archiv fur Protistenkunde* 124:271–282.

Pincemin, J. M., Gayol, P., & Salvano, P. 1982. *Pyrocystis lunula*: Dinococcide giving rise to a biflagellated thecate stage. *Archiv fur Protistenkunde* 125: 95–107.

Pomroy, A. J. 1989. Scanning electron microscopy of *Heterocapsa minima* sp. nov. (Dinophyceae) and its seasonal distribution in the Celtic Sea. *British Phycological Journal* 24:131–135.

Popovský, J., & Pfiester, L. A. 1990. Dinophyceae (Dinoflagellida). *In* "Süsswasserflora von Mitteleuropa" (H. Ettl, J. Gerloff, H. Heynig, & D. Mollerhauer, eds.), Brgründet von A. Pascher, Band 6, pp. 272. Gustav Fischer Verlag, Jena Stuttgart.

Posek, M. T., Johnson, A. H., & McMichael, K. L. 1980. "Scanning Electron Microscopy. A Student's Handbook," pp. 305. Ladd Research Industries, Inc.

Rampi, L., & Bernhard, M. 1980. "Chiave per la Determinazione delle Peridinee Pelagiche Mediterranee," pp. 193. Comitato Nazionale Energina Nucleare CNEN-RT/B10(80)8.

Reinecke, P. 1967. *Gonyaulax grindleyi* sp. nov.: A dinoflagellate causing a red tide at Elands Bay, Cape Province, in December 1966. *Journal of South African Botany* 33:157–160.

Rizzo, P. J. 1987. Biochemistry of the dinoflagellate nucleus. *In* "The Biology of Dinoflagellates" (F. J. R. Taylor, ed.),Botanical Monographs Vol. 21, pp. 143–173. Blackwell, Oxford.

Roberts, K. R., & Timpano, P. 1989. Comparative analyses of the dinoflagellate flagellar apparatus. 1. *Woloszynskia* sp. *Journal of Phycology* 25:26–36.

Roberts, K. R., Farmer, M. A., Schneider, R. M., & Lemoine, J. E. 1988. The microtubular cytoskeleton of *Amphidinium rhinocephalum* (Dinophyceae). *Journal of Phycology* 24: 544–553.

Rowan, R., & Powers, D. A. 1991. A molecular genetic classification of zooxanthellae and the evolution of animal–algal symbioses. *Science* 251:1348–1351.

Saunders, R. D., & Dodge J. D. 1984. An SEM study and taxonomic revison of some armoured sand-dwelling marine dinoflagellates. *Protistologica* 20:271–283.

Schiller, J. 1933. Dinoflagellatae (Peridineae) in monographischer Behandlung. I. Teil, Lieferung 3. *In* "Dr. L. Rabenhorst's Kryptogamen-Flora von Deutschland, Österreich und der Schweiz," pp. 617. Akademische Verlagsgesel- lschaft, Leipzig.

Schiller, J. 1937. Dinoflagellatae (Peridineae) in monographischer Behandlung. 2. Teil, Lieferung 4. *In* "Dr. L. Rabenhorst's Kryptogamen-Flora von Deutschland, Österreich und der Schweiz." pp. 589. Akademische Verlagsgesellschaft, Leipzig.

Schmidt, R. J., & Loeblich, A. R., III. 1979. Distribution of paralytic shellfish poisoning among Pyrrhophyta. *Journal of the Marine Biological Association of the United Kingdom* 59:479–487.

Schoenberg, D. A., & Trench, R. K. 1980. Genetic variation in *Symbiodinium* (= *Gymnodinium*) *microadriaticum* Freudenthal, and specificity in its symbiosis with marine invertebrates. II. Morphological variation in *S. microadriaticum. Proceedings of the Royal Society of London (B)* 207:429–444.

Silva, P. C. 1960. Remarks on algal nomenclature, III. *Taxon* 9:18–25.

Sournia, A. 1967a. Contribution a la connaissance des peridiniens microplanctoniques du Canal de Mozambique. *Bulletin du Musee National d'Histoire Naturelle* 39:417–438.

Sournia, A. 1967b. Le genre *Ceratium* (peridinien planctionique) dans le canal de Mozambique contribution a une revision mondiale. *Vie Milieu* 18:375–500.

Sournia, A. 1972. Une periode de poussees phytoplanctoniques pres de Nosy-be (Madagascar) en 1971. I. Especes rares ou nouvelles du phytoplancton. *Cahiers ORSTOM, Serie Oceanographique* 10:151–159.

Sournia, A. 1978. Phytoplankton manual. "Monographs on *Oceaongraphic Methodology*," No. 6, pp. 337. UNESCO. Paris.

Sournia, A. 1986. "Atlas du Phytoplankton Marin. Volume I: Introduction, Cyanophycées, Dictyochophycées, Dinophycées et Raphidophycées," pp. 219. Editions du Centre National de la Recherche Scientifique, Paris.

Soyer, M.-O. 1970. Les ultrastructures liées aux fonctions de relation chez *Noctiluca miliaris* S. (Dinoflagellata). *Zeitschrift Zellforsch* 104:29–55.

Spector, D. L. (ed.) 1984. "Dinoflagellates," pp. 545. Academic Press, Orlando.

Spero, H. J., & Moree, M. 1981. Phagotrophic feeding and its importance to the life cycle of the holozoic dinoflagellate, *Gymnodinium fungiforme*. *Journal of Phycology* 17:43–51.

Spero, H. J. 1982. Phagotrophy in *Gymnodinium fungiforme* (Pyrrhophyta): The peduncle as an organelle of ingestion. *Journal of Phycology* 18:356–360.

Steidinger, K. A. 1968. The genus *Gonyaulax* in Florida waters. I. Morphology and thecal development in *Gonyaulax polygramma* Stein, 1883. *Florida Board of Conservation, Division of Salt Water Fisheries, Marine Laboratory, Leaflet Series* 1:1–5.

Steidinger, K. A. 1971. *Gonyaulax balechii* sp. nov. (Dinophyceae) with a discussion of the genera *Gonyaulax* and *Heteraulacus*. *Phycologia* 10:183–187.

Steidinger, K. A. 1979. Collection, enumeration and identification of free-living marine dinoflagellates. *In* "Toxic Dinoflagellate Blooms" (D. L. Taylor and H. H. Seliger, eds.), pp. 435–442. Elsevier, New York.

Steidinger, K. A. 1983. A re-evaluation of toxic dinoflagellate biology and ecology. *In* "Progress in Phycological Research" (F. E. Round & D. J. Chapman, eds.), pp. 147–188. Elsevier, New York.

Steidinger, K. A. 1990. Species of the *tamarensis/catenella* group of *Gonyaulax* and the fucoxanthin derivative-containing gymnodinioids. *In* "Toxic Marine Phytoplankton" (E. Graneli, B. Sundstrom, L. Edler, & D. M. Anderson, eds.), pp. 11–16. Elsevier, New York.

Steidinger, K. A. 1993. Some taxonomic and biologic aspects of toxic dinoflagellates. *In* "Algal Toxins in Seafood and Drinking Water" (I. R. Falconer, ed.), pp. 1–28. Academic Press, London.

Steidinger, K. A., Babcock, C., Mahmoudi, B., Tomas, C., & Truby, E. 1989. Conservative taxonomic characters in toxic dinoflagellate species identification. *In* "Red Tides: Biology, Environmental Science, and Toxicology" (T. Okaichi, D. M. Anderson, & T. Nemoto, eds.), pp. 285–288. Elsevier, New York.

Steidinger, K. A., Burkholder, J. A., Glasgow, H. B., Jr., Hobbs, C. W., Garrett, J. K., Truby, E. W., Noga, E. J., & Smith, S. A. 1996. *Pfiesteria piscicida*, gen. et sp. nov. (Pfiesteriaceae, fam. nov., Dinophyceae), a new dinoflagellate with a complex life cycle and behavior. *Journal of Phycology* 32:157–164.

Steidinger, K. A., Truby, E. W., Garrett, J. K., & Burkholder, J. M. 1995. The morphology and cytology of a newly discovered toxic dinoflagellate. *In* "Harmful Marine Algal Blooms" (P. Lassus, G. Arzul, E. Erard, P. Gentien, & C. Marcaillou, eds.), pp. 83–88. Lavoisier, Cachan Cedex, France.

Steidinger, K. A., & Balech, E. 1977. *Scrippsiella subsalsa* (Ostenfield) comb. nov. (Dinophyceae) with a discussion on *Scrippsiella*. *Phycologia* 16:69–73.

Steidinger, K. A., & Cox, E. R. 1980. Free-living dinoflagellates. *In* "Phytoflagellates" (E. R. Cox, ed.), pp. 407–432. Elsevier/North Holland, New York.

Steidinger, K. A., & Davis, J. T. 1967. The genus *Pyrophacus*, with a description of a new form. *Florida Board of Conservation Marine Laboratory Leaflet Series* 1:1–8.

Steidinger, K. A., & Moestrup, Ø. 1990. The taxonomy of *Gonyaulax, Pyrodinium, Alexandrium, Gessnerium, Protogonyaulax*, and *Goniodoma*. *In* "Toxic Marine Phytoplankton" (E. Graneli, B. Sundstrom, L. Edler, & D. M. Anderson, eds.), pp. 522–523. Elsevier, New York.

Steidinger, K. A., & Vargo, G. A. 1988. Marine dinoflagellate blooms: Dynamics and impacts. *In* "Algae and Human Affairs" (C. A. Lembi & J. R. Waaland, eds.), pp. 373–401. Cambridge University Press, Cambridge.

Steidinger, K. A., & Williams, J. 1970. Dinoflagellates. "Memoirs of the Hourglass Cruises," Vol. 2, pp. 251. Florida Department of Natural Resources Marine Research Laboratory, St. Petersburg, FL.

Subrahmanyan, R. 1966. New species of Dinophyceae from Indian waters. I. The genera *Haplodinium* Klebs emend. Subrahmanyan and *Mesoporos* Lillick. *Phykos* 5:175–180.

Subrahmanyan, R. 1968. "The Dinophyceae of the Indian Seas. Part 1. Genus *Ceratium* Schrank," pp. 118. Marine Biological Association of India, Memoir II.

Subrahmanyan, R. 1971. "The Dinophyceae of the Indian Seas. Part 2. Family Peridiniaceae Schütt emend. Lindemann," pp. 334. Marine Biological Association of India, Memoir II.

Sweeney, B. M. 1978. Ultrastructure of *Noctiluca miliaris* (Pyrrhopyhta) with green flagellate symbionts. *Journal of Phycology* 14:116–120.

Swift, E., & Durbin, E. G. 1971. Similarities in the asexual reproduction of the oceanic dinoflagellates, *Pyrocystis fusiformis*, *Pyrocystis lunula*, and *Pyrocystis noctiluca*. *Journal of Phycology* 7:89–96.

Swift, E., & Wall, D. 1972. Asexual reproduction through a thecate stage in *Pyrocystis acuta* Kofoid, 1907 (Dinophyceae). *Phycologia* 11:57–65.

Tafall, B. F. O. 1942. Notas sobre algunos dinoflagelados planctonicos marinos de Mexico, con descripcion de nuevas especies. *Anales de la Escuela Nacional de Ciencias Biologicas* 2:435–447.

Tai, L., & Skosberg T. 1934. Studies on the Dinophysoidae, marine armored dinoflagellates of Monterey Bay, California. *Archiv fur Protistenkunde* 82:380–482.

Takayama, H., & Adachi, R. 1984. *Gymnodinium nagasakiense* sp. nov., a red-tide forming dinophyte in the adjacent waters of Japan. *Bulletin of the Plankton Society of Japan* 31:7–14.

Takayama, H. 1985. Apical grooves of unarmored dinoflagellates. *Bulletin of Plankton Society of Japan* 32:129–140.

Tangen, K., Brand, L. E., Blackwelder, P. L., & Guillard, R. R. L. 1982. *Thoracosphaera heimii* (Lohmann) Kamptner is a dinophyte: Observations on its morphology and life cycle. *Marine Micropaleontology* 7:193–212.

Taylor, D. L. 1971a. Taxonomy of some common *Amphidinium* species. *British Phycological Journal* 6:129–133.

Taylor, D. L. 1971b. Ultrastructure of the "zooxanthella," *Endodinium chattonii* in situ. *Journal of the Marine Biological Association of the United Kingdom* 51:227–234.

Taylor, F. J. R. 1963. *Brachydinium*, a new genus of the Dinococcales from the Indian Ocean. *Journal of South African Botany* 29:75–78.

Taylor, F. J. R. 1971. Scanning electron microscopy of thecae of the dinoflagellate genus *Ornithocercus*. *Journal of Phycology* 9:1–10.

Taylor, F. J. R. 1972. Unpublished observations on the thecate stage of the dinoflagellate genus *Pyrocystis* by the late C. A. Kofoid and Josephine Michener. *Phycologia* 11:47–55.

Taylor, F. J. R. 1973. Topography of cell division in the structurally complex dinoflagellate genus *Ornithocercus*. *Journal of Phycology* 9:1–10.

Taylor, F. J. R. 1975. Taxonomic difficulties in red tide and paralytic shellfish poison studies: the "*Tamarensis* complex" of *Gonyaulax*. *Environmental Letters* 9:103–119.

Taylor, F. J. R. 1976. Dinoflagellates from the International Indian Ocean Expedition. *Bibliotheca Botanica* 132:1–234.

Taylor, F. J. R. 1979. The toxigenic gonyaulacoid dinoflagellates. *In* "Toxic Dinoflagellate Blooms" (D. L. Taylor & H. H. Seliger, eds.), pp. 47–56. Elsevier/North Holland, New York.

Taylor, F. J. R. 1980. On dinoflagellate evolution. *Biosystems* 13:65–108.

Taylor, F. J. R. 1984. Toxic dinoflagellates: taxonomic and biogeographic aspects with emphasis on *Protogonyaulax*. *In* "Seafood Toxins" (E. P. Ragelis, ed.), pp. 77–97. American Chemical Society, Washington, DC.

Taylor, F. J. R. (ed.) 1987. "The Biology of Dinoflagellates," pp. 705. Blackwell, Oxford.

Thompson, R. H. 1950. A new genus and new records of freshwater Pyrrhophyta in the Desmokontae and Dinophyceae. *Lloydia* 13:277–299.

Throndsen, J. 1993. The planktonic marine flagellates. *In* "Marine Phytoplankton: A Guide to Naked Flagellates and Coccolithophorids" (C. R. Tomas, ed.) pp. 7–146. Academic Press, San Diego.

Toriumi, S., & Dodge, J. D. 1993. Thecal apex structure in the Peridiniaceae (Dinophyceae). *British Phycological Journal* 28:39–45.

Trench, R. K., & Blank, R. J. 1987. *Symbiodinium microadriaticum* Freudenthal, *S. goreauii* sp. nov., *S. kawagutii* sp. nov. and *S. pilosum* sp. nov.:gymnodinioid dinoflagellate symbionts of marine invertebrates. *Journal of Phycology* 23: 469–481.

Von Stosch, H. A. 1964. Zum problem der sexuellen fortflanzung in der Peridineengattung *Ceratium*. *Helgoländer Meeresuntersuchungen* 10:140–152.

Von Stosch, H. A. 1969a. Dinoflagellaten aus der Nordsee I. ›ber *Cachonina niei* Loeblich (1968), *Gonyaulax grindleyi* Reinecke (1967) und ein methode zur darstellung von Peridinnenpanzern. *Helgoländer Meeresuntersuchungen* 19:558–568.

Von Stosch, H. A. 1969b. Dinoflagellaten aus der Nordsee II. *Helgolandinium subglobosum* gen. et spec. nov. *Helgoländer Meeresuntersuchungen* 19:569–577.

Wall, D., & Dale, B. 1968. Modern dinoflagellate cysts and evolution of the Peridiniales. *Micropaleontology* 14:265–304.

Wall, D., & Dale, B. 1969. The "hystrichosphaerid" resting spore of the dinoflagellate *Pyrodinium bahamense* Plate, 1906. *Journal of Phycology*. 5:140–149.

Wall, D., & Dale, B. 1971. A reconsideration of living and fossil *Pyrophacus* Stein, 1883 (Dinophyceae). *Journal of Phycology* 7:221–235.

Wall, D. & Evitt, W. R. 1975. A comparison of the modern genus *Ceratium* Schrank, 1973, with certain Cretaceous marine dinoflagellates. *Micropaleontology* 21:14–44.

Watanabe, M. M., Inouye, I., Sawaguchi, T., & Chihara, M. 1990. *Lepidodinium viride* gen. et sp. nov. (Gymnodiniales, Dinophyta), a green dinoflagellate with a chlorophyll a- and b-containing endosymbiont. *Journal of Phycology* 26:741–751.

Whedon, W. F., & Kofoid, C. A. 1936. Dinoflagellata of the San Francisco region. I. On the skeletal morphology of two new species, *Gonyaulax catenella* and *G. acatenella*. *University of California Publications in Zoology* 41:25–34.

Yarranton, G. A. 1967. Parameters for use in distinguishing populations of *Euceratium* Gran. *Bulletin of Marine Ecology* 6:147–158.

Yuki, K., & Fukuyo, Y. 1992. *Alexandrium satoanum* sp. nov. (Dinophyceae) from Matoya Bay, central Japan. *Journal of Phycology* 28:395–399.

Yuki, K., & Yoshimatsu, S. 1989. New record of *Alexandrium insuetum* Balech (Dinophyceae) from Japan with some supplementary observations on thecal morphology. *Bulletin of Plankton Society of Japan* 36:121–126.

Zingmark, R. G. 1970. Sexual reproduction in the dinoflagellate *Noctiluca miliaris* Suriray. *Journal of Phycology* 6:122–126.

Zingone, A., & Montresor, M. 1988. *Protoperidinium parthenopes* sp. nov. (Dinophyceae), an intriguing dinoflagellate from the Gulf of Naples. *Cryptogamie Algologie* 9:117–125.

REFERENCES ADDED IN PROOF:

Elbrächter, M. 1994. Redescription of *Gymnodinium heterostriatum* Kofoid et Swezy 1921 (Dinophyceae). *Helgolander Meeresuntersuchungen*. 48:359–363.

Hansen, G. 1995. Analysis of the thecal plate pattern in the dinoflagellate *Heterocapsa rotundata* (Lohmann) comb. nov. [=*Katodinium rotundatum* (Lohmann) Loebilch III]. *Phycologia* 34:444–448.

Chapter 4

Introduction

Carmelo R. Tomas

For nearly two centuries, the identification of marine planktonic organisms has intrigued, inspired, and even entertained the natural scientist. Approached from the various disciplines of natural history, geology, botany, zoology, and microbiology, a steadily increasing body of knowledge has developed regarding the description and identification of marine phytoplankton species. Among these, the flagellated forms are the most varied and difficult to identify, partly due to the lack of adequate preservatives. Those species having "hard parts," for example mineralized scales, spines, silica skeletons, or coccoliths, are better documented, presumably because of the presence of these durable features. Species lacking these structures belong to the category of "unidentified flagellates," which has accompanied numerous species lists of marine samples.

In part, it is necessary to know how to observe these species and what features to note in properly identifying the flagellates. There are problems relating to observing living specimens, adequate preservation procedures, and diagnostic structures as seen in light and electron microscopy, and these are critical to proper identification.

One difficulty in coherently describing the planktonic marine flagellates is

Identifying Marine Phytoplankton
Copyright © 1993 by Academic Press, Inc. All rights of reproduction in any form reserved.

that two major divisions have evolved, separating this general group. The coccolithophorids, discovered in the last century, were originally described from fossil forms and later were discovered to have extant species. Terminology, classification, and systematics eventually evolved dealing with fossil and living forms and are now being modified to be consistent with biological reality. In botanical terms, however, the coccolithophorids are considered to be within two or three orders in the class *Prymnesiophyceae,* with the major criteria being the presence, even in rudimentary form, of a haptonema and coccoliths. This taxonomy has developed almost independently from that of the other marine flagellates.

With the exception of the dinoflagellates, the other marine flagellates have a less coherent history, with most species being described from living specimens by biologists examining seawater samples. In this general group, major advances were made during the twentieth century, and at present significant inroads are being made in their systematics and taxonomy. The diversity of this group is emphasized by the fact that the marine flagellates, as treated here, are spread among the two major algal divisions (Chromophyta and Chlorophyta) of the Eukaryophyta, in nine of ten algal classes, and three zooflagellate orders. Unlike the coccolithophorids, the taxonomy and systematics of this group are mediated by botanists and zoologists, who alternately have identified and described the same or similar species. In addition, evidence from recent studies with the electron microscope shows strong structural similarities between pigmented and nonpigmented flagellates, drawing into question the notion of separating flagellate species based on the presence or absence of pigments alone. Any modern systematic treatment of the flagellates will have to address this problem.

The algal groups in marine plankton presented here follow a classification scheme partially modified from that of Christensen (1966; see Chapter 2). Below is a key to suffixes.

-phyta	division
-phyceae	class
-ales	order
-aceae	family

PROTOKARYOTA—without a true nucleus

  Cyanophyta—containing *c*-phycocyanin, *c*-phycoerythrin

    Cyanophyceae

      Chroococcales

      Nostocales

EUKARYOTA—having a true nucleus

  (ACONTA—without flagella)

Rhodophyta—red algae; having *r*-phycoerythrin, r-phycocyanin
  Bangiophyceae
    Porphyridiales
(CONTOPHORA—with flagella)
  Chromophyta—having carotenoids (yellow/brown pigments)
    Cryptophyceae
      Cryptomonadales
    Dinophyceae—dinoflagellates
      Prorocentrales
      Dinophysales
      Peridiniales
      Phytodiniales
    Prymnesiophyceae–Haptophyceae—with haptonema
      Isochrysidales
      Coccolithophorales
      Prymnesiales
      Pavlovales
    Chrysophyceae
      Bicosoecales
      Ochromonadales
      Synurales
      Chrysosphaerales
      Sarcinochrysidales
    Dictyochophyceae
      Rhizochromulinales
      Pedinellales
      Dictyochales
    Bacillariophyceae—diatoms
      Biddulphiales—centric diatoms
      Bacillariales—pennate diatoms
    Raphidophyceae
      Chattonellales
  Chlorophyta—having chlorophyll *b*
    Euglenophyceae

Euglenales
Prasinophyceae
    Mamiellales
    Chlorodendrales
Chlorophyceae—green algae
    Volvocales
    Chlorococcales

The zooflagellates in marine plankton are classified as according to Lee et al. (1985), but the hierarchial level is raised to match the algal system. Below is a key to suffixes.

-a	phylum
-ea	class
-ida	order
-idae	family

Zoomastigophora
    Choanoflagellidea = Craspedophyceae—cells with collars
        Codonosigida
        Salpingoecida
        Acanthoecida
    Kinetoplastidea—flagella with paraxial rod
        Bodonida
    Ebriidea—having a silica skeleton
        Ebrida

The differences in how planktonic flagellates and coccolithophorids have been presented in the past is to some degree reflected in this volume. The planktonic flagellates are treated in a slightly different format from the coccolithophorids. This reflects in part the preference of different authors but also shows a major difference in how species were and are presently described. It is hoped that the users of this manual will not find this too disturbing but will find each presentation to be more consistent with each type of reference material, thus making it easier to compare the previous literature with that presented here.

Proper identification of species is becoming more and more important. New problems continuously arise regarding environmental changes and the resulting modifications of natural floras. The proper identification of species is essential for any discussion of species succession and its altering by various causes. The total concept of biodiversity and the challenge it poses rest heavily on being able to clearly identify species. Similarly, in the marine environment, phenomena such as blooms of harmful and/or noxious microalgae have become important on a global scale, and the proper identification of these species often has health implications. It is hoped that this volume will make the identification of these difficult species easier and stimulate researchers to further study this important component of plankton.

Chapter 5

The Planktonic Marine Flagellates

Jahn Throndsen

INTRODUCTION

Marine plankton mostly consists of unicellular, nonmotile cells from many classes of algae and bacteria, motile flagellates, and ciliates. These cells range from less than 1 μm to greater than 1 mm and can attain population abundances exceeding a million cells per liter during bloom periods.

The light microscope has been the most commonly used tool for qualitative and quantitative analysis of phytoplankton; however, success in studying these organisms in the past depended on the presence of a cell wall or an outer covering resistant to common fixing or preserving agents. Hence, most of our present knowledge of phytoplankton mainly concerns "preservable" species, such as diatoms (Bacillariophyceae), the dominating primary producers in the temperate and cold areas; coccolithophorids (Prymnesiophyceae *pro parte*), abundant in warm and tropical seas; blue-green algae (Cyanobacteria, Cyanophyta), important in tropical marine areas; and dinoflagellates (Dinophyceae *pro parte*) common in warmer seas and during warmer seasons of temperate areas. Most of these species preserve well and can be studied long after their collection.

Identifying Marine Phytoplankton

Less is known of the relative importance of planktonic "naked" flagellates, primarily due to the absolute requirement of examining living material and of having a familiarity with the characteristics spanning a number of classes and orders. A hint at the global importance of the planktonic flagellates is evident from the many phytoplankton counts of preserved material that consistently report a large component of "unidentified flagellates." Some indication of their abundance and impact on the marine ecosystem is also evident when flagellates (*Chrysochromulina polylepis; C. leadbeateri; Prymnesium parvum*) form toxic blooms.

Eukaryotic picoplankton are documented herein whenever adequate methods are used (Johnson & Sieburth, 1982; Joint & Pipe, 1984; Platt & Li, 1986; Throndsen, 1973, 1979); however, few picoflagellate species are known. Some of them, like *Micromonas pusilla*, are ubiquitous in the marine environment. These observations notwithstanding, the general knowledge of species composition, abundance, distribution, and physiology is still lacking.

GENERAL CONSIDERATIONS

The protistan plankton may be grouped by nonsystematic or ecologically important criteria, such as the ability to move, size, and their trophic mode. The following characteristics are those observed most often when making identifications.

Motility   Movement of these species within a water column depends upon the presence of a flagellum(a) or cilia. The motion produced by these organelles differs significantly from the gliding motion on solid substrates accomplished by benthic diatoms and cyanophyceaens using different locomotory structures. The ciliated protists constitute the well-defined phylum Ciliophora (Protozoa), whereas flagellated forms are scattered throughout many algal phyla as well as the protozoan phylum Sarcomastigophora. A flagellated stage may be a permanent or temporary part of a life cycle. Motility may have a great impact on the ability of a cell to react to environmental conditions as well as enhancing their nutrient uptake (Sommer, 1988). The type of locomotory appendage(s) (e.g., cilia, flagella) and the type of movement they produce are key characteristics in the correct identification of a species.

Size   A common criterion for grouping phytoplankton in a nonsystematic way is size. Lohmann (1909, 1911) introduced the term "Nannoplankton" for the cell fraction from 1 to 25 μm, while the term "Netzplankton" was commonly used for the cells held back by the finest silk number 20 net with a mesh size approximately 50 μm when wet. In addition to linear dimensions, Lohmann (1908) also tried to group plankton by specific cell volume. With today's modern filtration techniques, further separation into size fractions is possible, resulting in functional divisions such as those suggested by Sieburth et al. (1978), for example, picoplankton 0.2–2.0 μm, nanoplankton 2.0–20 μm, microplankton 20–200 μm, and mesoplankton 200 μm–20 mm. These general size definitions are followed in the present account, except for picoplankton, which is slightly modified to accommodate the smallest eukaryotic cells, the size of which often varies between 1.5 and 2.5 μm.

Most of the flagellates encountered in the plankton from open waters belong to the nanoplankton (2–20 μm). A number of dinoflagellates, however, will be in the microplankton or an even larger size group. A limited number of minute flagellates such as *Micromonas pusilla* and the eukaryotic nonmotile *Bathycoccus prasinos* are not uncommon in the picoplankton of oceanic waters, which are otherwise dominated by cyanophycean genera like *Synechococcus* and *Synechocystis*.

Trophic modes The form of nutrition or trophic mode is also a criterion used in identification of flagellates. Basically three trophic types are common in this diverse group.

Phototrophic: These forms rely solely on photosynthesis for their survival. These species are usually confined to the euphotic zone of the sea, where sufficient radiant energy can support cell growth by photosynthesis.

Heterotrophic: These organisms depend on the uptake of organic material from the sea, ingesting particles by phagotrophy or taking in soluble substances by pinocytosis. Choanoflagellates have a spectacular organelle, the collar, consisting of fine pseudopodia, which facilitates the uptake of minute food particles such as bacteria. In some Dinophyceae, digestion takes place outside the cell in a lobe of cytoplasm extruded to enclose the prey (Jacobson and Anderson, 1986).

Mixotrophic: This type of nutrition is common to flagellates in many algal classes. In the Chrysophyceae, mixotrophy is well documented for *Ochromonas* (Cole & Wynne, 1974). Uptake of organic particles in Prymnesiophyceae has been known since the first description of marine species of *Chrysochromulina* (Parke et al., 1955, 1956). Experimental evidence has not always been convincing, although it is well documented in some cases (see Green, 1991). The infrequent presence of nutrient vacuoles in some green algal flagellates may also indicate mixotrophy within Prasinophyceae (e.g., *Cymbomonas;* Throndsen, 1988).

The Problem of Phytoflagellates versus Zooflagellates

The presence or absence of chloroplasts is the most obvious criterion for distinguishing between phototrophic and heterotrophic flagellates. Fine structural evidence is presently redefining the criteria for modern flagellate systematics and questions the validity of chloroplast presence or absence in defining related species. Based on this evidence, many colorless (heterotrophic) flagellates appear to be related in evolutionary schemes to taxa (classes, orders) that also contain phototrophic species classified as algae. This is evident in the order Pedinellales (Dictyochophyceae) containing the phototrophic *Pseudopedinella* and the heterotrophic *Parapedinella*. With the exception of the choanoflagellates (Choanoflagellidea), very few of the known marine planktonic heterotrophs show zooflagellate characteristics exclusively among the Kinetoplastidea and Ebriidea.

ALGAL FLAGELLATE CHARACTERISTICS

Flagellates are found scattered throughout the algal system, and therefore any treatise on them as a group must include a number of taxa of higher ranks including classes, orders, and families, which often contain one or a few genera at the lower levels. Some genera are monospecific, whereas others such as *Chrysochromulina* and *Pyramimonas* may contain 50 or more species.

Modern systematics recognizes four classes in which the flagellate stage either is dominant or plays an important part in the life cycle of the species. The classes are Cryptophyceae, Dinophyceae, Prymnesiophyceae (= Haptophyceae), and Prasinophyceae. The first two classes were introduced early in this century based on observations using light microscopy, whereas the last two have recently been separated from the Chrysophyceae and the Chlorophyceae based on ultrastructural criteria.

The Flagellate Cell

The surface of flagellate cells may either be smooth, covered with scalelike structures composed wholly of organic material (Prymnesiophyceae, Prasinophyceae), calcified as in coccolithophorids, or silicified (Chrysophyceae). Silica skeletons are either found externally, as in the Dictyochophyceae, or internally, as in the Ebriidae.

Organic scales may cover the cell surface and/or the flagella in many species, especially in the Prasinophyceae and Prymnesiophyceae. These scales often show a pattern typical for the taxon, some typical for higher taxa such as class, order, family, or genus, and others characteristic for the single species. The scales are usually very difficult to observe in the light microscope, but some exceptions exist. Scales of *Apedinella,* for instance, are visible in bright field after cellulose staining, and *Chrysochromulina polylepis* scales may be seen in fluorescence microscopes using acridine orange.

Flagellates are also characterized by the presence of one or more flagella, with one, two, or four a. the most common number, and eight and 16 occasionally found in some *Pyramimonas* species (Prasinophyceae). The general flagellar anatomy is a fairly conservative feature, though details (e.g., of the transitional region) give important information on evolutionary trends. Variation in flagellar roots and external flagellar structures (scales and hairs) reflects the polyphyletic evolutionary affiliations of the flagellates. The flagellar root may be quite conspicuous, as in *Tetraselmis* (Prasinophyceae), where a broad cross-banded root is readily revealed in sectioned material, or it can consist of a single or a few microtubuli that have to be traced by tedious serial sectioning.

Paracrystalline structures may run alongside the axoneme, as in *Eutreptiella* (Euglenophyceae), or support a flagellar wing, as typical in Pedinellales (Dictyochophyceae) and Dinophyceae (transverse flagellum). Flagellar appendages called tripartite tubular hairs are typical of heterokont algae such as the Chrysophyceae. These hairs are produced in the perinuclear cavity, while other types of hairs are nontubular and are produced in vesicles derived from the Golgi complex.

Scale-covered flagella are typically found in the Prasinophyceae, where one, two, or three scale types may be present on the same flagellum. Flagellar scales may, however, also be found occasionally in Dinophyceae (*Oxyrrhis marina;* Clarke & Pennick 1972), Prymnesiophyceae (*Pavlova;* Van der Veer, 1969; Green, 1980), Dictyochophyceae (*Actinomonas;* Larsen, 1985; *Parapedinella;* Pedersen et al., 1986) and Cryptophyceae (*Chroomonas;* Santore, 1983).

Flagella may also serve for purposes other than swimming, as in *Bicosoeca* (Chrysophyceae), where the smooth flagellum is used for attaching the cell in the lorica, and as in *Cafeteria* (Chrysophyceae) for temporarily anchoring it to the substratum. In *Rhynchomonas* (Kinetoplastidea) the second flagellum is trailing.

Another threadlike organelle is the haptonema (Parke et al., 1955), typical of the class Prymnesiophyceae. It may be long and coiling upon irritation, as in *Chrysochromulina,* or short and noncoiling, as in *Prymnesium.* The haptonema may anchor the cell to the substratum, as reported by Leadbeater (1969), and the function of this organelle as a food-capturing device has been documented by Kawachi et al. (1991).

A trailing pseudopodium is easily observed in *Pseudopedinella* (Dictyochophyceae), whereas the thin filopodium anchoring *Pavlova* (Prymnesiophyceae) cells may be difficult to detect.

Also important in distinguishing flagellate cells are subsurface structures such as proteinaceous plates in the periplast of Cryptophyceae (Gantt, 1971), the amphiesma in Dinophyceae (Loeblich, 1970), and the pellicula in Euglenophyceae (Leedale, 1966). These structures may be revealed with staining under the light microscope, but for finer details must be studied with the electron microscope. Mitotic and meiotic characters are systematically important at higher taxonomic levels.

The microanatomy of the flagellate cell varies with its evolutionary background and hence is a valuable clue in modern taxonomy and classification. Important organelles in this context are mitochondria, which may have tubular or flat cristae (**Fig. 1**), and chloroplasts, which may have two to four enveloping

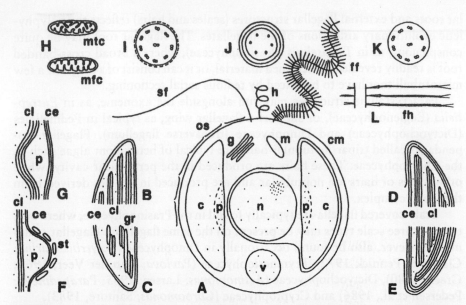

FIGURE 1 Fine-structure features of algal flagellates. (A) Simplified section of a generalized phytoflagellate cell with smooth (naked) flagellum sf, "flimmer" (with flagellar hairs) flagellum ff, haptonema h, organic scales os, and cell membrane cm. In the cell there are two chloroplasts c, one with an outer envelope in common with the nucleus n. Other organelles are the Golgi complex g, the mitochondrion m, and the vacuole v (which may contain chrysolaminaran). The pyrenoid p is part of the chloroplast. (B–E) Simplified sections through part of the chloroplast: (B) Cryptophyceae with double chloroplast envelope ce and lamellae cl with two thylakoids. (C) Chlorophyceae with a trend towards grana formation gr. (D) Euglenophyceae and Dinophyceae with three-layered lamellae and triple membranes enclosing the chloroplast. (E) Chrysophyceae with three-layered lamellae and four layers in the chloroplast envelope (the outer also enclosing the nucleus). (F, G) Two examples of pyrenoids: (F) cryptophycean pyrenoid with starch st; (G) submerged pyrenoid without starch. (H) Schematic section of mitochondrion with tubular cristae mtc and with flattened cristae mfc. (I, J) Schematic transverse section of flagellum; (I) with flagellar scales; (J) smooth. (K) Simplified section of haptonema. (L) Flagellar hairs fh on a flimmer flagellum. (Modified from Throndsen, 1980.)

membranes. The photosynthetic apparatus of the chloroplasts, the thylakoids and photosynthetic pigments, show systematic variations within the algae and phytoflagellates as well. The accumulation products in these flagellates, as in other algae, vary from fats and oils, which are both generally present, to specific carbohydrates like starch (Crypto-, Dino-, Prasino-, and Chlorophyceae), chrysolaminaran (Chryso- and Prymnesiophyceae), and paramylon (Euglenophyceae).

Reproduction and Life Cycles

Binary fission is the most common mode of vegetative reproduction in the planktonic flagellates. Division in phycoma stages of *Halosphaera, Ptero-*

sperma and *Pachysphaera* (Prasinophyceae) results in hundreds of motile cells from a single nonmotile one (Parke & Adams, 1961; Parke & den Hartog-Adams, 1965).

Isogamous sexual reproduction has been described in the classes Chrysophyceae, Prymnesiophyceae, and the Dinophyceae, where anisogamy is also known.

Most planktonic flagellates are known only as motile cells. Probably not all of these species lack other nonmotile stages, since they may not be recognized as the same species and may be described under a different name, as in the case of *Hymenomonas* and *Apistonema* (Rayns, 1962) in the Prymnesiophyceae. For the majority, the flagellated cell may be the dominant life stage, as indicated by their ability to reproduce by binary fission for 10–20 years in culture.

Many flagellates are part of a life cycle. In some cases the cycle consists of just one motile and one nonmotile stage, as in *Prymnesium* (Carter, 1937) and *Pycnococcus* (Guillard et al., 1991). Two distinct flagellated stages may also occur, as in *Dictyocha speculum*, where one is the well-known cell with the silica skeleton while the other is a naked spherical cell without a skeleton. In other cases, the life cycle may be more complex, as described for *Dicrateria* and *Isochrysis* (Parke, 1949), in which naked, nonmotile, coccoid and palmelloid stages are common in addition to the free-swimming flagellate.

Several genera form gelatinous colonial stages, as in *Phaeocystis*, or walled phycoma stages, as in *Pterosperma* and *Pachysphaera*, and are best known from these nonmotile stages, which are more commonly observed.

Cysts are a special stage encountered in many flagellate species, presumably facilitating survival through periods of poor conditions for growth. Cyst walls can vary in resistence to mechanical and chemical environmental stress. In *Prymnesium parvum*, cells within soft-walled cysts will survive in surface sediments for a period, after which the cyst degrades; this contrasts with the siliceous stomatocysts of the Chrysophyceae and the sporopollenin-containing cysts of some dinoflagellates, which readily fossilize and can be traced back in the geological record.

Occurrence and Distribution

Flagellates are found in all marine biotopes from the oligotrophic open oceans to eutrophic inshore waters, mud flats, and marshland ditches. Similar biotopes in geographically distant areas are often inhabited by the same morphological species.

Variation in morphological details are probably due more to recognized environmental factors rather than to geographical isolation of genotypes. This is particularly evident in *Chrysochromulina leadbeateri*, where scale type and pattern reported from Australian and Norwegian waters appear closer than those of the scales in North Pacific and North Atlantic specimens. On the other hand, differences in scale morphology in specimens identified as *Pyramimonas amylifera* (Prasinophyceae) eventually were discovered to be due to the confusion of two closely related species, *P. amylifera* and *P. propulsa* (Moestrup & Hill, 1991). Their distribution is clearly more related to biology than geography, since *P. propulsa* is found in Australian, Japanese, South African, and Mediterranean waters, while *P. amylifera* is confined to brackish waters of northern Europe and America.

Importance

The ecological importance of flagellates appears to be quite comprehensive and varies with the trophic roles they play in different ecosystems. As grazers of bacteria they are vital components of the microbial loop but also serve as food for ciliates and mesozoan larvae.

Virus Infections

Viruses may infect flagellates and are widely distributed among eukaryotic algae, as reviewed by Van Etten et al., 1991. It is even documented for the minute species *Micromonas pusilla* (Mayer & Taylor, 1979). While flagellates are ubiquitous, the abundance of viruses in the sea is probably variable (Bergh et al., 1989; Bratbak et al., 1990). In Japanese coastal and oceanic waters, viral particles were found in the order of 10^9 to 10^{10} per liter (Hara et al., 1991). Presently there is little information to evaluate the impact of viral infection on flagellate populations.

History

With the exception of armored dinoflagellates, the delicate vegetative cells of many plankton flagellates have left no traces in the sediment to be later found in the fossil record. Exceptions are some resistant walls of cysts, silicious skeletons of silicoflagellates (Dictyochophyceae; order Dictyochales), and calcified coccoliths of the coccolithophorids. This last group is best documented in the fossil record and can be traced back to the Cretaceous era.

Literature

Most "naked" flagellates cannot be identified from preserved material in the light microscope, hence most of the taxonomic information on this group comes from observing live cells almost exclusively from studies of coastal areas including brackish water localities. Classical exceptions from this are studies of the Kara Sea (Meunier, 1910), the Barents Sea (Wulff, 1916), the Atlantic Ocean (Lohmann, 1912), and the Adriatic Sea (Schiller, 1913, 1925).

For colorless flagellates from coastal areas the "Manual of the Infusoria" by W. Saville Kent (1880–1882), an extensive work for the time, still provides descriptions for many genera and species.

A single investigation that is still a major source of information is that published by Nellie Carter (1937) for a brackish water locality in England; many of the species described are valid for coastal and oceanic areas as well.

One of the most comprehensive surveys on naked algal flagellates is that of Walter Conrad, who presented brackish water flagellates in a series of papers (Conrad, 1926, 1938, 1939, 1941) ending with a more extensive publication on taxonomy, occurrence, and ecology of brackish water flagellates in Belgium (Conrad & Kufferath, 1954).

Mary Parke's (1949) description of several flagellates originally used in experiments as food for rearing oyster larvae (Bruce et al., 1940) introduced several of the common flagellates encountered in the sea in many parts of the world today. These included *Hemiselmis rufescens*, *Pyramimonas grossii*, *Dicrateria inornata*, and *Isochrysis galbana*. Further flagellate works stimulated by oyster cultivation experiments are those of R. W. Butcher (1952), "Contributions to our knowledge of the smaller marine algae," later followed by the important series "An introductory account of the smaller algae of British coastal waters" (Butcher 1959, 1961, 1967). The latter series is divided into Chlorophyceae (including Prasinophyceae), Euglenophyceae, and Cryptophyceae; despite inconsistencies with present-day taxonomy, this work still remains a useful basis for further studies.

Recent publications on algal protist genera by Sournia (1986) and Chrétiennot-Dinet (1990), together with Farr et al. (1979, 1986), constitute a taxonomic base at the genus level. The information at the species level, however, is still scattered throughout the literature.

The present account is meant to be a simple aid in identifying phytoflagellates that lack a characteristic cell wall but that may have firmly arranged surface coverings composed of organic scales. There is no "key," but the taxa of ranks higher than genus are presented in accordance with present-day systematics. All identifications indicated should be confirmed with the original description of the species in question. It should be noted that the species selected in the illustrations may not be typical in a particular region.

The general problem related to working with "naked" flagellates is the lack of adequate fixing agents, requiring the study of living material for a reliable identification in the light microscope. The proper identification of many species relies upon submicroscopic details revealed only by the electron microscope. When dominant species require electron microscopy for proper identification, materials may be prepared as described in advanced techniques (see Techniques).

"Naked" flagellates are met with in most biotopes, including oceanic and coastal waters as well as rock pools, sandy beaches, and mud flats. Some are heterotrophic while others are phototrophic. In this account, emphasis has been laid on the latter, mainly those found in the plankton. Natural material often has cell densities far too low for conducting the necessary steps for identification. In this instance the use of serial dilution cultures (SDC) increasing the concentrations of the most common species may be very helpful or required (see Techniques).

Following a brief description of the systematics used, a condensed outline of general characteristics for each algal group is presented. The overall systematic organization is that of Christensen (1962, 1966, 1980) but modified in accordance with Moestrup & Throndsen (1988; Moestrup, 1992).

FLAGELLATE TERMINOLOGY

The following is an abbreviated list of terms used in the descriptions of flagellate taxa. Some of these terms are not explained in the text and are thus presented here as an aid to the beginning researcher. For further details and explanation of terms see the glossary at the end of this chapter and the original reference cited with each species description. Abbreviations used are: a., adjective; n., noun; gen., genitive; Gr., Greek; L., Latin; G., German.

Acronematic (a.)—(Gr. *akron*, peak; Gr. *nema*, gen. *nematos*, thread) flagellum terminating in a narrow hair-like part.

Actinopod (n.)—(Gr. *aktis*, gen. *aktinos*, ray; Gr. *pus*, gen. *podos*, foot) ray-shaped pseudopodium.

Anisokont (a.)—(Gr. *anisos*, unlike; Gr. *kontos*, pole) having different flagella of unequal length and/or quality; usually one with flagellar hairs, the other smooth.

Auxotrophic (a.)—(Gr. *auxein*, to increase; Gr. *trophe*, food) photosynthetic organsms that need minor supply of organic substances like vitamins.

Axopodium (n.)—(Gr. *axon*, axle; Gr. *pus*, gen. *podos*, foot) thin, straight pseudopodium supported by microtubules.

Coccoid (a.)—(Gr. *kokkos*, grain) spherical cells with a firm outer wall, without flagella.

Ejectosome (n.)—(L. *ejectus*, thrown out; Gr. *soma*, body) ejectile organelle typical of cryptophyceans but also present in some prasinophyceans.

Extrusome (n.)—(L. *extrudere*, thrust out; Gr. *soma*, body) a membrane-bound organelle usually confined to the cell surface; mucoid content may be extruded to form threads.

Filopodium (n.)—(L. *filum*, thread; Gr. *pus*, gen. *podos*, foot) fine protoplasmic thread protruding from the cell surface, often trailing behind a swimming flagellate or anchoring the cell to the substratum.

Flimmer flagellum (n.)—(G. *Flimmer*, glimmer; L. *flagrum*, whip) flagellum with usually two rows of tripartite tubular hairs along the whole or part of the flagellar axis.

Heterodynamic (a.)—(Gr. *heteros*, different; Gr. *dynamis*, power) regarding flagellar movement, flagella of the same cell being used in different ways, for example, one pulling, the other trailing.

Heterokont (a.)—(Gr. *heteros*, different; Gr. *kontos*, pole, oar) flagella of the cell being different, one with tripartite flagellar hairs, the other smooth.

Homodynamic (a.)—(Gr. *homos*, one and the same; Gr. *dynamis*, power) flagellar movement: flagella of the same cell used (more or less synchronously) in the same mode.

Isokont (a.)—(Gr. *isos*, equal; Gr. *kontos*, pole, oar) the flagella of a cell being of the same kind and length; usually smooth, or with organic scales.

Lorica (n.)—wide cell investment of organic material (cellulose or chitin) impregnated with Mn, Fe, or Si, or made up from silica units as in choanoflagellates.

Mixotrophy—(L. *mixis*, mixing; Gr. *trophe*, nutrition) phototrophic organism taking up organic substances for supplementary nutrition, for example, by phagotrophy.

Monadoid (a.)—(Gr. monas, gen. *monados* unity) unicellular, often flagellated stage.

Mucocyst (n.)—(L. *mucus*, snot; Gr. *kystis*, bag) mucus-containing extrusome.

"Naked"—usually applied to characterize the lack of a cell wall.

Palmelloid (a.)—(like the Chlorophycean genus *Palmella*) a stage where the flagella are thrown off and cells proliferate within a mucilagenous matrix.

Paraxial rod (n.)—(Gr. *para*, beside; L. *axis*, axle) cross-striated structure running parallel with the axoneme in certain flagella.

Phototroph (n.)—(Gr. *phot*, light; Gr. *trophe*, food) organism using light as energy source to produce food.

Pyrenoid (n.)—(Gr. *pyren*, stone [of a fruit]) clear area of the chloroplast stroma naked (chryso-, prymnesiophyceans), with paramylon shield (euglenophyceans), or with starch shield (crypto-, dino-, chloro-, prasinophyceans).

Pyriform (a.)—(L. *pirum*, pear; L. *formis*, form) pear shaped.

Stigma (n.)—(Gr. *stigma*, dot) red or orange-red spot (microgranule or oil droplet) usually confined to the chloroplast, often identical to eyespot.

Stomatocyst (n.)—(Gr. *stoma*, gen. *stomatos*, mouth; Gr. *kystis*, bag) cyst, often silicified, with a narrow opening closed by a pectic plug.

Tractellum (n.)—(L. *tractare,* to handle, from L. *trahere,* to draw) flagellum pulling the cell.

Trichocyst (n.)—(Gr. *thrix,* gen. *trichos* hair; Gr. *kystis,* bag) ejectile organelle, found in raphidophyceans (and dinoflagellates).

PHYTOFLAGELLATE TAXONOMY

The morphology of the living cell is still the basis for flagellate taxonomy. At present, nearly all known genera may be identified by light microscopical criteria; however, species identification may be difficult and dependent on electron microscopy (EM) for verification. In some genera, this is imperative. The morphological properties of living cells, however, are still very important elements that are strongly influenced by the environment, making it necessary to know the cell in its natural condition. Thus any taxonomic study of flagellates will require the use of both types of observational techniques. Neither light nor electron microscopical observations alone can resolve the essentials of morphology needed.

Future identification of some flagellates may be completely dependent on EM analysis of specific morphological structures/features or be replaced by DNA-based identification of genomes. Until the latter becomes a general routine in plankton research, we will face a complex set of features to be used for flagellate identification.

At present, the valid publication of a taxon is based on a diagnosis that includes those features that distinguish one taxon from all other taxa. These diagnoses were traditionally based on morphological characters observed in the light microscope, but almost all modern diagnoses include submicroscopical characters as well. In some cases, the type illustration (usually the holotype) will be an electron micrograph.

In short, light microscopy at present offers a varying degree of possibility for identification. It is generally sufficient to identify class, order, family, and often the genus. At species level, light microscopy may also be sufficient; however, in many genera submicroscopical structures revealed by electron microscopy become increasingly more important.

PHYTOFLAGELLATE SYSTEMATICS

Phytoflagellates in this context will be restricted to species that are able to multiply in the flagellate stage, regardless of whether or not they have a nonmotile stage. In this functional definition, the swarmers of brown and green benthic macro algae are excluded, whereas *Phaeocystis, Halosphaera, Pterosperma,* and *Pachysphaera (Tasmanites)* are included.

Flagella-bearing species are common in all algal classes except Cyanophyceae, Rhodophyceae, Phaeophyceae, and Bacillariophyceae. Though the basic internal structure of the flagella is surprisingly uniform within the eukaryotic algae, the external morphology varies widely and reflects the polyphyletic status of the flagellate group: Flagella may be smooth (**Fig. 1A-Sf**; Prymnesiophyceae, Chlorophyceae), with special hair-like appendages. (**Fig.1A-ff, 1L-fh**; Chrysophyceae, Raphidophyceae, Euglenophyceae, Cryptophyceae, Xanthophyceae, Eustigmatophyceae), or with organic scales (**Fig. 1I**; Prasinophyceae). These are details difficult to observe in the light microscope, but the morphology is often reflected by the mode of use; hence a swimming flagellate may render valuable information also about flagellar morphology. [For flagellar fine structure see Moestrup (1982).]

Flagella and haptonemata are different organelles despite, in some cases, their similarity in appearance. The haptonema is a thread-like appendage, the end of which may stick to the substrate (*hapto*, to attach; *nema*, thread). Long haptonemata will coil when relaxed while short ones will not contract. Anatomically, the haptonema has six to eight longitudinal microtubules surrounded by three coaxial membranes (**Fig. 1J**).

Pigments, flagella and microanatomy are important features in modern algal systematics. The main divisions including flagellate species according to the system of Christensen (1962, 1966) are Chromophyta, lacking chlorophyll *b*, but with chlorophyll *a* and accessory pigments, and Chlorophyta with chlorophylls *a* and *b*:

Division	Chromophyta	Division	Chlorophyta
Class	Cryptophyceae	Class	Euglenophyceae
	Dinophyceae		Prasinophyceae
	Raphidophyceae		Chlorophyceae
	Chrysophyceae		
	Dictyochophyceae		
	Prymnesiophyceae		

The general characteristics of the classes most frequently encountered in the phytoplankton are presented next (letters refer to **Figs. 1** and **2**).

Division Chromophyta: Chlorophyll *a*, not *b*, yellowish-green, golden, blue, or red chloroplasts with lamellae (**Fig. 1B, D, E**); mitochondria with tubular cristae (**Fig. 1H-mtc**).

Class Cryptophyceae: (**Fig. 2A**, also see **Fig. 3**) Asymmetric cell shape, ejectosomes.

Class Dinophyceae: (**Fig. 2B**) Nucleus large, transverse + longitudinal flagellum.

Class Raphidophyceae: (**Fig. 2E**) Many chloroplasts, anterior + posterior pointing flagellum.

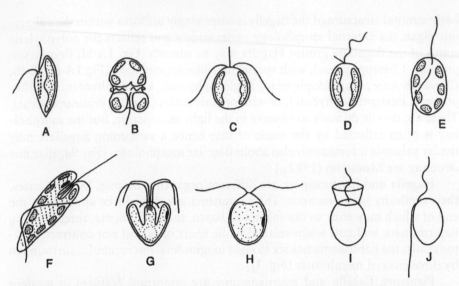

FIGURE 2 "Naked" flagellates from different classes in marine plankton. (A) Cryptophyceae; (B) Dinophyceae; (C) Prymnesiophyceae; (D) Chrysophyceae; (E) Raphidophyceae; (F) Euglenophyceae; (G) Prasinophyceae; (H) Chlorophyceae; (I) Choanoflagellidea; (J) Kinetoplastidea. (Modified from Throndsen, 1980.)

Class Chrysophyceae: (**Fig. 2D**) "Flimmer" flagellum +/− smooth
    flagellum.

Class Dictyochophyceae: "Flimmer"/winged flagellum, +/− radial
    symmetry, +/− external skeleton.

Class Prymnesiophyceae/Haptophyceae: (**Fig. 2C**) Two smooth
    flagella, +/− haptonema.

Division Chlorophyta: Chlorophyll *a* and *b*, green chloroplast with lamellae
or grana structure (**Fig. 1C**) mitochondria with flattened cristae (**Fig. 1H-mfc**).

Class Euglenophyceae: (**Fig. 2F**) Pellicula, flagellar canal with one or two
    emerging flagella, containing paramylon.

Class Prasinophyceae: (**Fig. 2G**) Flagella +/− cell covered with organic
    scales.

Class Chlorophyceae: (**Fig. 2H**) Smooth flagella, +/− cell wall.

Heterotrophic flagellates encountered in the plankton may be closely related to the algal flagellates and hence included in the preceding, or they may be distinctly zooflagellates.

Division Zoomastigophora (devoid of chloroplasts).

Class Choanoflagellidea: (Fig. 2I) One smooth flagellum + pseudopodial collar.

Class Kinetoplastidea: (Fig. 2J) Anisokont: flimmer flagellum +/− smooth flagellum.

Class Ebriidea: Inner silica skeleton.

Algal classes Xanthophyceae and Eustigmatophyceae (see Hibberd, 1981) are less common in marine phytoplankton and therefore are excluded here. Classes Bacillariophyceae and Phaeophyceae have flagellated swarmers only.

The variation in appearance of different species is often great within the flagellate classes, and a short account on each class may be useful.

CHROMOPHYTA

CRYPTOPHYCEAE

Systematic position: Class Cryptophyceae Fritsch 1927 (in West & Fritsch, 1927).

Characteristic Features

Cell shape—asymmetrical with furrow or depression.

Gullet or furrow—lined with two or more rows of ejectosomes (Fig. 3).

Flagella—two, originating at the end of the furrow/gullet, with two and one row of fine tubular hairs (EM).

Mode of swimming—heterodynamic.

Color—brown, green, red, or blue.

Chloroplasts—one or two, with double thylakoids, girdle lamella lacking (EM).

Pyrenoid—often with starch shield, which is contained inside chloroplast ER-membrane (EM).

Storage product—starch, which stains brownish with Lugol's (I_2 KI) solution.

Nutrition—phototrophic and heterotrophic species.

Distribution—plankton, all seasons.

Identification/Observation and Problems

Material preserved with formaldehyde may or may not show flagella, ejectosomes, and cell shape. With iodine (Lugol's), distinct flagella and cell outline can be seen but cell content is obscured by the stain.

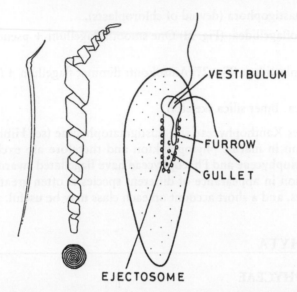

FIGURE 3 Cryptomonad cell morphology, with ejectosomes: whole cell to the right, a single ejectosome charged in front view at lower left, and released (partly and fully extended) at upper left.

Living material allows the most salient features to be observed in bright-field and phase-contrast microscopy. Details of the vestibulum/furrow region may be revealed by interference contrast, but scanning electron microscopy (SEM) may prove necessary. Ejectosomes may reveal their presence by making the cryptomonad produce sudden "jumps." Flagella are easily shed, and chloroplast color is often variable within the species.

Problems are often related to variability in cell shape and color within the species. Electron microscopy has revealed the fine-structural characteristics of several species, and modern taxonomy is based on combined use of SEM and advanced transmission electron microscopy (TEM) methods such as freeze-etching. However, these methods have been of limited use in practical routine identification.

Both colored (photosynthetic) and colorless (heterotrophic) cryptomonad species are found in the marine plankton. The photosynthetic species are included in three families, as follow.

Order Cryptomonadales Engler 1903.

Family Hilleaceae Butcher 1967: Cells devoid of ejectosomes, a longitudinal groove or furrow present.

Genus *Hillea* Schiller 1925: (**Fig. 4A**) Insufficiently known; *H. fusiformis* Schiller, *H. marina* Butcher.

Family Hemiselmidaceae Butcher (1967) *ex* Silva 1980: Cells with transverse furrow/gullet, with ejectosomes.

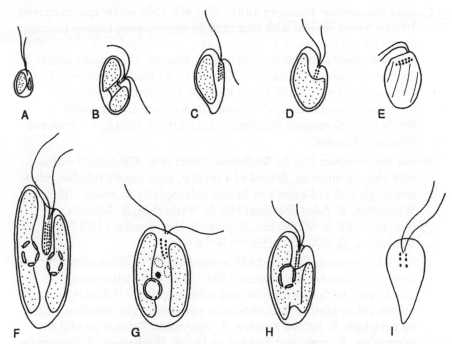

FIGURE 4 Common Cryptophyceae genera from marine plankton: (A) *Hillea* 2.5–25 μm; (B) *Hemiselmis* 4–8.5 μm; (C) *Plagioselmis* 5.5–8 μm; (D) *Rhinomonas* 5–10 μm; (E) *Goniomonas* 5–10 μm; (F) *Cryptomonas* 15–40 μm (freshwater); (G) *Chroomonas* 4–20 μm; (H) *Rhodomonas* 10–30 μm; (I) *Leucocryptos* 15–18 μm.

Genus *Hemiselmis* Parke 1949: (Fig. 4B) With transverse gullet, furrow lacking; chloroplast single, green, blue-green, red, or brown; *H. anomala* Butcher, *H. rotunda* Butcher, *H. rufescens* Parke, *H. virescens* Droop, *H. simplex* Butcher.

Family Cryptomonadaceae Ehrenberg 1831, Pascher 1913: Cells with longitudinal groove-gullet system, with ejectosomes.

Genus *Plagioselmis* Butcher 1967 *ex* Hill 1990: (**Fig. 4C**) Small, 6–10 μm, lacrymuloid cells with a longitudinal furrow; single chloroplast pink, red, orange, or yellow; single pyrenoid; heterodynamic flagella; *P. prolonga* Butcher, *P. punctata* Butcher.

Genus *Isoselmis* Butcher 1967: Like *Plagioselmis*, but with homodynamic flagella and two short rows of ejectosomes; *I. obconica* Butcher.

Genus *Falcomonas* Hill 1990: With longitudinal furrow; cells falcate, 6–10 μm, single chloroplast, green, blue-green; *F. daucoides* (Conrad & Kufferath) Hill.

Genus *Chroomonas* Hansgirg 1885: (**Fig. 4G**) Cells often approximately 10 μm, barrel shaped with two rows of ejectosomes, furrow lacking, branched tubular gullet present, chloroplast bilobed, color blue, blue-green, or green; central eyespot usually present; *C. africana* Meyer & Pienaar, *C. baltica* (Büttner) N. Carter, *C. diplococca* Butcher, *C. dispersa* Butcher, *C. collegionis* Butcher, *C. extensa* Butcher, *C. mesostigmatica* Butcher, *C. monococca* Butcher, *C. placoidea* Butcher, *C. plurococca* Butcher, *C. vectensis* N. Carter, *C. virescens* (Butcher) Butcher.

Genus *Rhinomonas* Hill & Wetherbee 1988: (**Fig. 4D**) Cells 5–10 μm, with rhinote anterior, devoid of a furrow, with simple tubular gullet, and single red, red-brown or brown chloroplast; *R. pauca* Hill & Wetherbee, *R. fulva* (Butcher) Hill & Wetherbee, *R. lateralis* (Butcher) Hill & Wetherbee, *R. fragarioides* (Butcher) Hill & Wetherbee, *R. reticulata* (Lucas) Novarino.

Genus *Rhodomonas* Karsten 1898 (emended Hill & Wetherbee 1989, = *Pyrenomonas* Santore 1984): (**Fig. 4H**) Cells ovoid, 10–30 μm, with short furrow and tubular gullet, bilobed red or red-brown chloroplast with conspicuous pyrenoid, faint striation visible on periplast; *R. baltica* Karsten, *R. abbreviata* Butcher *ex* Hill & Wetherbee, *R. atrorosea* Butcher *ex* Hill & Wetherbee, *R. chrysoidea* Butcher *ex* Hill & Wetherbee, *R. duplex* Hill & Wetherbee (=*Cryptomonas appendiculata* Butcher), *R. falcata* Butcher *ex* Hill & Wetherbee, *R. heteromorpha* Butcher *ex* Hill & Wetherbee, *R. maculata* Butcher *ex* Hill & Wetherbee, *R. marina* (Dangeard) Lemmermann, *R. salina* (Wislouch) Hill & Wetherbee.

Genus *Geminigera* Hill 1991: Cells 15–18 μm, ovoid with furrow and sack-like gullet, and single red or red-brown chloroplast with two conspicuous pyrenoids; *G. cryophila* (Taylor & Lee) Hill.

Genus *Proteomonas* Hill & Wetherbee 1986: Small faintly striated cells 6–9 μm or large 11–15 μm fusiform cells (alternating life stages), with a true furrow, gullet lacking, and single chloroplast pink, red, or red-orange, with single pyrenoid; *P. sulcata* Hill & Wetherbee.

Genus *Cryptomonas* Ehrenberg 1832: (**Fig. 4F**) Cells 15–40 μm, mostly ovoid with furrow and sack-like gullet, two brown, yellow-brown or yellow-green chloroplasts, each with a pyrenoid. Probably confined to fresh water only, marine species transferred to other genera.

Genus *Teleaulax* Hill 1991: Cells 12–15 μm with acute anterior and posterior, with long furrow, single red-brown or orange chloroplast with single pyrenoid; *T. acuta* (Butcher) Hill, *T. amphioxeia* (Conrad) Hill.

Genus *Storeatula* Hill 1991: Cells 20–25 μm, oval, laterally compressed without furrow, with long tubular gullet, single multilobed/reticulate red or red-brown chloroplast with single conspicuous pyrenoid; *S. major* Butcher *ex* Hill.

Genus *Leucocryptos* Butcher 1967: (**Fig. 4I**) Colorless, phagotrophic, without true furrow or gullet, with few ejectosomes present in ventral rows; *L. marina* (Braarud) Butcher.

Genus *Goniomonas* Stein 1878: (**Fig. 4E**) Cells 5–10 μm, truncate, laterally compressed, colorless, with anterioventral furrow, flagella inserted dorsally and ejectosomes in a transverse ring; *G. truncata* (Fresenius) Stein, *G. pacifica* Larsen & Patterson, *G. amphinema* Larsen & Patterson.

(Systematics according to Hill & Wetherbee, 1988, 1989, and D. R. A. Hill, personal communication, 1990.)

Selected Cryptophycean Species When referring to species measurements, please note that numbers given in decimals (e.g., 2.5 μm) are based on actual measurements while those given as fractions (e.g., flagella 1½ × cell length) are approximated measurements.

Hillea fusiformis (Schiller) Schiller 1925 (**Plate 1**)
Chlamydomonas fusiformis Schiller 1913
Cell length: 6–7 μm (Schiller, 1913), 8–10 μm (Schiller, 1925), 5–10 μm (Butcher, 1967).
Flagella: Two, twice the cell length.
Chloroplasts: Single, green to blue-green.
Characteristic features: Cells with a faint groove from apex to about the middle of the cell.
Distribution: Coastal; Mediterranean.

Hillea marina Butcher 1952 (**Plate 1**)
Cell length: 2.5 μm.
Flagella: Two, subequal, longer than the cell.
Chloroplasts: Single, dull yellow.
Characteristic features: Cells laterally flattened, with distinct pyrenoid.
Distribution: Coastal; Atlantic, Mediterranean.

Hemiselmis virescens Droop 1955 (**Plate 1**)
Cell length: 4.5–7 μm.
Flagella: Two, unequal.
Chloroplasts: Single, green.
Characteristic features: Cell not compressed, but with ventral depression.
Distribution: Coastal; Atlantic, Pacific.

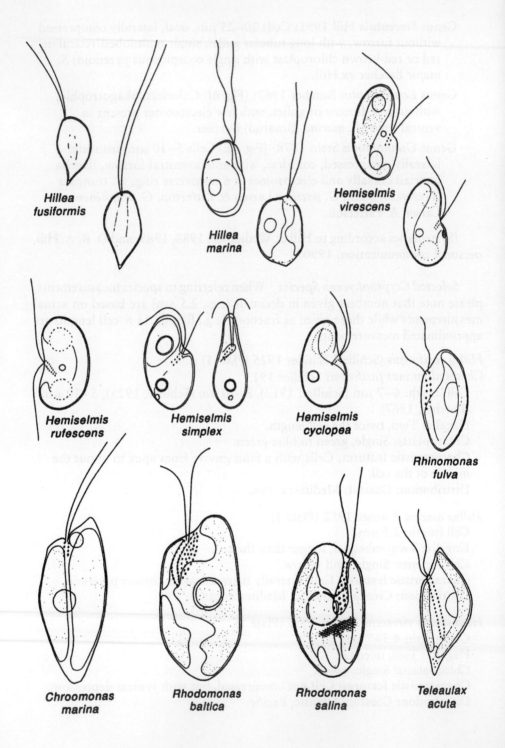

Hillea
fusiformis

Hillea
marina

Hemiselmis
virescens

Hemiselmis
rufescens

Hemiselmis
simplex

Hemiselmis
cyclopea

Rhinomonas
fulva

Chroomonas
marina

Rhodomonas
baltica

Rhodomonas
salina

Teleaulax
acuta

Hemiselmis rufescens Parke 1949 (Plate 1)
Cell length: 4–8.5 μm.
Flagella: Two, 1½ × cell length, and shorter.
Chloroplasts: Single, crimson.
Characteristic features: Cell bean-shaped with a "stigma" on the ventral surface.
Distribution: Coastal; Atlantic.

Hemiselmis simplex Butcher 1967 (Plate 1)
Cell length: 5–6.5 μm.
Flagella: Two, subequal.
Chloroplasts: Single, Paris green.
Characteristic features: Cell laterally compressed, pyriform in ventral view, with refractive, pale orange colored posterior body.
Distribution: Coastal; Atlantic, Pacific.

Hemiselmis cyclopea Butcher 1967 (Plate 1)
Cell length: 4.5–5.5 μm.
Flagella: Two, subequal.
Chloroplasts: Single, poppy red.
Characteristic features: Cell with distinct refractive body.
Distribution: Coastal; Atlantic, Pacific.

Rhinomonas fulva (Butcher) Hill & Wetherbee 1988 (Plate 1)
(*Chroomonas fulva* (Butcher) Butcher 1967)
Cryptochrysis fulva Butcher 1952
Cell length: 5–7 μm.
Flagella: Two, subequal, rather longer than the cell.
Chloroplasts: Single, bronze yellow.
Characteristic features: Cell with bluntly rhinote anterior, prominent central pyrenoid.
Distribution: Coastal; Atlantic.

Chroomonas marina (Büttner) Butcher 1967 (Plate 1)
Cryptomonas marina Büttner 1911
Cell length: 18–22 μm.
Flagella: Two.
Chloroplasts: Two, pyrenoid lacking.
Characteristic features: Cell scarcely compressed, with contractile vacuole.
Distribution: Coastal; Atlantic.

Plate 1 *Hillea fusiformis, H. marina, Hemiselmis virescens, H. rufescens, H. simplex, Chroomonas marina, Rhodomonas salina, Teleaulax acuta* based on type illustrations; *Hemiselmis cyclopea,* original from Mikawa Bay, Japan; *Rhinomonas fulva,* original from Misumi Bay, Japan; *Rhodomonas baltica* based on Zimmermann (1925).

Rhodomonas baltica Karsten *sensu* Zimmerman 1925 (**Plate 1**)
Cell length: 18–30 μm.
Flagella: Two, slightly shorter than the cell.
Chloroplasts: One or two, lobed, poppy red to olive yellow.
Characteristic features: Cell somewhat flattened, with short gullet lined
with three or four rows of ejectosomes.
Distribution: Coastal, oceanic; Baltic, Atlantic.

Rhodomonas salina (Wislouch) Hill & Wetherbee 1989 (**Plate 1**)
Cryptomonas salina Wislouch 1924
Cell length: 14–17 μm.
Flagella: Two.
Chloroplasts: Single, lobed.
Characteristic features: Cell with short but prominent gullet lined with
ejectosomes, pellicular striations stands out in phase contrast.
Distribution: Coastal; Black Sea, Pacific.

Teleaulax acuta (Butcher) Hill 1991 (**Plate 1**)
Cryptomonas acuta Butcher 1952
Cell length: 12–15 μm.
Flagella: Two, shorter than the cell.
Chloroplasts: One or two, red-brown to golden brown.
Characteristic features: Cell with prominent pyrenoid, acute posterior
with minute oscillating granules, several rows of ejectosomes.
Distribution: Coastal, offshore; Atlantic.

Leucocryptos marina (Braarud) Butcher 1967
Chilomonas marina (Braarud) Halldal 1953
Bodo marina Braarud 1935 (illustrated in **Fig. 4I**)
Cell length: 15–18 μm.
Flagella: Two, equal, longer than the cell.
Chloroplasts: Lacking.
Characteristic features: Cell pyriform with two ventral rows of
ejectosomes (5–18 in each row).
Distribution: Coastal, oceanic; Atlantic, Arctic, Pacific.

RAPHIDOPHYCEAE

Systematic position: Class Raphidophyceae Chadefaud *ex* Silva 1980 = Class
Chloromonadophyceae Papenfuss 1955.

Characteristic Features

Cell shape—ovoid to pyriform, asymmetric, more or less flattened. A more
or less pronounced flagellar groove may be present (*Olisthodiscus,
Heterosigma*).

Cell surface—apparently smooth, but may bear extrusomes (Heterosigma), the subsurface layer often with trichocysts (*Chattonella, Fibrocapsa*).

Flagella—two, one flimmer flagellum, pointing forward, the other also with hairs, often in a shallow ventral groove, pointing backward.

Mode of swimming—rapid movements of the anterior-pointing flagellum pull the cell, the posterior-pointing flagellum apparently less active.

Color—marine species yellow to yellowish-brown (with fucoxanthin, affinities to Chrysophyceae, freshwater species green, with pigment affinities to Xanthophyceae).

Chloroplasts—many discoid, with thylakoids in triplets and +/- girdle lamellae.

Pyrenoids—naked, submerged, on inside of chloroplasts (may be lacking).

Nutrition—only phototrophic species presently known.

Distribution—coastal and shallow marine or brackish waters, may cause toxic discolored water.

Identification/Observation and Problems

Preserved material is usually inadequate for identification, but may be useful for estimating cell concentration during blooms once the identity of the species has been confirmed.

Living material is a requisite for reliable identification of Raphidophycean species. Examination with light and electron microscopes may be needed to reveal internal structures.

Problems are due to the sensitivity of the cells under the microscope, where flagella are often quickly shed and surface details such as flagellar grooves become smooth or obscure, in addition to a wide variation in cell shape.

* Order Chattonellales ord. nov. (formal diagnosis: Cellula solitaria flagellis binis opposite directis natans. Chloroplasti numerosi fucoxanthino colorati, chlorophyllo b carentes): The order Chattonellales is based on *Chattonella subsalsa* Biecheler (1936, p. 80), and comprises genera with fucoxanthin as the main accessory pigment. These cells have many yellow- green chloroplasts with pyrenoids. Species lacking fucoxanthin, that is, those with bright green chloroplasts, are to be retained in the order Raphidomonadales Chadefaud (1960) with family Vacuolariaceae Luther (1899), Loeblich and Loeblich (1978), mainly restricted to fresh water.

  * Family Chattonellaceae fam. nov. (formal diagnosis: Familia Rhaphidophyceas complectens fucoxanthino inter carotenoides primario donatas, pyrenoidibus instructas): With chloroplasts containing fucoxanthin as the main accessory pigment, and with pyrenoids, based

---

* All nomenclatural novelties were validated in the original publication, Tomas, C. (ed.) Marine Phytoplankton: A Guide to Naked Flagellates and Coccólithophorids, Academic Press, 1993.

on *Chattonella subsalsa* Biecheler, B. 1936 Archives de zoologie experimentale et génerale 78, p. 80, comprising five marine genera.

Genus *Chattonella* Biecheler 1936: *C. subsalsa* Biecheler, *C. marina* (Subrahmanyan) Hara & Chihara, *C. antiqua* (Hada) Ono, *C. verruculosa* Hara & Chihara. The distinguishing morphological features are given in Hara & Chihara (1982), Fukuyo et al. (1990).

Genus *Olisthodiscus* N. Carter 1937: Cells strongly dorsiventrally compressed; *O. luteus* N. Carter, *O. magnus* Hulburt, *O. carterae* Hulburt.

Genus *Heterosigma* Hada 1968: Cells moderately compressed, with mucus bodies; *H. inlandica* Hada, *H. akashiwo* (Hada) Hada.

Genus *Fibrocapsa* Toriumi & Takano 1973: With trichocysts, especially in the posterior part; *F. japonica* Toriumi & Takano.

Genus *Oltmannsia* Schiller 1925: With "ribbon-shaped," apical flagella, tentatively included in Raphidophyceae; *O. viridis* Schiller.

Selected species of Raphidophyceae

Chattonella subsalsa Biecheler 1936 (Plate 2)
Cell length: 30–50 μm.
Flagella: Two, swimming and trailing flagellum, approximately equal to cell length.
Chloroplasts: Many, green (yellowish green).
Characteristic features: The cell is asymmetric in side view, and with pointed posterior end. Ejected mucocysts oboe shaped. Often regarded as synonymous with *Hornellia marina* Subrahmanyan (1954) e.g., Sournia (1986), but see Hara and Chihara (1982).
Distribution: Described from brackish areas rich in organic material; coastal; Mediterranean.

Olisthodiscus luteus N. Carter 1937 (Plate 2)
Cell length: 12–19 μm.
Flagella: two, swimming flagellum 1¼–1½ × cell length, trailing flagellum cell length or shorter.
Chloroplasts: Many, pale yellow.
Characteristic features: Cell exceedingly flattened, ventral side devoid of chloroplasts, not rotating when swimming (EM; Hara et al., 1985).
Distribution: Salt marshes; Europe, Japan.

Heterosigma inlandica Hada 1968 (Plate 2)
Cell length: 10–18 μm.
Flagella: Approximately equal to cell length, only one observed.
Chloroplasts: 10–20 Greenish brown, disc-shaped.

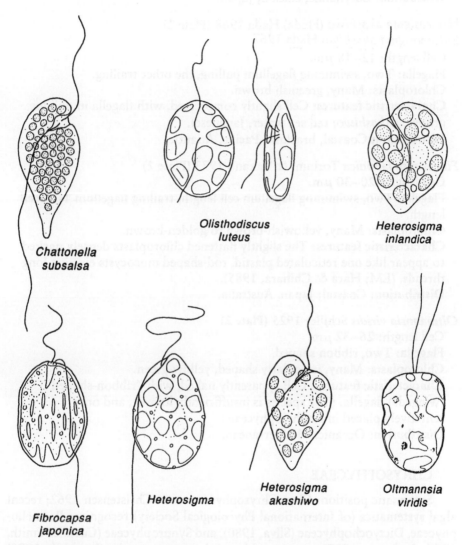

Plate 2 *Chattonella subsalsa, Olisthodiscus luteus, Heterosigma inlandica, H. akashiwo, Fibrocapsa japonica, Oltmannsia viridis* based on type illustrations; *heterosigma* sp. original from the Oslofjord, Norway.

Characteristic features: Cell more or less compressed with a branched (?) ventral groove, probably conspecific with *Heterosigma akashiwo*.
Distribution: Euryhaline, coastal; Japan.

Heterosigma akashiwo (Hada) Hada 1968 (Plate 2)
Entomosigma akashiwo Hada 1967
Cell length: 12–18 μm.
Flagella: Two, swimming flagellum pulling, the other trailing.
Chloroplasts: Many, greenish brown.
Characteristic features: Cell slightly compressed, with flagella in oblique groove. (*Akashiwo;* red sea water, Japanese).
Distribution: Coastal, brackish; Pacific, Atlantic.

Fibrocapsa japonica Toriumi & Takano 1973 (Plate 2)
Cell length: 20–30 μm.
Flagella: Two, swimming flagellum cell length, trailing flagellum 1¼ × cell length.
Chloroplasts: Many, yellowish-brown to golden-brown.
Characteristic features: The slightly flattened chloroplasts densely packed to appear like one reticulated plastid, rod-shaped mucocysts ejecting long threads. (EM; Hara & Chihara, 1985).
Distribution: Coastal; Japan, Australia.

Oltmannsia viridis Schiller 1925 (Plate 2)
Cell length: 26–32 μm.
Flagella: Two, ribbon shaped.
Chloroplasts: Many, irregularly shaped, yellow-green.
Characteristic features: Cell apparently naked, with "ribbon-shaped" (winged?) flagella. (The species is insufficiently known, and only tentatively placed in Raphidophyceae.)
Distribution: Oceanic: Mediterranean.

CHRYSOPHYCEAE

Systematic position: Class Chrysophyceae *sensu* Christensen 1962; recent algal systematics (of International Phycological Society) recognizes Pedinellophyceae, Dictyochophyceae (Silva, 1980), and Synurophyceae (Cavalier-Smith, 1986; Andersen, 1987) as separate classes. In zoosystematics (Lee et al., 1985) the order Chrysomonadida and order Silicoflagellida are treated separately.

Characteristic Features

Cell shape—often round or pyriform.

Cell covering—naked or covered by scales, which may be silicified, cellulose or chitin lorica present in some species.

Flagella—heterodynamic, flimmer flagellum + smooth flagellum.

Mode of swimming—the flimmer flagellum pointing forward, pulling the cell.

Silicified cysts—produced endogenously.

Color—yellow to golden brown.

Chloroplasts—one or two with triple thylakoids and girdle lamella (EM).

Storage product—chrysolaminaran (liquid β-1,3-glucan) to be stained *in vivo* by Brilliant Cresyl Blue (to produce rose color). Fat and oil may be present.

Nutrition—phototrophic, mixo- and heterotrophic species.

Distribution—Planktonic species found in coastal and oceanic areas but mostly confined to inshore waters, especially brackish sheltered areas. Benthic flagellate species and several multicellular species forming threads and simple thalli are present in marine and brackish environments.

Identification/Observation and Problems

Preserved material is usually inadequate unless the species can be identified by a cellulose lorica or silicified scales. Though iodine (Lugol's solution) may save the flagella, the cell shape is only occasionally preserved well enough for proper identification.

Living material offers the best possibilities for identification in the light microscope, and for some species, special life history stages are necessary for a reliable identification (e.g., cysts in many *Dinobryon* species).

Problems are that the marine species of the Chrysophyceae are still insufficiently known and thin sections in EM may be necessary for some identifications.

The class contains a variety of forms: walled and naked cells, monadoid, rhizopodoid, palmelloid, coccoid, as well as many cellular species. The "naked" flagellate species are included in the order Ochromonadales with two unequal flagella (Pedinellales with a single winged flagellum, and Dictyochales with an external silica skeleton have recently been included in a separate class Dictyochophyceae, Moestrup, 1992). Coccoid forms like *Meringosphaera* Lohmann, *Pelagococcus subviridis* Norris (in Lewin et al., 1977), and *Aureococcus anophagefferens* Hargraves & Sieburth (in Sieburth et al., 1988) have recently been shown to be important in marine areas (Throndsen & Kristiansen, 1988; Sieburth et al., 1988; Cosper et al., 1989). *Meringosphaera* Lohmann is included despite its nonflagellate character.

Order Bicosoecales Grassé 1926: Colorless cells, often attached by the smooth flagellum; a more or less prominent lip is used in food uptake.

Family Cafeteriaceae Moestrup 1992: Cells without lorica. See Plate 17 for illustrations.

Genus *Cafeteria* Fenchel & Patterson 1988: Cells sessile, fixed by the smooth flagellum; *C. roenbergensis* Fenchel & Paterson, *C. minuta* (Ruinen) Larsen & Patterson.

Genus *Pseudobodo* Griessman 1913: Cells rounded, wider than long, highly compressed, flagella inserted on the side; *P. tremulans* Griessmann, *P. minimus* Ruinen.

Family Bicosoecaceae Stein 1878: Cells with loricae, has also been treated as a separate zooflagellate class (Bicoecids; Dyer, 1990), but see Moestrup (1992).

Genus *Bicosoeca* James-Clark 1866: (Fig. 5C) With cell attached to lorica by the smooth ˉagellum; *B. maris* Picken, *B. mediterranea* Pavillard, *B. gracilipes* James-Clark, *B. griessmannii* (Griessmann) Bourrelly.

Order Ochromonadales Pascher 1910: Contains four flagellate families in marine environments.

Family Ochromonadaceae Lemmermann 1899: Cells are naked and free living.

Genus *Ochromonas* Wyssotski 1888: (Fig. 5A) Cells elongated, with one or two chloroplasts; *O. bourrellyi* Magne, *O. cosmopoliticus* Ruinen, *O. marina* Lackey, *O. minima* Throndsen, *O. mexicana* Norris (*O. crenata* Klebs, *O. oblonga* N. Carter, halotolerant species).

Genus *Sphaleromantis* Pascher 1910: Cells compressed, with two chloroplasts; for example, *S. marina* Pienaar, *S. subsalsa* Conrad, *S. alata* Conrad, *S. ochracea* Pascher.

Genus *Boekelovia* Nicolai & Baas-Becking 1935: Cells triangular, compressed, flagella inserted ventrally; *B. hooglandii* Nicolai & Baas-Becking.

Genus Spumella Cienkowski 1870 (= *Monas* O. F. Müller = *Heterochromonas* Pascher 1912); Colorless cells; *Spumella hovassei* (Fiatte & Joyon) Bourrelly.

Family Chromulinaceae Engler 1897 (1898): Cells with one flagellum visible with light microscopy.

Genus *Chromulina* Cienkowski 1870: (Fig. 5L) Spherical or elongated cells; *C. pleiades* Parke.

Family Chrysococcaceae Lemmermann 1899: Cells with "narrow" loricae.

Genus *Calycomonas* Lohmann 1908 (= *Codonomonas* Van Goor 1925b): (Fig. 5H) Heterotrophic; *C. gracilis* Lohmann, *C. cylindrica* (Conrad & Kufferath) Lund, *C. dilatata* (Conrad & Kufferath) Lund, *C.*

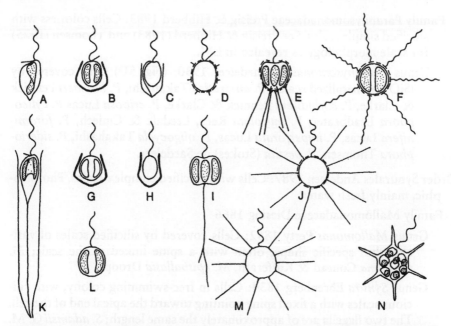

FIGURE 5   Chryso- and Dictyochophycean genera from marine plankton: (A) *Ochromonas* 3–10 μm; (B) *Pseudokephyrion* cell 5–7 μm, lorica 9–11 μm; (C) *Bicosoeca* cell 4–6 μm, lorica 10–30 μm; (D) *Paraphysomonas* 2–20 μm; (E) *Apedinella* 7.5–10 μm; (F) *Meringosphaera* 4–9 μm; (G) *Kephyrion* cell 5 μm, lorica 4.5–18 μm; (H) *Calycomonas* cell 3–8 μm, lorica 4–15 μm; (I) *Pseudopedinella* 3.5–8 μm; (J) *Parapedinella* 7–15 μm + actinopodia; (K) *Dinobryon* cell 7–11 μm, lorica 19–64 μm; (L) *Chromulina* 3–10 μm; (M) *Actinomonas* 6–9 μm + actinopodia; (N) *Dictyocha* 10–45 μm + spines. (Modified from Throndsen, 1980.)

globosa Lohmann, *C. vangoorii* (Conrad) Lund, *C. wulffii* Conrad & Kufferath (*C. ovalis* Wulff transferred to genus *Paulinella* Lauterborn; testate rhizopod: *P. ovalis* (Wulff) Johnson, Hargraves & Sieburth).

Genus *Kephyrion* Pascher 1911: (**Fig. 5G**) Photosynthetic, *K. cinctum* (Lackey) Bourrelly, *K. petasatum* Conrad, and other species in brackish water.

Family Dinobryaceae Ehrenberg 1834 (Dinobryonaceae Ehrenberg 1838): Cells with two flagella, in open cellulose loricae.

Genus *Dinobryon* Ehrenberg 1834: (**Fig. 5K**) Cells attached by a cytoplasmic strand in "deep" loricae, +/− colonies; *D. balticum* (Schütt) Lemmermann, *D. belgica* Meunier, *D. coalescens* Schiller, *D. faculiferum* (Willén) Willén, *D. porrectum* Schiller, *D. mediterraneum* Pavillard.

Genus *Pseudokephyrion* Pascher 1913: (**Fig. 5B**) Solitary forms in "short" loricae; *P. formosissimum* Conrad.

Family Paraphysomonadaceae Preisig & Hibberd 1983: Cells colorless with silicified simple scales. See Preisig & Hibberd (1983) and Thomsen (1975) for scale morphology as revealed in EM.

Genus *Paraphysomonas* de Saedeleer 1930: (**Fig. 5D**) Cells covered by (SiO$_2$) mineralized scales; *P. antarctica* Takahashi, *P. butcheri* Pennick & Clarke, *P. corbidifera* Pennick & Clarke, *P. cribosa* Lucas *P. cylicophora* Leadbeater, *P. faveolata* Rees, Leedale & Cmiech, *P. foraminifera* Lucas, *P. imperforata* Lucas, *P. oligocycla* Takahashi, *P. sideriophora* Thomsen, *P. vestita* (Stokes) de Saedeleer.

Order Synurales Andersen 1987: Cells with silicified complex scales. Phototrophic, mainly fresh water.

Family Mallomonadaceae Diesing 1866

Genus *Mallomonas* Perty 1852: Cells covered by silicified scales of variable, but specific shape, often with a spine linked to the scale; *M. subsalina* Conrad & Kufferath, *M. epithallatia* Droop.

Genus *Synura* Ehrenberg 1834: Cells in free-swimming colony, with silicious scales with a fixed spine pointing toward the apical end of the cell. The two flagella are of approximately the same length; *S. adamsii* G. M. Smith.

Order Chrysosphaerales Bourrelly 1957a: A predominating nonmotile stage (zoospores uniflagellated).

Family Aurosphaeraceae Schiller 1925: Cells with silicified scales and spines.

Genus *Meringosphaera* Lohmann 1902: (**Fig. 5F**) Cells spherical with radiating spines; *M. mediterranea* Lohmann, *M. tenerrima* Schiller 1925. See also Pascher (1932).

Family Chrysosphaeraceae Pascher 1914: Cells without spines.

Genus *Aureococcus* Hargraves & Sieburth (in Sieburth et al., 1988); *A. anophagefferens* Hargraves & Sieburth.

Genus *Pelagococcus* Norris 1977; *P. subviridis* Norris (in Lewin et al., 1977).

Order Sarcinochrysidales Gayral & Billard 1977: Flagellated swarmers.

Family Sarcinochrysidaceae Gayral & Billard 1977: Swarmers with trailing flagellum.

Genus *Sarcinochrysis* Geitler 1930; e.g., *S. marina* Geitler.

Selected Flagellate Species of Chrysophyceae and Dictyochophyceae Species naked or with organic scales:

Ochromonas oblonga N. Carter 1937 (Plate 3)
   Cell length: 8–10 μm.

Flagella: Two, 1–1 ½ and ⅓ × cell length.
Chloroplasts: Single, parietal with bright red eyespot.
Characteristic features: Cell posterior with dark granular mass probably
indicating phagotrophic nutrition.
Distribution: Littoral halotolerant species; Europe.

Ochromonas crenata Klebs 1893 (Plate 3)
Cell length: 12–20 μm.
Flagella: Two, the longer 2 × cell length.
Chloroplasts: Single, twisted.
Characteristic features: Cell surface warty due to mucocysts.
Distribution: Brackish water; Europe.

Ochromonas marina Lackey 1940 (Plate 3)
Cell length: 35 μm.
Flagella: Two, unequal.
Chloroplasts: Two, golden-brown.
Characteristic features: Cell with smooth anterior and warty posterior
surface.
Distribution: Coastal; northwest Atlantic.

Ochromonas cosmopoliticus Ruinen 1938 (Plate 3)
Cell length: 10–16 μm.
Flagella: Two, 1½ and ½ × cell length.
Chloroplasts: Single, yellow-brown?
Characteristic features: Cell with a prominent median vacuole, flagella
protruding from apical pointed part of the cell.
Distribution: Coastal (salines); Atlantic.

Ochromonas bourrellyi Magne 1954 (Plate 3)
Cell length: 7–8 μm.
Flagella: Two, the longer approximately cell length.
Chloroplasts: Two, yellow.
Characteristic features: Cells with one or more contractile vacuoles,
usually with large posterior chrysolaminaran vacuole.
Distribution: Inshore; Atlantic.

Ochromonas minima Throndsen 1969 (Plate 3)
Cell length: 3.5–6.5 μm.
Flagella: Two, 1–2 and ⅓ × cell length.
Chloroplasts: Single, yellow-brown to yellow-green.
Characteristic features: Cell laterally flattened, with twisted chloroplast.
Distribution: Coastal; Atlantic.

Sarcinochrysis marina Geitler 1930 (Plate 3)
Cell length: 6–7.5 μm.

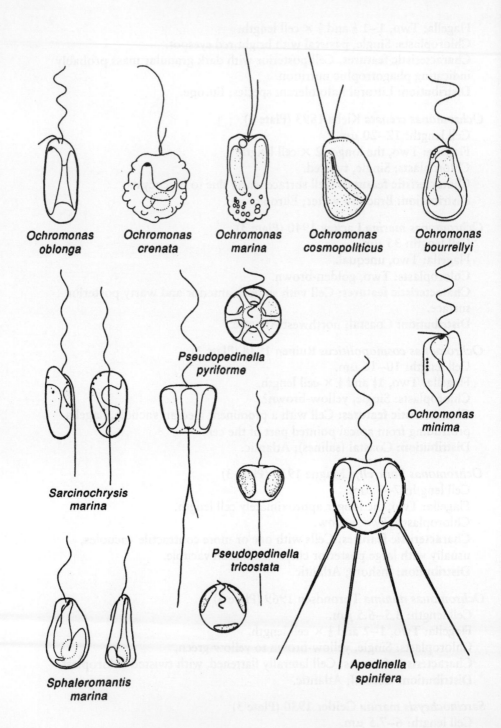

Ochromonas oblonga

Ochromonas crenata

Ochromonas marina

Ochromonas cosmopoliticus

Ochromonas bourrellyi

Sarcinochrysis marina

Pseudopedinella pyriforme

Ochromonas minima

Pseudopedinella tricostata

Sphaleromantis marina

Apedinella spinifera

Flagella: Two, 2–3 and 1¼–1½ × cell length (trailing).
Chloroplasts: Two, yellow-brown.
Characteristic features: Flagellated cells are swarmers of the palmelloid stage, and the shorter flagellum is trailing.
Distribution: Coastal; Atlantic.

Pseudopedinella pyriforme N. Carter 1937 (Plate 3)
Syn. *P. variabilis* Rouchijajnen 1968
Cell length: 5–8 μm.
Flagella: 3–5 × cell length.
Chloroplasts: Six, yellow-brown.
Characteristic features: Cell usually with a long trailing pseudopodium.
Distribution: Coastal, Oceanic; Atlantic, Arctic, Pacific, Mediterranean.

Pseudopedinella tricostata (Rouchijajnen) Thomsen 1988 (Plate 3)
Pedinella tricostata Rouchijajnen 1966
Cell length: 4–5½ μm.
Flagella: 10–12 μm.
Chloroplasts: Three, golden-brown.
Characteristic features: Cell with three chloroplasts only, originally described with axopodia protruding anteriorly.
Distribution: Coastal, oceanic; Black Sea, Pacific, Mediterranean.

Apedinella spinifera (Throndsen) Throndsen 1971 (Plate 3)
Apedinella radians (Lohmann) Campbell 1973
Meringosphaera radians Lohmann, 1908
Cell length: 6½–10 μm.
Flagella: 1–2 × cell length.
Chloroplasts: Six, yellow-brown.
Characteristic features: Cell covered with cellulose scales, four to nine (six) spiny scales easily recognizable in light microscope (phase contrast).
Distribution: Coastal; Atlantic, Pacific, Mediterranean, Arctic.

Sphaleromantis marina Pienaar 1976 (Plate 3)
Cell length: 5–7 μm.
Flagella: Two, 6 μm and 2 μm.
Chloroplast: Single, lobed, yellow-brown.
Characteristic features: Triangular cells, with two types of body scales and flagellar scales (EM).
Distribution: Coastal; Pacific.

Plate 3 *Ochromonas oblonga, O. marina, O. cosmopoliticus, O. bourrellyi, O. minima, Sarcinochrysis marina, Sphaleromonatis marina* based on type illustrations; *Ochromonas crenata* redrawn from Conrad (1926); *Pseudopedinella pyriforme, P. tricostata,* and *Apedinella spinifera,* originals.

Selected Chrysophycean Species with Lorica

Dinobryon balticum (Schütt) Lemmermann 1900 (**Plate 4**)
Dinodrendron balticum Schütt 1892
Syn. *Dinobryon pellucidum* Levander 1894
   Lorica length: 50–66 μm (basal), 32–35 μm (distal part of the colony).
   Lorica width: 3–5 μm.
   Characteristic features: Ochromonadoid cells in long narrow loricae,
   species often regarded as conspecific with *Dinobryon pellucidum*.
   Distribution: Oceanic; Baltic, Atlantic, Arctic.

Dinobryon coalescens Schiller 1925 (**Plate 4**)
   Lorica length: 50–60 μm.
   Characteristic features: Loricae coalescing, few individuals per colony.
   Distribution: Coastal; Mediterranean.

Dinobryon porrectum Schiller 1925 (**Plate 4**)
   Lorica length: 26–40 μm.
   Lorica width: 6–7 μm.
   Characteristic features: Neighboring lorica walls nearly parallel, small
   colonies.
   Distribution: Coastal; Mediterranean.

Dinobryon belgica Meunier 1910 (**Plate 4**)
   Lorica length: approximately 24 μm.
   Lorica width: approximately 7 μm.
   Characteristic features: Cells with short wide loricae.
   Distribution: Oceanic; Arctic.

Dinobryon faculiferum (Willén) Willén 1992 (**Plate 4**)
Dinobryon petiolatum sensu Willén 1963
   non *Dinobryon petiolatum* Dujardin 1841
   Lorica length: 65–85 μm.
   Lorica width: 4–4.8 μm.
   Characteristic features: Solitary, lorica with extremely long basal spine.
   Distribution: Coastal, oceanic; Baltic, Atlantic.

Bicosoeca maris Picken 1941 (**Plate 4**)
   Lorica length: 10 μm.
   Cell length: 4–6 μm.
   Flagella: Two, 10–25 μm and 6–10 μm.
   Chloroplasts: Lacking.
   Characteristic features: Lorica with a short pointed stalk.
   Distribution: Coastal; Atlantic.

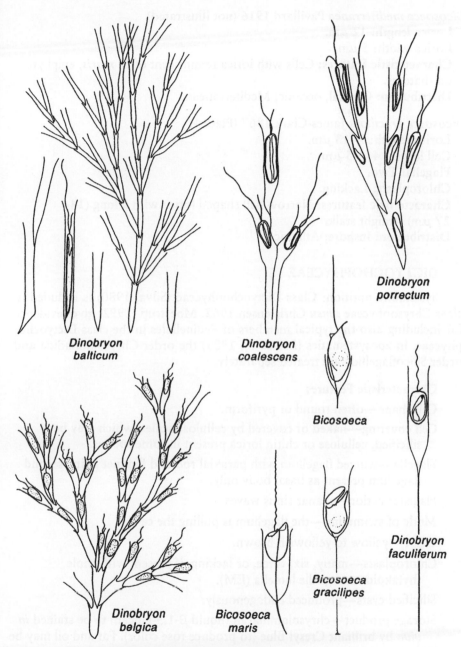

Plate 4 *Dinobryon coalescens, D. porrectum, D. belgica, Bicosoeca gracilipes, B. maris* based on type illustration; *Dinobryon balticum* redrawn from Lemmermann as reproduced in Starmach (1980); *Dinobryon faculiferum* and *Bicosoeca* sp., originals.

Bicosoeca mediterranea Pavillard 1916 (not illustrated)
Lorica length: 12 μm.
Lorica width: 5 μm.
Characteristic features: Cells with lorica reminiscent of *B. maris*, epiphyte on diatoms.
Distribution: Coastal, oceanic; Mediterranean.

Bicosoeca gracilipes James-Clark 1867 (**Plate 4**)
Lorica length: 9–13 μm.
Cell length: 4–6.5 μm.
Flagella: Two.
Chloroplasts: Lacking.
Characteristic features: Narrow bell shaped lorica with a long (14–27 μm) straight stalk.
Distribution: Inshore; Atlantic.

DICTYOCHOPHYCEAE

Systematic position: Class Dictyochophyceae (Silva, 1980) is included in class Chrysophyceae *sensu* Christensen 1962. Moestrup (1992) gives evidence for including also the typical members of Pedinellales in the class Dictyochophyceae. In zoosystematics (Lee et al., 1985) the order Chrysomonadida and order Silicoflagellida are treated separately.

Characteristic Features

Cell shape—often round or pyriform.

Cell covering—naked or covered by cellulose scales, which may be silicified, cellulose or chitin lorica present in some species.

Flagella—winged flagellum with paraxial rod and flimmer hairs, second flagellum present as basal body only.

Flagellar action—planar sinus waves.

Mode of swimming—the flagellum is pulling the cell.

Color—yellow to yellowish brown.

Chloroplasts—many, six, three, or lacking, if present with triple thylakoids and girdle lamella (EM).

Silicified cysts—produced endogenously.

Storage product—chrysolaminaran (liquid β-1,3-glucan) to be stained *in vivo* by brilliant Cresyl Blue (to produce rose color). Fat and oil may be present.

Nutrition—phototrophic, mixo- and heterotrophic species.

Distribution—planktonic species found in coastal and oceanic areas but mostly confined to inshore waters, especially brackish sheltered areas.

Benthic flagellate species and ameboid species are present in marine and brackish environments.

Identification/Observation and Problems

Preserved material is usually inadequate unless the species can be identified by a silica skeleton or organic scales. Though iodine (Lugol's solution) may save the flagellum, the typical cell shape is only occasionally kept for proper identification.

Living material offers the best possibilities for identification in the light microscope except when silica structures are present. Information on fine structure may be needed for validating the identification in some species.

The class contains a variety of forms, including naked cells or cells covered with organic scales, external silica skeletons, and ameboid and rhizopodoid species.

The "naked" flagellate species are included in the two orders Rhizochromulinales with naked zoospores and Pedinellales with radially symmetrical cells. The third order, Dictyochales, is characterized by an external silica skeleton, but naked cells also occur.

Order Rhizochromulinales O'Kelly & Wujek 1992: Rhizopodial amoeboid and flagellated cells.

Family Rhizochromulinaceae O'Kelly & Wujek 1992: Ameboid vegetative cells and naked zoospores.

Genus *Rhizochromulina* Hibberd & Chrétiennot 1979: Flagellated zoospores; *R. marina* Hibberd & Chrétiennot-Dinet.

Order Pedinellales Zimmermann, Moestrup & Hällfors 1984: radially symmetrical cells.

Family Pedinellaceae Pascher 1910: six, three, or no chloroplasts.

Genus *Apedinella* Throndsen 1971: (Fig. 5E) Organic scales, without trailing pseudopodium; *A. spinifera* (Throndsen) Throndsen/*A. radians* (Lohmann) Campbell. [The conspecificy of *A. radians* and *A. spinifera* is based on the assumption that Lohmann (1908) overlooked the flagellum when describing *Meringosphaera radians*.]

Genus *Pseudopedinella* N. Carter 1937: (Fig. 5I) Trailing pseudopodium; *P. pyriforme* N. Carter, *P. tricostata* (Rouchijajnen) Thomsen, *P. elastica* Skuja (brackish water).

Genus *Actinomonas* Kent 1880–1882: (Fig. 5M) Colorless, with single active flagellum, actinopods radiating in all directions, and a slender stalk that may attach the cell to a substratum; *A. mirabilis* Kent.

Genus *Ciliophrys* Cienkowski 1875: Colorless, spherical, with actinopods radiating in all directions, flagellum inactive in heliozoan stage; *C. infusionum* Cienkowski (*C. marina* Caullery).

Genus *Pteridomonas* Penard 1890: Colorless, with a single row of actinopods encircling the proximal part of the flagellum, slender stalk; *P. pulex* Penard.

Genus *Parapedinella* S. M. Pedersen & Thomsen in Pedersen, Beech, & Thomsen 1986: (Fig. 5J) Colorless, with slender radiating pseudopodia and organic scales; *P. reticulata* S. M. Pedersen, & Thomsen.

Order Dictyochales Haeckel 1894: External silica skeleton, one extant family. For an extensive illustrated review of extant (and fossil) species and varieties see Loeblich et al. (1968).

Family Dictyochaceae Lemmermann 1901: Cells with one flagellum (+ one very short in naked phase), many chloroplasts, identification based on external SiO_2 skeleton.

Genus *Dictyocha* Ehrenberg 1837: (Fig. 5N) *Dictyocha speculum* Ehrenberg, *D. fibula* Ehrenberg, *D. crux* Ehrenberg, *D. antarctica* Lohmann, *D. rhombus* Haeckel.

Genus *Octactis* Schiller 1925: basal ring wide, with eight spines; *O. octonaria* (Ehrenberg) Hovasse, probably conspecific with *O. pulchra* Schiller.

Genus *Mesocena* Ehrenberg 1843: *M. annulus* Haeckel, *M. hexagona* Haeckel, *M. quadrangula* Haeckel, *M. stellata* Haeckel.

Selected Chryso- and Dictyochophycean Species with Radiating Actinopods or Spines

Actinomonas mirabilis Kent 1880–82 (Plate 5)
Cell diameter: 6–9 μm.
Actinopodia: 1½–4–6 × cell diameter.
Flagella: One, 3–4 × cell length.
Chloroplasts: Lacking.
Characteristic features: Cell with actinopodia (1½–4–6 × cell diameter) radiating in all directions, a fairly long pedicel may be present.
Distribution: Neritic; Europe, cosmopolitan.

Ciliophrys infusionum Cienkowski 1876 (Plate 5)
Syn. *Ciliophrys marina* Caullery 1910
Cell diameter: 7–20 μm + actinopodia.
Actinopodia: Up to 50 μm long.
Flagella: One.
Chloroplasts: Lacking.
Characteristic features: Cell in heliozoan stage with actinopodia (3–4 × cell diameter) radiating in all directions, flagellum fairly inactive

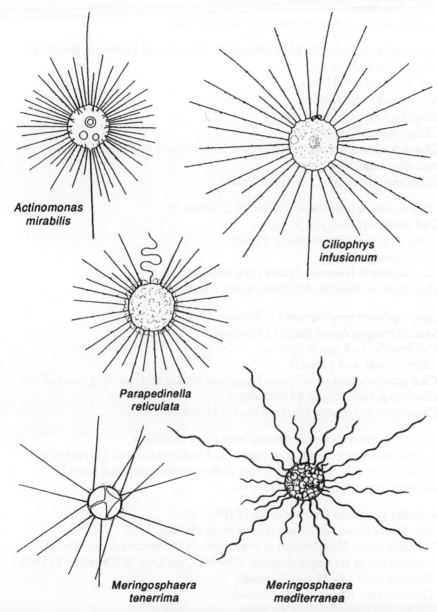

Plate 5 *Meringosphaera tenerrima* based on type illustration; *Actinomonas mirabilis, Ciliophrys infusionum* redrawn from Griessmann (1913); *Parapedinella reticulata* based on light micrograph in Pedersen et al. (1986); *Meringosphaera mediterranea* redrawn from Pascher (1932).

making a figure "8." Cell in flagellate phase without actinopods, with active flagellum.
Distribution: Littoral; North Sea, Mediterranean.

Parapedinella reticulata S. M. Pedersen & Thomsen in Pedersen, Beech, & Thomsen 1986 (**Plate 5**)
Cell diameter: 7–15 μm.
Actinopodia: Numerous.
Flagella: One, 17 μm.
Chloroplasts: Lacking.
Characteristic features: Cell covered with organic scales, actinopods radiating in all directions.
Distribution: Coastal; Atlantic, Pacific.

Meringosphaera tenerrima Schiller 1925 (**Plate 5**)
Cell diameter: 4–6 μm + spines.
Spine length: Approximately 18 μm.
Chloroplasts: Four.
Characteristic features: Spines thin, needle formed.
Distribution: Neritic; Mediterranean, Arctic.

Meringosphaera mediterranea Lohmann 1902 (**Plate 5**)
Syn. *Meringosphaera baltica* Lohmann
Cell length: 5–9 μm + spines.
Chloroplast: 3–6 parietal.
Characteristic features: Spines long, undulating and tapering toward the distal end, radiating in all directions.
Distribution: Neritic; Mediterranean, North Sea, Baltic.

Selected Dictyochophycean species with silica skeletons
These are *extremely variable* in shape, and hence the noted characteristic features may not be seen in all forms of the species; only one form is illustrated for each species, as follows.

Dictyocha speculum Ehrenberg 1839 (**Plate 6**)
Syn. *Distephanus speculum* (Ehrenberg) Haeckel 1887
[Generic name *Distephanus* is preoccupied; the species is provisionally retained under its original name. However, see Ling & Takahasi (1985).]
Skeleton size: 19–34 μm + spines.
Chloroplasts: Many, yellow-brown.

Plate 6 *Dictyocha antarctica, Octactis octonaria* forma *pulchra* based on type illustrations; *Dictyocha speculum, D. fibula, D. crux* and *D. staurodon* redrawn from Gemeinhardt (1930); *Octactis octonaria* from Lemmermann (1901); *Mesocena hexagona* from Schulz as reproduced in Gemeinhardt (1930).

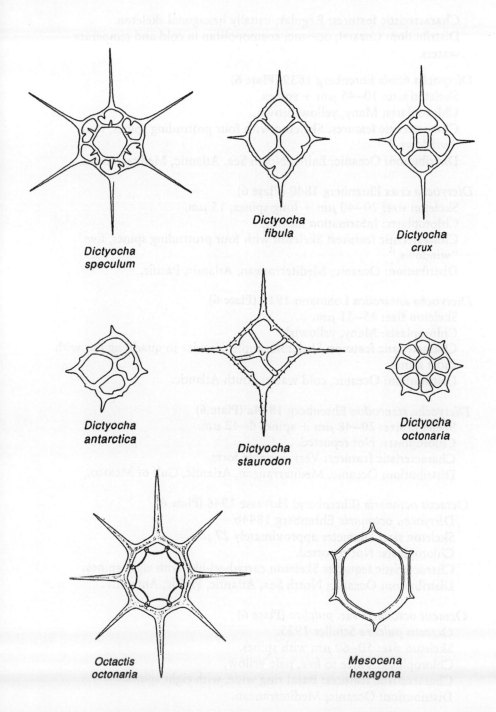

Dictyocha
speculum

Dictyocha
fibula

Dictyocha
crux

Dictyocha
antarctica

Dictyocha
staurodon

Dictyocha
octonaria

Octactis
octonaria

Mesocena
hexagona

Characteristic features: Regular, usually hexagonal skeleton.
Distribution: Coastal, oceanic; cosmopolitan in cold and temperate waters.

Dictyocha fibula Ehrenberg 1839 (Plate 6)
Skeleton size: 10–45 μm + spines.
Chloroplasts: Many, yellow-brown.
Characteristic features: Skeleton with four protruding spines, four "windows."
Distribution: Oceanic; Baltic, North Sea, Atlantic, Mediterranean.

Dictyocha crux Ehrenberg 1840 (Plate 6)
Skeleton size: 20–40 μm + long spines, 15 μm.
Chloroplasts: Information lacking.
Characteristic features: Skeleton with four protruding spines, five "windows."
Distribution: Oceanic; Mediterranean, Atlantic, Pacific.

Dictyocha antarctica Lohmann 1919 (Plate 6)
Skeleton size: 45–51 μm.
Chloroplasts: Many, yellowish.
Characteristic features: Skeleton simple, circular to quadrangular with short spines.
Distribution: Oceanic, cold water; South Atlantic.

Dictyocha staurodon Ehrenberg 1844a (Plate 6)
Skeleton size: 20–48 μm + spines 4–42 μm.
Chloroplasts: Not reported.
Characteristic features: Very variable form.
Distribution: Oceanic, Mediterranean, Atlantic, Gulf of Mexico.

Octactis octonaria (Ehrenberg) Hovasse 1946 (Plate 6)
  Dictyocha octonaria Ehrenberg 1844b
Skeleton size: Diameter approximately 27 μm.
Chloroplasts: Not reported.
Characteristic features: Skeleton cartwheel-like with eight spines.
Distribution: Oceanic; North Sea, Atlantic, Pacific, Antarctic.

Octactis octonaria var. *pulchra* (Plate 6)
  Octactis pulchra Schiller 1925.
Skeleton size: 50–60 μm with spines.
Chloroplasts: Three to five, pale yellow.
Characteristic features: Basal ring wide, with eight spines.
Distribution: Oceanic; Mediterranean.

Mesocena hexagona Haeckel 1887 (Plate 6)
  Skeleton size: 60–82 × 46–60 μm + spines 5–16 μm.
  Chloroplasts: Not reported.
  Characteristic features: Open ring with short spines.
  Distribution: Oceanic; Mediterranean.

PRYMNESIOPHYCEAE—HAPTOPHYCEAE (EXCLUSIVE OF COCCOLITHOPHORIDS)

Systematic position: Class Haptophyceae Christensen 1962, renamed Prymnesiophyceae Hibberd 1976, part of Chrysophyceae (e.g., in Bourrelly, 1957a).

Characteristic Features

Cell shape—spherical, round or flattened, elongated or saddle-shaped.

Cell covering—Organic scales covering the cell body surface are usually not seen in the light microscope; large spiny scales may be observed in the fluorescence microscope after staining (e.g., with acridine orange).

Haptonema—short or long thread-like organelle (sometimes used for anchoring the cell), may be protruding in the swimming direction. Long haptonemata coil when relaxed.

Flagella—two, most often smooth, but may have minute organic scales and/or tiny fibrillar hairs on the surface (EM).

Mode of swimming—homo- or heterodynamic, flagella pushing the cell.

Color—yellow-brown to golden-brown, may be pale.

Chloroplasts—one or two, with triple thylakoids, girdle lamella lacking (EM).

Pyrenoids—naked, on the inside of chloroplasts, embedded or stalked.

Storage products—chrysolaminaran (liquid β-1,3-glucan) is stained *in vivo* by brilliant cresyl blue to a rose color, paramylon (solid β-1,3-glucan, not stained) present in one order (Pavlovales).

Distribution—Planktonic: many unicellular species, single or in colonies recorded in neritic and oceanic waters; heterotrophic species not common. Benthic species: insufficiently known, but many are recorded from shallow areas.

Identification/Observation and Problems ("Naked" Forms)

Preserved material is usually inadequate, though some species (e.g., *Phaeocystis pouchetii*) may be identified. Haptonema and flagella may be kept in material fixed with osmic acid (OsO₄), but even so identification in the light microscope is difficult, unless dried shadowcast or stained preparations are studied in the EM for scale morphology.

Living material offers the best possibilities for identification in the light microscope to genus level, but identification to species most often relies on scale morphology (see earlier listing).

Light microscopy is often insufficient for a reliable identification beyond genus level. The species are often small, and many descriptions are based on scale morphology.

The class may be divided into four orders: Isochrysidales, devoid of haptonema or with a rudimentary one; Coccosphaerales, comprising most of the coccolithophorids; and Prymnesiales and Pavlovales, with haptonema.

Order Isochrysidales Pascher 1910: Contains two families with "naked" representatives (see also Green & Pienaar, 1977):

Family Gephyrocapsaceae Black 1971: Cells with two smooth flagella and devoid of haptonema.

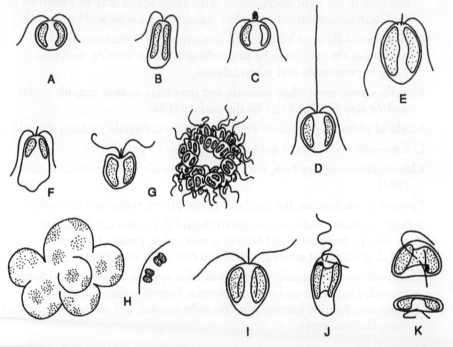

FIGURE 6   Prymnesiophycean genera from marine plankton: (A) *Dicrateria/Imantonia* 3–8 μm; (B) *Isochrysis* 5–6 μm; (C, D) *Chrysochromulina* 4–26 μm; (E) *Prymnesium* 6–18 μm; (F) *Platychrysis* 6–12 μm; (G) *Corymbellus* 8–11 μm, colony 200 μm; (H) *Phaeocystis* nonmotile in colony ≥2000–8000 μm; (I) *Phaeocystis* motile 4.5–8 μm; (J) *Pavlova* 4–10 μm; (K) *Diacronema* 3–6 μm. (F based on Carter, 1937; G based on Green, 1976a, modified as in Throndsen, 1980.)

Genus *Dicrateria* Parke 1949: (**Fig. 6A**) Naked cells; *D. gilva* Parke, *D. inornata* Parke.

Genus *Imantonia* Reynolds 1974: (**Fig. 6A**) Cells with organic scales; *I. rotunda* Reynolds.

Genus *Parachrysidalis* Hulburt 1965: ventrally incerted flagella; *P. estuariale* Hulburt.

Family Isochrysidaceae Pascher 1910: Cells with two smooth flagella and a rudimentary haptonema.

Genus *Isochrysis* Parke 1949: (**Fig. 6B**) *I. galbana* Parke.

Order Coccosphaerales: With calcified body scales or coccoliths is treated in a separate section: Coccolithophorids, also including the coccolith carrying genera of Isochrysidales.

Order Prymnesiales Papenfuss 1955: Cells covered by organic scales only.

Family Prymnesiaceae Conrad 1926: Cells with a short or long haptonema, the flagellate stages prevailing. [A key to the genera is found in Chrétiennot (1973).]

Genus *Chrysochromulina* Lackey 1939: (**Fig. 6C,D**) Coiling haptonema, homodynamic flagella, many species, to be identified by EM mostly, *Chrysochromulina parva* Lackey (FW); >45 marine species including *C. alifera* Parke & Manton, *C. camella* Leadbeater & Manton, *C. ephippium* Parke & Manton, *C. ericina* Parke & Manton, *C. hirta* Manton, *C. leadbeateri* Estep et al., *C. mantoniae* Leadbeater, *C. minor* Parke & Manton, *C. parkeae* Green & Leadbeater, *C. polylepis* Manton & Parke, *C. pringsheimii* Parke & Manton, *C. spinifera* (Fournier) Pienaar & Norris. [For a more extensive list of species with references see Estep et al. (1984).]

Genus *Prymnesium* Massart *ex* Conrad 1926: (**Fig. 6E**) Short haptonema, heterodynamic flagella; *P. czosnowskii* Starmach, *P. saltans* Massart (*ex* Conrad), *P. parvum* N. Carter, *P. patelliferum* Green, Hibberd, & Pienaar, *P. minutum* N. Carter, *P. calathiferum* Chang & Ryan, *P. zebrinum* Billard, *P. annuliferum* Billard.

Genus *Platychrysis* Geitler 1930: (**Fig. 6F**) Variable form; *P. pigra* Geitler, *P. pienaarii* Gayral & Fresnel, *P. simplex* Gayral & Fresnel, *P. neustophila* Norris [see Gayral & Fresnel (1983) for review].

Genus *Corymbellus* Green 1976a: (**Fig. 6G**) Motile cells in colonies; *C. aureus* Green.

Family Phaeocystaceae Lagerheim 1896: Cells with short haptonema, a palmelloid phase dominant (or most conspicuous).

Genus *Phaeocystis* Lagerheim 1893: (**Fig. 6H,I**) Nonmotile cells embedded in round or lobed jelly colonies, motile stage *Prymnesium-*

like; *P. pouchetii* (Hariot) Lagerheim, *P. globosa* Scherffel (see
Jahnke & Baumann, 1987), *P. antarctica* Karsten, *P. brucei* Mangin,
P. giraudyi (Derbés & Solier) Hamel, *P. scrobiculata* Moestrup.

Order Pavlovales Green 1976b comprises one family, cells with two different
flagella, one acting as a tractellum, storage products are polyphosphates
and paramylon like compounds. Cells devoid of scales, but with "skin,"
long flagellum with knob scales. A short haptonema as well as an eyespot
may be present.

Family Pavlovaceae Green 1976b: with the characteristics of the order. [A
key is found in Green (1980).]

Genus *Pavlova* Butcher 1952: (Fig. 6J) Flagella and haptonema inserted
subapically; *P. gyrans* Butcher, *P. salina* (N. Carter) Green
(*Nephrochloris salina* N. Carter), *P. lutheri* (Droop) Green
(*Monochrysis lutheri* Droop); see also Green (1976b, 1980).

Genus *Diacronema* Prauser 1958: (Fig. 6K) Flattened cells with laterally
inserted flagella; *D. vlkianum* Prauser.

Selected species of Prymnesiophyceae Lacking or with a Short Haptonema

Isochrysis galbana Parke 1949 (Plate 7)
  Cell length: 5–6 μm.
  Flagella: Two, approximately 7 μm.
  Chloroplasts: Single, yellow-brown.
  Characteristic features: Cell elongated, variable in shape, devoid of
  haptonema.
  Distribution: Coastal; Atlantic.

Dicrateria inornata Parke 1949 (not illustrated)
  Cell length: 3–5.5 μm.
  Flagella: Two, subequal, 7–9 μm.
  Chloroplasts: Two (four), yellow-brown.
  Characteristic features: Cell naked, may be confused with *Imantonia
  rotunda* in the LM, see later listing.
  Distribution: Coastal, oceanic; Atlantic.

Imatonia rotunda Reynolds 1974 (Plate 7)
  Cell length: 2–4 μm.
  Flagella: Two, 4½–7 μm.
  Chloroplasts: Two, yellow-brown.

Plate 7 *Isochrysis galbana, Platychrysis pigra, P. neustophila, Prymnesium annuliferum, P. zebrinum, Corymbellus aureus* based on type illustrations; *Imantonia rotunda, Prymnesium parvum,*
originals; *Prymnesium saltans* redrawn from Conrad (1941).

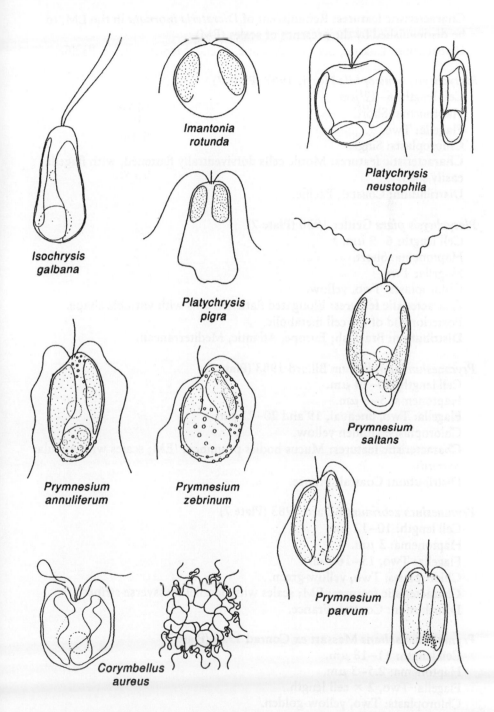

Imantonia rotunda

Platychrysis neustophila

Isochrysis galbana

Platychrysis pigra

Prymnesium saltans

Prymnesium annuliferum

Prymnesium zebrinum

Prymnesium parvum

Corymbellus aureus

Characteristic features: Reminiscent of *Dicrateria inornata* in the LM, to be distinguished by the presence of scales (EM).
Distribution: Coastal, oceanic; Atlantic, Arctic, Mediterranean, Pacific.

Platychrysis neustophila Norris 1967 (Plate 7)
   Cell length: 6–12 μm.
   Haptonema: Short.
   Flagella: Two.
   Chloroplasts: Single.
   Characteristic features: Motile cells dorsiventrally flattened, with flagella easily shed.
   Distribution: Coastal; Pacific.

Platychrysis pigra Geitler 1930 (Plate 7)
   Cell length: 6–9 μm.
   Haptonema: Short.
   Flagella: Two.
   Chloroplasts: Two, yellow.
   Characteristic features: Elongated flattened cells with variable shape. Posterior end of the cell metabolic.
   Distribution: Brackish; Europe, Atlantic, Mediterranean.

Prymnesium annuliferum Billard 1983 (Plate 7)
   Cell length: 10–14 μm.
   Haptonema: 5–6 μm.
   Flagella: Two, unequal, 19 and 20–22 μm.
   Chloroplasts: Golden yellow.
   Characteristic features: Mucus bodies prominent (EM; scales with annular pattern).
   Distribution: Coastal; France.

Prymnesium zebrinum Billard 1983 (Plate 7)
   Cell length: 10–12 μm.
   Haptonema: 3 μm.
   Flagella: Two, 15–16 μm.
   Chloroplasts: Two, yellow-green.
   Characteristic features: EM; scales with parallel transverse stripes.
   Distribution: Coastal; France.

Prymnesium saltans Massart *ex* Conrad 1926 (Plate 7)
   Cell length: 11–18 μm.
   Haptonema: 2.5–5 μm.
   Flagella: Two, 2 × cell length.
   Chloroplasts: Two, yellow-golden.

Characteristic features: Large species, jerky swimming movements, characteristic flagella.
Distribution: Brackish water; Europe.

Prymnesium parvum N. Carter 1937 (Plate 7)
Cell length: 8–10 μm.
Haptonema: ⅓ × cell length.
Flagella: Two, 1½–2 × cell length.
Chloroplasts: Two, golden yellow.
Characteristic features: Vibrating granules in the posterior part of the cell. EM necessary to distinguish the species from *P. patelliferum* (cf. Green, Hibberd, & Pienaar, 1982).
Distribution: Brackish, coastal; Atlantic, Mediterranean, Pacific.

Prymnesium patelliferum Green, Hibberd, & Pienaar 1982, not illustrated, see earlier.
Cell length: 6–12 μm.
Haptonema: 3–5 μm.
Flagella: Two, 10–14.5 μm.
Chloroplasts: Two, yellow-green to olive.
Characteristic features: Vibrating granules in the posterior part of the cell. Electron microscopy necessary to distinguish the species from *P. parvum* (cf. Green, Hibberd, & Pienaar, 1982).
Distribution: Coastal; Pacific, Atlantic.

Corymbellus aureus Green 1976a (Plate 7)
Cell length: 8–11 μm.
Haptonema: Noncoiling, 3 μm.
Flagella: Two, subequal.
Chloroplasts: Two, yellow-green.
Characteristic features: Cells in colonies up to 200 μm in diameter (scales to be checked in EM).
Distribution: Coastal, Oceanic; Atlantic, Mediterranean.

Selected Single and Colonial Prymnesiophycean Species

Phacocystis pouchetii (Hariot) Lagerheim 1893 (Plate 8)
Tetraspora pouchetii Hariot in Pouchet 1893
Motile stage:
Cell length: 4.5–8 μm.
Haptonema: Short.
Flagella: Two, 1½ × cell length.
Chloroplasts: Two, yellow-brown.
Nonmotile stage:
Cell length: 4–8 μm in colony.

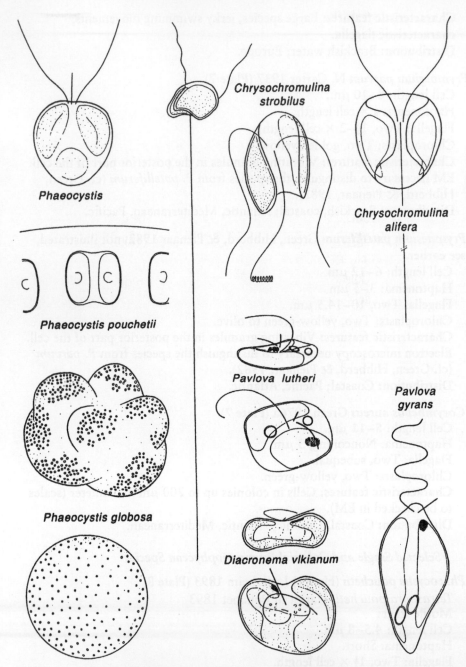

Plate 8 *Chrysochromulina strobilus, C. alifera, Pavlova lutheri* based on type illustrations; *Pavlova gyrans, Phaeocystis* motile, P. *pouchetti,* and P. *globosa,* originals; *Diacronema vlkianum* modified from Vlk (1938).

Haptonema: Lacking.
Flagella: Lacking.
Chloroplasts: Two, with pyrenoids.
Colony size: Up to 1.5–2 mm.
Characteristic features: Nonmotile stage in lobed gelatinous colonies, cells in groups (of four); motile stage *Prymnesium*-like, to be distinguished by scale morphology and filamentous investment pattern (five-pointed star— EM., Parke et al., 1971).
Distribution: Cold water species; Oceanic, coastal; Arctic, Antarctic, Atlantic, Pacific.

Phaeocystis globosa Scherffel 1899, 1900 (Plate 8)
Motile stage:
Cell length: 5 μm.
Haptonema: Haptonema short.
Flagella: Two, greater than cell length.
Chloroplasts: Two, yellow-brown.
Nonmotile stage:
Cell length: Approximately 7 μm in colony.
Haptonema: Lacking in colonial stage.
Flagella: Lacking.
Chloroplasts: Two, yellowish brown.
Colony size: Up to 2 mm, maximum 8 mm.
Characteristic features: Nonmotile stage in spherical smooth colonies without lobes up to 2 mm in diameter, cells evenly distributed along the surface of the colony; motile stage similar to *Phaeocystis pouchetii*.
Distribution: Temperate waters; Oceanic, coastal; North Sea.

Chrysochromulina strobilus Parke & Manton 1959 (in Parke et al., 1959) (Plate 8)
Cell length: 6–10 μm.
Haptonema: 12–18 × cell length.
Flagella: Two, 2–3 × cell length.
Chloroplasts: Two, golden brown.
Characteristic features: Cell dorsoventrally flattened and strongly curved with flagella and haptonema protruding from the ventral concave side, fairly long haptonema. Scales to be identified in EM.
Distribution: Oceanic, coastal; Atlantic.

Chrysochromulina alifera Parke & Manton 1956 (in Parke et al., 1956) (Plate 8)
Cell length: 6–10 μm.
Haptonema: 10–12 × cell length.
Flagella: Two, 2–2½ × cell length.
Chloroplasts: Two, golden brown.

Characteristic features: Cell saddle-shaped, two types of scales to be distinguished in EM.
Distribution: Oceanic: Atlantic.

Pavlova lutheri (Droop) Green 1975 (Plate 8)
Monochrysis lutheri Droop 1953.
Cell length: 6–10 μm.
Haptonema: Insignificant in the light microscope.
Flagella: Two, the longer 1–1½ × cell length, the shorter insignificant in the light microscope.
Chloroplasts: Two or three, olive/yellow.
Characteristic features: Cell concave, flattened, triangular to square in outline.
Distribution: Coastal; Baltic, Pacific.

Pavlova gyrans Butcher 1952 (Plate 8)
Cell length: 4–10 μm.
Haptonema: Insignificant in the light microscope.
Flagella: Two, the longer about 2 × cell length.
Chloroplasts: Two, dull yellow.
Characteristic features: Cell strongly metabolic.
Distribution: Coastal; Atlantic, Pacific.

Diacronema vlkianum Prauser 1958 (Plate 8)
Cell length: 4–8 μm.
Haptonema: Approximately 1 μm.
Flagella: Two, 7–10 μm and 6–9 μm.
Chloroplasts: Single, yellow-green to olive-green.
Characteristic features: Cell flattened with flagella protruding from the middle of the concave ventral side.
Distribution: Fresh water, marine; Europe.

Selected Species of Chrysochromulina

Chrysochromulina parkeae Green & Leadbeater 1972 (Plate 9)
Cell length: 10–26 μm.
Haptonema: 2.5–4.5 μm.
Flagella: Two, 8–20 μm.
Chloroplasts: Two, golden brown.
Characteristic features: Cell with posterior (and occasionally anterior) spiny scales (20–31 μm) visible in the light microscope, to be identified in the EM.
Distribution: Coastal, oceanic; Atlantic.

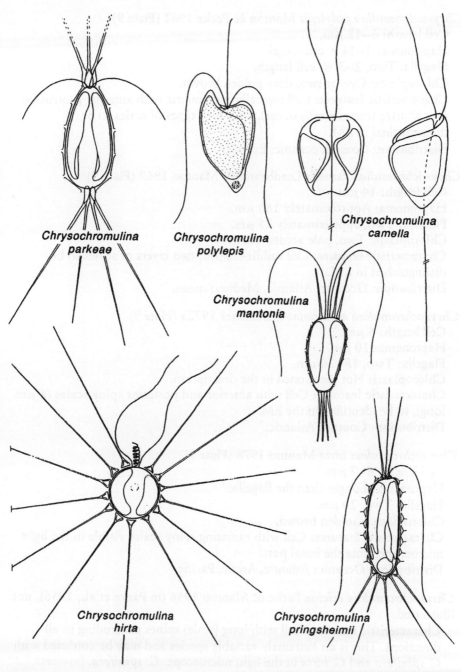

Chrysochromulina parkeae

Chrysochromulina polylepis

Chrysochromulina camella

Chrysochromulina mantonia

Chrysochromulina hirta

Chrysochromulina pringsheimii

Plate 9 *Chrysochromulina parkeae, C. camella, C. mantoniae, C. hirta, C. pringsheimii* based on type illustrations; *Chrysochromulina polylepis* modified from Manton & Parke (1962).

Chrysochromulina polylepis Manton & Parke 1962 (Plate 9)
Cell length: 6–12 μm.
Haptonema: 1–1½ × cell length.
Flagella: Two, 2–3 × cell length.
Chloroplasts: One or two, deep golden brown.
Characteristic features: Cell typically asymmetric with anterior protrusion (but shorter truncate cells occur), with four types of scales to be distinguished in EM.
Distribution: Coastal, oceanic; Europe.

Chrysochromulina camella Leadbeater & Manton 1969 (Plate 9)
Cell length: 14 μm.
Haptonema: Approximately 160 μm.
Flagella: Two, approximately 25 μm.
Chloroplasts: Two, pale golden brown.
Characteristic features: Cell saddle-shaped, two layers of scales to be distinguished in EM.
Distribution: Oceanic; Atlantic, Mediterranean.

Chrysochromulina mantoniae Leadbeater 1972a (Plate 9)
Cell length: 6 μm.
Haptonema: 10 μm.
Flagella: Two, 18–20 μm.
Chloroplasts: Not mentioned in the description.
Characteristic features: Cell with anterior and posterior spiny scales (6 μm long), to be identified in the EM.
Distribution: Coastal; Atlantic.

Chrysochromulina hirta Manton 1978 (Plate 9)
Cell length: 6–7 μm.
Haptonema: Longer than the flagella.
Flagella: Two, 20 μm.
Chloroplasts: Golden brown.
Characteristic features: Cell with radiating spiny scales visible in the light microscope, note the basal part.
Distribution: Oceanic; Atlantic, Arctic, Pacific.

Chrysochromulina ericina Parke & Manton 1956 (in Parke et al., 1956), not illustrated.
Characteristic features: Cell with long (scale) spines protruding in all directions. This is an extremely variable species and may be confused with *C. spinifera* and *C. hirta* in the light microscope. *C. spinifera*, however, has a nonmotile haptonema.
Distribution: Coastal; Europe.

Chrysochromulina pringsheimii Parke & Manton 1962 (Plate 9)
  Cell length: 14–20 μm.
  Haptonema: 1–2 × cell length.
  Flagella: Two, 1½–2 × cell length.
  Chloroplasts: Two (four), golden brown.
  Characteristic features: Cell with anterior and posterior spiny scales
  (12–20 μm), to be identified in the EM.
  Distribution: Coastal; Atlantic.

CHLOROPHYTA

EUGLENOPHYCEAE

Systematic position: Class Euglenophyceae Schoenichen 1925, Order Euglenida Bütschli 1884, Phylum Euglenophyta in Leedale, 1967.

Characteristic Features (Fig. 7)

Cell covering—pellicula, subsurface system of proteinacious interlocking
  bands running usually in spiral, giving the cells a striped pattern.

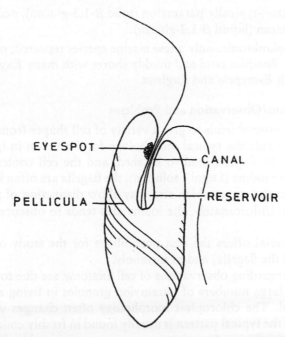

FIGURE 7 Euglenophycean cell morphology, with flagellar canal, reservoir, and flagella. Pellicular stripes surround the whole cell body.

Body metaboly—euglenoid movements (a swelling of the cell body running from posterior end of the cell) and more irregular changes in the cell shape.

Flagella—one, two, or four emergent, running from their bases in the reservoir through the canal. Nonemergent flagellum to the level of the eyespot.

Mode of swimming—homo- or heterodynamic.

Color—bright green in phototrophic forms.

Chloroplasts—one or many, reticulated, ribbon or disc shaped.

Pyrenoids—often with paramylon shields or clusters of paramylon granules.

Eyespot—orange or red, usually conspicuous, situated near the canal plasmalemma, separate from the chloroplasts.

Nucleus—large, with condensed chromosomes, often prominent, in the middle of posterior part or the cell.

Canal—apically or subapically, leading from the anterior cell surface to the reservoir.

Reservoir—an interior dilatation into which the contractile vacuole empties. A contractile vacuole is lacking in true marine species.

Storage products—typically paramylon (solid β-1,3-glucan), occasionally chrysolaminaran (liquid β-1,3-glucan).

Distribution—planktonic: only a few marine species reported, mainly *Eutreptiella*. Benthic: sand and muddy shores with many *Euglena*, salt marshes with *Eutreptia* and *Euglena*.

Identification/Observation and Problems

Preserved material tends to give a variety of cell shapes from which it may be difficult to decide the typical morphology for the taxon in question. With formaldehyde, the flagella tend to be shed, and the cell content is rendered rather pale. With iodine (Lugol's solution), the flagella are often kept and easily observed due to the stain, which also may allow distinction of pyrenoids and mucous bodies. Unfortunately, the iodine also tends to obscure the details of cell anatomy.

Living material offers the best possibilities for the study of body shape, mode of use of the flagella, and of metaboly.

Problems regarding observations of cell anatomy are due to the obscuring of features by large numbers of paramylon granules in living as well as preserved material. The chloroplast morphology often changes with the living condition, and the typical pattern is usually found in freshly collected material.

The phototrophic euglenophycean species encountered in the marine environment belong to the order Euglenales:

Order Euglenales Engler 1898: Two families with marine representatives.

Family Eutreptiaceae: Cells with two flagella emerging from the canal.

Genus *Eutreptia* Perty 1852: (**Fig. 8A**) Two equal (homodynamic?) flagella; *E. viridis* Perty, *E. pertyi* Pringsheim, *E. globulifera* van Goor, *E. lanowii* Steur.

Genus *Eutreptiella* de Cunha 1914: (**Fig. 8B**) Two subequal heterodynamic flagella; *E. marina* da Cunha, *E. gymnastica* Throndsen, *E. hirudoidea* Butcher, *E. braarudii* Throndsen, *E. eupharyngea* Moestrup & Norris, *E. cornubiense* Butcher, *E. elegans* (Schiller) Pascher, *E. dofleinii* (Schiller) Pascher, *E. pascheri* (Schiller) Pascher. See Walne et al. (1986) for comparative data on the species described for this genus.

Family Euglenaceae Dujardin 1841: Cells with one flagellum emerging from the canal.

Genus *Euglena* Ehrenberg 1838: (**Fig. 8C**) With many green chloroplasts; *E. proxima* Dangeard, *E. ascusformis* Schiller.

[Systematics according to Butcher (1967), but see also Leedale (1967).]

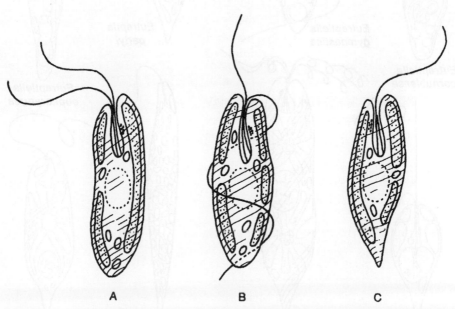

A   B   C

FIGURE 8 Euglenophycean (phototrophic) genera from marine plankton. (A) *Eutreptia* (20–120 μm; (B) *Eutreptiella* 17–90(115) μm; (C) *Euglena* 45–70 μm. (Modified from Throndsen, 1980.)

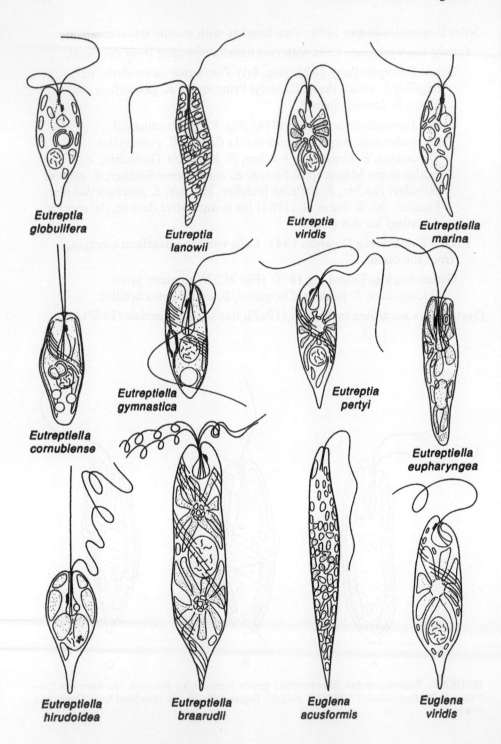

Eutreptia globulifera

Eutreptia lanowii

Eutreptia viridis

Eutreptiella marina

Eutreptiella cornubiense

Eutreptiella gymnastica

Eutreptia pertyi

Eutreptiella eupharyngea

Eutreptiella hirudoidea

Eutreptiella braarudii

Euglena acusformis

Euglena viridis

Selected Species of Euglenophyceae

Eutreptia globulifera van Goor 1925a (Plate 10)
Cell length: 20–30 μm.
Flagella: Two, equal (10–12 μm?).
Chloroplasts: Many, green.
Characteristic features: Many small (3–4 μm) chloroplasts, median globular body (pyrenoid with paramylon shields?), metabolic.
Distribution: Onshore; Atlantic.

Eutreptia lanowii Steur 1904 (Plate 10)
Cell length: 25–60 μm.
Flagella: Two, approximately cell length.
Chloroplasts: Many, green.
Characteristic features: Steur (1904) stated that the species is like *E. viridis*, but slightly smaller, with flagella of different thickness, and living in salt water.
Distribution: Coastal; Mediterranean.

Eutreptia viridis Perty 1852 (Plate 10)
Cell length: 49–66 μm.
Flagella: Two, approximately cell length.
Chloroplasts: Many, green.
Characteristic features: Single paramylon center with radiating chloroplasts, which may also appear as scattered discs, sharply pointed posterior end.
Distribution: Coastal; Atlantic.

Eutreptiella marina da Cunha 1914 (Plate 10)
Cell length: 40–50 μm.
Flagella: Two, unequal, cell length and ⅓ × cell length.
Chloroplasts: Many, green.
Characteristic features: Chloroplasts disc-shaped without pyrenoid, paramylon grains rod-shaped. Metaboly pronounced.
Distribution: Coastal; Atlantic.

Eutreptiella cornubiense Butcher 1961 (Plate 10)
Cell length: 12–50 μm.
Flagella: Two, unequal, cell length and ¾ × cell length.
Chloroplasts: Single, green.

Plate 10 *Eutreptia globulifera, E. lanowii, E. viridis, Eutreptiella marina, E. cornubiense, E. pertyi, E. eupharyngea, E. hirudoidea, Euglena ascusformis* based on type illustrations; *Euglena viridis, Eutreptiella braarudii, E. gymnastica* originals.

Characteristic features: Single reticulated chloroplasts with large double pyrenoid covered by two semilunar paramylon shields, posterior end rounded. Pronounced metaboly.
Distribution: Coastal; Atlantic.

Eutreptiella gymnastica Throndsen 1969 (Plate 10)
Cell length: 17–30 μm.
Flagella: Two, 20–32 μm and 8–13 μm.
Chloroplasts: Single, green.
Characteristic features: The reticulated chloroplast has a single pyrenoid covered by two opposed, flattened, bowl-shaped paramylon shields; chrysolaminaran may be present.
Distribution: Coastal; Atlantic.

Eutreptia pertyi Pringsheim 1953 (Plate 10)
Cell length: 45–80 μm.
Flagella: Two, approximately cell length.
Chloroplasts: Many, green.
Characteristic features: Single paramylon center with radiating chloroplasts, which may also appear as scattered discs, canal opening slightly subapically, with pronounced metaboly. Differs in size and shape from E. *viridis*.
Distribution: Coastal; Atlantic.

Eutreptiella eupharyngea Moestrup & Norris (In Walne et al., 1986) (Plate 10)
Cell length: 35–70 μm.
Flagella: Two, 60–80 μm, 20–24 μm.
Chloroplasts: Many, green, band-shaped.
Characteristic features: Large cell size combined with a very distinct eyespot, a deep canal-reservoir complex, and chloroplasts in two rosettes with paramylon centers.
Distribution: Neritic; North Atlantic, northeast Pacific.

Eutreptiella hirudoidea Butcher 1961 (Plate 10)
Cell length: 24–30 μm.
Flagella: Two, unequal, cell length and 3 × cell length.
Chloroplasts: Many, green.
Characteristic features: Six to ten leaf-like chloroplasts, prominent eyespot, cell shape clavate, pellicular stripes not apparent and with pronounced metaboly.
Distribution: Coastal; Atlantic.

Eutreptiella braarudii Throndsen 1969 (Plate 10)
Cell length: 64–115 μm.
Flagella: Two or four.
Chloroplasts: Many, green.

Characteristic features: Pellicula usually quite rigid with prominent broad stripes, chloroplasts radiating from two paramylon centers, and often with four flagella. Cell shape variable, euglenoid movement may occur.
Distribution: Cold water; Coastal; North Atlantic, Arctic.

Euglena acusformis Schiller 1925 (Plate 10)
 Cell length: 40–50 μm.
 Flagella: One, $\frac{2}{3}$ × cell length.
 Chloroplasts: Many, green.
 Characteristic features: Cell fusiform with many disc-shaped chloroplasts; may be conspecific with *E. proxima* Dangeard (see Butcher, 1967).
 Distribution: Coastal, oceanic; Mediterranean.

Euglena viridis Ehrenberg 1830 (in Ehrenberg, 1828–1831) (Plate 10)
 Cell length: 35–70 μm.
 Flagella: One, cell length.
 Chloroplasts: Many, green.
 Characteristic features: Single paramylon center or naked pyrenoid with radiating chloroplasts, pointed posterior end.
 Distribution: Psammobious; Atlantic shores.

PRASINOPHYCEAE

Systematic position: Class Prasinophyceae Moestrup & Throndsen 1988 (non Silva 1980), formalized from Christensen (1962, 1966).

Characteristic Features

Cell shape—quadrangular or bilaterally compressed, often with a depression where the flagella originate.

Cell covering—organic scales cover cell body and flagella, which may assemble to form a theca (e.g., *Tetraselmis*). Naked species also occur.

Flagella—one, two, four, eight (or 16) covered with minute scales and simple hairs, appear rather stiff and "thick."

Mode of swimming—hetero- or homodynamic, flagella pushing the cell.

Color—slightly olive-green.

Chloroplasts—one (or two) simple or lobed campanulate, or many disc-shaped (phycoma stages).

Eyespot—in the chloroplast, often present.

Ejectsomes—present in a subgenus of *Pyramimonas*.

Storage product—starch in shield around pyrenoid, and as stroma starch in the chloroplast.

Distribution—planktonic: coastal and oceanic, some species common and with a worldwide distribution. Benthic: as attached colonies on rock (*Prasinocladus*), as symbionts (*Tetraselmis* [*Platymonas*] *convoluta*).

Identification/Observation and Problems

Preserved material, whether fixed with iodine or formaldehyde, is most often inadequate for identification. Shadowcast whole cell mounts fixed with osmic acid will provide a reliable means for identification of most species in the electron microscope.

Living material offers a fairly good possibility for identification to genus level, and if you know the swimming pattern also, many species can be identified. Division stages will often reveal if the species is thecate or not.

Problems are often related to size, as many flagellated forms (e.g., *Micromonas*) are too small to be enumerated by routine counting techniques, and reliable identification depends on EM. Species known in a nonmotile stage only need to be studied at maturity for a safe identification.

Based on fine-structural evidence, present systematics recognizes two orders, the Mamiellales and Chlorodendrales, which comprise 13 genera in marine planktonic environments.

Order Pedinomonadales Moestrup 1991: Contains naked species with one flagellum (+ basal body of second flagellum). Accommodated as the single order of the class Pedinophyceae by Moestrup (1991).

Family Pedinomonadaceae Korshikov 1938

Genus *Resultor* Moestrup 1991: (Fig. 9A) Laterally flattened cells carrying one long flagellum creating jumping movements; single marine species *R. mikron* (Throndsen) Moestrup.

Order Mamiellales Moestrup 1984: Contains species lacking the typical inner layer scales (EM), one family.

Family Mamiellaceae Moestrup 1984: Usually with spider's web type of scales (EM).

Genus *Dolichomastix* Manton 1977: With one type of body scales; *D. eurylepidea* Manton, *D. lepidota* Manton, *D. nummulifera* Manton.

Genus *Mamiella* Moestrup 1984: (Fig. 9F) Two slightly subequal heterodynamic flagella and two types of body scales; *M. gilva* (Parke & Rayns) Moestrup.

Genus *Mantoniella* Desikachary 1972: (Fig. 9C) Very unequal flagella and two types of body scales; *M. squamata* (Manton & Parke) Desikachary, *M. antarctica* Marchant.

Genus *Bathycoccus* Eikrem & Throndsen 1990: Nonmotile with one type of scales only; *B. prasinos* Eikrem & Throndsen.

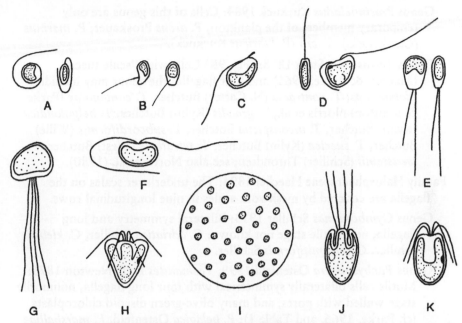

FIGURE 9 Prasinophycean genera from marine plankton. (A) *Resultor* 1.5–2.5 μm; (B) *Micromonas* 1–3 μm; (C) *Mantoniella* 3–5 μm; (D) *Nephroselmis* 3.5–6.5 μm; (E) *Pseudoscourfieldia* 3–5 μm; (F) *Mamiella* 4–6.5 μm; (G) *Pachysphaera/Pterosperma/Cymbomonas* 7–12–16 μm; (H) *Pyramimonas* 4.5–16–35 μm; (I) *Halosphaera* phycoma stage ≥800 μm; (J) *Halosphaera* motile 10–28 μm; (K) *Tetraselmis* 4.5–25 μm. (Modified from Throndsen, 1980.)

Family Pycnococcaceae Guillard 1991 (in Guillard et al., 1991): With affinities to Mamiellaceae; vegetative stage with solitary, walled coccoid cells.

   Genus *Pycnococcus* Guillard 1991: Coccoid olive-green cells, flagellate stage with single flagellum; *P. provasolii* Guillard.

Order Chlorodendrales Fritsch 1917: Flagella and cell body covered by an underlayer of square or diamond-shaped scales.

Family Chlorodendraceae Oltmanns 1904: Flagellar underlayer scales covered by small (usually) rod-shaped scales, apparently in 24 double rows.

   Genus *Nephroselmis* Stein 1878: (**Fig. 9D**) Flattened cells with two heterodynamic flagella; *N. astigmatica* Pienaar, *N. fissa* Lackey, *N. marina* Schiller, *N. minuta* (N. Carter) Butcher, *N. pyriformis* (N. Carter) Ettl, *N. rotunda* (N. Carter) Fott.

   Genus *Pseudoscourfieldia* Manton 1975: (**Fig. 9E**) Two homodynamic flagella; *P. marina* (Throndsen) Manton.

Genus *Prasinocladus* Kuckuck 1984: Cells of this genus are only
temporary members of the plankton; *P. ascus* Proskauer, *P. marinus*
(Cienkowski) Waern (P. lubricus Kuckuck).

Genus *Tetraselmis* Stein 1878: (Fig. 9K) Cells with "scale theca"
(Manton & Parke, 1965) and four flagella; the genus may include
Platymonas; T. contracta (N. Carter) Butcher, *T. convolutae* (Parke
& Manton) Norris et al., *T. gracilis* (Kylin) Butcher, *T. helgolandica*
(Kylin) Butcher, *T. inconspicua* Butcher, *T. subcordiformis* (Wille)
Butcher, *T. suecica* (Kylin) Butcher, *T. tetrathele* (West) Butcher, *T.*
wettsteinii (Schiller) Throndsen; see also Norris et al. (1980).

Family Halosphaeraceae Haeckel 1894: The underlayer scales on the
flagella are covered by meshwork scales in nine longitudinal rows.

Genus *Cymbomonas* Schiller 1913: Bilateral symmetry and long
flagella, nonmotile stage not known; *C. adriatica* Schiller, *C. klebsii*
Schiller, *C. tetramitiformis* Schiller.

Genus *Pachysphaera* Ostenfeld 1899, *Tasmanites* E. T. Newton 1875:
Motile cells bilaterally symmetrical with four long flagella, nonmotile
stage walled with pores, and many olive-green discoid chloroplasts
(cf. Parke, 1966, and Table 1); *P. pelagica* Ostenfeld, *P. marshalliae*
Parke.

Genus *Pterosperma* Pouchet 1893: Motile cells bilaterally symmetrical
with four long flagella, nonmotile stage walled, with wings (alae),

TABLE 1   Prasinophycean Genera with Phycoma Stages

	Genus		
Characteristic	*Pachysphaera* *(Tasmanites)*	*Pterosperma*	*Halosphaera*
Cell wall[a]	+ pores 1 or 2 sizes	+/− pores + alae[b]	+/− punctae
Size (mature)	100–175 μm	25–100 μm + alae	190–800 μm
Chloroplasts	many disc-shaped	disc-shaped	disc, irregular
Color	olive-green	greenish-yellow/ golden brown	pale yellow– deepish green
Pyrenoid starch[c]	+	+	+
Motile symmetry	bilateral	bilateral	quadrilateral
More information	Parke et al. (1978)	Boalch & Mommaerts (1969)	
Described species	2	16	4

[a] Stains yellow with iodine (Lugol's solution; see Table 3).
[b] Alae = wings, hyaline wing-like protrusions from the cell wall.
[c] Stains purple red with iodine (see above).

and many discoid greenish yellow to golden brown chloroplasts. Nonmotile (phycoma) stage predominant? Sixteen species described (cf. Parke et al., 1978); *P. rotondum* Pouchet, *P. citriforme* Parke, *P. polygonum* Ostenfeld, *P. cristatum* Schiller, *P. cuboides* Gaarder, *P. dictyon* (Jørgensen) Ostenfeld, *P. eurypteron* Parke, *P. inornatum* Parke, *P. marginatum* Gaarder, *P. michaelsarsii* (Gaarder) Parke & Boalch, *P. moebii* (Jørgensen) Ostenfeld, *P. nationalis* Lohmann, *P. parallellum* Gaarder, *P. porosum* Parke, *P. undulatum* Ostenfeld, *P. vanhoeffenii* (Jørgensen) Ostenfeld.

Genus *Halosphaera* Schmitz 1878: (Fig. 9I,J) *Pyramimonas*-like motile stage (chloroplast with two or four pyrenoids, EM; Manton et al., 1963, Parke & den Hartog-Adams, 1965), nonmotile stage spherical with +/− punctate wall and many more or less irregular disc formed chloroplasts, pale yellowish to deepish green. Nonmotile stage predominant; *H. minor* Ostenfeld, *H. parkeae* Boalch & Mommaerts, *H. russellii* Parke, *H. viridis* Schmitz. (A key to the species is given in Boalch & Mommaerts, 1969; see also Table 2.)

Genus *Pyramimonas* Schmarda 1850: (Fig. 9H) Quadrilateral symmetry and pyramidoidal cells with four flagella. It comprises approximately 50 species most of them reported from marine plankton; *P. adriaticus* Schiller (*P. lunata* Inouye, Hori, & Chihara, 1983), *P. amylifera* Conrad, *P. disomata* McFadden et al. (= *P. disomata* Butcher), *P. grossii* Parke, *P. moestrupii* McFadden, *P. nansenii* Braarud, *P. nephroidea* McFadden, *P. obovata* N. Carter, *P. octociliata* N. Carter, *P. octopus* Moestrup & Aa Kristiansen, *P. olivacea* N. Carter, *P. orientalis* McFadden et al. (= *P. orientalis* Butcher), *P. pisum* Conrad & Kufferath, *P. plurioculata* Butcher, *P.*

TABLE 2 Comparative Data of *Halosphaera* Species, Based on Boalch & Mommaerts (1969)

Characteristic	*H. viridis*	*H. minor*	*H. russellii*	*H. parkeae*
Motile				
Size	20–28 μm	14–28 μm	12–22.5 μm	10–16 μm
Flagella length	2 × cell	1.5–2 × cell	2 × cell	1.5 × cell
Chloroplast color	pale green	deepish green[b]	pale yellowish green	pale green
Nonmotile				
Size at maturity	400–800 μm	190–235 μm	300–500 μm	200–550 μm[a]
Chloroplast size	<8 μm	3–5 μm	3–6 μm	9 μm
Rosettes	256/512	64	128/256	128/256/512

[a] May be divided into *H. parkeae* f. *parkeae*, 400–550 μm; *H. parkeae* f. *minuta*, 200–250 μm.
[b] *Sensu* Boalch & Mommaerts (1969).

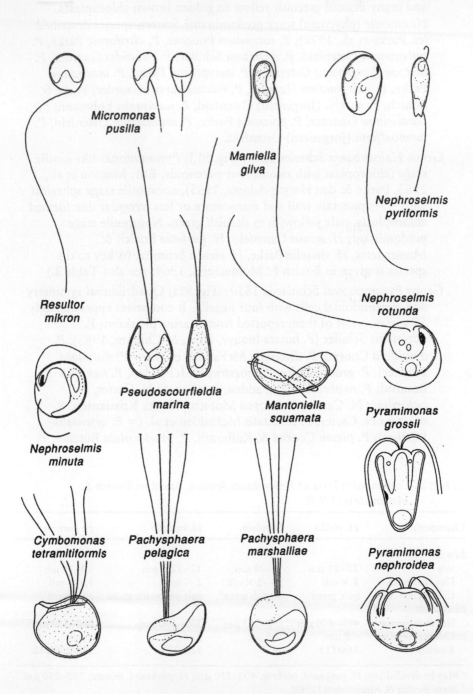

Micromonas
pusilla

Mamiella
gilva

Nephroselmis
pyriformis

Resultor
mikron

Nephroselmis
rotunda

Nephroselmis
minuta

Pseudoscourfieldia
marina

Mantoniella
squamata

Pyramimonas
grossii

Cymbomonas
tetramitiformis

Pachysphaera
pelagica

Pachysphaera
marshalliae

Pyramimonas
nephroidea

propulsa Moestrup & Hill, *P. pseudoparkeae* Pienaar & Aken, *P. virginica* Pennick. The identification of most of the species has to be confirmed by EM of the scale cover. (The genus may be divided into subgenera based on submicroscopic structures; cf. McFadden et al., 1986).

Genus *Micromonas* Manton & Parke 1960: (Fig. 9B) Naked cells carrying one short flagellum; single species *M. pusilla* (Butcher) Manton & Parke is tentatively placed in Prasinophyceae.

The genera *Pachysphaera* (motile Fig. 9G), *Pterosperma,* and *Halosphaera* are best known from the phycoma or nonmotile stage (Table I).

Selected Species of Prasinophyceae (Including Pedinophyceae)

Resultor mikron (Throndsen) Moestrup 1991 (Plate 11)
Pedinomonas mikron Throndsen 1969
Cell length: 1.5–2.5 μm.
Flagella: Single, 7–12 μm.
Chloroplasts: Single, green.
Characteristic features: Cell flattened, often with jerky movements.
Distribution: Coastal; Atlantic.

Micromonas pusilla (Butcher) Manton & Parke 1960 (Plate 11)
Chromulina pusilla Butcher 1952
Cell length: 1–3 μm.
Flagella: Single, short, but with terminal "hair" approximately 3 μm.
Chloroplasts: Single, green.
Characteristic features: Extremely small, but with a characteristic swimming pattern, flagellum with terminal "hair" may be observed in phase contrast.
Distribution: Coastal, oceanic; Atlantic, Mediterranean, Arctic, Pacific.

Mamiella gilva (Parke & Rayns) Moestrup 1984 (Plate 11)
Nephroselmis gilva Parke & Rayns 1964.
Cell length: 4–6.5 μm.
Flagella: Two, 2½–3½ × cell length.
Chloroplasts: Yellowish green.

Plate 11 *Nephroselmis minuta, Mantoniella squamata, Pachysphaera marshalliae* (motile), *Pyramimonas grossii, P. nephroidea* based on type illustrations; *Resultor mikron, Mamiella gilva, Micromonas pusilla, Nephroselmis pyriformis, N. rotunda,* originals; *Pseudoscourfieldia marina* modified from Moestrup & Throndsen (1988); *Pachysphaera pelagica* Ostenfeld 1899 (motile phase) redrawn from Parke (1966); *Cymbomonas tetramitiformis* Schiller modified from Schiller (1913).

Characteristic features: Cell bean-shaped, with two very long flagella, scale pattern (EM) essential for species identification.
Distribution: Coastal, oceanic; North Sea, Mediterranean, Pacific.

Nephroselmis pyriformis (N. Carter) Ettl 1982 (Plate 11)
Bipedinomonas pyriformis N. Carter 1937.
Cell length: 5–7 µm.
Flagella: Two, 4–5 and 1½–2 × cell length.
Chloroplasts: Single, green.
Characteristic features: Cell compressed, heterodynamic mode for flagellar action is also reflected at rest.
Distribution: Coastal; Atlantic, Arctic, Pacific.

Nephroselmis rotunda (N. Carter) Fott 1971 (Plate 11)
Bipedinomonas rotunda N. Carter 1937.
Cell length: 6–8 µm.
Flagella: Two, 4–5 and 1½–2 × cell length.
Chloroplasts: Single, green.
Characteristic features: Cell more rounded than in *N. pyriformis*.
Distribution: Coastal; Atlantic, Pacific.

Nephroselmis minuta (N. Carter) Butcher 1959 (Plate 11)
Heteromastix minuta N. Carter 1937.
Cell length: 3.5 µm.
Flagella: Two, subequal.
Chloroplasts: Single, green.
Characteristic features: Cell compressed, chloroplast with conspicuous pyrenoid.
Distribution: Coastal; Atlantic.

Pseudoscourfieldia marina (Throndsen) Manton 1975 (Plate 11)
Scourfieldia marina Throndsen 1969.
Cell length: 3–5 µm.
Flagella: Two, of different lengths.
Chloroplasts: Single, green.
Characteristic features: Cell flattened, with two homodynamic flagella, chloroplast devoid of eyespot.
Distribution: Coastal, oceanic; Atlantic.

Mantoniella squamata (Manton & Parke) Desikachary 1972 (Plate 11)
Micromonas squamata Manton & Parke 1960.
Cell length: 3–5 µm.
Flagella: Two, 3–3½ × cell length, and short.
Chloroplasts: Single, green.

Characteristic features: Cell shape with the looped long flagellum is quite typical, but identification should be checked in EM.
Distribution: Coastal, oceanic; Atlantic, Arctic.

Cymbomonas tetramitiformis Schiller 1913 (Plate 11)
Cell length: 13.5–16 μm.
Flagella: Four, 1½–2 × cell length or longer.
Chloroplasts: Single, green, obliquely bell-shaped.
Characteristic features: The cell has an horseshoe-shaped ridge surrounding the apical depression from which the four fairly long flagella are protruding. Pyrenoid and starch shield not always evident. [For scale morphology see Throndsen (1988).]
Distribution: Coastal; Mediterranean, Atlantic, Pacific.

Pachysphaera pelagica Ostenfeld 1899 (Plate 11) (motile phase)
Motile phase:
Cell length: 9–12 μm.
Flagella: Four, 6–9 × cell length.
Chloroplasts: Single, pale yellowish green.
Characteristic features: Cell dorsoventrally flattened with extremely long flagella.
Distribution: Oceanic; Atlantic.

Pachysphaera marshalliae Parke 1966 (Plate 11) (motile phase)
Motile phase:
Cell length: 7–9 μm.
Flagella: Four, 4–6 × cell length.
Chloroplasts: Single, pale yellowish green.
Characteristic features: Cell nearly spherical with long flagella.
Distribution: Oceanic; Atlantic, Mediterranean.

Pyramimonas grossii Parke 1949 (Plate 11)
Cell length: 4–8 μm.
Flagella: Four, slightly longer than the cell.
Chloroplasts: Single, green.
Characteristic features: Cell pyramidal to rounded, chloroplast lobed with single small eyespot, ejectosomes present.
Distribution: Coastal, oceanic; Arctic, Atlantic, Pacific, Mediterranean, probably cosmopolitan.

Pyramimonas nephroidea McFadden 1986 in McFadden et al. (1986) (Plate 11)
Cell length: 8–9 μm.
Cell width: 9–10 μm.
Flagella: Four, slightly shorter than the cell.
Chloroplasts: Single, green.

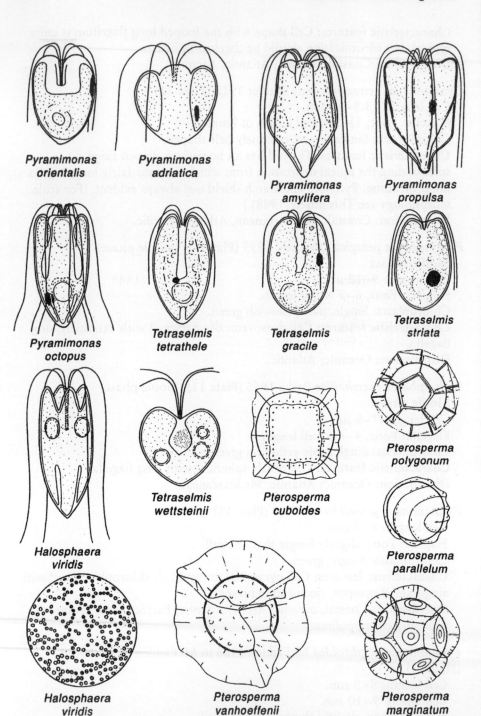

Pyramimonas
orientalis

Pyramimonas
adriatica

Pyramimonas
amylifera

Pyramimonas
propulsa

Pyramimonas
octopus

Tetraselmis
tetrathele

Tetraselmis
gracile

Tetraselmis
striata

Halosphaera
viridis

Tetraselmis
wettsteinii

Pterosperma
cuboides

Pterosperma
polygonum

Pterosperma
parallelum

Halosphaera
viridis

Pterosperma
vanhoeffenii

Pterosperma
marginatum

Characteristic features: Cell broader than wide, chloroplast bell-shaped with lobes, and double eyespot.
Distribution: Coastal; Australia, New Zealand.

Selected Prasinophycean Species with Four Flagella or with a Characteristic Phycoma Stage

Pyramimonas orientalis McFadden, Hill, & Wetherbee 1986 (Plate 12)
Pyramimonas orientalis Butcher 1959.
Cell length: 4–6 μm.
Flagella: Four, cell length.
Chloroplasts: Single, green.
Characteristic features: Cell shape ovoid pyramidal with bluntly rounded posterior, chloroplast with four short lobes and a parietal cup-shaped pyrenoid, usually with a single eyespot.
Distribution: Coastal; Atlantic, Arctic, Mediterranean, Pacific.

Pyramimonas adriaticus Schiller 1913 (Plate 12)
Cell length: 6–12 μm.
Cell width: 7–10 μm.
Flagella: Four, twice the cell length.
Chloroplasts: Single, green with four lobes.
Characteristic features: The luneate form, cells being broader than long (probably conspecific with *P. lunata* Inouye; Hori & Chihara, 1983).
Distribution: Coastal; Mediterranean, Pacific.

Pyramimonas amylifera Conrad 1939 (Plate 12)
Cell length: 18–20 μm.
Flagella: Four or eight, 1½ × cell length.
Chloroplasts: One, olive to yellow.
Characteristic features: Conspicuous starch bodies in old cells. Often assumed to be conspecific with *A. propulsum* Butcher, but see *Pyramimonas propulsa* Moestrup & Hill.
Distribution: Coastal; Atlantic, Pacific, Mediterranean.

Pyramimonas propulsa Moestrup & Hill 1991 (Plate 12)
cfr. *Asteromonas propulsum* Butcher 1959.
Cell length: 18–35 μm.

Plate 12 *Pyramimonas orientalis, P. adriatica, P. amylifera, P. propulsa, P. octopus, Tetraselmis striata, T. wettsteinii, Pterosperma cuboides, P. parallelum, Halosphaera viridis* phycoma based on type illustrations; *Pterosperma vanhoeffenii, P. polygonum,* and *P. marginatum* based on light micrographs in Parke *et al.* (1978); *Halosphaera viridis* motile, modified from Parke & Adams (1965); *Tetraselmis gracilis, T. tetrathele* redrawn from Butcher (1959).

Flagella: Eight, approximately cell length.
Chloroplasts: Single, yellow-green.
Characteristic features: Cell with pronounced wings, permanently with 8
flagella (see also *Pyramimonas amylifera* Conrad, scale morphology
indicate that 2-3 species may be distinguished); six to eight starch
granules.
Distribution: Coastal; Atlantic, Mediterranean.

Pyramimonas octopus Moestrup & Kristiansen 1987 (in Moestrup, Hori, &
Kristiansen, 1987) (Plate 12)
Cell length: 16–20 µm.
Flagella: Eight, longer than the cell.
Chloroplasts: Single, lobed.
Characteristic features: Chloroplast with four large anterior starch grains,
and with two prominent starch shields lateral to the basal pyrenoid, with
8 flagella (16 during division).
Distribution: Psammophilic species; Denmark.

Tetraselmis tetrathele (G. S. West) Butcher 1959 (Plate 12)
Platymonas tetrathele G. S. West 1916.
Cell length: 10–16 µm.
Flagella: Four, ½–¾ × cell length.
Chloroplasts: Bright green.
Characteristic features: Compressed, with deep and wide four-lobed apical
furrow.
Distribution: Salt marshes, tidal pools; Europe.

Tetraselmis gracilis (Kylin 1935) Butcher 1959 (Plate 12)
Platymonas gracilis Kylin 1935.
Cell length: 8–12 µm.
Flagella: Four, cell length.
Chloroplasts: Yellow-green.
Characteristic features: Cell slightly compressed, with four anterior lobes;
pyrenoid large, subbasal, with U-shaped starch shield; median eyespot,
large, red-orange.
Distribution: Salt marshes, tidal pools; Europe.

Tetraselmis striata Butcher 1959 (Plate 12)
Cell length: 6.5–8 µm.
Flagella: Four, ¾–1 × cell length.
Chloroplasts: Yellow-green.
Characteristic features: Cell with marked longitudinal rows of granules,
and posterior eyespot.
Distribution: Coastal; England.

Tetraselmis wettsteinii (Schiller) Throndsen 1988 (in Throndsen & Zingone 1988) (Plate 12)
  Carteria wettsteinii Schiller 1913.
  Cell length: 11–17 μm.
  Flagella: Four, approximately cell length.
  Chloroplasts: Single, green.
  Characteristic features: Cell heart-shaped, chloroplast with two to four pyrenoids. In culture the cells extrude a substance that gives the medium a violet color.
  Distribution: Coastal; Mediterranean.

Halosphaera viridis Schmitz 1878 (Plate 12) (motile)
  Motile stage:
  Cell length: 20–28 μm.
  Flagella: Four, 2 × cell length.
  Chloroplasts: Pale green.
  Characteristic features: Cell body rounded with a narrow cylindrical apical depression.
  Distribution: Coastal, oceanic; Mediterranean, west Atlantic.
  Phycoma stage: Spherical cells, up to 400–800 μm, with many light green chloroplasts; see Table 2.

Selected Pterosperma Species, Phycoma Stages Only

Pterosperma cuboides Gaarder 1954 (Plate 12)
  Cell diameter: 18–50 μm.
  Chloroplasts: many, green (C. Billard, personal communication 1994).
  Characteristic features: Alae dividing the cell surface into six quadrates.
  Distribution: Oceanic; Atlantic.

Pterosperma polygonum Ostenfeld (in Ostenfeld & Schmidt, 1902) (Plate 12)
  Cell diameter: 30 μm.
  Chloroplasts: Not specified.
  Characteristic features: Wall with pores, alae in a pentagonal pattern (*P. cristatum* lack pores).
  Distribution: Oceanic; Atlantic, Red Sea.

Pterosperma cristatum Schiller 1925 (Plate 12)
  [Not illustrated, see *P. polygonum;* for EM see Inouye et al. (1990).]
  Cell diameter: 10–12 μm.
  Chloroplasts: Many, yellow-green.
  Characteristic features: Wall without evident pores, alae in a pentagonal pattern (*P. polygonum* with pores).
  Distribution: Coastal, oceanic; Mediterranean, Pacific, Antarctic?

Pterosperma parallelum Gaarder 1938 (Plate 12)
Cell diameter: 9–20 μm.
Chloroplasts: Single?
Characteristic features: Wall with low curved parallel alae.
Distribution: Coastal, oceanic; Arctic, Antarctic, Pacific.

Pterosperma vanhoeffenii (Jørgensen) Ostenfeld 1899 (Plate 12)
Pterosphaera venhöffenii Jørgensen 1900.
Cell diameter: 47–53 μm + alae.
Chloroplasts: Many, yellowish green.
Characteristic features: Alae few and pronounced, conspicuous species.
Distribution: Oceanic, coastal; Atlantic.

Pterosperma marginatum Gaarder 1954 (Plate 12)
Cell diameter: 15 μm + alae.
Chloroplasts: Not reported.
Characteristic features: Protruding alae encircling a pronounced pore in
the wall.
Distribution: Oceanic, coastal; Atlantic.

For further information on *Pterosperma*–phycoma stages of different
species, see Parke et al. (1978).

CHLOROPHYCEAE

Systematic position: Class Chlorophyceae *sensu* Christensen 1962.

Characteristic Features

Cell shape—often rounded or ovoid, may be lobed (e.g., *Brachiomonas*).

Cells—naked or with cellulose wall.

Flagella—one, two, four (or eight), smooth (or with tomentum).

Mode of swimming—homodynamic, flagella pushing the cell.

Color—bright green.

Chloroplast—(in flagellates) one, parietal or campanulate, lobed or
reticulated.

Eyespot—in chloroplast.

Pyrenoid—in chloroplast, with starch shield.

Storage product—starch surrounding pyrenoid, and elsewhere in the
chloroplast stroma.

Distribution—planktonic species found in coastal waters near and on
shore, rock pools.

Identification/Observation and Problems

Preserved material fixed with formaldehyde or iodine may be adequate for identification of some species, but the iodine tends to obscure the cell contents, especially when a large amount of stroma starch is present.

Living material offers the best possibilities for the identification in the light microscope. Electron microscopy is often of limited importance unless embedding and sectioning techniques are applied.

Problems are often related to determining whether a cell wall is present or not. Applying a hypertonic medium to a living preparation to cause the cell to retract from the cell wall, observing if the species divides within the mother cell wall, or staining for cellulose with $ZnCl_2 + H_2SO_4$ (Jane, 1942) are among some of the methods used.

Most chlorophycean flagellates found in marine plankton are included in two families of the order Volvocales.

Order Volvocales Oltmanns 1904

Family Dunaliellaceae Christensen 1967: Cells naked.

Genus *Dunaliella* Teodoresco 1905: (Fig. 10A) Two flagella; *D. salina* (Dunal) Teodoresco, *D. tertiolecta* Butcher.

Genus *Collodictyon* H. J. Carter 1865: Two or three flagella, *C. sphaericum* Norris.

Genus *Oltmannsiellopsis* Chihara & Inouye (in Chihara et al., 1986) (= *Oltmannsiella sensu* Hargraves & Steele, 1980): Two "double" flagella, cells laterally attached in "flat" groups/colonies of two to four specimens; *O. viridis* (Hargraves & Steele) Chihara & Inouye, *O. geminata* Inouye & Chihara, *O. unicellularis* Inouye & Chihara.

Family Chlamydomonadaceae G. M. Smith 1920: Cells with wall.

Genus *Brachiomonas* Bohlin 1897: (Fig. 10D) "Lobed" cells; *B. submarina* Bohlin, *B. simplex* Hazen (rock pool plankton).

Genus *Carteria* Diesing 1866: (Fig. 10C) Four flagella; *C. marina* Wulff.

Genus *Chlamydomonas* Ehrenberg 1834: (Fig. 10B) With "papilla" and two flagella; very many species, some marine e.g. *C. coccoides* Butcher, *C. euryale* Lewin, *C. pulsatilla* Wollenweber, *C. reginae* Ettl & Green, *C. quadrilobata* N. Carter.

Genus *Oltmannsiella* Zimmermann 1930: Two flagella, 4(1-8) cells attached laterally to make "flat" groups/colonies; *O. lineata* Zimmermann (The species is almost identical to the Dunaliellacean species *Oltmannsiellopsis,* but according to Zimmermann (1930) the cell divide within the mother cell wall, hence the present genus/species appears to belong to the family Chlamydomonadaceae).

(Systematics according to Parke & Burrows, 1976.)

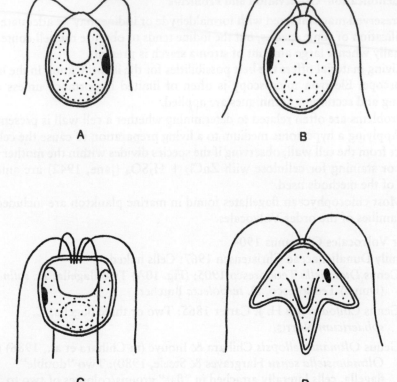

FIGURE 10   Chlorophycean flagellate genera from marine plankton. (A) *Dunaliella* 5–24 μm; (B) *Chlamydomonas* 4.5–29 μm; (C) *Carteria* 7–32 μm; (D) *Brachiomonas* 15–48 μm. (Modified from Throndsen, 1980.)

Selected Species of Chlorophyceae   Selected naked and walled species with two or four flagella:

Dunaliella maritima Massjuk 1973 (Plate 13)
　　Cell length: 7–12 μm.
　　Flagella: Two, 1½–2 × cell length.
　　Chloroplasts: Yellowish green.

Plate 13   *Asteromonas gracilis* redrawn from Ruinen (1938); *Brachiomonas submarina*, original; *Brachiomonas simplex*, *Chlamydomonas pulsatilla*, *Dunaliella maritima*, and *D. salina*, based on Butcher (1959); *Pyramichlamys vectensis* based on Carter (1937).

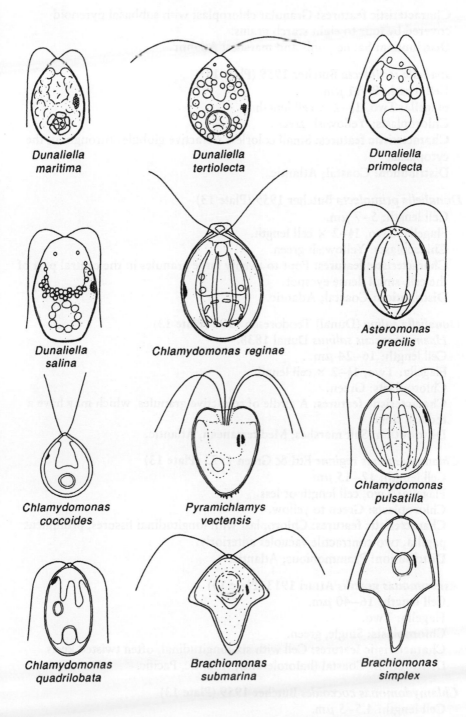

Dunaliella
maritima

Dunaliella
tertiolecta

Dunaliella
primolecta

Dunaliella
salina

Chlamydomonas reginae

Asteromonas
gracilis

Chlamydomonas
coccoides

Pyramichlamys
vectensis

Chlamydomonas
pulsatilla

Chlamydomonas
quadrilobata

Brachiomonas
submarina

Brachiomonas
simplex

Characteristic features: Granular chloroplast with subbasal pyrenoid covered by four to eight starch grains.
Distribution: Saline pools and marshes; Atlantic.

Dunaliella tertiolecta Butcher 1959 (Plate 13)
Cell length: 9–11 μm.
Flagella: Two, 1½–2 × cell length.
Chloroplasts: Yellowish green.
Characteristic features: Small colorless refractive globules throughout the cytoplasm.
Distribution: Coastal; Atlantic.

Dunaliella primolecta Butcher 1959 (Plate 13)
Cell length: 5–7 μm.
Flagella: Two, 1½–2 × cell length.
Chloroplasts: Yellowish green.
Characteristic features: Four to twenty large granules in the central part of the cell, small dense eyespot.
Distribution: Coastal; Atlantic.

Dunaliella salina (Dunal) Teodoresco 1905 (Plate 13)
Haematococcus salinus Dunal 1838.
Cell length: 16–24 μm.
Flagella: Two, 1½–2 × cell length.
Chloroplasts: Green.
Characteristic features: A girdle of refractive granules, which may have a reddish tinge.
Distribution: Salt marshes; Mediterranean, Atlantic.

Chlamydomonas reginae Ettl & Green 1973 (Plate 13)
Cell length: 12–15 μm.
Flagella: Two, cell length or less.
Chloroplasts: Green to yellow.
Characteristic features: Chloroplast with longitudinal fissures, prominent papilla, two contractile vacuoles anteriorly.
Distribution: Psammobious; Atlantic.

Asteromonas gracilis Artari 1913 (Plate 13)
Cell length: 16–40 μm.
Flagella: Two.
Chloroplasts: Single, green.
Characteristic features: Cell with six longitudinal, often twisted keels.
Distribution: Coastal (halotolerant); Atlantic, Pacific.

Chlamydomonas coccoides Butcher 1959 (Plate 13)
Cell length: 4.5–5 μm.

Flagella: Two (easily shed).
Chloroplasts: Yellow-green.
Characteristic features: Small species with a relatively thick cell wall, a papilla lacking, devoid of contractile vacuole (see also *C. minuta* Schiller 1925).
Distribution: Coastal; Atlantic.

Pyramichlamys vectensis (Kufferath) Ettl & Ettl 1959 (Plate 13)
Carteria vectensis Kufferath (in Conrad & Kufferath, 1954).
Cell length: 7 μm.
Flagella: Four, 1–1½ × cell length.
Chloroplasts: Green.
Characteristic features: Walled species with apical depression and pointed posterior with a few spicules.
Distribution: Salt marshes; Atlantic.

Chlamydomonas pulsatilla Wollenweber 1926 (Plate 13)
Cell length: 15–29 μm.
Flagella: Approximately cell length.
Chloroplasts: Green.
Characteristic features: Large species with a fimbriate chloroplast, four contractile vacuoles.
Distribution: Rock pools; Atlantic.

Chlamydomonas quadrilobata N. Carter 1937 (Plate 13)
Cell length: 12 μm.
Flagella: Two.
Chloroplasts: Green.
Characteristic features: Very elongated cell (length/width 2–2.5) with four-lobed chloroplast.
Distribution: Salt marshes; Atlantic.

Brachiomonas submarina Bohlin 1897 (Plate 13)
Cell length: 15–40 μm.
Flagella: Two.
Chloroplasts: Green with single pyrenoid.
Characteristic features: More or less pronounced "arms" protruding laterally in four directions.
Distribution: Coastal (rock pools); Atlantic.

Brachiomonas simplex Hazen 1922 (Plate 13)
Cell length: 30–48 μm.
Flagella: Two.
Chloroplasts: Green.
Characteristic features: Devoid of protruding "arms."
Distribution: Brackish water; Europe.

Species of the Oltmannsiella–Oltmannsiellopsis *complex*

Oltmannsiella lineata Zimmermann 1930 (Plate 14)
Cell length: 15–22 µm.
Flagella: Two(?), 20–30 µm.
Chloroplasts: Green with eyespot.
Characteristic features: Cells, one to eight, but commonly four attach laterally to form "flat" groups/colonies. The relationship to *Oltmannsiellopsis viridis* Chihara & Inouye, from which it differs by the two flagella being single, is discussed.
Distribution: Coastal; Mediterranean.

Oltmannsiellopsis viridis (Hargraves & Steele) Chihara & Inouye (in Chihara et al., 1986) (Plate 14)
Oltmannsiella virida Hargraves & Steele 1980
Cell length: 10–30 µm.
Flagella: Two, "double" flagella, approximately cell length.
Chloroplasts: Green, cup-shaped with two to four lobes, with eyespot.

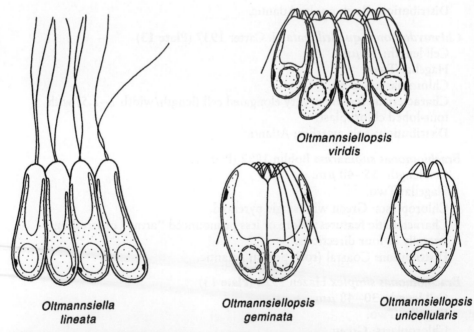

Oltmannsiellopsis
viridis

Oltmannsiella
lineata

Oltmannsiellopsis
geminata

Oltmannsiellopsis
unicellularis

Plate 14   *Oltmannsiella lineata* redrawn from Zimmermann (1930); *Oltmannsiellopsis geminata,* *O. unicellularis,* and *O. viridis* redrawn from Chihara & Inouye et al. (1986).

Characteristic features: Cells laterally attached in "flat" groups/colonies of four specimens, flagella originate in a depression.
Distribution: Coastal; East Atlantic, Japan.

Oltmannsiellopsis geminata Inouye & Chihara (in Chihara et al., 1986)
(Plate 14)
  Cell length: 10.5–15 μm.
  Flagella: Two, "double" flagella.
  Chloroplasts: Green, dorsolateral with eyespot.
  Characteristic features: Cells laterally attached in "flat" groups/colonies of two specimens, may be a variety of *O. viridis.*
  Distribution: Coastal; Japan.

Oltmannsiellopsis unicellularis Inouye & Chihara (in Chihara et al., 1986)
(Plate 14)
  Cell length: 9–14 μm.
  Flagella: Two, "double" flagella.
  Chloroplasts: Green, cup-shaped with eyespot.
  Characteristic features: Single cells, eyespot in anterior half of the cell, may form gelatinous, palmelloid stages.
  Distribution: Coastal; Japan.

ZOOFLAGELLATES (PHYLUM ZOOMASTIGOPHORA)

In protozoan systematics (e.g., according to Lee et al., 1985), the phytoflagellates are included in one class, Phytomastigophorea, the classes of the phytosystematics just described being treated at the level of orders. The zooflagellates are included in the class Zoomastigophorea Calkins. The systematics used here generally follow Lee et al. (1985), though the rank of the suprageneric taxa of the choanoflagellates has been raised to a level comparable to that found in the algal system: Phylum Zoomastigophora (class Zoomastigophorea Calkins).

Three orders/classes with flagellate members are frequently found in the marine plankton:

Choanoflagellidea—with collared cells.

Kinetoplastida—"naked" cells, flagella with paraxial rod.

Ebriida—with internal siliceous skeleton.

CHOANOFLAGELLIDEA

Systematic position: Class choanoflagellidea (order Choanoflagellida, class Zoomastigophorea, subphylum Mastigophora, phylum Sarcomastigophora in Lee et al., 1985).

The class contains almost invariably colorless species (one possible exception being *Stylochromonas minuta* Lackey). For some time these species were included in the algal system (e.g., as class Craspedophyceae; Christensen, 1962), but based on their microanatomy they are now regarded as true zooflagellates (Leadbeater, 1972b; Hibberd, 1976).

Characteristic Features

Cell shape—oval with a single flagellum encircled at the base by a collar composed by fine pseudopodia.

Cell covering—many species enveloped in a lorica composed of silica rods, costal strips; taxonomy is based almost solely on the lorica morphology (Fig. 11).

Flagellum—smooth, pushing the cell or producing a water current with particles hitting the outside of the collar.

Mode of swimming—active swimming not pronounced, but may occur in small species (e.g., *Monosiga micropelagica* Throndsen).

Distribution—common in coastal as well as oceanic areas, and from polar to tropical waters. At present about 60 species are known.

Identification/Observation and Problems

Living material is required for most naked species to give accurate observations of cell shape and size when using phase or interference contrast micro-

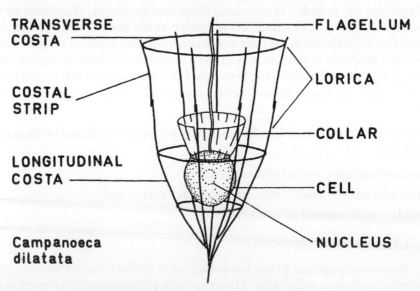

FIGURE 11 Choanoflagellate morphology with simple lorica terminology. (From Throndsen, 1974).

scopy. The loricate species are usually identified from the dimensions and construction of the loricae, and the most relevant method is to study dry mounts in the light (or electron) microscope (Thomsen, 1982).

Preserved material may be studied with phase contrast to reveal loricae of choanoflagellates, and some of the nonloricate species may fix in a characteristic way (e.g., *Monosiga marina*). Dry preparations are effective in revealing lorical structure in phase or interference contrast.

Problems are related to the delicate loricate structures and apparently low sinking rate, which render these forms easy to miss or overlook during routine observations on preserved material. Staining with iodine (Lugol's solution) may overcome this problem, but may obscure details important for a reliable identification.

Order Choanoflagellida Kent 1880: Three families which are tentatively raised to the rank of order in the overview below (only most common genera included):

Order Codonosigida/Family Codonosigidae Kent 1880: Cells naked or with an investment invisible in the LM.

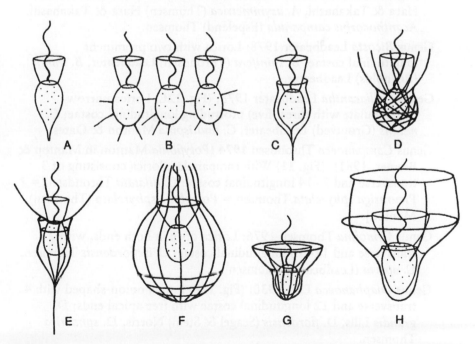

FIGURE 12 Choanoflagellate genera from marine plankton: (A) *Monosiga* 3–11 μm; (B) *Desmarella* approximately 6 μm; (C) *Salpingoeca* cell 3–6 μm, theca 3.5–9 μm; (D) *Acanthocorbis* cell 4–5, lorica 14–16 μm; (E) *Calliacantha* cell 9–22 μm; (F) *Diaphanoeca* cell 3–8 μm, lorica 10–38 μm; (G) *Parvicorbicula* cell 7–8 μm, lorica 10–18 μm; (H) *Pleurasiga* cell 5–12 μm, lorica 10–23 μm. (Modified from Throndsen, 1980.)

Genus *Monosiga* Kent 1882: (**Fig. 12A**) Solitary, free-swimming cells;
   M. marina Grøntved, *M. micropelagica* Throndsen.

Genus *Desmarella* Kent 1882: (**Fig. 12B**) Cells in "loose" colonies; *D. moniliformis* Kent.

Genus *Proterospongia* Kent 1882: Cells in "jelly" colonies; *P. nana* (Braarud) Tuzet, *P. dybsoeensis* (Grøntved) Loeblich III.

Genus *Stylochromonas* Lackey 1940: With chloroplasts and short stalk: *S. minuta* Lackey.

Order Salpingoecida/Family Salpingoecidae Kent 1880: Cells with a
   hyaline close-fitting theca, visible in the light microscope.

Genus *Salpingoeca* James-Clark 1867: (**Fig. 12C**) *S. inquillata* Kent.

Order Acanthoecida/Family Acanthoecidae Norris 1965: Cells surrounded
   by often wide loricae consisting of costae constructed from costal strips
   of silica (see **Fig. 11**).

Genus *Acanthocorbis* Hara & Takahashi 1984: (**Fig. 12D**) Lorica
   basket shaped with 8–17 protruding longitudinal costae; *A. unguiculata* (Thomsen) Hara & Takahashi, *A. apoda* (Leadbeater) Hara & Takahashi, *A. asymmetrica* (Thomsen) Hara & Takahashi. *Acanthocorbis campanula* (Espeland) Thomsen.

Genus *Bicosta* Leadbeater 1978: Lorica with two prominent
   longitudinal costae; *B. spinifera* (Throndsen) Leadbeater, *B. minor* (Reynolds) Leadbeater.

Genus *Calliacantha* Leadbeater 1978: (**Fig. 12E**) Lorica narrow
   companulate with three (five) protruding longitudinal costae; *C. natans* (Grøntved) Leadbeater, *C. multispina* Manton & Oates.

Genus *Campanoeca* Throndsen 1974 (*Polyfibula* Manton in Manton &
   Bremer, 1981): (**Fig. 11**) With campanulate lorica consisting of 3 transverse and 7–14 longitudinal costae; *C. dilatata* Throndsen [= ? *Pleurasiga sphyrelata* Thomsen = *Polyfibula sphyrelata* (Thomsen) Manton].

Genus *Crinolina* Thomsen 1976: Lorica open in both ends, with 2
   transverse and 12–16 longitudinal costae; *C. isefiordensis* Thomsen, *C. aperta* (Leadbeater) Thomsen.

Genus *Diaphanoeca* Ellis 1930: (**Fig. 12F**) Lorica onion-shaped with 4
   transverse and 12 longitudinal costae with free apical ends; *D. grandis* Ellis, *D. fiordensis* (Scagel & Stein) Norris, *D. sphaerica* Thomsen.

Genus *Parvicorbicula* (Meunier 1910) Deflandre 1960: (**Fig. 12G**)
   Lorica funnel-shaped with 2 transverse and 4–10 longitudinal costae; *P. socialis* (Meunier) Deflandre, *P. quadricostata* Throndsen.

Genus *Pleurasiga* Schiller 1925: (**Fig. 12H**) Lorica with three transverse and seven longitudinal costae; *P. reynoldsii* Throndsen, *P. minima* Throndsen (*P. orculaeformis* Schiller to be recollected).

Genus *Stephanoeca* Ellis 1930: Lorica two-chambered with transverse and longitudinal costae; *S. diplocostata* Ellis (including *S. pedicellata* Leadbeater), *S. complexa* (Norris) Throndsen, *S. degans* (Norris) Throndsen.

Selected species of Choanoflagellidea

Monosiga marina Grøntved 1952 (**Plate 15**)
Cell length: 6–11 μm.
Cell width: 3.5–5.5 μm.
Characteristic features: Cell pyriform is preserved material.
Distribution: Temperate; Oceanic; North Atlantic.

Desmarella moniliformis Kent 1878 (**Plate 15**)
Cell length: 6 μm.
Characteristic features: Naked cells with thin protruding pseudopodial threads, in series of two to four cells.
Distribution: Coastal shallow waters: North Atlantic, North East Pacific?

Salpingoeca inquillata Kent 1880–1882 (**Plate 15**)
Cell investment length: 10 μm.
Characteristic features: Cell enclosed in a goblet-shaped hyaline investment.
Distribution: Coastal shallow waters; North Atlantic.

Acanthocorbis unguiculata (Thomsen) Hara & Takahashi 1984 (**Plate 15**)
Acanthoecopsis unguiculata Thomsen 1973.
Lorica length: 12—14 μm.
Lorica width: 11 μm.
Characteristic features: Basal part of lorica with irregularly arranged costa, number of protruding spines 12–14.
Distribution: Temperate, cold; Coastal; Europe, Japan.

Calliacantha natans (Grøntved) Leadbeater 1978 (**Plate 15**)
Salpingoeca natans Grøntved 1956.
Lorica length: 17–26 μm.
Costae: Two transverse, six longitudinal, three spines.
Characteristic features: The three curved spines protruding anteriorly also constitute the first costal ring. The remaining costae and the posterior spine are less pronounced.
Distribution: Coastal to offshore?; Arctic, North Atlantic, Pacific.

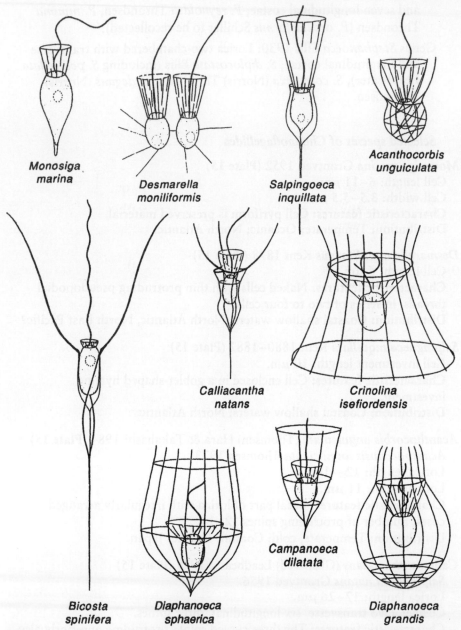

Plate 15 *Acanthocorbis unguiculata, Bicosta spinifera, Calliacantha natans, Campanoeca dilatata, Desmarella moniliformis, Diaphanoeca grandis, D. spherica, Monosiga marina,* and *Salpingoeca inquillata* from Throndsen (1974); *Crinolina isefiordensis* redrawn from Thomsen (1976).

Crinolina isefiordensis Thomsen 1976 (Plate 15)
Lorica length: 25–30 μm.
Lorica width: 20–30 μm.
Costae: Two transverse, 15–16 longitudinal.
Characteristic features: Lorica with wide anterior and posterior opening,
the cell apparently suspended by the rim of the collar from the anterior
transverse costa, which is situated at the constricted part of the lorica.
Distribution: Temperate; Coastal; North Atlantic, South Pacific.

Bicosta spinifera (Throndsen) Leadbeater 1978 (Plate 15)
Salpingoeca spinifera Throndsen 1970a.
Lorica length: 50–80 μm.
Costae: Two longitudinal.
Characteristic features: A large species, the pronounced longitudinal
spines fairly easy to observe.
Distribution: Temperate, cold; Oceanic, coastal; Arctic, North Atlantic,
Pacific.

Diaphanoeca sphaerica Thomsen 1982 (Plate 15)
Syn. *Campanoeca pedicellata* (Leadbeater) Throndsen 1976.
Lorica length: 24–35 μm.
Lorica width: 20 μm.
Costae: Three transverse, 14 longitudinal.
Characteristic features: Lorica barrel-shaped with pointed posterior, cell
located in the basal part of the lorica. May form colonies.
Distribution: Coastal; North Atlantic, Baltic.

Campanoeca dilatata Throndsen 1974 (Plate 15)
Lorica length: 19 μm including stalk.
Lorica width: 10 μm.
Costae: Three transverse, seven longitudinal.
Characteristic features: Lorica bell-shaped, with costal joints in the
anterior part fairly distinct in the LM.
Distribution: Coastal; Europe.

Diaphanoeca grandis Ellis 1930 (Plate 15)
Lorica length: 24–38 μm (30 μm; Ellis, 1930).
Lorica width: 20 μm.
Costae: 4 transverse, 13 longitudinal.
Characteristic features: Lorica onion-shaped with longitudinal costa
protruding from the anterior transverse costal ring, from which the cell
appears to be suspended by the distal part of the pseudopodial collar.
Distribution: Apparently worldwide in coastal areas.

Loricate Choanoflagellate Species (Plate 16)

Parvicorbicula socialis (Meunier) Deflandre 1960 (Plate 16)
Lorica length: 10 μm.
Lorica width: 12 μm.
Costae: Two transverse, 10 longitudinal.
Characteristic features: Anteriorly wide, funnel-shaped loricae, may form colonies with lateral contact rim to rim.
Distribution: Coastal, oceanic; Arctic, Antarctic, North Atlantic, Mediterranean (*Parvicorbicula socialis* var. *pavillardi*).

Parvicorbicula quadricostata Throndsen 1970b (Plate 16)
Lorica length: 18 μm.
Lorica width: 24 μm.
Costae: Three transverse, four longitudinal.
Characteristic features: Lorica funnel-shaped with circular anterior costa, the other transverse costae being quadrangular.
Distribution: Temperate; Oceanic, coastal; Atlantic.

Pleurasiga minima Throndsen 1970b (Plate 16)
Lorica length: 10 μm.
Lorica width: 10 μm.
Costae: Two transverse, 7–10 longitudinal.
Characteristic features: Similar to *P. reynoldsii* (in LM), but smaller.
Distribution: Cold, temperate, tropical; Oceanic; Atlantic, Pacific.

Pleurasiga reynoldsii Throndsen 1970b (Plate 16)
Lorica length: 23 μm.
Lorica width: 22 μm.
Costae: Three transverse, seven longitudinal.
Characteristic features: The barrel/funnel-shaped lorica appears to have a basal chamber with an investment surrounding the cell; the number of longitudinal costae is reduced to four in this part.
Distribution: Temperate, tropical; Oceanic; North Atlantic, Pacific.

Pleurasiga orculaeformis Schiller 1925 (Plate 16)
Lorica length: 16–18 μm.
Lorica width: 14 μm.
Costae: Three transverse, seven longitudinal.
Characteristic features: Lorica with a barrel-shaped anterior and conical basal part in which the cell is located. The present illustration is modified from Schiller (1925), assuming that the original drawing was made from a microscope image at extremely small aperture (to facilitate high contrast).
Distribution: Oceanic; Mediterranean.

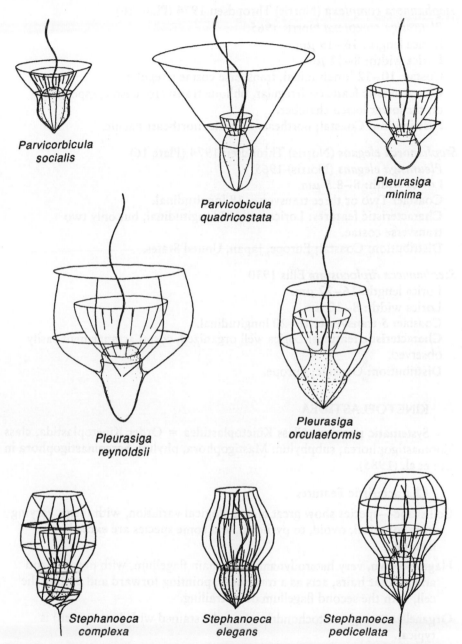

Plate 16 *Parvicorbicula quadricostata, P. socialis, Pleurasiga minima, P. reynoldsii, Stephanoeca complexa, S. elegans,* and *S. pedicellata* from Throndsen (1974); *Pleurasiga orculaeformis* modified from Schiller (1925).

Stephanoeca complexa (Norris) Throndsen 1974 **(Plate 16)**
　Pleurasiga complexa Norris 1965.
　Lorica length: 16–19 μm.
　Lorica width: 8–11 μm.
　Costae: 10–12 longitudinal, transverse costae irregular.
　Characteristic features: Irregular, oblique transverse costae, especially
　on the lower lorica chamber.
　Distribution: Coastal; northeast Atlantic, northeast Pacific.

Stephanoeca elegans (Norris) Throndsen 1974 **(Plate 16)**
　Pleurasiga elegans (Norris) 1965
　Lorica width: 8–8.5 μm.
　Coastae: Two or three transverse, 18 longitudinal.
　Characteristic features: Lorica with 18 longitudinal, but only two
　transverse costae.
　Distribution: Coastal; Europe, Japan, United States.

Stephanoeca diplocostata Ellis 1930
　Lorica length: 16–22 μm.
　Lorica width: 8–10 μm.
　Coastae: 5 transverse, 16–20 longitudinal.
　Characteristic features: Lorica well organized, costal arrangement easily
　observed.
　Distribution: Oceanic; Europe.

KINETOPLASTIDEA

Systematic position: Class Kinetoplastidea = Order Kinetoplastida, class
Zoomastigophorea, subphylum Mastigophora, phylum Sarcomastigophora in
Lee et al. (1985).

Characteristic Features

Cell shape—species show great morphological variation, with forms varying
　from spherical, ovoid, to pyriform, and some species are exceedingly
　flattened.
Flagella—two, very heterodynamic; the main flagellum, with paraxial rod
　and tubular hairs, acts as a tractellum, pointing forward and pulling the
　cell, with the second flagellum often trailing.
Organelles—single mitochondrion (may be stained with Janus green) is
　typical.
Distribution—naked zooflagellate species are most often reported from
　coastal and littoral areas. The apparent lack of species identified in
　offshore waters probably reflects inadequate preservation methods. The
　fairly high numbers for "flagellates and monads" frequently reported

from oceanic areas certainly include, in varying proportions, members of naked zooflagellates.

Identification/Observation and Problems

Living material is a prerequisite for identification in the light microscope. Preserved material using any of the common preservatives are generally inadequate for identification except on rare occasions.

Problems are mostly related to the variable shape, and the general lack of knowledge on morphological details for identification. Further, when such details have been described they are usually revealed at the EM level.

Order Bodonida Hollande 1952: Free-living with marine representatives.

Family Bodonidae Bütschli 1884: Cells naked, free-swimming with two heterodynamic flagella.

Genus *Bodo* Ehrenberg 1832: Anterior swimming flagellum and a trailing flagellum; *B. saltans* Ehrenberg, *B. caudatus* Dujardin, *B. curvifilus* Griessmann, *B. edax* Klebs, *B. parvulus* Griessmann, *B. underboolensis* Ruinen.

Genus *Phyllomitus* Stein 1883; *P. yorkeënsis* Ruinen.

Genus *Pleurostomum* Namyslovski 1913: Cells with lateral cytostome; *P. gracile* Namyslovski, *P. caudatum* Namyslovski, *P. salinum* Namyslovski, *P. flabellatum* Ruinen.

Genus *Rhynchomonas* Klebs 1893: Cells elongated, somewhat compressed, with trailing long flagellum, short flagellum adpressed to a proboscis-like structure pointing forward; *R. nasuta* (Stokes) Klebs, *R. mutabilis* Griessmann.

Family Jacobidae Patterson 1990: Two apical flagella, one curved pointing forward, the other bending backward in a ventral groove.

Genus *Jacoba* Patterson 1990: With free-living species; *J. libera* (Ruinen) Patterson.

Incertae sedis

Genus *Metromonas* Larsen & Patterson 1990: Cells flattened, leaf-like, with one or two posterior flagella; *M. simplex* (Griessmann) Larsen & Patterson.

Genus *Telonema* Griessmann 1913: Cells rigid with two posterior flagella; *T. subtilis* Griessmann.

Selected Colorless Species of Kinetoplastidea and Chrysophyceae

Bodo parvulus Griessmann 1913 (Plate 17)
Cell length: 3–4 μm.
Cell width: 4–5 μm.

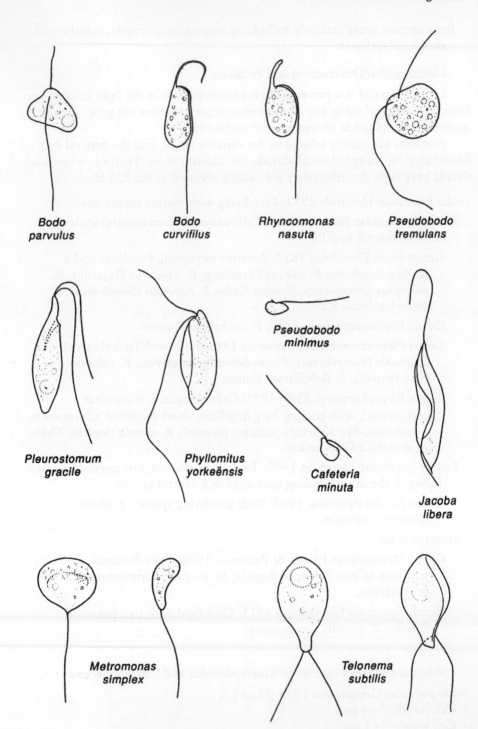

Bodo
parvulus

Bodo
curvifilus

Rhyncomonas
nasuta

Pseudobodo
tremulans

Pseudobodo
minimus

Pleurostomum
gracile

Phyllomitus
yorkeënsis

Cafeteria
minuta

Jacoba
libera

Metromonas
simplex

Telonema
subtilis

Flagella: Two, swimming flagellum cell length, trailing flagellum 2 × cell length.
Characteristic features: Cell often triangular with the shorter flagellum pulling the cell.
Distribution: Coastal or littoral; Europe.

Bodo curvifilus Griessmann 1913 (Plate 17)
Cell length: 4–7 μm.
Cell width: 2–4 μm.
Flagella: Two, swimming flagellum (proboscis type) cell length, trailing flagellum 2 × cell length.
Characteristic features: The cell is often bean-shaped and somewhat compressed, the fairly thick anterior flagellum having a peculiar sort of searching movement.
Distribution: Littoral; Europe.

Rhynchomonas nasuta (Stokes) Klebs 1893 (Plate 17)
Heteromita nasuta Stokes 1888.
Cell length: 5–8 μm.
Cell width: 2–3 μm.
Flagella: Two; short "swimming" flagellum pointing forward closely adpressed to a proboscis equal to cell length, trailing flagellum 2 × cell length.
Characteristic features: The relatively prominent proboscis with the adpressed anterior flagellum makes searching movements; the trailing flagellum is much thinner.
Distribution: Euryhaline, coastal; Europe, Japan, Arctic (Greenland), probably cosmopolitan.

Pseudobodo tremulans Griessmann 1913 (Plate 17)
Cell length: 4–5(8) μm.
Cell width: 5–7 μm.
Flagella: Two, anterior 4 × cell length, posterior half as long.
Characteristic features: Cell broader than long, flagella inserted on the side, free-swimming.
Distribution: Coastal; Atlantic, Mediterranean, Pacific.

Pleurostomum gracile Namyslovski 1913 (Plate 17)
Cell size: 8–14 μm.

Plate 17   *Bodo parvulus, B. curvifilus, Pseudobodo tremulans, Metromonas simplex*, and *Telonema subtilis* based on type illustrations (Griessmann, 1913); *Pleurostomum gracile, Phyllomitus yorkëensis, Pseudobodo minimus, Cafeteria minuta,* and *Jacoba libera* based on type illustrations (Ruinen, 1938); *Rhynchomonas nasuta*, original.

Cell width: 2–4 μm.
Flagella: Two, homodynamic.
Characteristic features: Cell spindle-shaped, with a subapical, lateral cytostome.
Distribution: Brackish, coastal; Baltic, India, Australia, Mediterranean.

Phyllomitus yorkeënsis Ruinen 1938 (Plate 17)
Cell length: 14–22 μm.
Cell width: 4–6 μm.
Flagella: Two, swimming flagellum ½–1 × cell length, trailing flagellum up to 2 × cell length.
Characteristic features: Cells lancet-shaped, with a ventral groove.
Distribution: Littoral; Australia.

Pseudobodo minimus Ruinen 1938 (Plate 17)
Cell length: 2 μm.
Cell width: 1 μm.
Flagella: Two, swimming flagellum 5 × cell length, trailing flagellum ½ × cell length.
Characteristic features: Cells very small, pointed ovoid, with flagella inserted in a faint groove.
Distribution: Salinas; Portugal, Australia.

Cafeteria minuta (Ruinen) Larsen & Patterson 1990 (Plate 17)
Pseudobodo minutus Ruinen 1938.
Cell length: 1.5–2.5 μm.
Cell width: 1–1.5 μm.
Flagella: Two, swimming flagellum 3–5 × cell length, trailing flagellum 1–1½ × cell length.
Characteristic features: Cells are pyriform to ovoid, with a long swimming flagellum and a very short trailing flagellum.
Distribution: Salinas; India.

Jacoba libera (Ruinen) Patterson 1990 (Plate 17)
Cryptobia libera Ruinen 1938.
Cell length: 8–16 μm.
Cell width: 3–5 μm.
Flagella: Two, swimming flagellum is cell length, trailing flagellum 1½ × cell length.
Characteristic features: Cells spindle-shaped, the trailing flagellum running along the rim of the undulating membrane.
Distribution: Salinas: India, Australia; Oceanic: Kattegat, Atlantic, Sargasso Sea (also reported from deep-water sediment: Atlantic).

Metromonas simplex (Griessmann) Larsen & Patterson 1990 (Plate 17)
   Phyllomonas simplex Griessmann 1913.
   Cell length: 6–8 μm.
   Cell width: 5–8 μm.
   Flagella: One, posterior.
   Characteristic features: Cell rounded oval, ventral side invaginated, often
   with creeping movement.
   Distribution: Littoral, coastal; North Sea.

Telonema subtilis Griessmann 1913 (Plate 17)
   Cell length: 6–8 μm.
   Flagella: Two, cell length.
   Characteristic features: Body shape fairly firm, flagella pointing backward
   during swimming.
   Distribution: Coastal; Europe, Japan.

   EBRIIDEA

   Systematic position: Class Ebriidea = Order Ebriida, Class Zoomastigophorea, Subphylum Mastigophora, Phylum Sarcomastigophora in Lee et al.
(1985).

   Characteristic Features
Cells—naked with two flagella and an internal siliceous skeleton.
Distribution—cold or temperate waters; Pacific Ocean, Barents Sea, Baltic.

   Identification/Observation and Problems
   Preserved material is adequate for identification since it is based on the
silica structures. Living material is usually not necessary for identification.
   Problems are mainly related to the great variation in the silica structures.

Order Ebriida Poche 1913: Single order with extant representatives.
   Family Ebriidae (Lemmermann) Deflandre 1950 (Ebriaceae Lemmermann
      1901)
      Genus *Ebria* Borgert 1891: Convex silica skeleton; *E. antiqua* Schulz,
         E. tripartita (Schumann) Lemmermann.
   Family Ebriopsidae Deflandre 1950.
      Genus *Hermesinum* Zacharias 1906: Pointed silica skeleton; *H.
         adriaticum* Zacharias, *H. platense* Frenguelli.

   Selected Species of Ebriidea
Ebria tripartita (Schumann) Lemmermann 1899 (Fig. 13)
   Dictyocha tripartita Schumann 1867.

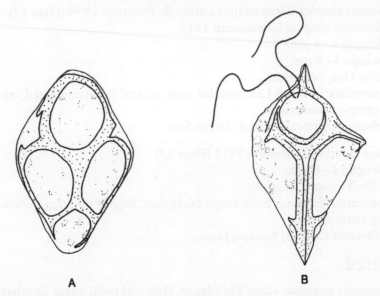

FIGURE 13 Ebriidean genera/species from marine plankton: *Ebria tripartita* 30–40 μm; *Hermesinum adriaticum* 45–50 μm. (Based on Hovasse as reproduced in Chadefaud, 1960.)

Skeleton size: 30–40 μm.
Flagella: Two.
Characteristic features: To be identified by its skeleton.
Distribution: Coastal, oceanic; Baltic, Mediterranean, East Pacific.

Hermesinum adriaticum Zacharias 1906 (**Fig. 13**)
Cell length: 45–50 μm.
Flagella: Two.
Characteristic features: Stenohaline warm water organism (Hargraves & Miller, 1974).
Distribution: Coastal, oceanic; Mediterranean, Black Sea, (inshore from Woods Hole to Rio de Janeiro).

TECHNIQUES

Flagellate studies will be of two kinds: the static ones (i.e., looking at dead cells) and dynamic ones (studying living and often motile cells).

Flagellate Studies—Living versus Fixed Material

In general, the living cell will offer the best possibilities for the identification of "naked" flagellates in the light microscope. Valuable information may be obtained from fixed material, especially after staining to emphasize specific organelles or inclusions. But this will only add to the basic knowledge gained by observation of the active living specimens.

Most of the naked flagellate species known at present are described from the light microscope, and only occasionally from preserved material. Some species, like many of *Chrysochromulina* (Prymnesiophyceae), are known only from preserved material investigated in the electron microscope. This may be sufficent to separate species on specific structural details such as scale morphology, but the knowledge of the appearance of individual species will be rather limited.

Presently, only light microscopy allows the observation of living cells. The criteria for identification of nearly all genera and most of the species known today are based on light microscopy. It is still imperative to master this instrument when dealing with identification of flagellates. In some genera, species identification relies on submicroscopic details, but cell morphology, size, flagellar length, and haptonema observed in the light microscope often help in delimiting the number of possible species. Therefore, even in taxa that cannot be fully identified with light microscopy, any observable features should be noted. If a sample is to be forwarded for electron microscopical analysis, data on observations from light microscopy should be submitted together with the material for further study.

Light Microscopy

A good-quality, properly aligned light microscope generally offers adequate possibilities for observation of flagellate cells. In bright-field microscopy cell shape and size as well as color can be observed. By combining bright-field illumination with staining procedures, information on size and structure of the nucleus as well as the reserve products (starch, paramylon, chrysolaminaran, fat/oil) can be obtained. Adding facilities for phase contrast (lenses and condenser) greatly improves the possibilities for studying flagella, haptonemata, and delicate external structures like loricae of choanoflagellates. The differential contrast can improve the observation of morphology and anatomy, provided the preparations are thin enough.

Fluorescence microscopy, usually combined with selective staining, offers the possibility to study chloroplast pigments and reveal differences of taxonomical importance. Fluorescence observations alone seldom give sufficient information for identification at the species level, but pigments and fluorescent inclusions will be revealed. Combined with other morphological observations, those made with a fluorescent microscope may render decisive evidence to distinguish between species.

The light microscopist studying living cells should note the mode of swimming along with shape and morphology of the cells to give some indication as to the type of flagella it possesses. The planar sinuoidal waves of heterokont flimmer flagella clearly reflect the morphology of the pulling main flagellum. The haptonema (typical of Prymnesiophyceae) are much more delicate than the flagella. Its action and also its true proportions will only be observed in healthy cells. The color, provided bright-field illumination is used as well as the cell proportions, is important. Observation of living cells will also reveal details like swimming behavior and food uptake relative to their ecology.

PREPARING SAMPLES FOR OBSERVATION

With a few exceptions, the material to be analyzed has to be living, and if there is no bloom of the species of interest, some method of concentration is necessary. Common methods for concentrating are:

1. Plankton nets (cf. Tangen, 1978), mesh size 5–50 μm, will concentrate the material during the sampling in the sea.

2. Membrane filters with "pore" size 0.45–0.8 μm will also retain nanoplankton flagellates, but may destroy the most delicate forms. The sample may be filtered directly onto a filter and resuspended in filtered sea water.

3. Centrifuge used at moderate speed (300–1000 × g, approximately 1500 –3000 rpm at a radius 10 cm) will concentrate flagellates in 20–40 min.

4. Crude cultures will increase the cell number for part of the phytoflagellate community. Should be frequently subsampled.

5. Dilution cultures will give material from a larger part of the phytoflagellate community, and also give the possibility for estimating cell numbers.

Culture methods have the advantage that relatively large amounts of material of each species can be produced; however, the disadvantage may be the selectivity due to different viability and competition between the flagellate species under the culture conditions.

Examination of the material is preferably done in a compound light microscope, aligned for critical illumination and equipped with phase contrast or interference contrast and bright-field optics. A good-quality 100× objective is indispensable, and a calibrated ocular micrometer is a necessity. Observations on swimming behavior are preferably made at a medium total magnification (200–400×), whereas observation on cell morphology and anatomy should be made at 1000–1200× as well. If necessary, cell motion may be slowed down by adding a drop of uranylacetate solution (1 g uranyl acetate + 15 g sea water). Many "naked" flagellates will be intact for only a few minutes under the microscope, even when a heat-absorbing filter is used on the microscope lamp. Drawings and measurements of the cells in question are therefore important for further identification work. Another consequence of the short "microscope life" of "naked" cells is that it is very important to know what to look for. The following outline lists those common characters needed for routine identification.

What to Look For

I. Motility
    A. Flagella
        1. Mode of use
            a. Pulling the cell: "flimmer" flagella
                Chrysophyceae
                Raphidophyceae
                Dictyochophyceae
                Prymnesiophyceae (*Pavlova*)

            b. Pushing the cell: smooth flagella or flagella with scales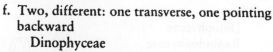
                Prymnesiophyceae
                Chlorophyceae
                Prasinophyceae

        2. Number and type
            a. One pulling
                Dictyochophyceae
            b. One pushing
                Prasinophyceae
                (Choanoflagellates)

            c. Two, equal, pushing
                Prymnesiophyceae
                Chlorophyceae

            d. Two, different: one pulling, one smooth
                Chrysophyceae
                Raphidophyceae

            e. Two, different: one pulling, one trailing
                Chrysophyceae
                Kinetoplastida

            f. Two, different: one transverse, one pointing backward
                Dinophyceae

            g. Four or more, equal
                Chlorophyceae
                Prasinophyceae

    B. Haptonema (Prymnesiophyceae)
        1. Short, limited movement
            Prymnesium
            Phaeocystis (flagellated stage)
            Pavlova

2. Long, coiling
       Chrysochromulina

C. Pseudopodia
    1. Blunt or pointed; for cell movement
        Amoeba

    2. Thin threadlike; trailing
        Pseudopedinella

    3. Very thin threadlike, anchoring the cell
        Pavlova

II. Cell shape
     A. Symmetry
         1. Bilateral; many species
         2. Radial
             Dictyochophyceae
     B. Asymmetrical
         Cryptophyceae

III. Chloroplasts
      A. Number
          1. Many
              Dinophyceae
              Raphidophyceae
              Dictyochophyceae
              Euglenophyceae

          2. One or two
              Cryptophyceae
              Prymnesiophyceae
              Chrysophyceae
              Prasinophyceae
              Chlorophyceae

　　　　3. Lacking
　　　　　　Choanoflagellates
　　　　　　Kinetoplastida
　　　　　　Ebriida
　　　　　　Chrysophyceae
　　　　　　Dictyochophyceae
　　B. Color
　　　　1. Yellow-green
　　　　2. Yellow-brown
　　　　3. Blue-red
IV. Special features
　　A. Furrow
　　　　1. Transverse + longitudinal
　　　　　　Dinophyceae

　　　　2. Single longitudinal, oblique or transverse
　　　　　　Cryptophyceae
　　　　　　Raphidophyceae

　　B. Eyespot/stigma
　　　　1. Within the chloroplast
　　　　　　Chrysophyceae
　　　　　　Chlorophyceae
　　　　　　Prasinophyceae

　　　　2. Outside the chloroplast
　　　　　　Dinophyceae
　　　　　　Euglenophyceae
　　　　　　Eustigmatophyceae

　　C. Channel or gullet
　　　　Euglenophyceae
　　　　Cryptophyceae
　　D. Ejectile structures
　　　　1. Ejectosomes
　　　　　　Cryptophyceae
　　　　2. Trichocysts
　　　　　　Dinophyceae

E. Periplast stripes
    Euglenophyceae
    Dinophyceae

F. Euglenoid movements
    Euglenophyceae
G. Pseudopodial collar
    Choanoflagellates

H. Skeleton
    1. Internal
        Ebriidea

    2. External
        Dictyochophyceae

I. Nucleus
    1. Inconspicuous (most common)
    2. Large, conspicuous
        Dinophyceae
        Raphidophyceae
        Euglenophyceae

V. Storage Products
    A. Solid
        1. Starch
            Cryptophyceae
            Dinophyceae
            Prasinophyceae
            Chlorophyceae

        2. Paramylon
            Euglenophyceae
            Prymnesiophyceae (*Pavlova*)
    B. Fluid
        1. Chrysolaminarin
            Chrysophyceae
            Prymnesiophyceae

        2. Oil (common among classes)

VI. Cell wall
   A. Lacking (cell division in flagellated stage), all classes

   B. Present
      1. Single (cell division in a nonmotile phase inside mother cell wall/theca)
         Chlorophyceae
         Prasinophyceae

      2. Divided (+/− transverse and longitudinal furrow)
         Dinophyceae

VII. Lorica
   A. Lacking (or not visible: common)
   B. Present
      1. Close-fitting
         Chrysophyceae (*Chrysococcus*)

      2. Wide
         Chrysophyceae (*Dinobryon, Bicosoeca*)
      3. Basket-shaped
         Choanoflagellates
         (Acanthoecaceae)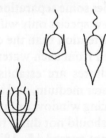

After careful observation of the living material, additional preparations stained and/or preserved in various ways may give further information: for example, Lugol staining/fixation will facilitate counting of flagella and measurement of their length, and starch will be revealed. Other stains are indicated in Table 3.

TABLE 3   Useful Stains[a] for Flagellate Identification Work

Lugol's solution (*sensu* Jane, 1942): 6 g KI, 4 g I$_2$, 100 ml H$_2$O
Brilliant Cresyl Blue (see, e.g., Butcher, 1959): in sea water
Uranyl acetate: 1 g uranylacetate in 15 ml sea water.
Sudan black: Saturated solution in 70% C$_2$H$_5$OH
Sudan blue: Saturated solution in 50% C$_2$H$_5$OH
Osmic acid (extremely dangerous): 200 mg OsO$_4$, 10 ml H$_2$O

[a] All stains to be added at the edge of the cover glass to diffuse into the preparation. Add sample to stain (one drop of each) on the microscope slide before mounting the cover glass to produce instant fixation (e.g., with Lugol's solution to facilitate counting and measuring flagella). For iron acetocarmine, acridine orange (with epifluoroscence), or chlor-zinciodide (Schultze's solution), consult textbooks on cytological methods.

CULTIVATION FOR IDENTIFICATION

The "naked" flagellates often show a great variation in shape, and hence it is important to examine many specimens of the same species when a reliable identification is desired. It is also a general problem that the "microscope stage life" of each specimen is fairly short, thus preventing long examination, especially when the species is encountered for the first time. Only occasionally will the natural concentration of cells be high enough to fulfill this requirement.

Cultivation is an effective method for increasing the concentration of single species. Two methods used are crude cultures and serial dilution cultures. Crude cultures depend on the growth dynamics in the flagellate community present, and unless the original sample is monospecific, they will progress through a succession of species. These cultures need frequent attention. The serial dilution method includes some separation of the components of the original community, and one or a few species only will be present in each culture. These cultures need less frequent attention than the crude cultures, but more cultures (up to 25) are to be examined from each water sample.

Crude cultures are established by mixing a seawater inoculum with a suitable seawater medium, and leaving the culture in light of moderate intensity (e.g., north facing window or fluorescent light tubes at a distance of 20–30 cm). Temperature should not differ too much from that of the sea, and in temperate areas should not exceed 15–18°C. Since "weed species" will rapidly overgrow all other species, these cultures have limited life and should be frequently subsampled and examined in order to obtain the optimal amount of information from crude cultures (as already discussed).

Dilution cultures are set up by stepwise dilution of the inoculum, and adding subsamples of it to (3–10 parallel) tubes containing culture medium (**Fig. 14**). The quantitative principle of the method is to determine the smallest

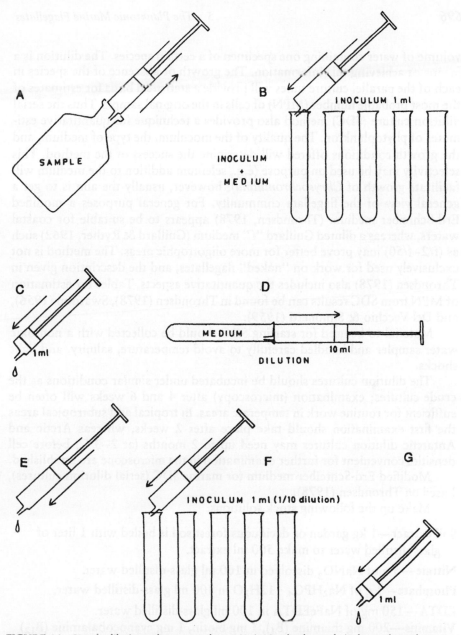

FIGURE 14 Simple dilution culture procedure: (A) A 10-ml subsample is drawn from the water sample by means of a syringe with excentric opening. (B) One milliliter inoculum is given to each of five culture tubes containing approximately 9 ml of culture medium. (C) The syringe is emptied except for the last 1 ml. (D) Nine milliliters of medium is sucked into the syringe at a rate appropriate for mixing the contents. (E) One milliliter is expelled and (F) 1 ml is given to each of five culture tubes, the inoculum representing 100 μl of the original sample. (G) The syringe is emptied except for the last 1 ml, and steps D–F are repeated to give the next dilution step (inoculum = 10 μl of the original sample). (From Throndsen, 1978.)

volume of water containing one specimen of a certain species. The dilution is a means of achieving this information. The growth or abscence of the species in each of the parallel culture tubes will provide a statistical basis for estimates of the most probable number (MPN) of cells in the original sample. Thus the serial dilution culture (SDC) method also provides a technique for quantitative estimates of phytoplankton. The quality of the inoculum, the type of medium, and the growth conditions offered will determine the success of the method. This selectivity may be used on purpose (e.g., selenium addition to the medium will facilitate growth of *Chrysochromulina*); however, usually the aim is to get a general view of the flagellate community. For general purposes a modified Erd-Schreiber medium (Throndsen, 1978) appears to be suitable for coastal waters, whereas a diluted Guillard "f" medium (Guillard & Ryther, 1962) such as (f/2–f/50) may prove better for more oligotrophic areas. The method is not exclusively used for work on "naked" flagellates, and the description given in Throndsen (1978) also includes the quantitative aspects. Tables for estimation of MPN from SDC results can be found in Throndsen (1978), Swaroop (1956), and Del Vecchio & Simonetti (1959).

Material to be used for starting SDC should be collected with a nontoxic water sampler and handled carefully to avoid temperature, salinity, and light shocks.

The dilution cultures should be incubated under similar conditions as the crude cultures; examination (microscopy) after 4 and 6 weeks will often be sufficient for routine work in temperate areas. In tropical and subtropical areas the first examination should take place after 2 weeks, whereas Arctic and Antarctic dilution cultures may need up to 2 months (at 2–3°C) before cell densities convenient for further examination in the microscope are established.

Modified Erd-Schreiber medium for marine SDC (serial dilution cultures), based on Throndsen (1978):

Make up the following stock solutions:

Soil extract—1 kg garden or deciduous forest soil is boiled with 1 liter of glass-distilled water to make 500 ml extract.

Nitrate—5 g of $NaNO_3$ dissolved in 100 ml glass-distilled water.

Phosphate—1 g of $Na_2HPO_4 \cdot 12H_2O$ in 100 ml glass-distilled water.

EDTA—150 mg of NaFeEDTA in 100 ml glass-distilled water.

Vitamins—200 mg thiamine (B_1), 1 mg biotin, 1 mg cyanocobalamine (B_{12}) in 1 liter glass-distilled water. The vitamin solution may be subdivided into convenient volumes (2–5 ml), sterilized, and stored frozen to be added aseptically for each batch of media.

To make 1 liter Erd-Schreiber medium, use seawater, preferably from the same water mass/locality as the inoculum:

1000 ml	Seawater
12.0 ml	Soil extract
1.0 ml	Nitrate
1.0 ml	Phosphate
0.5 ml	EDTA
2.0 ml	Vitamins

Pasteurize the medium at 80°C for 15 min, and cool it to incubation temperature in running water. If possible add the vitamins aseptically after cooling.

For other media including Guillard's "f" medium (Guillard & Ryther, 1962), consult McLachlan (1973).

PREPARATION OF SAMPLES FOR FURTHER STUDIES

The present guide is based on light microscopy only, and for many species electron microscopy may be necessary for a reliable identification. Surface structures as organic scales, silicified scales, and coccoliths are more or less easily observed in the light microscope; however, their submicroscopical structure may be crucial for identification. With adequate preparation, material may be sent to experts for further investigation. It should be noted, however, that for species without scales, loricae, or a characteristic cell wall such as *Pelagococcus subviridis*, embedded material sectioned for electron microscopy may be needed. For such species it is generally best to send a culture.

Preserving Material

Material for electron microscopy analysis may be either preserved in fluid for transportation, or prepared desiccated on EM grids ready for investigation after staining or shadow casting. In the latter case, it is important to rinse the grids with distilled water (pH must be checked, see below), as salts will cause corrosion of copper grids and ruin the preparations. A final check under the light microscope will reveal if there are cells present.

Unmineralized Scales

Organic scales are usually difficult to detect in the light microscope, but heavy staining with acridine orange may reveal their presence in the fluorescence microscope. Information is then usually limited to show the outline of the scales.

The scale structure may be preserved in such a way that EM experts can give further advice on the identification. Organic scales are usually stained with osmic acid (OsO_4) and uranyl acetate, or shadowed to reveal details of their structural pattern. The latter technique is particularly helpful when structures are different on the two sides of the scale. A good electron microscope and an apparatus for shadowing the preparations with heavy metals are necessary for this work.

To prepare the material for EM work one needs some basic equipment and chemicals. The preparation procedure requires some skill, and may be hazardous since glutaraldehyde as well as osmic acid are very toxic substances requiring particular precautions.

Depending on the laboratory to receive the sample, different procedures may be followed: Samples for direct EM preparations to be shadowed may be fixed with 0.5–2% glutaraldehyde; samples for stained preparations are better fixed with 2% osmic acid. In some cases even material fixed with Lugol's solution may prove adequate for species identification. To exploit the possibilities without wasting time and material, it is necessary to establish contact with the specialized laboratories in advance.

Silicified Structures

Silicified scales are resistent to acids and are therefore relatively easy to prepare for electron microscopy. Their gross morphology may be revealed in the light microscope upon oxidizing (by heat or by chemicals) away the organic cell material. Alkaline solutions will dissolve silica structures.

Choanoflagellate loricae consist of fine tubular silicified costae, which are delicate and need to be handled almost like organic scales. The costae may be observed also by the light microscope using dried and rinsed material on cover glasses inverted and mounted on slides (Thomsen, 1982), but details in costal morphology will require preparation for electron microscopy.

Calcified Structures

Scanning electron microscopy usually offers the best possibilities for identifying coccolithophorids by their coccolith morphology. The material may be collected on Nucleopore filters or sedimented in drops on thin glass chips covered by polylysine. The latter compound facilitates adhesion and thus prevents the cells from being washed off during rinsing.

Coccoliths are easily soluble even in weakly acid solutions, and hence also the desalting part of the preparation may cause problems; note that the pH of distilled water lowers as CO_2 is absorbed upon distillation.

SPECIFIC PROBLEMS TO AVOID

It is important that any workable preparation is *clean from dust,* so take particular care to use clean equipment and solutions.

Marine material has to be desalted before the final preparations are ready for insertion in the electron microscope. Otherwise the salt crystals will disturb the image. This "rinsing" step is important. The problems will differ with the substructure in question; organic scales may be washed off the cells, coccoliths may be dissolved in slightly acidic water, and silica structure in alkaline water. The salts may be washed out of the material either before the EM preparation is made by dialyzing the fixed sample against distilled water (Hara et al., 1991) or during the preparation procedure by carefully rinsing the grids of dried material with distilled water (Moestrup & Thomsen, 1980).

COMMON FLAGELLATE SYNONYMS

Acanthocorbis unguiculata (Thomsen) Hara & Takahashi 1984
= *Acanthoecopsis unguiculata* Thomsen 1973

Apedinella spinifera (Throndsen) Throndsen 1971
= *Pseudopedinella spinifera* Throndsen 1969
= *Apedinella radians* (Lohmann) *sensu* Campbell 1973
= ?*Meringosphaera radians* Lohmann 1908

Bicosta spinifera (Throndsen) Leadbeater 1978
= *Salpingoeca spinifera* Throndsen 1970a

Bilucifera cryophila (Taylor & Lee) Hill 1991
= *Cryptomonas cryophila* Taylor & Lee 1971

Cafeteria minuta (Ruinen) Larsen & Patterson 1990
= *Pseudobodo minutus* Ruinen 1938

Calliacantha natans (Grøntved) Leadbeater 1978
= *Salpingoeca natans* Grøntved 1956

Calycomonas Lohmann 1908
= *Codonomonas* Van Goor 1925

Calycomonas dilatata (Conrad & Kufferath) Lund 1960
= *Codonomonas dilatata* Conrad & Kufferath 1954

Calycomonas ovalis Wulff 1916 (and 1919)
transferred to *Paulinella*; testate rhizopod, see Johnson et al. (1988)

Calycomonas vangoorii (Conrad) Lund 1960
= *Codonomonas vangoorii* Conrad 1938

Campanoeca Throndsen 1974
= ?*Polyfibula* Manton & Bremer 1981

Chattonella antiqua (Hada) Ono 1980 (in Ono & Takano 1980)
= *Hemieutreptia antiqua* Hada 1974

Chattonella marina (Subrahmanyan) Hara & Chihara 1982
= *Hornellia marina* Subrahmanyan 1954

Chroomonas baltica (Büttner) N. Carter 1937
= *Cyanomonas baltica* Büttner 1911

Chroomonas marina (Büttner) Butcher 1967
= *Cryptomonas marina* Büttner 1911

Chroomonas virescens (Butcher) Butcher 1967
= *Cryptochrysis virescens* Butcher 1952

Chrysochromulina spinifera (Fournier) Pienaar & Norris 1979
= *Chrysocampanula spinifera* Fournier 1971

Ciliophrys infusionum Cienkowski 1875
= *Ciliophrys marina* Caullery 1910

Diaphanoeca sphaerica Thomsen 1982
= *Campanoeca pedicellata* (Leadbeater) Throndsen 1974
= *Stephanoeca pedicellata* Leadbeater *sensu* Thomsen 1976

Dictyocha speculum Ehrenberg 1837
= *Distephanus speculum* Haeckel 1897

Dinobryon balticum (Schütt) Lemmermann 1900
= *Dinobryon pellucidum* Levander 1894

Dinobryon faculiferum Willén 1992
= *Dinobryon petiolatum* Willén 1963

Eutreptiella da Cunha 1914
= *Gymnastica* Schiller 1925

Eutreptiella dofleinii (Schiller) Pascher 1927
= *Gymnastica dofleinii* Schiller 1925

Eutreptiella pascheri (Schiller) Pascher 1927
= *Gymnastica elegans* Schiller 1925

Heterosigma akashiwo (Hada) Hada 1968
= *Entomosigma akashiwo* Hada 1967

Leucocryptos marina (Braarud) Butcher 1967
= *Chilomonas marina* (Braarud) Halldal 1953
= *Bodo marina* Braarud 1935

Mamiella gilva (Parke) Moestrup 1984
= *Nephroselmis gilva* Parke in Parke & Rayns 1964

Metromonas simplex (Griessmann) Larsen & Patterson 1990
= *Phyllomons simplex* Griessmann 1913

Micromonas pusilla (Butcher) Manton & Parke 1960
= *Chromulina pusilla* Butcher 1952

Mantoniella squamata (Manton & Parke) Desikachary 1972
= *Micromonas squamata* Manton & Parke 1960

Nephroselmis minuta (N. Carter) Butcher 1959
= *Heteromastix minuta* N. Carter 1937

Nephroselmis pyriformis (N. Carter) Ettl 1982

= *Heteromastix pyriformis* (N. Carter) Manton in Parke & Dixon 1968
= *Bipedinomonas pyriformis* N. Carter 1937

Nephroselmis rotunda (N. Carter) Fott 1971
= *Heteromastix rotunda* (N. Carter) Manton 1964
= *Bipedinomonas rotunda* N. Carter 1937

Oltmannsiellopsis Chihara & Inouye 1986
= *Oltmannsiella sensu* Hargraves & Steele 1980

Oltmannsiellopsis viridis (Hargraves & Steele) Chihara & Inouye 1986
= *Oltmannsiella virida* Hargraves & Steele 1980

Parvicorbicula socialis (Meunier) Deflandre 1960
= *Corbicula socialis* Meunier 1910

Pavlova lutheri (Droop) Green 1975
= *Monochrysis lutheri* Droop 1953

Pavlova salina (N. Carter) Green 1976b
= *Nephrochloris salina* N. Carter 1937

Polyfibula sphyrelata (Thomsen) Manton 1981 in Manton & Bremer (1981)
= *Pleurasiga sphyrelata* Thomsen 1973

Pseudopedinella tricostata (Rouchijajnen) Thomsen 1988
= *Pedinella tricostata* Rouchijajnen 1966

Pseudoscourfieldia marina (Throndsen) Manton 1975
= *Scourfieldia marina* Throndsen 1969

Pyramimonas adriaticus Schiller 1913
= ?*Pyramimonas lunata* Inouye, Hori, & Chihara 1983

Pyramimonas disomata McFadden, Hill, & Wetherbee 1986
= *Pyramimonas disomata* Butcher 1959

Pyramimonas orientalis McFadden, Hill, & Wetherbee 1986
= *Pyramimonas orientalis* Butcher 1959

Pyramimonas propulsa Moestrup & Hill 1991
= Cfr. *Asteromonas propulsum* Butcher 1959

Resultor mikron (Throndsen) Moestrup 1991
= *Pedinomonas mikron* Throndsen 1969

Rhinomonas fragarioides (Butcher) Hill & Wetherbee 1988
= *Chroomonas fragaroides* Butcher 1967

Rhinomonas fulva (Butcher) Hill & Wetherbee 1988
= *Cryptochrysis fulva* Butcher 1952
= *Chroomonas fulva* (Butcher) Butcher 1967

Rhinomonas lateralis (Butcher) Hill & Wetherbee 1988
= *Cryptochrysis lateralis* Butcher 1952
= *Chroomonas lateralis* (Butcher) Butcher 1967

Rhinomonas reticulata (Lucas) Novarino 1990
= *Rhodomonas reticulata* (Lucas) Hill & Wetherbee 1989
= *Pyrenomonas reticulata* (Lucas) Santore 1986
= *Cryptomonas reticulata* Lucas 1968

Rhodomonas Karsten 1898
= *Pyrenomonas* Santore 1984

Rhodomonas duplex Hill & Wetherbee 1989
(= *Cryptomonas appendiculata sensu* Butcher 1967 *non sensu* Schiller 1957)

Rhodomonas marina (Dangeard) Lemmermann 1900
= *Cryptomonas marina* Dangeard 1892

Rhodomonas salina (Wislouch) Hill & Wetherbee 1989
= *Chroomonas salina* Butcher 1967
= *Cryptomonas salina* Wislouch 1924

Spumella Cienkowski 1870
= *Monas* O. F. Müller 1773
= *Heterochromonas* Pascher 1912

Stephanoeca complexa (Norris) Throndsen 1974
= *Pleurasiga complexa* Norris 1965

Stephanoeca diplocostata Ellis 1930
= including *S. pedicellata* Leadbeater 1972b

Stephanoeca elegans (Norris) Throndsen 1974
= *Pleurasiga elegans* Norris 1965

Teleaulax acuta (Butcher) Hill 1991
= *Cryptomonas acuta* Butcher 1952

Tetraselmis von Stein 1878
= *Platymonas* G. S. West 1916; for most recombinations see Butcher 1959

Tetraselmis convolutae (Parke & Manton) Norris et al. 1980
= *Platymonas convolutae* Parke & Manton 1967

Tetraselmis wettsteinii (Schiller) Throndsen 1988
= *Carteria wettsteinii* Schiller 1913

Thalassomonas Butcher 1959
= *Mantoniella* Desikachary 1972; *pro parte* see Manton & Parke (1960)

GLOSSARY

The following list explains terms and words as they are used in the descriptions of the taxa. Some of these terms have been presented in the introductory list of terminology.

Abbreviations: a., adjective; dim., diminutive; gen., genitive; n., noun; D., Dutch; G., German; Gr., Greek; I., Italian; L., Latin; OE., Old English.

Acronematic (a.)—(Gr. *akron*, peak; Gr. *nema*, gen. *nematos*, thread) flagellum terminating in a narrow hair-like part.

Actinopod (n.)—(Gr. *aktis*, gen. *aktinos*, ray; Gr. *pus*. gen. *podos*, foot) ray-shaped pseudopodium.

Adnate (a.)—(L. *adnatus*, congenitally attached) for spines not protruding freely.

Amoeboid (a.)—(Gr. *amoibe*, change) like single-celled organism with continously variable shape.

Anisokont (a.)—(Gr. *anisos*, unlike; Gr. *kontos*, pole, oar) the flagella of the cell being different in length and/or quality.

Apical (a.)—(Gr. *apex*, gen. *apicis*, top) related to what is defined as the top of the cell.

Asymmetric (a.)—(Gr. prefix *a*, without; Gr. *symmetria*, commensurateness) lacking symmetry.

Autotrophic (a.)—(Gr. *autos*, self; Gr. *trophe*, food) growing with light energy and inorganic nutrients only.

Auxotrophic (a.)—(Gr. *auxein*, to increase; Gr. *trophe*, food) photosynthetic organisms that need minor supply of organic substances like vitamins.

Axis (n.)—(Gr. *axon*, axle) an imaginary line in the middle of the cell (e.g., length axis).

Axoneme (n.)—(Gr. *axon*, axle; Gr. *nema*, thread) the group of microtubules (9+2 structure) constituting the active part of the flagellum.

Axopodium (n.)—(Gr. *axon*, axle; Gr. *pus*. gen. *podos*, foot) thin, straight pseudopodium supported by microtubules.

Bilateral (a.)—(L. *bis*, two; L. *lateralis*, on the side) similar on two sides (e.g., about symmetry).

Brackish water (a. + n.)—(D. *brak*, salty) mixture of fresh water and saltwater; mostly used for salinities below 25–30‰.

Canal (n.)—(L. *canalis*) used for the tubular passage into the reservoir in euglenophyceans.

Cell membrane (n.)—(L. *cella*, room; L. *membrana*, skin) the outer membrane of the cell.

Chrysolaminaran—liquid β-1,3-glucan typically found in Chryso- and Prymnesiophyceae.

Coccoid (a.)—(Gr. *kokkos*, grain) cell with a firm outer wall, without flagella.

Cosmopolitan (a.)—(Gr. *kosmopolites*, citizen of the world) species being present in same type of biotypes in all parts of the world.

Costa (n.)—(L. *costa*, rib) in choanoflagellates, the ribs of the lorica.

Costal strip (n.)—(L. *costa*, rib) in choanoflagellates, the tubular silica units making up the costa.

Crista (n.)—(L. *crista*, crest) the invaginations of the inner mitochondrial membrane.

Diagnosis (n.)—(Gr. *diagnosis*, distinguishing) a short description of a taxon making it distinguishable from all other (described) taxa.

Dorsal (a.)—(L. *dorsum*, back) pertaining to the back of a cell when a ventral side can be defined (cf. ventral).

Ejectosome (n.)—(L. *ejectus*, thrown out; Gr. *soma*, body) ejectile organelle typical of cryptophyceans, but also present in prasinophyceae.

Endogene (a.)—(Gr. *endon*, within; Gr. *genesis*, origin) process taking place inside the cell.

Exogene (a.)—(Gr. *ex*, out of; Gr. *genesis*, origin) process related to the cell, but taking place outside the cell membrane.

Extrusome (n.)—(L. *extrudere*, thrust out; Gr. *soma*, body) a membrane-bound organelle usually confined to the cell surface; mucoid content may be extruded to form threads.

Falcate (a.)—(L. *falx*, sickle) scythe or sickle shaped.

Filopodium (n.)—(L. *filum*, thread; Gr. *pus* gen. *podos*, foot) fine protoplasmic thread protruding from the cell surface; trailing behind a swimming flagellate or anchoring the cell to the substratum.

Flagellum (n.)—(L. *flagrum*, whip) the thread-like swimming organelle of flagellates.

Flimmer flagellum (n.)—(G. *Flimmer*, glimmer; L. *flagrum*, whip) flagellum with usually two rows of tripartite tubular hairs along the whole or part of the flagellar axis.

Formvar (n.)—material for producing very thin films for supporting material in the electron microscope.

Grid (for EM) (n.)—circular, thin, finely meshed plate approximately 2.5 mm in diameter for support of formvar film with dried material to be studied in the electron microscope.

Gubernaculum (n.)—(L. *gubernaculum*, rudder) trailing flagellum (term not commonly used).

Gullet (n.)—(L. *gula*, throat) a sack-like invagination from the cell surface of cryptophyceans.

Heterodynamic (a.)—(Gr. *heteros*, different; Gr. *dynamis*, power) about flagellar movement, flagella of the same cell being used in different ways (e.g., one pulling, the other trailing).

Heterokont (a.)—(Gr. *heteros*, different; Gr. *kontos*, pole, oar) the flagella of the cell being different: one with tripartite flagellar hairs, the other smooth.

Heterotrophy (n.)—(Gr. *heteros*, different; Gr. *trophe*, food) nonphotosynthetic organisms, using organic material for their nutrition.

Homodynamic (a.)—(Gr. *homos*, one and the same; Gr. *dynamis*, power) flagellar movement: flagella of the same cell used (more or less synchronously) in the same mode.

Hypertonic (a.)—(Gr. *hyper*, over, above; Gr. *tonos*, tension) the osmotic pressure being higher in the medium than in the cell; living cells will shrink.

Hypotonic (a.)—(Gr. *hypo*, under; Gr. *tonos*, tone) the osmotic pressure of the cell is higher than in the medium; the cells will blow up and rupture.

Isokont (a.)—(Gr. *isos*, equal; Gr. *kontos*, pole, oar) the flagella of a cell being of the same kind and length; usually smooth, or with organic scales.

Lacrymuloid (a.)—(L. *lacrimula*, little tear) tear-shaped.

Lamella (n.)—(L. *lamella*, dim. of *lamina*, plate, platelet) stack of (two, three, or a few) thylakoids kept together.

Lateral (a.)—(L. *lateralis*, on the side) at the side.

Lorica (n.)—(L. *lorica*, cuirass) wide cell investment of organic material (cellulose or chitin) impregnated with Mn, Fe, or Si, or made up from silica units as in choanoflagellates.

Marine (a.)—(L. *mare*, sea) pertaining to seawater, from coastal or oceanic areas.

Mesoplankton (Gr. *mesos*, middle; Gr. *planktos*, drifting) plankton organisms in the size range 200–2000 μm.

Microplankton (n.)—(Gr. *mikros*, small; Gr. *planktos*, drifting) plankton organisms in the size range 20–200 μm.

Microtubule (n.)—(Gr. *mikros*, small; L. *tubulus*, pipe) hollow (tubular) cytoplasmic structures constituting part of the cellular endoskeleton, and flagellar axoneme and roots.

Mixotrophy (n.)—(Gr. *mixis*, mixing; Gr. *trophe*, food) phototrophic organism taking up organic substances for supplementary nutrition (e.g., by phagotrophy).

Monadoid (a.)—(Gr. *monas*, gen. *monados*, unity) unicellular, often flagellated stage.

Mucocyst (n.)—(L. *mucus*, snot; Gr. *kystis*, bag) mucus-containing extrusome.

"Naked"—usually applied to characterize the lack of a cell wall.

Nanoplankton (n.)—(Gr. *nanos, nannos,* dwarf; Gr. *planktos,* drifting) plankton organisms in the size range 2–20 μm.

Neritic (a.)—(Gr. *nerites,* pertaining to Nereus) coastal.

Oligotrophic (a.)—(Gr. *oligos,* few; Gr. *trophe,* food) pertaining to availability of nutrients, low production areas.

Palmelloid (a.)—(like the volvocales genus *Palmella*) a stage where the flagella are thrown off, and cells proliferate within a mucilaginous matrix.

Paracrystalline (a.)—(Gr. *para,* beside; Gr. *krystallinos,* like clear ice) structurally strictly organized pattern in electron-dense organic material.

Paramylon (n.)—(Gr. *para,* beside; Gr. *amylon,* starch) solid β-1,3-glucan typically found in euglenoids.

Paraxial rod (n.)—(Gr. *para,* beside; Gr. *axis,* axle; OE. *rodd,* rod) cross-striated structure running parallel with the axoneme in certain flagella.

Pellicula (n.)—(L. *pellicula* from *pellis,* skin) a layer of interlocking proteinaceous ribbons located next to the cell membrane in euglenophycean cell.

Periplast (n.)—(Gr. *peri,* around; Gr. *plastos,* formed) any membranes other than the cell membrane surrounding the cell body.

Phagotroph (n.)—(Gr. *phagein,* to eat; Gr. *trophos,* food) cell taking up food particles.

Photoautotroph (n.)—(Gr. *phos,* gen. *photos,* light; Gr. *autos,* self; Gr. *trophe,* food) organism needing only inorganic nutrient and light for growth.

Phototroph (n.)—(Gr. *phot,* light; Gr. *trophe,* food) organism using light as energy source.

Picoplankton (n.)—(I. *piccolo,* tiny; Gr. *planktos,* drifting) plankton organisms in the size range 0.2–2 μm.

Pinocytosis (n.)—(Gr. *pinein,* drinking) uptake of liquid into vacuoles in the cell.

Plasmalemma (n.)—(Gr. *plasma,* formed; Gr. *lemma,* shell) cell membrane.

Pleuronematic (a.)—(Gr. *pleura,* side; Gr. *nema,* hair) flagellum with two opposite rows of hairs.

Proboscis (n.)—(Gr. *proboskis,* elephant's trunk) permanent movable organelle considerably thicker than the flagellum (e.g., *Rhynchomonas*).

Psammobious, psammophilic (a.)—(Gr. *psammos,* sand) pertaining to sand (e.g., species living in the interstitial water of sandy beaches).

Pseudopodial collar (n.)—(Gr. *pseudes*, false; Gr. *pus*, gen. *podos*, foot; L. *collum*, neck) a ring of thin pseudopodia making a wide funnel around the flagellum in choanoflagellates.

Pseudopodium (n.)—(Gr. *pseudes*, false; Gr. *pus*, gen. *podos*, foot) a temporary protoplasmic protrusion from the cell.

Pyrenoid (n.)—(Gr. *pyren*, stone [of a fruit]) clear area of the chloroplast stroma naked (chryso-, prymnesiophyceans), with paramylon shield (euglenophyceans), or with starch shield (crypto-, dino-, chloro-, prasinophyceans).

Pyriform (a.)—(L. *pyrum*, pear; L. *formis*, form) pear shaped.

Reservoir (n.)—(L. *reservare*, retain) the sack-like dilatation beyond the canal in euglenophyceans.

Rhinote (a.)—(Gr. *Rhis* gen. *rhinos*, nose) nose-shaped.

Rhizopodial (a.)—(Gr. *rhiza*, root; Gr. *pus.* gen. *podos*, foot) rootlike.

Salinas (n.)—(L. *salinae*, saltworks) hypersaline areas.

Saprotroph (n.)—(Gr. *sapros*, rotten; Gr. *trophe*, food) organism living on decaying organic matter.

Scale theca (n.)—a close-fitting cell investment composed of minute organic scales merged to form a wall-like structure (*Tetraselmis*).

Stigma (n.)—(Gr. *stigma*, dot) red or orange-red spot (microgranule or oil droplet) usually confined to the chloroplast, often identical to eyespot.

Stipitate (a.)—(L. *stipes*, gen. stipitis, stalk) with a stalk.

Stomatocyst (n.)—(Gr. *stoma*, gen. *stomatos*, mouth; Gr. *kystis*, bag) cyst, often silicified, with a narrow opening closed by a pectic plug.

Striated strand (n.)—a paracrystalline structure of some flagella running parallel to the axoneme.

Subapical (a.)—located behind the top of the cell.

Symmetrical (a.)—(Gr. *symmetria*, commensurateness) morphology similar on both sides of a plane through the cell (bilateral symmetry) or around the cell axis (radial symmetry).

Taxon (n.)—(Gr. *tasso, taxis*, arrange, arrangement) a systematical unit the level of which is often indicated by the suffix; -phyta, division; -phyceae, class; -ales, order; -aceae, family, in the botanical nomenclature. Can also refer to genus and species.

Theca (n.)—(Gr. *theke*, case) a close-fitting investment of the cell (in dinoflagellates also used about the amphiesma when cellulose is extruded).

Thylakoid (n.)—(Gr. *thylakos*, pouch) the flattened invagination of the inner chloroplast membranes in which the photosynthetic pigments are deposited/located; the units of the lamella.

Tractellum (n.)—(L. *tractare*, to handle, from L. *trahere*, to draw) flagellum pulling the cell.

Trichocyst (n.)—(Gr. *thrix*, gen. trichos hair; Gr. *kystis*, bag) ejectile organelle, found in raphidophyceans (and dinoflagellates).

Tripartite hair (n.)—(L. *tri*, three; L. *partitus*, divided) flagellar hair consisting of three parts: the basis, shaft, and terminal hair(s).

Trophic mode (n.)—(Gr. *trophe*, food; L. *modus*, manner) the way of nutrition.

Tubular hair (n.)—(L. *tubus*, pipe; OE. hære, hair) a hollow type of proteinacious structure constituting the shaft of the tripartite hairs found in heterokont algal flagellates.

Ventral (a.)—(L. *venter*, belly) often applied to the side of the cell where subapical or median flagella are protruding. When flagella are apical, the ventral side may be impossible to determine in the light microscope.

INDEX OF FLAGELLATE TAXA

Synonyms in boldface.

Acanthocorbis, 673, 674
 A. apoda, 674
 A. asymmetrica, 674
 A. campanula, 674
 A. unguiculata, 674, 675, 699
Acanthoecida, 674
Acanthoecaccae, 693
Acanthoecopsis
 A. unguiculata, 675, 699
Actinomonas, 595, 619, 627
 A. mirabilis, 627, 628, 629
Apedinella, 594, 619, 627
 A. spinifera, 622, 623, 627, 699
 A. radians, 623, 627, 699
Apistonema, 597
Asteromonas
 A. gracilis, 667, 668
 A. propulsum, 661, 701
Aureococcus, 620
 A. anophagefferens, 617, 620
Aurosphaeraceae, 620

Bacillariophyceae, 587, 591, 602, 605
Bathycoccus, 652
 B. prasinos, 593, 652
Bicosoecaceae, 618
Bicosoecales, 587, 617
Bicosoeca, 595, 619, 625, 693
 B. gracilipes, 618, 625, 626
 B. griessmannii, 618
 B. maris, 618, 624, 625, 626
 B. mediterranea, 618, 626
Bicosta, 674
 B. minor, 674
 B. spinifera, 674, 676, 677, 699
Bilucifera
 B. cryophila, 699
Bipedinomonas
 B. phyriformis, 658
 B. rotunda, 658, 701
Bodo, 681
 B. saltans, 681
 B. caudatus, 681

B. curvifilus, 681, 682, 683
B. edax, 681
B. marina, 612, 700
B. parvulus, 681, 682
B. underboolensis, 681
Bodonida, 681
Bodonidae, 681
Boekelovia, 618
  B. gallica, 618
  B. hooglandii, 618
Brachiomonas, 664, 665, 666
  B. simplex, 665, 667, 669
  B. submarina, 665, 667, 669
  Cafeteriaceae, 618
Cafeteria, 595, 618
  C. roenbergensis, 618
  C. minuta, 618, 682, 684, 699
Calliacantha, 673, 674
  C. natans, 674, 675, 676, 699
  C. multispina, 674
Calycomonas, 618, 619, 699
  C. dilatata, 699
  C. ovalis, 699
  C. vangoorii, 699
Campanoeca, 674, 699
  C. dilatata, 674, 676, 677
  C. pedicellata, 677, 700
Carteria, 665, 666
  C. marina, 665
  C. vectensis, 669
  C. wettsteinii, 663, 702
Chattonellaceae, 613
Chattonellales, 587, 613
Chattonella, 613, 614
  C. antiqua, 614, 699
  C. marina, 614, 699
  C. subsalsa, 613, 614, 615
Chilomonas
  C. marina, 612, 700
Chlamydomonadaceae, 665
Chlamydomonas, 665, 666
  C. coccoides, 665, 667, 668
  C. euryale, 665
  C. fusiformis, 609
  C. minuta, 669
  C. pulsatilla, 665, 667, 669
  C. reginae, 665, 667, 668
  C. quadrilobata, 665, 667, 669
Chlorodendraceae, 653

Chlorodendrales, 588, 652, 653
Chloromonadophyceae, 612
Chlorophyceae, 588, 594, 596, 599, 603,
  604, 664, 666, 689, 690, 692, 693
Chlorophyta, 586, 587, 603, 604, 645
Choanoflagellidea, 588, 594, 604, 605,
  671, 673, 675
Chromophyta, 586, 587, 603, 605
Chromulinaceae, 618
Chromulina, 618, 619
  C. pleiades, 618
  C. pusilla, 657, 700
Chroomonas, 595, 607, 608
  C. africana, 608
  C. baltica, 608, 699
  C. diplocca, 608
  C. dispersa, 608
  C. collegionis, 608
  C. extensa, 608
  C. fragaroides, 701
  C. fulva, 611, 701
  C. lateralis, 701
  C. marina, 610, 611, 700
  C. mesostigmatica, 608
  C. placoidea, 608
  C. plurococca, 608
  C. salina, 702
  C. vectensis, 608
  C. virescens, 608, 700
Chrysocampanula spinifera, 700
Chrysochromulina, 593, 594, 595, 634,
  635, 642, 687, 690, 696
  C. parva, 635
  C. alifera, 635, 640, 641
  C. camella, 635, 643, 644
  C. ephippium, 635
  C. ericina, 635, 644
  C. hirta, 635, 643, 644
  C. leadbeateri, 592, 597, 635
  C. mantonia, 635, 643, 644
  C. minor, 635
  C. parkeae, 635, 642, 643
  C. polylepis, 592, 594, 635, 643,
    644
  C. pringsheimii, 635, 643, 645
  C. spinifera, 635, 644, 700
  C. strobilus, 640, 641
Chrysococcaceae, 618
Chrysococcus, 693

Chrysophyceae, 587, 593, 594, 595, 596, 597, 603, 604, 613, 616, 617, 620, 626, 681, 689, 670, 671, 672, 673
Chrysosphaeraceae, 620
Chrysosphaerales, 587, 620
Ciliophrys, 627
 C. infusionum, 627, 628, 629, 700
 C. marina, 627, 628, 700
Coccosphaerales, 634, 635
Codonomonas, 618, 699
 C. gracilis, 618
 C. cylindrica, 618
 C. dilatata, 618, 699
 C. globosa, 619
 C. vangoorii, 619, 699
 C. wulffii, 619
 C. ovalis, 619
Codonsigida, 673
Codonosigidae, 673
Collodictyon, 665
 C. sphaericum 665
Corbicula
 C. socialis, 701
Corymbellus, 634, 635
 C. aureus, 635, 637, 639
Craspedophyceae, 672
 Crinolina, 674
 C. isefiordensis, 674, 676, 677
 C. aperta, 674
Cryptobia
 C. libera, 684
Cryptochrysis
 C. fulva 611, 701
 C. lateralis, 701
 C. virescens, 700
Cryptomonadaceae, 607
Cryptomonadales, 606
Cryptomonas, 607, 608
 C. acuta, 612, 702
 C. appendiculata, 608, 702
 C. cryophila, 699
 C. marina, 611, 700, 702
 C. reticulata, 702
 C. salina, 612, 702
Cryptophyceae, 594, 595, 596, 599, 603, 604, 605, 607, 690, 691, 692
Cyanomonas
 C. baltica, 699

Cymbomonas, 593, 653, 654
 C. adriatica, 654
 C. klebsii, 654
 C. tetramitiformis, 654, 656, 659
Desmarella, 673, 674
 D. moniliformis, 674, 675, 676
Diacronema, 634, 636
 D. vlkianum, 636, 640, 642
Diaphanoeca, 673, 674
 D. grandis, 674, 676, 677
 D. fiordensis, 674
 D. pedicellata, 677
 D. sphaerica, 674, 676, 677, 700
Dicrateria, 597, 634, 635
 D. inornata, 599, 635, 636, 638
 D. gilva, 635
Dictyocha, 619, 628
 D. antarctica, 628, 631, 632
 D. crux, 628, 631, 632
 D. fibula, 628, 631, 632
 D. octonaria 631, 632
 D. rhombus, 628
 D. speculum, 597, 628, 630, 631, 700
 D. staurodon, 631, 632
 D. tripartita, 685
Dictyochaceae, 628
Dictyochales, 587, 598, 617, 627, 628, 587, 594, 595, 598, 603, 604, 616, 617, 620, 626, 689, 690, 691, 692
Dinobryon, 617, 619, 693
 D. balticum, 619, 624, 625, 700
 D. belgica, 619, 624, 625
 D. coalescens, 619, 624, 625
 D. faculiferum, 619, 624, 625, 700
 D. mediterraneum, 619
 D. pellucidum, 624, 700
 D. petiolatum, 624, 700
 D. porrectum, 619, 624, 625
Dinobryaceae, 619
 Dinodendron
 D. balticum, 624
Dinophyceae, 587, 591, 593, 594, 595, 596, 597, 603, 604, 689, 690, 691, 692, 693
Distephanus, 630
 D. speculum, 630, 700
Dolichomastix, 652
 D. eurylepidea, 652

D. lepidota, 652
D. nummulifera, 652
Dunaliella, 665, 666
   D. maritima, 666, 667
   D. primolecta, 667, 668
   D. salina, 665, 667, 668
   D. tertiolecta, 665, 667, 668
Dunaliellaceae, 665
Ebria, 685
   E. antiqua, 685
   E. tripartita, 685, 686
Ebriaceae, 685
Ebriidea, 588, 594, 605, 671, 685, 691, 692
Ebriopsidae, 685
Entomosigma
   E. akashiwo, 616, 700
Euglena, 646, 647
   E. ascusformis, 647, 648, 651
   E. proxima, 647, 651
   E. viridis, 648, 651
Euglenaceae, 647
Euglenales, 588, 646, 647
Euglenida, 645
Euglenophyceae, 587, 595, 596, 599, 603, 604, 645, 649, 690, 691, 692
Euglenophyta, 645
Eustigmatophyceae, 603, 605, 691
Eutreptia, 646, 647
   E. pertyi, 647, 648, 650
   E. globulifera, 647, 648, 649
   E. lanowii, 647, 648, 649
   E. viridis, 647, 648, 649, 650
Eutreptiaceae, 646
Eutreptiella, 595, 646, 647, 700
   E. braarudii, 647, 648, 650
   E. dofleinii, 647, 700
   E. cornubiense, 647, 648, 649
   E. elegans, 647
   E. eupharyngea, 647, 648, 650
   E. gymnastica, 647, 648, 650
   E. hirudoidea, 647, 648, 650
   E. marina, 647, 648, 649
   E. pascheri, 647, 700
Falcomonas, 607
   F. daudoides, 607
Fibrocapsa, 613, 614
   F. japonica, 614, 615, 616

Gephyrocapsaceae, 634
Germinigera, 608
   G. cryophila, 608
Goniomonas, 607, 609
   G. truncata, 609
   G. pacifica, 609
   G. amphinema, 609
Gymnastica, 700
   G. dofleinii, 700
   G. elegans, 700
Haematococcus
   H. salinus, 668
Halosphaera, 596, 602, 653, 654, 655, 657
   H. minor, 655
   H. parkeae, 655
   H. russellii, 655
   H. viridis, 655, 660, 663
Halosphaeraceae, 654
Haptophyceae, 594, 633
Hemieutreptia
   H. antiqua, 699
Hemiselmis, 607
   H. cyclopea, 610, 611
   H. rotunda, 607
   H. rufescens, 599, 607, 610, 611
   H. simplex, 607, 610, 611
   H. virescens, 607, 609, 610
Hemiselmidaceae, 606
Hermesinum, 685
   H. adriaticum, 685, 686
   H. platense, 685
Heterochromonas, 618, 702
Heteromastix
   H. minuta, 658, 700
   H. pyriformis, 701
   H. rotunda, 701
Heteromita
   H. nasuta, 683
Heterosigma, 612, 613, 614, 615
   H. akashiwo, 614, 615, 616, 700
   H. inlandica, 614, 615
Hillea, 606, 607
   H. fusiformis, 606, 609, 610
   H. marina, 606, 609, 610
Hilleaceae, 606
Hornellia
   H. marina, 614, 699

Hymenomonas, 597
Imantonia, 634, 635
  I. rotunda, 635, 636, 637
Isochrysis, 597, 634, 635
  I. galbana, 599, 635, 636, 637
Isochrysidaceae, 635
Isochrysidales, 587, 634, 635
Isoselmis, 607
  I. obconica, 607
  Jacoba, 681
  J. libera, 681, 682, 684
  Kephyrion, 619
  K. cinctum, 619
  K. petasatum, 619
Kinetoplastidea, 589, 594, 595, 604,
    605, 671, 680, 681, 689, 690
Leucocryptos, 607, 608
  L. marina, 608, 612, 700
  Mallomonas, 620
  M. subsalina, 620
  M. epithallatia, 620
Mallomonadaceae, 620
Mamiella, 652, 653
  M. gilva, 652, 656, 657, 700
Marmiellaceae, 652
Mamiellales, 588, 652
Mantoniella, 652, 653, 702
  M. antarctica, 652
  M. squamata, 652, 656, 658, 700
Meringosphaera, 617, 619, 620
  M. baltica, 630
  M. mediterranea, 620, 629, 630
  M. radians, 623, 627, 699
  M. tenerrima, 620, 629, 630
Mesocena, 628
  M. annulus, 628, 631, 632
  M. hexagona, 628, 631, 633
  M. quadrangula, 628, 631, 632
  M. stellata, 628, 631, 632
Metromonas, 681
  M. simplex, 681, 682, 685, 700
Micromonas, 652, 653, 657
  M. pusilla, 592, 593, 598, 655, 657,
    700
  M. squamata, 658, 700
Monas, 618, 702
Monochrysis
  M. lutheri, 636, 642, 701

Monosiga, 673, 674
  M. marina, 673, 674, 675, 676
  M. micropelagica, 672, 674
  Nephrochloris
  N. salina, 636, 701
Nephroselmis, 653
  N. astigmatica, 653
  N. fissa, 653
  N. gilva, 657, 700
  N. marina, 653
  N. minuta, 653, 656, 658, 700
  N. pyriformis, 653, 656, 658, 700
  N. rotunda, 653, 656, 658
  Ochromonas, 593, 618, 619
  O. bourrellyi, 618, 621, 622
  O. cosmopoliticus, 618, 621, 622
  O. crenata, 618, 621, 622
  O. marina, 618, 621, 622
  O. minima, 618, 621, 622
  O. mexicana, 618
  O. oblonga, 618, 620, 622
Ochromonadaceae, 618
Ochromonadales, 587, 618, 620
Octactis, 628
  O. octonaria, 628, 631, 632
  O. pulchra, 628, 632
Olisthodiscus, 612, 614
  O. luteus, 614, 615
  O. magnus, 614
  O. carterea, 614
Oltmannsia, 614
  O. viridis, 614, 616
Oltmannsiella 665, 670, 701
  O. virida, 670, 701
  O. lineata, 665, 670
Oltmannsiellopsis, 665, 670, 701
  O. geminata, 665, 670, 671
  O. unicellularis, 665, 670, 671
  O. viridis, 665, 670, 671, 701
  Pachysphaera, 597, 602, 653, 654, 657
  P. pelagica, 654, 656, 659
  P. marshalliae, 654, 656, 659
Parachrysidalis, 635
  P. estuariale, 635
Parapedinella, 594, 595, 619, 628
  P. reticulata, 628, 629, 630
Paraphysomonas, 619, 620
  P. antarctica, 620

P. butcheri, 620
P. corbidifera, 620
P. cribosa, 620
P. cylicophora, 620
P. faveolata, 620
P. foraminifera, 620
P. imperforata, 620
P. sideriophora, 620
P. vestita, 620
Paraphysomonadaceae, 620
Parvicorbicula, 673, 674
   P. socialis, 674, 678, 679, 701
   P. socialis var. *pavillardi,* 678
   P. socialis var. *socialis* 674
   P. quadricostata, 674, 678, 679
Paulinella, 619, 699
   P. ovalis, 619
Pavlova, 595, 634, 636, 689, 690, 692
P. gyrans, 636, 640, 642
   P. lutheri, 636, 640, 642, 701
   P. salina, 636, 701
Pavlovaceae, 636
Pavlovales, 587, 634, 636
Pedinella
   P. tricostata, 623, 701
Pedinellaceae, 627
Pedinellales, 587, 594, 595, 617, 626, 627
Pedinomonadaceae, 652
Pedinomonas
   P. mikron, 658, 701
Pedinomonadales, 652
Pedinophyceae, 652, 658
Pelagococcus, 620
   P. subviridis, 617, 620, 697
Phaeocystaceae, 635
Phaeocystis, 597, 602, 634, 635, 640, 689
   P. antarctica, 636
   P. brucei, 636
   P. globosa, 636, 610, 641
   P. giraudyi, 636
   P. pouchetii, 633, 636, 639, 640, 641
   P. scrobiculata, 636
Phaeophyceae, 602, 605
Phyllomitus, 681
   P. yorkeënsis, 681, 682, 684
Phyllomonas

P. simplex, 685, 700
Plagioselmis, 607
   P. prolonga, 607
   P. punctata, 607
Platychrysis, 634, 635
   P. pigra, 635, 637, 638
   P. pienaarii, 635
   P. simplex, 635
   P. neustophila, 635, 637, 638
Platymonas, 654, 702
   P. convoluta, 652, 702
   P. gracilis, 662
   P. tetrathele, 662
Pleurasiga, 673, 674
   P. complexa, 680, 702
   P. elegans, 702
   P. minima, 675, 678, 679
   P. orculaeformis, 675, 678, 679
   P. pedicellata, 679
   P. reynoldsii, 675, 678, 679
   P. sphyrelata, 674, 701
Pleurostomum, 681
   P. gracile, 681, 682, 683
   P. caudatum 681
   P. salinum 681
   P. flabellatum 681
Polyfibula, 674, 699
   P. sphyrelata, 674, 701
Prasinocladus, 652, 654
   P. ascus, 654
   P. marinus, 654
   P. lubricus, 654
Prasinophyceae, 588, 593, 594, 595, 596, 597, 598, 603, 604, 651, 657, 689, 690, 692, 693
Proteomonas, 608
   P. sulcata, 608
Proterospongia, 674
   P. nana, 674
   P. dybsoeensis, 674
Prymnesiaceae, 635
Prymnesiales, 587, 634, 635
Prymnesiophyceae, 586, 587, 591, 593, 594, 595, 596, 597, 598, 633, 636, 687, 688, 689, 690, 692
Prymnesium, 595, 597, 634, 635, 641, 689
   P. annuliferum, 635, 637, 638

Prymnesium (continued)
  P. calathiferum, 635
  P. czosnowskii, 635
  P. saltans, 635, 637, 638
  P. parvum, 592, 597, 635, 637, 639
  P. patelliferum, 635, 639
  P. minutum, 635
  P. zebrinum, 635, 637, 638
Pseudobodo, 618
  P. tremulans, 618, 682, 683
  P. minimus, 618, 682, 684
  P. minutus, 684, 699
Pseudokephyrion, 619
  P. formosissimum, 619
Pseudopedinella, 594, 595, 619, 627,
  690
  P. elastica, 627
  P. pyriforme, 622, 623, 627
  P. spinifera, 699
  P. tricostata, 622, 623, 627, 701
  P. variabillis, 623
Pseudoscourfieldia, 653
  P. marina, 653, 656, 658, 701
Pteridomonas, 628
  P. pulex, 628
Pterosperma, 596, 602, 653, 654, 657,
  661, 662
  P. citriforme, 655
  P. cristatum, 655, 661
  P. cuboides, 655, 660, 661
  P. dictyon, 655
  P. eurypteron, 655
  P. inornatum, 655
  P. marginatum, 655, 660, 664
  P. michaelsarsii, 655
  P. moebii, 655
  P. nationalis, 655, 660
  P. parallellum, 655, 664
  P. polygonum, 655, 660, 663
  P. porosum, 655
  P. rotondum, 655
  P. undulatum, 655
  P. vanhoeffenii, 655
Pterosphaera venhoffenii, 664
Pycnococcaceae, 653
Pycnococcus, 597, 653
  P. provasolii, 653
Pyramichlamys

P. vectensis, 667, 669
Pyramimonas, 594, 651, 655
  P. adriatica, 655, 660, 661, 701
  P. lunata, 655, 661, 701
  P. amylifera, 597, 598, 655, 660, 661,
  662
  P. disomata, 655, 701
  P. grossii, 599, 655, 656, 659
  P. moestrupii, 655
  P. nansenii, 655
  P. nephroidea, 655, 656, 659
  P. obovata, 655
  P. octociliata, 655
  P. octopus, 655, 660, 662
  P. olivacea, 655
  P. orientalis, 655, 660, 661, 701
  P. pisum, 655
  P. plurioculata, 655
  P. propulsa, 598, 655, 660, 661, 701
  P. pseudoparkeae, 655
  P. virginica, 655
Pyrenomonas, 594, 608, 653, 702
  P. reticulata, 702
  Raphidomonadales, 613
Raphidophyceae, 587, 603, 604, 612,
  614, 616, 689, 690,692
Resultor, 652, 653
  R. mikron, 652, 656, 657, 701
Rhinomonas, 607, 608
  R. pauca, 608
  R. fulva, 608, 610, 611, 701
  R. lateralis, 608, 701
  R. fragarioides, 608, 701
  R. reticulata, 608, 702
Rhizochromulina, 627
  R. marina, 627
Rhizochromulinaceae, 627
Rhizochromulinales, 587, 627
Rhodomonas, 607, 608, 702
  R. baltica 608, 610, 612
  R. abbreviata, 608
  R. atrorosea, 608
  R. chrysoidea, 608
  R. duplex, 608, 702
  R. falcata, 608
  R. heteromorpha, 608
  R. maculata, 608
  R. marina, 608, 702

R. reticulata, 702
R. salina, 608, 610, 612, 702
Rhynchomonas, 595, 681
  R. nasuta, 681, 682, 683
  R. mutabilis, 681
Salpingoeca, 588, 673, 674
  S. inquillata, 674, 675, 676
  S. natans, 675, 699
  S. spinifera, 677, 699
Salpinogoecida, 674
Salpingoecidae, 674
Sarcinochrysidaceae, 620
Sarcinochrysidales, 587, 620
Sarcinochrysis, 620
  S. marina, 621, 622
Scourfieldia
  S. marina, 658, 701
Sphaleromantis, 618
  S. marina, 618, 622, 623
  S. subsalsa, 618
  S. alata, 618
  S. ochracea, 618
Spumella, 618, 702
  S. hovassei, 618
Stephanoeca, 675
  S. diplocostata, 675, 680, 702
  S. complexa, 675, 679, 680, 702
  S. elegans, 679, 680, 702
  S. pedicellata, 675, 679, 700, 702
Storeatula, 609
  S. major, 609

Stylochromonas, 674
  S. minuta, 672, 674
Synura, 620
  S. adamsii, 620
Synurales, 587, 620
Synurophyceae, 616
Tasmanites, 602, 654
Teleaulax, 608
  T. acuta, 608, 610, 612, 702
  T. amphioxeia, 608
Telonema, 681
  T. subtilis, 681, 682, 685
Tetraselmis, 595, 651, 653, 654, 702
  T. contracta, 654
  T. convolutae, 652, 654, 702
  T. gracile, 654, 660, 662
  T. helgolandica, 654
  T. inconspicua, 654
  T. striata, 660, 661
  T. subcordiformis, 654
  T. suecica, 654
  T. tetrathele, 654, 660, 662
  T. wettsteinii, 654, 660, 663, 702
Tetraspora
  T. poucheti, 639
Thalassomonas, 702
Vacuolariaceae, 613
Volvocales, 588, 665
Xanthophyceae, 603, 605, 613
Zoomastigophora, 588, 605, 671, 680, 685

REFERENCES

Andersen, R. A. 1987. Synurophyceae classis nov., a new class of algae. *American Journal of Botany* 74:337–353.

Artari, A. 1913. Zur Physiologie der Chlamydomonaden. Versuche und Beobachtungen an *Chlamydomonas ehrenbergii* Gorosch. und verwandten Formen. *Jahrbuch für wissenschaftliche Botanik* 52:410–466.

Bergh, O., Børsheim, K. Y., Bratbak, G., & Heldal, M, 1989. High abundance of viruses found in aquatic environments. *Nature (London)* 340:467–468.

Biecheler, B. 1936. Sur une Chloromonadine nouvelle d'eau saumâtre *Chattonella subsalsa* n. gen., n. sp. *Archives de zoologie experimentale et générale* 78:79–83.

Billard, C. 1983. *Prymnesium zebrinum* sp. nov. et *P. annuliferum* sp. nov., deux nouvelles espèces apparentées a *P. parvum* (Prymnesiophyceae). *Phycologia* 22:141–151.

Black, M. 1971. The systematics of coccoliths in relation to the palaeontological record. *In* "The Micropalaentology of Oceans" (B. M. Funnel & W. R. Riedel, eds.), pp. 611–624. Cambridge University Press, Cambridge.

Boalch, G. T., & Mommaerts, J. P. 1969. A new punctate species of *Halosphaera*. *Journal of the Marine Biological Association of the United Kingdom* 49:129–139.

716 5 The Planktonic Marine Flagellates

Bohlin, K. 1897. Zur Morphologie und Biologie einzelliger Algen. *Öfversigt af Kongliga Vetenskaps-Akademiens Förhandligar* 54/1897(9):507–529.

Borgert, A. 1891. Über die Dictyochiden, insbesondere über *Distephanus speculum*, sowie Studien an Phaeodarien. *Zeitschrift für Wissenschaftliche Zoologie* 51:629–676.

Bourrelly, P. 1957a. Recherches sur les Chrysophycées, morphologi, phylogenie, systematique. *Revue algologique Memoires Hors-Serie.* 1:1–412.

Bourrelly, P. 1957b. Note systématique sur quelques algues microscopiques des cuvettes supra-littorales de la région de Dinard. *Bulletin du Laboratoire maritime de Dinard* 43:111–118.

Braarud, T. 1935. The "Øst" expedition to the Denmark Strait 1929. II. The phytoplankton and its conditions of growth. *Hvalrådets Skrifter* 10:1–173.

Bratbak, G., Heldal, M., Norland, S., & Thingstad, T. F. 1990. Viruses as partners in spring bloom microbial trophodynamics. *Applied and Environmental Microbiology* 45:1400–1405.

Bruce, J. R., Knight, M., & Parke, M. W. 1940. The rearing of oyster larvae on an algal diet. *Journal of the Marine Biological Association of the United Kingdom* 24:337–374.

Butcher, R. W. 1952. Contributions to our knowledge of the smaller marine algae. *Journal of the Marine Biological Association of the United Kingdom* 31:175–191.

Butcher, R. W. 1959. An introductory account of the smaller algae of British coastal waters. Part I: Introduction and Chlorophyceae. Fisheries Investigations, Series 4, London.

Butcher, R. W. 1961. An introductory account of the smaller algae of British coastal waters. Part 8: Euglenophyceae = Euglenineae. Fisheries Investigations, Series 4, London.

Butcher, R. W. 1967. An introductory account of the smaller algae of British coastal waters. Part 4: Cryptophyceae. Fisheries Investigations, Series 4, London.

Bütschli, O. 1884 (1883–1887). Mastigophora. *Dr. H. G. Bronn's Klassen und Ordnungen des Thier-Reichs, wissenschaftlich dargestellt in Wort und Bild,* Leipzig, I:785–864.

Büttner, J. 1911. Die farbigen Flagellaten des Kieler Hafens. *Wissenschaftliches Meeresunter-suchungen Neue Folge, Abteilung Kiel* 12:119–133.

Campbell, P. H. 1973. Studies on the brackish water phytoplankton. Sea Grant Publications, University of North Carolina, UNC.SG.73.07.

Carter, H. J. 1865. On the fresh- and saltwater Rhizopoda of England and India. *Annual Magazine of Natural History* 15(88):277–293.

Carter, N. 1937. New or interesting algae from brackish water. *Archiv für Protistenkunde* 90:1–68.

Caullery, M. 1910. Sur un protozoaire marin du genre *Ciliophrys* Cienkowski (*C. Marina* n. s. p.). *Association Française Pour l'avancement des sciences, Compte Rendu de la 38ᵉ session, Lille* 1909:708–709.

Cavalier-Smith, T. 1986. The kingdom Chromista: Origin and systematics. *Progress in Phycological Research* 4:309–347.

Chadefaud, M. 1960. Les vegetaux non vasculaires (Cryptogamie). *In* "Traité de Botanique Systématique," Tome I (M. Chadefaud & L. Emberger, eds.) pp. 1–1018. Masson & Cie, Paris.

Chihara, M., Inouye, I. & Takahata, N. 1986. *Oltmannsiellopsis,* a new genus of marine flagellate (Dunaliellaceae, Chlorophyceae). *Archiv für Protistenkunde* 132:313–324.

Chrétiennot, M. J. 1973. The fine structure and taxonomy of *Platychrysis pigra* Geitler (Haptophyceae). *Journal of the Marine Biological Association of the United Kingdom* 53:905–914.

Chrétiennot-Dinet, M. J. 1990. Chlorarachniophycées, Chlorophycées, Chrysophycées, Cryptophycées, Euglénophycées, Eustigmatophycées, Prasinophycées, Prymnésiophycées, Rhodophycées, Tribophycées. *In* "Atlas du phytoplancton marin," vol. 3 (A. Sournia, ed.), pp. 1–261. Editions du C.N.R.S., Paris.

Christensen, T. 1962, 1966. "Botanik," Vol. 2 ("Systematisk Botanik"), No. 2, Alger. Munksgaard, København.

Christensen, T. 1967. Two new families and some new names and combinations in the algae. *Blumea* 15:91–94.

Christensen, T. 1980. "Algae, A Taxonomic Survey." AiO Tryk, Odense.

Cienkowski, L. 1870. Ueber Palmellaceen und einige Flagellaten. *Archiv für Mikroskopische Anatomie* 7:421–436.

Cienkowski, L. 1875. Ueber einige Rhizopoden und verwandte organismen. *Archiv für Mikroskopische Anatomie* 12:15–50.

Clarke, K. J., & Pennick, N. C. 1972. Flagellar scales in *Oxyrrhis marina* Dujardin. *British Phycological Journal* 7:357–360.

Cole, G. T., & Wynne, M. J. 1974. Endocytosis of *Microcystis aeruginosa* by *Ochromonas danica*. *Journal of Phycology* 10:397–410.

Conrad, W. 1926. Recherches sur les Flagellates de nos eaux saumâtres. *Archiv für Protistenkunde* 56:167–231.

Conrad, W. 1938. Notes protistologiques III. Chrysomonadines interéssantes du nannoplankton saumâtre. *Bulletin du Musée royal d'Histoire naturelle de Belgique* 14(29):1–7.

Conrad, W. 1939. Notes protistologiques, XI.—Sur *Pyramidomonas amylifera*, n. sp. *Bulletin du Musée royale d'histoire naturelle de Belgique.* 15:1–9.

Conrad, W. 1941. Notes protistologiques, XXI.—Sur les Chrysomonadines à trois fouets. Aperçu synoptique. *Bulletin du Musée royal d'Histoire naturelle de Belgique* 17(45):1–16.

Conrad, W., & Kufferath, H. 1954. Recherches sur les eaux saumâtres des environs de Lilloo. 2. Partie descriptive. *Memoires Institut de sciences naturelles de Belgique* 127:1–346.

Cosper, E. M., Bricelj, V. M., & Carpenter, E. J. (eds.) 1989. Novel phytoplankton blooms. *In* "Coastal and Estuarine Studies," Vol. 35. Springer-Verlag, Berlin.

da Cunha, A. M. 1914. Contribucao paro o conhecimento da faune de Protozoarios do Brazil II. *Memorias do Instituto Oswaldo Cruz, Brazil* 6(3):169–179.

Dangeard, P. 1892. Sur un *Cryptomonas marina*. *Le Botaniste* 3:32.

Deflandre, G. 1950. Sur l'evolution des Ébriédiens. Interprétation du genre *Ebriopsis*. *Comptes Rendus des séances de Academie des Sciences* 230:1683–1685.

Deflandre, G. 1960. Sur la présence de *Parvicorbicula* n. g. *socialis* (Meunier) dans le plancton de l'Antarctique (Terre Adélie). *Revue algologique nouvelle serie* 5:183–188.

Del Vecchio, V., & D'Arca Simonetti, A. 1959. Il metodo del Most Probable Number (M.P.N.), e la sua importanza nella colimetria delle aque, con particolare riguardo a quelle destinate ad uso potabile. *Nuovi Annali d'Igiene e Microbiologia* 10(6):441–481.

de Saedeleer, H. 1930. Notules systématiques. VI. *Physomonas*. *Annales Protistologie* 2(4):177–178.

Desikachary, T. V. 1972. Notes on Volvocales. *Indian Current Science* 41:445–447.

Diesing, K. M. 1866. Revision der Protohelminthen. Abteilung: Mastigophoren. *Sitzungsberichte der Akademie der Wissenschaften zu Wien* 52:298–332.

Droop, M. R. 1953. On the ecology of flagellates from some brackish and freshwater rockpools of Finland. *Acta Botanica Fennica* 51:1–52.

Droop, M. R. 1955. Some new supra-littoral protista. *Journal of the Marine Biological Association of the United Kingdom* 34:233–245.

Dujardin, F. 1841. "Histoire Naturelle des Zoophytes. Infusoires." Librairie encyclopédique de Roret, Paris.

Dunal, F. 1838. Extrait d'un mémoire de M. F. Dunal sur les Alguers qui colorent en rouge certaines eaux des marais salans méditerranéens. *Annales des sciences naturelles. Botanique 2 Ser.* 9:172–175.

Dyer, B. D. 1990. Phylum Zoomastigina, Class Bicoecids. *In* "Handbook of Protoctista" (L. Margulis, J. O. Corliss, M. Melkonian, & D. J. Chapman, eds.), pp. 191–193. Jones and Bartlett, Boston.

Ehrenberg, C. G. 1828–1831. Animalia evertebrata. *In* "P. C. Hemprich & C. G. Ehrenberg, Symbolae physicae." Pars zoologica. Berlin. 10 pls. (1928), unpaginated text (1831).

Ehrenberg, C. G. 1832. Über die Entwicklung und Lebensdauer der Infusionsthiere. *Abhandlungen der Deutschen Akademie der Wissenschaften zu Berlin* 1831:1–154.

Ehrenberg, C. G. 1834. Dritter Beitrag zur Erkenntnis grosser Organisationen in der Richtung des Kleinsten Raumes. *Abhandlungen der Akademie der Wissenschaften zu Berlin* 1833: 145–336.

Ehrenberg, C. G. 1837. Eine briefliche Nachricht des Hrn. Agassiz in Neuchatel über den ebenfalls aus mikroskopischen Kisel-Organismen gebildeter Polirschifer von Oran in Africa. *Bericht der Königlichen Akademie der Wissenschaften, Berlin (Physiche Klasse)* 1836:59–61.

Ehrenberg, C. G. 1838. *Die Infusionsthierchen als vollkommene Organismen: Ein Blick in das Tiefere organische Leben der Natur.* Leipzig 1837:1–548.

Ehrenberg, C. G. 1839. Über die Bildung der Kreidefelsen und des Kreidemergels durch unsichtbare Organismen. *Königliche Akademie der Wissenschaften, Berlin, Abhandlunge* 1838: 59–148.

Ehrenberg, C. G. 1840. Blätter von ihm selbst ausgeführter Zeichnungen von ebenso vielen Arten. *Berichte über die Verhandlungen der Königliche Akademi der Wissenschaften, Berlin* 1839:197–219.

Ehrenberg, C. G. 1843. Verbreitung und Einfluss des mikroskopischen Lebens in Süd- und Nord-Amerika. *Abhandlungen der Akademie der Wissenschaften zu Berlin* 1841:291–446.

Ehrenberg, C. G. 1844a. Mittheilung über zwei neue Lager von Gebirgsmassen aus Infusorien als Meeres-Absatz in Nord-Amerika und eine Vergleichung derselben mit den organischen Kreide-Gebilden in Europa und Afrika. *Berichte über die Verhandlungen der Königliche Preussische Akademie der Wissenschaften, Berlin* 1842:57–97.

Ehrenberg, C. G. 1844b. Resultate seiner Untersuchungen der ihm von der Südpolreise des Capitain Ross, so wie von den Herren Schayer und Darwin zugekommenen Materialien über das Verhalten des kleinsten Lebens in den Oceanen und den grössten bisher zugänglichen Tiefen des Weltmeeres vor. *Berichte über die Verhandlungen der Königliche Preussische Akademie der Wissenschaften, Berlin* 1843:182–207.

Eikrem, W., & Throndsen, J. 1990. The ultrastructure of *Bathycoccus* gen. nov. and *B. prasinos* sp. nov., a nonmotile picoplanktonic alga (Chlorophyta, Prasinophyceae) from the Mediterranean and Atlantic. *Phycologia* 29(3):344–350.

Ellis, W. N. 1930. Recent researches on the Choanoflagellata (Craspedomonadines) (freshwater and marine) with description of new genera and species. *Annales de la Société royale zoologique de Belgique* 60:49–88.

Engler, A. 1897(1898). "Syllabus der Pflanzenfamilien." Gebr. Bornträger, Berlin.

Engler, A. 1903. "Syllabus der Pflanzenfamilien," 3rd ed. Gebr. Bornträger, Berlin.

Estep, K. W., Davis, P. G., Hargraves, P. E., & Sieburth, J. M. 1984. Chloroplast containing microflagellates in natural populations of North Atlantic nanoplankton, their identification and distribution; including a description of five new species of *Chrysochromulina* (Prymnesiophyceae). *Protistologica* 20:613–634.

Ettl, H. 1982. Taxonomische Namensänderungen und Neubeschreibungen unter den Phytomonadina. *Nova Hedwigia* 35:731–736.

Ettl, H., & Ettl, D. 1959. Zur Kenntnis der Klasse Volvophyceae, II. (Neue oder wenig bekannte Chlamydomonadalen). *Archiv für Protistenkunde* 104:51–112.

Ettl, H., & Green, J. C. 1973. *Chlamydomonas reginae* sp. nov. (Chlorophyceae), a new marine flagellate with unusual chloroplast differentiation. *Journal of the Marine Biological Association of the United Kingdom* 53:975–985.

Farr, E. R., Leussink, J. A., & Stafleu, F. A. 1979. "Index Nominum Genericorum (Plantarum)," Vols. I–III. Bohn, Scheltema & Holkema, Utrecht.

Farr, E. R., Leussink, J. A., & Zijlstra, G. 1986. "Index Nominum Genericorum (Plantarum), Supplementum I." Bohn, Scheltema & Holkema, Utrecht.

Fenchel, T., & Patterson, D. J. 1988. *Cafeteria roenbergensis* nov. gen., nov. sp., a heterotrophic microflagellate from marine plankton. *Marine Microbial Food Webs* 3:9–19.

Fott, B. 1971. Taxonomische Übertragungen und Namensänderungen unter den Algen IV. *Chlorophyceae* und *Euglenophyceae*. *Preslia* 43:289–303.

Fournier, R. O. 1971. *Chrysocampanula spinifera* gen. et sp. nov., a new marine haptophyte from the Bay of Chaleur, Quebec. *Phycologia* 10:89 92.

Fritsch, F. E. 1917. Contributions to our knowledge of the freshwater algae of Africa. 2. *Annals of the South African Museum* 9:483–611.

Fukuyo, Y., Takano, H., Chihara, M., & Matsuoka, K. (eds.) 1990. "Red Tide Organisms in Japan." Uchida Rokakuho, Tokyo.

Gaarder, K. R. 1938. Phytoplankton studies from the Tromsø district 1930–31. *Tromsø museums årshefter, naturhistorisk avdeling nr. 11* 55(1):1–160 + plates.

Gaarder, K. R. 1954. Coccolithineae, Silicoflagellatae, Pterospermataceae and other forms from the "Michael Sars" North Atlantic deep-sea expedition 1910. Report on the Scientific Results of the "Michael Sars" North Atlantic Deep-sea Expedition 1910, Vol. II, 4:1–20.

Gantt, E. 1971. Micromorphology of the periplast of *Chroomonas* sp. (Cryptophyceae). *Journal of Phycology* 7:177–184.

Gayral, P., & Billard, C. 1977. Synopsis du nouvel order des Sarcinochrysidales (Chrysophyceae). *Taxon* 26(2-3):241–245.

Gayral, P., & Fresnel, J. 1983. *Platychrysis pienaarii* sp. nov. et. *P. simplex* sp. nov. (Prymnesiophyceae): Description et ultrastructure. *Phycologia* 22:29–45.

Geitler, L. 1930. Ein grünes Filarplasmodium und andere neue Protisten. *Archiv für Protistenkunde* 69:615–636.

Grassé, P. P. 1926. Contribution à l'étude des Flagellés parasites. *Archives de zoologie experimentale et générale* 65:345–602.

Green, J. C. 1975. The fine structure and taxonomy of the haptophycean flagellate *Pavlova lutheri* (Droop) comb. nov. (*Monochrysis lutheri* Droop). *Journal of the Marine Biological Association of the United Kingdom* 55:785–793.

Green, J. C. 1976a. *Corymbellus aureus* gen. et sp. nov., a new colonial member of the Haptophyceae. *Journal of the Marine Biological Association of the United Kingdom* 56:31–38.

Green, J. C. 1976b. Notes on the flagellar apparatus of *Pavlova mesolychnon* Van der Veer and on the status of *Pavlova* Butcher and related genera within the Haptophyceae. *Journal of the Marine Biological Association of the United Kingdom* 56:595–602.

Green, J. C. 1980. The fine structure of *Pavlova pinguis* Green and a preliminary survey of the order Pavlovales (Prymnesiophyceae). *British Phycological Journal* 15:151–191.

Green, J. C. 1991. Phagotrophy in prymnesiophyte flagellates. *In* "The Biology of Free-living Heterotrophic Flagellates," Systematics Association Special Vol. 45 (D. J. Patterson & J. Larsen, eds.), pp. 401–414. Clarendon Press, Oxford.

Green, J. C., & Leadbeater, B. S. C. 1972. *Chrysochromulina parkeae* sp. nov. (Haptophyceae) a new species recorded from S.W. England and Norway. *Journal of the Marine Biological Association of the United Kingdom* 52:469–474.

Green, J. C., & Pienaar, R. N. 1977. The taxonomy of the order Isochrysidales (Prymnesiophyceae) with special reference to the genera *Isochrysis* Parke, *Dicrateria* Parke and *Imantonia* Reynolds. *Journal of the Marine Biological Association of the United Kingdom* 57:7–17.

Green, J. C., Hibberd, D. J., & Pienaar, R. N. 1982. The taxonomy of *Prymnesium* (Prymnesiophyceae) including a description of a new cosmopolitan species, *P. patellifera* sp. nov., and further observations on *P. parvum* N. Carter. *British Phycological Journal* 17:363–382.

Griessmann, K. 1913. Über marine Flagellaten. *Archiv für Protistenkunde* 32:1–78.

Grøntved, J. 1952. Investigations on the phytoplankton in the southern North Sea in May 1947. *Meddelelser fra Kommissionen for Danmarks Fiskeri og Havundersøkelser, Serie Plankton* 5(5):1–49.

Grøntved, J. 1956. Planktological contributions II. Taxonomical studies in some Danish coastal localities. *Meddelelser fra Kommissionen for Danmarks Fiskeri og Havundersøkelser, Serie Plankton, Ny Serie* 1(12):1–13.

Guillard, R. R. L., & Ryther, J. H. 1962. Studies of marine planktonic diatoms. I. *Cyclotella nana* Hustedt and *Detonula confervaceae* (Cleve) Gran. *Canadian Journal of Microbiology* 8: 229–239.

Guillard, R. R. L., Keller, M. D., O'Kelly, C. J., & Floyd, G. L. 1991. *Pycnococcus provasolii* gen. et sp. nov., a coccoid prasinoxanthnin-containing phytoplankter from the western north Atlantic and Gulf of Mexico. *Journal of Phycology* 27:39–47.

Hada, Y. 1967. Protozoan plankton of the Inland Sea, Setonaikai. I. The Mastigophora. *Bulletin of Suzugamine Women's College of Natural Sciences* 13:1–26.

Hada, Y. 1968. Protozoan plankton of the Inland Sea, Setonaikai. II. The Mastigophora and Sarcodina. *Bulletin of Suzugamine Women's College of Natural Sciences* 14:1–28.

Hada, Y. 1974. The flagellata examined from polluted water of the Inland Sea, Setonaikai. *Bulletin of the Plankton Society of Japan* 20:20–33.

Haeckel, E. 1887. Report on the Radiolaria collected by H.M.S. Challenger. Report on the scientific results of the voyage of H.M.S. Challenger during the years 1873–1876. *Zoology* 18:(2 vol text, 1–1803 pp., + 1 vol plates, 1–140).

Haeckel, E. 1894. "Systematische Phylogenie der Protisten und Pflanzen." Georg Reimer, Berlin.

Halldal, P. 1953. Phytoplankton investigations from weather ship M in the Norwegian Sea, 1948–49. *Hvalrådets Skrifter* 38:5–91.

Hansgirg, A. 1885. Anhang zu meiner Abhandlung "Über den Polymorphismus der Algen." *Botanische Centralblatt* 23(8),34:229–233.

Hara, S., & Takahashi, E. 1984. Re-investigation of *Polyoeca dichotoma* and *Acanthoeca spectabilis* (Acanthoecidae: Choanoflagellida). *Journal of the Marine Biological Association of the United Kingdom* 64:819–827.

Hara, Y., & Chihara, M. 1982. Ultrastructure and taxonomy of *Chattonella* (Class Raphidophyceae) in Japan. *Japanese Journal of Phycology* 30:47–56.

Hara, Y., & Chihara, M. 1985. Ultrastructure and taxonomy of *Fibrocapsa japonica* (Class Raphidophyceae). *Archiv für Protistenkunde* 130:133–141.

Hara, Y., Inouye, I., & Chihara, M. 1985. Morphology and ultrastructure of *Olisthodiscus luteus* (Raphidophyceae) with special reference to the taxonomy. *Botanical Magazine, Tokyo* 98:251–262.

Hara, S., Terauchi, K., & Koike, I. 1991. Abundance of viruses in marine waters: Assessment by epifluorescence and transmission electron microscopy. *Applied and Environmental Microbiology* 57:2731–2734.

Hargraves, P. E., & Steele, R. L. 1980. Morphology and ecology of *Oltmannsiella virida*, sp. nov. (Chlorophyceae: Volvocales). *Phycologia* 19:96–102.

Hargraves, P. E., & Miller, B. T. 1974. The Ebridian flagellate *Hermesinum adriaticum* Zach. *Archiv für Protistenkunde* 116:280–284.

Hazen, T. E. 1922. The phylogeny of the genus *Brachiomonas*. *Bulletin of the Torrey Botanical Club* 49:75–92.

Hibberd, D. J. 1976. The ultrastructure and taxonomy of the Chrysophyceae and Prymnesiophyceae (Haptophyceae): A survey with some new observations on ultrastructure of the Chrysophyceae. *Botanical Journal of the Linnean Society* 72:55–80.

Hibberd, D. J. 1981. Notes in the taxonomy and nomenclature of the algal classes Eustigmatophyceae and Tribophyceae (synonym Xanthophyceae). *Botanical Journal of the Linnean Society* 82:93–119.

Hibberd, D. J., & Chrétiennot, M. J. 1979. The ultrastructure and taxonomy of *Rhizochromulina marina* gen. et. sp. nov., an amoeboid marine chrysophyte. *Journal of the Marine Biological Association of the United Kingdom* 59:179–193.

Hill, D. R. A. 1990. *Chroomonas* and other blue-green cryptomonads. *Journal of Phycology* 27:133–145.

Hill, D. R. A. 1991. A revised circumscription of *Cryptomonas* (Cryptophyceae) based on examination of Australian strains. *Phycologia* 30:170–188.

Hill, D. R. A., & Wetherbee, R. 1986. *Proteomonas sulcata* gen. et sp. nov. (Cryptophyceae), a cryptomonad with two morphologically distinct and alternating forms. *Phycologia* 25:521–543.

Hill, D. R. A., & Wetherbee, R. 1988. The structure and taxonomy of *Rhinomonas pauca* gen. et sp. nov. (Cryptophyceae). *Phycologia* 27:335–365.

Hill, D. R. A., & Wetherbee, R. 1989. A reappraisal of the genus *Rhodomonas* (Cryptophyceae). *Phycologia* 28:143–158.

Hollande, A. 1952. Ordre de Bodonides (Bodonidea ord. nov.). *In* "Traité de Zoologie" (P. P. Grassé, ed.), Vol. 1(1), pp. 669–693. Masson & Cie, Paris.

Hovasse, R. 1946. Flagellés à squelette siliceux: Silicoflagellés et Ebriidés provenant du plancton recueilli au cours des campagnes scientifiques du prince Albert 1er de Monaco (1885–1912). Résultats des Campagnes Scientifiques Albert Ier, fascicule 107. Impremerie National de Monaco.

Hulburt, E. M. 1965. Flagellates from brackish waters in the vicinity of Woods Hole, Massachusetts. *Journal of Phycology* 1:87–94.

Inouye, I., Hori, T., & Chihara, M. 1983. Ultrastructure and taxonomy of *Pyramimonas lunata*, a new marine species of the Class Prasinophyceae. *Japanese Journal of Phycology* 31:238–249.

Inouye, I., Hori, T., & Chihara, M. 1990. Absolute configuration analysis of the flagellar apparatus of *Pterosperma cristatum* (Prasinophyceae) and consideration of its phylogenetic position. *Journal of Phycology* 26:329–344.

Jacobson, D. M., & Anderson, D. M. 1986. Thecate heterotrophic dinoflagellates: Feeding behavior and mechanisms. *Journal of Phycology* 22:249–258.

Jahnke, J., & Baumann, M. E. M. 1987. Differentiation between *Phaeocystis pouchetii* (Har.) Lagerheim and *Phaeocystis globosa* Scherffel. I. Colony shapes and temperature tolerances. *Hydrobiological Bulletin* 21:141–147.

James-Clark, H. 1866. Conclusive proofs of the animality of the Ciliate Sponges and their affinities with the Infusoria flagellata. *American Journal of Science 2. Series* 1866:320–324.

James-Clark, H. 1867. On the Spongiæ ciliatæ as Infusoria flagellata. *Annals and Magazine of Natural History* 1:133–142, 188–215, 250–264.

Jane, F. W. 1942. Methods for the collection and examination of fresh-water algae, with special reference to flagellates. *Journal of the Quekett Microscopical Club, Series 4,* 1:217–229.

Johnson, P. W., & Sieburth, J. M., 1982. In-situ morphology and occurrence of eucaryotic phototrophs of bacterial size in the picoplankton of estuarine and oceanic waters. *Journal of Phycology* 18:318–327.

Johnson, P. W., Hargraves, P. E., & Sieburth, J. M. 1988. Ultrastructure and ecology of *Calycomonas ovalis* Wulff, 1919, (Chrysophyceae) and its redescription as a testate rhizopod, *Paulinella ovalis* n. comb. (Filosea: Euglyphina). *Journal of Protozoology* 35:618–626.

Joint, I. R., & Pipe, R. K. 1984. An electron microscope study of a natural population from the Celtic Sea. *Marine Ecology Progress Series* 20:113–118.

Jørgensen, E. 1900. Protophyten und Protozoën im Plankton aus der norwegischen Westküste. *Bergens museums Årbog* 1899(6):1–110.

Karsten, G. 1898. *Rhodomonas baltica* N. g. et sp. *Wissenschaftliches Meeresuntersuchungen, Abteilung Kiel, Neue Folge* 3:15–16.

Kawachi, M., Inouye, I., Maeda, O., & Chihara, M. 1991. The haptonema as a food-capturing device: observations on *Chrysochromulina hirta* (Prymnesiophyceae). *Phycologia* 30:563–573.

Kent, W. S. 1878. Notes on the embryology of sponges. *Annual Magazine of natural History, 2 Series* 5:139–156.

Kent, W. S. 1880–1882. "A Manual of Infusoria" Vols. I–III. David Bogue, London.

Klebs, G. 1893. Flagellatenstudien. II. *Zeitschrift für Wissenschaftliche Zoologie* 15:353–445.

Korshikov, A. 1923. Protochlorinae, eine neue Gruppe der grünen Flagellaten. *Russ. Arch. Protistologie* 2:148–169.

Korshikov, A. 1938. Volvocineae. *In* "Vyznachnyk prisnovodnykh vodorostej URSR," IV (J. V. Roll, eds.), pp. 1–183. Akademia Nauk. URSR, Kiev.

Kuckuck, P. 1894. Bemerkungen zur marinen Algenvegetation von Helgoland. *Wissenschaftliche Meeresuntersuchungen, Neue Folge* 1(1):225–263.

Kylin, H. 1935. Über *Rhodomonas, Platymonas* und *Prasinocladus. Kungliga fysiografiska Sällskapets i Lund Förhandlingar* 5(22):1–13.

Lackey, J. B. 1939. Note on plankton flagellates from the Scioto River. *Lloydia* 2:128–143.

Lackey, J. B. 1940. Some new flagellates from the Woods Hole Area. *American Midland Naturalist* 23:463–471.

Lagerheim, G. 1893. *Phaeocystis*, nov. gen., grundadt på *Tetraspora Poucheti* Har. *Botaniska notiser* 1893:12–33.

Lagerheim, G. 1896. Ueber *Phaeocystis poucheti* (Har.) Lagerh., eine Plankton-Flagellate. *Öfversigt af Kongliga Vetenskaps-Akademiens Förhandlingar* 53:277–288.

Larsen, J., 1985. Ultrastructure and taxonomy of *Actinomonas pusilla*, a heterotrophic member of the Pedinellales (Chrysophyceae). *British Phycological Journal* 20:341–355.

Larsen, J., & Patterson, D. J. 1990. Some flagellates (Protista) from tropical marine sediments. *Journal of Natural History* 24:801–937.

Leadbeater, B. S. C. 1969. Observations on the Haptophyceae. *British Phycological Journal* 4:214–221.

Leadbeater, B. S. C. 1972a. Fine-structural observations on six new species of *Chrysochromulina* (Haptophyceae) from Norway with preliminary observation on scale production in *C. microcylindra* sp. nov. *Sarsia* 49:65–80.

Leadbeater, B. S. C. 1972b. Ultrastructural observation on some marine choanoflagellates from the coast of Denmark. *British Phycological Journal* 7:195–211.

Leadbeater, B. S. C. 1978. Renaming of *Salpingoeca sensu* Grøntved. *Journal of the Marine Biological Association United Kingdom* 58:511–515.

Leadbeater, B. S. C., & Manton, I. 1969. *Chrysochromulina camella* sp. nov. and *C. cymbium* sp. nov., two relatives of *C. strobilus* Parke and Manton. *Archiv für Mikrobiologie* 68:116–132.

Lee, J. J., Hutner, S. H. & Bowee, E. C. (eds.) 1985. "An Illustrated Guide to the Protozoa." Society of Protozologists. Allen Press, Lawrence, Kansas.

Leedale, G. F. 1966. *Euglena*: A new look with the electron microscope. *Advance of Science* 17:22–37.

Leedale, G. F. 1967. "Euglenoid Flagellates." Prentice-Hall, Englewood Cliffs, N.J.

Lemmermann, E. 1899. Ergebnisse einer Reise nach dem Pazifik. (H. Schauinsland 1896/97) Planktonalgen. *Abhandlungen der naturwissenschaftliches Verein von Bremen* 16:313–398.

Lemmermann, E. 1900. Beiträge zur Kenntnis der Planktonalgen XI. Die Gattung *Dinobryon* Ehrenb. *Berichten der Deutsche Botanische Gesellschaft* 18:500–524.

Lemmermann, E. 1901. Silicoflagellate. Ergebnisse einer Reise nach dem Pazifik. H. Schauinsland 1896/97. *Berichte der Deutschen botanischen Gesellschaft* 19:247–271.

Levander, K. M. 1894. Materialen zur Kenntnis der Wasserfauna in der Umgebung von Helsingfors, mit besonderer Berücksichtigung der Meeresfauna. I. Protozoa. *Acta Societatis pro Fauna et Flora Fennica* 12(2):1–115.

Lewin, J., Norris, R. E., Jeffrey, S. W., & Pearson, B. E. 1977. An aberrant chrysophycean alga *Pelagococcus subviridis* gen. nov. et sp. nov. from the North Pacific Ocean. *Journal of Phycology* 13:259–266.

Ling, H. Y., & Takahashi, K. 1985. The silicoflagellate genus *Octactis* Schiller 1925: A synonym of the genus *Distephanus. Micropaleontology* 31:76–81.

Loeblich, A. R. III. 1970. The amphiesma or dinoflagellate cell covering. *Proceedings of the North American Paleontological Convention*, vol. 2, Chicago, 1969 (E. L. Yochelson, ed.), pp. 867–929. Allen Press, Lawrence, Kansas.

Loeblich, A. R., & Loeblich, L. A. 1978. Division Chloromonadophyta. *In* "Handbook of Microbiology 2." (A. I. Laskin & H. A. Lechevalier, eds.), pp. 375–380. CRC Press, Boca Raton, Fla.

Loeblich, A. R. III, Loeblich, L. A., Tappan, H., & Loeblich, A. R. Jr. 1968. Annotated index of fossil and recent silicoflagellates and ehridians with descriptions and illustrations ot validly proposed taxa. *Geological Society of America, Memoire* 106.

Lohmann, H. 1902. Neue Untersuchungen über den Reichtum des Meeres an Plankton und über die Brauchbarkeit der verschiedenen Fangmethoden. *Wissenschaftliches Meeresuntersuchungen Abteilung Kiel, Neue Folge* 7:1–87.

Lohmann, H. 1908. Untersuchungen zur Festellung des vollständigen Gehaltes des Meeres an Plankton. *Wissenschaftliches Meeresuntersuchungen Abteilung Kiel, Neue Folge* 10:129–370.

Lohmann, H. 1909. Die Gehäuse und Gallertblasen der Appendicularien und ihre Bedeutung für die Erforshung des Lebens im Meer. *Verhandlungen der Deutschen zoologischen Gesellschaft* 1909:200–239.

Lohmann, H. 1911. Über das Nannoplankton und die Zentrifugierung kleinster Wasserproben zur Gewinnung desselben in lebendem Zustande. *Internationale Revue der gesamten Hydrobiologie und Hydrographie* 4:1–38.

Lohmann, H. 1912. Untersuchungen über das Pflanzen- und Tierleben der Hochsee im Atlantischen Ozean währen der Ausreise der "Deutschland." *Sitzungsberichte der Gesellschaft naturforschender Freunde, Berlin.* 1912(2a):23–54.

Lohmann, H. 1919. Die Bevölkerung des Ozeans mit Plankton nach den Ergebnissen der Zentrifugenfränge während der Ausreise der "Deutschland" 1911. *Archiv für Biontologie* 4:1–617.

Lucas, I. A. N. 1968. Three species of the genus *Cryptomonas* (Cryptophyceae). *British Phycological Bulletin* 3:535–541.

Lund, J. W. G. 1960. Concerning *Calycomonas* Lohmann and *Codonomonas* Van Goor. *Nova Hedwigia* 1:423–429.

Luther, A. 1899. Ueber *Chlorosaccus* eine neue Gattung der Süsswasseralgen, nebst einigen Bemerkungen zur Systematik verwandter Algen. *Bihang Kongliga Svenska Vetenskapsakademiens Handlingar* 24-III(13):1–22.

Magne, F. 1954. Les Chrysophycées marines de la station biologique de Roscoff. *Revue générale de botanique* 61:389–416.

Manton, I. 1977. *Dolichomastrix* (Prasinophyceae) from Arctic Canada, Alaska and South Africa: A *Scourfieldia caeca* (Korsch.) Belcher et Swale. *Archiv für Protistenkunde* 117:358–368.

Manton, I. 1977. *Dolichomastrix* (Prasinophyceae) from Artic Canada, Alaska and South Africa: A new genus of flagellates with scaly Flagella. *Phycologia* 16:427–438.

Manton, I. 1978. *Chrysochromulina hirta* sp. nov., a widely distributed species with unusual spines. *British Phycological Journal* 13:3–14.

Manton, I., & Bremer, G. 1981. Observations on lorica structure and aspects of replication in the *Pleurasiga sphyrelata* Thomsen complex (= *Polyfibula* spp., gen. n.), (Choanoflagellata). *Zoologica Scripta* 10:273–291.

Manton, I., & Parke, M. 1960. Further observations on small green flagellates with special reference to possible relatives of *Chromulina pusilla* Butcher. *Journal of the Marine Biological Association of the United Kingdom* 39.275–298.

Manton, I., & Parke, M. 1962. Preliminary observations on scales and their mode of origin in *Chrysochromulina polylepis* sp. nov. *Journal of the Marine Biological Association of the United Kingdom* 42:565–578.

Manton, I., & Parke, M. 1965. Observations on the fine structure of two species of *Platymonas* with special reference to flagellar scales and the mode of origin of the theca. *Journal of the Marine Biological Association of the United Kingdom.* 45:743–754.

Manton, I., Oates, K., & Parke, M. 1963. Observations on the fine structure of the *Pyramimonas* stage of *Halosphaera* and preliminary observations on three species of *Pyramimonas*. *Journal of the Marine Biological Association of the United Kingdom* 43:225–238.

Massjuk, N. P. 1973. "Morfologia, sistematika, ekologia, geograficeskoje rasprostramenie roda Dunaliella Teod." Naukova Dumka, Kiev.

Mayer, J. A., & Taylor, F. J. R. 1979. A virus which lyses the marine nanoflagellate *Micromonas pusilla. Nature (London)* 281:299–301.

McFadden, G. I., Hill, D. R. A., & Wetherbee, R. 1986. A study of the genus *Pyramimonas* (Prasinophyceae) from southeastern Australia. *Nordic Journal of Botany* 6:209–234.

McLachlan, J. 1973. Growth media—Marine. *In* "Handbook of Phycological Methods; Culture Methods & Growth Measurements" (J. Stein, ed.), pp. 25–51. Cambridge University Press, Cambridge.

Meunier, A. 1910. Microplankton des mers de Barents et de Kara. *Duc d'Orléans Campagne Arctique de 1907, Bruxelles.*

Moestrup, Ø. 1982. Flagellar structure in algae: A review, with new observations particularly on the Chrysophyceae, Phaeophyceae (Fucophyceae), Euglenophyceae, and Rickertia. *Phycologia* 21:427–528.

Moestrup, Ø. 1984. Further studies on *Nephroselmis* and its allies (Prasinophyceae). II. *Mamiella* gen. nov. (Mamiellales ord. nov.). *Nordic Journal of Botany* 4:109–121.

Moestrup, Ø. 1991. Further studies of presumedly primitive green algae, including the description of Pedinophyceae class. nov. and *Resultor* gen. nov. *Journal of Phycology* 27:119–133.

Moestrup, Ø. 1992. Current status of chrysophyte "splinter groups": Synurophytes, pedinellids, silicoflagellates. *Proceeding from the Third International Chrysophyte Symposium* (C. D. Sandgren, J. Small, & J. Kristiansen, eds.), in press.

Moestrup, Ø., & Thomsen, H. A. 1980. Preparation of shadow-cast whole mounts. *In* "Handbook of Phycological Methods; Developmental & Cytological Methods" (E. Gantt, ed.), pp. 386–390. Cambridge University Press, Cambridge.

Moestrup, Ø., & Throndsen, J. 1988. Light and electron microscopical studies on *Pseudoscourfieldia marina*, a primitive scaly green flagellate (Prasinophyceae) with posterior flagella. *Canadian Journal of Botany* 66:1415–1434.

Moestrup, Ø., & Hill, D. R. A. 1991. Studies on the genus *Pyramimonas* (Prasinophyceae) from Australian and European waters: *P. propulsa*, sp. nov. and *P. mitra*, sp. nov. *Phycologia* 30:534–546.

Moestrup, Ø., Hori, T., & Kristiansen, Aa. 1987. Fine structure of *Pyramimonas octopus* sp. nov., an octoflagellated benthic species of *Pyramomonas* (Prasinophyceae), with some observations on its ecology. *Nordic Journal of Botany* 7:339–352.

Müller, O. F. 1773. *Vermium terrestrium et fluviatilum seu animalium infusoriorum, helminthicorum et testaceorum, non marinorum suecincta historia. Heineck et Faber typis Martini Hallager, Havniae et Lipsiae* 1(1):30+.

Newton, E. T. 1875. On "tasmanite" and Australian "white coal." *Geological Magazine, Series 2* 2(8):337–342.

Namyslovski, B. 1913. Über unbekannte Mikroorganismen aus dem Innern des Salzbergwerkes Wieliczka. *Bulletin international Academia Science Cracow, Serie B* 3/4:88–104.

Nicolai, F. E., & Baas-Becking, L. G. M. 1935. Einige Notizen über Salzflagellaten. *Archiv für Protistenkunde* 85:319–328.

Norris, R. E. 1965. Neustonic marine Craspedomonadales (choanoflagellates) from Washington and California. *Journal of Protozoology* 12:589–602.

Norris, R. E. 1967. Microalgae in enrichment cultures from Puerto Penasco, Sonora, Mexico. *Bulletin of Southern California Academy of Science* 66:233–250.

Norris, R. E., Hori, T., & Chihara, M. 1980. Revision of the genus *Tetraselmis* (Class Prasinophyceae). *Botanical Magazine Tokyo* 93:317–339.

Novarino, G. 1990. Observations on *Rhinomonas reticulata* comb. nov. and *R. reticulata* var. *eleniana* var. nov. (Cryptophyceae), with comments on the genera *Pyrenomonas* and *Rhodomonas. Nordic Journal of Botany* 11:243–252.

O'Kelly, C. J., & Wujek, D. E. 1992. Status of the Chrysamoebiales (Chrysophyceae): observations on *Chrysamoeba pyrenoidifera*, *Rhizochromulina marina* and *Lagynion delicatulum*. *Proceeding from the Third International Chrysophyte Symposium* (C. D. Sandgren, J. Small, & J. Kristiansen, eds.), in press.

Oltmanns, F. 1904. "Morphologie und Biologie der Algen," 1 Band. G. Fischer, Jena.

Ono, C., & Takano, H. 1980. *Chattonella antiqua* (Hada) comb. nov., and its occurrence on the Japanese coast. Bulletin of the Tokai Regional Fisheries Research Laboratory 102,93–100.

Ostenfeld, C. H. 1899. Note. *In* "Iagtagelser over overfladevandets temperatur, saltholdighed og plankton paa islandske og grønlandske skibsrouter i 1898" (M. Knudsen & C. H. Ostenfeld, eds.), p. 93. Bianco Lunos Kgl. Hof-Bogtrykkeri (F. Dreyer), Kjøbenhavn.

Ostenfeld, C. H., & Schmidt, J. 1902. Plankton fra det Røde Hav og Adenbugten. *Videnskapelige Meddelelser fra den naturhistoriske Forening i Kjøbenhaven for Aaret* 1901:141–182.

Papenfuss, G. F. 1955. Classification of the algae. *In* "A Century of Progress in the Natural Sciences 1853–1953," pp. 115–224. California Academy of Sciences, San Francisco.

Parke, M. 1949. Studies on marine flagellates. *Journal of the Marine Biological Association of the United Kingdom* 28:255–286.

Parke, M. 1966. The genus *Pachysphaera* (Prasinophyceae). *In* "Some Contemporary Studies in Marine Science" (H. Barnes, ed.), pp. 555–563. Allen and Unwin, London.

Parke, M., & Adams, I. 1961. The *Pyramimonas*-like motile stage of *Halosphaera viridis* Schmitz. *Bulletin of the Research Council of Israel* 10D:94–100.

Parke, M., & Burrows, E. M. 1976. Chlorophyceae. Checklist of British marine algae—Third revision (M. Parke & P. S. Dixon, eds.). *Journal of the Marine Biological Association of the United Kingdom* 56:566–570.

Parke, M., & den Hartog-Adams, I. 1965. Three species of *Halosphaera*. *Journal of the Marine Biological Association of the United Kingdom* 45:537–557.

Parke, M., & Dixon, P. S. 1968. Check-list of British marine algae—Second revision. *Journal of the Marine Biological Association of the United Kingdom* 48:783–832.

Parke, M., & Manton, I. 1962. Studies of marine flagellates. VI. *Chrysochromulina pringsheimii* sp. nov. *Journal of the Marine Biological Association of the United Kingdom* 42:391–404.

Parke, M., & Rayns, D. G. 1964. Studies on marine flagellates. VII. *Nephroselmis gilva* sp. nov. and some allied forms. *Journal of the Marine Biological Association United Kingdom* 44:209–217.

Parke, M., Manton, I., & Clarke, B. 1955. Studies of marine flagellates. II. Three new species of *Chrysochromulina*. *Journal of the Marine Biological Association of the United Kingdom* 34:579–609.

Parke, M., Manton, I., & Clarke, B. 1956. Studies of marine flagellates. III. Three further species of *Chrysochromulina*. *Journal of the Marine Biological Association of the United Kingdom* 35:387–414.

Parke, M., Manton, I., & Clarke, B. 1959. Studies of marine flagellates. V. Morphology and microanatomy of *Chrysochromulina strobilus* sp. nov. *Journal of the Marine Biological Association of the United Kingdom* 38:169–188.

Parke, M., Green, J. C., & Manton, I. 1971. Observation on the fine structure of zooids of the genus *Phaeocystis* (Haptophyceae). *Journal of the Marine Biological Association of the United Kingdom* 51:927–941.

Parke, M., Boalch, G. T., Jowett, R., & Harbour, D. S. 1978. The genus *Pterosperma* (Prasinophyceae): Species with a single equatorial ala. *Journal of the Marine Biological Association of the United Kingdom* 58:239–276.

Pascher, A. 1910. Chrysomonaden aus dem Hirschberger Grossteiche. *Monographien und Abhandlungen zur Internationalen Revue der gesamten Hydrobiologie und Hydrographie* 1:1–66.

Pascher, A. 1911. Über der Beziehungen der Cryptomonaden zu den Algen. *Berichte der Deutschen Botanischen Gesellschaft* 29:193–203.

Pascher, A. 1912. Über Rhizopoden und Palmellastadien bei Flagellaten (Chrysomonaden), nebst einer Übersicht über die braunen Flagellaten. *Archiv für Protistenkunde* 25:153–200.

Pascher, A. 1913. "Die Süsswasser-Flora Deutschlands, Osterreichs und der Schweiz. 2. Flagellatae II." G. Fischer, Jena.

Pascher, A. 1914. Über Flagellaten und Algen. *Berichte der Deutschen botanischen Gesellschaft* 32:136–160.

Pascher, A. 1927. Neue oder wenig bekannte Flagellaten. XVII. *Arch für Protistenkunde* 58:577–598.

Pascher, A. 1932. Zur Kenntnis mariner Planktonten I: *Meringosphaera* und ihre Verwandten. *Archiv für Protistenkunde* 77:195–218.

Patterson, D. J. 1990. *Jacoba libera* (Ruinen, 1938), a heterotrophic flagellate from deep ocean sediments. *Journal of the Marine Biological Association of the United Kingdom* 70:381–393.

Pavillard, J. 1916. Flagellés nouveaux, épiphytes des diatomées pélagiques. *Comptes rendus hebdomadaire des Séances d l'Academie des Sciences, Paris* 163:65–68.

Pedersen, S. A., Beech, P. L., & Thomsen, H. A. 1986. *Parapedinella reticulata* gen et sp. nov. (Chrysophyceae) from Danish waters. *Nordic Journal of Botany* 6:507–513.

Penard, E. 1890. Über einige neue oder wenig bekannte Protozoën. *Jahrbücher des Nassauischen Vereins für Naturkunde* 43:73–91.

Perty, M. 1852. "Zur Kenntnis kleinster Lebensformen nach Bau, Funktionen, Systematik mit specialverzeichniss in der Schweiz beobachteten." Jent & Reinert, Bern.

Picken, L. E. R. 1941. On the Bicoecidae: A family of colourless flagellates. *Philosophical Transactions of the Royal Society, Series B* 230:451–473.

Pienaar, R. N. 1976. The microanatomy of *Sphaleromantis marina* sp. nov. (Chrysophyceae). *British Phycological Journal* 11:83–92.

Pienaar, R. N., & Norris, R. E. 1979. The ultrastructure of the flagellate *Chrysochromulina spinifera* (Fournier) comb. nov. (Prymnesiophyceae) with special reference to scale production. *Phycologia* 18:99–108.

Platt, T., & Li, W. K. W. (eds.) 1986. Photosynthetic picoplankton. *Canadian Bulletin of Fisheries and Aquatic Sciences* 214:1–583.

Poche, F. 1913. Das System der Protozoa. *Archiv für Protistenkunde* 33:125–321.

Pouchet, G. 1893. Histoire Naturelle. Voyage de "La Manche" à l'Ile Jan Mayen et au Spitzberg (juillet–aôut 1892). *Nouvelles Archives des Missions Scientifiques et Littéraires* 5(10):155–217.

Prauser, H. 1958. *Diacronema vlkianum*, eine neue chrysomonade. *Archiv für Protistenkunde* 103:117–128.

Preisig, H. R., & Hibberd, D. J. 1983. Ultrastructure and taxonomy of *Paraphysomonas* and related genera 3. *Nordic Journal of Botany* 3:695–723.

Pringsheim, E. G. 1953. Salzwasser-Eugleninen. *Arch für Mikrobiologie* 18:149–164.

Rayns, D. G. 1962. Alternation of generations in a coccolithophorid, *Cricosphaera carterae* (Braarud & Fagerl.) Braarud. *Journal of the Marine Biological Association of the United Kingdom* 42:481–484.

Reynolds, N. 1974. *Imantonia rotunda* gen. et sp. nov., a new member of the Haptophyceae. *British Phycological Journal* 9:429–434.

Rouchijajnen, M. I. 1966. Duae species novae e chrysophytis mobilibus Maris Nigri. *Academia scientiarum URSS, Novitates systematical plantarum non vascularium* 1966:10–15.

Rouchijajnen, M. I. 1968. De specie nov chrysophytorum e Mare Nigro. *Academia scientiarum URSS, Novitates systematicae plantarum non vascularium* 1968:8–11.

Ruinen, J. 1938. Notizen über Salzflagellaten. II. Über die verbreitung der Salzflagellaten. *Archiv für Protistenkunde* 90:210–258.

Santore, U. J. 1983. Flagellar and body scales in the Cryptophyceae. *British Phycological Journal* 18:239–248.

Santore, U. J. 1984. Some aspects of taxonomy in the Cryptophyceae. *New Phytologist* 98:627–646.

Santore, U. J. 1986. The ultrastructure of *Pyrenomonas heteromorpha* comb. nov. (Cryptophyceae). *Botanica Marina* 29:75–82.

Scherffel, A. 1899. *Phaeocystis globosa* n. sp. *Berichte der deutschen botanischen Gesellschaft* 17:317–318.

Scherffel, A. 1900. *Phaeocystis globosa* nov. spec. nebst einigen Betrachtungen Über die Phylogenie niederer, unbesondere brauner Organismen. *Wissenschaftliche Meeresuntersuchungen Abteilung Helgoland N.S.* 4:1–29.

Schiller, J. 1913. Vorläufige Ergebnisse der Phytoplankton untersuchung auf den Fahrten S.M.S. "Najade" in der Adria. II. Flagellaten und Chlorophyceen. *Sitzberichtung der Königlichen Akademie der Wissenschaften Wien, mathematisch-naturwissenschaftilche Klasse* 122:621–630.

Schiller, J. 1925. Die planktontischen Vegetationen des adriatischen Meeres. B. Chrysomonadina, Heterokontae, Cryptomonadina, Eugleninae, Volvocales. 1. Systematischer Teil. *Archiv für Protistenkunde* 53:59–123.

Schiller, J. 1957. Untersuchungen an den planktischen Protophyten des Neusiedlersees 1950-1954, II. Teil. *Wissenschaftliche Arbeiten aus dem Burgenland* 18:1–44.

Schmarda, L. K. 1850. Neue Formen der Flagellaten. *Denkschrift der Akademie der Wissenschaften, Wien* 1:9–14.

Schmitz, F. 1878. *Halosphaera*, eine neue Gattung grüner Algen aus dem Mittelmeer. *Mitteilungen der zoologishes Station Neapel* 1:67–92.

Schoenichen, W. 1925. "Eyferth's Einfachste Lebensformen des Tier- und Pflanzenreiches. 5 Auflage, Band 1, Spaltpflanzen, Geisselinge, Algen, Pilze." Bermühler, Berlin.

Schumann, J. 1867. Preussischen Diatomeen. *Schriften der Königlichen physikalisch-ökonomischen Gesellschaft zu Königsberg* 8:37–68.

Schütt, F. 1892. Das Pflanzenleben der Hochsee. *Ergebnisse der im Atlantischen Ozean Planktonexpedition der Humboldt-Stiftung* 1889 1A:241–314.

Sieburth, J. M., Johnson, P. W., & Hargraves, P. E. 1988. Ultrastructure and ecology of *Aureococcus anophagefferens* gen. et sp. nov. (Chrysophyceae): The dominant picoplankter during a bloom in Narragansett Bay, Rhode Island, summer 1985. *Journal of Phycology* 24:416–425.

Silva, P. C. 1980. Names of classes and families of living algae. *Regnum Vegetabile* 103:1–156.

Smith, G. M. 1920. Phytoplankton of the Inland Lakes of Wisconsin. *Bulletin of the Wisconsin Geology and Natural History Survey* 57:1–243.

Sommer, U. 1988. Some relationships in phytoflagellate motility. *Hydrobiologia* 161:125–131.

Sournia, A. 1986. "Atlas du phytoplancton marin, volume 1. Cyanophycées, Dictyochophycées, Dinophycèes & Raphidophycées." Editions du C.N.R.S., Paris.

Starmach, K. 1980. Chrysophyta I, Chrysophyceae—Zlotowiciowce. *Flora Slodkowodna Polski* 5:1–776.

Stein, F. 1878. "Der Organismus der Infusionsthiere. III. Der organismus der Flagellaten I." Wilhelm Engelmann, Leipzig.

Stein, F. 1883. "Der Organismus der Infusionsthiere. III. Der organismus der Flagellaten II." Wilhelm Engelmann, Leipzig.

Steur, A. 1904. Über eine Euglenoide (*Eutreptia*) aus dem Canale grande von Triest. *Archiv für Protistenkunde* 3:126–137.

Stokes, A. C. 1888. Notices of new infusoria flagellata from American fresh waters. *Journal of the Royal microscopical Society* 1888:698–704.

Subrahmanyan, R. 1954. On the life-history and ecology of *Hornellia marina* gen. et sp. nov., (Chloromonadineae), causing green discoloration of the sea and mortality among marine organisms off the Malabar Coast. *Indian Journal of Fisheries* 1:182–203.

Swaroop, S. 1956. Estimation of bacterial density of sea water samples. *Bulletin of World Health Organization* 14:1089–1107.

Tangen, K. 1978. Nets. *In* "Phytoplankton Manual" (A. Sournia, ed.), pp. 50–58. UNESCO Monographs on Oceanographic Methodology no. 6, Paris.

Taylor, D. L., & Lee, C. C. 1971. A new cryptomonad from Antarctica: *Cryptomonas cryophila* sp. nov. *Archiv für Mikrobiologie* 75:269–280.

Teodoresco, E. C. 1905. Organisation et développement du *Dunaliella*, nouveau genre de Volvocacée-Polyblépharidée. *Beihäft zu Botanische Centralblatt* 18(1):215–232.

Thomsen, H. A. 1973. Studies on marine choanoflagellates I. Silicified choanoflagellates of the Isefjord (Denmark). *Ophelia* 12:1–26.

Thomsen, H. A. 1975. An ultrastructural survey of the Chrysophycean genus *Paraphysomonas* under natural conditions. *British Phycological Journal* 10:113–127.

Thomsen, H. A. 1976. Studies on marine choanoflagellates II. Fine structural observations on some silicified choanoflagellates from the Isefjord (Denmark), including the description of two new species. *Norwegian Journal of Botany* 23:33–51.

Thomsen, H. A. 1982. Planktonic choanoflagellates from Disko Bugt, West Greenland with a survey of the nanoplankton of the area. *Meddelelser om Grønland, Bioscience* 8:1–35.

Thomsen, H. A. 1988. Ultrastructural studies of the flagellate and cyst stages of *Pseudopedinella tricostata* (Pedinellales, Chrysophyceae). *British Phycological Journal* 23:1–16.

Throndsen, J. 1969. Flagellates of Norwegian coastal waters. *Nytt Magasin for Botanikk* 16:161–216.

Throndsen, J. 1970a. *Salpingoeca spinifera* sp. nov., a new plankton species of the Craspedophyceae recorded in the Arctic. *British Phycological Journal* 5:87–89.

Throndsen, J. 1970b. Marine planktonic Acanthoecaceans (Craspedophyceae) from Arctic waters. *Nytt Magasin for Botanikk* 17:103–111.

Throndsen, J. 1971. *Apedinella* gen. nov. and the fine structure of *A. spinifera* (Throndsen) comb. nov. *Norwegian Journal of Botany* 18:47–64.

Throndsen, J. 1973. Phytoplankton occurrence and distribution in stations sampled during the SCOR WG 15 cruise to the Caribbean Sea, Pacific Ocean and Sargasso Sea in May 1970, based on direct cell counts. S.C.O.R. W.G. 15. Data Report SCOR Discoverer Expedition, May 1970. SIO Ref. 73–16, University of California, D 1–28. San Diego.

Throndsen, J. 1974. Planktonic choanoflagellates from North Atlantic waters. *Sarsia* 56:95–122.

Throndsen, J. 1978. The dilution culture method. In "Phytoplankton Manual" (A. Sournia, ed.), pp. 218–224. UNESCO Monographs on Oceanographic Methodology no. 6, Paris.

Throndsen, J. 1979. The significance of ultraplankton in marine primary production. *Acta Botanica Fennica* 110:53–56.

Throndsen, J. 1980. Bestemmelse av marine nakne flagellater [Identification of marine naked flagellates]. *Blyttia* 38:189–207.

Throndsen, J. 1988. *Cymbomonas* Schiller (Prasinophyceae) reinvestigated by light and electron microscopy. *Archiv für Protistenkunde* 136:327–336.

Throndsen, J., & Kristiansen, S. 1988. Nanoplankton communities in Haltenbanken waters during the FOH oil pollution experiment, July–August, 1982. *Sarsia* 73:71–74.

Throndsen, J., & Zingone, A. 1988. *Tetraselmis wettsteinii* (Schiller) Throndsen comb. nov. and its occurrence in golfo di Napoli. *Giornale Botanico Italiano* 122:227–235.

Toriumi, S., & Takano, H. 1973. *Fibrocapsa*, a new genus in Chloromonadophyceae from Atsumi Bay, Japan. *Bulletin of Tokai Regional Fisheries Research Laboratory* 76:25–35.

Van der Veer, J. 1969. *Pavlova mesolychnon* (Chrysophyta), a new species from the Tamar estuary, Cornwall. *Acta Botanica Neerlandica* 18:496–510.

Van Etten, J. L., Lane, L. C., & Meints, R. H. 1991. Virus and viruslike particles of eukaryotic algae. *Microbiological Reviews* 55:586–620.

van Goor, A. C. J. 1925a. Die Euglenineae des holländischen Brackwassers mit besonderer Berücksichtigung ihrer Chromatophoren. *Recueil des travaux botaniques néerlandais* 22:275–291.

van Goor, A. C. J. 1925b. Über einige bemerkenswerte Flagellaten der holländischen Gewässer. *Recueil des travaux botaniques néerlandais* 22:315–319.

Vlk, W. 1938. Über den Bau der Geissel. *Archiv für Protistenkunde* 90:448–488.

Walne, P. L., Moestrup, Ø., Norris, R. E., & Ettl, H. 1986. Light and electron microscopical studies of *Eutreptiella eupharyngea* sp. nov. (Euglenophyceae) from Danish and American waters. *Phycologia* 25:109–126.

West, G. S. 1916. Algological notes. 18–23. *Journal of Botany* 54:1–10.

West, G. S., & Fritsch, F. E. 1927. "A Treatise on the British Freshwater Algae." Cambridge University Press, Cambridge.

Willén, T. 1963. Notes on Swedish plankton algae. *Nova Hedwigia* 6:39–56.

Willén, T. 1992. *Dinobryon faculiferum*, a new name for *Dinobryon petiolatum* (Chrysophyceae: Dinobryaceae). *TAXON*, 42:62–63.

Wislouch, S. 1924. Pryczynek do biologji solnisk i genezy szlamów leczniczych na Krymie (Beitrage zur biologie und Entstehung von Heilschlamm der Salinen der Krim). *Acta Societatis Botanicorum Poloniae* 2:99–129. (In Polish, German summary.)

Wollenweber, W. 1926. Viervakuolige Chlamydomonaden. *Berichte der Deutschen botanischen Gesellschaft* 44:52–59.

Wulff, A. 1916. Ueber das Kleinplankton der Barentssee. *Arbeiten der Deutschen wissenschaftlichen Kommission für die internationale Meeresforschung. A. Aus dem Laboratorium für internationale Meeresforschung in Kiel. Biologische Abteilung* 28:95–125.

Wulff, A. 1919. Ueber das Kleinplankton der Barentssee. *Wissenschaftliche Meeresuntersuchungen, Neue Folge, Abteilung Helgoland* 13(1):95–125.

Wyssotski, A. 1888. Mastigophora i Rhizopoda, najdennyja v Vejsovom i Repnom ozerach. *Travaux de la Société des naturalistes à la l'Université Impériale de Kharkow/Trudy Obsc. Isp. Prir. Imp. Kharkovsk. Univ. (1887)* 21:119–140.

Zacharias, O. 1906. Eine neue Dictyochide aus dem Mittelmeer, *Hermesinum adriaticum* n. g., n. sp. *Archiv Hydrobiologie und Planktonkunde* 1:394–398.

Zimmermann, B., Moestrup, Ø., & Hällfors, G. 1984. Chrysophyte or heliozoon: Ultrastructural studies on a cultured species of *Pseudopedinella* (Pedinellales ord. nov.), with comments on species taxonomy. *Protistologica* 20:591–612.

Zimmermann, W. 1925. Helgoländer Meeresalgen I–VI. Beiträge zur Morphologie, Physiologie und Oekologie der Algen. *Wissenschaftliches Meeresuntersuchungen der Kommission zur wissenschaftlichen Untersuchung der deutschen Meere, Abteilung Helgoland, Neue Folge* 16(1):1–25.

Zimmermann, W. 1930. Neue und wenig bekannte Kleinalgen von Neapel I–V. *Zeitschift für Botanik* 23:419–442.

REFERENCE ADDED IN PROOF:

Sieburth, J. McN, Smetachek, V. & Lenz, J. 1978. Pelagic ecosystem structure: heterotrophic compartments of the plankton and their relationship to plankton sizes. Limnology and Oceanography 23:1256–1263.

Chapter 6

Modern Coccolithophorids

Berit R. Heimdal

INTRODUCTION

The Living Cell

Coccolithophorids are biflagellate (**Fig. 1**) or coccoid unicells, whose longest dimensions rarely exceed 30 μm and are most often < 10 μm. They may be spherical, ovoid to oval, elongated cylinders, or spindle-shaped cells tapering gradually toward one or both ends (**Fig. 2**). The flagellar apparatus and intracellular fine structure of motile coccolithophorids resemble the typical prymnesiophyte cell in most characters. They usually contain two parietal chloroplasts including a pyrenoid. In *Emiliania huxleyi* (Lohmann) Hay & Mohler, the main carotenoid pigment is shown to be 19'-hexanoyloxy-fucoxanthin (Arpin et al., 1976), which was also recorded in the prymnesiophytes *Corymbellus aureus* and *Phaeocystis pouchetii* (Nelson & Wakeham, 1989) and in some species of the dinoflagellate genus *Gyrodinium* (Tangen & Bjørnland, 1981). The chlorophyll composition of *E. huxleyi* is also distinctive, and recently Jeffrey & Wright (1987) described a new chlorophyll c designated as chlorophyll c_3 in this species. The presence of a phytol-substituted chlorophyll c with chromatogra-

Identifying Marine Phytoplankton
Copyright © 1993 by Academic Press, Inc. All rights of reproduction in any form reserved.

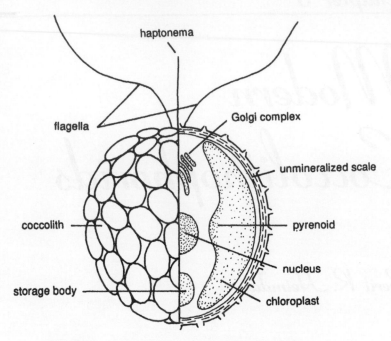

FIGURE 1   Structure of a motile coccolithophorid cell.

phic properties quite different from those of the chlorophylls c_2 and c_3 but similar to those of chlorophyll a was documented for the first time by Nelson & Wakeham (1989). The main reserve substance in the Prymnesiophyceae is a water-soluble carbohydrate resembling chrysolaminarin (Christensen, 1980). A study of the lipid composition of *E. huxleyi* and other prymnesiophycean algae (Volkman et al., 1981) suggests that comparisons of lipid profiles can give useful taxonomic information (Marlowe et al., 1984). Furthermore, there is evidence that coccolithophorids produce large quantities of volatile organic sulfur compounds, which can eventually pass to the atmosphere (Andreae, 1986; Turner et al., 1988).

Coccoliths

The most distinctive feature of coccolithophorids is the outer covering of small regular calcareous plates or coccoliths. The coccoliths consist of calcium carbonate, primarily in the crystalline form of calcite. In nitrogen-limited cultures of *E. huxleyi*, calcite was largely replaced by aragonite and vaterite (Wilbur & Watabe, 1963). Since each species of coccolithophorid exhibits its

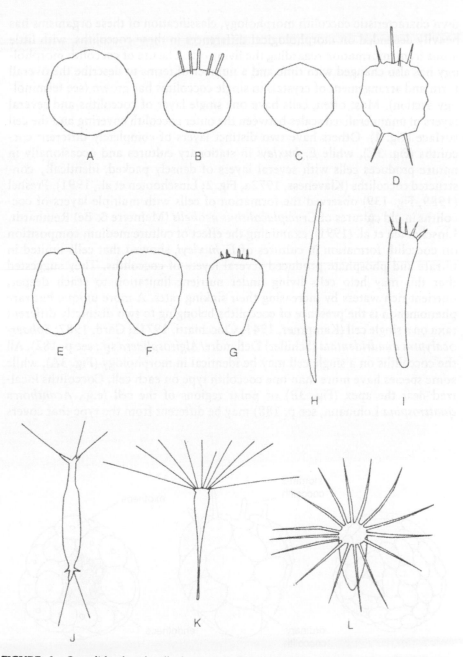

FIGURE 2 Coccolithophorid cell shapes: (A) spherical; (B) subspherical; (C) polygonal;
(D) oval; (E) ellipsoidal; (F) ovate; (G) pyriform; (H) fusiform; (I) clavate; (J) cylindrical;
(K) caudate; (L) cuneate.

own characteristic coccolith morphology, classification of these organisms has heavily depended on morphological differences in these coccoliths, with little input from information regarding the living cell. The use of coccolith morphology has also changed with time and a number of terms to describe the overall form and arrangement of crystals in single coccoliths has grown (see terminology section). Most often, cells have one single layer of coccoliths and several layers of unmineralized scales between the outer coccolith covering and the cell surface (**Fig. 1**). Others have two distinct layers of completely different coccoliths (**Fig. 3C**), while *E. huxleyi* in stationary cultures and occasionally in nature produces cells with several layers of densely packed, identical!, constructed coccoliths (Klaveness, 1972a, Fig. 2; Linschooten et al., 1991). Fresnel (1989, Fig. 139) observed the formation of cells with multiple layers of coccoliths in old cultures of *Cruciplacolithus neohelis* (McIntyre & Bé) Reinhardt. Linschooten et al. (1991), examining the effect of culture medium composition on coccolith formation in cultures of *E. huxleyi,* showed that cells limited in nitrate and phosphate produced several layers of coccoliths. They suggested that this may help cells living under nutrient limitation to reach deeper, nutrient-rich waters by increasing their sinking rates. A more unique but rare phenomenon is the presence of coccoliths belonging to two distinctly different taxa on a single cell (Kamptner, 1941; Clocchiatti, 1971a; Gard, 1987; *Sphaerocalyptra quadridentata* (Schiller) Deflandre/*Algirosphaera* sp., see p. 182). All the coccoliths on a single cell may be identical in morphology (**Fig. 3A**), while some species have more than one coccolith type on each cell. Coccoliths localized near the apex (**Fig. 3B**) or polar regions of the cell (e.g., *Acanthoica quattrospina* Lohmann, see p. 188) may be different from the type that covers

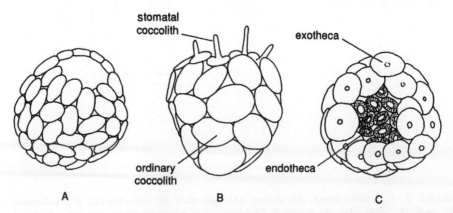

FIGURE 3 Various types of coccospheres: (A) monomorphic; (B) dimorphic; (C) dithecate.

the rest of the cell. In yet other species, coccoliths having a central spine are intermixed with coccoliths without such an extension (e.g., *Rhabdosphaera claviger* Murray & Blackman, see p. 218). All the coccoliths on a single cell may also be identical in morphology even if they are two different sizes, such as the micro- and macrococcoliths of *Umbellosphaera irregularis* Paasche and *U. tenuis* (Kamptner) Paasche (see p. 226).

The varying degree of organic control on crystal growth in coccoliths is the basis for the subdivision into holococcoliths and heterococcoliths (Black, 1963). Holococcoliths (**Fig. 4**) are composed of regularly packed calcite crystals of rhombohedral or hexagonal prismatic shapes. In heterococcoliths (**Figs. 5 and 6**), the crystals are also rhombohedral but the shape may be modified so that their crystal faces and angles become partly or wholly suppressed. The genus *Turrisphaera* exhibits an unusual type of calcification in which small hexagonal crystallites are reticularly arranged within an organic matrix (Manton et al., 1976b).

The majority of coccoliths presently described are of the heterococcolith type, which are more robust than the smaller, loosely packed crystalline holococcoliths. Therefore, it is expected that the delicate holococcoliths are rare in bottom sediments, resulting in an underestimate of abundance from the fossil record. In spite of this, a few holococcoliths are known from sediments as old as the Early Cretaceous (Tappan, 1980). In the Indian Ocean, Red Sea, Mediterranean Sea, and North Atlantic Ocean, Kleijne (1991) recently showed that holococcolithophorids can form a significant part of the summer flora.

Observations of *E. huxleyi* from different water masses suggest that temperature is an important factor controlling calcification (McIntyre et al., 1970; Burns, 1977), with cold water species being the most heavily calcified. This view was not confirmed from culture experiments using *E. huxleyi*, which indicated that temperature alone could not explain the natural variability (Watabe & Wilbur, 1966) and that the amount of calcium carbonate deposited did not appreciably vary with temperature (Paasche, 1968). Investigations of arctic and subarctic nanoplankton (Thomsen, 1980a) showed the frequent occurrence of coccolithophorids with a normal morphology but having uncalcified coccoliths. Since the physiological mechanism(s) that determine the mineralization process in coccolithophorids is not well understood, it is difficult to evaluate whether the arctic observations are due to environmental conditions or are an artifact.

Coccoliths normally consist of an organic base plate scale with a calcified rim. The coccolith-forming stage of *E. huxleyi* is atypical in that it lacks both the base plate scale in the coccoliths and the uncalcified organic body scales (Klaveness, 1972b). Its motile stage has uncalcified scales only. Recent observations of coccolith production by *E. huxleyi* indicated, however, that there is a thin organic plate between the calcite rhombohedra at the onset of calcification (Westbroek et al., 1989). This plate can be observed in the coccolith vesicle

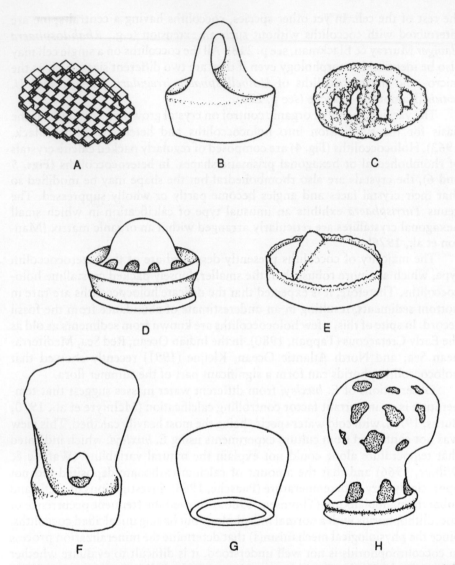

FIGURE 4 Various types of holococcoliths (not drawn to scale): (A) crystallolith; (B) zygolith with a single bridge; (C) zygolith with three parallel bridges, distal surface view; (D) areolith; (E) laminolith with a distal rim and a transverse ridge on the distal surface; (F) helladolith with a pored distal petaloid extension; (G) calyptrolith; (H) gliscolith. Part A redrawn from Gaarder & Markali (1956); B, D, F, H redrawn from Norris (1985).

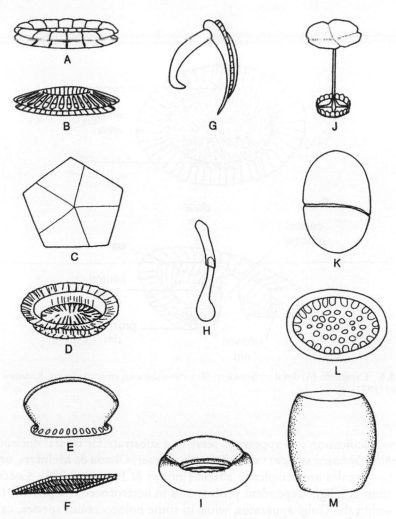

FIGURE 5 Various types of heterococcoliths (not drawn to scale): (A) cricolith; (B) placolith; (C) pentalith; (D) caneolith; (E) cyrtolith; (F) scapholith; (G) ceratolith; (H) two links of an arm of osteoliths; (I) modified caneolith; (J) pappolith; (K) lepidolith; (L) cribrilith; (M) lopadolith. Parts A, B redrawn from Braarud et al. (1952); D redrawn from Halldal & Markali (1954b); E redrawn from Halldal & Markali (1954a); F redrawn from Reinhardt (1972); G redrawn from Farinacci (1971); J redrawn from Tangen (1972).

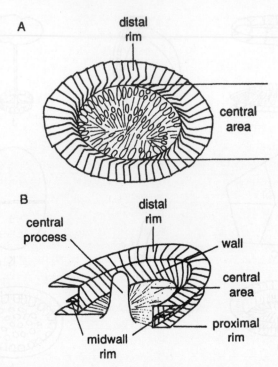

FIGURE 6   Caneolith: (A) distal surface view; (B) section showing structural units. Redrawn from Farinacci (1971).

prior to calcification and appears to serve as a substrate for crystal nucleation. In *Umbilicosphaera sibogae* var. *foliosa* (Kamptner) Okada & McIntyre, uncalcified body scales are completely absent (Inouye & Pienaar, 1984). Coccolith production is a light-dependent process and in heterococcolithophorids takes place within the Golgi apparatus, while in some holococcolith species, calcification occurs externally (Rowson et al., 1986).

Reproduction and Life Cycles

Coccolithophorids multiply vegetatively by binary fission. In *Pleurochrysis carterae* (Braarud & Fagerland) Christensen, the life cycle involves alternation of a benthic filamentous phase with either scale-bearing swarmers or flagellated cells bearing coccoliths. Ploidy remains somewhat uncertain during the various stages, although the work of Rayns (1962) indicated that the coccolith-bearing form is diploid and the benthic form is haploid. The life history study of *E. huxleyi* conducted by Klaveness (1972b) reported naked, scale-bearing, and coccolith-bearing cells, each capable of indefinite vegetative reproduction. In

Cruciplacolithus neohelis nonmotile coccolith-bearing cells under some culture conditions produce biflagellate cells bearing the same type of coccoliths as nonmotile cells (Fresnel, 1989). The position of these biflagellated cells in the life history of *C. neohelis* is unknown. Some species may have hetero- or holococcoliths in different stages of their life cycle, and this has caused some confusion. In both *Calcidiscus leptoporus* (Murray & Blackman) Loeblich & Tappan and *Coccolithus pelagicus* (Wallich) Schiller, nonmotile cells have heterococcoliths while their motile holococcolith-bearing stages were originally described as separate species of the genus *Crystallolithus* (see remarks on pp. 175 and 194; also Parke & Adams, 1960; Rowson et al., 1986; Kleijne, 1991). Each cell type can reproduce by binary fission when kept in culture. Examination of a number of cells from the heterococcolithophorid genus *Papposphaera* Tangen and the holococcolithophorid genus *Turrisphaera* Manton, Sutherland, & Oates, led to the conclusion that these coccolithophorids have a heteromorphic life history somewhat similar to that of *Coccolithus pelagicus/Crystallolithus hyalinus* Gaarder & Markali (Thomsen et al., 1991). Since the genus *Papposphaera* has priority over *Turrisphaera*, all species of the latter genus were made synonyms of *Papposphaera* species. Similar relationships were also indicated for *Pappomonas* Manton & Oates/*Trigonaspis* Thomsen and *Calciarcus* Manton, Sutherland, & Oates/*Wigwamma* Manton, Sutherland, & Oates (Thomsen et al., 1991). Since the identities of these last combinations were somewhat uncertain, their taxonomic revisions must await further analysis with the scanning electron microscope and possibly further culture studies. Under favorable conditions, coccolith-covered cells of some species can produce calcareous cysts (Kamptner, 1937a).

History of Coccolithophorids

The first evidence for the presence of coccolithophorids came from isolated fossil coccoliths observed by Ehrenberg (1836) in Cretaceous rocks that were thought to be of inorganic origin. Later, the term coccolith was proposed by Huxley (1858) for the calcareous remains he found in ocean sediments, and Wallich (1861) used the term coccosphere for the coccolith-covered globules he observed in sediment samples. Their algal nature, however, was not demonstrated until 1898, when Murray & Blackman noted that coccospheres from the North Atlantic contained yellow-green plastids. These authors were unaware of the earlier description of the fresh-water coccolithophorid *Hymenomonas roseola* by von Stein (1878) and the nature of its wall structure, which Lohmann (1902) documented as consisting of small acid-soluble plates that formed the cell wall of this species. It was Conrad (1914) who realized that these were the same as those plates found in fossil sediments called coccoliths.

At present, 200 living species of coccolithophorids belonging to nearly 70 genera have been described. These species can occur in great abundances or "blooms," which cause discoloration of our modern seas, while in past geologic time they played an important role in the formation of calcareous deposits. The most abundant living species, *Emiliania huxleyi*, was reported at 115×10^6 cells per liter on the southwestern Norwegian coast where the sea had a milky green color (Berge, 1962). Holligan et al. (1983) found densities up to 8.5×10^6 cells per liter of this species at the edge of the continental shelf between France and Ireland. The water in that region was noted as having a pale turquoise color. At one station, integrated values for the upper 60 m had $>3 \times 10^{11}$ cells of *E. huxleyi* and 40 g calcium carbonate per square meter. For a bloom covering 7200 km^2, as seen with remote sensing, they estimated that there were 7.2×10^4 tons of calcite in the upper 60 m of water. The total annual calcite production for the northwest European shelf edge by coccolithophorids was estimated to be at least an order of magnitude greater (Holligan et al., 1983). *Coccolithus pelagicus* also was noted to be important for the extraction of carbon dioxide in the open ocean, especially at high latitudes (Knappertsbusch, 1991), with estimates of the coccolith carbonate production from $0.7 \text{ g} \cdot \text{m}^{-2} \cdot \text{yr}^{-1}$ at 40.6°N to $33.1 \text{ g} \cdot \text{m}^{-2} \cdot \text{yr}^{-1}$ at 53°N. Recent coccolithophorid populations, therefore, are likely to be major contributors to transfer of carbon between the atmosphere and the ocean-bottom sediments. Extensive calcium carbonate oozes on the ocean floor and massive limestone deposits containing coccoliths as their principal components from the Mesozoic and Cenozoic suggest that they were even more important in the past.

Occurrence and Distribution

Most living coccolithophorids inhabit tropical or subtropical offshore waters. Although predominantly oceanic in distribution, a few species are neritic, and some are found indifferently in both regions (Smayda, 1958). The most abundant living species, *E. huxleyi*, is normally considered to be an oceanic species but may also be very important in neritic areas (Birkenes & Braarud, 1952), consistent with its euryhaline and eurythermal nature. Representatives of the genera *Cricosophaera*, *Hymenomonas*, *Ochrosphaera*, *Pleurochrysis*, and *Cruciplacolithus* consistently occur in near shore waters. *Hymenomonas roseola* is the only representative of the coccolithophorids common in fresh water (Schiller, 1930), and *Balaniger balticus* Thomsen & Oates may be among the very limited number of coccolithophorids that prefer brackish water (Thomsen & Oates, 1978). *Emiliania huxleyi* and *Coccolithus pelagicus* were also reported from polar waters where minimum temperatures fall below freezing (Braarud, 1979; Heimdal, 1983). Detailed investigations on nanoplankton from arctic and antarctic areas revealed that identical cold-water-adapted coccolithophorids occurred at both poles (Manton & Oates, 1975; Manton &

Sutherland, 1975; Manton et al., 1976a, 1976b, 1977, Thomsen, 1980a, 1980b, 1980c, 1980d; Thomsen et al., 1988). This, in addition to the observations of typically "arctic" species in the antarctic Weddell Sea, suggests that, in geological terms, exchange of biological material between poles has taken place relatively recently.

When the vertical distribution is considered, data from the Pacific (Okada & Honjo, 1973; Reid, 1980) show that some species were confined to the upper layers and some to the lower layers of the water column. Between these layers was an admixture of species from above and below. According to Okada & Honjo (1973), *Florisphaera profunda* Okada & Honjo and *Thorosphaera flabellata* Halldal & Markali were strictly confined to the deep photic zone in the same way as Jordan et al. (1991) described for *Turrilithus latericioides* gen. et sp. nov. from the North Atlantic and the Mediterranean.

Literature

Historically, relative to the other phytoplankton groups, especially diatoms and armored dinoflagellates, the study of extant coccolithophorids has received relatively little attention. In recent years, however, greater use of electron microscopy has resulted in significant advances in the understanding of the morphology of this group. In addition, an increasing awareness of the importance of coccolithophorids in sedimentation of biogenic carbon and lipids, in paleoclimate studies, and added interest in physiology and ecology of these species have all contributed to a proliferation of new literature. A number of these papers on living species deserve mention: Holococcolithophorids from the Indian Ocean, Red Sea, Mediterranean Sea, and North Atlantic Ocean (Kleijne, 1991); *Atlas du phytoplankton marin. Prymnésiophycées* (Chrétiennot-Dinet, 1990); McIntyre & Bé (1967a), Okada & McIntyre (1977), Heimdal & Gaarder (1980, 1981), and Mostajo (1985) for the Atlantic; Norris (1984, 1985) and Friedinger & Winter (1987) for the Indian Ocean; Okada & Honjo (1970, 1973, 1975), Okada & McIntyre (1977), Nishida (1979), and Reid (1980) for the Pacific Ocean; Okada & Honjo (1975), Winter et al. (1979), and Okada (1992) for the marginal seas along the western Pacific Ocean and the Red Sea; Borsetti & Cati (1972, 1976, 1979) and Rampi & Bernhard (1981) for the Mediterranean; Gaarder & Hasle (1971) for the Gulf of Mexico; Throndsen (1972) for the Caribbean Sea; Hallegraeff (1984) for Australian waters; and Manton et al. (1976b, 1977), Thomsen (1981), and Thomsen et al. (1988) for the Arctic and Antarctic. Because of the scattered nature of the literature and the fact that many of the publications only describe a few taxa or deal only with a specific geographical area, nonspecialists are advised to start with introductory coccolithophorid books such as Schiller (1930) or Kamptner (1941).

GENERAL CHARACTERISTICS

Form—The majority of species are biflagellate or coccoid unicells.

Size—Most with a diameter <30 μm, many <10 μm.

Cell covering—Calcified (calcium carbonate, calcite) scales, the *coccoliths*, lying over uncalcified scales formed in the Golgi apparatus. *Heterococcoliths* are composed of assemblages of morphologically complex and diverse calcium carbonate elements. *Holococcoliths* are made up of one type of calcite (often rhombohedral) crystals. In species of heterococcolithophorids the calcium carbonate is deposited internally during coccolith formation. Holococcolith calcification takes place outside the plasmalemma. The cell wall of benthic stages where this is known (e.g., *Pleurochrysis carterae*) consists of organic scales, but they do not produce coccoliths.

Flagella—Motile forms have two smooth flagella, anteriorly inserted, equal or subequal in length, with the usual 9 + 2 pattern of axoneme microtubular organization.

Haptonema—This structure is present in some motile forms. In some coccolithophorids the only evidence for the haptonema is the presence of the basal structures within the cell, the emergent portion being absent.

Internal structures—One or two chloroplasts are present, each having a single pyrenoid enclosed in an envelope consisting of four membranes. The chloroplast lamellae are composed of three thylakoids with no girdle lamella. Each cell has several mitochondria with tubular cristae. The Golgi complex occupies a characteristic position between the nucleus and the kinetids. The plastid and pyrenoid are surrounded by endoplasmic reticulum confluent with the nuclear envelope. The nucleus itself is usually located close to the plastids.

Pigments—Chlorophyll *a* and *c* together with a number of carotenoids are the major pigments present. In *Emiliania huxleyi*, fucoxanthin is replaced by the fucoxanthin derivate 19′-hexanoyloxyfucoxanthin (Hibberd, 1980), which was first isolated from this species (Arpin et al., 1976).

Storage products—The major deposits consist of a water-soluble carbohydrate similar or identical to chrysolaminarin (Christensen, 1980), and lipids.

Reproduction—Vegetative reproduction occurs by binary fission, mostly longitudinal, or by repeated divisions of the mother cell within the crust of coccoliths and release of motile or nonmotile daughter cells. Three planktonic species, *Calcidiscus leptoporus*, *Coccolithus pelagicus*, and *Emiliania huxleyi*, are shown to have a heteromorphic life history involving a nonmotile heterococcolith-bearing form as well as motile flagellated phases. The motile forms of *Calcidiscus leptoporus* and *Coccolithus pelagicus* have holococcoliths (Parke & Adams, 1960; Rowson et al., 1986; Kleijne, 1991) and were originally described as species of the genus *Crystallolithus* (pp. 175, 194), while the flagellate stage of *E. huxleyi* has unmineralized scales only (Klaveness, 1972b). Similarly, *Calciarcus*, *Pappomonas*, and *Papposphaera* species were shown to have a heteromorphic life history involving both flagellate hetero- and holococcolithophorid stages (Thomsen et al., 1991). Nonmotile cells of *Cruciplacolithus neohelis* under some conditions in culture produced coccolith-bearing flagellates with the same type of coccoliths as the nonmotile cells (Fresnel, 1989). The position of these cells in the life history is unknown. The life cycle of *Pleurochrysis carterae* involves alternation between a benthic filamentous phase with either scale-bearing swarmers or flagellated coccolith-bearing cells.

Resting spores—Observed by Kamptner (1937a).

Distribution—Most modern coccolithophorids live in tropical or subtropical offshore waters. The work of Manton et al. (1976b, 1977) and Thomsen (1980a) revealed species living in very cold waters as well. The most abundant living species, the cosmopolitan *E. huxleyi,* is normally oceanic, but may also be neritic in its distribution (Birkenes & Braarud, 1952). It has a particularly wide ecological range, occurring in waters either elevated or reduced in salinity and temperature. *Hymenomonas roseola* is the only representative of the coccolithophorids common in fresh water (Schiller, 1930), and *Balaniger balticus* Thomsen & Oates is among the very limited number of coccolithophorids that prefer brackish water (Thomsen & Oates, 1978). Representatives of the genera *Cricosphaera*, *Hymenomonas*, *Ochrosphaera*, *Pleurochrysis*, and *Cruciplacolithus* consistently occur in near shore waters.

Ecological Importance—Coccolithophorids are often more abundant than other phytoplankton groups in warm, oligotrophic waters and may occasionally bloom in boreal waters (Raymont, 1980). Data from remote sensing revealed that such blooms cover many thousands of square kilometers, but few attempts have been made to investigate their dynamics or determine their importance with respect to carbon removal from surface waters and their influence on the air–sea exchange of carbon dioxide. So far, our knowledge of coccolithophorid blooms is largely based on blooms of *E. huxleyi* in the North Atlantic. Satellite studies indicated that they occur most frequently, both in oceanic and offshore coastal waters, at mid latitudes (45–65°N) during spring and summer seasons (Holligan & Groom, 1986). In Norwegian coastal waters, *E. huxleyi* blooms causing water discoloration were reported with some regularity (Berge, 1962). During the early summer of 1978, surface waters on Rockall Bank, 350 km west of Scotland, had exceptionally high populations of *Coccolithus pelagicus* (in excess of 10^7 cells $\cdot$ L^{-1}), reflecting the effects of upwelling on the shallower portion of the bank (Milliman, 1980). At low latitudes, elevated coccolithophorid populations occurred in upwelling areas (Dupouy & Demarcq, 1987; Mitchell-Innes & Winter, 1987; Kleijne, 1990), where they appeared to prefer warmer, "mature" water, while diatoms dominate in "young" upwelled water. In South African waters the only reported occurrence of a coccolithophorid bloom causing water discoloration was due to *Gephyrocapsa oceanica* Kamptner (Grindley & Taylor, 1970). This event took place after a diatom bloom in mature upwelled water during the austral summer when water temperatures were within the range of 14.8 to 17.4°C. *Gephyrocapsa oceanica* disappeared when the temperature increased.

A few species of coccolithophorids were reported living attached to centric diatoms (Lohmann, 1912; Gaarder & Hasle, 1962; Okada & McIntyre, 1977; Hallegraeff, 1984) in a possible symbiotic relationship. Gaarder & Hasle (1962, p. 148), analyzing samples from the equatorial Pacific Ocean, interpreted the attachment of coccolithophorid cells and single coccoliths on diatoms as a mere agglutination, while Okada & McIntyre (1977), describing another type of coccolithophorid/diatom association from the Pacific, indicated an actual symbiosis between the coccolithophorid *Crenalithus sessilis* (Lohmann) Okada & McIntyre and the centric diatom *Thalassiosira* sp.

Geological Age—Coccoliths and fossilized remnants of other calcareous organisms are generally abundant in rocks of Jurassic to the present (Tappan, 1980). Rare occurrences were noted for the Paleozoic and Triassic, but they increased rapidly in both diversity and abundance in the Jurassic and Cretaceous, when their remains accumulated as extensive coccolith limestone and chalks.

TERMINOLOGY AND MORPHOLOGY

The terminology used in this review is mainly based upon the publications of Braarud et al. (1955), Halldal & Markali (1955), Hay et al. (1966), McIntyre & Bé (1967a), Farinacci (1971), and Okada & McIntyre (1977). As the understanding of coccoliths has changed during the past two decades, the terminology has also changed, and the readers should be aware of a new guide to coccolith terminology presently in course of preparation (Young, 1991). In the text, terms that occur in the most frequently used books and also those more recently defined terms based on electron microscope observations are used. Because many of the terms come from Latin or Greek, a few of the more common terms and their English equivalents are also listed in the glossary (p. 248). Derivations of generic names are considered outside the scope of this work and are not discussed here.

Abbreviations: a., adjective; e.g., for example (Latin *exempli gratia*); Gr., Greek; L., Latin.

Antapical (a)—(Gr. *anti*, opposite; L. *apex*, summit) pertaining to region opposite apex.

Appendix—long distal growth from the basal plate of the coccolith.

Apex—(L. *apex*, summit) the terminal point or end of a cell or organism; the end in the direction of movement of a flagellate.

Apical (a.)—at a tip or summit.

Aragonite—mineral, made of calcium carbonate, like calcite, but differing from calcite in having orthorhombic crystals, less distinct cleavage and greater density. See Calcite.

Areolith—dome-shaped holococcolith with an aerolate interior comprised of thickened ridges of calcite elements (**Fig. 4D**).

Bar—bridge, e.g. in *Coccolithus pelagicus* (see **Plate 4**) and *Gephyrocapsa oceanica* (see **Plate 5**).

Base plate scale—organic basal plate of a coccolith upon which the calcified elements are deposited.

Calcareous (a.)—containing calcium, usually in the form of calcium carbonate.

Calcite—mineral, like aragonite, composed of calcium carbonate, but with hexagonal crystallization. See Aragonite.

Calicalith—(L. *calix*, chalice) tube-like coccolith with an open distal end.

Calyptrolith—(Gr. *kalyptra*, covering) calotte (cap)-shaped holococcolith having the form of an open cap or basket, e.g., coccoliths of *Calyptrosphaera oblonga* (**Fig. 4G**). Calyptroliths consist mostly of a single layer of crystallites, except for rims, ridges, or knobs of additional layers as in the genera *Calyptrolithina* and *Calyptrolithophora* (see **Plate 2**) which bear microcrystals in groups of various shapes superposed on its distal surface.

Caneolith—elliptical discoid heterococcolith, in a basket-like arrangement with a central area or bottom of lamellae, and a simple or complex girdle bordered above and below petaloid elements (**Figs. 5D and 6**).

Caudate (a.)—(L. *cauda*, tail) having a tail (**Fig. 2K**).

Centripetal (a.)—(L. *centrum*, center; *petere*, to seek) developing from the outside toward the center.

Ceratolith—horseshoe-shaped coccolith (**Fig. 5G**).

Clavate (a.)—(L. *clava*, club) club-shaped, thickened at one end (**Fig. 2I**).

Clonal culture—a population descended from one individual.

Clone—a group of individuals propagated by mitosis from a single ancestor.

Coccoid (a.)—(Gr. *kokkos*, berry; *eidos*, form) spherical or globose in form.

Coccolith—(Gr. *kokkos*, berry; *lithos*, stone) minute scale on cell surface of some prymnesiophytes (coccolithophorids); encrusted with calcium carbonate, usually in the crystalline form of calcite; often abundant as fossil remains of coccolithophorids in chalk.

Coccolith case—a cell covering of coccoliths in which the coccoliths hold together to form an intact shell of scales.

Coccolithophorid—a unicellular organism classified within the Prymnesiophyceae, which bears variously scultpured calcareous scales (coccoliths) upon its cell surface.

Coccosphere—a cell covering of coccoliths, which may not necessarily be spherical.

Corona—(L. *corona*, crown) crown or crown-shaped structure.

Cribrate (a.)—(L. *cribrum*, sieve) sieve-like, perforated.

Cribrilith—(L. *cribrum*, sieve) strainer-shaped discolith with a perforated central area and a rim composed of lamellae (**Fig. 5L**) *Pontosphaera discopora* (see **Plate 6**).

Cricolith—(Gr. *krikos*, ring) elliptical heterococcolith with elements arranged in a simple ring on a base plate scale (**Fig. 5A**).

Crystallolith—disciform holococcolith, consisting of regular rhombohedrons of calcite deposited on the distal surface of an organic scale (**Fig. 4A**).

Cuneate (a.)—(L. *cuneatus*, wedge-shaped) triangular with the acute angle toward the base (**Fig. 2L**).

Cyrtolith—(Gr. *kurtos*, curved) heterococcolith with a convex disc-like shape; cyrtoliths may have various types of centers that protrude distally from the cell. Calyptroform cyrtolith: Central area of coccolith having a cap or sacculiform shape (**Fig. 5E**), e.g., *Algirosphaera* spp. (see **Plate 3**). Salpingiform cyrtolith, with trumpetlike central process, e.g., *Discosphaera tubifer* (see **Plate 5**). Styliform cyrtolith, with styliform central process, e.g., polar coccoliths of *Acanthoica quattrospina* (see **Plate 3**) and intermixed with coccoliths lacking processes in *Rhabdosphaera claviger* (see **Plate 6**).

Dimorphism (a. dimorphic)—coccolith case bearing two types of coccoliths (**Fig. 3B**).

Discoid—(Gr. *diskos*, disc; *eidos*, form) flat and circular, disc-shaped.

Discolith—(Gr. *diskos*, disc) an ellipsoidal coccolith, usually with a raised margin and a central pattern that is different in its arrangement of elements from the margin, opening distally; in some species the margin is very high, forming a vase- or barrel-shaped coccolith, e.g., *Scyphosphaera apsteinii* (see **Plate 6**).

Distal—side away from cell.

Dithecate (a.)—double-layered coccolith case (**Fig. 3C**); distal layer monomorphic, proximal layer dimorphic; coccoliths of both layers different.

Element—any of the basic parts of which a coccolith is constructed, discernable by the bounding suture lines.

Endotheca—the proximal layer of coccoliths.

Endothecal (a.)—from the proximal layer.

Envelope—a continuous outer case ("skin" in Manton & Leedale, 1963), covering the scales and coccoliths in, e.g., *Calyptrosphaera sphaeroidea* and *Crystallolithus hyalinus*.

Exotheca—the distal layer of coccoliths.

Exothecal (a.)—from the distal layer.

Flosculolith—(L. *flosculus*, little flower) holococcolith consisting of a distally widening tube with a vaulted roof that partially closes the distal tube opening.

Fragariolith—stomatal coccolith of *Anthosphaera* species with a large distal, single-layered, leaf-like process.

Fusiform (a.)—(L. *fusus*, spindle; *forma*, shaped) tapering gradually at both ends (**Fig. 2H**).

Gliscolith—holococcolith with a circular to elliptical basal ring, a distal neck of the same diameter as the ring, and a distal inflated, globular section (**Fig. 4H**); internally the gliscolith is hollow, the inflated distal part of the coccolith, and sometimes the stalk, being penetrated by pores.

Helicoid placolith/helicolith—lopodolith. Placolith with helical shape, e.g., *Helicosphaera* spp. (see **Plate 5**). See Placolith.

Helladolith—holococcolith, with a double-layered leaf-like appendix (**Fig. 4F**).

Heterococcolith—heterococcoliths are built of elements of different forms; the crystals are basically rhombohedral, but continue to grow so that their characteric crystal face becomes somewhat modified. See Holococcolith.

Holococcolith—coccolith formed solely of calcite microcrystals held together by an organic matrix; the calcium carbonate is deposited as uniform rhombohedral or hexagonal crystals showing little modification. See Heterococcolith.

Imbrication pattern—overlapping pattern.

Imbrication, dextral—in distal view each element overlaps the one to the right when viewed from the center of the coccolith.

Imbrication, sinistral—in distal view each element overlaps the one to the left when viewed from the center of the coccolith.

Inclination, clockwise—suture inclined to the right. See Suture line.

Inclination, counterclockwise—suture inclined to the left. See Suture line.

Laminolith—(L. *lamina,* plate) laminated disc-shaped holococcolith, with or without perforations (**Fig. 4E**).

Lanceolate (a.)—(L. *lanceola,* little lance) lance-shaped.

Lepidolith—thin ellipsoidal disc-shaped coccolith (**Fig. 5K**), e.g., the surface coccoliths of *Thorosphaera flabellata* (**Plate 7**).

Lopadolith—barrel-shaped coccolith, opening distally (**Fig. 5M**).

Lopodolith—see Helicoid placolith.

Macrococcolith—coccolith with an almost circular proximal disc and a funnel-shaped upper part, e.g., *Umbellosphaera irregularis* and *U. tenuis* (see **Plate 7**).

Microcococcolith—coccolith with an oval proximal disc, upper part as a very short tube with an upper brim not necessarily smaller than macrococcolith, e.g., *Umbellosphaera irregularis* and *U. tenuis* (see **Plate 7**).

Monomorphic (a.)—coccolith case bearing one type of coccoliths (**Fig. 3A**).

Monospecific (a.)—belonging to a single species; a monospecific bloom consists of a single species.

Ordinary coccolith—see **Fig. 3B** and Stomatal coccolith.

Osteolith—(Gr. *osteon,* bone) femur-shaped coccolith (**Fig. 5H**), built up of lamellae, e.g., the appendage coccoliths of *Ophiaster hydroideus* (see **Plate 6**).

Pappolith—(L. *pappus,* pappus) heterococcolith having upright marginal elements on basal disc and in which the central area is partially covered by a single layer of elements arranged in rows or bars. A spine-like process may emerge from the central area. Styliform pappoliths, with styliform central process having a distal end of different shape. See diagrams of coccoliths with appendages in *Papposphaera lepida* (**Fig. 5J**) and two *Pappomonas* species (Manton & Sutherland 1975, Figs. 11 and 12).

Pentalith—coccolith of five calcite units arranged radially in a single plane (**Fig. 5C**).

Placolith—heterococcolith composed of an upper and a lower shield of radial elements interconnected by a tube (**Fig. 5B**).

Polar coccolith—coccolith type occurring only in the polar regions.

Polarizing microscope—a microscope using polarized light for examining the crystalline structure of specimens held in between polarizing and analyzer lenses. Mineral deposits are detected by alterations in the path of polarized light.

Polymorphism (a. polymorphic)—coccolith case bearing more than two types of coccoliths.

Proximal—side facing the cell.

Pyriform (a.)—(L. *pyrum*, pear; *forma*, shape) pear-shaped (**Fig. 2G**).

Rhabdolith—(Gr. *rhabdos*, rod) heterococcolith having a basal, circular to ellipsoidal disc and an elevated central region extending distally from the coccosphere. See Styliform cyrtolith.

Rhombic scale—scale that takes the form of a parallelogram.

Scale—organic or mineralized structure of specific architecture deposited on the cell surface.

Scapholith—diamond-shaped heterococcolith with a central area of transverse lamellae, which usually mesh in the center (**Fig. 5F**).

Stomatal coccolith—coccolith belonging to the stomatal (flagellar) area (**Fig. 3B**). See Ordinary coccolith.

Suture—general morphological term referring to a seam or a furrow between adjacent parts.

Suture line—line of adhesion between the elements in, e.g., the shields of a placolith. See Inclination.

Tremalith—a calcified or potentially calcified scale rim with more or less vertical sides.

Vaterite—crystalline form of calcium carbonate.

Wall—that portion of a caneolith that forms a girdle (**Fig. 6B**).

Whorl—a crown-like appendage of arms made of modified coccoliths, e.g., *Calciopappus caudatus* (see **Plate 3**), *Halopappus adriaticus* and *Michaelsarsia elegans* (see **Plate 5**), and *Ophiaster hydroideus* (see **Plate 6**).

Zygolith—zygoform holococcolith having one or several bridges across the central tube (**Fig. 4, B and C**).

PROBLEMS IN STUDYING RECENT COCCOLITHOPHORIDS

The methods developed for the collection, preservation, and identification of coccolithophorids were discussed in the "Phytoplankton Manual" by Sournia (1978), to which references are made for details. Identification problems are in most cases related to the small size of the cells and their individual coccoliths. Electron microscopic (EM) examination is thus essential for elucidating the more detailed analysis of the coccolith microstructure. During recent decades, the study of coccolithophorids was also conducted with the aid of scanning electron microscopy (SEM). In fact, most coccolithophorids can only be satisfactorily described through a combined use of conventional light microscopy (LM), SEM, and transmission electron microscopy (TEM). In order to link the LM and EM observations used in defining a species and thus avoid taxonomic uncertainties, it is recommended that the same specimen be studied under both the light and the electron microscopes whenever possible. A good quality light microscope with an oil immersion objective of 100× as well as a polarizing microscope, for analyzing coccoliths, should be used. For observation of general cell morphology, various types of shallow-focus light microscopes, having bright-field , interference, and phase-contrast microscopy, are quite valuable. Good results can also be obtained by unconventional use of regular bright-field systems, which can improve contrast.·

Identification and preservation procedures will differ according to the ultimate goal of the study. Taxonomic and ecological studies can be made after preservation and rinsing of samples in buffered distilled water, which prevents rapid dissolution of the coccoliths in aqueous solutions. In vivo observations, as well as ultrastructural and physiological studies, must be conducted on fresh material. Such studies may be done with natural samples or laboratory cultures. So far, however, only a few species of coccolithophorids have been cultured and maintained for any length of time. With media specifically designed for coccolithophorids, this may be less of a problem. Methods for sampling and isolation of coccolithophorids are generally the same as for other microalgae (see Sournia, 1978). For the majority of species in nature, single-cell isolation is probably the best technique, especially as most coccolithophorids are usually found in small numbers. The micropipette method may, however, be very difficult with such small cells. Many coccolithophorids will not tolerate excessive manipulation experienced during the cleaning process. In these cases, it may be necessary to transfer them directly into a small volume of the growth medium or use an indirect isolation method such as Throndsen's (1978) serial dilution technique. The indirect methods have the disadvantage, however, that one cannot always relate the morphology of cells grown in culture to that of the original cell isolated from the natural sample.

CLASSIFICATION

Living coccolithophorids are mainly classified on the basis of morphology of the cells and of their coccoliths (Schiller, 1930; Kamptner, 1941; Deflandre, 1952; Parke & Green, 1976; Christensen, 1980; Norris, 1984, 1985; Young, 1987; Chrétiennot-Dinet, 1990; Jordan & Young, 1990; Kleijne & Jordan, 1990; Kleijne, 1991). With increasing knowledge of life cycles and cellular ultrastructure, this is likely to change (Parke & Adams, 1960; Hibberd, 1976,

1980; Inouye & Pienaar, 1984, 1988; Rowson et al., 1986). So far, only a small fraction of the known species have been studied extensively in culture. At present, our knowledge of the critical characteristics is very fragmentary and is too limited to provide a base for another classification system.

The taxonomic system for the Prymnesiophyceae most commonly followed at present is that of Parke & Green (1976), which includes four orders within the class. In their scheme, the coccolithophorids are distributed between the Isochrysidales and the Coccosphaerales, the latter order being a heterogeneous assemblage that will have to be modified as the structure and the life histories of the different species are better understood. Recent work on members of the order Isochrysidales, including organisms that differ morphologically and structurally in a number of important ways, also indicates that this order is not a natural taxonomic unit. The distribution of hydrocarbons, fatty acids, and carotenoids within these species can provide additional information of taxonomic value to help clarify relationships (Marlowe et al., 1984; Jordan, 1991).

Tappan (1980), who extended the classification scheme of Parke & Green (1976) to accommodate the fossil forms, separated the coccolithophorids into six orders, while Christensen (1980), Fresnel (1989), and Chrétiennot-Dinet (1990) have chosen to include all of them in a single order. This may lead to different views on the systematic position and relationships of some species. For example, the genus *Hymenomonas* placed in the order Isochrysidales by Tappan (1980), based on the shape of its coccoliths and the elements composing them, is included in the order Coccolithophorales Schiller (1926) by Fresnel (1989) and Chrétiennot-Dinet (1990). Thus, care must be exercised when referring to the different taxonomic schemes.

The sole reliance on the morphology of coccoliths and cells as the major criteria for the present distribution into orders, families, and genera, has limitations rendering such schemes as *very preliminary*. For example, in accordance with Tappan (1980) and others, all genera and species of holococcolithophorids are included in one single family, the Calyptrosphaeraceae (Boudreaux & Hay, 1969). This is so even if some of its members could be transferred to other families characterized by heterococcoliths. Presently, *Calciarcus, Calcidiscus, Coccolithus, Pappomonas,* and *Papposphaera* species are reported to show such an alteration between holococcolith and heterococcolith stages in their life cycle (Rowson et al., 1986; Kleijne, 1991; Thomsen et al., 1991). Should future studies of other species currently not cultured find similar results, the strict separation based on coccolith type will have to be reevaluated. It should be remembered, however, that coccolith morphology will still remain an important identification character for the different stages in the life cycle of species.

In the absence of a comprehensive revision, the following outline for the classification and arrangement of genera makes use of information from Parke & Green (1976), Tappan (1980), Norris (1983, 1984, 1985), and Fresnel

(1989). An attempt has been made to assign some of the newly described genera having lightly calcified organic scales, (i.e., *Balaniger, Quaternariella, Trigonaspis,* and *Turrisphaera*) to positions in this scheme, even if it is possible they may not be closely related to each other or to the other types of holococcolithophorids. They perhaps should not be placed in the Calyptrosphaeraceae, but in one or several independent families. At present there is, however, insufficient data to compare these genera in detail, and any taxonomic conclusion must be postponed until more material has been studied. With the exception of the family Calyptrosphaeraceae (holococcolithophorids), the families, genera, and species are listed in alphabetical order.

OUTLINE FOR CLASSIFICATION AND ARRANGEMENT OF GENERA

I. Order Isochrysidales Pascher 1910

　　Motile cells with two acronematic flagella of equal or subequal length, haptonema rudimentary or absent.

　　A. Family Gephyrocapsaceae Black 1971

　　　　Naked prymnesiophytes, with organic scales or placoliths, motile cells lacking a haptonema.

　　　　Crenalithus Roth 1973

　　　　Emiliania Hay & Mohler 1967 (in Hay et al., 1967)

　　　　Gephyrocapsa Kamptner 1943

　　B. Family Hymenomonadaceae Senn 1900

　　　　Motile cells with organic scales and cricolith cover and visible haptonema; may have benthic filamentous stage alternating with motile one.

　　　　Cricosphaera Braarud 1960

　　　　Hymenomonas von Stein 1878, emend. Gayral & Fresnel 1979

　　　　Pleurochrysis Pringsheim 1955, emend. Gayral & Fresnel 1983

Remarks: The genus *Cricosphaera* has been regarded as a synonym of *Hymenomonas,* but was reinstated as distinct by Gayral & Fresnel (1976). In Fresnel (1989) and Fresnel & Billard (1991) the two genera *Cricosphaera* and *Pleurochrysis* are placed in the family Pleurochrysidaceae Fresnel & Billard.

II. Order Coccosphaerales Haeckel 1894

　　Coccolith-bearing cells, some genera with a polymorphic life cycle, flagella and haptonema recorded in some genera.

　　A. Family Calyptrosphaeraceae Boudreaux & Hay 1969

　　　　Motile cell with holococcoliths; some may be an alternate stage of other taxa.

　　　　Anthosphaera Kamptner 1937b, emend. Kleijne 1991

　　　　Balaniger Thomsen & Oates 1978

　　　　Calyptrolithina Heimdal 1982

　　　　Calyptrolithophora Heimdal 1980 (in Heimdal & Gaarder, 1980)

Calyptrosphaera Lohmann 1902
Corisphaera Kamptner 1936
Crystallolithus Gaarder & Markali 1956, emend. Gaarder 1980 (in Heimdal & Gaarder, 1980)
Daktyletra Gartner 1969 (in Gartner & Bukry, 1969)
Gliscolithus Norris 1985
Helladosphaera Kamptner 1936
Homozygosphaera Deflandre 1952 (in Grassé, 1952)
Periphyllophora Kamptner 1936
Quaternariella Thomsen 1980a
Sphaerocalyptra Deflandre 1952 (in Grassé, 1952)
Syracolithus (Kamptner) Deflandre 1952 (in Grassé, 1952)
Trigonaspis Thomsen 1980b
Turrisphaera Manton, Sutherland, & Oates 1976
Zygosphaera Kamptner 1936, emend. Heimdal 1982

B. Family Braarudosphaeraceae Deflandre 1947
Cell enclosed in a cover of 12 pentaliths fitted together in a regular dodecahedron.
Braarudosphaera Deflandre 1947

C. Family Calciosoleniaceae Kamptner 1937b
Cylindrical or fusiform cells with scapholiths; may have needle-like spines at one or both ends.
Anoplosolenia Deflandre 1952 (in Grassé, 1952)
Calciosolenia Gran 1912 (in Murray & Hjort, 1912)

D. Family Ceratolithaceae Norris 1965
Cells without flagella; coccoliths of two different types, a single ceratolith that engirdles the protoplast and many ring-shaped coccoliths, adhering together to form a sphere that encloses the protoplast and its surrounding horseshoe-shaped coccolith. Several cells, each with a ceratolith, may be present within a single sphere.
Ceratolithus Kamptner 1950

E. Family Coccolithaceae Poche 1913
Spherical cell with placoliths.
Calcidiscus Kamptner 1950
Coccolithus Schwarz 1894
Cruciplacolithus Hay & Mohler 1967 (in Hay et al., 1967)
Oolithotus Reinhardt 1968 (in Cohen & Reinhardt, 1968)
Umbilicosphaera Lohmann 1902

F. Family Deflandriaceae Black 1968, emend. Norris 1983
Coccoliths as pappoliths, which may or may not be styliform.
Pappomonas Manton & Oates 1975
Papposphaera Tangen 1972

Remarks: Jordan & Young (1990) recommended that the genera *Pappomonas* and *Papposphaera* should be transferred to a new family Papposphaeraceae since the Deflandriaceae is extinct and the coccoliths of the family differ in microstructure from that of the Papposphaeraceae.

G. Family Halopappaceae Kamptner 1928
    Ordinary coccoliths as caneoliths, variously modified coccoliths around apical opening.
    Calciopappus Gaarder & Ramsfjell 1954, emend. Manton & Oates 1983
    Halopappus Lohmann 1912
    Michaelsarsia Gran 1912, emend. Manton, Bremer, & Oates 1984
    Ophiaster Gran 1912, emend. Manton & Oates 1983

H. Family Helicosphaeraceae Black 1971
    Ellipsoidal cell with apical opening left by the cover of helicoform placoliths spirally arranged.
    Helicosphaera Kamptner 1954

I. Family Pontosphaeraceae Lemmermann 1908
    Coccoliths as cribriliths.
    Discolithina Loeblich & Tappan 1963
    Pontosphaera Lohmann 1902
    Scyphosphaera Lohmann 1902

J. Family Rhabdosphaeraceae Lemmermann 1908
    Coccoliths as cyrtoliths.
    Acanthoica Lohmann 1903
    Algirosphaera Schlauder 1945, emend. Norris 1984
    Discosphaera Haeckel 1894
    Palusphaera Lecal 1965a, emend. Norris 1984
    Rhabdosphaera Haeckel 1894

K. Family Syracosphaeraceae Lemmermann 1908
    Coccoliths as caneoliths.
    Alisphaera Heimdal 1973
    Caneosphaera Gaarder 1977 (in Gaarder & Heimdal, 1977)
    Coronosphaera Gaarder 1977 (in Gaarder & Heimdal, 1977)
    Deutschlandia Lohmann 1912, emend. Gaarder 1981 (in Heimdal & Gaarder, 1981)
    Syracosphaera Lohmann 1902, emend. Gaarder 1977 (in Gaarder & Heimdal, 1977)
    Umbellosphaera Paasche 1955, emend. Gaarder 1981 (in Heimdal & Gaarder, 1981)

L. *Incertae sedis*
    Calciarcus Manton, Sutherland, & Oates 1977
    Florisphaera Okada & Honjo 1973
    Thorosphaera Ostenfeld 1910
    Wigwamma Manton, Sutherland, & Oates 1977

SYSTEMATIC DESCRIPTIONS

The generic and species descriptions presented here have been standardized as much as possible and are concerned with details observable both under the light and electron microscopes. Some genera include a large variety of forms; others (e.g., *Periphyllophora*, p. 177) are monospecific. References are made to the original description and to other publications that contain good illustrations of the species as interpreted by this author. The figures accompanying the text are examples, and in genera having variable forms, several different species may be illustrated. There exists a considerable literature on the taxonomy, morphology, and distribution of coccolithophorids (indexed by Loeblich & Tappan, 1966, 1968, 1969, 1970a, 1970b, 1971, 1973, and continued in the *INA Newsletter* from 1979), and this literature should be consulted for further information and comparison. For more information about the holococcolithophorids, readers are referred to the recent papers of Kleijne (1991) and Kleijne et al. (1991), which include the descriptions for four new genera, *Calicasphaera*, *Flosculosphaera*, *Poricalyptra*, and *Poritectolithus*, with two new types of holococcoliths, calicaliths and flosculoliths.

HOLOCOCCOLITHOPHORIDS

Coccosphere consisting of holococcoliths (see p. 163).

		Coccolith types	
Genus/coccolith case	Coccolith type	Ordinary	Stomatal
Monomorphic			
Calyptrosphaera	Calyptrolith		
Crystallolithus	Crystallolith		
Daktylethra	Areolith		
Gliscolithus	Gliscolith		
Homozygosphaera	Zygolith		
Periphyllophora	Helladolith		
Syracolithus	Laminolith		
Dimorphic			
Anthosphaera		Calyptrolith	Fragariolith
Calyptrolithina		Calyptrolith	Zygolith
Calyptrolithophora		Calyptrolith	Calyptrolith
Corisphaera		Zygolith	Zygolith
Helladosphaera		Zygolith	Helladolith
Sphaerocalyptra		Calyptrolith	Calyptrolith
Zygosphaera		Laminolith	Laminolith

Monomorphic Species

Coccosphere with one type of coccoliths.

Genus Balaniger Thomsen & Oates 1978

   Type: *Balaniger balticus* Thomsen & Oates 1978.
   Small cells having two flagella and a short haptonema.
   The cell surface is covered with organic scales supporting calcified
   structures shaped as hollow pyramids.

Balaniger balticus Thomsen & Oates

Reference: Thomsen & Oates, 1978, pp. 773–779, Plates 1–3.

   Saddle-shaped cells, 3.1–4.7 μm long, 3.6–6.0 μm wide, with two
   20–26 μm long flagella and a 7.0–8.8 μm long haptonema. The cell
   surface has rimmed organic scales 0.4–0.5 μm long and 0.3–0.35 μm
   wide, with radiating ridges on the proximal surface and concentric rings
   on the distal surface. Scales are generally covered by a single layer of small
   calcified structures, shaped as hollow 0.10–0.15 μm high pyramids with
   triangular bases (sides ~ 0.10–0.15 μm) and faceted or rounded apices.

Distribution: Brackish water species, western Baltic.

Genus *Calyptrosphaera* Lohmann 1902

Type: *Calyptrosphaera globosa* Lohmann 1902.

   Cells spherical or ovate with cap-shaped coccoliths (calyptroliths).
   Stomatal coccoliths may be higher than the ordinary ones and may carry a
   short distal spine.

Calyptrosphaera oblonga Lohmann (Plate 1)

References: Lohmann, 1902, p. 135, Plate 5, Figs. 43–46; Schiller, 1930,
pp. 222–224, Text-Fig. 108; Halldal & Markali, 1955, p. 8, Plate 1;
Gaarder & Hasle, 1971, p. 529, Fig. 5a,b; Heimdal & Gaarder, 1980, p. 3,
Plate 1, Figs. 4 and 5; Reid, 1980, p. 164, Plate 6, Figs. 9 and 10, Plate 7,
Fig. 1; Hallegraeff, 1984, p. 242, Fig. 52a,b; Norris, 1985, p. 628, Fig. 36;
Kleijne, 1991, p. 28, Plate 3, Figs. 3 and 4.

   Cells are spherical, ovoid or pyriform in shape, and have two flagella,
   17–28 μm long, covered with more than 100 coccoliths, but were also
   reported in the 10 μm range (Reid, 1980) with about 40–50 calyptroliths.

Plate 1   *Calyptrosphaera oblonga* Lohmann, single coccolith: (a) surface view; (b) broad lateral
view; (c) vertical longitudinal section; redrawn from Kamptner (1941); a–c not drawn to scale.
Calyptrosphaera sphaeroidea Schiller. *Crystallolithus hyalinus* Gaarder & Markali. *Daktylethra
pirus* (Kamptner) Norris: (a) ordinary coccolith; (b) stomatal coccolith; a, b not drawn to scale.
Periphyllophora mirabilis (Schiller) Kamptner, single coccolith: (a) distal surface view; (b) narrow
lateral view; (c) broad lateral view; a–c redrawn from Kamptner (1941), not to scale. *Syracolithus
dalmaticus* (Kamptner) Loeblich & Tappan. *Syracolithus quadriperforatus* (Kamptner) Gaarder,
single coccolith: (a) surface view; (b) vertical longitudinal section; (c) broad lateral view; redrawn
from Kamptner (1941); a–c not drawn to scale. Scale bar = 5 μm.

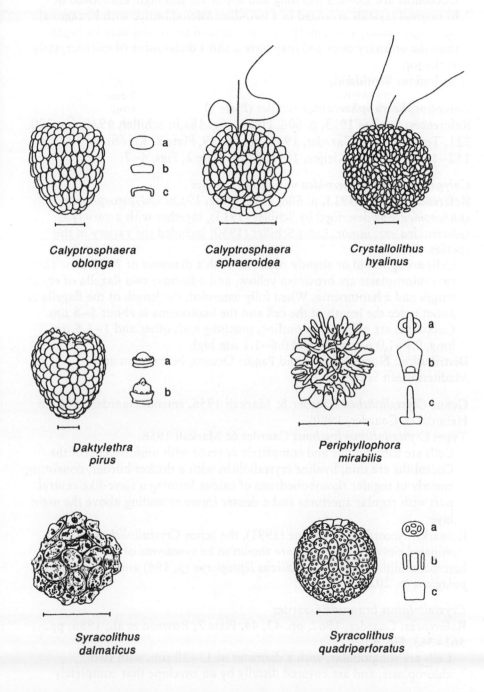

Calyptrosphaera
oblonga

Calyptrosphaera
sphaeroidea

Crystallolithus
hyalinus

Daktylethra
pirus

Periphyllophora
mirabilis

Syracolithus
dalmaticus

Syracolithus
quadriperforatus

Coccoliths are 2.0–2.5 μm long and about 1.5 μm high, composed of hexagonal crystals arranged in a two-dimensional lattice with hexagonal openings, usually monomorphic, but stomatal coccoliths may be higher than the ordinary ones and may have a short distal spine of microcrystals at the top.
Distribution: Worldwide.

Calyptrosphaera sphaeroidea Schiller (Plate 1)
References: Schiller 1913, p. 606, Plate 3, fig. 18a,b; Schiller, 1930, pp. 220, 221, Text-Fig. 104; Gaarder, 1962, pp. 38, 39, Plate 2; Klaveness, 1973, pp. 152–158, Figs. 1–8; Kleijne, 1991, p. 28, Plate 2, Figs. 4–7.

Calyptrosphaera sphaeroidea var. *minor* Schiller
Reference: Schiller, 1913, p. 606, Plate 3, Fig. 19a,b. *Calyptrosphaera sphaeroidea* was described by Schiller (1913), together with a variety C. *sphaeroidea* var. *minor*. Later Schiller (1930) included the variety in the species.
  Cells are spheroid or slightly pyriform with a diameter of 7–12 μm. The two chloroplasts are brownish yellow, and cells have two flagella of equal length and a haptonema. When fully extended, the length of the flagella is about twice the length of the cell and the haptonema is about 5–8 μm. Coccoliths are elliptical in outline, touching each other and 1–1.5 μm long, 0.9–1.0 μm wide, and 0.6–1.1 μm high.
Distribution: North Atlantic and Pacific Oceans, Norwegian and Mediterranean Seas.

Genus *Crystallolithus* Gaarder & Markali 1956, emend. Gaarder 1980 (in Heimdal & Gaarder, 1980)
Type: *Crystallolithus hyalinus* Gaarder & Markali 1956.
  Cells are subglobular and completely covered with uniform coccoliths. Coccoliths are thin, hyaline crystalloliths with a thicker border, consisting entirely of regular rhombohedrons of calcite forming a sieve-like central part with regular apertures and a denser frame extending above the main layer.
Remarks: According to Kleijne (1991), the genus *Crystallolithus* is eliminated since all its species are shown to be synonyms of the heterococcolithophorids *Calcidiscus leptoporus* (p. 194) and *Coccolithus pelagicus* (p. 202).

Crystallolithus braarudii Gaarder
References: Gaarder, 1962, pp. 43, 44, Plate 7; Rowson et al., 1986, pp. 361–363, Fig. 4.
  Cells are subglobular, with a diameter of 13–20 μm, with two chloroplasts, and are covered distally by an envelope that completely

covers the cell except where it is penetrated by the flagella and haptonema. Coccoliths are monomorphic holococcoliths (crystalloliths), having a length 2.0–3.0 μm and breadth 1.4–1.9 μm. Most coccoliths have a double layer of crystals located at the edge of the coccolith base plate and a perforated central area of one crystal thickness where the crystals form an ellipse, which is linked to the outer edge of the coccolith by radial lines of crystals. There may be more extra crystals superposed on the central area.

Distribution: Northern Atlantic and Pacific Oceans, Mediterranean Sea.

Crystallolithus hyalinus Gaarder & Markali (Plate 1)
References: Gaarder & Markali, 1956, pp. 1–4, Plate 1; Parke & Adams, 1960, pp. 264–274, Plates 1–4 [referred to as the motile *"Crystallolithus"* phase of *Coccolithus pelagicus* (Wallich) Schiller]; Nishida, 1979, Plate 3, Fig. 1b,c (referred to as the motile phase of *C. pelagicus*); Heimdal & Gaarder, 1980, Plate 2, Fig. 13.

Cells are globose to subglobose having a diameter of 8–18 μm, with two or four golden brown chloroplasts. There are two equal to subequal smooth flagella, 1.5–2.5 times the cell diameter in length, and a short haptonema, 0.5–0.75 of the cell diameter when fully extended. The flagella and haptonema arise close together at one pole. Coccoliths are about 2 μm long and 1.3–1.5 μm broad elliptical crystalloliths embedded in an envelope that varies considerably in thickness. Coccoliths are composed of microcrystals, cleavage rhombohedrons of calcite, which form an elliptic main layer and an extra peripheral row. Crystals in the main layer are arranged in a regular pattern forming a strainer-like plate having mostly rhombohedral apertures. Extra crystals may also be superposed distally on the central part.

Distribution: Atlantic and Northern Pacific Oceans.

Remarks: Parke & Adams (1960) have shown that the motile *Crystallolithus hyalinus* and the nonmotile *Coccolithus pelagicus* are different stages in the life cycle of a single species. The name *C. pelagicus* should be used for both the motile and the nonmotile phases since it has priority over *Crystallolithus hyalinus*. Rowson et al. (1986), using cultures, found evidence that *C. braarudii* is a third stage in the life cycle of *Coccolithus pelagicus*. Thus, both *Crystallolithus braarudii* and *C. hyalinus* fall within the limits of one single species and are synonyms. To acknowledge this morphological variation of *Coccolithus pelagicus*, Kleijne (1991) proposed the names *C. pelagicus* f. *braarudii* (Gaarder) and *C. pelagicus* f. *hyalinus* (Gaarder and Markali).

Crystallolithus rigidus Gaarder
References: Gaarder, 1980, in Heimdal & Gaarder, 1980, pp. 6, 7, Plate 2, Figs. 10–12; Norris, 1985, p. 630, Figs. 13 and 37; Kleijne, 1991,

p. 17, Plate 4, Figs. 4–6 (referred to as *Calcidiscus leptoporus* f. *rigidus*).
Cells are subglobular with a diameter of 11–14 μm. Coccoliths are
randomly arranged, irregularly elliptical crystalloliths, 1.9–2.6 μm in
length, length/breadth ratio 1.3–1.8, height 0.3 μm. Coccoliths have two
layers of crystals arranged in a regular pattern with six-sided perforations
present over the central part, but the second layer is not continuous and
consists in some coccoliths of single crystals that are not connected with
each other. The frame is made of three closely set crystal rings.
Distribution: North Atlantic and Indian Oceans, Mediterranean Sea.
Remarks: Kleijne (1991) showed that the holococcolith-bearing *C. rigidus* is
the motile stage in the life cycle of *Calcidiscus leptoporus* (Murray &
Blackman) Loeblich & Tappan (p. 194), which implies that *C. leptoporus*
and *Crystallolithus rigidus* are synonyms. To class this morphological
variation she proposed the name *Calcidiscus leptoporus* f. *ridigus*
(Gaarder).

Genus *Daktylethra* Gartner 1969 (in Gartner & Bukry, 1969)
Type: *Daktylethra punctulata* Gartner 1969 (fossil).
   Monomorphic coccolith case consisting of areoliths, dome-shaped
   holococcoliths with an areolate interior comprised of thickened ridges of
   calcite elements.

Daktylethra pirus (Kamptner) Norris (**Plate 1**)
Basionym: *Calyptrosphaera pirus* Kamptner (Kamptner, 1937b, p. 304,
Plate 16, Figs. 21–23). References: Kamptner, 1941, pp. 78, 98, Plate 2,
Figs. 17–19; Throndsen, 1972, pp. 53, 54, Figs. 2–9; Borsetti & Cati,
1976, p. 211, Plate 13, Figs. 1–3; Reid, 1980, p. 164, Plate 7, Figs. 2
and 3).
Synonym: *Calyptrolithophora pirus* (Kamptner) Hallegraeff (Hallegraeff,
1984, p. 244, Fig. 54).
References: Norris, 1985, p. 631, Figs. 10, 38, 39; Kleijne, 1991, pp. 28, 29,
Plate 3, Figs. 5 and 6, Plate 4, Fig. 1.
   Cells ovoid to pear-shaped, 18–26 μm long and 15–22 μm wide, with a
   5–8 μm wide stomatal opening. Coccoliths are 2.3–3.0 × 2.1 × 2.0 μm
   high areoliths, with a distinct collar at the base of the dome-shaped cap
   and approximately eight to twelve pores at the junction of the cap and the
   basal collar; in lateral view the pores may be obscured by the raised
   marginal collar. On the coccolith dome, one or two large pores sometimes
   are present as thin-walled circular areas (interior areolae), but open spaces
   between crystallites are not present. The coccoliths around the stomatal
   opening have a short conical extension protruding from the central area
   and are somewhat higher than the ordinary ones.
Distribution: North Atlantic, Pacific and Indian Oceans, Mediterranean Sea,
Australian waters, Caribbean Sea.

Genus *Gliscolithus* Norris 1985
Type: *Gliscolithus amitakarenae* Norris 1985.
Monospecific genus.
  Coccosphere is monomorphic, comprised of gliscoliths, holococcoliths consisting of a circular to elliptical proximal ring with an inflated, hollow distal part penetrated by large pores. Both the neck and the globular part of the coccoliths are comprised of a single layer of hexagonal elements.

Gliscolithus amitakarenae Norris
References: Norris, 1985, p. 630, Figs. 11, 12, 40, 42; Kleijne, 1991, p. 29, Plate 5, Figs. 1 and 2.
  Monomorphic coccosphere consisting of 2.7–4.0 μm high gliscoliths with large pores in the inflated part and sometimes in the stalk; proximal ring 1.0–2.0 × 1.0–1.5 μm; the distal section, diameter 1.5–2.0 μm, sometimes with a small distal protuberance.
Distribution: Mediterranean Sea, Indian Ocean.

Genus *Homozygosphaera* Deflandre 1952 (in Grassé, 1952)
Type: *Homozygosphaera spinosa* (Kamptner) Deflandre 1952.
  Some cells have been observed with two flagella and a short haptonema. Spherical to ellipsoidal coccolith case with monomorphic zygoliths. Stomatal coccoliths may be somewhat larger than the ordinary ones. The number of bridges spanning the central tube distally seems to vary between one crossing the narrower dimension of the elliptical coccolith and several arches meeting in the center of the tube opening, leaving open space or pores between the bridges and the tube rim.

Homozygosphaera triarcha Halldal & Markali
References: Halldal & Markali, 1955, p. 9, Plate 4; Borsetti & Cati, 1972, p. 404, Plate 50, Fig. 2; Okada & McIntyre, 1977, p. 32, Plate 13, Fig. 12; Nishida, 1979, Plate 18, Fig. 3a,b; Norris, 1985, p. 636, Figs. 25 and 49; Kleijne, 1991, pp. 31, 33, Plate 5, Figs. 5 and 6.
  Ellipsoidal cells with diameters around 8 μm. Coccoliths are 1.6–2.4 μm long and up to 3.0 μm high zygoliths that consist of an oval proximal tube carrying three arches that join distally, leaving three lateral pores, to form the long characteristic towering central area.
Distribution: Worldwide.

Genus *Periphyllophora* Kamptner 1936
Type: *Periphyllophora mirabilis* (Schiller) Kamptner 1937b.
Monospecific genus.
  Monomorphic ellipsoidal or ovoid coccosphere consisting of helladoliths having the bridge area expanded into a petaloid form with a forked base and a pore on each basal piece.

Periphyllophora mirabilis (Schiller) Kamptner (**Plate 1**)
Basionym: *Calyptrosphaera* (?) *mirabilis* Schiller (Schiller, 1925, p. 34, Plate 3, Fig. 31, 31a).
References: Kamptner, 1937b, p. 309, Plate 17, Figs. 39, 40; Kamptner, 1941, pp. 92, 108, Plate 12, Figs. 129–131; Halldal & Markali, 1955, p. 9, Plate 3; Borsetti & Cati, 1972, pp. 404, 405, Plate 51, Fig. 1a,b; Nishida, 1979, Plate 20, Fig. 3a,b; Hallegraeff, 1984, p. 242, Fig. 49; Norris, 1985, pp. 621, 636, Figs. 26 and 50; Kleijne, 1991, p. 33, Plate 14, Figs. 1 and 2.
   Ellipsoidal or ovate cells having a diameter of 7–12 μm. Coccoliths consist of helladoliths having an oval proximal tube, 1.5–2.3 μm long and 1.3 μm high, carrying a 2 μm broad and 2.0–2.5 μm high leaf-like extension composed of two layers of elements and arranged normal to the long axis of the basal tube. The base of the petaloid structure is forked, with a pore on each basal piece.
Distribution: Worldwide.

Genus *Syracolithus* (Kamptner) Deflandre 1952 (in Grassé, 1952)
Type: *Syracolithus dalmaticus* (Kamptner) Loeblich & Tappan 1963.
   Spherical cells covered by monomorphic laminoliths. Pores may be present, penetrating the entire coccolith, but in some the basal plate may close the proximal side.

Syracolithus dalmaticus (Kamptner) Loeblich & Tappan (**Plate 1**)
Basionym: *Syracosphaera dalmatica* Kamptner (Kamptner, 1927, p. 178, Fig. 2). **Reference:** Kamptner, 1941, pp. 81, 104, Plate 4, Figs. 46–48.
Synonyms: *Homozygosphaera wettsteinii* (Kamptner) Halldal & Markali (Okada & McIntyre, 1977, pp. 32, 34, Plate 13, Fig. 8); *Corysphaera* sp. (Nishida, 1979, Plate 18, Fig. 1); *Homozygosphaera wettsteinii* (Kamptner) (Winter et al., 1979, p. 212, Plate 5, Fig. 6); *Homozygosphaera halldalii* Gaarder, (invalid), validated in Gaarder, 1983, p. 50; (Kleijne, 1991, p. 37).
References: Loeblich & Tappan, 1963, p. 193; Norris, 1985, pp. 637, 638, Fig. 8; Kleijne, 1991, p. 37, Plate 7, Fig. 1.
   Ellipsoidal cells having a diameter of 11–19 μm, with a 4–5 μm wide flagellar area. Two flagella and a short haptonema are present as are the characteristic organic scales. Elliptical coccoliths are 1.9–3.4 μm long and 1.0–1.4 μm high laminoliths with four to seven pores surrounding an elevated central structure that often is terminated by two distal nodules. Central structure in coccoliths surrounding the flagellar area are sometimes higher than in the other coccoliths.
Distribution: North Atlantic, Pacific and Indian Oceans, Mediterranean Sea.

Syracolithus quadriperforatus (Kamptner) Gaarder (**Plate 1**)
Basionym: *Syracosphaera quadriperforata* Kamptner (Kamptner, 1937b, p. 302, Plate 15, Figs. 15 and 16). Reference: Kamptner, 1941, pp. 81, 82, Plate 4, Fig. 49, Plate 5, Figs. 50 and 51.
Synonym: *Homozygosphaera quadriperforata* (Kamptner) Gaarder (Gaarder, 1962, pp. 48–50, Text-Fig. 2c,d, Plate 12; Borsetti & Cati, 1976, p. 222, Plate 16, Figs. 7–10; Okada & McIntyre, 1977, pp. 30, 32, Plate 13, Fig. 9).
References: Gaarder, 1980, in Heimdal & Gaarder, 1980, pp. 10, 12; Norris, 1985, p. 638, Figs. 9, 42, 51, 52; Kleijne, 1991, pp. 37, 38, Plate 7, Figs. 3 and 4.
Spherical cells with a diameter of 12–15 μm and a 3–4 μm wide circular or elliptical flagellar area from which two flagella emerge. Coccoliths consist of laminoliths approximately 2.3–2.6 × 1.6 μm wide and up to 1.5 μm high having four to seven large pores. Distally projecting knob-like extensions between the pores give the surface an irregular appearance.
Distribution: Worldwide.

Dimorphic Species

Coccospheres having two different kinds of holococcoliths. Ordinary and stomatal coccoliths are either of two different types, or of a single type with different shapes.

Genus *Anthosphaera* Kamptner 1937b, emend. Kleijne, 1991
Type: *Anthosphaera fragaria* Kamptner 1937b, emend. Kleijne 1991.
Dimorphic coccosphere. Ordinary coccoliths are calyptroform. Basal part of stomatal coccoliths (fragarioliths; Kleijne, 1991) are similarly constructed, with a distal part drawn out in a single layered leaf-like extension.

Anthosphaera fragaria Kamptner, emend. Kleijne
Synonyms: *Helladosphaera fragaria* (Kamptner) Gaarder (Gaarder, 1962, pp. 47, 48, Plate 11; Norris, 1984, p. 38); *Calyptrolithina fragaria* (Kamptner) Norris (Norris, 1985, p. 625).
References: Kamptner, 1937b, p. 304, Plate 15, Fig. 20; Kamptner, 1941, pp. 86, 106, Plate 9, Figs. 89 and 90; Kleijne, 1991, p. 42, Plate 8, Figs. 3–6.
Cells are nearly spherical, having a diameter of 4.7–6.5 μm, and up to 3.7 μm wide flagellar area with two flagella. Ordinary coccoliths are 1.0–1.3 μm long and 0.5 μm high calyptroliths, having a regularly perforated marginal rim. Stomatal coccoliths carry a large petaloid process with rounded off angles, approximately 2.0–2.5 μm high.
Distribution: Mediterranean Sea.

Remarks: Gaarder (1962) regarded both the ordinary coccoliths and the stomatal coccoliths of *A. fragaria* as zygoliths and transferred *A. fragaria* to the genus *Helladosphaera*. Norris (1985) characterized the ordinary coccoliths as being calyptroform and the stomatal coccoliths as helladoliths, thus placing this species in *Calyptrolithina*, while Kleijne (1991) emended the generic description of *Anthosphaera* to species having stomatal coccoliths that are fragarioliths.

Genus *Calyptrolithina* Heimdal 1982
Type: *Calyptrolithina divergens* (Halldal & Markali) Heimdal 1982.
The coccolith case is subspherical and dimorphic with a distinct flagellar area. Ordinary coccoliths are calyptroliths, while stomatal coccoliths are zygoliths with a high pointed bridge at a right angle to the long axis of coccolith.
Remarks: This genus is characterized by having calyptroform ordinary coccoliths and zygoform stomatal coccoliths (Heimdal, 1982). Norris (1985) considered that species having stomatal helladoliths also should be included in this genus and transferred several other species originally described in other genera to *Calyptrolithina*.

Calyptrolithina divergens (Halldal & Markali) Heimdal
Basionym: *Zygosphaera divergens* Halldal & Markali (Halldal & Markali, 1955, p. 8, Plate 2). References: Okada & McIntyre, 1977, p. 36, Plate 12, Fig. 1a,b; Winter et al., 1979, Plate 5, Fig. 10; Reid, 1980, p. 166, Plate 7, Fig. 10; emend. Heimdal, 1980, in Heimdal & Gaarder, 1980, p. 12, Plate 3, Fig. 24a,b.
References: Heimdal, 1982, p. 54; Kleijne, 1991, p. 45, Plate 10, Figs. 1–3.
The dimorphic subspherical coccolith case has length 5.5–8.0 μm, breadth 5.5–6.0 μm, and flagellar area about 3.0 μm wide. Ordinary coccoliths are calyptroliths with the proximal part consisting of a basal ring that bears a tube widening distally. The central part of the coccolith is highly vaulted, surrounded at the base by distal crystal rings of a tube with a length of 1.3–2.3 μm, length/breadth ratio 1.3–1.4, and total height 0.6–1.2 μm. The proximal ring, tube, and central part are formed by a single layer of microcrystals arranged in the usual hexagonal pattern, second crystal ring from distal margin of tube with a regular row of holes, most easily seen in distal view. Stomatal coccoliths are zygoliths with high pointed bridge normal to long axis of coccolith, length 1.1–1.6 μm, total height 1.5–1.8 μm.
Distribution: North Atlantic and Pacific Oceans, Norwegian Sea, Mediterranean Sea.

Calyptrolithina multipora (Gaarder) Norris (**Plate 2**)
Basionym: *Corisphaera multipora* Gaarder 1980, in Heimdal & Gaarder (1980, pp. 3, 4, Plate 1, Fig. 7).

Synonyms: *Syracolithus dalmaticus* (Kamptner) [Borsetti & Cati, 1972, p. 399, Plate 43, Fig. 1a,b (non 1c)]; *Helladosphaera dalmatica* (Kamptner) Okada & McIntyre (Okada & McIntyre, 1977, pp. 28, 29, Plate 12, Figs. 3–6); *Helladosphaera* sp. (Nishida, 1979, Plate 19, Fig. 3a,b).
References: Norris, 1985, p. 625, Fig. 31; Kleijne, 1991, p. 46, Plate 10, Figs. 5 and 6.

Cells are subspherical with a diameter of 13.9–22.5 μm, covered with characteristic organic scales. Ordinary coccoliths are calyptroliths with a length of 2.3–3.4 μm, length/breadth ratio 1.3–1.5, height about 1.5 μm, consisting of a proximal tube one crystal thick composed of approximately 13 densely packed crystal rings and a flat distal part, which is perforated. The uppermost ring has an elevated central part with two complete or partially complete concentric rings of regular six-sided pores. A central knob consisting of closely set crystals reaches well above the coccolith top in lateral view. Stomatal coccoliths are zygoliths, with a proximal tube of about the same height as in the ordinary coccoliths, and crossed by a rather broad bridge slightly pointed at the top. The bridge has about the same height as the tube and extends down to slightly below the tube top.
Distribution: Northern Atlantic, Pacific and Indian Oceans, Mediterranean Sea, Gulf of Mexico.

Genus *Calyptrolithophora* Heimdal 1980 (in Heimdal & Gaarder, 1980)
Type: *Calyptrolithophora papillifera* (Halldal) Heimdal 1980.

Coccosphere is ellipsoidal to pear-shaped, showing dimorphism of calyptroliths. Ordinary coccoliths are calyptroliths with a low, elliptical, proximal tube and nearly flat distal part, with or without superimposed microcrystals in groups of various shapes. Stomatal coccoliths have a similar proximal tube with a highly convex upper portion. Microcrystals are regular rhombohedrons of calcite.
Remarks: This genus was described by Heimdal (in Heimdal & Gaarder, 1980) to contain three species that had been previously included in the genus *Sphaerocalyptra* Deflandre. In contrast to the previous genus being characterized by having campanulate calyptroliths, the coccoliths of *Calyptrolithophora* have straight sides slightly wider at the top.

Calyptrolithophora gracillima (Kamptner) Heimdal
Basionym: *Calyptrosphaera gracillima* Kamptner (Kamptner 1941, pp. 77, 98, Plate 1, Figs. 13–16).
Synonyms: *Syracolithus catilliferus* (Kamptner) (Borsetti & Cati, 1972, pp. 398, 399, Plate 40, Fig. 2a,b); *Sphaerocalyptra gracillima* (Kamptner) Throndsen (Throndsen, 1972, pp. 54–56, Figs. 10–15; Nishida, 1979, Plate 17, Fig. 4a,b).
References: Heimdal, 1980, in Heimdal & Gaarder, 1980, p. 2; Norris, 1985, p. 626.

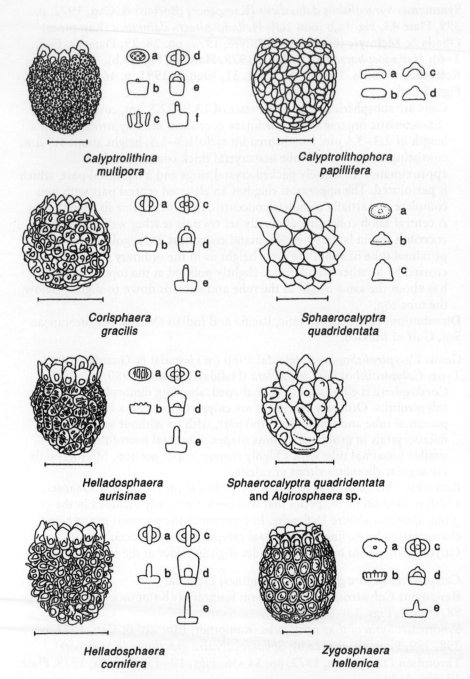

*Calyptrolithina
multipora*

*Calyptrolithophora
papillifera*

*Corisphaera
gracilis*

*Sphaerocalyptra
quadridentata*

*Helladosphaera
aurisinae*

Sphaerocalyptra quadridentata
and *Algirosphaera* sp.

*Helladosphaera
cornifera*

*Zygosphaera
hellenica*

Ovoid to pear-shaped cells having a length from 9.5 to 18.0 μm, breadth 9–16 μm, flagellar area of 3.8 μm wide, and dimorphic coccoliths. Ordinary coccoliths about 2 μm long and 0.7 μm high with a central knob-like protrusion of crystals. Stomatal coccoliths deviate from the ordinary ones in having a highly convex central area.

Distribution: Pacific and Indian Oceans, Mediterranean Sea, Australian waters, Caribbean Sea.

Calyptrolithophora papillifera (Halldal) Heimdal (**Plate 2**)

Basionym: *Calyptrosphaera papillifera* Halldal (Halldal, 1953, p. 48, Fig. 14).

Synonym: *Sphaerocalyptra papillifera* (Halldal) Halldal (Halldal, 1954, p. 123; Borsetti & Cati, 1972, p. 398, Plate 40, Fig. 1a,b; Okada & McIntyre, 1977, p. 34, Plate 11, Figs. 1 and 2; Nishida, 1979, Plate 23, Fig. 3a,b).

References: Heimdal, 1980, in Heimdal & Gaarder, 1980, pp. 2, 3, Plate 1, Figs. 2 and 3; Norris, 1985, p. 626, Fig. 34; Kleijne, 1991, p. 50, Plate 12, Figs. 1 and 2.

Pear-shaped to spherical cells are 9–20 μm long, with a flagellar area of 4–6 μm wide. Ordinary coccoliths are calyptroliths, 2.0–2.5 μm long and about 0.5 μm high consisting of three or four concentric rings of microcrystals with no ornamentation on the flat distal surface. The highly convex stomatal coccoliths have parallel rows of microcrystals normal to the long axis of the coccolith on its distal surface.

Distribution: Worldwide.

Plate 2 *Calyptrolithina multipora* (Gaarder) Norris, ordinary coccolith: (a) distal surface view; (b) broad lateral view; (c) vertical longitudinal section. Stomatal coccolith: (d) distal surface view; (e) narrow lateral view; (f) broad lateral view; a–f redrawn from Kamptner (1941), not to scale. *Calyptrolithophora papillifera* (Halldal) Heimdal, ordinary coccolith: (a) vertical longitudinal section; (b) lateral view. Stomatal coccolith: (c) vertical longitudinal section; (d) lateral view; redrawn from Halldal (1953). *Corisphaera gracilis* Kamptner, ordinary coccolith: (a) distal surface view; (b) broad lateral view. Stomatal coccolith: (c) distal surface view; (d) narrow lateral view; (e) broad lateral view; a–e redrawn from Kamptner (1941), not to scale. *Helladosphaera aurisinae* Kamptner, ordinary coccolith: (a) distal surface view; (b) broad lateral view. Stomatal coccolith: (c) distal surface view; (d) narrow lateral view; (e) broad lateral view; a–e redrawn from Kamptner (1941), not to scale. *Helladosphaera cornifera* (Schiller) Kamptner, ordinary coccolith: (a) distal surface view; (b) broad lateral view. Stomatal coccolith: (c) distal surface view; (d) narrow lateral view; (e) broad lateral view; a–e redrawn from Kamptner (1941), not to scale. *Sphaerocalyptra quadridentata* (Schiller) Deflandre, ordinary coccolith: (a) distal surface view; (b) broad lateral view. Stomatal coccolith: (c) broad lateral view; a–c redrawn from Kamptner (1941), not to scale. *Sphaerocalyptra quadridentata* (Schiller) Deflandre/*Algirosphaera* sp., dimorphic coccosphere, redrawn from Kamptner (1941). *Zygosphaera hellenica* Kamptner, ordinary coccolith: (a) distal surface view; (b) broad lateral view. Stomatal coccolith: (c) distal surface view; (d) narrow lateral view; (e) broad lateral view; redrawn from Kamptner (1941), a–e not to scale. Scale bar = 5 μm.

Genus *Corisphaera* Kamptner 1936
Type: *Corisphaera gracilis* Kamptner 1937b.
  Dimorphic coccosphere consisting of zygoliths with stomatal coccoliths as
  enlarged zygoliths with an extended bridge.
Remarks: Norris (1985) drew attention to the resemblance of the stomatal
coccoliths to helladoliths and transferred several species originally described
in *Corisphaera* to *Helladosphaera*. In the present study, species bearing
zygoform ordinary coccoliths and stomatal coccoliths that are helladoliths
are placed in *Helladosphaera*.

***Corisphaera gracilis* Kamptner (Plate 2)**
Synonym: *Helladosphaera gracilis* (Kamptner) Norris (Norris, 1985, pp.
631, 633).
References: Kamptner, 1937b, p. 307, Plate 16, Figs. 33–35; Kamptner,
1941, pp. 90, 107, Plate 11, Figs. 113–116; Gaarder, 1962, p. 41, Plate 4;
Okada & McIntyre, 1977, p. 28, Plate 13, Fig. 3; Heimdal & Gaarder,
1980, p. 3, Plate 1, Fig. 6a,b; Kleijne, 1991, p. 52, Plate 12, Figs. 3–5; non
Corisphaera gracilis Kamptner (Borsetti & Cati, 1976, p. 218, Plate 15, Fig.
9); non *C*. aff. *gracilis* Kamptner (Borsetti & Cati, 1976, p. 218, Plate 15,
Figs. 10 and 11).
  Spherical, ovoid or berry-shaped dimorphic coccosphere 7.0–9.5 μm long,
  6.0–7.2 μm wide, having a flagellar area of 2.3–3.0 μm wide. Coccoliths
  consisting of zygoliths, with an oval proximal tube, about 1.3 μm long
  and 0.6 μm wide. The low bridge of ordinary coccoliths apparently does
  not reach above the tube top, while the petaloid bridge in stomatal
  coccoliths is considerably higher.
Distribution: North Atlantic, Pacific and Indian Oceans, Mediterranean Sea.

Genus *Helladosphaera* Kamptner 1936
Type: *Helladosphaera cornifera* (Schiller) Kamptner 1937b.
  Coccosphere characterized by its ordinary zygoform coccoliths and
  helladoform stomatal coccoliths with a petaloid expansion consisting of
  two layers of elements.
Remarks: Norris (1985) considered *Corisphaera* and *Helladosphaera* very
closely related and merged these two genera into one, *Helladosphaera* (see
remarks on *Corisphaera*, mentioned above).

***Helladosphaera aurisinae* Kamptner (Plate 2)**
Synonym: *Poricalyptra aurisinae* (Kamptner) Kleijne (Kleijne, 1991, p. 61).
References: Kamptner, 1941, p. 91, Plate 11, Figs. 121–124; Gaarder, 1962,
pp. 44–46, Plates 8 and 9; Borsetti & Cati, 1972, p. 403, Plate 49, Fig.
1a,b; Okada & McIntyre, 1977, p. 28, Plate 13, Fig. 7; Reid, 1980, p. 166,
Plate 7, Figs. 5 and 6).

Coccolith case is ovoid with the widest diameter approximately of 8 μm and a flagellar area 3.8 μm wide. Ordinary coccoliths, about 2.0–2.4 μm long and 1.0–1.2 μm high zygoliths with three (rarely two) parallel bridges crossing the narrow dimension of the elliptical coccolith. The central bridge may be thicker than the two lateral ones, most often with an elevated ridge. Stomatal helladoliths consisting of a shorter, thicker and nearly circular tube with a leaf-like distal extension, made up of two layers of elements.

Distribution: Northern Atlantic and Indian Oceans, Mediterranean Sea.

Helladosphaera cornifera (Schiller) Kamptner (**Plate 2**)
Basionym: *Syracosphaera cornifera* Schiller (Schiller, 1913, pp. 602, 603, Plate 2, Fig. 13).
References: Kamptner, 1937b, p. 308, Plate 17, Figs. 36–38; Kamptner, 1941, p. 91, Plate 12, Figs. 125–128; Borsetti & Cati, 1972, p. 404, Plate 49, Fig. 2a,b; Okada & McIntyre, 1977, p. 28, Plate 13, Figs. 4 and 5; Reid, 1980, p. 166, Plate 7, Figs. 7–9; Norris, 1985, p. 633, Fig. 45; Kleijne, 1991, p. 57, Plate 14, Figs. 3–6.
Cells ovoid having a size of 4.9–6.9 × 4.9–6.4 μm, and a 3.0–4.4 μm wide flagellar area and two flagella. Ordinary coccoliths are zygoliths consisting of an oval tube, about 1.5 μm long, 1.0 μm wide, and 0.5 μm high, with a 0.7–0.8 μm high thin bridge. The shape and angle of the bridge may vary among individuals. About 12 stomatal helladoliths with a maximum height of approximately 2.5 μm and a distal triangular portion of 1.5 μm wide surround the flagellar area.
Distribution: Northern Atlantic, Pacific and Indian Oceans, Mediterranean Sea, Australian waters, Belize Lagoon (British Honduras).

Genus *Sphaerocalyptra* Deflandre 1952 (in Grassé 1952)
Type: *Sphaerocalyptra quadridentata* (Schiller) Deflandre 1952.
Coccosphere consisting of two types of campanulate calyptroliths with a tapered shape, the height of the bell differing only between ordinary and stomatal coccoliths.
Remarks: The tapered campanulate coccoliths of *Sphaerocalyptra* differ distinctly from the cap-shaped coccoliths of *Calyptrosphaera* (p. 172). For comparison with *Calyptrolithophora*, see remarks on p. 181.

Sphaerocalyptra quadridentata (Schiller) Deflandre (**Plate 2**)
Basionym: *Calyptrosphaera quatridentata* Schiller (Schiller 1913, p. 607, Plate 3, Figs. 20 and 21); corrected to *C. quadridentata* (Schiller, 1925, p. 33). Reference: Kamptner, 1941, pp. 78, 99, Plate 2, Figs. 20–23).
References: Deflandre, 1952, in Grassé, 1952, p. 452, Fig. 350B; Borsetti & Cati, 1972, p. 398, Plate 41, Fig. 1; Winter et al., 1979, Plate 5, Fig. 9; Kleijne, 1991, p. 65, Plate 17, Fig. 3.

Slightly oval coccosphere having a length of 6–9 μm. Coccoliths consisting of 1.4–2.0 μm long calyptroliths having a distinct rim at the base of a tapered conical central area, which may be pointed or have a thickened, distal ridge. Heights of the ordinary and stomatal coccoliths are 0.6–1.2 μm and 1.5–2.2 μm, respectively.

Distribution: Northern Atlantic, Mediterranean Sea.

Genus *Zygosphaera* Kamptner 1936, emend. Heimdal 1982
Type: *Zygosphaera hellenica* Kamptner 1937b (type designated by Loeblich & Tappan, 1963, p. 194).

Dimorphic coccosphere characterized by its ordinary coccoliths being laminoliths and the stomatal zygoform laminoliths, with a high transverse ridge. Pores may be present.

***Zygosphaera bannockii* (Borsetti & Cati) Heimdal**
Basionym: *Sphaerocalyptra bannockii* Borsetti & Cati (Borsetti & Cati, 1976, p. 212, Plate 13, Figs. 4–6).
Synonym: *Laminolithus bannockii* (Borsetti & Cati) Heimdal (Heimdal, 1980, in Heimdal & Gaarder, 1980, p. 8, Plate 2, Fig. 18).
References: Heimdal, 1982, p. 53; Norris, 1985, p. 640, Figs. 6 and 56; Kleijne, 1991, p. 67, Plate 18, Fig. 1.

Subspherical cells having a diameter approximately 6–7 μm. Ordinary coccoliths are elliptical laminoliths having a length of 1.0–1.3 μm, length/breadth ratio 1.3–1.6, and height excluding the raised part about 0.3 μm. The flat distal surface is divided into two parts by a short, 0.4 μm high ridge, comprised of one layer of crystals running nearly at a right angle to the long axis of the coccolith. Stomatal coccoliths are 1.0–1.2 μm high and have a perforation on each side of the high, pointed ridge.

Distribution: North Atlantic and Indian Oceans, Mediterranean Sea.

***Zygosphaera hellenica* Kamptner (Plate 2)**
Synonyms: *Helladosphaera* (?) sp. 1 & *H.* sp. 2 (Borsetti & Cati, 1979, pp. 160, 161, Plate 17, Figs. 3–8; *Laminolithus hellenicus* (Kamptner) Heimdal (Heimdal, 1980, in Heimdal & Gaarder, 1980, p. 8, Plate 3, Figs. 19–21).
References: Kamptner, 1937b, p. 306, Plate 16, Figs. 27–29; Reid, 1980, pp. 166, 168, Plate 8, Figs. 1 and 2; Heimdal, 1982, p. 53; Norris, 1985, p. 639, Fig. 57; Kleijne, 1991, p. 69, Plate 18, Figs. 3–5.

The subspherical coccolith case has a diameter about 15 μm. Ordinary coccoliths consist of elliptical laminoliths, either unperforated or with a perforation on one or both sides of a central knob of microcrystals. Coccolith length is 1.9–2.4 μm, with a length/breadth ratio of 1.2–1.7, and height excluding the top of 0.5–0.6 μm. The coccoliths are usually comprised of nine layers of microcrystals, the third from the bottom with a more open structure, leaving a regular row of holes visible in lateral

view. Stomatal coccoliths have a homologously built proximal part, usually unperforated with pronounced grooves on each side of a transverse ridge. Their height including the ridge is 1.4–1.9 μm.
Distribution: North Atlantic, Pacific and Indian Oceans, Mediterranean Sea.
Remarks: This species seems to be closely related to *Z. debilis* Kamptner, with which it overlaps in the size of the cells and the coccoliths (Kamptner, 1941, pp. 87, 88). According to Kamptner *Z. hellenica* is more rigid than *Z. debilis*, but combination of these two species into one should be reconsidered after more specimens have been studied.

Zygosphaera marsilii (Borsetti & Cati) Heimdal
Basionym: *Sphaerocalyptra marsilii* Borsetti & Cati (Borsetti & Cati, 1976, pp. 212, 213, Plate 13, Figs. 7–10).
Synonyms: ?*Zygosphaera amoena* Kamptner (Kamptner, 1937b, p. 305, Plate 16, Figs. 24–26); *Corisphaera gracilis* Kamptner (Borsetti & Cati, 1976, p. 218, Plate 15, Fig. 9); *C.* aff. *gracilis* Kamptner (Borsetti & Cati, 1976, p. 218, Plate 15, Figs. 10 and 11); *Laminolithus marsilii* (Borsetti & Cati) Heimdal (Heimdal, 1980, in Heimdal & Gaarder, 1980, pp. 8, 10, Plate 3, Figs. 22 and 23); *Zygosphaera amoena* Kamptner (Norris, 1985, pp. 639, 640, Fig. 55).
References: Heimdal, 1982, p. 53; Kleijne, 1991, p. 69, Plate 18, Fig. 6.
Subspherical cells with diameters approximately 6–8 μm, and a flagellar area about 5 μm wide from which two flagella emerge. Ordinary coccoliths are elliptical laminoliths, 1.4–1.6 μm long, length/breadth ratio of 1.2–1.3, and 0.3–0.4 μm high. These coccoliths usually have seven layers of microcrystals with a central collection of microcrystals mostly in the shape of an irregular ridge on the distal surface. Disc sides show the characteristic hexagonal pattern. Solid stomatal coccoliths, total height 1.0–1.4 μm, have shallow grooves on each side of the high ridge.
Distribution: Northern Atlantic and Indian Oceans, Mediterranean Sea.

Comparison of *Z. bannockii*, *Z. hellenica*, and *Z. marsilii*
The cells of *Z. hellenica* are definitely larger than those of the other two species. The ordinary coccoliths of *Z. bannockii* differ from those of *Z. hellenica* in having a distinct ridge, usually unconnected with the rim on its distal surface, while the collections of microcrystals on the distal surface of *Z. hellenica* are knob-like and in *Z. marsilii* somewhat intermediate. The stomatal coccoliths of *Z. bannockii* are perforated at the base of the ridge, but not in the case of *Z. hellenica* and *Z. marsilii*. *Zygosphaera hellenica* differs also from the two other species in showing a regular row of pores near the base of the coccoliths.
In all three species the microcrystals forming the distal layer of the ordinary coccoliths show a fairly regular pattern of concentric rings. This arrangement

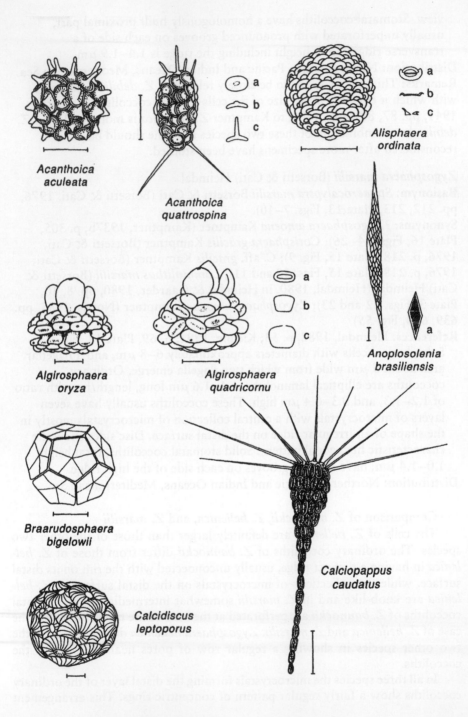

Acanthoica
aculeata

a
b
c

Acanthoica
quattrospina

Alisphaera
ordinata

a
b
c

Anoplosolenia
brasiliensis

a

Algirosphaera
oryza

a
b
c

Algirosphaera
quadricornu

Braarudosphaera
bigelowii

Calciopappus
caudatus

Calcidiscus
leptoporus

may be a secondary feature and is also found in some other species, as for example in *Calyptrolithophora gracillima* (synonym *Sphaerocalyptra gracillima;* see Throndsen, 1972, p. 56, Fig. 15). It is possible that *Z. amoena* and *Z. marsilii* should be treated as one species (Norris, 1985), but I prefer to retain them as separate until further studies have clarified their similar characteristics.

HETEROCOCCOLITHOPHORIDS

Coccolith case consisting of heterococcoliths (see p. 163).

Genus *Acanthoica* Lohmann 1903
Type: *Acanthoica coronata* Lohmann 1903.
Spherical to ellipsoidal dimorphic coccolith case having no stomatal area. Styliform cyrtoliths (rhabdoliths) are present on one or both poles of the coccolith case. Ordinary coccoliths simple cyrtoliths with open spaces between radiating elements of the central area.

Acanthoica aculeata Kamptner (Plate 3)
References: Kamptner, 1941, pp. 76, 97, Plate 1, Figs. 1 and 2; Gaarder & Hasle, 1971, p. 523, Fig. 2; Throndsen, 1972, pp. 56, 57, Figs. 16–19; Borsetti & Cati, 1976, p. 209, Plate 12, Fig. 1; Nishida, 1979, Plate 13, Fig. 3.
Spherical to ovoid cell having a diameter of 7.6–9.5 μm and covered by elliptical coccoliths. Ordinary coccoliths are elliptical cyrtoliths, usually slightly constricted in the middle, 2.3–3.5 μm long, with a length/breadth ratio of 1.2–1.7, and a number of lamellae (34–52) in the peripheral ring of the central area. About five rhabdoliths with a 1.2 μm long spine surround the flagellar pore.
Distribution: North Atlantic, Pacific and Indian Oceans, Mediterranean Sea

Acanthoica maxima Heimdal
References: Heimdal, 1981, in Heimdal & Gaarder, 1981, p. 39, Plate 1, Figs. 1 and 2; Norris, 1984, pp. 37, 38.

Plate 3 *Acanthoica aculeata* Kamptner. *Acanthoica quattrospina* Lohmann, ordinary coccolith: (a) distal surface view; (b) vertical longitudinal section. Pole coccolith: (c) vertical longitudinal section; a–c redrawn from Kamptner (1941), not to scale. *Algirosphaera oryza* Schlauder, redrawn from Norris (1984). *Algirosphaera quadricornu* (Schiller) Norris, ordinary coccolith: (a) distal surface view; (b) broad lateral view. Stomatal coccolith: (c) broad lateral view; a–c redrawn from Kamptner (1941), not to scale. *Alisphaera ordinata* (Kamptner) Heimdal, single coccolith: (a) distal surface view; (b) broad lateral view; (c) narrow lateral view; a–c not drawn to scale. *Anoplosolenia brasiliensis* (Lohmann) Deflandre: (a) single coccolith, not drawn to scale. *Braarudosphaera bigelowii* (Gran & Braarud) Deflandre. *Calcidiscus leptoporus* (Murray & Blackman) Loeblich & Tappan. *Calciopappus caudatus* Gaarder & Ramsfjell. Scale bar = 5 μm.

Dimorphic oval coccolith case about 23 μm in length without pole spines. Ordinary coccoliths are 2.9–4.2 μm long simple cyrtoliths with a length/ breadth ratio of 1.3–2.0, and a number of lamellae (40–56) in the peripheral ring of the central area. The central part is distinctly delimited and seems to be formed by simple overlapping of lamellae rising toward the center where a pore is present. Polar coccoliths, present at both poles, apparently have a smaller disc than the ordinary coccoliths and an appendix consisting of a 9.0–18.0 μm long spine composed of spirally arranged lamellae.

Distribution: North Atlantic and Indian Oceans.

Acanthoica quattrospina Lohmann (**Plate 3**)
Synonym: *Acanthoica acanthifera* Lohmann (Kamptner, 1941, pp. 76, 98, Plate 1, Figs. 5–8; Halldal & Markali, 1955, p. 16, Plate 19, Figs. 1 and 2; Borsetti & Cati, 1972, pp. 397, 398, Plate 39, Fig. 1).
References: Lohmann, 1903, p. 68, Plate 2, Figs. 23 and 24; Halldal & Markali, 1955, pp. 15, 16, Plate 18, Figs. 1 and 2; Borsetti & Cati, 1972, p. 398, Plate 39, Fig. 2; Nishida, 1979, Plate 13, Fig. 1.
  Cells are subspherical to oval in outline, 7.6–13.2 μm long, 6.0–9.5 μm wide, and with two flagella. The length of ordinary coccoliths is 1.1– 2.4 μm, with a length/breadth ratio of 1.0–1.4, and number of lamellae (20–33) in the peripheral ring of the central area. Polar coccoliths are present at both poles, with a maximum spine length of approximately 10 μm.
Distribution: Norwegian Sea, North Atlantic, Pacific and Indian Oceans, Mediterranean Sea, Australian waters.

Comparison of *A. aculeata, A. maxima,* and *A. quattrospina*

Acanthoica maxima differs from *A. aculeata* in the shape and size of the cell, and in having polar coccoliths at both poles of the coccolith case. It is separated from *A. quattrospina* by having a larger coccolith case consisting of a greater number of coccoliths. Furthermore, the ordinary coccoliths of *A. maxima* differ markedly from those of the latter species in being much larger and longer, elliptical in outline and in having about twice the number of lamellae in the basal plate.

Genus *Algirosphaera* Schlauder 1945, emend. Norris 1984
Type: *Algirosphaera oryza* Schlauder 1945 (type designated by Loeblich & Tappan, 1963, p. 191).
  Spherical to ellipsoidal cells with distinct flagellar areas and dimorphic sacculiform cyrtoliths. The basal plate of the coccoliths has a broad to narrow elliptical outline, open spaces between the elements radiating from the protuberance and a solid marginal ring. In stomatal coccoliths, the

sacculiform elevation of the central area is modified into a petaloid form. In ordinary coccoliths, the elevations have a similar shape but are not so high and often have an infolding at the distal end, giving a labiatiform appearance, with one or two distal pores.

Algirosphaera oryza Schlauder (Plate 3)
Synonyms: *Anthosphaera oryza* (Schlauder) Gaarder (Gaarder, 1971, in Gaarder & Hasle, 1971, pp. 523, 529, Fig. 4a–e; Reid, 1980, p. 157, Plate 3, Fig. 10); *Anthosphaera quadricornu* (Schiller) (Borsetti & Cati, 1972, p. 403, Plate 48, Fig. 1); *Anthosphaera quatricornu* (Schiller) (Winter et al., 1979, Plate 2, Fig. 7).
References: Schlauder, 1945, p. 23, Plate 5, Fig. 19; Norris, 1984, p. 38.
   Flattened, ellipsoidal coccospheres have a diameter about 13 μm, with two rows of equatorial coccoliths projecting approximately at right angles to each other giving the coccosphere a "coronate" profile (Gaarder & Hasle, 1971, Fig. 4a). The basal plate of the ordinary coccoliths is about 2.2 μm long, with an elevation up to 2.4 μm high. The inflated elevation of the stomatal coccoliths has a maximum width of 2.7 μm at its broadest point and is up to 4.6 μm high.
Distribution: Atlantic, Pacific and Indian Oceans, Mediterranean Sea, Australian waters.

Algirosphaera quadricornu (Schiller) Norris (Plate 3)
Basionym: *Syracosphaera quatricornu* Schiller [Schiller, 1914, p. 6, Plate 2, Fig. 19 ("18"); corrected to *S. quadricornu* (Schiller, 1925, p. 22)].
Synonyms: *Anthosphaera quadricornu* (Schiller) Halldal & Markali (Halldal & Markali, 1955, p. 17, Plate 21); *Anthosphaera robusta* (Lohmann) Kamptner *pro parte* (Kamptner, 1941, pp. 86, 87, Plate 9).
Reference: Norris, 1984, p. 38.
   Ellipsoidal cells having a diameter of 10–15 μm, with two flagella.
   Ordinary coccoliths are 1.9–2.5 μm long and 1.8–2.0 μm high cyrtoliths.
Distribution: Worldwide.

Algirosphaera robusta (Lohmann) Norris
Basionym: *Syracosphaera robusta* Lohmann (Lohmann, 1902, pp. 133, 135, Plate 4, Figs. 34 and 35).
Synonym: *Anthosphaera robusta* (Lohmann) Kamptner *pro parte* (Kamptner, 1941, pp. 86, 87, Plate 9; Halldal & Markali, 1954a, pp. 117, 118, Plate 1; Gaarder & Hasle, 1971, p. 529, Fig. 4f,g).
Reference: Norris, 1984, p. 38.
   Spherical cells having a diameter approximately 9–13 μm, with two flagella. Ordinary coccoliths consist of oval base plates 1.9–2.5 μm long and 1.1–1.3 μm wide with a 1.5 μm high elevation.
Distribution: Worldwide.

Comparison of *A. oryza, A. quadricornu,* and *A. robusta*

The shape of the coccosphere of *A. robusta* and *A. quadricornu* varies from spherical to flattened ellipsoidal (Lohmann, 1902; Schiller, 1930; Kamptner, 1941). Specimens of *A. quadricornu* are often considerably flattened like Halldal & Markali's Fig. 1 (1955, Plate 21). This character is even more pronounced in *A. oryza,* which in addition shows the "coronate" profile (Gaarder & Hasle, 1971, Fig. 4a). Ordinary coccoliths of *A. oryza* and *A. quadricornu* are indistinguishable even under an electron microscope, where as those of *A. robusta* are markedly lower than those of *A. quadricornu* (Halldal & Markali, 1955), giving the cells of this species a rather low-walled appearance compared to the other two species. In distal view, the outline of the ordinary coccoliths in *A. quadricornu* is elliptical rather than nearly parallel long sides as in *A. robusta.* The stomatal coccoliths of the three species can hardly be distinguished.

Genus *Alisphaera* Heimdal 1973
Type: *Alisphaera ordinata* (Kamptner) Heimdal 1973.
> Spherical to subspherical cells having two flagella and covered by uniform elliptical coccoliths wedged together in approximately regular meridian rows with the short coccolith axes in a polar direction. The proximal part of the coccolith consists of an elliptical disc attached to a short tube. This tubular part widens into a distal rim split in a larger flattened part and a smaller obliquely raised flange or tooth-like structure on one side. Regularly placed nodules occur at the inner periphery of distal rim opposite the raised area. The central area has a median longitudinal slit with irregular zig-zag borders.

Alisphaera capulata Heimdal
Reference: Heimdal, 1981, in Heimdal & Gaarder, 1981, pp. 39, 40, Plate 1, Figs. 3 and 4.
> Coccolith case is subspherical, and about 10 μm in diameter. The proximal coccolith disc is about 1.3 μm long and 0.6 μm wide with a coccolith tube 0.3 μm high. Maximum height of the tongue-shaped, raised part of the distal rim is 0.8 μm and the flattened slightly convex part is approximately 0.3 μm wide. The number of elements in the flattened region is approximately 20. The raised portion has a slight curve to the left if viewed from coccolith center. In distal view, no mark of distinction between tube and tongue-shaped protrusion can be observed from inside. Outside, the latter has a distinct narrow ledge near base.

Distribution: North Atlantic, Mediterranean Sea.

Alisphaera ordinata (Kamptner) Heimdal (Plate 3)
Basionym: *Acanthoica ordinata* Kamptner (Kamptner, 1941, pp. 76, 98, 113, Plate 1, Figs. 3 and 4).
Synonym: *Arisphaera ordinata* (Kamptner) Heimdal (Nishida, 1979, Plate 15, Fig. 2).

Reference: Heimdal, 1973, pp. 74, 75, Figs. 6–8.
Subspherical cells are 6.9–10.0 μm long and 7.8–10.0 μm wide, with two flagella. Coccoliths are 1.7–2.0 μm long, 0.9 μm wide, and 0.7–0.8 μm high, including the raised part of the distal rim. Height of the tubular wall is 0.2 μm. The proximal disc is about 1.4 μm long and 0.6 μm wide and the width of proximal rim is 0.1 μm. Both flattened and raised portions of the distal rim are approximately 0.5 μm wide with the flange spreading out peripherally, giving an impression of overlapping flattened part when viewed distally. When observed from above, no marked distinction can be made between the raised part of distal rim and tube continuing into the rim. In lateral view a faint proximal ridge is seen.
Distribution: North Atlantic and Pacific Oceans, Mediterranean Sea.

Alisphaera unicornis Okada & McIntyre
Synonym: *Arisphaera unicornis* Okada & McIntyre (Nishida, 1979, Plate 15, Fig. 1).
Reference: Okada & McIntyre, 1977, p. 18, Plate 6, Figs. 7 and 8.
Spherical to subspherical coccospheres having a diameter of 7.3–12.0 μm are composed of more than 100 overlapping coccoliths. Coccoliths have one side of the distal rim broader than the other, extending distally into a pointed projection, which is parallel to longer axis of coccolith. Proximal rim is much narrower than any part of the distal one. Coccolith length is 1.3–2.7 μm and width 0.8–1.9 μm exclusive of the beak-like protrusion.
Distribution: Pacific Ocean.

Comparison of *A. capulata, A. ordinata,* and *A. unicornis*
The cells of *A. unicornis* might be larger than those of *A. capulata* and *A. ordinata*. Furthermore, the coccoliths of *A. unicornis* differ from those of *A. ordinata* by having a beak-shaped projection on its distal rim, which has a continuous periphery, while the raised part of the distal rim of *A. capulata* is tongue-shaped and somewhat in between in *A. ordinata*.

Genus *Anoplosolenia* Deflandre 1952 (in Grassé 1952)
Type: *Anoplosolenia brasiliensis* (Lohmann) Deflandre 1952.
Cells are spindle-shaped with long, gradually tapering ends, which never bear spines. Coccoliths consist of rhombohedral heterococcoliths with a central area of parallel lamellae that usually mesh together in the center (scapholiths).

Anoplosolenia brasiliensis (Lohmann) Deflandre (Plate 3)
Basionym: *Cylindrotheca brasiliensis* Lohmann (Lohmann, 1919, p. 187, Fig. 56).
References: Deflandre, 1952, in Grassé, 1952, p. 458, Figs. 356D, 356E; Halldal & Markali, 1955, pp. 14, 15, Plate 16; Gaarder & Hasle, 1971, p. 523, Figs. 3a, b, c; Borsetti & Cati, 1972, p. 409, Plate 56, Fig. 1;

Manton & Oates, 1985, pp. 466–469, Plates 1 and 2, Figs. 1–7.
　　Cells are 70–100 μm long and 4–5 μm wide. The rhomboidal coccoliths covering the whole cell in a single layer become narrower but not shorter toward the ends. Coccoliths are 4.0–7.0 μm long, have a short diagonal axis of 2.0–2.5 μm, and height 0.2–0.3 μm.
Distribution: Atlantic, Indian and Pacific Oceans, Mediterranean Sea, and Australian waters.

Genus *Braarudosphaera* Deflandre 1947
Type: *Braarudosphaera bigelowii* (Gran & Braarud) Deflandre 1947.
　　Cells are covered by 12 pentaliths fit together in a regular pentagonal dodecahedron. The flat relatively thick pentaliths touch each other by the margin.

Braarudosphaera bigelowii (Gran & Braarud) Deflandre (Plate 3)
References: Deflandre, 1947, pp. 439–441, Figs. 1–5; Borsetti & Cati, 1972, p. 410, Plate 57, Fig. 3; Nishida, 1979, Plate 16, Fig. 1.
　　Cells with a diameter of 12–16 μm and coccoliths of 6–8 μm.
Distribution: Norwegian Sea, Atlantic and Pacific Oceans.
Remarks: The absence of an opening in coccospheres of *B. bigelowii* may suggest that these cells are cysts, but none have yet been germinated in culture. According to Lefort (1972), nonmotile coccoid cells of *B. magnei* Lefort have been observed to alternate with motile biflagellate cells, indicating that the two cell types are interconvertible.

Genus *Calcidiscus* Kamptner 1950
Type: *Calcidiscus leptoporus* (Murray & Blackman) Loeblich & Tappan 1978.
　　Spherical to subspherical coccosphere with coccoliths consisting of tightly interlocked, circular placoliths, which are formed of an upper and a lower shield connected by a tube.
Remarks: Kleijne (1991) found evidence that one species of this genus *Calcidiscus leptoporus* f. *leptoporus* and the holococcolith bearing *Crystallolithus rigidus* (p. 175) are two stages in the life cycle of one and the same species, and included this information in the description of the genus *Calcidiscus*.

Calcidiscus leptoporus (Murray & Blackman) Loeblich & Tappan (Plate 3)
Basionym: *Coccosphaera leptopora* Murray & Blackman (Murray & Blackman, 1898, pp. 430, 439, Plate 15, Figs. 1–7).
Synonyms: *Coccolithophora leptopora* (Murray & Blackman) Lohmann (Lohmann, 1902, p. 137, Plate 5, Figs. 52, 61–64); *Coccolithus leptoporus* (Murray and Blackman) Schiller (Schiller, 1930, pp. 100, 101, 245, Figs. 9a,b, 10, 121, 122; Kamptner, 1941, p. 94, Plate 13, Figs. 137–139);

Cyclococcolithus leptoporus (Murray & Blackman) Kamptner (Kamptner, 1954, p. 23, Fig. 20; Hasle, 1960, Plate 1, Figs. 3, 4, Plate 3, Figs. 1 and 2, McIntyre & Bé, 1967a, p. 569, Plate 7; Gaarder & Hasle, 1971, pp. 529, 533, Fig. 7; Nishida, 1979, Plate 4, Fig. 1; Hallegraeff, 1984, p. 233, Fig. 6).
Reference: Loeblich & Tappan, 1978, p. 1391.
  Nonmotile coccosphere: Spherical to subspherical coccospheres having a diameter of 22–28 μm. Coccoliths consist of circular placoliths, with a distal shield 6.3–11.3 μm in diameter and a proximal shield 4.4–8.5 μm. Both shields are composed of 16–31 petaloid elements with dextral imbrication in the distal shield and sinistral imbrication in the proximal shield. In distal view, the suture lines between elements are easily seen with a good quality light microscope. For an account of the motile stage, described independently as the holococcolithophorid *Crystallolithus rigidus* Gaarder (see p. 175).
Distribution: Worldwide.

Genus *Calciopappus* Gaarder & Ramsfjell 1954, emend. Manton & Oates 1983
Type: *Calciopappus caudatus* Gaarder & Ramsfjell 1954.
  Narrowly conical to cylindrical cells having a slightly contracted apex with a depression and usually a thin antapical process. The two flagella are equally long but the haptonema is very short. Two chloroplasts are located in the broadest part of the cell. Ordinary coccoliths are elliptical, nearly flat, narrow rimmed, incomplete caneoliths; they are longitudinally oriented and arranged in close hexagonal packing at times suggesting coaxial rings. An apical ring of spines (modified coccoliths) is located at the anterior end of the cell. Each spine, distributed along the rim of the apical depression, has a split base and is attached to a whorl of ring-shaped coccoliths, which have a finger-like projection toward the center of the whorl. Also present in the apical ring are coccoliths of the same basic structure as ordinary coccoliths, but bearing a central rod.
Remarks: For comparison of the genera *Calciopappus, Halopappus, Michaelsarsia,* and *Ophiaster* see p. 217 and Table 1.

Calciopappus caudatus Gaarder & Ramsfjell (Plate 3)
References: Gaarder & Ramsfjell, 1954, p. 155, Fig. 1; Gaarder et al., 1954, pp. 3–9, Figs. 1 and 2, Plates 1–4; Nishida 1979, Plate 10, Fig. 1; Manton & Oates, 1983, pp. 452–455, Plate 7, Plate 8, Figs. 38–42, Hallegraeff, 1984, p. 239, Fig. 37.
  The cell length without spines is 26–36 μm, with a greatest breadth of 3.5–4.0 μm, and an antapical process <1 μm broad. Ordinary coccoliths are approximately 2 μm long and 1 μm wide with a central area consisting of 28–36 lamellae running nearly parallel and somewhat

TABLE 1 Morphology of *Calciopappus (C)*, *Halopappus (H)*, *Michaelsarsia (M)*, and *Ophiaster (O)*

	C	H	M	O
Cell spherical to ovoid			+	+
Cell conical, elliptical or oblong	+	+		
Ordinary coccoliths placed at random			+	+
Ordinary coccoliths placed in hexagonal close packing sometimes suggesting coaxial rings, long axis mainly parallel to that of the cell	+	+		
Stomatal coccoliths like ordinary ones with high pointed central protrusion	+[a]			+
Stomatal coccoliths rhomboidal with low tube-like central protrusion			+	
Stomatal coccoliths rhomboidal with low mound-like central protrusion			+	
Apically placed whorl of arms:	+	+	+	
Ring-shaped coccolith wide-walled with bottom partially covered by flat bands, finger-like protrusion toward whorl-center and arm in one part	+			
Ring shaped coccolith wide-walled with apparently open bottom, arm with three links		+	+	
Antapically placed whorl of arms, proximal coccolith specially constructed, twisted, arm with numerous links				+

[a] Observed in *C. rigidus*.

obliquely to the nearly straight sides of the coccoliths. The number of slender apical spines is 5–14 with lengths from 20–30 μm.

Distribution: Norwegian Sea, Atlantic and Pacific Oceans.

Calciopappus rigidus Heimdal

Reference: Heimdal, 1981, in Heimdal & Gaarder, 1981, pp. 42, 44, Plate 2, Figs. 5–8.

Cells have a stiff, slender, cone-shaped, tetramorphic coccolith case. Ordinary coccoliths are arranged in coaxial rings with the long axis running parallel to long axis of cell. They have a length of 1.0–1.6 μm and length/breadth ratio of 1.4–1.8. The central area is formed of 16–25 μm flat lamellae arranged at approximately right angles to the long axis of the coccolith, and a solid flat central part. An apical whorl is formed by subcircular, overlapping coccoliths with a central opening partly filled by one or more flat bands, each with a finger-like projection toward center of the whorl. Between these coccoliths is a third type with the basic structure of an ordinary coccolith and a central rod of considerable length. A fourth coccolith type with a bayonet-like distal part and split base with flattened appendage is attached to the proximal side of the whorl coccoliths.

Distribution: North Atlantic and Indian Oceans, Mediterranean Sea.

Remarks: In preserved material, cells of *C. rigidus* are often found without the whorl and the attachment to the cell of the coccoliths bordering the apical area, and the connections between these coccoliths seem to be rather loose.

Comparison of *C. caudatus* and *C. rigidus*

The ordinary coccoliths of *C. rigidus* differ from those of *C. caudatus* in being smaller and having a lesser number of central lamellae running at nearly right angles to the long axis of the coccoliths. In most of the ordinary coccoliths of *C. caudatus* examined so far, the central lamellae run somewhat obliquely to the nearly straight sides of the coccoliths (Gaarder et al., 1954; Nishida, 1979, Plate 10, Fig. 1). As already pointed out by these authors, this may be due to a disarrangement during the preservation or isolation processes. The apical spines of *C. rigidus* resemble those of the type, and the architectures of the subcircular whorl coccoliths of *C. caudatus* and *C. rigidus* are also almost identical. But in contrast to *C. rigidus,* the *C. caudatus* specimens examined so far seem to lack the incomplete caneoliths bearing a central rod. The *C. caudatus* specimens studied by Gaarder et al. (1954) were, however, poorly preserved and it is difficult, on the basis of the transmission electron micrographs, to decide whether more than one kind of coccoliths with a spiny process was present on the cell. Even Nishida's seemingly well-preserved specimens failed to demonstrate this type of coccolith.

Genus *Calciosolenia* Gran 1912 (in Murray & Hjort, 1912)
Type: *Calciosolenia murrayi* Gran 1912.
Cells are long, cylindrical, enveloped by scapholiths, with apical end more constricted and elongated than cone-shaped antapical end, both ends with one or more long spines. These cells have two flagella.

Calciosolenia murrayi Gran (Plate 4)
Synonyms: *Calciosolenia sinuosa* Schlauder (Halldal & Markali, 1955, p. 15, Plate 17; *C.* aff. *murrayi* Gran (Manton & Oates, 1985, pp. 469–471, Plate 4).
References: Gran, 1912, in Murray & Hjort, 1912, p. 332; Gaarder & Hasle, 1971, p. 529, Fig. 3d,e.
Cells are 50–75 μm long and 4–5 μm wide. Coccoliths consist of scapholiths 2.5–3.5 μm long, 1.0–1.3 μm wide, and 0.2 μm high, with a central area of alternate band-shaped elements arranged at a right angle to the long axis as in *Anoplosolenia brasiliensis* (see p. 193).
Distribution: Atlantic and Pacific Oceans, Mediterranean Sea, Australian waters.

Comparison of *A. brasiliensis* and *C. murrayi*

Calciosolenia murrayi is characterized by having a cylindrical shape with a neck-like constriction at the apical end and a cone-shaped antapical end, both terminating with spines. The cylindrical shape and terminal spines are easily

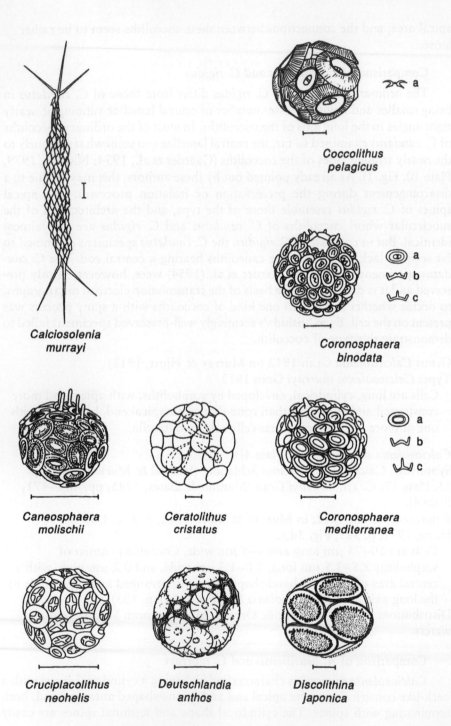

Calciosolenia murrayi

Coccolithus pelagicus

Coronosphaera binodata

Caneosphaera molischii

Ceratolithus cristatus

Coronosphaera mediterranea

Cruciplacolithus neohelis

Deutschlandia anthos

Discolithina japonica

recognized with the light microscope. In contrast, *A. brasiliensis* has a spindle-shaped coccolith case tapering gradually at both ends into long horns and generally longer coccoliths with greater number of lamellae. The cells of *C. murrayi* appear more robust and easier to preserve than those of *A. brasiliensis*.

Genus *Caneosphaera* Gaarder 1977 (in Gaarder & Heimdal, 1977)
Type: *Caneosphaera halldalii* (Gaarder) Gaarder 1977.
Spherical to subspherical cells having a distinct flagellar area. The coccolith case consists of a single layer, and stomatal coccoliths have a central rod. Ordinary and stomatal coccoliths are complete caneoliths, elliptical in outline with a broad distal rim and narrow proximal rim. The bottom plate has a peripheral ring of narrow radial lamellae fused along a central line or connected to a more or less complex central field. The coccolith wall is approximately at a right angle to the bottom and forms a continuous tube.

Caneosphaera halldalii (Gaarder) Gaarder
Basionym: *Syracosphaera halldalii* Gaarder 1971, in Gaarder & Hasle (1971, p. 536).
Synonym: *Syracosphaera mediterranea* Lohmann (Halldal & Markali, 1954b, p. 330, Fig. 2).
Reference: Gaarder, 1977, in Gaarder & Heimdal, 1977, pp. 64, 66, Plate 6, Figs. 35–39.
Cell diameter is about 10 μm. Coccoliths are 2.0–3.3 μm long, have a length/breadth ratio of 1.3–1.5 and height of 0.4–0.6 μm. The length of the stomatal rod is approximately 1.1 μm, while the breadth of distal rim about 0.3 μm. The slightly concave coccolith bottom consists of 22–30 flat lamellae, arranged at right angles to the elliptical outline. The wall is formed of alternating broad and narrow elements. Each bottom rib corresponds to a broad element in the wall and to the center of an element in the upper rim. The uppermost part of the wall may or may not bear a varying number of finger-like centripetal protrusions. These protrusions may differ in length and be more or less irregularly spaced,

---

Plate 4. *Calciosolenia murrayi* Gran. *Caneosphaera molischii* (Schiller) Gaarder. *Ceratolithus cristatus* Kamptner. *Coccolithus pelagicus* (Wallich) Schiller: (a) single coccolith, vertical section, not to scale. *Coronosphaera binodata* (Kamptner) Gaarder, ordinary coccolith: (a) distal surface view; (b) vertical longitudinal section. Stomatal coccolith: (c) vertical longitudinal section; a–c redrawn from Kamptner (1941), not to scale. *Coronosphaera mediterranea* (Lohmann) Gaarder, ordinary coccolith: (a) distal surface view; (b) vertical longitudinal section. Stomatal coccolith: (c) vertical longitudinal section; a–c redrawn from Kamptner (1941), not to scale. *Cruciplacolithus neohelis* (McIntyre & Bé) Reinhardt. *Deutschlandia anthos* Lohmann. *Discolithina japonica* Takayama. Scale bar = 5 μm.

but their number always seems to be less than the bottom lamellae and
distal rim elements.
Distribution: North Atlantic and Pacific Oceans, Mediterranean Sea,
Australian waters.

Caneosphaera molischii (Schiller) Gaarder (**Plate 4**)
Basionym: *Syracosphaera Molischi* Schiller (Schiller, 1925, p. 21, Fig. K).
Synonyms: *Syracosphaera corrugis* Okada & McIntyre (Okada & McIntyre,
1977, p. 21, Plate 8, Figs. 3 and 6); *S. molischii* Schiller (Okada &
McIntyre, 1977, p. 24, Plate 8, Figs. 4 and 5); *S. elatensis* Winter (Winter et
al., 1979, p. 207, Plate 3, Figs. 11–13).
References: Gaarder, 1977, in Gaarder & Heimdal, 1977, pp. 66, 68, Plate
7, Figs. 40–46, Plate 8, Figs. 47–49; Nishida, 1979, Plate 7, Fig. 2;
Gaarder, 1981, in Heimdal & Gaarder, 1981, pp. 44, 46, Plate 3;
Hallegraeff, 1984, p. 242, Fig. 47.

Spherical to subspherical cells with a diameter of 4.5–9.0 μm, coccoliths
1.2–3.3 μm long, a length/breadth ratio of 1.1–1.7, and height of
0.2–0.3 μm. The number of bottom lamellae varies from 18–35. These
cells have two flagella. Stomatal coccoliths are usually smaller and more
circular in outline than ordinary ones, with a 1.1–1.8 μm high protrusion
that can be rounded or flattened at right angles to the short coccolith axis
or even bear four wings along its axis. Maximal breadth of the flattened
protrusion is 0.4–0.7 μm. Ordinary coccoliths are very similar to those of
C. halldalii except for the central area, where an excess precipitate of
calcite may form a more or less intricate and variable pattern, even in a
single cell. The distal rim has a breadth of 0.3–0.8 μm and is constructed
of the same number of elements as those in the bottom, and presumably
as those in the wall. On the distal side the border lines between the single
elements are most often obscured by an excess of calcite precipitate in the
radially placed ribs. The tooth-like protrusions from the inner side of the
coccolith wall, present in some specimens of *C. molischii,* seem to extend
farther down the wall than the corresponding finger-like protrusions in *C.
halldalii.*
Distribution: Norwegian Sea, Atlantic, Indian and Pacific Oceans,
Mediterranean Sea, Australian waters.
Remarks: According to Okada & McIntyre (1977), Nishida (1979), Winter
et al. (1979), and Heimdal & Gaarder (1981), *C. molischii* may occur
bearing some deviating coccoliths, most probably placed around the flagellar
opening. They are elliptical in outline with an asymmetrical outer rim
consisting of about 20–30 elements and with a fan-like widening on one
side along the long axis (Heimdal & Gaarder, 1981). Okada & McIntyre
(1977) described these cells as dithecate, which would mean having two
complete layers of coccoliths. *Syracosphaera halldalii* and *S. molischii* were

transferred to the genus *Caneosphaera* by Gaarder & Heimdal (1977) on the basis of the coccosphere being monothecate and the coccoliths lacking an intermediate continuous or beaded mid wall rim. Recently Jordan & Young (1990) found the reliability of the generic description of *Caneosphaera* uncertain and they therefore proposed that the species of *Caneosphaera* be transferred back to *Syracosphaera*. However, judging from the illustrations hitherto published and from my own observations it seems probable that the external coccoliths of *C. molischii* are restricted to the apical area. Until better illustrations of these coccoliths and particularly their arrangement on the cell are available, I therefore wish to retain the species in the genus *Caneosphaera*. As to the type of the genus, *C. halldalii* and its form *C. halldalii* f. *dilatata* Heimdal, so few specimens have hitherto been observed that the possibility of their bearing external coccoliths is still questionable and I therefore hesitate to revise the diagnosis of the genus at present.

Genus *Ceratolithus* Kamptner 1950
Type: *Ceratolithus cristatus* Kamptner 1950.
Cells without flagella are engirdled by horseshoe-shaped coccoliths (ceratoliths). Coccoliths are somewhat asymmetrical in form, with one side being slightly shorter than the other. The cell with its engirdling coccolith is closed within a sphere of circular to oval ring-shaped coccoliths, which overlap. Several cells, each with a ceratolith, may be present within a single sphere.

***Ceratolithus cristatus* Kamptner (Plate 4)**
Synonym: *Ceratolithus telesmus* Norris (Jordan & Young, 1990, pp. 15, 16).
References: Kamptner, 1950, p. 154; Kamptner, 1954, pp. 43–45, Figs. 44 and 45; Norris, 1965, pp. 19–21, Plate 11, Figs. 1–4, Plate 12; Borsetti & Cati, 1976, p. 224, Plate 17, Figs. 1–8; Nishida, 1979, Plate 13, Fig. 2.
Elliptical cells are 15–18 μm long, 12–15 μm wide, with four chloroplasts, and have a small, usually centrally located nucleus. Coccoliths of the ceratolith type are 10–36 μm long. Ring-shaped coccoliths are circular to oval having a diameter of 7–9 μm, with thickened rims.
Distribution: Atlantic, Indian, and Pacific Oceans, Mediterranean Sea.
Remarks: According to Jordan & Young (1990), there are considerable variations in ceratolith morphology with continuous transition between the *C. cristatus* and *C. telesmus* morphotypes. They have therefore proposed that the status of *C. telesmus* should be changed to that of a variety, *C. cristatus* var. *telesmus* (Norris) Jordan & Young.

Genus *Coccolithus* Schwarz 1894
Type: *Coccolithus oceanicus* Schwarz 1894.

Spherical to subspherical cells have a nonmotile stage bearing monomorphic coccoliths consisting of tightly interwedged placoliths (heterococcoliths) with both shields constructed from imbricating radial segments. In one species the motile stage bears holococcoliths.

Coccolithus pelagicus (Wallich) Schiller (**Plate 4**)
Basionym: *Coccosphaera pelagica* Wallich (Wallich, 1877, p. 348, Plate 17, Figs. 1, 2, 5, 11d, 16).
References: Schiller, 1930, p. 246, Fig. 123, a, c, and d; Kamptner, 1954, p. 20, Figs. 14–16; McIntyre & Bé, 1967a, pp. 569, 570, Plate 8; McIntyre et al., 1967, p. 11, Plate 4, Fig. A, B; Okada & McIntyre, 1977, p. 6, Plate 1, Figs. 5, 9, 10; Nishida, 1979, Plate 3, Fig. 1; Heimdal & Gaarder, 1981, p. 48; Samtleben & Schröder, 1990, p. 37, Plate 1, Fig. 8.

Nonmotile coccosphere: Nonmotile *Coccolithus pelagicus* cells are spherical to subspherical and up to 32 μm in diameter. Coccoliths consist of placoliths, oval in plan view, with wide shields convex distally, concave proximally and a narrow elliptical central pore. A bar made of rhombic elements crosses the central pore at right angles to the long axis of the pore. Sometimes this bar is situated near the bottom of the coccolith tube and hence is not always clearly seen. Each element of the distal shield overlaps the one to the right when viewed from the center of the coccolith, and the imbrication is dextral, while the proximal shield is smaller and the sutures inclined clockwise. The length of distal shield is 6.8–10.9 μm, its breadth 5.3–9.1 μm, and element counts for both shields vary from 36 to 44.

For a description of the other life cycle stages, described independently as the holococcolithophorids *Crystallolithus braarudii* Gaarder and *C. hyalinus* Gaarder & Markali, see pp. 174 and 175.

Distribution: According to McIntyre & Bé (1967a), *Coccolithus pelagicus* is a cold-water species limited to subarctic and cold temperate waters in the North Atlantic. I have also found occasional coccospheres of this species in samples from the Mediterranean, and Nishida (1979) reported it as a common species in the cold current around New Zealand.

Genus *Coronosphaera* Gaarder 1977 (in Gaarder & Heimdal, 1977)
Type: *Coronosphaera mediterranea* (Lohmann) Gaarder 1977.

Spherical to subspherical cells, which have distinct flagellar areas and are covered by a single layer of dimorphic coccoliths. Ordinary and stomatal coccoliths are elliptical in outline, with a similar basic structure. The proximal rim is extremely narrow, and the bottom plate consists of a peripheral ring of radial lamellae connected to a solid central field of varying structure. The wall is compact, with an intermediate centripetal widening, apparently composed of tightly compressed flat elements of variable size and height. Peripheral elements end in a crown shaped wall.

Coronosphaera binodata (Kamptner) Gaarder (**Plate 4**)
Basionym: *Syracosphaera mediterranea* Lohmann var. *binodata* Kamptner (Kamptner, 1927, p. 178, Fig. 3).
Synonyms: *Coccolithophora pelagica* (Wallich) Lohmann (Schiller 1925, p. 36, Fig. T); *Syracosphaera binodata* (Kamptner) Kamptner (Kamptner, 1937b, p. 300, Plate 14, Figs. 7–9; Kamptner, 1941, p. 82, Plate 5, Figs. 55–57, Plate 6, Fig. 59).
References: Gaarder, 1977, in Gaarder & Heimdal, 1977, p. 62, Plate 5, Figs. 27–32; Nishida, 1979, Plate 6, Fig. 2.
  Cells are mostly spherical, 15.0–25.0 μm long and 15.0–18.5 μm wide. Coccoliths are 2.7–4.2 μm long, have a length/breadth ratio of 1.3–1.5 and height of 1.1–1.3 μm. The number of bottom lamellae varies from 33 to 49. The length of the stomatal rod is 1.2–1.6 μm. The central field of ordinary coccoliths bears two pointed knobs distinguishable in the light microscope, especially when using oblique light.
Distribution: Atlantic and Pacific Oceans, Mediterranean Sea.

Coronosphaera maxima (Halldal & Markali) Gaarder
Basionym: *Syracosphaera maxima* Halldal & Markali (Halldal & Markali, 1955, p. 11, Plates 8, 9).
References: Gaarder, 1977, in Gaarder & Heimdal, 1977, p. 62, Plate 5, Figs. 33, 34; Reid, 1980, p. 160, Plate 5, Figs. 1–3.
  Spherical to ovoid cells, which vary from 26 to 53 μm in size. Ordinary coccoliths, 3.0–6.0 μm long, have a length/breadth ratio of 1.3–1.6 and height of 0.7–1.4 μm. The bottom plate has 60–80 thin and narrowly spaced lamellae and a central structure parted in two or more sections in some broad elliptical coccoliths. Stomatal coccoliths, about 10 in number, are smaller than the ordinary ones and are extremely narrow, with a length/breadth ratio up to 2.0, and 1.0–2.7 μm long central rod.
Distribution: Atlantic and Pacific Oceans.

Coronosphaera mediterranea (Lohmann) Gaarder (**Plate 4**)
Basionym: *Syracosphaera mediterranea* Lohmann (Lohmann, 1902, pp. 133, 134, Plate 4, Figs. 31, 31a, 32).
Synonyms: *Syracosphaera tuberculata* Kamptner (Kamptner, 1937b, p. 302, Plate 15, Figs. 17–19; Kamptner, 1941, pp. 86, 106, Plate 8, Figs. 85–87, Plate 9, Fig. 88; Borsetti & Cati, 1972, p. 402, Plate 47, Fig. 2); *Syracosphaera pulchra* Lohmann (Halldal & Markali, 1954b, p. 332, Fig. 4); *Syracosphaera pulchroides* Halldal & Markali (Halldal & Markali, 1955, pp. 10, 11).
References: Gaarder, 1977, in Gaarder & Heimdal, 1977, pp. 60, 62, Plate 4, Figs. 21–26; Nishida, 1979, Plate 6, Fig. 1; Hallegraeff, 1984, p. 242, Fig. 48a, b, c.

Biflagellate, spherical to ellipsoidal cells are 13.0–17.0 μm long (13.0–24.0 μm in Hallegraeff, 1984), and 13.0–16.0 μm wide. Coccoliths consist of a single layer of dimorphic coccoliths 3.0–4.2 μm long, with a length/breadth ratio of 1.3–1.6, and height of 0.9–1.4 μm. The number of bottom lamellae varies from 46 to 54 and the stomatal rod is 1.2–2.0 μm long. The central field of ordinary coccoliths appears to be a structureless mass composed of two flattened parts connected by a transverse keel placed at or near the center.

Distribution: Worldwide.

Comparison of *C. maxima, C. binodata,* and *C. mediterranea*

The cells of *C. maxima* are larger than those of the other two species, and they appear to have rather low walls in relation to both coccolith and cell size. The extremely small and narrow stomatal coccoliths also distinguish this species from the others. The specimens of *C. binodata* usually exceed those of *C. mediterranea* in size and give the impression of being more opaque. Also, the coccoliths seem to be relatively higher in *C. binodata* than in *C. mediterranea*. In proximal view, ordinary coccoliths of *C. binodata* and *C. mediterranea* cannot be distinguished, while in distal view the coccoliths of *C. binodata* are easily recognized by the two separate knobs at the bottom. Even if the central field of coccoliths of *C. maxima* are similar to that of *C. mediterranea*, they may be readily identified in proximal view by the numerous, closely spaced, narrow bottom lamellae.

Genus *Crenalithus* Roth 1973
Type: *Crenalithus doronicoides* Roth 1973 (fossil).

Coccosphere of spherical to subspherical shape with coccoliths consisting of tightly interwedged placoliths, which are oval to elliptical in outline. The placoliths have two shields, approximately equal in size and a large central pore. Proximal area of the central pore is closed by a plate constructed of interconnected rods with open spaces between forming a reticulate grid or by interlocking rods with no open spaces forming a solid plate.

Crenalithus parvulus Okada & McIntyre
References: Okada & McIntyre, 1977, pp. 6, 7, Plate 2, Figs. 1 and 2; Heimdal & Gaarder, 1981, p. 48, Plate 4, Fig. 17.

Coccospheres are spherical to subspherical with a diameter of 3.0–3.8 μm with approximately 10–25 coccoliths of the placolith type. The distal shield is 1.5–2.0 μm long, 1.2–1.6 μm wide, and composed of 20–30 elements.

Distribution: North Atlantic and Pacific Oceans.

Remarks: According to Jordan & Young (1990), many of the species previously described as *Crenalithus* belong in the genus *Reticulofenestra* described by Hay et al. (1966). They therefore proposed the new combination *R. parvula* (Okada & McIntyre) Jordan & Young.

Genus *Cricosphaera* Braarud 1960
Type: *Cricosphaera elongata* (Droop) Braarud 1960 (type designated by Chrétiennot-Dinet, 1990, p. 99).
Cells are elongate to globular with two equally long flagella and a short haptonema. They contain monomorphic coccoliths of the cricolith type. Benthic stages are unkown.

Genus *Cruciplacolithus* Hay & Mohler 1967 (in Hay et al., 1967)
Type: *Cruciplacolithus tenuis* (Stradner) Hay & Mohler 1967 (fossil).
Spherical to subspherical cells, which have a nonmotile stage bearing monomorphic coccoliths. The coccoliths are tightly interwedged elliptical placoliths having a cruciform structure of intergrown rhombic crystals that partially cover the central pore. A coccolith-bearing, motile stage is known for one species.

Cruciplacolithus neohelis (McIntyre & Bé) Reinhardt (**Plate 4**)
Basionym: *Coccolithus neohelis* McIntyre & Bé (McIntyre & Bé, 1967b, pp. 369–371, Fig. 1).
References: Reinhardt, 1972, p. 52, Fig. 54; Fresnel, 1989, pp. 111–129, Figs. 135–166.
Nonmotile coccosphere: Cells are spherical to subspherical, with a diameter of 7.5–9.0 μm. Elements of the larger distal shield are dextrally imbricate with a pronounced suture pattern toward the margin. In the smaller proximal shield, the slightly dextrally imbricate elements are flat, broad, and with uneven margins. Arms of the central cross-like structure are oriented along the major and minor axes of the central pore. The length of the distal shield is 2.2–3.2 μm, breadth 1.8–2.6 μm, and average coccolith height is 0.4 μm. The number of elements in the distal shield varies from 19 to 28.
Motile stage: Biflagellate coccolith-bearing cells without any visible haptonema are about 10 × 6 μm in size. Flagella are approximately 7 μm long. Internal structure and coccolith type are similar to the nonmotile stage. Reproduction occurs by division in both nonmotile and motile stages. The exact position of the flagellated cells in life history is unknown.
Distribution: Littoral, Atlantic and Pacific Oceans.

Genus *Deutschlandia* Lohmann 1912, emend. Gaarder 1981 (in Heimdal & Gaarder, 1981)
Type: *Deutschlandia anthos* Lohmann 1912.
Cells are spherical to subspherical and have two flagella. Coccolith case is

dithecate, with a dimorphic endotheca and monomorphic exotheca.
Endothecal coccoliths consist of incomplete caneoliths, elliptical in outline
and touching each other. Stomatal coccoliths have a central rod.
Exothecal coccoliths are circular cyrtoliths, slightly concave proximally,
with a wide, flat marginal part and a narrow central part raising distally
into a hollow cone. Discs overlap each other to form a complete sphere,
which is often open apically.

Deutschlandia anthos Lohmann (Plate 4)

Synonym: *Syracosphaera variabilis* (Halldal & Markali) Okada & McIntyre
(Okada & McIntyre, 1977, p. 27, Plate 9, Figs. 7 and 8; Nishida, 1979,
Plate 8, Fig. 1; Winter et al., 1979, p. 210, Plate 4, Figs. 7 and 8).
References: Lohmann, 1912, p. 47, Fig. 10, 1, 3 (not 2); Reid, 1980, p. 156,
Plate 2, Figs. 5 and 6; Heimdal & Gaarder, 1981, pp. 48, 50, 51, Plate 5,
Figs. 23–26; Hallegraeff, 1984, p. 236, Fig. 25.

Cells are mostly spherical with an endothecal sphere of 7.4–10.0 μm in
diameter and two equally long flagella emerging from a narrow flagellar
opening having a length about one and a half times the endothecal
diameter. The endothecal sphere is composed of approximately 30–50
incomplete caneoliths, 1.7–3.0 μm long, with a length/breadth ratio of
1.3–1.5, and stomatal rods of 1.2–1.6 μm long. The number of bottom
lamellae varies from 16 to 34. Exothecal coccoliths are circular cyrtoliths,
3.0–5.5 μm in diameter, with a wide peripheral collar composed of
17–21 flat radiating elements. A distally raised hollow cone is constructed
of eight specially shaped elements with eight trapezoidal openings at their
base.
Distribution: Atlantic and Pacific Oceans, Mediterranean Sea, Australian
waters.
Remarks: According to Jordan & Young (1990), the reliability of the
description of *Deutschlandia* is still uncertain. They therefore proposed that
C. anthos be transferred back to the genus *Syracosphaera* (pp. 224) and
proposed the new combination *S. anthos* (Lohmann) Jordan & Young.

Genus *Discolithina* Loeblich & Tappan 1963

Type: *Discolithina vigintiforata* (Kamptner ex Deflandre) Loeblich &
Tappan 1963 (fossil).

Cells are subspherical to spherical covered by large elliptical cribriliths
having a smooth periphery and touching each other.

Discolithina japonica Takayama (Plate 4)

Synonym: *Pontosphaera japonica* (Takayama) Burns (Burns, 1973, p. 154,
Plate 2, Figs. 10–13; Hallegraeff, 1984, p. 236, Fig. 23).
References: Takayama, 1967, pp. 189, 190, Plates 9 and 10; Okada &

McIntyre, 1977, p. 15, Plate 6, Fig. 3; Reid, 1980, p. 156, Plate 2, Figs. 7–9.

Spherical to subspherical coccospheres are 17.5–21.0 μm along the longer axis. Coccospheres consist of approximately 15 cribriliths, about 7 μm long × 5 μm wide × 0.4 μm high, with numerous small holes in the central part of the distal surface and circular depressions on the distal margin.

Distribution: North Atlantic and Pacific Oceans, Australian waters.

Remarks: The holotype is a fossil coccolith, which has a longitudinal slit in the central area. The study of recent material so far shows some specimens with a slit, while others lack this structure. Also, the size and number of holes in the central part of the disc vary greatly. Coccoliths of this species resemble the discoliths of *Scyphosphaera apsteinii* Lohmann (p. 224), which has randomly placed larger pores in small numbers inside the smooth border of the disc and lack the circular depressions on the distal margin.

Genus *Discosphaera* Haeckel 1894
Type: *Discosphaera tubifer* (Murray & Blackman) Ostenfeld 1900.

Spherical cells have uniform salpingiform cyrtoliths with a trumpet-shaped appendix considerably longer than coccolith base plate diameter. Circular to elliptical base plates have marginal rings of flat lamellae connected to a homogeneous inner center by separated flattened elements leaving small open spaces between radiating elements inside each marginal ring. A pore in the center of the plate represents the attachment area of the appendix.

Discosphaera tubifer (Murray & Blackman) Ostenfeld (Plate 5)
References: Ostenfeld, 1900, p. 200; McIntyre & Bé, 1967a, p. 566, Plate 1; Gaarder & Hasle, 1971, p. 533, Fig. 8; Nishida, 1979, Plate 12, Fig. 1; Hallegraeff, 1984, p. 236, Figs. 34–36.

Cells without processes have diameters of 7–14 μm, and coccoliths similar to those in *Acanthoica* (p. 189) in most characteristics, the main difference being the salpingiform process of the uniform coccoliths in *Discosphaera*. The diameter of the coccolith base plate is 1.6–2.3 μm, and the number of elements in the marginal ring of the base plate are 24–30. The length of the appendix is 4–9 μm and the diameter of bell 2.5–4.2 μm.

Distribution: Worldwide.

Genus *Emiliania* Hay & Mohler 1967 (in Hay et al., 1967)
Type: *Emiliania huxleyi* (Lohmann) Hay & Mohler 1967.
Monospecific genus.

This coccolithophorid has three cell types: nonmotile spherical coccolith-bearing cells enclosed in a covering of oval placoliths, completely naked

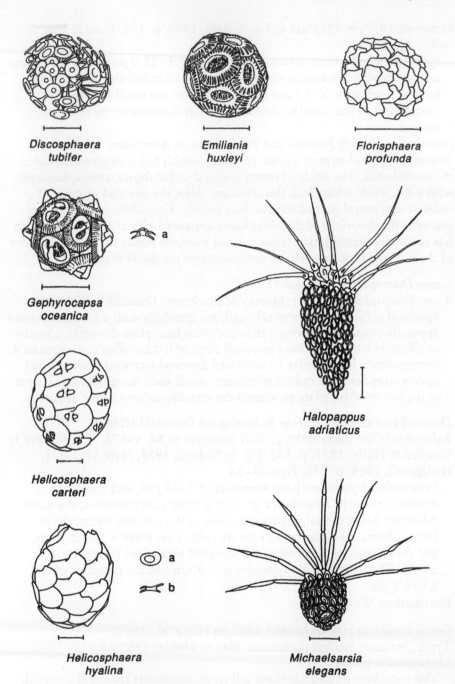

Discoscphaera
tubifer

Emiliania
huxleyi

Florisphaera
profunda

Gephyrocapsa
oceanica

Halopappus
adriaticus

Helicosphaera
carteri

Helicosphaera
hyalina

Michaelsarsia
elegans

cells, and scaly motile cells with flagella, each type capable of independent vegetative propagation. Placoliths have shields equal in size with a solid or reticulate central pore. The distal shield is formed by T-shaped elements with interlocking ends at the margin. The shape and development of the proximal shield elements are variable.

Emiliania huxleyi (Lohmann) Hay & Mohler (Plate 5)
Basionym: *Pontosphaera huxleyi* Lohmann (Lohmann, 1902, pp. 129, 130, Plate 4, Figs. 1–9, Plate 6, Fig. 69).
Synonyms: *Hymenomonas huxleyi* (Lohmann) Kamptner (Kamptner, 1930, pp. 155, 159); *Coccolithus huxleyi* (Lohmann) Kamptner (Kamptner, 1943, p. 43); McIntyre & Bé, 1967a, pp. 568, 569, Plate 5, Fig. D, Plate 6); *Gephyrocapsa huxleyi* (Lohmann) Reinhardt (Reinhardt, 1972, p. 89).
References: Hay & Mohler, 1967, in Hay et al., 1967, p. 447, Plates 10 and 11, Figs. 1 and 2; Borsetti & Cati, 1972, p. 405, Plate 51, Fig. 2; Okada & Honjo, 1973, Plate 1, Figs. 1–3; Nishida, 1979, Plate 1, Fig. 1; Winter et al., 1979, pp. 200, 201, Plate 1, Figs. 2–5; Hallegraeff, 1984, p. 233, Figs. 8–12.

Nonmotile coccosphere: Cells are spherical to subspherical and 5–10 μm in diameter. Coccoliths consist of partially interlocked placoliths. In stationary cultures and sometimes in nature cells can form multiple layers of coccoliths. Coccoliths average 3.5 μm long × 2.9 μm wide, but may vary considerably even on one coccosphere. Observations of *E. huxleyi* from different water masses suggest that the number of elements, the development of the shields, and the proximal covering of the central tube are related to water temperature. Specimens from warm waters have both shields formed of long T-shaped elements, which are fused at their outermost ends, and a delicate reticulate grid covering the proximal end of the central tube. There seems to be a gradual increase in calcification in colder waters. The additional calcite thickens the individual elements so that the spaces between them are gradually narrowed, and extends the wall of the central tube. In the most heavily calcified specimens the elements become so broad and close that they form a nearly solid shield, and the proximal covering of the central tube consists of a solid plate of many interlocking rods. In some heavily calcified specimens the extra

---

Plate 5. *Discosphaera tubifer* (Murray & Blackman) Ostenfeld. *Emiliania huxleyi* (Lohmann) Hay & Mohler. *Florisphaera profunda* Okada & Honjo. *Gephyrocapsa oceanica* Kamptner: (a) single coccolith, vertical section, not drawn to scale. *Halopappus adriaticus* Schiller. *Helicosphaera carteri* (Wallich) Kamptner. *Helicosphaera hyalina* Gaarder, single coccolith: (a) distal surface view; (b) broad lateral view; a, b not drawn to scale. *Michaelsarsia elegans* Gran. Scale bar = 5 μm.

calcite may cover the central tube completely. All gradations may be found between these extremes.

Distribution: This cosmopolitan species is the most ubiquitous coccolithophorid species in today's oceans. Two recent papers (von Bleijswijk et al., 1991; Young & Westbroek, 1991) based on coccolith morphology and immunological properties of the coccolith polysaccharide show, however, that there are distinct types of *E. huxleyi* that may differ in their physiology and ecology. Taxonomically, the status of these types is still uncertain and their extablishment as subtaxon of *E. huxleyi* must await future work (Young & Westbroek, 1991).

Remarks: *Emiliana huxleyi* has an internal structure that agrees well with the general concepts of a prymnesiophycean cell. But unlike the other coccolithophorids studied so far, *E. huxleyi* has no base plate scale, although scale-bearing forms have been observed as an alternative phase in life cycle (Klaveness, 1972b). Hence Klaveness & Paasche (1979) indicated that *E. huxleyi* may be more closely related to certain genera of noncalcified flagellates than to the more typical coccolithophorids.

It is obvious from the list of synonyms that the taxonomic history of *E. huxleyi* has been rather turbulent, and its systematic position may still be changed. I agree with Hay et al. (1967) that the segments of the distal shield are strongly imbricate in *Coccolithus*, while in *Emiliania* the distal shield is formed of T-shaped elements separated from each other except at the margins, where they are fused. In *Gephyrocapsa* the elements of both shields are slighly imbricate. Previous descriptions of *Gephyrocapsa* spp., however, are based almost exclusively on light and electron microscopy of nonreplicated or unshadowed wild material, and there is no information on the presence or absence of scales in this genus. Awaiting this, I hesitate to adopt the generic name *Gephyrocapsa* proposed by Reinhardt (1972).

Genus *Florisphaera* Okada & Honjo 1973
Type: *Florisphaera profunda* Okada & Honjo 1973.
> Coccospheres are subspherical with a shallow dome top. Polygonal plate-shaped coccoliths overlap like tiles from the base, forming a rosette when spread open in apical view. In lateral view, the upper half of the sphere is steplike and increases in height toward center.

Florisphaera profunda Okada & Honjo (Plate 5)
Synonym: Coccolithophorid sp. 1 (Throndsen, 1972, p. 59, Figs. 29–32).
References: Okada & Honjo, 1973, pp. 373, 374, Plate 2, Figs. 4 and 5; Borsetti & Cati, 1976, p. 225, Plate 18, Figs. 5 and 6; Okada & McIntyre, 1977, p. 36; Nishida, 1979, Plate 16, Fig. 3; Reid, 1980, p. 168, Plate 8, Figs. 3 and 4.
> Coccospheres have an apical axis length ranging from 2.7 to 7.5 µm and diameter from 3.7 to 8.5 µm. The covering consists of approximately 30 to more than 100 coccoliths. The quadrangular coccoliths are

approximately 2.0 μm long × 1.5 μm wide and have sides that slope slightly inward toward the base ending roughly parallel. The thickness varies from approximately 0.05 to 0.25 μm, with the maximum at the basal end. The lower end is angled at the corners and one side has a small protuberance, while the upper edge has two blunt projections, giving it a wavy appearance.

Distribution: North Atlantic and Pacific Oceans.

Remarks: *Florisphaera profunda* var. *elongata* Okada & McIntyre (1977, p. 36, and validated in Okada & McIntyre, 1980) differs from this species in having 4.5 to 12.0 μm high and 8.3 to 13.2 μm wide coccospheres and larger coccoliths (length 1.7–7.8 μm, width 1.3–5.0 μm). Their thickness is similar to that of *F. profunda*. Kleijne (1991) indicates that *F. profunda* var. *elongata* and an unidentified holococcolithphorid type B may be two different stages in the life cycle of one and the same species.

Genus *Gephyrocapsa* Kamptner 1943

Type: *Gephyrocapsa oceanica* Kamptner 1943.

Spherical to subspherical cells, which have tightly interlocked elliptical placoliths consisting of two shields connected by a short tube. The proximal side of the central tube is covered by a reticulate grid, while the distal side is spanned by a bridge usually formed by two diametrically opposite plates placed at different angles to the long axis of the coccolith. Bridge plates may be overlapping, offset, touching, or separated and one or both may be lacking. The elements of the shields are slightly imbricate, dextral in the distal and sinistral on the proximal shield.

Gephyrocapsa caribbeanica Boudreaux & Hay

Synonym: *Gephyrocapsa oceanica* Kamptner (McIntyre & Bé, 1967a, Plate 9, Fig. A).

Reference: Boudreaux & Hay, 1967, in Hay et al., 1967, p. 447, Plates 12 and 13, Figs. 1–4.

Coccoliths of this species are similar in construction to those of *G. oceanica* (p. 212) with the exception of the bridge having an angle of about 30–35° with the long axis of the coccolith. In *G. oceanica* the bridge is set at approximately a right angle to the long axis of the coccolith.

Distribution: North Atlantic and Pacific Oceans.

Gephyrocapsa ericsonii McIntyre & Bé

References: McIntyre & Bé, 1967a, p. 571, Plate 10, Plate 12, Fig. B; Borsetti & Cati, 1979, p. 158, Plate 14, Figs. 1 and 2; Nishida, 1979, Plate 2, Fig. 3; Winter et al., 1979, p. 201, Plate 1, Fig. 6; Reid, 1980, p. 154, Plate 1, Fig. 5; Hallegraeff, 1984, p. 233, Figs. 13 and 14.

The coccosphere is spherical to subspherical with a covering consisting of approximately 16–21 tightly interwedged coccoliths. Maximal diameters range from 3.1 to 4.1 μm. Oval to elliptical placoliths are similar in basic construction to those of *G. oceanica* (see later) with the exception of the thinner blade-like bridge elements. Both high and low arched bridges have been observed (Okada & McIntyre, 1977; Hallegraeff, 1984). The distal shield is composed of 30–34 elements, 1.9–2.2 μm in length and 1.4–1.7 μm wide. The height of the coccoliths averages 0.4 μm, and the bridge reaches from 0.5 to 0.7 μm above the distal shield.

Distribution: North Atlantic and Pacific Oceans, Australian waters.

Remarks: Coccoliths of this species differ from those of *G. crassipons* Okada & McIntyre (1977, p. 10, Plate 2, Figs. 5 and 6) by having thin bridge plates consisting of a single element, instead of a short thick bridge consisting of two plates, each of which is formed by several elements extending from the collar, which partly closes the distal opening of the central tube. In *G. crassipons* the bridge arises at a low angle to the coccolith surface (Reid, 1980), and is markedly diagonal to the long axis of the coccolith, while in *G. ericsonii* the bridge is oriented at a slight diagonal from the long axis of the coccolith.

Gephyrocapsa oceanica Kamptner (**Plate 5**)

References: Kamptner, 1943, p. 45; Halldal & Markali, 1955, p. 18, Plates 23 and 24; Hasle, 1960, Plate 2, Figs. 3–5; Gaarder & Hasle, 1971, p. 533, Fig. 6d–f; Okada & McIntyre, 1977, pp. 10, 11, Plate 3, Figs. 3–9; Nishida, 1979, Plate 2, Fig. 1; Winter et al., 1979, p. 201, Plate 1, Figs. 7–9; Reid, 1980, p. 154, Plate 1, Fig. 6; Hallegraeff, 1984, p. 233, Figs. 15–18.

Spherical cells are 5–15 μm in diameter with coccoliths 3.5–5.5 μm long and 2.4–4.2 μm wide. The size of the central area, the length and thickness of the bridge, and the development of a collar around the distal opening of the central tube show considerable variations (Okada & McIntyre, 1977, p. 10). The bridge, oriented at approximately right angles to the long axis of the coccolith, is usually formed by a pair of plates, but Okada & McIntyre (1977, Plate 3, Figs. 7–9) also observed specimens having more than two bridge plates.

Distribution: Worldwide.

Remarks: Although coccospheres of *G. oceanica* are usually found as isolated separate individuals, they sometimes form clusters (Okada & Honjo, 1970, Plate 2, Fig. 3; Burns, 1977, Figs. 28 and 29; Nishida, 1979, Plate 2, Fig. 1b; Winter et al., 1979, Plate 1, Fig. 8; Hallegraeff, 1984, Fig. 17).

Gephyrocapsa ornata Heimdal

References: Heimdal, 1973, pp. 71, 74, Figs. 1–5; Okada & McIntyre, 1977, p. 11, Plate 3, Figs. 1 and 2; Nishida. 1979, Plate 2, Fig. 2.

Cells are spherical with a diameter of 3.3–4.2 μm exclusive of protrusions. Coccoliths consist of elliptical placoliths, tightly interwedged

and about 0.4 μm high exclusive of protrusions. The distal shield is slightly wider than the proximal one and is composed of 30–36 flattened segments, concurrent or with narrow slits between each segment but with a continuous periphery. Length of the distal shield averages 2.3 μm, while the width is 1.9 μm. Central area is radially perforated, about 1.0 μm long and 0.6 μm wide. The central tube is crossed on the distal side by a narrow bridge consisting of two diametrically opposite thin plates of varying shape. Plates generally overlap and are sometimes separated or missing. In distal view, the bridge axis is displaced clockwise at low angle to long axis of the coccolith. The maximum height of bridge above distal shield is 1.5 μm. The central tube widens outwardly into a ring of flattened, irregularly formed, tooth-like protrusions reaching 0.3–0.6 μm above the distal shield. Coccoliths number from 14 to 18.
Distribution: Atlantic and Pacific Oceans.

Gephyrocapsa protohuxleyi McIntyre
References: McIntyre, 1970, pp. 188, 189, Fig. 1; Winter et al., 1978, pp. 295–298, Plate 1, Fig. 1, Figs. 1–6; Winter et al., 1979, p. 201, Plate 1, Fig. 11; Reid, 1980, pp. 154, 155, Plate 1, Fig. 7; Heimdal & Gaarder, 1981, p. 52, Plate 4, Figs. 20 and 21.

Cells are spherical with a diameter about 5 μm with tightly interwedged placoliths with a large elliptical central area and a bridge displaced clockwise about 30° from the longest dimension when viewed from the distal surface. Specimens having coccoliths with or without a bridge on the same cell have been observed (Heimdal & Gaarder, 1981, Plate 4, Fig. 20). The orientation of the bridge is constant, while its thickness and height above the distal shield varies in warm water and cold water forms. As with *Emiliania huxleyi* (p. 209), this species shows definite ecological variations in structure of the shield structure and the proximal covering of the central tube. The distal shield size is 2.2–3.4 × 2.2–2.6 μm in low-latitude coccoliths and 2.9–3.4 × 2.2–2.6 μm in high-latitude forms. Considerable variation in the size of the coccoliths was observed even on one coccosphere (Heimdal & Gaarder, 1981, Plate 4, Fig. 21).
Distribution: North Atlantic and Pacific Oceans, Mediterranean Sea.
Remarks: *Gephyrocapsa protohuxleyi* was described from cores by observations made under the transmission electron microscope by McIntyre (1970). Based on its stratigraphic position and coccolith structure, he concluded that *G. protohuxleyi* may be a phyletic link between *E. huxleyi* and the *Gephyrocapsa* complex, and an index fossil for the Pleistocene. Later, living specimens of the same species were recorded from the Gulf of Elat (Winter et al., 1978, 1979).

Genus *Halopappus* Lohmann 1912
Type: *Halopappus vahseli* Lohmann 1912.
Cells have two equally long flagella and a haptonema. Cell shape is long

conical, elliptical or oblong, with an apical depression surrounding the flagellar bases and apical crown or jointed appendages approximately the same length as the cell body. Each appendage consists of three strongly modified, linearly attached, elongated coccoliths with apparently vacant centers. Each is supported at its base by a ring-shaped coccolith possessing a relatively wide calcified rim and an apparently vacant center bridged by an unmineralized, patternless membrane. The link coccoliths themselves appear with an unmineralized, centrally located, fragile reticulum replacing the membrane of the other coccolith types. Ordinary coccoliths are incomplete caneoliths arranged in a hexagonal close packing pattern with their long axes mainly parallel to that of the cell. The apical depression is lined with similar but smaller rhomboidal coccoliths, each carrying a short, distally directed central tube.

Halopappus adriaticus Schiller, emend. Manton, Bremer, & Oates (Plate 5)
Synonyms: *Syracosphaera corii* Schiller (Schiller, 1925, p. 20, Plate 1, Fig. 15; Kamptner, 1941, p. 83, Plate 5, Fig. 58, Plate 6, Figs. 60 and 61);
Michaelsarsia adriaticus (Schiller) Manton, Bremer, & Oates (Manton et al., 1984, p. 198).
References: Gaarder & Hasle, 1971, p. 533, Figs. 5c,d; Nishida, 1979, Plate 10, Fig. 3; Reid, 1980, p. 160, Plate 5, Figs. 4 and 5; Heimdal & Gaarder, 1981, pp. 52, 54, 56, Plate 6; Hallegraeff, 1984, p. 239, Fig. 38; Manton et al., 1984, pp. 191–194, Plates 5–8.
Cells are conical, elliptical, or almost oblong, 12.0–29.0 μm long, about 9 μm wide with 8–17 slender, three-linked appendages each up to 25 μm long. These appendages have linked centers partly occluded by thin intrusive crystallites attached to the inner edge of the rim. The two proximal links of the appendages are about the same length, 6.7–9.0 μm, the distal one being slightly shorter with a pointed end. Ring-shaped coccoliths supporting the appendages are similar in size and structure to those of *Michaelsarsia elegans* (see p. 216), except for the presence of short fringing crystallites reducing the unoccluded central area by 50% or more. Ordinary coccoliths are 1.8–2.4 μm long, with a length/breadth ratio of 1.5–1.8 with 24–30 bottom ribs placed approximately parallel to the periphery and fused together along the long axis of the coccolith to a central ridge. The smaller rhomboid coccoliths surrounding the apical depression have a lower wall and about 20 bottom ribs. Their central tube reaches somewhat above the top of the wall.
Distribution: Atlantic, Indian, and Pacific Oceans, Mediterranean Sea, Australian waters.
Remarks: Manton et al. (1984) drew attention to the resemblance of the genera *Halopappus* and *Michaelsarsia* (for comparison see pp. 214 and 216 and Table 1). They recommended that *Michaelsarsia* should also include *H.*

adriaticus and proposed the new combination *M. adriaticus* (Schiller) Manton, Bremer, & Oates. In my opinion its position in *Michaelsarsia* may, however, still be questionable, and I therefore propose to retain it in *Halopappus*.

Genus *Helicosphaera* Kamptner, 1954
Type: *Helicosphaera carteri* (Wallich) Kamptner 1954.
Coccospheres are ellipsoidal with an apical opening. Coccoliths consist of long elliptical helicoform placoliths (helicoliths) arranged in spiral around the long axis of cell and wedged in one another by the flanges but with wings free.

Helicosphaera carteri (Wallich) Kamptner (Plate 5)
Synonyms: *Coccolithus carteri* (Wallich) Kamptner (Kamptner, 1941, pp. 93, 94, 111, 112, Plate 12, Fig. 134, Plate 13, Figs. 135 and 136; *Helicopontosphaera kamptneri* Hay & Mohler (Hay et al., 1967, p. 448, Plates 10 and 11, Fig. 5).
References: Kamptner, 1954, pp. 21, 23, Figs. 17–19; Gaarder, 1970, pp. 114, 117, Fig. 2e,f; Gaarder & Hasle, 1971, p. 533, Fig. 9, a, b, and e; Borsetti & Cati, 1972, p. 405, Plate 52, Figs. 1 and 2; Nishida, 1979, Plate 9, Fig. 4; Hallegraeff, 1984, p. 233, Figs. 19 and 20.
Coccosphere ranges up to 25 μm along the longer axis with a length/breadth ratio of 1.3–1.7 and an apical area about 6 μm wide. Coccoliths, 16–38 in number, are 6.4–7.5 μm long and 4.4–5.5 μm wide and have a central area with two large pores.
Distribution: Atlantic and Pacific Oceans, Mediterranean Sea, Australian waters.
Remarks: In agreement with Clocchiatti (1969), Gaarder (1970), and Jafar & Martini (1975), the present author refrains from adopting the name *Helicopontosphaera kamptneri* proposed for this species by Hay & Mohler (in Hay et al., 1967) and on the same premises.

Helicosphaera hyalina Gaarder (Plate 5)
References: Gaarder, 1970, pp. 113–119, Figs. 1a–g, 2a–d; Gaarder & Hasle, 1971, p. 533, Fig. 9c,d,f; Borsetti & Cati, 1972, p. 406, Plate 52, Figs. 3 and 4; Nishida, 1979, Plate 9, Fig. 1.
Cells are 12–22 μm long, 11–18 μm wide, and have a length/breadth ratio of 1.1–1.5. Coccoliths, 5.2–7.6 μm long, have a nearly homogeneous central area.
Distribution: Atlantic and Pacific Oceans, Mediterranean Sea.

Comparison of *H. carteri* and *H. hyalina*

Intact specimens of *H. carteri* and *H. hyalina* are rather easily distinguished in water mounts under bright-field microscopy. The two separate pores in the central part of the *H. carteri* coccoliths are distinct especially when using oblique light.

Genus *Hymenomonas* von Stein 1878, emend. Gayral & Fresnel 1979
Type: *Hymenomonas roseola* von Stein 1878.

Cells may be motile or nonmotile, covered with identical tremalith coccoliths in the form of a short tube with more or less straight sides and a jagged distal top.

Genus *Michaelsarsia* Gran 1912, emend. Manton, Bremer, & Oates 1984
Type: *Michaelsarsia elegans* Gran 1912, emend. Manton, Bremer, & Oates, 1984.

Cells are spherical to ovoid with two equally long flagella, haptonema, and an apical crown of jointed appendages of about the same length as the cell body. As in *Halopappus* (p. 214), each appendage consists of three links of strongly modified, elongated coccoliths supported by a ring-shaped coccolith at the base. The central slit of the arm links is, however, broader and lacks the intrusive crystals found in *Halopappus*. Ordinary coccoliths are incomplete caneoliths with a raised, rather wide central area. Small rhomboidal coccoliths lining the apical depression have a solid conical protrusion.

Michaelsarsia elegans Gran, emend. Manton, Bremer, & Oates (**Plate 5**)
Synonyms: *Michaelsarsia*? (Borsetti & Cati, 1976, pp. 216, 217, Plate 15, Figs. 1, 2, 7, 8); Winter et al. (1979, p. 207, Plate 3, Fig. 8); *Michaelsarsia* sp. (Nishida, 1979, Plate 10, Fig. 2).
References: Gran, 1912, in Murray & Hjort, 1912, p. 332; Heimdal & Gaarder, 1981, pp. 56, 58, Plate 7; Hallegraeff, 1984, p. 239, Fig. 39; Manton et al., 1984, pp. 187–191, 198, Plates 1–4).

Cells are spherical to ovoid, sometimes pointed antapically, 9.0–11.0 μm long, with a length/breadth ratio of 1.2–1.7, and 12–21 three-linked appendages each up to 20 μm long. Ordinary coccoliths are 1.2–2.6 μm long, have a length/breadth ratio of about 1.4, with 20–28 narrow bottom ribs radially arranged and a somewhat raised, wide central area. The apical depression surrounding the flagellar bases has a converging row of smaller, low-walled rhomboid coccoliths with 20–25 bottom ribs, which fuse in the center to a slightly pointed mound somewhat higher than the wall. As in *Halopappus*, the ring-shaped coccoliths of *M. elegans*, 2.6–3.0 μm in diameter, show angular concentric rings. The diameter of the unobstructed central area, nearly half that of the whole coccolith, is wider than in *Halopappus*.

Distribution: Atlantic Ocean, Mediterranean Sea, Australian waters.
Remarks: As previously stressed by, for example, Manton et al. (1984), the most characteristic feature of *Michaelsarsia* and *Halopappus* is, as in intact specimens of *Calciopappus* (p. 195), the crown of apical appendages. While

each of the spines in *Calciopappus* is a single, modified coccolith, each of the appendages in *Halopappus* and *Michaelsarsia* consists of three differently modified coccoliths, which are united linearly. In all three genera, modified ring shaped coccoliths distributed along the rim of the apical depression support the bases of the appendages. In both *H. adriaticus* and *M. elegans* the crown of appendages is easily separated from the cell proper. In *M. elegans* the individual parts of the appendages also seem to be easily separated from each other. Relatively often we have observed individuals deficient in one or more arm links (Heimdal & Gaarder, 1981, Plate 7, Figs. 36 and 37). This fact may lead to the assumption that some of the previously described *Michaelsarsia* species need to be reinvestigated. For comparison of *Calciopappus*, *Halopappus*, *Michaelsarsia*, and *Ophiaster*, see Table 1.

Genus *Oolithotus* Reinhardt 1968 (in Cohen & Reinhardt, 1968)
Type: *Oolithotus fragilis* (Lohmann) Reinhardt 1972.
  Cells are spherical with circular to semicircular placoliths, which are convex distally, concave proximally, with both shields connected by an asymmetrically placed central tube.

***Oolithotus fragilis* (Lohmann) Reinhardt (Plate 6)**
Basionym: *Coccolithophora fragilis* Lohmann (Lohmann, 1912, pp. 49, 54, Fig. 11).
Synonyms: *Coccolithus fragilis* (Lohmann) Schiller (Schiller, 1930, pp. 243–245, Figs. 43 and 120); *Cyclococcolithus fragilis* (Lohmann) Deflandre (in Deflandre & Fert, 1954, p. 151, Plate 6, Figs. 1–3); Hasle, 1960, Plate 3, Figs. 3–7; McIntyre & Bé, 1967a, p. 570, Plate 9, Fig. C); *Cyclococcolithina fragilis* var. A (Okada & Honjo, 1973, Plate 2, Fig. 2).
References: Reinhardt, 1972, p. 89; Okada & McIntyre, 1977, p. 11, Plate 4, Fig. 3; Nishida, 1979, Plate 5, Fig. 3.
  Coccospheres are spherical to subspherical up to 22–30 μm along the longer axis. Coccoliths consist of approximately 50 to >100 tightly interlocked placoliths ranging from 5.4 to 10.9 μm along their longer axis. Coccoliths have straight suture lines, usually no holes in the central area, and an element count of 12–18. The proximal shield is smaller than the distal shield but usually larger than half of the coccolith diameter.
Distribution: Atlantic and Pacific Oceans, Mediterranean Sea.

***Oolithotus fragilis* subsp. *cavum* Okada & McIntyre**
References: Okada & McIntyre, 1977, pp. 11, 12, Plate 4, Figs. 4 and 5; Winter et al., 1979, p. 206, Plate 2, Fig. 2; Nishida, 1979, Plate 5, Fig. 4; Hallegraeff, 1984, p. 233, Fig. 7.
  Coccospheres are spherical to subspherical and 4–8 μm along the longer axis with approximately 15–30 tightly interlocked placoliths 2.4–5.5 μm long. Circular to semicircular coccoliths have a small proximal shield

**Oolithotus
fragilis**

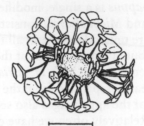

**Papposphaera
lepida**

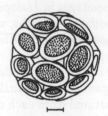

**Pontosphaera
discopora**

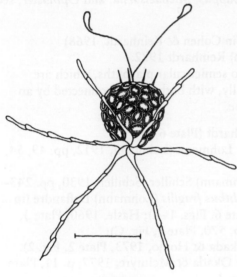

**Ophiaster
hydroideus**

**Pleurochrysis
carterae**

**Pontosphaera
syracusana**

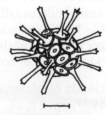

**Rhabdosphaera
claviger**

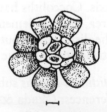

**Scyphosphaera
apsteinii**

usually less than half the size of the distal shield, which has one small hole
in the central area and curved suture lines. Element count varies from 11
to 15. On the proximal side the central area has one or two small holes.
Distribution: North Atlantic and Pacific Oceans, Australian waters.

Genus *Ophiaster* Gran 1912, emend. Manton & Oates 1983
Type: *Ophiaster hydroideus* (Lohmann) Lohmann 1913, emend. Manton &
Oates, 1983.
Spherical to ovoid cells have two flagella and a short haptonema.
Ordinary coccoliths are normal elliptical discs, while stomatal coccoliths
bear a central spine nearly equal to the coccolith in ength. At the
posterior pole of the cell there is a star-like appendage of elongated
transformed coccoliths (osteoliths) linked by narrow ends into flexible
arms attached to the cell by differently shaped proximal links. Small
unmineralized scales are present beneath the ordinary coccoliths, together
with unmineralized membranes individually attached to the proximal sides
of the coccoliths.

***Ophiaster formosus* Gran *sensu* Gaarder, emend. Manton & Oates**
References: Gran, 1912, p. 331, Fig. 239, 2; Gaarder, 1967, p. 185, Text-
Fig. 1A, Plate 1, Fig. C, Plate 3, Figs. B and E; Winter et al., 1979, p. 206,
Plate 3, Fig. 6; Manton & Oates, 1983, p. 460.
Cells can have a diameter of 4.5–10.5 μm but are commonly found
between 8 and 10 μm. Flagella can be as long as 20 μm. The haptonema
is shorter than that of *O. hydroideus*. Star-like appendage of 5–10 band-
like arms each with 12–14 links has swollen joints in lateral view.
Ordinary and stomatal coccoliths as well as proximal arm links (4.5–
6.0 μm) are as in *O. hydroideus* (see later); intermediary flat arm links,
3.5–4.5 μm long, with a breadth/length ratio of about 1/3 are not
constricted centrally. Distal links may appear with two longer diverging
thorns.
Distribution: Probably worldwide.

***Ophiaster hydroideus* (Lohmann) Lohmann, emend. Manton & Oates**
(Plate 6)
Basionym: *Meringosphaera hydroidea* Lohmann (Lohmann, 1903, p. 69).

Plate 6. *Oolithotus fragilis* (Lohmann) Reinhardt. *Papposphaera lepida* Tangen. *Pontosphaera discopora* Schiller. *Ophiaster hydroideus* (Lohmann) Lohmann, redrawn from Gaarder (1967). *Pleurochrysis carterae* (Braarud & Fagerland) Christensen. *Pontosphaera syracusana* Lohmann. *Rhabdosphaera claviger* Murray & Blackman. *Scyphosphaera apsteinii* Lohmann. Scale bar = 5 μm.

References: Lohmann, 1913, p. 152, Fig. 9; Halldal & Markali, 1955, pp. 18, 19, Plate 25; Gaarder, 1967, pp. 184, 185, Text-Fig. 1C, Plate 1, Figs. A and B, Plate 2, Fig. A, Plate 3, Fig. A; Okada & McIntyre, 1977, p. 19, Plate 10, Fig. 13; Winter et al., 1979, pp. 206, 207, Plate 3, Fig. 7; Manton & Oates, 1983, pp. 441–446, 460, Plates 1–4.

Cells have a diameter varing from 3.5 to 6.5 μm, but commonly found at 5–6 μm and having flagella up to 15 μm long and a haptonema. Star-like appendage consists of 5–11 cord-like arms each with six to nine links, which detach easily and have swollen joints in apical and lateral view. Ordinary coccoliths are 0.9–1.8 μm long incomplete caneoliths with 10–30, commonly up to 25, radial lamellae fused to form a solid central area varying in width. A slender central spine in stomatal coccoliths is up to 1.5 μm high. Intermediary arm links are approximately 2.4–3.0 μm long and 0.3–0.5 μm wide, with concave sides and a minimum breadth/length ratio of about 1/6. Proximal links are about 4.5 μm long, twisted and enlarged at the proximal end mediating attachment to the cell. The distal arm link ends in a terminal thorn.

Distribution: Probably worldwide.

Remarks: For comparison with *Calciopappus, Halopappus,* and *Michaelsarsia,* see **Table 1.**

Genus *Palusphaera* Lecal 1965a, emend. Norris 1984
Type: *Palusphaera vandeli* Lecal 1965a, emend. Norris 1984.

Coccospheres are monomorphic with styliform cyrtoliths having a circular outline. No open spaces are present on coccolith disc.

***Palusphaera vandeli* Lecal, emend. Norris**
Synonym: *Rhabdosphaera longistylis* Schiller (Heimdal & Gaarder, 1981, p. 58).
References: Lecal, 1967, pp. 318–320, Text-Fig. 13, Figs. 19 and 20; Norris, 1984, p. 35, Figs. 1F,G, 9, 10.

Spherical coccosphere has a diameter of 19–30 μm. Coccoliths consist of approximately 15–20 styliform cyrtoliths. The diameter of the basal coccolith disc is approximately 1.5–3.0 μm with a narrow central spine having a maximal length of 10 μm.

Distribution: North Atlantic and Indian Oceans, Mediterranean Sea.

Remarks: *Palusphaera vandeli* is a distinctly different species from those placed in the genus *Rhabdosphaera,* which have dimorphic coccoliths (see p. 222).

Genus *Pappomonas* Manton & Oates 1975
Type: *Pappomonas flabellifera* Manton & Oates 1975.

Small cells that have two flagella and a short haptonema. Coccoliths comprised of styliform pappoliths with a conspicuous central appendage around the flagellar pole and ordinary pappoliths without central processes.

Genus *Papposphaera* Tangen 1972
Type: *Papposphaera lepida* Tangen 1972.

Cells are spherical with two equal flagella two or three times the cell diameter in length and a haptonema. Coccosphere is composed of monomorphic styliform pappoliths uniformly distributed over the cell. The basal disc of each coccolith is elliptical to circular in outline with a marginal ring of elements, usually standing almost perpendicular to the floor of the disc. Each coccolith has a centrally placed appendix with distal end of a different shape.

Papposphaera lepida Tangen (Plate 6)
References: Tangen, 1972, pp. 171–177, Figs. 1–11, Text-Figs. 12 and 13; Manton & Oates, 1975, pp. 94, 96, Figs. 1–4; Okada & McIntyre, 1977, p. 38, Plate 6, Fig. 4; Norris, 1983, Fig. 1D.

The coccosphere, $11-16$ μm in diameter, has a cell body of $4.5-7.0$ μm and is sometimes bilaterally flattened with slightly smaller coccoliths on flattened sides. Hyaline coccoliths (50–100) are uniformly distributed on cell surface. The basal part of coccolith ($0.8-1.4$ μm long) is elliptical to subcircular with a marginal ring from $0.20-0.35$ μm high. The coccolith appendage consists of a $2.0-4.0$ μm long delicate rod with four decurrent ridges on the proximal part of the rod, which diverge at the bottom plate forming a distinct cross. At the top of the appendage, a wide funnel-shaped structure having a diameter of $1.1-2.1$ μm is formed by four flattened lobes, mostly having shallow incisions.
Distribution: Worldwide.

Genus *Pleurochrysis* Pringsheim 1955, emend. Gayral & Fresnel 1983
Type: *Pleurochrysis scherffelii* Pringsheim 1955.

This genus comprises species that may occur as unicellular motile or coccoid forms or short filaments. Motile coccolith-bearing cells are spherical, oval, or elongate with two flagella and a short haptonema. Coccoliths consist of monomorphic cricoliths. Benthic stage is pseudofilamentous or pseudoparenchymatic.

Pleurochrysis carterae (Braarud & Fagerland) Christensen, (Plate 6)
Basionym: *Syracosphaera carterae* Braarud & Fagerland (Braarud & Fagerland, 1946, p. 4, Figs. a–h).
Synonyms: *Syracosphaera Brandti* Schiller (Carter, 1937, p. 37, Fig. 3); *Hymenomonas carterae* (Braarud & Fagerland) Braarud (Braarud, 1954, p. 3); *Cricosphaera carterae* (Braarud & Fagerland) Braarud (Braarud, 1960, pp. 211, 212).

Reference: Christensen, 1978, p. 68.
  Motile coccosphere: Cells have a diameter of 12–17 μm. Coccoliths
  consist of cricoliths measuring about 1.3 μm × 2.0 μm.
Distribution: Described from brackish water in England; widely distributed,
described only from inshore waters.

Genus *Pontosphaera* Lohmann 1902
Type: *Pontosphaera syracusana* Lohmann 1902 (type designated by
Loeblich & Tappan, 1963, p. 193).
  Spherical cells are completely covered by monomorphic cribriliths.

***Pontosphaera discopora* Schiller (Plate 6)**
References: Schiller, 1930, pp. 186, 187, Fig. 6; Halldal & Markali, 1955,
p. 19, Plate 27; Gaarder & Hasle, 1971, p. 536, Fig. 10a–d; Nishida, 1979,
Plate 11, Fig. 4a; Hallegraeff, 1984, p. 233, Fig. 22.
  Coccospheres of 17–28 μm in diameter have cribriliths 6.0–10.0 μm long
  and 4.5–8.0 μm wide. The cribriliths are composed of a low funnel-
  shaped rim and a central area with perforations arranged more or less
  regularly in concentric lines following the elliptical outline of the
  coccolith.
Distribution: Atlantic and Pacific Oceans, Mediterranean Sea, Australian
waters.

***Pontosphaera syracusana* Lohmann (Plate 6)**
References: Schiller, 1930, pp. 187, 188, Figs. 5 and 22; Gaarder & Hasle,
1971, p. 536, Fig. 10e,f; Nishida, 1979, Plate 11, Fig. 2.
  Cells normally have a diameter from 22 to 30 μm but are occasionally
  found to be only 15 μm. Cribriliths are 12–15 μm long, consisting of a
  perforated bottom plate with randomly placed circular pores and a funnel-
  shaped rim up to 5 μm high.
Distribution: Atlantic and Pacific Oceans, Mediterranean Sea.

Genus *Rhabdosphaera* Haeckel 1894
Type: *Rhabdosphaera claviger* Murray & Blackman 1898 (type designated
by Hay & Towe, 1962, p. 504).
  Cells are spherical with covering of styliform or claviform cyrtoliths and
  cyrtoliths having no central process. Spined coccoliths are not restricted to
  polar regions of the coccosphere but are intermixed with spineless
  coccoliths. Coccoliths completely enclose the cell.

***Rhabdosphaera claviger* Murray & Blackman (Plate 6)**
Synonym: *Rhabdosphaera stylifera* Lohmann (Schiller, 1930, p. 250, Fig.
129; Kamptner, 1941, p. 96, Plate 15, Figs. 148 and 149; McIntyre & Bé,
1967a, p. 567, Plate 4).
References: Gaarder & Hasle, 1971, p. 536, Fig. 11; Borsetti & Cati, 1972,
pp. 407–409, Plate 15, Figs. 2–6; Nishida, 1979, Plate 12, Fig. 2;

Hallegraeff, 1984, p. 236, Figs. 32 and 33.
The dimorphic coccosphere is 10–12 µm in diameter with cyrtoliths that
are 2.7–3.4 µm long and 1.7–2.5 µm wide at the basal plate. Some
cyrtoliths carry a 3.6–5.0 µm high appendix ending in four vanes around
a short central spine.
Distribution: Worldwide.
Remarks: According to workers on recent material (McIntyre & Bé, 1967a;
Gaarder & Hasle, 1971; Borsetti & Cati, 1972; Kling, 1975; Hallegraeff,
1984; Norris, 1984), *R. claviger* and *R. stylifer* are conspecific. McIntyre &
Bé (1967a) also found intergradation between the *stylifer* and the broad club
shaped *claviger* type appendix in sediment samples. Kleijne & Jordan (1990)
agreed that these should be regarded as intraspecific taxa but thought that
the *claviger* and *stylifer* morphotypes were distinct enough to be separated
as varieties. Clocchiatti (1971b) and Perch-Nielsen (1971), working with
fossil material, maintain, however, the separation on specific level, while
Cohen & Reinhardt (1968) and Boudreaux & Hay (1969) follow McIntyre
& Bé when studying core samples. The specific epithet *claviger*, being a
noun, should not be changed by the gender of the genus (Kamptner, 1967).

Rhabdosphaera longistylis Schiller
Synonym: *Palusphaera vandeli* Lecal (Heimdal & Gaarder, 1981, p. 58).
References: Schiller, 1925, p. 40, Plate 4, Fig. 40; Schiller, 1930, pp. 251,
252, Fig. 130; Borsetti & Cati, 1972, p. 409, Plate 55, Fig. 1; Okada &
McIntyre, 1977, p. 17, Plate 5, Fig. 6; Winter et al., 1979, Plate 2, Fig. 11;
Norris, 1984, p. 34.
The dimorphic spherical coccospheres have a diameter of 20–25 µm.
Coccoliths consist of 50 to >100 of styliform type intermixed with
coccoliths without processes. The basal disc is 1.8–2.3 µm in length and
1.4–1.8 µm in width, and the spine is 40–50 µm long.
Distribution: North Atlantic, Indian, and Pacific Oceans, Mediterranean Sea.
Remarks: Unlike *Palusphaera vandeli* (p. 220), which has only styliform
coccoliths, almost half of the total coccoliths of *R. longistylis* are spineless.

Genus *Scyphosphaera* Lohmann 1902
Type: *Scyphosphaera apsteinii* Lohmann 1902.
Spherical cells are covered with disciform ordinary coccoliths (cribriliths)
and large barrel-shaped coccoliths (lopadoliths).

Scyphosphaera apsteinii Lohmann (Plate 6)
References: Schiller, 1930, p. 195, Figs. 36 and 75a; Gaarder, 1970, Figs.
4e,f, 6; Gaarder & Hasle, 1971, p. 536, Fig. 12b–d; Borsetti & Cati, 1972,
p. 399, Plate 41, Fig. 3, Plate 42, Figs. 1 and 2; Nishida, 1979, Plate 11, Fig.
1; Hallegraeff, 1984, p. 236, Fig. 24.
Cells have a diameter of 20–25 µm, excluding lopadoliths. Coccoliths
consist of cribriliths about 10–13 µm long with a smooth border and a
perforated central area and lopadoliths with a web-like pattern of

transverse and longitudinal ridges on the surface, 7–17 µm long (bottom plate) and 16–25 µm high.

Distribution: Atlantic, Indian, and Pacific Oceans, Mediterranean Sea.

Remarks: For comparison of the discoliths of this species with those of the monomorphic species *Discolithina japonica*, see p. 206.

Genus *Syracosphaera* Lohmann 1902, emend. Gaarder 1977 (in Gaarder & Heimdal, 1977)

Type: *Syracosphaera pulchra* Lohmann 1902 (type designated by Loeblich & Tappan, 1963, p. 193).

Cells are variable in outline with a distinct flagellar area. Dithecate coccospheres have both layers consisting of heterococcoliths that are elliptical in outline, differing in shape but resembling each other structurally. The exotheca is monomorphic while the endotheca is dimorphic. Endothecal coccoliths are complete caneoliths with an intermediate continuous or beaded midwall rim. Stomatal coccoliths bear a central rod. Exothecal coccoliths are distally convex with central depression.

Remarks: Jordan and Young (1990) and Jordan (1991) proposed that the species of *Caneosphaera* (p. 199) and *Deutschlandia anthos* (p. 206) be transferred to *Syracosphaera*, resulting in a much wider morphological variation for this genus than described by Gaarder (in Gaarder & Heimdal, 1977). Kleijne (1991) indicates a possible relationship between a *Syracosphaera* sp. and a species of the holococcolithophorid genus *Zygosphaera* (p. 186). Further observations are, however, needed to verify this.

***Syracosphaera histrica* Kamptner (Plate 7)**

Synonyms: *Syracoosphaera nodosa* (Kamptner) (Okada & Honjo, 1970, p. 21, Plate 1, Figs. 1 and 2); *S.* aff. *pirus* Halldal & Markali (Borsetti & Cati, 1972, p. 401, Plate 47, Fig. 1).

References: Kamptner, 1941, pp. 84, 104, Plate 6, Figs. 65–68; Borsetti & Cati, 1972, p. 400, Plate 44, Fig. 3; Gaarder & Heimdal, 1977, pp. 55, 56, Plate 2; Okada & McIntyre, 1977, p. 22, Plate 8, Fig. 12; Nishida, 1979, Plate 7, Fig. 1.

The ovoid coccolith case is 11–20 µm long inclusive of stomatal rods and 9.5–12.5 µm broad. Endotheca consists of approximately 25–70 densely set coccoliths. The base of the endothecal coccoliths is composed of a peripheral ring of evenly spaced, short, femur-shaped elements of equal length. Inside this ring there is a mound of more wedge-shaped elements with slightly pointed ends directed toward the center. The innermost elements are raised and joined to form a short protuberance with an irregularly toothed top, reaching up to or slightly above the low wall in ordinary coccoliths. Elements of the proximal part of the wall, equal in number to the ring elements in the bottom, are broad at the base, tongue-

shaped at the top with a bead-like knob. The ring of beads in profile looks like a midwall rim. Distal wall elements start as tongues between proximal elements and broaden above the beads bending outward at the top to form a continuous distal rim. Endothecal coccoliths are 2.0–4.7 μm in length, have a length/breadth ratio of 1.3–1.6 and height 0.4–0.8 μm. The length of stomatal rod is 1.0–2.0 μm, and the number of lamellar elements varies from 20–40. Structure of exothecal coccoliths resembles a flat-bottomed shallow basket with the bottom oriented distally as in the exothecal coccoliths of *S. pulchra* (see later). The length of exothecal coccoliths is 2.1–4.3 μm, with a length/breadth ratio of 1.1–1.5 and height 0.4–0.5 μm. The number of lamellae varies from 23 to 34.
Distribution: Atlantic and Pacific Oceans, Mediterranean Sea.

Syracosphaera pirus Halldal & Markali (Plate 7)
Synonym: *Syracosphaera prolongata* Gran ex Lohmann (Throndsen, 1972, pp. 57–59, Figs. 22–28; Okada & McIntyre, 1977, p. 26, Plate 7, Figs. 2 and 3; Winter et al., 1979, p. 210, Plate 4, Fig. 4).
References: Halldal & Markali, 1955, pp. 11, 12, Plate 10; Gaarder & Heimdal, 1977, pp. 56, 58, Plate 3; Okada & McIntyre, 1977, p. 26, Plate 9, Figs. 10 and 11; Nishida, 1979, Plate 10, Fig. 4.
Cells are pear-shaped to caudate measuring 12–70 μm along longer axis, 8–10 μm at its broadest part and with two flagella. Endothecal coccoliths (1.4–3.6 μm long) have a continuous midwall rim and a length/breadth ratio of 1.2–1.5. The central area of these coccoliths consists of 18–48 radially arranged lamellae and a blunt-toothed central protrusion that often exceeds the height of the wall, giving the cell a knobby appearance. Ordinary endothecal coccoliths are 0.2–0.5 μm high, exclusive of the central process. As in *S. pulchra,* the rather long central rod of the stomatal coccoliths (maximum length about 3.0 μm) is usually forked at the tip. Exothecal coccoliths, lying with concave side toward the cell, are circular to semicircular in outline, about 3 μm in diameter, and consist of approximately 25–40 radial lamellae and a conical central process.
Distribution: Atlantic Ocean, Mediterranean and Caribbean Seas.

Syracosphaera prolongata Gran ex Lohmann, emend. Heimdal & Gaarder (Plate 7)
Synonym: ?*Syracosphaera* sp. (Okada & McIntyre, 1977, Plate 7, Fig. 4); non *Syracosphaera prolongata* Gran ex Lohmann (Throndsen, 1972, pp. 57–59, Figs. 22–28; Okada & McIntyre, 1977, p. 26, Plate 7, Figs. 2 and 3; Winter et al., 1979, p. 210, Plate 4, Fig. 4).
References: Gran, 1912, in Murray & Hjort, 1912, p. 332, Fig. 239 (invalid, no description); Lohmann, 1913, p. 161, Fig. 16d,e; Schiller, 1930, pp. 215, 216, Fig. 100; Heimdal & Gaarder, 1981, pp. 60, 62, Plate 10, Figs. 48–50.

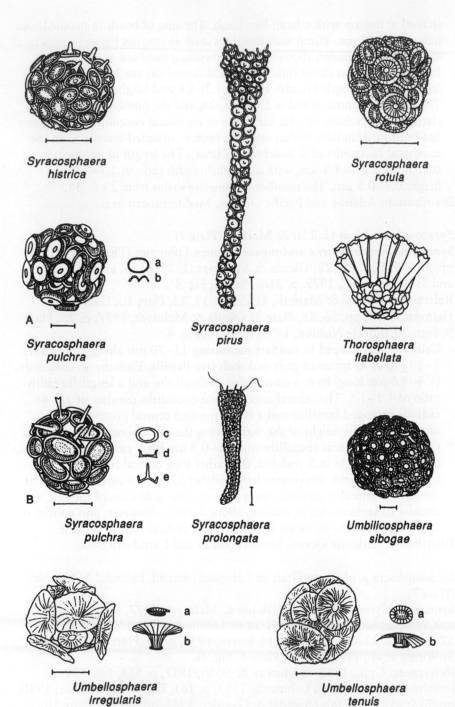

Syracosphaera
histrica

Syracosphaera
rotula

A
Syracosphaera
pulchra

a
b

Syracosphaera
pirus

Thorosphaera
flabellata

B
Syracosphaera
pulchra

c
d
e

Syracosphaera
prolongata

Umbilicosphaera
sibogae

Umbellosphaera
irregularis

a
b

Umbellosphaera
tenuis

a
b

Endotheca: The coccolith case is elongated and cone-shaped, 35–45 μm long, about 7 μm at its broadest part, and with the short axis of ordinary coccoliths running parallel to the long axis of cell. These cells have two flagella. The length of the coccoliths is 2.0–2.8 μm with a length/breadth ratio of 1.3–1.5. The height of ordinary coccoliths is about 0.3 μm and that of the stomatal coccoliths up to 2.5 μm inclusive of rod, which is forked at the tip. Ordinary and stomatal coccoliths have a continuous midwall rim, appearing not as wide as the distal and proximal ones. The central area of ordinary coccoliths consists of about 35 lamellae, which form a mound with no spine.

Distribution: North Atlantic, Mediterranean Sea.

Remarks: According to the descriptions presented here, both *S. pirus* and *S. prolongata* produce long tube-like coccolith cases first described as *S. prolongata* by Gran (1912). Unlike the ordinary endothecal coccoliths of *S. pirus,* those of *S. prolongata* have a swollen central area rather than a central process. This feature is easily distinguished when using phase contrast, but is also distinct with bright-field illumination. The description here of *S. prolongata* is, however, based on a few specimens lacking the exotheca. A more critical treatment of this species and its relation to *S. pirus* must therefore obviously await the examination of additional cells. The stomatal coccoliths of *S. pirus, S. prolongata,* and *S. pulchra* show a high degree of conformity, while the rod is less distinctly forked at the top in *S. histrica.*

Syracosphaera pulchra Lohmann (Plate 7)

References: Kamptner, 1941, pp. 85, 86, 105, 106, Plate 7, Figs. 77 and 78, Plate 8, Figs. 79–84; Gaarder & Heimdal, 1977, p. 55, Plate 1; Okada & McIntyre, 1977, p. 27, Plate 10, Figs. 11 and 12; Nishida, 1979, Plate 6, Fig. 3; Hallegraeff, 1984, p. 239, Fig. 46; Inouye & Pienaar, 1988, pp. 207–216, Figs. 1–15.

Plate 7. *Syracosphaera histrica* Kamptner; note central raised part of endothecal coccoliths. *Syracosphaera pulchra* Lohmann: (A) Coccosphere with endothecal and exothecal coccoliths. Exothecal coccolith: (a) distal surface view; (b) vertical longitudinal section. (B) Coccosphere with endothecal coccoliths only. Ordinary coccolith: (c) distal surface view; (d) vertical longitudinal section. Stomatal coccolith: (e) vertical longitudinal section. Parts a–c, e redrawn from Kamptner (1941), not to scale. *Syracosphaera pirus* Halldal & Markali. Endothecal coccolith case. *Syracosphaera prolongata* Gran ex Lohmann. *Syracosphaera rotula* Okada & McIntyre. Note the two large exothecal coccoliths. *Thorosphaera flabellata* Halldal & Markali. *Umbellosphaera irregularis* Paasche: (a) micrococcolith; (b) macrococcolith; a, b redrawn from Markali & Paasche (1955), not to scale. *Umbellosphaera tenuis* (Kamptner) Paasche: (a) micrococcolith, distal surface view; (b) macrococcolith; a, b redrawn from Markali & Paasche (1955), not to scale. *Umbilicosphaera sibogae* (Weber-von Bosse) Gaarder. Scale bar = 5 μm.

A spherical to pyriform coccolithophorid which measures 12–39 × 12–18 μm and has two almost equally long (34–51 μm) flagella and a conspicuous haptonema (15–28 μm). A wide range in shape and size of endothecal coccoliths occurs with a maximum/minimum length ratio varying up to 2.3 on the same cell, with stomatal coccoliths usually being among the smallest ones observed. Ordinary coccoliths are 4.5–7.0 μm long, 1.2–1.4 μm high, with radially orientated lamellar elements arranged on an organic base plate scale and two rings of alternating elements (for shape and arrangement of these elements see Inouye & Pienaar, 1988, p. 209, Text-Fig. 15a), having a continuous midwall rim. Stomatal coccoliths are identical to ordinary ones except for a central rod forked at the tip extending up to 3.0 μm high. Endothecal and exothecal coccoliths differ in shape but have similar basic structures. Exothecal dome-shaped coccoliths have no associated organic base plate and can vary on the same cell from a nearly circular to an elliptical outline. The length/breadth ratio can reach 1.6, while the length varies from 3.4 to 7.2 μm and the height from 0.9 to 1.5 μm.

Distribution: Worldwide.

Remarks: Some new observations of ultrastructure and coccolith morphology obtained from unialgal cultures (Inouye & Pienaar, 1988) indicate that *S. pulchra* is more closely related to members of the Prymnesiales than to many of the coccolithophorids. The production of two different types of coccoliths, endothecal coccoliths with an organic base plate and exothecal coccoliths without, is of particular interest. Manton & Leedale (1969) defined a coccolith as an organic scale with a calcified rim. Calcification of such coccoliths occurs on the associated organic base plate within a Golgi vesicle as shown for an endothecal coccolith of *S. pulchra* by Inouye & Pienaar (1988, Fig. 18), while exothecal coccoliths not associated with organic scales suggest that the calcified rim also can be produced independently. The same authors indicate that cells possessing coccoliths with associated base plate scales are more primitive than those not possessing such scales because, within the Prymnesiophyceae, the uncalcified organic scales are much more widely distributed than coccoliths.

Syracosphaera rotula Okada & McIntyre (Plate 7)

References: Okada & McIntyre, 1977, p. 27, Plate 9, Figs. 9 and 12; Winter et al., 1979, Plate 4, Fig. 6; Heimdal & Gaarder, 1981, Plate 9, Figs. 46 and 47.

The dithecate cells are spherical to subspherical and have a diameter from 5.4 to 7.2 μm. Endothecal coccoliths are 1.2–2.4 μm long, 0.9–2.0 μm wide, and have 15–25 lamellae. Exothecal coccoliths are circular in outline with a diameter of 3.0–3.5 μm, having a narrow marginal rim with a smooth central portion connected by 20–25 radially arranged elements.

Distribution: North Atlantic and Pacific Oceans, Mediterranean Sea.

Genus *Thorosphaera* Ostenfeld 1910
Type: *Thorosphaera elegans* Ostenfeld 1910.
Spherical cells have oval lepidoliths and elongate tubular coccoliths with open distal ends and closed proximal ends. The tube coccoliths increase in width distally.

Thorosphaera flabellata **Halldal & Markali (Plate 7)**
References: Halldal & Markali, 1955, p. 19, Plate 26; Okada & Honjo, 1973, Plate 2, Fig. 3; Borsetti & Cati, 1976, p. 214, Plate 14, Figs. 5–7; Nishida, 1979, Plate 15, Fig. 3; Reid, 1980, p. 170, Plate 8, Figs. 9–11, Plate 9, Figs. 1 and 2; Heimdal & Gaarder, 1981, Plate 5, Fig. 27.
The coccosphere has a diameter of approximately 8 μm, excluding the tubular appendages. Coccoliths consist of lepidoliths, 1.5–2.0 μm long, 1.0–1.4 μm wide, and 0.1 μm high. The distal side of these coccoliths has a transverse furrow from which the two parts incline slightly upward. Tubular coccoliths have a crown-shaped distal end with a bottom length of 1.6 μm, top length of 2.0 μm, and height between 6.0 and 12.0 μm.
Distribution: Atlantic and Pacific Oceans, Mediterranean Sea.
Remarks: It appears that the tubular coccoliths of *T. flabellata* bear some resemblance to the tower-shaped coccoliths of *Turrilithus latericioides* described from the deep photic waters of the North Atlantic by Jordan et al. (1991). Further observations are, however, needed to understand the taxonomic position of both these genera.

Genus *Umbellosphaera* Paasche 1955, in Markali & Paasche, 1955 emend. Gaarder 1981 (in Heimdal & Gaarder, 1981)
Type: *Umbellosphaera tenuis* (Kamptner) Paasche 1955 (in Markali & Paasche, 1955).
Cells are spherical or slightly oval with two types of caneoliths: micrococcoliths on the cell surface and macrococcoliths, with walls increasing in height and width located between or outside micrococcoliths. The largest macrococcoliths delimit the coccosphere externally.

Umbellosphaera corolla **(Lecal) Gaarder**
Basionym: *Syracosphaera (Syracolithus) corolla* Lecal (Lecal, 1965b, p. 252, Plate 1, Figs. 1–4, corr. in Loeblich & Tappan 1968, p. 591).
Synonym: *Syracosphaera corolla* Lecal (in the text to Lecal, 1965b, Plate 1, Fig. 2; Okada & McIntyre, 1977, p. 20, Plate 8, Figs. 1 and 2; Nishida, 1979, Plate 6, Fig. 4; Winter et al., 1979, Plate 3, Fig. 9).
Reference: Gaarder, 1981, in Heimdal & Gaarder, 1981, pp. 62, 64, Plate 11, Figs. 52–57.
Spherical to oval coccosphere having a diameter from 8.0–15.0 μm and two flagella. Micrococcoliths are 2.7–4.0 μm long, have a length/breadth ratio of about 1.5, are composed of 20–29 elements, and have walls with regular rings of somewhat blunt spines near the base. Macrococcoliths are

2.9–6.0 μm long, have a length/breadth ratio of 1.3–1.9, and a basically identical architecture.

Distribution: Atlantic and Pacific Oceans, Mediterranean Sea.

Remarks: The most conspicuous deviation from *U. irregularis* and *U. tenuis* lies in the ring of spines near the wall bases in both macro- and microcôccoliths of *U. corolla.*

Umbellosphaera irregularis Paasche (Plate 7)
References: Paasche, 1955, in Markali & Paasche, 1955, p. 97, Plates 3–6; McIntyre & Bé, 1967a, p. 567, Plate 2; Nishida, 1979, Plate 12, Fig. 4; Reid, 1980, p. 158, Plate 4, Figs. 4 and 5.

The protoplast of this species varies from 6 to 10 μm. The complete coccosphere has a diameter of 10–20 μm, with an angular appearance due to shape of macrococcoliths. Micrococcoliths are elliptical with a greater diameter of 1.6–3.6 μm and distal part of 0.2 μm high with about 20 radiating elements. Large macrococcoliths are almost circular in outline, and smaller ones are more elliptical with a delicate smooth funnel-shaped structure measuring 4.6–10.0 μm at the widest part. The funnel height never exceeds 4 μm or two-fifths of the coccolith diameter. The number of radiating elements varies between 15 and 25.

Distribution: Atlantic and Pacific Oceans, Mediterranean Sea, Australian waters.

Remarks: Kleijne (1991) indicated that the heterococcolith-bearing *U. irregularis* and the holococcolith-bearing *Flosculosphaera calceolariopsis* Jordan & Kleijne may represent different stages in the life cycle of one species.

Umbellosphaera tenuis (Kamptner) Paasche (Plate 7)
Basionym: *Coccolithus tenuis* Kamptner (Kamptner, 1937b, pp. 311, 312, Plate 17, Figs. 41 and 42).
References: Paasche, 1955, in Markali & Paasche, 1955, pp. 96, 97, Plates 1 and 2; McIntyre & Bé, 1967a, pp. 566, 567, Plate 3; Borsetti & Cati, 1972, pp. 406, 407, Plate 53, Fig. 3, Plate 54, Figs. 1 and 2; Nishida, 1979, Plate 12, Fig. 3; Winter et al., 1979, p. 206, Plate 3, Figs. 1 and 2; Reid, 1980, p. 158, Plate 4, Figs. 6 and 7.

The protoplast of this species has a diameter of 7–10 μm. The complete coccosphere is 10–16 μm in diameter with a rather smooth surface due to partly overlapping convex distal rims of macrococcoliths. Micrococcoliths are elliptical with a greater diameter of 2.6–3.0 μm and very low distal part consisting of about 20 elements. Macrococcoliths are slightly elliptical with a greater diameter of 4.4–8.1 μm and the height of umbrella-shaped distal portion never exceeding one-fourth of its diameter. The number of radial elements is about 25. Coarse irregular radial furrows occur on the distal surface.

Distribution: Atlantic and Pacific Oceans, Mediterranean Sea, Australian waters.

Genus *Umbilicosphaera* Lohmann 1902
Type: *Umbilicosphaera sibogae* (Weber-van Bosse) Gaarder 1970.
The cell is spherical or oval and completely covered with overlapping circular or elliptical placoliths of a single type having an open central tube.

Umbilicosphaera hulburtiana Gaarder
References: Gaarder, 1970, pp. 121, 122, Figs. 7, 8a–d, 9a,b; Okada & McIntyre, 1977, p. 12, Plate 3, Fig. 12; Nishida, 1979, Plate 2, Fig. 4; Hallegraeff, 1984, p. 231, Fig. 4.
Cells have a spherical to ellipsoidal outline and are 8.5–28.0 μm long with a width of 8.5–24.0 μm and a length/width ratio of 1.0–1.2. Coccoliths consist of partly interlocked, broad elliptical placoliths. The size of the coccoliths (2.6–5.3 μm) and the width of the central tube can vary greatly.
Distribution: North Atlantic and Pacific Oceans, Mediterranean Sea, Australian waters.
Remarks: Small cells of *U. hulburtiana* may easily be confused with *Emiliania huxleyi* (p. 209) in water mounts at lower magnifications in LM. However, they are usually less hyaline with a more even circumference, and in profile the collar button shape of the coccoliths is more pronounced.

Umbilicosphaera sibogae (Weber-van Bosse) Gaarder (**Plate 7**)
Basionym: *Coccosphaera sibogae* Weber-van Bosse (Weber-van Bosse, 1901, pp. 137, 140, Plate 17, Figs. 1 and 2).
Synonyms: *Umbilicosphaera mirabilis* Lohmann (Lohmann, 1902, pp. 139, 140, Plate 5, Fig. 66, 66a; Schiller, 1930, p. 249, Fig. 126); *Cyclococcolithus mirabilis* (Lohmann) Kamptner (Kamptner, 1954, pp. 24, 25, Figs. 21–23).
References: Gaarder, 1970, pp. 122–126, Figs. 8e, 9c, 9d; Kling, 1975, p. 10, Plate 1, Fig. 2; Borsetti & Cati, 1976, pp. 223, 224, Plate 18, Figs. 3 and 4; Okada & McIntyre, 1977, p. 13, Plate 4, Fig. 2; Nishida, 1979, Plate 5, Fig. 2; Winter et al., 1979, Plate 2, Fig. 3; Reid, 1980, p. 155, Plate 2, Figs. 1 and 2; Hallegraeff, 1984, p. 231, Fig. 5a.
Spherical or oval coccospheres vary from 15.0 to 50.0 μm in size. The number of partly overlapping coccoliths comprising a coccosphere varies from 40 to more than 100. Circular placoliths have a larger open central tube. The distal shield measuring 3.2–6.2 μm is equal to or slightly narrower than the proximal shield and has straight suture lines in the inner one-third and a zig-zag pattern of suture lines in outer part.

Distribution: Atlantic, Indian, and Pacific Oceans, Mediterranean Sea, Australian waters.

Umbilicosphaera sibogae var. *foliosa* (Kamptner) Okada & McIntyre
Basionym: *Cyclococcolithus foliosus* Kamptner (Kamptner, 1963, pp. 167, 168, Plate 7, Fig. 38).
References: Okada & McIntyre, 1977, p. 13, Plate 4, Fig. 1; Nishida, 1979, Plate 5, Fig. 1; Reid, 1980, p. 155, Plate 2, Figs. 3 and 4; Hallegraeff, 1984, p. 231, Fig. 5b; Inouye & Pienaar, 1984, pp. 358, 361, Figs. 1–14.
Spherical to subspherical coccospheres have a diameter from 10.0 to 18.0 μm with a protoplast size of 6.3–9.5 μm. Coccoliths consist of approximately 15–30 tightly interlocked circular placoliths. The diameter of the distal shield is 4.8–8.3 μm and the diameter of the central tube is 2.0–2.6 μm at the proximal end and 2.4–2.7 at the distal end. Several hook-like spines protrude upward from the lowest part of the central tube.
Distribution: Atlantic and Pacific Oceans, Australian waters.
Remarks: *Umbilicosphaera sibogae* differs from this form by having a larger coccosphere consisting of a far greater number of coccoliths. In addition, the type form has a relatively larger central tube without spines and coccoliths with a rather smooth continuation of the suture lines on the distal shield, instead of sharply offset suture.

COMMON COCCOLITHOPHORID SYNONYMS

Acanthoica quattrospina Lohmann 1903
= *Acanthoica acanthifera* Lohmann *sensu* Kampter 1941

Algirosphaera oryza Schlauder 1945
= *Anthosphaera oryza* (Schlauder) Gaarder (in Gaarder & Hasle, 1971)

Algirosphaera quadricornu (Schiller) Norris 1984
= *Anthosphaera quadricornu* (Schiller) Halldal & Markali 1955
= *Syracosphaera quadricornu* Schiller 1914

Algirosphaera robusta (Lohmann) Norris 1984
= *Anthosphaera robusta* (Lohmann) Kamptner 1941
= *Syracosphaera robusta* Lohmann 1902

Alisphaera ordinata (Kamptner) Heimdal 1973
= *Acanthoica ordinata* Kampter 1941

Anoplosolenia brasiliensis (Lohmann) Deflandre (in Grassé, 1952)
= *Cylindrotheca brasiliensis* Lohmann 1919

Anthosphaera fragaria Kamptner 1937b
= *Helladosphaera fragaria* (Kamptner) Gaarder 1962

Calcidiscus leptoporus (Murray & Blackman) Loeblich & Tappan 1978
= *Coccolithophora leptopora* (Murray & Blackman) Lohmann 1902
= *Coccolithus leptoporus* (Murray & Blackman) Schiller 1930
= *Coccosphaera leptopora* Murray & Blackman 1898
= *Cyclococcolithus leptoporus* (Murray & Blackman) Kamptner 1954

Calciosolenia murrayi Gran (in Murray & Hjort, 1912)
= *Calciosolenia sinuosa* Schlauder *sensu* Halldal & Markali 1955

Calyptrolithina divergens (Halldal & Markali) Heimdal 1982
= *Zygosphaera divergens* Halldal & Markali 1955

Calyptrolithina multipora (Gaarder) Norris 1985
= *Corisphaera multipora* Gaarder (in Heimdal & Gaarder, 1980)

Calyptrolithophora gracillima (Kamptner) Heimdal (in Heimdal & Gaarder, 1980)
= *Calyptrosphaera gracillima* Kamptner 1941
= *Sphaerocalyptra gracillima* (Kamptner) Throndsen 1972
= *Syracolithus catilliferus* (Kamptner) Borsetti & Cati 1972

Calyptrolithophora hasleana (Gaarder) Heimdal (in Heimdal & Gaarder, 1980)
= *Corisphaera hasleana* Gaarder 1962
= *Sphaerocalyptra hasleana* (Gaarder) Kling 1975

Calyptrolithophora papillifera (Halldal) Heimdal (in Heimdal & Gaarder, 1980)
= *Calyptrosphaera papillifera* Halldal 1953
= *Sphaerocalyptra papillifera* (Halldal) Halldal 1954

Caneosphaera halldalii (Gaarder) Gaarder (in Gaarder & Heimdal, 1977)
= *Syracosphaera halldalii* Gaarder (in Gaarder & Hasle, 1971)

Caneosphaera molischii (Schiller) Gaarder (in Gaarder & Heimdal, 1977)
= *Syracosphaera corrugis* Okada & McIntyre *sensu* Heimdal & Gaarder 1981
= *Syracosphaera elatensis* Winter *sensu* Heimdal & Gaarder 1981
= *Syracosphaera Molischi* Schiller 1925

Ceratolithus cristatus Kamptner 1950
= *Ceratolithus telesmus* Norris *sensu* Jordan & Young 1990

Coccolithus pelagicus (Wallich) Schiller 1930
= *Coccosphaera pelagica* Wallich 1877
= *Crystallolithus braarudii* Gaarder *sensu* Rowson et al. 1986
= *Crystallolithus hyalinus* Gaarder & Markali *sensu* Parke & Adams 1960

Coronosphaera binodata (Kamptner) Gaarder (in Gaarder & Heimdal, 1977)
= *Syracosphaera binodata* (Kamptner) Kamptner 1937b

Coronosphaera maxima (Halldal & Markali) Gaarder (in Gaarder & Heimdal, 1977)

= *Syracosphaera maxima* Halldal & Markali 1955

Coronosphaera mediterranea (Lohmann) Gaarder (in Gaarder & Heimdal, 1977)
= *Syracosphaera mediterranea* Lohmann 1902
= *Syracosphaera pulchroides* Halldal & Markali 1955
= *Syracosphaera tuberculata* Kamptner 1937b

Cruciplacolithus neohelis (McIntyre & Bé) Reinhardt 1972
= *Coccolithus neohelis* McIntyre & Bé 1967b

Daktylethra pirus (Kamptner) Norris 1985
= *Calyptrolithophora pirus* (Kamptner) Hallegraeff 1984
= *Calyptrosphaera pirus* Kamptner 1937b

Deutschlandia anthos Lohmann 1912
= *Syracosphaera variabilis* (Halldal & Markali) Okada & McIntyre *sensu* Heimdal & Gaarder 1981

Discolithina japonica Takayama 1967
= *Pontosphaera japonica* (Takayama) Burns 1973

Emiliania huxleyi (Lohmann) Hay & Mohler (in Hay et al., 1967)
= *Coccolithus huxleyi* (Lohmann) Kamptner, 1943
= *Gephyrocapsa huxleyi* (Lohmann) Reinhardt 1972
= *Hymenomonas huxleyi* (Lohmann) Kamptner 1930
= *Pontosphaera huxleyi* Lohmann 1902

Halopappus adriaticus Schiller 1914
= *Michaelsarsia adriaticus* (Schiller) Manton, Bremer, & Oates 1984
= *Syracosphaera corii* Schiller *sensu* Heimdal & Gaarder 1981

Helicosphaera carteri (Wallich) Kamptner 1954
= *Coccolithus carteri* (Wallich) Kamptner 1941
= *Helicopontosphaera kamptneri* Hay & Mohler (in Hay et al., 1967)

Helladosphaera cornifera (Schiller) Kamptner 1937b
= *Syracosphaera cornifera* Schiller 1913

Homozygosphaera tholifera (Kamptner) Halldal & Markali 1955
= *Calyptrosphaera tholifera* Kamptner 1941

Oolithotus fragilis (Lohmann) Reinhardt 1972
= *Coccolithophora fragilis* Lohmann 1912
= *Coccolithus fragilis* (Lohmann) Schiller 1930
= *Cyclococcolithus fragilis* (Lohmann) Deflandre (in Deflandre & Fert, 1954)

Periphyllophora mirabilis (Schiller) Kamptner 1937b
= *Calyptrosphaera mirabilis* Schiller 1925

Rhabdosphaera claviger Murray & Blackman 1898
= *Rhabdosphaera stylifera* Lohmann *sensu* Gaarder & Hasle 1971

Sphaerocalyptra quadridentata (Schiller) Deflandre (in Grassé, 1952)
= *Calyptrosphaera quadridentata* Schiller 1913

Syracolithus catilliferus (Kamptner) Deflandre (in Grassé, 1952)
= *Calyptrosphaera catillifera* (Kamptner) Gaarder 1962
= *Syracosphaera catillifera* Kamptner 1937b

Syracolithus dalmaticus (Kamptner) Loeblich & Tappan 1963
= *Homozygosphaera halldalii* Gaarder *sensu* Kleijne 1991
= *Homozygosphaera wettsteinii* (Kamptner) Halldal & Markali *sensu*
   Norris 1985
= *Syracosphaera dalmatica* Kamptner 1927

Syracolithus quadriperforatus (Kamptner) Gaarder (in Heimdal & Gaarder,
   1980)
= *Homozygosphaera quadriperforata* (Kamptner) Gaarder 1962
= *Syracosphaera quadriperforata* Kamptner 1937b

Umbellosphaera corolla (Lecal) Gaarder (in Heimdal & Gaarder, 1981)
= *Syracosphaera corolla* Lecal 1965b

Umbellosphaera tenuis (Kamptner) Paasche (in Markali & Paasche, 1955)
= *Coccolithus tenuis* Kamptner 1937b

Umbilicosphaera sibogae (Weber-van Bosse) Gaarder 1970
= *Coccosphaera sibogae* Weber-van Bosse 1901
= *Cyclococcolithus mirabilis* (Lohmann) Kamptner 1954
= *Umbilicosphaera mirabilis* Lohmann 1902

Zygosphaera bannockii (Borsetti & Cati) Heimdal 1982
= *Laminolithus bannockii* (Borsetti & Cati) Heimdal (in Heimdal &
   Gaarder, 1980)
= *Sphaerocalyptra bannockii* Borsetti & Cati 1976

Zygosphaera hellenica Kamptner 1937b
= *Laminolithus hellenicus* (Kamptner) Heimdal *sensu* Heimdal 1982

Zygosphaera marsilii (Borsetti & Cati) Heimdal 1982
= *Laminolithus marsilii* (Borsetti & Cati) Heimdal (in Heimdal & Gaarder,
   1980)
= *Sphaerocalyptra marsilii* Borsetti & Cati 1976
= *Zygosphaera amoena* Kamptner *sensu* Norris 1985

INDEX OF COCCOLITHOPHORID TAXA

Acanthoica, 754, 773
  A. aculeata, 772, 773
  A. maxima, 774
  A. quattrospina, 735, 746, 772, 774
Algirosphaera, 734, 746, 750, 754, 766, 774
  A. oryza, 772, 774, 775, 776
  A. quadricornu, 772, 775
  A. robusta, 775, 776
Alisphaera, 754, 776
  A. capulata, 776, 777
  A. ordinata, 772, 776, 777
  A. unicornis, 777
Anoplosolenia, 753, 777
  A. brasiliensis, 772, 777, 781, 783
Anthosphaera, 747, 755, 763
  A. fragaria, 763
  Balaniger, 752, 756
  B. balticus, 740, 743, 756
Braarudosphaera, 753, 778
  B. bigelowii, 772, 778
  Calciarcus, 739, 751
Calcidiscus, 751, 753, 778
  C. leptoporus, 738, 742, 743, 760, 772, 778
Calciopapppus, 754, 779, 780, 801, 804
  C. caudatus, 749, 772, 779, 781
  C. rigidus, 780, 781
Calciosolenia, 753, 781
  C. murrayi, 781, 782
Calicasphaera, 754
Calyptrolithina, 745, 752, 755, 764
  C. divergens, 764
  C. multipora, 764, 766
Calyptrolithophora, 745, 752, 755, 765
  C. gracillima, 765, 773
  C. papillifera, 765, 766, 767
Calyptrosphaera 747, 757, 758
  C. oblonga, 745, 756, 757
  C. sphaeroidea, 747, 757, 758
Caneosphaera, 754, 783, 785, 808
  C. halldalii, 783
  C. molischii, 782, 784
Ceratolithus 753, 785
  C. cristatus, 782, 785
  C. telesmus, 785
Coccolithus, 751, 753, 785
  C. pelagicus, 739, 740, 742, 743, 744, 745, 759, 782, 786

Corisphaera, 753, 755, 768
  C. gracilis, 766, 768, 771
Coronosphaera, 754, 786
  C. binodata, 782, 787, 788
  C. maxima, 787, 788
  C. mediterranea, 782, 786, 787, 788
Crenalithus, 752, 788
  C. parvulus, 788
Cricosphaera, 740, 743, 752, 789
Cruciplacolithus, 740, 743, 753, 789
  C. neohelis, 734, 739, 743, 782, 789
Crystallolithus, 739, 743, 753, 755, 758
  C. braarudii, 758, 786
  C. hyalinus, 739, 747, 757, 758
  C. rigidus, 758, 759, 778, 779
  Daktylethra, 753, 755, 760
  D. pirus, 757, 760
Deutschlandia, 754, 789
  D. anthos, 782, 789, 790, 808
Discolithina,754, 790
  D. japonica, 782, 790, 808
Discosphaera, 754, 791
  D. tubifer, 791, 792
Emiliania, 791
  E. huxley 731, 732, 735, 738, 740, 742, 743, 752, 791, 792, 793, 794, 796, 815
Florisphaera, 754, 794
  F. profunda, 741, 792, 794
  F. profunda var. elongata
Flosculosphaera, 755
  F. calceolariopsis, 814
Gephyrocapsa, 752, 795
  G. caribbeanica, 795
  G. ericsonii, 795
  G. oceanica, 744, 745, 792, 795, 796
  G. ornata, 796
  G. protohuxleyi, 796
Gliscolithus, 753, 755, 761
  G. amitakarenae, 761
Halopappus, 754, 780, 798, 799, 800, 801, 804
  H. adriaticus, 749, 792, 798, 801
Helicopontoshaera, 799
  H. kamptneri, 799
Helicosphaera, 747, 754, 799
  H. carteri, 792, 799, 800
  H. hyalina, 792, 799, 800
Helladosphaera, 753, 755, 768

H. *aurisinae*, 766, 768
H. *cornifera*, 766, 768, 769
Homozygosphaera, 753, 755, 761
　H. *triarcha*, 761
Hymenomonas, 740, 743, 751, 752, 800
　H. *roseola*, 739, 743, 800
　Michaelsarsia, 754, 780, 792, 799,
　　800, 801, 804
　M. *elegans*, 749, 798, 800, 801
　Oerosphaera, 740, 743
Oolithotus, 753, 801
　O. *fragilis*, 801, 802
　O. *fragilis*, subsp. *cavum*, 801
Ophiaster, 754, 780, 801
　O. *formosus*, 803
　O. *hydroideus*, 748, 749, 802, 803
Palusphaera, 804
　P. *vandeli* 804
Pappomonas, 739, 751, 753, 804
　P. *fabellifera*, 804
Papposphaera, 739, 743, 751, 753, 805
　P. *lepida*, 748, 802, 805
Periphyllophora, 753, 755, 761
　P. *mirabilis*, 757, 761, 762
Pleurochrysis, 740, 743, 752, 805
　P. *carterae*, 738, 743, 802, 805
　P. *scherffelii*, 805
Pontosphaera, 754, 806
　P. *discopora*, 747, 802, 806
　P. *syracusana*, 802, 806
Poricalyptra, 755
Poritectolithus, 755
Quaternariella, 752, 753
Rhabdosphaera, 754, 806
　R. *claviger*, 735, 746, 802, 806, 807

R. *longistylis*, 804, 807
Scyphosphaera, 754, 807
　S. *apsteinii*, 745, 791, 807, 808
Spaerocalyptra, 753, 755, 769
　S. *quadridentata*, 734, 766, 769
Syracolithus, 753, 755, 762
　S. *dalmaticus*, 757, 762, 764
　S. *quadriperforatus*, 757, 763
Syracosphaera, 754, 808
　S. *histrica*, 808, 810, 812
　S. *pirus*, 809, 810, 811
　S. *prolongata*, 809, 810, 811
　S. *pulchra*, 787, 808, 809, 810, 811,
　　812
　S. *rotula*, 810, 813
Thorosphaera, 753, 754, 813
　T. *flabellata*, 741, 748, 810, 813
Trigonaspis, 739, 752
Turrilithus, 751
　T. *latericioides*, 751, 813
Turrisphaera, 735, 739, 752–753
Umbellosphaera, 753, 754, 813
　U. *corolla*, 814
　U. *irregularis*, 735, 748, 810, 814
　U. *tenuis*, 735, 784, 810, 813, 814
Umbilicosphaera, 815
　U. *hulburtiana*, 815
　U. *sibogae*, 810, 815, 816
　U. *sibogae*, var. *foliosa*, 738, 816
Wigwamma, 739, 754
Zygosphaera, 753, 755, 770–771,
　808
　Z. *bannockii*, 770, 771
　Z. *hellenica*, 766, 770, 771
　Z. *marsilii*, 771, 773

REFERENCES

Andreae, M. O. 1986. The ocean as a source of atmospheric sulphur compounds. *In* "The role of air-sea exchange in geochemical cycling" (P. Buat-Menard, ed.), pp. 331–362. Reidel, Dordrect, The Netherlands.

Arpin, N., Svec, W. A., & Liaaen-Jensen, S. 1976. New fucoxanthin-related carotenoids from *Coccolithus huxleyi. Phytochemistry* 15:529–532.

Berge, G. 1962. Discoloration of the sea due to *Coccolithus huxleyi* "bloom." *Sarsia* 6:27–40.

Birkenes, E., & Braarud, T. 1952. Phytoplankton in the Oslo Fjord during a "*Coccolithus huxleyi* summer." *Avhandlinger utgitt av Det Norske Videnskaps-Akademi i Oslo I. Matematisk-Naturvidenskapelig Klasse* 2: 3–23.

Black, M. 1963. The fine structure of the mineral parts of the Coccolithophoridae. *Proceedings of the Linnean Society of London* 174:41–46.

Black, M. 1968. Taxonomic problems in the study of coccoliths. *Palaeontology* 11:793–813.

Black, M. 1971. The systematics of coccoliths in relation to the palaeontological record. *In* "The Micropalaentology of Oceans" (B. M. Funnel & W. R. Riedel, eds.), pp. 611–624. Cambridge University Press, Cambridge.

Borsetti, A. M., & Cati, F. 1972. Il nannoplancton calcareo vivente nel Tirreno centro-meridionale. *Giornale di Geologia Serie 2a* 38:395–452.

Borsetti, A. M., & Cati, F. 1976. Il nannoplancton calcareo vivente nel Tirreno centro-meridionale, Parte II. *Giornale di Geologia Serie 2a* 40:209–240.

Borsetti, A. M., & Cati, F. 1979. Il nannoplancton calcareo vivente nel Tirreno centro-meridionale, Parte III. *Giornale di Geologia Serie 2a* 43:157–174.

Boudreaux, J. E., & Hay, W. W. 1969. Calcareous nannoplankton and biostratigraphy of the late Pliocene-Pleistocene-Recent sediments in the Submarex cores. *Revista Española de Micropaleontologia* 1:249–292.

Braarud, T. 1954. Coccolith morphology and taxonomic position of *Hymenomonas roseola* Stein and *Syracosphaera carterae* Braarud & Fagerland. *Nytt Magasin for Botanikk* 3: 1–4.

Braarud, T. 1960. On the coccolithophorid genus *Cricosphaera* n. gen. *Nytt Magasin for Botanikk* 8:211–212.

Braarud, T. 1979. The temperature range of the non-motile stage of *Coccolithus pelagicus* in the North Atlantic region. *British Phycological Journal* 14:349–352.

Braarud, T., & Fagerland, E. 1946. A coccolithophoride in laboratory culture *Syracosphaera carterae* n. sp. *Avhandlinger utgitt av Det Norske Videnskaps-Akademi i Oslo I. Matematisk-Naturvidenskapelig Klasse* 2:1–10.

Braarud, T., Deflandre, G., Halldal, P., & Kamptner, E. 1955. Terminology, nomenclature, and systematics of the Coccolithophoridae. *Micropaleontology* 1:157–159.

Braarud, T., Gaarder, K. R., Markali, J., & Nordli, E. 1952. Coccolithophorids studied in the electron microscope. Observations on *Coccolithus Huxleyi* and *Syracosphaera Carterae. Nytt Magasin for Botanikk* 1:129–134.

Burns, D. A. 1973. Structural analysis of flanged coccoliths in sediments from the South West Pacific Ocean. *Revista Española de Micropaleontologia* 5:147–160.

Burns, D. A. 1977. Phenotypes and dissolution morphotypes of the genus *Gephyrocapsa* Kamptner and *Emiliania huxleyi* (Lohmann). *New Zealand Journal of Geology and Geophysics* 20: 143–155.

Carter, N. 1937. New or interesting algae from brackish water. *Archiv für Protistenkunde* 90: 1–69.

Chrétiennot-Dinet, M.-J. 1990. Chlorarachniophycées, Chlorophycées, Chrysophycées, Cryptophycées, Euglénophycées, Eustigmatophycées, Prasinophycées, Prymnésiophycées, Rhodophycées et Tribophycées. *In* "Atlas du phytoplancton marin" (A. Sournia, ed.). Editions du C.N.R.S., Paris.

Christensen, T. 1978. Annotations to a textbook of phycology. *Botanisk Tidsskrift* 73:65–70.

Christensen, T. 1980. "Algae. A Taxonomic Survey," 1. Odense, AiO Tryk.

Clocchiatti, M. 1969. Contribution à l'étude de *Helicosphaera carteri* (Wallich) Kamptner (Coccolithophoridae). *Revue de Micropaléontologie* 12:75–83.

Clocchiatti, M. 1971a. Sur l'existence de coccosphères portant des coccolithes de *Gephyrocapsa oceanica* et de *Emiliania huxleyi* (Coccolithophoridés). *Compte rendu de l'Académie des sciences, Paris* 273:318–321.

Clocchiatti, M. 1971b. Remarques sur quelques rhabdolithes de Méditerranée. *Cahiers de Micropaléontologie* (2)9:1–8.

Cohen, C. L. D., & Reinhardt, P. 1968. Coccolithophorids from the Pleistocene Caribbean deep-sea Core CP-28. *Neues Jahrbuch für Geologie und Paläontologie. Abhandlungen* 131:289–304.

Conrad, W. 1914. Contributions à l'étude des flagellates. III. La morphologie et la nature des enveloppes chez *Hymenomonas roseola* Stein et *H. coccolithophora* Massart et Conrad, nov. spec., et les Coccolithophoridae. *Annales de Biologie Lacustre, Bruxelles* 7:155–164.

Deflandre, G. 1947. *Braarudosphaera* nov. gen., type d'une famille nouvelle de Coccolithophoridés actuels à éléments composites. *Compte rendu de l'Académie des sciences, Paris* 225:439–441.

Deflandre, G. 1952. Classe des Coccolithophoridés (Coccolithophoridae Lohmann, 1902). *In* "Traité de Zoologie" (P.-P. Grassé, ed.), Vol. 1, pp. 439–470. Masson Cie., Paris.

Deflandre, G., & Fert, C. 1954. Observations sur les coccolithophoridés actuels et fossiles en microscopie ordinaire et électronique. *Annales de Paléontologie* 40:115–176.

Dupouy, C., & Demarcq, H. 1987. CZCS as an aid for understanding modalities of the phytoplankton productivity during upwelling off Senegal. *Advances in Space Research* 7:63–71.

Ehrenberg, C. G. 1836. Bemerkungen über feste mikroskopische, anorganische Formen in den erdigen und derben Mineralien. *Preussische Akademie der Wissenschaften Berlin: Bericht über die zur Bekanntmachung geeigneten Verhandlungen der Königlich-Preussischen Akademie der Wissenschaften Berlin* 1836:84–85.

Farinacci, A. 1971. Round table on calcareous nannoplankton. *Proceedings of the II Plantonic Conference, Roma 1970* (A. Farinacci, ed.), pp. 1343–1360. Edizioni Tecnoscienza, Roma.

Fresnel, J. 1989. Les coccolithophorides (Prymnesiophyceae) du littoral: Genres: *Cricosphaera, Pleurochrysis, Cruciplacolithus, Hymenomonas* et *Ochrosphaera*. Ultrastructure, cycle biologique, systematique. These de Doctorat d'etat Sciences, Université de Caen.

Fresnel, J., & Billard, C. 1991. *Pleurochrysis placolithoides* sp. nov. (Prymnesiophyceae), a new marine coccolithophorid with remarks on the status of cricolith-bearing species. *British Phycological Journal* 26:67–80.

Friedinger, P. J. J., & Winter, A. 1987. Distribution of modern coccolithophore assemblages in the south-west Indian Ocean off Southern Africa. *Journal of Micropaleontology* 6:49–56.

Gaarder, K. R. 1962. Electron microscope studies on holococcolithophorids. *Nytt Magasin for Botanikk* 10:35–51.

Gaarder, K. R. 1967. Observations on the genus *Ophiaster* Gran (Coccolithineae). *Sarsia* 29:183–192.

Gaarder, K. R. 1970. Three new taxa of Coccolithineae. *Nytt Magasin for Botanikk* 17:113–126.

Gaarder, K. R. 1983. Validation of *Homozygosphaera halldalii*. *INA Newsletter, Proceedings of the International Nannoplankton Association* 5:50.

Gaarder, K. R., & Hasle, G. R. 1962. On the assumed symbiosis between diatoms and coccolithophorids in *Brenneckella. Nytt Magasin for Botanikk* 9:145–149.

Gaarder, K. R., & Hasle, G. R. 1971. Coccolithophorids of the Gulf of Mexico. *Bulletin of Marine Science* 21:519–544.

Gaarder, K. R., & Heimdal, B. R. 1977. A revision of the genus *Syracosphaera* Lohmann (Coccolithineae). *"Meteor" Forschungsergebnisse Reihe D* 24:54–71.

Gaarder, K. R., & Markali, J. 1956. On the coccolithophorid *Crystallolithus hyalinus* n. gen., n. sp. *Nytt Magasin for Botanikk* 5:1–5.

Gaarder, K. R., & Ramsfjell, E. 1954. A new coccolithophorid from northern waters. *Calciopappus caudatus* n. gen., n. sp. *Nytt Magasin for Botanikk* 2:155–156.

Gaarder, K. R., Markali, J., & Ramsfjell, E. 1954. Further observations on the coccolithophorid *Calciopappus caudatus. Avhandlinger utgitt av Det Norske Videnskaps-Akademi i Oslo I. Matematisk- Naturvidenskapelig Klasse* 1:1–9.

Gard, G. 1987. Observation of a dimorphic coccosphere. *Abhandlungen der Geologischen Bundesanstalt* 39:85–87.

Gartner, S., & Bukry, D. 1969. Tertiary holococcoliths. *Journal of Paleontology* 43:1213–1221.

Gayral, P., & Fresnel, J. 1976. Nouvelles observations sur deux Coccolithophoracées marines: *Cricosphaera roscoffensis* (P. Dangeard) comb. nov. et *Hymenomonas globosa* (F. Magne) comb. nov. *Phycologia* 15:339–355.

Gayral, P., & Fresnel, J. 1979. Révision du genre *Hymenomonas* Stein à propos de l'étude comparative de deux Coccolithacées: *Hymenomonas globosa* (Magne) Gayral et Fresnel et *Hymenomonas lacuna* Pienaar. *Revue Algologique N.S.* 14:117–125.

Gayral, P., & Fresnel, J. 1983. Description, sexualité et cycle de développement d'une nouvelle Coccolithophoracée (Prymnesiophyceae): *Pleurochrysis pseudoroscoffensis* sp. nov. *Protistologica* 19:245–261.

Gran, H. H. 1912. Pelagic plant life. *In* "The Depths of the Ocean" (J. Murray & J. Hjort, eds.), (pp. 307–386). Macmillan, London.

Grassé, P.-P. 1952. "Traité de zoologie. Anatomie, systématique, biologie. 1(1). Phylogénie. Protozoaires: Généralités. Flagellés." Masson & Cie, Paris.

Grindley, J. R., & Taylor, F. J. R. 1970. Factors affecting plankton blooms in False Bay. *Transactions of the Royal Society of South Africa* 39:201–210.

Haeckel, E. 1894. "Systematische Phylogenie der Protisten und Pflanzen." Reimer, Berlin.

Halldal, P. 1953. Phytoplankton investigations from weather ship M in the Norwegian Sea, 1948–49 (Including observations during the "Armauer Hansen" cruise July 1949). *Hvalrådets Skrifter* 38:1–91.

Halldal, P. 1954. Comparative observations on coccolithophorids in light and electron microscopes and their taxonomical significance. *Huitième Congrès International de Botanique Paris 1954. Rapports et communications parvenus avant le Congrès a la Section* 17:122–124.

Halldal, P., & Markali, J. 1954a. Morphology and microstructure of coccoliths studied in the electron microscope. Observations on *Anthosphaera robusta* and *Calyptrosphaera papillifera*. *Nytt Magasin for Botanikk* 2:117–119.

Halldal, P., & Markali, J. 1954b. Observations on coccoliths of *Syracosphaera mediterranea* Lohm., *S. pulchra* Lohm., and *S. molischi* Schill. in the electron microscope. *Extrait du Journal du Conseil International pour l'Exploration de la Mer* 19:329–336.

Halldal, P., & Markali, J. 1955. Electron microscope studies on coccolithophorids from the Norwegian Sea, the Gulf Stream and the Mediterranean. *Avhandlinger utgitt av Det Norske Videnskaps-Akademi i Oslo I. Matematisk-Naturvidenskapelig Klasse* 1:1–30.

Hallegraeff, G. M. 1984. Coccolithophorids (calcareous nanoplankton) from Australian waters. *Botanica Marina* 27:229–247.

Hasle, G. R. 1960. Plankton coccolithophorids from the Subantarctic and Equatorial Pacific. *Nytt Magasin for Botanikk* 8:77–88.

Hay, W. W., & Towe, K. M. 1962. Electronmicroscopic examination of some coccoliths from Donzacq (France). *Ecologae Geologicae Helvetiae* 55:497–517.

Hay, W. W., Mohler, H. P., & Wade, M. E. 1966. Calcareous nannofossils from Nal'chik (northwest Caucasus). *Ecologae Geologicae Helvetiae* 59:379–399.

Hay, W. W., Mohler, H. P., Roth, P. H., Schmidt, R. R., & Boudreaux, J. E. 1967. Calcareous nannoplankton zonation of the Cenozoic of the Gulf Coast and Caribbean-Antillean area, and transoceanic correlation. *Transactions of the GulfCoast Association of Geological Societies* 17:428–480.

Heimdal, B. R. 1973. Two new taxa of recent coccolithophorids. *"Meteor" Forschungsergebnisse Reihe D* 13:70–75.

Heimdal, B. R. 1982. Validation of the names of some species of *Zygosphaera* Kamptner. *INA Newsletter, Proceedings of the International Nannoplankton Association* 4:52–56.

Heimdal, B. R. 1983. Phytoplankton and nutrients in the waters north-west of Spitsbergen in the autumn of 1979. *Journal of Plankton Research* 5:901–918.

Heimdal, B. R., & Gaarder, K. R. 1980. Coccolithophorids from the northern part of the eastern central Atlantic. 1. Holococcolithophorids. *"Meteor" Forschungsergebnisse Reihe D* 32:1–14.

Heimdal, B. R., & Gaarder, K. R. 1981. Coccolithophorids from the northern part of the eastern central Atlantic. 2. Heterococcolithophorids. *"Meteor" Forschungsergebnisse Reihe D* 33:37–69.

Hibberd, D. J. 1976. The ultrastructure and taxonomy of the Chrysophyceae and Prymnesiophyceae (Haptophyceae): A survey with some new observations on the ultrastructure of the Chrysophyceae. *Botanical Journal of the Linnean Society* 72:55–80.

Hibberd, D. J. 1980. Prymnesiophytes (= Haptophytes). *In* "Phytoflagellates." (E. R. Cox, ed.), Developments in Marine Biology 2:273–317. Elsevier North Holland, Amsterdam.

Holligan, P. M., & Groom, S. B. 1986. Phytoplankton distribution along the shelf break. *Proceedings of the Royal Society of Edinburgh* 88B:239–263.

Holligan, P. M., Viollier, M., Harbour, D. S., Camus, P., & Champagne-Philippe, M 1983. Satellite and ship studies of coccolithophore production along a continental shelf edge. *Nature* (London) 304:339–342.

Huxley, T. H. 1858. Appendix A. *In* "Deep Sea Soundings in the North Atlantic Ocean between Ireland and Newfoundland" (J. Dayman, ed.), pp. 63–68. H. M. Stationary Office, London.

Inouye, I., & Pienaar, R. N. 1984. New observations on the coccolithophorid *Umbilicosphaera sibogae* var. *foliosa* (Prymnesiophyceae) with reference to cell covering, cell structure and flagellar apparatus. *British Phycological Journal* 19:357–369.

Inouye, I., & Pienaar, R. N. 1988. Light and electron microscope observations of the type species of *Syracosphaera, S. pulchra* (Prymnesiophyceae). *British Phycological Journal* 23:205–217.

Jafar, S. A., & Martini, E. 1975. On the validity of the calcareous nannoplankton genus *Helicosphaera*. *Senckenbergiana Lethaea* 56:381–397.

Jeffrey, S. W., & Wright, S. W. 1987. A new spectrally distinct component in preparations of chlorophyll *c* from the microalga *Emiliania huxleyi* (Prymnesiophyceae). *Biochimica et Biophysica Acta* 894:180–188.

Jordan, R. W. 1991. Problems in the taxonomy and terminology of living coccolithophorids. *INA Newsletter, Proceedings of the International Nannoplankton Association* 13:52–53.

Jordan, R. W., & Young, J. R. 1990. Proposed changes to the classification system of living coccolithophorids. *INA Newsletter, Proceedings of the International Nannoplankton Association* 12(1):15–18.

Jordan, R. W., Knappertsbusch, M., Simpson, W. R., & Chamberlain, A. H. L. 1991. *Turrilithus latericioides* gen. et sp. nov., a new coccolithophorid from the deep photic zone. *British Phycological Journal* 26:175–183.

Kamptner, E. 1927. Beitrag zur Kenntnis adriatischer Coccolithophoriden. *Archiv für Protistenkunde* 58:173–184.

Kamptner, E. 1928. Über das System und die Phylogenie der Kalkflagellaten. *Archiv für Protistenkunde* 64:19–43.

Kamptner, E. 1930. Die Kalkflagellaten des Süsswassers und ihre Beziehungen zu jenen des Brackwassers und des Meeres. *Internationale Revue der gesamten Hydrobiologie und Hydrographie* 24:147–163.

Kamptner, E. 1936. Über die Coccolithineen der Südwestküste von Istrien. *Anzeiger der Akademie der Wissenschaften, Wien* 73:243–247.

Kamptner, E. 1937a. Über Dauersporen bei marinen Coccolithineen. *Akademie der Wissenschaften in Wien* 146:67–76.

Kamptner, E. 1937b. Neue und bemerkenswerte Coccolithineen aus dem Mittelmeer. *Archiv für Protistenkunde* 89:279–316.

Kamptner, E. 1941. Die Coccolithineen der Südwestküste von Istrien. *Annalen des Naturhistorischen Museums in Wien* 51:54–149.

Kamptner, E. 1943. Zur Revision der Coccolithineen Spezies *Pontosphaera huxleyi* Lohm. *Anzeiger der Akademie der Wissenschaften, Wien.* 80:43–49.

Kamptner, E. 1950. Über den submikroskopischen Aufbau der Coccolithen. *Anzeiger Österreichische Akademie der Wissenschaften* 87:152–158.

Kamptner, E. 1954. Untersuchungen über den Feinbau der Coccolithen. *Archiv für Protistenkunde* 100:1–90.

Kamptner, E. 1963. Coccolithineen-Skelettreste aus Tiefseeablagerungen des Pazifischen Ozeans. *Annalen des Naturhistorischen Museums in Wien* 66:139–204.

Kamptner, E. 1967. Kalkflagellaten-Skelettreste aus Tiefseeschlamm des Südatlantischen Ozeans. *Annalen des Naturhistorischen Museums in Wien* 71:117–198.

Klaveness, D. 1972a. *Coccolithus huxleyi* (Lohmann) Kamptner. 1. Morphological investigations on the vegetative cell and the process of coccolith formation. *Protistologica* 8:335–346.

Klaveness, D. 1972b. *Coccolithus huxleyi* (Lohm.) Kamptn. II. The flagellate cell, aberrant cell types, vegetative propagation and life cycles. *British Phycological Journal* 7:309–318.

Klaveness, D. 1973. The microanatomy of *Calyptrosphaera sphaeroidea*, with some supplementary observations on the motile stage of *Coccolithus pelagicus*. *Norwegian Journal of Botany* 20:151–162.

Klaveness, D., & Paasche, E. 1979. Physiology of coccolithophorids. In "Biochemistry and Physiology of Protozoa" (M. Levandowsky & S. H. Hunter, eds.), second edition, vol. 1, pp. 191–213. Academic Press, London.

Kleijne, A. 1990. Distribution and malformation of extant calcareous nannoplankton in the Indonesian seas. *Marine Micropaleontology* 16:293–316.

Kleijne, A. 1991. Holococcolithophorids from the Indian Ocean, Red Sea, Mediterranean Sea and North Atlantic Ocean. *Marine Micropaleontology* 17:1–76.

Kleijne, A., & Jordan, R. W. 1990. Proposed changes to the classification system of living coccolithophorids. 2. *INA Newsletter, Proceedings of the International Nannoplankton Association* 12(2):13.

Kleijne, A., Jordan, R. W., & Chamberlain, A. H. L. 1991. *Flosculosphaera calceolariopsis* gen. et sp. nov. and *F. sacculus* sp. nov., new coccolithophorids (Prymnesiophyceae) from the N. E. Atlantic. *British Phycological Journal* 26:185–194.

Kling, S. A. 1975. A lagoonal coccolithophore flora from Belize (British Honduras). *Micropaleontology* 21:1–13.

Knappertsbusch, M. 1991. Estimates of coccolith-carbonate export production of modern calcareous nannoplankton in the North Atlantic. *INA Newsletter, Proceedings of the International Nannoplankton Association* 13:53–54.

Lecal, J. 1965a. A propos des modalités d'élaboration des formations épineuses des Coccolithophoridés. *Protistologica* 1:63–70.

Lecal, J. 1965b. Coccolithophorides littoraux de Banyuls. *Vie et Milieu* 16(1-B):251–270.

Lecal, J. 1967. Le nannoplancton des Côtes d'Israël. *Hydrobiologia* 29:305–387.

Lefort, F. 1972. Quelques caractères morphologiques de deux espèces actuelles de *Braarudosphaera* (Chrysophycées, Coccolithophoracées). *Botaniste* 55:81–93.

Lemmermann, E. 1908. Flagellate, Chlorophyceae, Coccosphaerales und Silicoflagellatae. In "Nordisches Plankton. Botanischer Teil" (K. Brandt & C. Apstein, eds.), pp. 1–40. Lipsius and Tischer, Kiel and Leipzig.

Linschooten, C., van Bleijswijk, J. D. L., van Emburg, P. R., de Vrind, J. P. M., Kempers, E. S., Westbroek, P., & de Vrind-de Jong, E. W. 1991. Role of the light-dark cycle and medium composition on the production of coccoliths by *Emiliania huxleyi* (Haptophyceae). *Journal of Phycology* 27:82–86.

Loeblich, A. R., Jr., & Tappan, H. 1963. Type fixation and validation of certain calcareous nannoplankton genera. *Proceedings of the Biological Society of Washington* 76:191–196.

Loeblich, A. R., Jr., & Tappan, H. 1966. Annotated index and bibliography of the calcareous nannoplankton. *Phycologia* 5:81–216.

Loeblich, A. R., Jr., & Tappan, H. 1968. Annotated index and bibliography of the calcareous nannoplankton 2. *Journal of Paleontology* 42:584–598.

Loeblich, A. R., Jr., & Tappan, H. 1969. Annotated index and bibliography of the calcareous nannoplankton 3. *Journal of Paleontology* 43:568–588.

Loeblich, A. R., Jr., & Tappan, H. 1970a. Annotated index and bibliography of the calcareous nannoplankton, 4. *Journal of Paleontology* 44:558–574.

Loeblich, A. R., Jr., & Tappan, H. 1970b. Annotated index and bibliography of the calcareous nannoplankton 5. *Phycologia* 9:157–174.

Loeblich, A. R., Jr., & Tappan, H. 1971. Annotated index and bibliography of the calcareous nannoplankton 6. *Phycologia* 10:315–339.

Loeblich, A. R., Jr., & Tappan, H. 1973. Annotated index and bibliography of the calcareous nannoplankton 7. *Journal of Paleontology* 47:715–759.

Loeblich, A. R., Jr., & Tappan, H. 1978. The coccolithophorid genus *Calcidiscus* Kamptner and its synonyms. *Journal of Paleontology* 52:1390–1392.

Lohmann, H. 1902. Die Coccolithophoridae, eine Monographie der Coccolithen bildenden Flagellaten, zugleich ein Beitrag zur Kenntnis des Mittelmeerauftriebs. *Archiv für Protistenkunde* 1:89–165.

Lohmann, H. 1903. Neue Untersuchungen über den Reichtum des Meeres an Plankton und über die Brauchbarkeit der vershiedenen Fangmethoden. Zugleich auch ein Beitrag zur Kenntniss des Mittelmeerauftriebs. *Wissenschaftliche Meeresuntersuchungen herausgegeben von der Kommision zur wissenschaftlichen Untersuchung der Deutschen Meere in Kiel und der Biologischen Anstalt auf Helgoland* N.F. 7:1–87.

Lohmann, H. 1912. Untersuchungen über das Pflanzen- und Tierleben der Hochsee. *Veröffentlichungen des Instituts für Meereskunde an dem Universität. Berlin*, N.F. 1:1–92.

Lohmann, H. 1913. Über Coccolithophoriden. *Verhandlungen der Deutschen zoologischen Gesellschaft* 23:143–164.

Lohmann, H. 1919. Die Bevölkerung des Ozeans mit Plankton nach den Ergebnissen der Zentrifugen-fänge während der Ausreise der "Deutschland" 1911. *Archiv für Biontologie. Berlin* 4(3):1–617.

Manton, I., & Leedale, G. F. 1969. Observations on the microanatomy of *Coccolithus pelagicus* and *Cricosphaera carterae*, with special reference to the origin and nature of coccoliths and scales. *Journal of the Marine Biological Association of the United Kingdom* 49:1–16.

Manton, I., & Oates, K. 1975. Fine-structural observations on *Papposphaera* Tangen from the southern hemisphere and on *Pappomonas* gen. nov. from South Africa and Greenland. *British Phycological Journal* 10:93–109.

Manton, I., & Oates, K. 1983. Nanoplankton from the Galapagos Islands: two genera of spectacular coccolithophorids (*Ophiaster* and *Calciopappus*), with special emphasis on unmineralized periplast components. *Philosophical Transactions of the Royal Society of London B* 300:435–462.

Manton, I., & Oates, K. 1985. Calciosoleniaceae (coccolithophorids) from the Galapagos Islands: Unmineralized components and coccolith morphology in *Anoplosolenia* and *Calciosolenia*, with a comparative analysis of equivalents in the unmineralized genus *Navisolenia* (Haptophyceae = Prymnesiophyceae). *Philosophical Transactions of the Royal Society of London B* 309:461–477.

Manton, I., & Sutherland, J., 1975. Further observations on the genus *Pappomonas* Manton et Oates with special reference to *P. virgulosa* sp. nov. from West Greenland. *British Phycological Journal* 10:377–385.

Manton, I., Bremer, G., & Oates, K. 1984. Nanoplankton from the Galapagos Islands: *Michaelsarsia elegans* Gran and *Halopappus adriaticus* Schiller (coccolithophorids) with special reference to coccoliths and their unmineralized components. *Philosophical Transactions of the Royal Society of London B* 305:183–199.

Manton, I., Sutherland, J., & McCully, M. 1976a. Fine structural observations on coccolithophorids from South Alaska in the genera *Papposphaera* Tangen and *Pappomonas* Manton and Oates. *British Phycological Journal* 11:225–238.

Manton, I., Sutherland, J., & Oates, K. 1976b. Arctic coccolithophorids: Two species of *Turrisphaera* gen. nov. from West Greenland, Alaska, and the Northwest Passage. *Proceedings of the Royal Society of London B* 194:179–194.

Manton, I., Sutherland, J., & Oates, K. 1977. Arctic coccolithophorids: *Wigwamma arctica* gen. et sp. nov. from Greenland and arctic Canada, W. *annulifera* sp. nov. from South Africa and S. Alaska and *Calciarcus alaskensis* gen. et sp. nov. from S. Alaska. *Proceedings of the Royal Society of London B* 197:145–168.

Markali, J., & Paasche, E. 1955. On two species of *Umbellosphaera*, a new marine coccolithophorid genus. *Nytt Magasin for Botanikk* 4:95–100.

Marlowe, I. T., Green, J. C., Neal, A. C., Brassell, S. C., Eglinton, G., & Course, P. A. 1984. Long chain (n-C_{37}-C_{39}) alkenones in the Prymnesiophyceae. Distribution of alkenones and other lipids and their taxonomic significance. *British Phycological Journal* 19:203–216.

McIntyre, A. 1970. *Gephyrocapsa protohuxleyi* sp. n., a possible phyletic link and index fossil for the Pleistocene. *Deep-Sea Research* 17:187–190.

McIntyre, A., & Bé, A. W. H. 1967a. Modern coccolithophoridae of the Atlantic Ocean. 1. Placoliths and cyrtoliths. *Deep-Sea Research* 14:561–597.

McIntyre, A., & Bé, A. W. H. 1967b. *Coccolithus neohelis* sp. n., a coccolith fossil type in contemporary seas. *Deep-Sea Research* 14:369–371.

McIntyre, A., Bé, A. W. H., & Preikstas, R. 1967. Coccoliths and the Pliocene-Pleistocene boundary. *Progress in Oceanography* 4:3–25.

McIntyre, A., Bé, A. W. H., & Roche, M. B. 1970. Modern Pacific coccolithophorida: A paleontological thermometer. *Transactions of the New York Academy of Sciences* 32:720–731.

Milliman, J. D. 1980. Coccolithophorid production and sedimentation, Rockall Bank. *Deep-Sea Research* 27:959–963.

Mitchell-Innes, B. A., & Winter, A. 1987. Coccolithophores: a major phytoplankton component in mature upwelled waters off the Cape Peninsula, South Africa in March, 1983. *Marine Biology* 95:25–30.

Mostajo, E. L. 1985. Nanoplancton calcareo del Oceano Atlanticio sur. *Revista Española de Micropaleontologia* 17:261–280.

Murray, G., & Blackman, V. H. 1898. On the nature of the coccospheres and rhabdospheres. *Philosophical Transactions of the Royal Society B* 190:427–441.

Murray, J., & Hjort, J. 1912. "The Depths of the Ocean. A General Account of the Modern Science of Oceanography Based Largely on the Scientific Researches of the Norwegian Steamer *Michael Sars* in the North Atlantic." MacMillan & Co., London.

Nelson, J. R., & Wakeham, S. G. 1989. A phytol-substituted chlorophyll *c* from *Emiliania huxleyi* (Prymnesiophyceae). *Journal of Phycology* 25:761–766.

Nishida, S. 1979. Atlas of Pacific nannoplanktons. *NOM (News of Osaka Micropaleontologists) Special paper* 3:1–31.

Norris, R. E. 1965. Living cells of *Ceratolithus cristatus* (Coccolithophorineae). *Archiv für Protistenkunde* 108:19–24.

Norris, R. E. 1983. The family position of *Papposphaera* Tangen and *Pappomonas* Manton & Oates (Prymnesiophyceae) with records from the Indian Ocean. *Phycologia* 22:161–169.

Norris, R. E. 1984. Indian Ocean nanoplankton. I. Rhabdosphaeraceae (Prymnesiophyceae) with a review of extant taxa. *Journal of Phycology* 20:27–41.

Norris, R. E. 1985. Indian Ocean nannoplankton. II. Holococcolithophorids (Calyptrosphaeraceae, Prymnesiophyceae) with a review of extant genera. *Journal of Phycology* 21:619–641.

Okada, H. 1992. Biogeographic control of modern nannofossil assemblages in surface sediments of Ise Bay, Mikawa Bay and Kumano-Nada, off coast of central Japan. *Memorie de Scienze Geologiche già Memorie degli Instituti di Geologia e Mineralogia dell'Università di Padova*. XLIII:431–449.

Okada, H., & Honjo, S. 1970. Coccolithophoridae distributed in southwest Pacific. *Pacific Geology* 2:11–21.

Okada, H., & Honjo, S. 1973. The distribution of oceanic coccolithophorids in the Pacific. *Deep-Sea Research* 20:355–374.

Okada, H., & Honjo, S. 1975. Distribution of coccolithophores in marginal seas along the Western Pacific Ocean and in the Red Sea. *Marine Biology* 31:271–285.

Okada, H., & McIntyre, A. 1977. Modern coccolithophores of the Pacific and North Atlantic Oceans. *Micropaleontology* 23:1–55.

Okada, H., & McIntyre, A. 1979. Seasonal distribution of modern coccolithophores in the Western North Atlantic Ocean. *Marine Biology* 54:319–328.

Okada, H., & McIntyre, A. 1980. Validation of *Florisphaera profunda* var. *elongata. INA Newsletter, Proceedings of the International Nannoplankton Association* 2:81.

Ostenfeld, C. H. 1900. Über *Coccosphaera. Zoologischer Anzeiger* 23:198–200.

Ostenfeld, C. H. 1910. *Thorosphaera*, eine neue Gattung der Coccolithophoriden. *Bericht der Deutschen botanischen Gesellschaft* 28:397–400.

Paasche, E. 1968. Biology and physiology of coccolithophorids. *Annual Review of Microbiology* 22:71–86.

Parke, M., & Adams, I. 1960. The motile (*Crystallolithus hyalinus* Gaarder & Markali) and non-motile phases in the life history of *Coccolithus pelagicus* (Wallich) Schiller. *Journal of the Marine Biological Association of the United Kingdom* 39:263–274.

Parke, M., & Green, J. C. 1976. Haptophyta. Haptophyceae. Check-list of British marine algae— third revision (M. Parke & P. S. Dixon). *Journal of the Marine Biological Association of the United Kingdom* 56:551–555.

Pascher, A. 1910. Chrysomonaden aus dem Hirschberger Grossteiche. *Monographien und Abhandlungen zur Internationalen Revue der gesamten Hydrobiologie und Hydrographie* 1:1–66.

Perch-Nielsen, K. 1971. Durchsicht tertiärer Coccolithen. *Proceedings of the II Planktonic Conference, Roma 1970* (A. Farinacci, ed.), pp. 939–980. Edizioni Tecnoscienza, Roma.

Poche, F. 1913. Das System der Protozoa. *Archiv für Protistenkunde* 33:125–321.

Pringsheim, E. G. 1955. Kleine Mitteilungen über Flagellaten und Algen. 1. Algenartige Chrysophyceen in Reinkultur. *Archiv für Mikrobiologie* 21:401–410.

Rampi, L., & Bernhard, M. 1981. Chiave per la determinazione delle coccolithoforidee mediterranee. Comitato Nazionale Energià Nucleare.

Raymont, J. E. G. 1980. "Plankton and Productivity in the Oceans. 1. Phytoplankton," 2nd ed. Pergamon Press, Oxford.

Rayns, D. G. 1962. Alternation of generations in a coccolithophorid, *Cricosphaera carterae* (Braarud & Fagerl.) Braarud. *Journal of the Marine Biological Association of the United Kingdom* 42:481–484.

Reid, F. M. H. 1980. Coccolithophorids of the North Pacific Central Gyre with notes on their vertical and seasonal distribution. *Micropaleontology* 26:151–176.

Reinhardt, P. 1972. "Coccolithen. Kalkiges Plankton seit Jahrmillionen." Die Neue Brehm-Bücherei, A. Ziemsen Verlag, Wittenberg, Lutherstadt.

Roth, P. H. 1973. Calcareous nannofossils. Leg. 17, Deep Sea Drilling Project. *In* "Initial Reports of the Deep Sea Drilling Project" (L. E. Winterer et al., eds.), vol. 17, pp. 695–795. Government Printing Office, Washington, D.C.

Rowson, J. D., Leadbeater, B. S. C., & Green, J. C. 1986. Calcium carbonate deposition in the motile (*Crystallolithus*) phase of *Coccolithus pelagicus* (Prymnesiophyceae). *British Phycological Journal* 21:359–370.

Samtleben, C., & Schröder, A. 1990. Coccolithophoriden Gemeinschaften und Coccolithen-Sedimentation im Europäischen Nordmeer. *Berichte aus dem Sonderforschungsbereich 313 "Sedimentation im Europäischen Nordmeer"* 25:1–52.

Schiller, J. 1913. Vorläufige Ergebnisse der Phytoplankton-Untersuchungen auf den Fahrten S.M.S. "Najade" in der Adria 1911/1912. 1. Die Coccolithophoriden. *Sitzungsberichte der Königlichen Akademie der Wissenschaften in Wien. Mathematisch-naturwissenschaftliche Klasse* 122(1):597–617.

Schiller, J. 1914. Bericht über Ergebnisse der Nannoplanktonuntersuchungen anlässlich der Kreu-
    zungen S.M.S. "Najade" in der Adria. *Internationale Revue der gesamten Hydrobiologie und
    Hydrographie Biologisches Supplement* 6(4):1–15.
Schiller, J. 1925. Die planktontischen Vegetationen des adriatischen Meeres. A. Die Coccolitho-
    phoriden-Vegetation in den Jahren 1911–14. *Archiv für Protistenkunde* 51:1–130.
Schiller, J. 1926. Über Fortpflanzung, geissellose Gattungen und die Nomenklatur der Coccolitho-
    phoraceen nebst Mitteilung über Copulation bei *Dinobryon*. *Archiv für Protistenkunde*
    53:326–342.
Schiller, J. 1930. Coccolithineae. *Kryptogamen-Flora von Deutschland, Österreich und der
    Schweiz* 10:89–273. Akademische Verlagsgesellschaft, Leipzig.
Schlauder, J. 1945. Recherches sur les flagellés calcaires de la baie d'Alger. Diplôme Faculté des
    Sciences, Université d'Alger.
Schwarz, E. H. L. 1894. Coccoliths. *Annals and Magazine of Natural History, London, Series 6*
    14:341–346.
Senn, G. 1900. Chrysomonadineae. *In* "Die natürlichen Pflanzenfamilien," Part 1 (A. Engler & K.
    Prantl, eds.), pp. 151–167. Wilhelm Englemann, Leipzig.
Smayda, T. J. 1958. Biogeographical studies of marine phytoplankton. *Oikos* 9:158–191.
Sournia, A. 1978. "Phytoplankton Manual." UNESCO Monographs on Oceanographic Method-
    ology 6. UNESCO, Paris.
Takayama, T. 1967. First report on nannoplankton of the upper Tertiary and Quaternary of the
    southern Kwanto region, Japan. *Jahrbuch der Geologischen Bundesanstalt (Wien)* 110:169–
    198.
Tangen, K. 1972. *Papposphaera lepida*, gen. nov., n.sp., a new marine coccolithophorid from
    Norwegian coastal waters. *Norwegian Journal of Botany* 19:171–178.
Tangen, K., & Bjørnland, T. 1981. Observations on pigments and morphology of *Gyrodinium
    aureolum* Hulburt, a marine dinoflagellate containing 19'-hexanoyloxyfucoxanthin as the
    main carotenoid. *Journal of Plankton Research* 3:389–401.
Tappan, H. 1980. Haptophyta, coccolithophores, and other calcareous nannoplankton. *In* "The
    Paleobiology of Plant Protists" (H. Tappan, ed.), pp. 678–803. W. H. Freeman, San Fran-
    cisco.
Thomsen, H. A. 1980a. *Wigwamma scenozonion* sp. nov. (Prymnesiophyceae) from West Green-
    land. *British Phycological Journal* 15:335–342.
Thomsen, H. A. 1980b. Two species of *Trigonaspis* gen. nov. (Prymnesiophyceae) from West
    Greenland. *Phycologia* 19:218–229.
Thomsen, H. A. 1980c. *Quaternariella obscura* gen. et sp. nov. (Prymnesiophyceae) from West
    Greenland. *Phycologia* 19:260–265.
Thomsen, H. A. 1980d. *Turrisphaera polybotrys* sp. nov. (Prymnesiophyceae) from West Green-
    land. *Journal of the Marine Biological Association of the United Kingdom* 60:529–537.
Thomsen, H. A. 1981. Identification by electron microscopy of nanoplanktonic coccolithophorids
    (Prymnesiophyceae) from West Greenland, including the description of *Papposphaera sarion*
    sp. nov. *British Phycological Journal* 16:77–94.
Thomsen, H. A., & Oates, K. 1978. *Balaniger balticus* gen. et sp. nov. (Prymnesiophyceae) from
    Danish coastal waters. *Journal of the Marine Biological Association of the United Kingdom*
    58:773–779.
Thomsen, H. A., Buck, K. R., Coale, S. L., Garrison, D. L., & Gowing, M. M. 1988. Nanoplankto-
    nic coccolithophorids (Prymnesiophyceae, Haptophyceae) from the Weddell Sea, Antarctica.
    Nordic Journal of Botany Section of Phycology 8:419–436.
Thomsen, H. A., Østergaard, J. B., & Hansen, L. E. 1991. Heteromorphic life histories in arctic
    coccolithophorids (Prymnesiophyceae). *Journal of Phycology* 27:634–642.
Throndsen, J. 1972. Coccolithophorids from the Caribbean Sea. *Norwegian Journal of Botany*
    19:51–60.

Throndsen, J. 1978. The dilution culture method. *In* "Phytoplankton Manual," UNESCO Monographs on Oceanographic Methodology 6 (A. Sournia, ed.), pp. 218–224. UNESCO, Paris.

Turner, S. M., Malin, G., & Liss, P. S. 1988. The seasonal variation of dimethyl sulfide and dimethylsulfoniopropionate concentrations in nearshore waters. *Limnology and Oceanography* 33:364–375.

van Bleijswijk, J., van der Wal, P., Kempers, R., Veldhuis, M., Young, J. R., Muyzer, G., de Vrind-de Jong, E., & Westbroek, P. 1991. Distribution of two types of *Emiliania huxleyi* (Prymnesiophyceae) in the northeast Atlantic region as determined by immunofluorescence and coccolith morphology. *Journal of Phycology* 27:566–570.

Volkman, J. K., Smith, D. J., Eglinton, G., Forsberg, T. E. V., & Corner, E. D. S. 1981. Sterol and fatty acid composition of four marine haptophycean algae. *Journal of the Marine Biological Association of the United Kingdom* 61:509–527.

von Stein, F. R. 1878. "Der Organismus der Flagellaten," vol. 1. Wilhelm Engelmann, Leipzig.

Wallich, G. C. 1861. Remarks on some novel phases of organic life, and on the boring powers of minute annelids, at great depths in the sea. *Annals and Magazine of Natural History, Series 3* 8:52–58.

Wallich, G. C. 1877. Observations on the coccosphere. *Annals and Magazine of Natural History, Series 4* 19:342–350.

Watabe, N., & Wilbur, K. M. 1966. Effects of temperature on growth, calcification, and coccolith form in *Coccolithus huxleyi* (Coccolithineae). *Limnology and Oceanography* 11:567–575.

Weber-van Bosse, A. 1901. Études sur les algues de l'Archipel Malaisien. 3. Note préliminaire sur les résultats algologiques de l'éxpédition du Siboga. *Annales du Jardin botanique de Buitenzorg* 17 (ser. 2, vol. 2): 126–141.

Westbroek, P., Young, J. R., & Linschooten, K. 1989. Coccolith production (biomineralization) in the marine alga *Emiliania huxleyi*. *Journal of Protozoology* 36:368–373.

Wilbur, K. M., & Watabe, N. 1963. Experimental studies in calcification in molluscs and the alga *Coccolithus huxleyi*. *Annals of the New York Academy of Sciences* 109:82–112.

Winter, A., Reiss, Z., & Luz, B. 1978. Living *Gephyrocapsa protohuxleyi* McIntyre in the Gulf of Elat ('Aqaba). *Marine Micropaleontology* 3:295–298.

Winter, A., Reiss, Z., & Luz, B. 1979. Distribution of living coccolithophore assemblages in the Gulf of Elat ('Aqaba). *Marine Micropaleontology* 4:197–223.

Young, J. 1987. Higher classification of coccolithophores. *INA Newsletter, Proceedings of the International Nannoplankton Association* 9:36–38.

Young, J. 1991. Terminology workshop. *INA Newsletter, Proceedings of the International Nannoplankton Association* 13:90.

Young, J. R., & Westbroek P. 1991. Genotypic variation in the coccolithophorid species *Emiliania huxleyi*. *Marine Micropaleontology* 18:5–23.

GLOSSARY

English	French	German
aperture	orifice m	Apertur f
apical spine	épine apicale f	apikaler Dorn m
arm	bras m	Arm m
bar	bare f	Balken m
base	base f	Basis f
blade	feuillet m	Blatt n
body	corps m	Körper m
branch	branche f	Zweig m
central area	aire centrale f	Zentralfeld n
central process	hampe centrale f	Zentralfortsatz m
central structure	structure centrale f	Zentralstruktur f
collar	collier m	Kragen m
column	colonne f	Säule f
cover plates	lames recouvrantes f	Deckplatten f
crystals	cristaux m	Kristalle m
cycle	cycle m	Zyklus m
depression	dépression f	Vertiefung f
disc	collerette f	Flansch m
distal	distal	distal
element	élément m	Element n
groove	sillon m	Rinne f
hight	hauter f	Höhe f
hole	trou m	Loch n
hook	croc m	Haken m
horseshoe-shaped	en fer à cheval	hufeisenförmig
interray area	aire interradiale f	Zwischenareal n
keel	crête f	Kiel m
knob	bouton m	Knopf m
lateral	latéral	lateral
median axis	axe médian m	Mittelachse f
median suture	suture médiane f	Mediansutur f
node	nodule f	Knoten m
opening	ouverture f	Öffnung f
perforation	perforation f	Perforation f
proximal	proximal	proximal
ray	rayon m	Strahl m
ridge	acrête f	Grat m
rim	anneau m	Kranz m
segment	segment m	Segment n
shield	disque m	Scheibe f
spur	éperon m	Sporn m
strut	contrefort m	Stütze f
suture	suture m	Sutur f
wall	paroi f	Wand f
width	largeur f	Weite f

f, feminine; m, masculine; n, neuter.
After Farinacci, 1971, pp. 1258, 1259.

Italian	Russian	Spanish
apertura f	апертура f	salida f
spina apicale f	апикальный отросток m	espina apical f
braccio m	рука f	brazo m
sbarra f	перемычка f	barra f
base f	базис m	base f
lama f	пластина f	hoja f
corpo m	тело n	cuerpo m
ramo m	ветвь f	ramo m
area centrale f	центральное поле n	área central f
formazione centrale f	стержень m	continuación central f
struttura centrale f	центральная структура f	structura central f
collare m	воротничок m	cuello m
colonna f	столбик m	columna f
lamelle ricoprenti	покровные пластинки f	láminas recubridoras f
cristalli m	кристаллы m	cristales m
giro m	цикл m	ciclo m
depressione f	углубление n	depresión f
disco m	диск m	disco m
distale	дистальный	distal
elemento m	элемент m	elemento m
solco m	желоб m	surco m
altezza f	высота f	altura f
fossetta f	ямка f	hoyuela f
uncino m	крючочек m	garifo m
a ferro di cavallo	подковообразный	forma de herradura
area interradiale f	межлучевая арея f	área interradial f
carena f	киль m	quilla f
bottone m	бугор m	botón m
laterale	латеральный	lateral
asse mediano m	медианная ось f	eje mediano m
sutura mediana f	медианный шов m	sutura mediana f
nodo m	бугорок m	nudo m
foro m	отверстие n	abertura f
perforazione f	перфорация f	perforación f
prossimale	проксимальный	proximal
raggio m	луч m	rayo m
cresta f	кромка f	cresta f
anello m	краевой ободок m	anillo m
segmento m	сегмент m	segmento m
scudo m	щиток m	escudo m
sperone m	шпора f	espolón m
contrafforte m	опора f	soporte m
sutura f	шов m	sutura f
parete f	стенка f	pared f
larghezza f	ширина f	largura f

Index

Aberrant cell 157, 313, 333
Acid
  Cleaned material 62, 63, 65, 79, 91,
    101, 111, 271, 289, 294, 329,
    335
  Cleaned mounted valves 111, 112 ,
    331
  Cleaning 40, 45, 335
Acid soluble plates 739
Acridine orange 594, 633, 694, 697
Acrobase 403
Acronematic flagella 600, 703, 752
Actinopod 600, 627, 628, 630, 703
Adnate 703
Adriatic 37, 96
Advanced techniques 599
Advalvar half 234
Africa 131, 249, 257
Agglutination 744
Air diatom mounts 7, 107, 112, 165,
  168, 278
Air-sea exchange of carbon dioxide 743
Alexandria harbor 497, 499
Algae
  accumulation products 596
  classification 23, 28, 750, 755
  eukaryotic 598
  heterokont 595
Algal Systematics 603
Alternating life stages 608
Alternation 697, 743
Alternative phase 794
Alveolus(i) 17, 18, 30, 33, 34
  Openings 33, 34
Ambush preditor 550
Amphiesma 338, 401
Amphisterol 390
Anastomosing costae 88

Anisogomy 471, 597
Anisogamous(y) 8, 10, 390
Anisokont 600, 605, 703
Annular pattern 638
Annulus 17, 23, 31, 36, 41, 42, 44, 47,
  53, 54, 56, 63, 74, 77, 79, 87, 117,
  120, 121, 124
Ansula(e) 226, 230, 231
Antapex 391
Antapical 743, 781
Antarctic 45, 71, 95, 97, 126, 140, 147,
  189, 240
Antiligula(e) 16, 144
Apical Axis 14, 15, 24, 235, 238, 243,
  244, 246–248, 251-263, 257, 261,
  266, 267, 269, 271, 273, 275, 276,
  280, 282, 284, 286, 290, 293, 294,
  286, 297, 308, 309, 311, 312, 316,
  324, 327, 328, 329, 330, 338, 391,
  392
  Heteropolar 296, 297, 299, 301, 303,
    307
  Isopolar 296, 298, 299, 300, 302
Apical 14, 242, 267, 703, 745
  area 781, 785, 799
  axis 794
  crown 798
  depression 659, 663, 669, 779, 798,
    800, 801
  end 674, 781
  field 241, 242, 253
  groove 403
  opening 754, 799
  plane 14, 312
  pole 240
  pore 245, 247, 249, 251
  pore complex (APC) 240, 247, 252,
    406

Apical (*continued*)
  pore field 24, 240-242, 244–248, 251, 252
  ring 779
  slit field 240, 241, 243, 246
  spine 253, 255–260, 262–265, 267, 338, 780, 781
  view 794, 804
  whorl 780
Apiculus(i) 101
Aplanospore 510, 520
Appendage coccoliths 748
Appendix 745, 747, 774, 791, 805, 807
Aragonite 732, 745
Araphid diatom species 7, 10, 26, 241, 253
Arctic 45, 69, 71, 74, 75, 82, 87, 95, 98, 215, 246, 247, 262, 269, 387, 471, 475,
  lakes 498
  material 82
  plankton 241
  sea ice 122, 278
Areola(e) 17, 18, 19, 30–35, 37, 42, 43, 46, 47, 50, 51, 53, 56–59, 61, 62, 64, 68, 74, 79, 82–84, 86, 87, 103, 104, 116–122, 132, 135, 144, 171, 231, 240, 244, 248, 263, 254, 257, 259–263, 265–267, 330, 331
  endochiastic 107
  hexagonal 51, 59, 62, 63, 69, 74, 80, 82, 92, 94
  loculate 17, 18, 30, 46, 57, 92, 97, 113, 117, 159, 253
  pentagonal 101, 105
  poroid 17, 18, 57, 142, 157, 160, 165, 167, 169, 181, 230, 233, 247
Areolate 406, 760
Areolated 31, 36
  sector 51, 54, 52, 113
  valves 82
Areolation 39, 43, 51, 56, 58, 59, 61, 62, 63, 65, 66, 69, 70, 71, 73, 77, 79, 82, 83, 84, 86, 87, 137, 139, 141, 177, 181, 245, 257, 264, 306

bifurcate 66
eccentric 50, 70, 71, 83, 86
fasciculate 19, 41, 45, 46, 51, 58, 61, 62, 69, 70, 71, 73, 79, 82, 86, 87, 88, 98, 106, 113, 117, 119, 120, 122, 124
furcate 113
linear 50, 71, 76, 86, 124
lineatus type 59, 62, 63, 77
radial 19, 41, 44, 51, 54, 61, 63, 66, 69, 73, 87, 98, 109, 112, 117, 118, 124–126, 128, 406, 409
rosette–larger areolae 98, 101, 104, 105, 107, 109
sublinear 83, 86, 124
tangential 50, 59, 62, 70, 77, 84, 87
Argentina 37, 39, 59, 490, 492, 500
Areolith 736, 745, 755, 760
Arm links 800, 801, 803
Armored 405, 406
  dinoflagellates 598, 741
  dinokont 405, 523
Asymmetric 612, 644, 690, 703, 785
Athecate 405
Atlantic Ocean 39, 40, 43, 74, 189, 453, 503, 504, 509, 512, 517, 519, 521, 541
  north 147, 246, 471, 500
  south 71, 257, 497, 530
  U.S. 490
Attachment pore 406
Austral summer 744
Australia 229, 230, 231, 232, 492, 497
  plankton 236
  waters 131, 447, 449
Autotrophic 388, 390, 703
Auxospore 8, 9, 11, 23, 93
  envelope 23
  formation 8, 9, 49
  lateral 201
  terminal 159
Auxotrophic 390, 600, 703
Axoneme 595, 601, 703–705, 707
  microtubular organization 742
Axopodium 600, 623, 703

Bacilliform 276
Bacteria 71, 74, 89, 224, 269, 591, 593, 598
Baltic Sea 71, 74, 89, 224, 269
Band(s) 14, 16, 17, 66, 70, 92, 96, 102, 103, 155, 157, 159, 160, 161, 163, 165, 166, 167, 168, 171, 175, 208, 209, 241, 246, 268, 275, 294, 295, 305, 306, 312, 329, 335, 338
  areolae 151, 238
  closed 16, 306
  connecting 17
  half 16, 93, 96
  hyaline, 17, 178
  intercalary 14, 17, 295, 307
  open 16, 110, 244, 303, 307
  structure 151, 156, 307
Bar 745, 748, 786
Barents Sea 140, 500
Baring Sea 499
Basal 186, 244
  body 626, 652
  chamber 30, 678
  coccolith disc 804
  collar 760
  disc 748, 749, 805, 807
  membrane 230
  pole 240
  pyrenoid 662
  ring 628, 632, 747, 764
  siliceous layer 17, 18, 30
  spine 624
  structures 742
Base plate Scale 745, 746, 794
Bay
  of Arcachon 500
  of Fundy 307, 494
  of Izmir 497
  of Naples 494
  of Salerno 492
Belize 425
Benthic 646, 652
  cyst 411, 415, 416
  diatoms 592
  filamentous phase 738, 743, 752, form 738
  macroalgae 602

species 627, 633
stages 742, 789, 805
Betacarotine 7
Biarcuate 286
Biflagellate 520, 519, 731, 739, 742, 778, 788, 789
Bifurcate 66
Bifurcation 188, 189
Bilabiate process 232, 226, 235, 236
Bilateral symmetry 654, 690, 707
Binary fission 7, 416, 596, 597, 738, 739, 742
Biodiversity 589
Biogenic carbon 741
Biogeographical 5
  provinces 28
Biotin 696
Biotopes 7, 597, 599
  marine 7,
  freshwater 7
  planktonic 7
  benthic 7
  epiphytic 7
  epizoic 7
  endozoic 7
  endophytic 7
Bipolar 20, 24, 25, 238
  elevations 20, 24, 169
  symmetry 7, 169, 178
Biseriate 17
Blooms 589, 592, 740, 743, 744, 748
Bluegreen rodlets 445, 453
Body metaboly 646
Boreal waters 743
Brackish Waters 85, 71, 75, 220, 223, 224, 619, 621, 684, 703, 740, 743, 756, 806
  species 33
Brazil 157
Bremerhaven 183
Bridge
  elements 796
  plates 795, 796
Brightfield
  Illumination 687, 688, 811
  microscopy 605, 687, 750, 800
Brilliant Cresyl Blue 617, 626, 633, 694

British Isles 449, 463, 464, 500

Calicalith 745, 755
Calcareous
  cysts 699
  organisms 744
  plates 734
Calcification 735, 793, 812
Calcified
  rim 798, 812
  scales 635, 742
  structures 698, 756
Calcite 732, 742, 745, 746, 748, 759,
  760, 765, 793, 794
  precipitate 784
  production 740
  rhombohedra 735, 758, 759
Calcium carbonate 732, 740, 742, 745,
  746, 747
Calcofluor 554
California 131, 184, 218, 248, 257, 307
Calotte (cap) shaped 745
Calyptoform 763, 764
Canada 307, 490, 494
Canal 645, 646, 647, 650, 703, 707
  plasmalemma 646
  raphe 290, 291, 294, 295, 303, 305,
    311, 315, 327, 329
  reservoir complex 650
  wall 295, 207, 327, 328, 329, 330,
    331
Caneolith 737, 738, 745, 749, 754,
  779, 781, 783, 790, 798, 800,
  804, 808, 813
Canopy 482
Cape of Good Hope 140
Cape Verde Islands 305
Capitate 207, 220, 246, 252, 330, 331
Caribbean Sea 488, 499
Carbon removal 743
Carina 88, 459
Carotenoids (see Pigments) 587, 731,
  742, 751
Caudate 733
Cell
  anatomy 646, 688

content 605, 646, 665,
covering 633, 645, 651, 672, 742
diminution 8, 9
densities 599, 696
Investment 601, 705, 707
investment length 675
motion 688, 690
surface 594, 600, 613, 646, 704, 734,
  746, 749, 756, 805, 813
type 739, 778
wall 591, 599, 604, 654, 665, 669,
  693, 697, 706, 739, 742
Cellular
  endoskeleton 468, 631, 632, 705
  ultrastucture 750
Cellulose 601, 616, 665, 705, 707
  lorica 617, 619, 626
  scales 623, 626
  staining 594, 665
  wall 664
Central
  areola 19, 23, 54, 56, 57, 59, 62, 77,
    87, 105,
  cavity 59,
  clustered processes 7, 31, 86
  concavity 63
  depression 193, 228, 232, 326, 329,
    808
  fibula(e) 268, 291
  field 783, 786, 787, 788
  innerspace 291, 296–298, 300, 302,
    303, 308–310, 313, 315, 317,
    319–321, 323, 328, 391, 330,
    331
  lenticular opening 37
  nodule 267, 268, 273, 275, 282, 284,
    286, 287, 291, 299, 313, 319,
    324, 330, 331
  pore 105, 267
  processes 738, 805, 806, 809, 811
  pyrenoid 210, 267, 276, 277, 278
  raphe ending 267, 291, 305, 327
  rod 779, 780, 781, 783, 787, 809,
    812
  rosette 19, 104, 105, 107, 109
  space 291
  spine 803, 804, 807

Central (*continued*)
  strutted processes 31, 35, 39, 40, 46,
    47, 50, 53, 54, 56, 59, 61, 62, 63,
    68, 69, 70, 71, 73, 75, 79, 80, 81,
    82, 83, 84, 86, 87
  thread(s) 86
  tube 793, 795, 798, 801, 815, 816
Central larger innerspace 291
Centric diatoms 7, 8, 9, 10, 11, 14,
    23–25, 27–29, 39, 123, 587, 744
Centrifuge 688
Centripetal 783, 786
Cenozoic 740
Ceratolith 737, 745, 753, 785
  morphology 785
Chalks, 744
Chain(s)
  axis 189, 195, 196, 199, 201, 203,
    204, 207, 209, 213, 315, 216,
    219, 220, 276
  close Set 160, 163, 165, 167, 169
  formation 10, 13, 31, 75, 86, 238
  helical 90, 160, 161, 163, 193, 209,
    210
  irregular 271
  loose 41
  motile 308
  margin 190, 210
  spiral 163, 168, 211, 213, 241
  star shaped 20, 241
  stepped 20, 270, 291, 305, 308, 310,
    324
  tight 31, 34, 93
  zig-zag shaped 20, 244, 249
Chamber 88
Chesapeake Bay 499
Chile 39, 71, 88, 248, 249, 492
Chitin 601, 705
  loricae 616, 626
Chloral hydrate 553
Chlorophyll (*see* Pigments) 7, 587, 603,
    604, 731, 742
Chloroplast 27, 30, 36, 43, 44, 48, 49,
    59, 63, 84, 89, 90, 91, 95, 96, 98,
    99, 101,104, 105,106, 107, 109,
    113–116, 230
  color 606, 655, 691

morpholoty 646
pigments 687
stroma 601, 664, 707
Choanoflagellates 593, 594, 601, 671,
    672, 673, 678, 687, 689, 691, 692,
    693, 698, 704, 705, 707
Cholesterol 390
Chromatographic qualities 732
Chrysolaminarin 7, 596, 617, 626, 633,
    646, 650, 687, 692, 703, 732, 742
  vacuole 621
Chuckchi Sea 499
Ciliated protists 592
Cingulum(a) 14, 96, 116, 391, 392,
  circular 403
  displaced 404
  median 405
  postmedian 405
  premedian 405
Circular placoliths 778, 779
Clasper(s) 20, 142, 143, 144, 146, 159,
    161
Clasping otarium(a)
Classification 595, 750, 751, 752
Clavate 262, 650 733, 745
Claviform cyrtoliths 806
Cleaning process 750
Cleavage rhombohedrons 759
Cleptochloroplasts 444, 445, 550
Clone 745
Cluster 50, 65, 73, 86
Coaxial
  membranes 603
  rings 779, 780
Coccoid 411
  stage 597, 617
  unicells 731, 741
Coccolith
  appendage 705
  axes 776, 784
  base plate diameter 791
  bearing form 738
  case 746, 747, 755, 761, 764, 769,
    770, 773, 774, 776, 783, 790,
    809, 810, 811
  center 776
  covered globules 739

Coccolith (*continued*)
  diameter 801
  disc 804
  formation 734, 742
  microstructure 750
  morphology 698, 734, 751, 794,
    812
  polysaccharide 794
  production 738
  structure 797
  terminology 744
  tube 786
  type 751, 755, 780, 789, 798
  vesicle 735
  wall 783, 784
Coccolithophorid
  bloom 743, 744
  diatom association 744
  cold water species 740
Coin shaped 48, 109
Cold water 28, 239, 243, 266, 267, 269,
    275, 289, 305, 316, 323
  northern 28, 33, 36, 43, 59, 71, 86,
    88, 89, 93
  southern 28, 43, 63, 71, 75, 80, 86,
    88, 110, 115
  species 28
Collar 88, 224, 588, 593, 672, 677,
    760, 796
  button shaped 815
Collar-like intercalary bands 16, 143
Collar shaped segments 16
Colonies 7, 20, 45, 54, 56, 57, 58, 61,
    63, 84, 96, 130, 203, 211, 221, 222,
    244, 253–255, 275, 278, 264, 275,
    293, 305, 321, 332
  bundle-shaped 240, 253
  fan-shaped 253, 257
  inseparable 20
  motile 308
  mucilage 54, 55, 56, 61, 63, 69, 634,
    635, 640, 641
  ovoid 267
  radiating 240, 263, 266
  separable 20, 45
  sheetlike 40
  spherical 193, 221

  stellate 240, 245, 247, 253
  stepped 307, 324
Colony formation 62, 266, 323, 333
Columbia 131
Combined fixative 552
Competition 688
Compound light microscope 688
Conical central process 809
Conopea 295, 307
Contiguous area 142, 143, 160
Continuous periphery 797
Contractile vacuole 611, 621, 646, 668,
    669
Copula(e) 14, 70, 142, 144, 146
Core samples 807
Corona 88, 89, 746
Costa(e) 30, 88, 178, 188, 191, 192,
    227, 229, 230
  Interareolar 231
  radial 30
Costal
  joints 677
  morphology 698
  ring 675, 677
  strip 672, 674, 704
Costate 672, 674, 675, 677, 678, 680,
    698, 704
Costate ocellus 115, 169, 170, 172, 175
Cretaceous 598, 739, 744
Cribate 746
Cribrillith 737, 746, 754, 790, 791, 806,
    808
Cribrum(a) 17, 19, 20, 30, 84, 97, 103,
    104, 105, 106, 107, 113, 117
Cricolith 737, 746, 752, 789, 806
Cristae 595, 704
Critical Illumination 688
Cross venation 79
Crossbanded root 595
Cross striated structure 601, 706
Cruciform structure 789
Crystal
  faces and angles 747
  nucleation 738
  orthorhombic 745
Crystallolith 736, 755, 758, 760
Cultivation for identification 694

Culture 597, 697, 732, 743, 751, 778
  clonal 745
  conditions 688
  crude 688, 694, 696
  medium composition 734
  serial dilution 600, 694, 696
  stationary 734, 793
  studies 739
Cuneate 733, 746
Cuneiform indentation 423
Cyanobacteria symbiont 423, 436
Cyanocobalamine (B12) 696
Cyrtolith 737, 746, 754, 773–775, 790,
  806, 807
Cytosome 681, 684
Cysts 597, 598, 601, 617, 707, 739, 778
Cytological methods 94
Cytoplasm 668
Cytosome 463

Danish waters 157, 184, 529
  Wadden Sea 445
Daughter Cells 742
Decaying Organic matter 741
Decussating arc 19, 29, 98, 103, 104,
  105, 109
Deep photic zone 741
Denticulate 407
Desmokont 391
Diatoxanthin 7
Dextral 779, 786, 795
Dextrally imbricate 789
Diagnostic Structures 585
Dialyzing 699
Diatom 5–71, 587, 591, 625, 741, 744,
  atlas 6
  frustule 6–8, 14, 15, 27, 334, 744
  morphology 13, 14, 33, 334
  resting spore 11, 333
  systematics 23
  system 11
  taxonomic revisions 338, 339
  terminology 13, 14, 28, 30, 88, 98,
  240
Diadinoxanthin 7
Diatoxanthin 7

Diatotepic layers 7
Differential contrast (DIC) 687
Dilution cultures 688, 94, 696
Dimethyl cholesterol 390
Dimorphism (dimorphic) 746, 747, 755,
  763, 764, 765, 773, 774,
  coccoliths 767, 786, 788, 804
  coccospheres 734, 763, 768, 770, 807
  endotheca 790
  sacculiform cyrtoliths 774
Dinoflagellates 591, 593, 597, 598, 602,
  707, 708, 731, 741
Dinokont 391, 400
  unarmored 444
Dinospore 460
Dinosterol 390
Diploid 8, 9, 411, 738
Diplonts 9
Disciform 746, 807
Discoid 40, 41, 46, 48, 83, 54, 63, 70,
  83, 91, 104, 105, 107, 109, 113, 46
Discolith 746, 791, 808
Distal
  arm link 804
  growth 745
  margin 764, 791
  nodules 762
  pores 775
  rim 736, 738, 776, 777, 783, 784,
    809, 811, 815
  spine 756, 758
Dithecate 734, 747, 784, 790, 808, 813
Dodecahedron 753
Domoic acid 307
Dormancy 11
Dorsal 303, 304, 400
Double flagella 665, 670, 671
Double thylakoids 605
Drum shaped 48, 113
Dry mounts 673
Dutch inshore waters 224

Eccentric 84, 115
  areolation 57, 70, 71, 83
  canal raphe 327
  convexity 110

Eccentric (*continued*)
    structure 51, 62, 84
Ecology 599, 688, 741, 794
Ecosystems 598
Ectocrines 389
Ejectile organelle 600, 602, 691, 704,
    708
Ejectosome 600, 603, 605, 606, 607,
    608, 609, 612, 651, 659, 691, 704
Electron micrographs 6, 19, 21, 49, 64,
    78, 85, 256, 259, 265, 326
Electron microscopy 5, 13, 20, 30, 35,
    39, 41, 45, 56, 57, 59, 63, 66, 77,
    78, 80, 83, 89, 92, 98, 103, 111,
    113, 271, 335
Electron microscope grids 697, 704
Elliptical
    valves 135, 177, 247
    coccolith 761, 762, 769, 773, 787
    cribriliths 790
    cyrtoliths 773
    discoid heterococcolith 745
    laminolith 770
    placoliths 789
    proximal ring 761
Embedding techniques 336, 665
Encystment 411
Endocyst 69
Endogenous 11, 12, 56, 66, 72, 73, 115,
    704
Endoplasmic rerticulum 742
Endophytic 7
Endosymbiotic 423, 436, 461
Endothecal 747
    coccoliths 790, 808, 809, 810, 811,
        812, 813,
    diameter 790
    sphere 790
Endozoic 7
English Channel 37, 229, 281, 529
Envelope 742, 747, 758, 759
Enveloping membranes
Epicingula 14, 16
Epicone 392
Epifluorsecence 694
Epiphyte 241, 246, 247, 324, 625
Episome 392

Epitheca(e) 7, 10, 14, 13, 16, 145, 153
Epivalve 13, 14, 16, 191
Eocene 7
Epizoic 7
Erd Schreiber medium 696
Euglenoid movements 646, 651, 692
Eukaryotic
    algae 598, 603
    picoplankton 592
Euphotic zone 593
Europe 447, 500
Euryhaline 683, 740
Eurythermal nature 740
Evolutionary
    background 595
    schemes 594
    trends 594
Exogene 704
Exotheca 734, 747, 790, 808, 811
Exothecal coccoliths 790, 809, 810,
    812, 813
Exotoxin 550
External coccolith
    morphology 603
    skeleton 604, 627, 692
Extrusome 600, 601, 613, 704, 705
Eyespot 601, 608, 621, 636, 645, 646,
    650, 651, 658, 659, 661, 664, 668,
    670, 671, 691, 707

Faeroe Islands 246
Falcate 607, 704
Fascia 178, 181, 183
Fascicle 29, 118, 119, 120, 122, 123,
    208
Fasciculate 19, 20, 50, 51, 58, 61, 62,
    66, 69, 70, 71, 73, 79, 117, 120,
    121, 128
    pattern 57, 98
    striae 19, 29, 98, 268, 271, 283, 284,
        291–295, 297, 300, 302, 303,
        305, 306, 308–313, 315–317,
        319, 320, 320, 321, 323, 324,
        326, 327, 328, 329, 330, 331
Fats 596, 617, 626, 687
Fatty acids 751

Filaments 219, 307
Filamentous investment 641
Filopodium 595, 600, 704
Filtration techniques 592
Fimbriate 105, 230, 257, 669
Fine tubular hairs 605
Fixed material 686, 687
Fixing agents 591, 599
Fissipariety 226
Flagellum (ar) 387, 390
  area 763, 764, 767, 769, 771, 774,
    783, 786, 808
  axis 600, 704
  canal 604
  flimmer 596, 600, 604, 605, 613,
    617, 626, 688, 689, 704
  hairs 596, 600, 703
  microanatomy 595
  morphology 603
  movements 600, 704, 705
  pore 400, 773
  roots 595
  scales 595, 596, 623
Flagellate
  taxonomic revisons 613, 614
  taxonomy 602
  terminology 600
Flagellated
  male gametes 7, 10
  stage 10
Flanges 408, 799
  ventral 405
Flat crstae 595, 596, 604
Florida 131, 497, 515
Flosculolith 747, 755
Fluorescence microscopy 594, 633, 687,
  697
Fluorescent inclusions 687
Foramen (ina) 17, 18, 20, 30, 84,
  92, 97, 103, 113, 117, 253,
  254, 256, 261, 263, 265,
  238
Formaldehyde (Formalin) 334, 552,
  605, 652, 665
Formvar 704
Foot pole 240, 241, 243, 244, 254, 255,
  257, 264-267

Fossilized remnants 744
Fossil
  coccolith 739, 791
  material 807
  record 598
  remains 746
  sediments 739
  species 628
Fragariolith 747, 755, 763, 764
Franz Josef Land 122
Free swimming 597, 620, 681, 683
Freeze etching 606
French Atlantic Coast 497
Freshwater 613, 620, 642, 703, 739,
  740, 743
  coccolithophorid 739
Frustule(s) 6, 7, 10, 14–17, 27, 101,
  155, 323, 238, 284, 289, 291, 294,
  305, 328, 335
  discoid 100, 103
  cylindrical 102, 293
  coin shaped 103, 107
  wedge shaped 103, 109
Fucoxanthin (see pigments) 7, 390, 613,
  742
Fultoportula(e) 22, 30
Furrow 171, 605, 606, 607, 608, 609,
  691, 693, 749
Fused setae 185, 186, 190, 191
Fusiform cells 269, 271, 276, 289,
  290, 293, 294, 307, 308,
  311–313, 315, 608, 651,
  733, 753

Gametes 7, 10, 415
Gametogenesis 9
Gelatinous masses 54, 57, 634, 635,
  640, 641
Gelatinous
  colonial stages 597
  colonies 641
  palmelloid stages 671
Generitype 27, 116, 184
Genotype 597
Genomes 602
Genuflexed 137, 139, 179

Geographical
  Areas 6, 82
  isolation 597
Geological age 7
Geotaxis 389
Germination 11, 13
Girdle 14–16
  band 16, 17, 92, 96, 144, 219, 225,
    226, 275, 293
  broad view 14, 15, 116, 163, 178,
    179, 181, 183, 186, 194, 201,
    202, 204, 210, 211, 213, 217,
    227
  lamella 7, 605, 613, 617, 626, 633,
    742
  narrow view 14, 15, 211, 213, 237
  morphology 10, 16, 40
  segments 14, 16, 106
  structure 8, 93, 307
Gliscolith 736, 747, 755, 761
Gluteraldehyde 552, 698
Glycerine jelly mounts 553
Glycocalyx 444
Golgi apparatus 595, 596, 742
Grana
  formation 596
  structure 604
Granular chloroplasts 668
Greece 492
Greenland 184
Groove gullet system 607
Growth 696, 706,
  conditions 696
  medium 750
  dynamics 694
Gubernaculum 704
Guillard's F medium 696, 697
Gulf
  of California 433, 221, 251, 492
  of Carpentaria 231
  of Mexico 37, 179, 189, 229, 251,
    257, 447, 492, 499, 509, 524
  of Naples 224, 234, 526, 542
  of Salerno 492, 494, 521
  of Tailand 112, 231, 257, 490, 494,
    497, 500
  of Townsville 112

stream warm core rings 131
Gullet 605, 606, 607, 608, 609, 612,
  691, 704

Habitat
  epiphytic 23,
  freshwater 23, 29, 33, 290
  marine 23, 33, 390
  planktonic 23, 390
Hair like appendages 603
Halotolerant species 618, 668
Halozoic 389
Haploid 8, 9, 415
Haplont(ic) 387, 390
Haptonema 586, 587, 595, 596, 603,
  604, 633, 634, 635, 636, 638,
  639, 641, 642, 644, 645, 687,
  688, 689, 742, 752, 756, 758,
  759, 761, 762, 779, 789, 798,
  800
Head pole 240, 242, 243, 254, 256,
  257, 264, 240, 242–244, 254–257,
  264, 265
Helicoform placoliths 754, 799
Helicoid placoliths/helicolith 747, 748
Helictoglossa(e) 267, 284, 287, 288,
  289
Helicoid placoliths/helicolith 747, 748
Heliozoan stage 627, 628
Helladoform stomatal coccoliths 768
Helladolith 736, 747, 755, 761, 762,
  764, 768
Helgoland 183
Hemolytic 390
Hepatotoxic 390
Heterococcolithophorids 738, 739, 742,
  743, 773
Heterococcoliths 737, 739, 742, 745,
  749, 751, 773, 777, 786, 808
Heterodynamic 601, 605, 607, 617,
  633, 635, 646, 647, 651, 652, 653,
  658, 680, 681, 704
Heterokont 595, 601, 687, 705, 708
Heteromorphic life history 739, 742,
  743
Heteromorphy 10

Heteropolar 254, 261, 262, 263, 266, 292, 296, 297, 299, 303, 308, 310, 315
Heteropolarity 262, 266, 297, 299, 301, 307
Heterothallic 390, 593, 594, 599, 604, 605, 606, 617, 618, 626
Heterotrophic 388, 389, 390, 550
Heterovalvate 10, 13, 31, 96, 193, 195, 196, 199, 224, 269
Heterovalvy 91, 111, 178
Hexagonal crystallization 745
Hexamin 334
19 Hexanoyloxyfucoxanthin 390, 732
High latitude forms 797
Histones 388
Holococcolithohorids 739, 741, 743, 752, 755, 779, 786, 795, 808
Holococcoliths 736, 738, 739, 742, 743, 745, 747, 748, 752, 755, 759, 760, 761, 763, 778, 786, 814
  calcification 742
  disciform 746
Holotype 267, 331, 602, 791
Homodynamic 601, 607, 633, 635
Homothallic 390
Horns 105, 169, 176, 177, 184, 225, 236, 238, 407, 470–482, 536, 538, 540
Hyaline 20, 122, 124, 125, 178
  area 98, 101, 103, 105, 106, 113, 119, 120
  band 17
  central area 20, 101, 106, 112
  coccoliths 805
  investment 675
  lines 98, 101, 103, 104, 105, 106, 107, 109, 110, 112
  margin 37, 111
  rays 132, 133–140
  ring 125, 128
  spaces 98
  valve 293
Hydroxymate siderophores 389
Hypersaline areas 707
Hypertonic medium 665
Hypnozygotes 415, 416

Hypocingulum(a) 14, 16
Hypocone 400
Hyposome 400
Hypotheca 7, 14, 16, 145, 153, 191, 400
Hypotonic 705
Hypovalve 11, 14, 16, 191, 209
Hystrichospherid 487

Iberian peninsula 490, 497, 500
Iceland 499
Imbricate 794, 795
imbricating radial segments 786
imbrication 143, 408
  dextral 747, 779, 786
  pattern 747
  senestral 747
Immobilization 552
Immunological properties 794
Inclination 748
Index fossil 797
Indian Ocean 36, 39, 43, 128, 131, 133, 177, 249, 251, 257, 281, 306, 513
Innerspace 291, 268, 291, 297, 302, 303,
Intercellular spaces 227, 228, 229, 235
Interconnected threads 31
Interference contrast 606, 672, 673, 688
intergrown rhombic crystals 789
interlocking proteinaceous ribbons 706
Interlocking ridges 89
  sibling valves 96
interlocking rods 788, 793
Interstitial meshes 101, 104, 105, 107, 109
interstitial water 706
Interstria 10, 17, 33, 34, 74, 98, 100, 101, 106, 251, 278, 284, 286, 292, 294, 295, 297, 299, 300, 302, 303–313, 315, 310, 321, 323, 324, 327, 329, 331
investment 673, 678, 705, 707
Iodine (*see* Lugol's Solution) 652, 654, 665, 673
Iron acetocarmine 694
Isogamous 10, 390, 597

Isogamy 8, 10
isokont 601, 705
Isolated fossil coccolith 739
Isolation of coccolithophorids 750
Isopolar 261, 263, 266, 267, 292, 297,
    299, 300, 302
Isotype 331
Isovalvate 195, 196, 199, 201, 202, 224,
    271

Japan 37, 65, 75, 86, 896, 157, 234,
    324, 447, 449, 490, 492, 494, 497,
    500, 550
Java 177, 257
Janus green 680
Junction zone 100, 106
Jurassic 7, 744

Kamchatka peninsula 492, 499
Kara Sea 122, 499
Keel 289, 291, 307, 668, 788
    punctum(e) 291
Kiel Bay 157
Kinetids 742
King's Bay 323
Kofoidian system 410, 412–414
Korea 490, 494, 497, 500

Labiate Process 19, 22, 236, 238, 241,
    242, 244–248, 251–254, 256, 265, 338
    shaped 144, 146, 155, 187, 188, 192,
    218
    structure 20, 144, 146, 155, 157, 218
Labiatiform appearance 775
Lacrymuloid cells 607
Lamella(e) 7, 596, 603, 604, 705, 707,
    708, 742, 745, 746, 748, 749, 773,
    774, 777, 780,   781, 783,
    784, 787, 788, 790, 791, 809, 811,
    813
Lamellar elements 809, 812
Laminolith 736, 748, 755, 762, 763,
    770

Lanceolate 748
Lectotype 246, 252, 307
Lepidoliths 737, 748, 813
left 400
Life cycles 7, 9, 13, 101, 411, 592, 594,
    596, 597, 738, 743, 750, 751, 759,
    760, 778, 786, 794, 795, 814,
Life histories 617, 738, 739, 742, 743,
    751, 789
Ligula(e) 16, 110, 144
Ligulate 96, 160
Limestone 740, 744
Lineatus structure 59, 62, 63
Linking
    spine 20, 178, 179, 181
    structure 43, 45
    thread 31, 35
Lipids 732, 741, 742
Lists 408
Lobes 17, 18, 188
Lobopodea 550
Longitudinal
    fissures 668
    Interstriae 278
Lopadolith 748, 807, 808
Lopodolith 737, 747, 748,
Lorica 595, 601, 618, 619, 624, 625,
    672, 675, 677, 678, 680, 687, 693,
    697, 698, 704, 705
    chamber 680
    length 624, 625, 675, 677, 678, 670
    width 624, 625, 675, 677, 678, 680
Lorical structure 673
Loricate species 673, 678
Low latitude coccoliths 797
Lugol's solution 334, 552, 605, 617,
    627, 646, 654, 673, 693, 694, 698
Luneate form 661

Macrorimportula(e) 98, 99
Macrolabiate process 104
Margin
    non-raphe bearing 328
Marginal
    bilabiate process 229, 236
    chambers 18, 33, 34

Marginal (*continued*)
  labiate processes 31, 33, 35, 36, 39,
    79, 97, 98, 105, 111
  ribs 19, 55, 63
  ridge 20, 57, 62,
  ring 19, 57, 116, 128, 130, 131, 132
  rows 68, 256, 259, 261, 263, 265
  spines 89, 91, 93, 96, 182, 253, 263,
    264, 265, 266, 267, 338
Marine plankton(ic) 23, 33, 117, 188
  diatom 6, 11, 23, 28, 29, 54, 92, 93,
    123
  genus(era) 6, 23, 45, 93
  Nanoplankton 39
  species 28, 33, 35, 271, 272, 273,
    276, 290, 294
Median
  transapical plane 296, 303, 308
Mediterranean Sea 131, 133, 155, 257,
    449, 464, 466, 468, 509, 512, 513
Meiosis 552
Melville Bay 229, 231
Megacytic zone 533
Melanosome 459
Membraneous
  costa(ae) 88, 178
Meshwork 75
Methyl cellulose 552
Microgranule 601, 707
Micropipette method 750
MIcroplankton 593, 705
Microscopy
  electron 5, 6, 13, 20, 39, 41, 56,
    57, 59, 63, 66, 77, 78, 80,
    83, 89, 92, 97, 105, 111, 113,
    316, 337
  fluorescence 594, 633, 687, 697
  interference 606, 672, 673, 688, 750
  phase contrast 606, 612, 623, 657,
    672, 673, 687, 688, 750, 811
  light (brightfield) 5, 6, 14, 19, 27,
    34, 36, 37, 39, 41, 43, 44, 45,
    58, 49, 53, 56, 57, 62, 63, 65,
    66, 73, 74, 77, 79, 80, 83, 84,
    93, 98, 100, 103, 104, 105,
    107, 109, 112, 334, 336,
    337

transmission 606, 750, 781, 797
Microtubules 595, 600, 603, 703, 705
Mid wall rim 738, 785, 808, 809, 811,
    812
Minute oscillating granules 612
Mitochondria 7, 595, 596, 603, 604,
    680, 742
  tubular type 7
Mitosis 13, 745
Mitotic characters 595
Mixotrophic 388, 526
Mobile biflagellate stage 641, 663, 752,
    779, 786, 789
Mode of swimming 605, 613, 617, 626,
    633, 646, 651, 664, 672
Mode of flagellar use 603, 689
Monomorphic 747, 748, 755, 758,
    759
  coccolith case 755, 760
  coccoliths 786, 789
  coccosphere 734, 761, 804
  cribrilliths 806
  cricoliths 805
  exotheca 790, 808
  laminoliths 762
  styliform pappoliths 805
  zygolith 761
Monospecific genus 31, 36, 115,
    183, 234, 243, 248, 267, 269,
    271, 273, 284, 286, 290, 442,
    594, 694, 748, 755, 761
Monothecate 785
Monotypic 93
  family 95, 116
  genera 27, 96, 97, 503
Morphotypes 807
Morphological
  adaptation 10
  character(s) 46, 190, 313
  dissimilarity 51
  distinct structures 10, 53, 54
  variation 48, 333
Morphotype 390
Most probable number (MPN) 696
Mother cell 665, 693, 742
Motility/movement 7, 689 592
Mucilaginous matrix 601, 706

Mucilage 20, 30, 31, 46, 54, 58, 61, 69,
   73, 89
  colonies 54, 55, 56, 61, 63, 69, 634,
      635, 640, 641
  pads 20, 89, 275
  stalk 241
  tubes 276
Mucoid
  contents 602
  halo 451
  production 419
  strands 501, 526
Mucocyst 601, 614, 616, 621, 705,
Mucus bodies 614, 638, 646,
Mud flats 178, 597, 599
Multiserial endosymboses 389
Multiseriate 17
Myxotrophy (*see* Trophic modes) 593,
   601, 617, 626, 705

Naked species 651, 652, 666, 672, 680,
Nannoplankton 592, 593, 688, 706,
   740
Narragansett Bay 252
Naviculoid 269, 271
Nematocysts 401, 444, 457, 459, 460
Netzplankton 592
Neurotoxin 307, 390
New Guinea 231
New Zealand 179, 229, 447, 492
Nitrate 696, 697, 734
Nitrogen limited cultures 734, 794
Nominate variety 223, 224, 230
Nonloricate species 673
Nonmotile coccoid cells 778
Nonmotile coccosphere 779, 786, 789,
   793
Nontoxic water sampler 696
Norman River 232
North Atlantic
  ocean 39, 147, 284, 323, 452
  plankton 323
  waters 40, 75, 110, 121
North European coastal waters 181
North Pacific 306
North Queensland

North Sea 168, 179, 189, 229, 249, 323,
   449, 488, 529
North West Africa
Northern
  cold water 28, 33, 36, 43, 59, 71, 86,
      88, 84, 93, 247
  hemisphere 66, 179
  temperate region 34, 36, 43, 59, 71,
      93
  variety 66, 68,
Norway 65, 168, 500,
Norwegian
  coastal waters 40, 43, 128, 278, 499,
      278
  fjords 95, 278
  sea 40, 130, 140, 323
Noxious microalgae 589
Nuclear envelopes 742
Nucleolus 388
Nucleosome 388
Nucleus 387, 586, 596, 603, 646. 672,
   687, 692. 742, 785
  dinokaryon 456
  pycnotic 8
Nutrient limitation 734
Nutrient rich waters 734
Nutrient uptake 592
Nutrient vacuoles 593
Nutrition 70, 593, 601, 605, 613, 617,
   626, 705, 708

Oblique light 800
Observational techniques 602
Ocellus(i) 163, 169, 178, 181, 236, 401,
   444, 459, 460
Ocellulus(i) 178, 181
Occluded process 19, 30–32, 36, 37,
   41, 46–49, 55, 58, 61, 63, 66,
   73, 77
Ochromonadoid cell 624
Offshore waters 680, 740, 743
Oil droplet 601, 707
Oils 596, 617, 626, 687, 692
Oligotrophic waters 597, 696, 743,
Oogamy 8, 10
Oozes 740

Organic
  base plate scale 812
  extrusions 30
  layer 7
  matrix 747
  membrane 8
  threads 30, 43, 46, 47
  scales 594, 596, 599, 601, 603, 604,
    627, 628, 630, 633, 635, 651,
    697, 698, 705, 707, 746, 752,
    756, 762, 765, 812
Orientation 797
Orthorhombic crystals 745
Oslofjord 65, 66, 89, 168
Osmic acid 633, 652, 694, 697, 698
Osmophilic material 444, 455
Osteoliths 737, 748, 803
Otarium(a) 142, 144–147, 150, 151,
    155, 156
  clasping 142, 143
  narrow 131
  pointed 130
Ovate 269

Pacific 39, 63, 128, 306, 453, 456, 503,
    504, 512, 513, 516, 521
  central 251, 306, 529
  north 251, 471, 490, 492, 530
  south 257
Paleoclimate studies 741
Paleozoic 744
Palmelloid stages 597, 617, 623, 635,
    706,
  papilla 668, 669
Paracrystalline structures 595, 706, 707
Paramylon 596, 604, 633, 636, 646,
    687, 692, 706,
  center 649, 650
  grains/granules 646, 649
  shield 601, 646, 649, 650, 707
Paraxial rods 588, 601, 626, 671, 680,
    706
Parent cell 7, 10, 11, 12
  frustule 7, 11
  theca 7, 10, 11
Parietal 621, 630, 661, 664, 731

Partially interlocking placoliths 793, 815
Pectic plug 601, 707
Pedicel 628
Peduncle 388, 444, 487, 527
Pellicle 388, 400
Pellicula 595, 604, 645, 651, 706
Pellicular striations 612, 645, 650
Pennate diatoms 7, 8, 10, 11, 14, 23–27,
    240, 241, 271, 587
Pentagonal dodecahedron 768
Pentalith 737, 748, 753, 778
Pentasters 468, 469
Peridinin 390
Perinuclear cavity 595
Peripheral
  collar 790
  elements 786
  ring 773, 774, 783, 786, 809,
Periplast 595, 608, 706
Periplast stripes 692
Perforated layer 17, 57
Perforation 17, 20, 230, 234, 295
Peru 499
Petaloid 736, 745, 761, 763, 768, 775,
    779
Phagocytic 390, 457, 470, 487
Phagocytosis 444
Phagotroph(ic) 388, 463, 521, 538, 607,
    621, 706
Phagotrophy 593, 601, 705
Phantom dinoflagellate 550
Phase contrast microscopy 606, 612,
    623, 657, 672, 673, 687, 688, 750,
    811
Philippines 490, 494, 497, 500
Photosynthetic symbionts 423, 436
Photoautotroph 706
Phototroph 601, 706
Phototrophic 593, 594, 599, 601, 605,
    613, 617, 620, 626, 646, 705
Phyobiliproteins 390
Phycocyanin 587
Phycoerythrin 587
Phycoma stages 596, 597, 651, 653,
    655, 657, 661, 664
Phyletic link 797
Phytoflagellates 599, 602, 671, 688

Phytol 731
Pigments 7, 586, 603, 687, 732, 742
  accessory pigments 603, 613
  affinities 613
  carotinoids 587, 732, 742, 751
  chlorophyll 587, 603, 604, 732, 742
  fucoxanthin 613, 742
  19' Hexanoyloxy-fucoxanthin 742
  phycocyanin 587
  phycoerythrin 587
  yellow/brown pigments 587
Pilus(i) 7, 178, 179, 183
Pilus valve 178
Pinocytosis 593, 706
Pit 407
Placolith 737, 747, 749, 752, 753, 778, 779, 786, 788, 793, 795, 797, 801, 803, 816
Planar sinus waves 626
Plane of division 14
Plankton nets 688
Planktonic
  diatoms 6, 11, 29, 190, 334
  genera 168
  species 28, 236, 238, 269, 272, 276, 281, 305
Planospores 415, 520
Planozygote 415, 416
Plasmalemma 226, 706, 742
Plastids 739, 742
Plate
  apical 406
  apical closing 410
  antapical 406, 410
  antapical closing 410
  anterior 410
  anterior closing plate 482
  first apical 407
  homologous 407
  intercalary 407, 410
  periflagellar 401, 419
  presingular 410
  postsingular 410
  transitional 410
Plate formula 410
Plate overlap 408
Pleistocene 797

Pleomorphic 451
Pleura(e) 17
Pleurax 231, 232, 336, 338
Pleuronematic 706
Ploidy 738
Polar
  coccoliths 746, 748, 774
  direction 776
  ice 294
  regions 6, 59, 63, 65, 142, 176, 179, 231, 295, 748, 806
  seas 137, 281
  species 6, 142, 221
  waters 672, 740
Polarity 23, 25, 263, 296
Polarizing microscope 748, 749
Poles 774, 803
Polygonal plate shaped coccoliths 794
Polylysine 698
Polymorphic life cycle 752
Polymorphism 749
Pore(s) 14, 17, 169, 188, 204, 240, 241, 243, 249, 251, 330
  attachment 406
  apical pore complex (APC) 406
  apical pore field 24, 240–242, 244–248, 251, 252
  cone shaped 418
  satellite 14
Poroid 17, 18, 30, 116, 130, 131, 157, 139, 181, 230, 232, 292, 295, 296, 297, 299, 300, 302, 303, 307, 309, 310, 312, 313, 317, 319, 321, 324, 329–331
Portugal 131, 257, 497,
  Atlantic coast 229
  Guinea 229
Precipitate of calcite 784
Preservation 43
  formaldehyde 605, 652, 665
  preservatives 681
  procedures and methods 680, 750
Primary
  resting spore valve 13, 66
  valve 31, 66, 68, 82
Proboscus 142, 158, 159, 681, 683, 706

Process(es) 8–10, 17
  base 147, 150, 157, 236
  bilabiate 22, 226, 227, 228, 229, 232,
    233, 235, 236
  central 31, 32, 36, 42, 48, 50, 53, 54,
    59, 61, 62, 63, 65, 66, 68, 69, 70,
    71, 73, 79, 84, 86, 87, 88, 113,
    233
  clustered 31, 32, 50, 68, 160
  labiate 17, 19–24
  occluded 19, 22, 77
  strutted 22
  subcentral 195, 196, 255
  tubes 160, 161
  tubular 10, 22, 161, 178, 179, 181,
    182
Proteinaceous locking band 645
Proteinaceous plate 595
Protestin plankton 592
Protoplasts 753, 814, 816
Protozoan systematics 671
Protuberance 107, 191, 198, 209, 219,
    224
  conical 192, 204
Proximal
  area 788
  arm link 803
  coccolith disc 776
  covering 793, 797
  elements 809
  end(s) 793, 804, 813
  layer 747
  links 798, 803, 804
  ridge 777
  rim 738, 777, 783, 786, 811
  ring 761, 764
  shield 779, 786, 795, 801, 803, 816
  shield elements 793
  side 762, 780, 795, 803
  tube 761, 762, 764, 765, 768, 776,
  view 788
Prymnesiophyte cell 731
Psammobius 651, 668, 706
Psammophilic species 662, 706
Pseudocolony 457
Pseudocellus(i) 169, 236
Pseudofilamentous 805

Pseudoloculus(i) 88, 92
Pseudonodulus 88, 115–123, 128, 130,
    131, 191, 339
Pseudoparenchymatic 805
Pseudopodia 593, 595, 599, 623, 627,
    628, 672, 690, 707
Pseudopodial 584
  collar 605, 677, 692, 707
  threads 675
Pseudoraphe 240
Pseudoseptum(a) 16, 178, 181, 183,
    293, 297, 305, 306
Punctum(a) 289
Punctate wall 655
Pyramidoidal cells 655
Pusule 401
Pyrenoids 16, 27, 116, 130, 209, 210,
    235, 275, 281, 596, 601, 605, 607,
    608, 609, 611, 612, 613, 633, 641,
    646, 649, 650, 651, 654, 655, 656,
    657, 658, 659, 661, 662, 663, 664,
    668, 669, 706, 732, 742
Pyriform 611, 612, 616, 626, 675, 680,
    684, 707, 733, 749, 756, 758, 812

Quadrilateral symmetry 655
Quadangular coccoliths 795

Radial
  furrows 815
  lamellae 783, 786, 804, 809
  rays 74, 75
  rows 51, 54, 57, 61, 69, 73, 74, 84,
    98, 106, 107, 109, 111, 113, 232
Radially oriented lamellar elements 812
Radially placed ribs 784
Radially symmetrical cells 627
Radial symmetry 604, 690, 707
Radiant energy 593
Raphe 7, 10, 23, 24–26, 240, 244–272,
    278, 281, 283, 284, 286, 290, 291,
    292, 294, 295, 297, 300, 306, 308,
    323, 326, 328, 329, 330, 338
  biarcuate 286
  canal 268, 291

Raphe (*continued*)
  central pores 276, 277, 278
  curvature 282
  eccentric 282, 291, 295, 308, 324
  ending 267, 291, 326, 329
  fins 272, 273, 283, 284, 286, 287,
      289, 290
  keel 283, 291, 293
  ridge 283, 287, 291
  slit 267, 268, 291, 328, 329
  sternum 26, 268, 269, 280, 282
  structure 293, 329
  system 267, 291, 295, 305, 306, 307
Raphid pennate diatoms 7, 26, 267
Refractive
  body 611
  granules/globules 668
Remote sensing 740, 743
Reproduction 390, 596, 738, 742, 789
Researve products 687
Resevoir 645, 646, 703, 707
Resting cell 12, 30, 45, 53, 69
Resting spore 10, 11, 743
  endogenous 11, 12, 56, 66, 72, 73,
      115, 168, 186
  epitheca 10, 13
  exogenous 11, 12, 32, 36, 56, 66, 150
  formation 10, 11-13, 35, 69, 89, 239,
      333
  germination 11, 13
  morphology 11, 31, 66
  parent cells 209, 211, 213
  semiendogenous 11, 12, 31, 32, 53,
      56, 66, 82, 89, 90, 94, 170, 171
  spines 210, 219, 220
  thecae 13
  valve 13, 31, 36, 68, 72, 96, 204, 207,
      210, 211, 219, 220, 221, 224,
      300
Reticulate 33, 254, 265, 267, 408
  central pore 793
  grid 788, 793, 795
Reticulated 616, 646, 650, 664
Reticulum 798
Rhabdolith 749, 773
Rhinote 608, 611, 707
Rhizopodial 627, 707

Rhizopodoid 617, 627
Rhombic elements 786
Rhombic scale 749
Rhombohedral 742, 747, 759
Rhombohedral heterococcoliths 777
Rhombohedrons 746, 758, 765
Rhomboidal coccoliths 778, 780, 798
Rhomboidal valves 321
Ria de Vigo 497
Ribbons 20
Ribs 17, 30, 31, 32, 35, 36, 37, 46, 48,
      63, 69, 70, 74, 77, 79, 230, 302,
      303, 312, 408
  radial 30, 31, 36, 57, 70, 79, 82
Ridge 17, 20, 91, 272, 286, 287, 291
  ventral 405
Rimoportula 17, 22
Rock pools 599, 664, 665, 669
Rostrum(a) 294
Rostrate 184, 246, 324
  apices 83, 249, 309,
  ends 251, 278, 280, 281, 294, 312,
      327
Rudimentary 31
  cells 31
  form 586
  valves 13, 72

Sacculiform elevation 775
Sacculiform shape 746
Salinas 684, 707
Saline pools 668
Salinity 696, 703, 743
Salpingiform cyrtolith 746, 791
Salpingiform process 791
Sand dwellwers 463, 547, 550
Sand flats 178
Salt marshes 614, 646, 662, 668, 669
Saprotroph 707
Scale
  bearing swarmers 738, 743
  covered flagella 595
  morphology 597, 620, 634, 641, 659,
      662, 687
  pattern 658
  type 595, 597

Scanning electron microscope 606, 698, 739, 750
Scapholith 737, 749, 753, 777, 781
Schultze's solution 694
Sea ice 28, 89, 115, 247, 295
Secondary
  resting spore valve 13, 207
  valve 31, 66, 82, 204, 206, 207, 210, 211, 218, 220, 223, 324
Sediment samples 739, 807
Selective staining 687
Semilunar paramylon shields 650
Separation valves 10, 20, 43, 90, 91, 183
Septate 70, 306
Septum(a) 16, 17, 242, 244, 306, 338
Serial dilution cultures 600, 694, 696
Sessile 618
Seta(ae) 20, 184, 186, 188, 189, 190, 191, 195, 196, 199, 201, 203, 206, 209, 211, 213, 219, 221–223, 275, 332, 334, 338
  fused 188, 189, 190, 191, 199, 203, 204, 205, 206, 211
  fused hypovalvar 193, 211, 213
  inner 186, 189, 190, 193, 196, 206, 207, 218, 219
  intercalary 185, 193, 198, 205, 206, 210, 213, 215, 216
  sibling 186, 190, 191, 201, 203, 204, 211
  terminal 183, 185, 186, 188, 189, 190, 192, 193, 196, 198, 199, 204, 206, 207, 209, 210, 213, 216, 218, 219, 220
Sexual reproduction 7, 9, 23, 333
Shadow casting 634, 697
  whole cell mounts 652
Shallow focus microscope 750
Short fringing crystallites 798
Sibling
  cells 215
  setae 190, 201, 203, 204, 211
  valves 45, 91, 96, 115
Sigmoid 74
Silica rods 672

Silica skeletons 585, 588, 594, 597, 598, 605, 617, 627, 628, 630, 671, 685
Silica structures 627, 685, 698, 699
Siliceous
  cell wall 7, 13, 17, 18, 91, 93, 111, 157, 238, 289, 295
  costae 227
  external extensions 92
  flaps 201
  frustule 11, 27
  layer 17, 18, 30, 226
  outgrowths 96
  ribs 77, 169
  skeleton 468
  spines 96, 116, 206, 207, 213, 234
  strip 17
  stomatocysts 597
  structures 10, 17, 23, 283, 333
  theca 8
  valve 20, 23, 35, 51, 84, 87, 105, 155
  wall 20, 334, 310, 320, 334, 133–135
Silicified
  bar 261, 262
  cysts 617, 626
  frustules 11
  interspaces 202
  resting cells 53
  resting spores 53, 66, 82, 115, 200
  scales 617, 620, 697, 698
  strips 329
Silurian 387
Sinestral 795
Sinestral imbrication 779
Single cell isolation 750
Sinking rates 673, 734
Slits 243, 267, 269
Skagerrak 75, 168
Sodium hypochlorite 553
Sodium thiosulfate 552
Soil extract 696, 697
Solitary free swiming cells 674
Southern
  Atlantic Ocean 17
  cold water region 28, 43, 65, 71, 75, 80, 86, 88, 110, 115, 239
  hemisphere 66
  ocean 6, 43, 140, 147, 159

Southern (*continued*)
   temperate region 34
   variety 68, 121
South Africa 184, 492, 494
South Pacific Ocean 63
Soviet Republic 500
Species
   composition 592
   level 599, 602
   measurements 609
   succession 588
Specific cell volume 592
Spicules 669
Spines 17, 20, 22, 62, 88, 89, 93, 94,
   95, 96, 97, 178, 179, 191, 193, 206,
   218, 223, 224, 232, 239, 251, 253,
   254, 257, 262, 266, 409, 620, 628,
   632, 633, 644, 675, 677, 703, 753,
   773, 774, 777, 779, 781, 801, 803,
   807, 811, 814, 816
   anterior winged 419
   coccolith 806
   dichotomously branching 220
   length 774
Spinulae 101
Spinules 57
Spiny
   setae 206, 215
   resting spores 207, 210
   valves 209
Spineless coccolith 806
Spiny process 781
Spiralling rows 98, 107
Spore 239
   germination 11, 13
   paired 191
   hypovalve 11
   valves 13, 72
Sporopollenin 597
Squash technique 553–554
Stain(ing) 595, 605, 633, 646, 687, 693,
   694, 697
Stalk 624, 626, 627, 628, 664, 674,
   677, 707, 747, 761
Starch 596, 605, 651, 654–665, 687,
   692, 693
Starch bodies 661

Starch granules/grains 662, 668
Starch shield 601, 605, 651, 659, 662,
   664, 707
Stationary cultures 734, 793
Stationary stage 307
Stauros 267, 268, 272, 273
   pseudo-stauros 284
Stepped chains 20
Stepped elevated valve face 103, 109
Sternum 25, 26, 240, 242, 244–249,
   252–254, 256, 259, 261, 263, 269,
   278, 280, 283, 338
Sterols 390
Stigma 601, 611, 691, 707
Stipitate 707
Stomatal
   coccolith 734, 748, 749, 756, 758,
     761, 763–772, 774, 780, 782,
     783, 784, 786, 787, 788, 790,
     803, 808, 809, 810, 811, 812
   helladoliths 764
   opeinings 760
   rod 783, 787, 788, 790, 809
   zygoform laminoliths 770
Stomatocysts 601, 707
Storage product(s) 605, 617, 626, 633,
   636, 646, 651, 664, 692, 742
Stria(e) 17, 19, 29, 89, 117, 119, 120,
   157, 240, 248, 251–253, 269, 275,
   280, 282, 284, 292, 293, 297, 300,
   306, 307, 309, 311, 316, 317, 319,
   324, 326, 327, 328, 329, 330, 409
   biseriate 17, 301, 312, 313, 328, 328
   convergent 241
   fasciculate 29
   incomplete 98, 101, 104, 105, 106,
     109, 112
   inserted 98
   lineate 240, 241
   longditudinal 241, 276, 280, 283
   marginal 263
   multiseriate 17, 312
   oblique 281, 282, 283, 328, 329
   parallel 29, 240, 246, 249, 252, 280,
     281, 290, 293, 296, 297
   punctate valve 29, 41, 98, 286, 287,
     289, 317

Stria(e) (*continued*)
  radial 29
  radiate 240, 247–249, 280
  radiating 244, 252
  reinforced 287
  spiraling 41
  tangential 29
  transapical 29, 243, 244, 246,
    247, 269, 276, 280, 284,
    290
  transverse 276, 280, 281, 282,
    283, 286, 289, 292, 293,
    297, 299, 302, 303, 306,
    328, 329
  uniseriate 17, 241, 247, 301
Striation pattern 156, 157, 272
Struts 30, 254, 261
Strutted process(es) 14, 19, 20, 28–37,
    41–50, 53–59, 61–66, 68–71,
    73–75, 79–84, 86–88
  arrangement 34, 55, 58, 71
  central 31, 35, 39, 40, 46, 47, 50, 53,
    54, 56, 59, 60, 62, 63, 68–71, 73,
    75, 79–84, 86, 87
  operculate 30
  trifultate 20, 30
  trumpet shaped 73
  tube 14, 22, 30, 35, 36, 44, 45
Styliform 584
  central process 746, 748
  coccoliths 807
  cyrtolith 746, 749, 804, 806
  pappoliths 748, 804,
  type 807
Subapical 707, 708
Sub basal pyrenoid 668
Subcircular whorl coccoliths 781
Subfusiform 442
Submicroscopic characters 599, 602
Sudan black 694
Sudan blue 694
Sulcal list 429
Sulcus 391, 401
Surf zone 241, 243
Suture 409, 749, 786,
  line 747, 749, 779, 801, 803, 816
  pattern 789

Swarmers 602, 605, 620, 623, 738
Swedish coastal waters 157
Swimming
  behaviour 688
  pattern 652, 657
Symbionts 652
Symbiosis 744
Symbiotic relationship 744
Symmetry 23, 25, 169, 178, 690, 703
Systematics 586, 594, 599, 600, 647,
    652, 665, 671
  algal 603, 616
  phytoflagellate 602
Systematic descriptions 755
Systematic position 605, 612, 616, 626,
    633, 645, 651, 664, 671, 680,
    685, 751, 794
Tabular 244
  array 293
Tabulation 410
Taiwan 500
Tangential
  areola walls 70
  areolation 59, 62, 63, 87
  rows 50, 51, 59, 62, 63, 77, 84, 87
  striae 29
Tangentially undulated 33, 74
Tasmanian waters 449, 492
Taxonomic
  base 599
  character 33
  history 794
  importance 687
  levels 5
  positions 115, 198, 204, 813
  problems 28, 41, 43, 89, 96
  revisions 739
  schemes 751
  studies 602, 750
  systems 750
Taxonomic index (revisions and novel
    names)
  diatoms 338, 339
  flagellates 613, 614
Taxonomy 586, 595, 599, 602, 606, 755
Terminal
  nodule 267, 278, 284

Terminal (*continued*)
  setae 188, 189, 192, 193, 196, 198,
    199, 204, 206, 207, 209, 210,
    213, 218, 219, 220
  thorn 43, 45, 188, 192, 206, 229,
    804
  valves 43, 45, 188, 192, 206, 229
Terminology 702, 744
Tetramorphic coccolith case 780
Texas 179
Theca 7, 8, 10, 11, 13, 103, 111, 387,
    388, 390, 674, 693, 707
Thecate 652
Thecal vesicles 390, 401, 444
Thiamine 696
Threads 30, 31, 35, 40, 43, 46, 47, 51,
    53–59, 61, 63, 65, 66, 69, 70, 74,
    80, 84, 86, 87, 336, 338
Thylakoids 7, 596, 605, 603, 617, 626,
    633, 705, 708, 742
Tidal pools 662
Tintinnid 201, 202, 203
Tomentum 664
Torsion 235
Townsville 231, 232
Toxic (harmful)
  blooms 592
  diatoms 54, 307
  dinoflagellates 420, 424, 425, 429,
    431, 433, 434, 439, 445, 446,
    447, 449, 490, 492, 494, 497,
    499, 500, 501, 503, 515, 516,
    550
  flagellates 614, 615, 639, 640, 641,
    643, 644,
  substances 698
Tracery of siliceous costae 227
Tractellum 602, 636, 680, 708
Trailing pseudopodium 595
Transapical
  axis 14, 15
  interstriae 271
  plane 14, 15
  pseudosepta 293
Transmission electron microscope 606,
    749, 781, 797
Transverse

flagellum 595, 603
groove
  interstriae 278, 299, 302, 319
  striae 299, 328, 329
Tremalith 749
Triassic 744
Trichocyst 402, 602, 613, 614, 691, 708
Trifultate 20
Tripartite flagellar hairs 601, 705
Tripartite tubular hairs 595, 600, 704
Triradiate 269
Trondheimsfjord 218
Trophic modes/roles 592, 593, 598, 708
  phototrophic 593, 594, 599, 601,
    605, 613, 617, 620, 646, 705
  heterotrophic 593, 594, 599, 604,
    605, 606, 617, 618, 626, 633
  mixotrophic 593, 601, 617, 626, 705,
Tropical marine areas 253, 272, 331,
    387, 591, 696
Tropical waters 672, 740, 743
Truncate proboscis 223
Trypan blue 553
Tube coccoliths 813
Tubular
  appendages 813
  cristae 595, 596, 603, 642
  hair 605, 680, 708
  silica units 704
  wall 777
Type 226
  localities 331
Tychoplanktonic species 89, 417, 420,
    423
Tyrrhenian
  brackish water lagoons 224
  sea 521

Unarmored 403, 405
Uncalcified coccliths 732, 735
Uncalcified organic body scales 735
Undulating membrant 684
Unialgal cultures 812
Unidentified flagellates 585, 592
Uniform elliptical coccoliths 776
Uniform salpingform cyrtoliths 746, 791

Unmineralized
  patternless membrane 798
  scales 697, 734, 803
Unipolar 24, 25
Uniseriate striae 17
Unoccluded central area 798
Unperforate margin 266
Unperforated areas 20, 98, 101, 106
Uranyl acetate 552, 688, 694, 697

Valvar plane 14, 15, 17, 220, 225, 234
Valve(s) 14–22, 403
  apex 88, 89, 144, 147, 150, 151,
    155, 156, 159, 246–249,
    252–254, 261–263, 266,
    271, 282, 286, 295–300,
    302, 309, 311, 315, 317,
    321, 326, 327, 328
  areolae 41, 54, 61, 65, 68, 109,
    73, 77, 82, 86, 92, 103, 111,
    123, 128, 130, 159, 175, 183,
    230, 231–233, 234, 238, 244,
    248, 249, 250, 305
  areolation 53, 121, 142, 125, 127,
    130, 132, 282
  conical 155, 137, 159
  contour 151, 155
  depression 65, 70, 94, 103, 104, 106
  elliptical 117, 113, 171, 175, 238,
    147, 248, 250, 271, 272, 273,
    281, 282, 296, 297
  face 14, 183
  face areola(e) 231, 252
  face areolation 234
  face structure 121, 184
  gross morphology 14–20, 142, 144,
    297, 299, 303, 326, 328
  hirsute
  intercalary 10, 43, 90, 91, 96, 186,
    188, 192
  mantle 14–20,
  membrane 290
  morphological structures 112
  morphology 10–20, 41, 62, 132, 135,
    137, 139, 141, 159
  morphotypes 10

  ornamentation 7
  outgrowths 184, 225
  pole(s) 209, 211, 216, 218, 219,
    220, 221, 223, 225, 246, 286,
    287, 289, 290, 291, 293, 296,
    297, 303, 315
  primary resting spore 13, 66
  process 137, 160, 165, 181
  protuberance 209, 219, 224, 338
  ribs 57, 58, 160, 165, 167, 302,
    303
  ridge 284, 286, 287, 289
  secondary resting spore 13
  semilanceolate 303
  striae 169, 240, 247, 296, 287, 295,
    312, 324
  striation 240, 245, 251, 272, 277,
    278, 282, 283, 287, 294, 299,
    304, 315, 319, 327, 328
  subconical 159, 160
  symmetry 142
  wall 17, 30, 33, 46, 92, 178, 238,
    239, 306, 328
Vacuoles 596, 621, 706
Valvocopula 306
Vane like structures 290
Vaulted valves 284
Vegetative
  cells 598, 627
  growth 8–10, 416, 596, 597, 738
  reproduction 596, 738, 742
  stage 653
  valves 32, 35, 223, 284
Velum(a) 17, 177, 253, 254, 256, 259,
    263, 265, 266, 267, 338, 267, 464,
    466,
Ventral 303, 305, 401
  pore 410
Vermiculate 411, 424, 433
Venation 79
Venezuela 500
Vertical distribution 741
Vesicles 595, 812
Vestibulum region 605
Viral particles 598
Vitamins 600, 696, 697, 703
Volatile organic sulfur comp. 732

Wall structure 707, 739
Water column 592, 741
Water discoloration 743, 744
Water mounts 800
Warts 33
Weakly silicified valves 35, 87, 105, 191,
   207, 224
Wedge shaped 59, 103, 109, 110, 159,
   165
West Indies 447
Whorl 749, 779, 780
   coccoliths 749, 779, 780
Wing 40, 63, 169, 171, 408,

Xanthophyll 390
X plate 411

Young upwelled water 744

Yucatan 229

Zooflagellates 594, 604, 618, 671, 672,
   680, 681
Zoospores 620, 627
Zoosystematics 616, 626
Zygoform
   holococcoliths 749
   ordinary coccoliths 768
   stomatal coccolith 764
Zygolith 736, 749, 755, 761, 764, 765,
   768, 769
Zig zag
   chains 20, 238, 244, 249
   colonies 247, 253
   line 87
   pattern 36
Zooxanthellae 461
Zygote 8, 9

Printed and bound by CPI Group (UK) Ltd, Croydon, CR0 4YY

03/10/2024

01040426-0008

MULTISCALE HYDROLOGIC REMOTE SENSING

Perspectives and Applications